Student & Parent

One-Stop Internet ~~Resources~~

D1037268

Log on to

www.mathmatters3.com

Online Study Tools

- Extra Examples
- Self-Check Quizzes
- Chapter Assessment Practice
- Standardized Test Practice

Online Resources

- Math*Works* Careers
- Links to Chapter Themes
- Multilingual Glossary

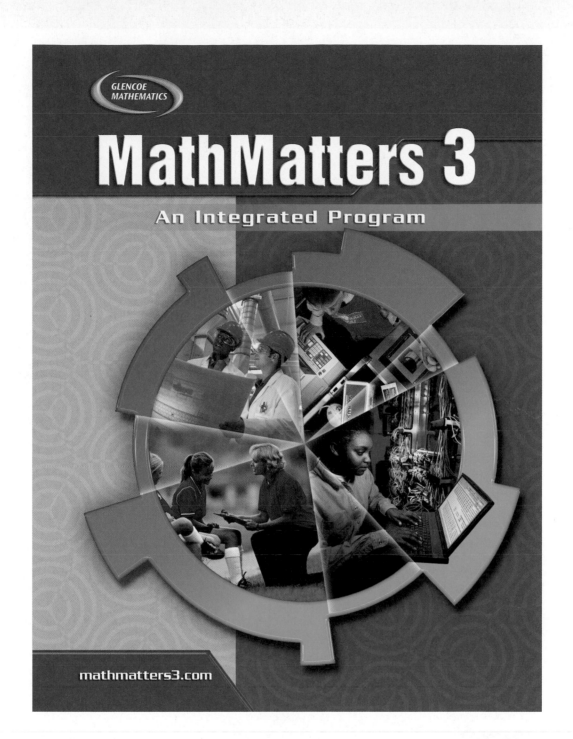

GLENCOE
MATHEMATICS

MathMatters 3

An Integrated Program

mathmatters3.com

Lynch

Olmstead

McGraw Hill Glencoe

New York, New York
Columbus, Ohio
Chicago, Illinois
Peoria, Illinois
Woodland Hills, California

Send all inquiries to:
Glencoe/McGraw-Hill
8787 Orion Place
Columbus, OH 43240-4027

ISBN: 978-0-07-868178-3
MHID: 0-07-868178-2

MathMatters 3 Student Edition

5   6   7   8   9   10   058/111   13   12   11   10   09   08   07

Contents in Brief

Chicha Lynch currently teaches Honors Advanced Algebra II at Marin Catholic High School in Kentfield, California. She is a graduate of the University of Florida. She was a state finalist in 1988 for the Presidential Award for Excellence in mathematics teaching. Currently, Ms. Lynch is a participating member of the National Council of Teachers of Mathematics as well as a long-time member of California Math Council North.

Eugene Olmstead is a mathematics teacher at Elmira Free Academy in Elmira, New York. He earned his B.S. in Mathematics at State University College at Geneseo in New York. In addition to teaching high school, Mr. Olmstead is an instructor for T^3, Teachers Teaching with Technology, and has participated in writing several of the T^3 Institutes. In 1991 and 1992, Mr. Olmstead was selected as a state finalist for the Presidential Award for Excellence in mathematics teaching.

Reviewers and Consultants

These educators reviewed every chapter and gave suggestions for improving the effectiveness of the mathematics instruction.

Tamara L. Amundsen
Teacher
Windsor Forest High School
Savannah, Georgia

Kyle A. Anderson
Mathematics Teacher
Waiakea High School
Milo, Hawaii

Murney Bell
Mathematics and Science
    Teacher
Anchor Bay High School
New Baltimore, Michigan

Fay Bonacorsi
High School Math Teacher
Lafayette High School
Brooklyn, New York

Boon C. Boonyapat
Mathematics Department
    Chairman
Henry W. Grady High School
Atlanta, Georgia

Peggy A. Bosworth
Retired Math Teacher
Plymouth-Canton High School
Canton, Michigan

Sandra C. Burke
Mathematics Teacher
Page High School
Page, Arizona

Jill Conrad
Math Teacher
Crete Public Schools
Crete, Nebraska

Nancy S. Cross
Math Educator
Merritt High School
Merritt Island, Florida

Mary G. Evangelista
Chairperson, Mathematics
    Department
Grove High School
Garden City, Georgia

Timothy J. Farrell
Teacher of Mathematics and
    Physical Science
Perth Amboy Adult School
Perth Amboy, New Jersey

Greg A. Faulhaber
Mathematics and Computer
    Science Teacher
Winton Woods High School
Cincinnati, Ohio

Leisa Findley
Math Teacher
Carson High School
Carson City, Nevada

Linda K. Fiscus
Mathematics Teacher
New Oxford High School
New Oxford, Pennsylvania

Louise M. Foster
Teacher and Mathematics
    Department Chairperson
Frederick Douglass High School
Altanta, Georgia

Darleen L. Gearhart
Mathematics Curriculum
    Specialist
Newark Public Schools
Newark, New Jersey

Faye Gunn
Teacher
Douglass High School
Atlanta, Georgia

Dave Harris
Math Department Head
Cedar Falls High School
Cedar Falls, Iowa

Barbara Heinrich
Teacher
Wauconda High School
Wauconda, Illinois

Margie Hill
District Coordinating Teacher
    Mathematics, K-12
Blue Valley School District
    USD229
Overland Park, Kansas

Suzanne E. Hills
Mathematics Teacher
Halifax Area High School
Halifax, Pennsylvania

Robert J. Holman
Mathematics Department
St. John's Jesuit High School
Toledo, Ohio

Eric Howe
Applied Math Graduate Student
Air Force Institute of Technology
Dayton, Ohio

Daniel R. Hudson
Mathematics Teacher
Northwest Local School District
Cincinnati, Ohio

Susan Hunt
Math Teacher
Del Norte High School
Albuquerque, New Mexico

Todd J. Jorgenson
Secondary Mathematics
   Instructor
Brookings High School
Brookings, South Dakota

Susan H. Kohnowich
Math Teacher
Hartford High School
White River Junction, Vermont

Mercedes Kriese
Chairperson, Mathematics
   Department
Neenah High School
Neenah, Wisconsin

Kathrine Lauer
Mathematics Teacher
Decatur High School
Federal Way, Washington

Laurene Lee
Mathematics Instructor
Hood River Valley High School
Hood River, Oregon

Randall P. Lieberman
Math Teacher
Lafayette High School
Brooklyn, New York

Scott Louis
Mathematics Teacher
Elder High School
Cincinnati, Ohio

Dan Lufkin
Mathematics Instructor
Foothill High School
Pleasanton, California

Gary W. Lundquist
Teacher
Macomb Community College
Warren, Michigan

Evelyn A. McDaniel
Mathematics Teacher
Natrona County High School
Casper, Wyoming

Lin McMullin
Educational Consultant
Ballston Spa, New York

Margaret H. Morris
Mathematics Instructor
Saratoga Springs Senior High
   School
Saratoga Springs, New York

Tom Muchlinski
Mathematics Resource Teacher
Wayzata Public Schools
Plymouth, Minnesota

Andy Murr
Mathematics
Wasilla High School
Wasilla, Alaska

Janice R. Oliva
Mathematics Teacher
Maury High School
Norfolk, Virginia

Fernando Rendon
Mathematics Teacher
Tucson High Magnet
   School/Tucson Unified
   School District #1
Tucson, Arizona

Candace Resmini
Mathematics Teacher
Belfast Area High School
Belfast, Maine

Kathleen A. Rooney
Chairperson, Mathematics
   Department
Yorktown High School
Arlington, Virginia

Mark D. Rubio
Mathematics Teacher
Hoover High School—GUSD
Glendale, California

Tony Santilli
Chairperson, Mathematics
   Department
Godwin Heights High School
Wyoming, Michigan

Michael Schlomer
Mathematics Department Chair
Elder High School
Cincinnati, Ohio

Jane E. Swanson
Math Teacher
Warren Township High School
Gurnee, Illinois

Martha Taylor
Teacher
Jesuit College Preparatory School
Dallas, Texas

Cheryl A. Turner
Chairperson, Mathematics
   Department
LaQuinta High School
LaQuinta, California

Linda Wadman
Instructor, Mathematics
Cut Bank High School
Cut Bank, Montana

George K. Wells
Coordinator of Mathematics
Mt. Mansfield Union
   High School
Jericho, Vermont

TABLE OF CONTENTS

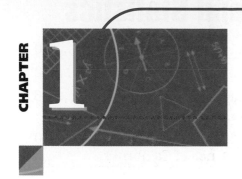

CHAPTER 1

Theme: Language and Communication

Math*Works* Careers
Cryptographer 15
Cashier 33

Applications
advertising 36
astronomy 40
business 41
chemistry 39
communication 9, 29
construction 23
finance 8, 21, 22
food service 28
geography 13
language 13, 19, 40
recreation 22
retail 9
sales 37
science 28, 37
sewing 31
temperature 13, 23, 28

Standardized Test Practice
Multiple Choice 46
Short Response/Grid In 47
Extended Response 47

Essential Mathematics

Page 19

Essential Algebra and Statistics

Theme: News Media

Math*Works* Careers
Environmental Journalist 61
Transcriptionist 81

Applications
advertising 74, 92
biology 59
business 69, 93
education 84
engineering 58
entertainment 89
finance 55, 69, 75
health 86
journalism 85
marketing 84
newspaper 53, 55
news media 69, 79, 84, 88
photography 65
recreation 74
reporting 88
retail 64
sales 79
sports 82, 85, 92
test taking 83
television 55, 59
temperature 63
weather 85

Standardized Test Practice
Multiple Choice 98
Short Response/Grid In 99
Extended Response 99

Page 54

Geometry and Reasoning

Theme: Geography

Math*Works* Careers
Cross-Country Bus Driver 113
Cartographer 133

Applications
architecture 111, 119
art 116, 127
carpentry 107, 135
city planning 125
drafting 129
geography 107, 130
mapmaking 117
navigation 121
scheduling 126
sports 130
surveying 121, 135
tiling 110
travel 138

Standardized Test Practice
Multiple Choice 144
Short Response/Grid In 145
Extended Response 145

Page 101

Triangles, Quadrilaterals, and Other Polygons

Theme: Art and Design

Math*Works* Careers
Jeweler 159
Animator 177

Applications
animation 154
architecture 162, 171, 174, 184
art 171, 174, 183, 191
bridge building 153, 163, 184
construction 156, 173, 175
design 160, 185
engineering 157
food service 159
physics 166
recreation 181
sports 181
stage design 189
surveying 179

Standardized Test Practice
Multiple Choice 196
Short Response/Grid In 197
Extended Response 197

Page 156

5

Theme: Lost Cities of Ancient Worlds

Math*Works* Careers
Equipment Operator 211
Archaeologist 229

Applications
archaeology 204, 206, 208, 214, 221, 226, 232, 233
architecture 223
art 217, 223, 227
astronomy 227, 232
carpeting 216
engineering 202
games 213, 214
history 205
manufacturing 224, 231
packaging 224, 233
recreation 209, 217
sports 208, 226, 227
stage design 208
weather 215

Standardized Test Practice
Multiple Choice 238
Short Response/Grid In 239
Extended Response 239

Page 217

Linear Systems of Equations

Theme: Manufacturing Industry

Math*Works* Careers
Precision Assembler 253
Engineering Technician 273

Applications
agriculture 285
budgeting 279
business 275
cartography 250
civics 267
community service 270
electronics 249
engineering 284
entertainment 269
farming 270
finance 247, 261, 266, 270, 275
fitness 257
health 279
income tax 260, 261
manufacturing 245, 250, 261,
270, 277, 282, 285
packaging 266
product design 255
real estate 257
recreation 266
shipping 265
small business 285
space 271
temperature 256

Standardized Test Practice
Multiple Choice 290
Short Response/Grid In 291
Extended Response 291

Page 253

Similar Triangles

Theme: Photography

Math*Works* Careers
Police Photographer 305
Photographic Processor 325

Applications
architecture 303, 306, 323
art 299, 311
business 298
construction 308
engineering 307
framing 303
investing 298
model building 308, 321
photography 301, 303, 308, 313, 319, 327
photo processing 297
real estate 298, 323
recreation 297
retail 298
satellite photography 307
scale models 317
surveying 312, 327

Standardized Test Practice
Multiple Choice 332
Short Response/Grid In 333
Extended Response 333

Page 293

Theme: Amusement Parks

Math*Works* Careers
Construction Supervisor 347
Aerospace Engineer 367

Applications
amusement park design 362
animation 344
art 340, 344, 351, 354
business 350, 372
computer graphics 344
cryptography 359
encryption 363, 365
engineering 353
food concessions 373
food distribution 373
game development 371
graphics design 349, 354
inventory 364
manufacturing 360
park admissions 365
population 361
recreation 339
ride design 370
ride management 343
sales 373
souvenir sales 359, 373
ticket sales 362

Standardized Test Practice
Multiple Choice 378
Short Response/Grid In 379
Extended Response 379

Transformations

Page 362

Probability and Statistics

Theme: Sports

Math*Works* Careers
Dietician 391
Physical Therapist 411

Applications
business 394
card games 386, 387, 395
education 387
fitness 408
games 386, 393, 394, 395
hiring 404
history 399
manufacturing 406, 409
marketing 388
office work 404
photography 395
programming 389
recreation 384
sales 407
scheduling 398
sports 385, 386, 387, 389, 392,
    394, 395, 397, 398, 403, 404,
    405, 407, 408, 409, 412
surveys 398
test taking 388, 414
transportation 387
travel 402
weather 387

Standardized Test Practice
Multiple Choice 420
Short Response/Grid In 421
Extended Response 421

Page 381

Right Triangles and Circles

Theme: Architecture

Math*Works* Careers
Construction and Building
Inspector 435
Landscape Architect 453

Applications
architecture 426, 431, 436,
449, 455, 457
art 451, 457
construction 429, 432, 439, 456
design 457
home repair 432
inventions 438
landscape architecture 440
math history 429
navigation 443
plumbing 439
road planning 438
small business 429
surveying 443, 451
urban planning 446

Standardized Test Practice
Multiple Choice 462
Short Response/Grid In 463
Extended Response 463

Page 422

Polynomials

Theme: Consumerism

Math*Works* Careers
Brokerage Clerks 477
Actuaries 497

Applications
advertising 473
archaeology 475
art 471, 495
boating 509
chemistry 501
construction 475, 485, 501, 509
design 491
income 470
landscaping 475, 480, 490, 495
manufacturing 479, 489, 493
marketing 474
packaging 469, 483, 508
payroll 474
product development 485, 499
sales 491
sculpture 480
sewing 485
shipping 471, 491
small business 484, 501, 507
transportation 471, 475, 484
travel 509

Standardized Test Practice
Multiple Choice 514
Short Response/Grid In 515
Extended Response 515

Page 464

Quadratic Functions

Theme: Gravity

Math*Works* Careers
Pilots 529
Air Traffic Controllers 549

Applications
aeronautics 536, 542, 543
archaeology 547
art 537
astronomy 527
business 527
geology 521
physics 522, 523, 532, 533, 537, 543, 551
science 535, 550
skydiving 543
small business 522
space exploration 545
sports 533, 537, 546, 547

Standardized Test Practice
Multiple Choice 556
Short Response/Grid In 557
Extended Response 557

Page 516

Advanced Functions and Relations

Theme: Astronomy

Math*Works* Careers
Payload Specialist 571
Astronomer 589

Applications
architecture 565
astronomy 563, 575, 577, 582, 585, 591
baked goods 582
biology 582
business 596
catering 582
chemistry 603
communications 577, 592
computer design 592
design 564
earnings 582
earthquakes 603
energy 568
farming 596
food prices 581
investments 596
magnetism 587
nutrition 596
oceanography 576
physics 582, 585
population 595, 596
postage 582
real estate 596
sales 582
satellite communications 567
science 564
sound 603
space exploration 581
sports 565
technology 596
travel 585
vehicle ownership 596

Standardized Test Practice
Multiple Choice 608
Short Response/Grid In 609
Extended Response 609

Page 576

14

Theme: Navigation

Math*Works* Career
Commercial Fisher 633

Applications

Standardized Test Practice

Trigonometry

Student Handbook 643

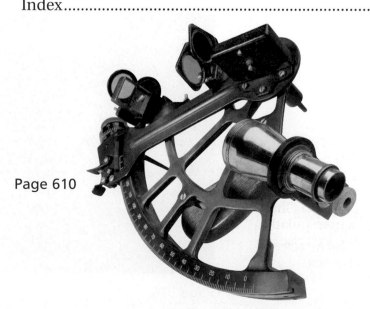

Page 610

How to Use Your MathMatters Book

Welcome to *MathMatters*! This textbook is different from other mathematics books you have used because *MathMatters* combines mathematics topics and themes into an integrated program. The following are recurring features you will find in your textbook.

Chapter Opener This introduction relates the content of the chapter and a theme. It also presents a question that will be answered as part of the ongoing chapter investigation.

Are You Ready? The topics presented on these two pages are skills that you will need to understand in order to be successful in the chapter.

Build Understanding The section presents the key points of the lesson through examples and completed solutions.

Try These Exercises Completing these exercises in class are an excellent way for you to determine if you understood the key points in the lesson.

Practice Exercises These exercises provide an excellent way to practice and apply the concepts and skills you learned in the lesson.

Extended Practice Exercises Critical thinking, advanced connections, and chapter investigations highlight this section.

Mixed Review Exercises Practicing what you have learned in previous lessons helps you prepare for tests at the end of the year.

MathWorks This feature connects a career to the theme of the chapter.

Problem Solving Skills Each chapter focuses on one problem-solving skill to help you become a better problem solver.

Look for these icons that identify special types of exercises.

 CHAPTER INVESTIGATION Alerts you to the on-going search to answer the investigation question in the chapter opener.

 WRITING MATH Identifies where you need to explain, describe, and summarize your thinking in writing.

 MANIPULATIVES Shows places where the use of a manipulative can help you complete the exercise.

 TECHNOLOGY Notifies you that the use of a scientific or graphing calculator or spreadsheet software is needed to complete the exercise.

 ERROR ANALYSIS Allows you to review the work of others or your own work to check for possible errors.

Essential Mathematics

THEME: Language and Communication

Translating ideas into language that others can understand has always been a challenge. You may be surprised to learn that mathematics is a language. People living 4000 years ago could solve difficult algebra and geometry problems, but they could not communicate their ideas to others because many of the symbols for mathematics had not been invented. Today, people in many careers use mathematical language to communicate their ideas.

- **Cryptographers** (page 15) encrypt important information so that messages sent electronically can be kept private. They must be able to apply special step-by-step mathematical processes called algorithms.

- **Cashiers** (page 33) use communication and math skills to help customers find the cost of purchases, apply discounts, and find the best value for their money. Cashiers must be able to use estimation and mental math skills to guard against errors in their work.

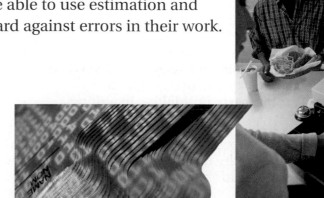

Math Online

mathmatters3.com/chapter_theme

The Science of Communication
Percent of Letters in Written English

E	13%	D	4%	G	1.5%
T	9%	L	3.5%	W	1.5%
A	8%	C	3%	V	1%
O	8%	M	3%	J	0.5%
N	7%	U	3%	K	0.5%
R	6.5%	F	2.5%	X	0.5%
I	6.5%	P	2%	Q	0.3%
S	6%	Y	2%	Z	0.2%
H	5.5%	B	1.5%		

Data Activity: Cryptology–The Science of Secret Communication

Use the table for Questions 1–3.

1. A secret message was encoded by replacing the letters of the alphabet with symbols. The message contains 2200 symbols. How many times would you expect the symbol representing the letter E to occur?

2. In any large passage, what percent of the letters would you expect to be vowels, excluding Y?

3. Which letter appears 20 times more often than the letter Q?

CHAPTER INVESTIGATION

A cipher is a secret method of writing. Many ciphers substitute numbers or symbols for the letters in a message. The person who receives the message must have a key in order to read it.

Working Together

Create a cipher using a number line. Then encode a short message, five or six sentences in length. Use the table above and logic to decode the message. Can you discover the key? Use the Chapter Investigation icons to guide your group.

1 Are You Ready?

Refresh Your Math Skills for Chapter 1

The skills on these two pages are ones you have already learned. Stretch your memory and complete the exercises. For additional practice on these and more prerequisite skills, see pages 654–661.

ADDITION AND SUBTRACTION

Add or subtract.

1. 8563 +9476	**2.** 6905 −4381	**3.** 2765 +7949	**4.** 8614 −2030
5. 7158 +6235	**6.** 8190 −5766	**7.** 5374 +7928	**8.** 6000 −4173
9. 24.86 +13.92	**10.** 58.43 −27.86	**11.** 30.247 +64.892	**12.** 71.056 −38.173
13. 4.638 +8.469	**14.** 4.76 −3.0892	**15.** 5.958 +8.0537	**16.** 6 −1.8428

17. $1\frac{7}{8} + 3\frac{3}{4}$ **18.** $5\frac{1}{3} - 2\frac{5}{6}$ **19.** $3\frac{5}{12} + 7\frac{3}{16}$ **20.** $8\frac{1}{15} - 4\frac{7}{8}$

MULTIPLICATION AND DIVISION

Multiply or divide. Give remainders in whole-number division. In decimal exercises, round to the nearest hundredth. Reduce fractions to lowest terms.

21. 6438 × 45	**22.** $32\overline{)1874}$	**23.** 5038 × 73	**24.** $24\overline{)5964}$
25. 397 ×482	**26.** $17\overline{)1394}$	**27.** 604 ×295	**28.** $56\overline{)7000}$
29. 3.48 × 2.5	**30.** $2.8\overline{)7.56}$	**31.** 6.143 × 0.25	**32.** $1.8\overline{)53.92}$
33. 7.641 × 0.03	**34.** $0.05\overline{)9.765}$	**35.** 5.05 ×0.0076	**36.** $8.08\overline{)62}$

37. $3\frac{2}{3} \times 5\frac{1}{8}$ **38.** $9\frac{7}{12} \div 2\frac{1}{4}$ **39.** $4\frac{3}{4} \times 6\frac{13}{16}$ **40.** $10\frac{1}{6} \div 2\frac{5}{8}$

SOLVING WORD PROBLEMS

In real life, problems are not always presented in numerical form. More often you have to decide what kind of answer you are looking for and how to go about finding it.

Remember the 5-Step Plan: Read - Plan - Solve - Answer - Check.

41. Keshawn bought a CD for $12.95 and a package of batteries for $3.79. How much change should he receive from a $20 bill?

42. Amie belongs to a book club. This month she ordered 3 books at $9.95 each. Shipping and handling for the 3 books came to $5.95. What is Amie's book bill this month?

43. Mr. Sanders pays $575 rent for his apartment. This month he also paid $57.82 for electricity, $12.86 for gas, $48.86 for telephone service, and $30 for garbage pickup. What were his total expenses this month?

44. Emily has a cat and a dog. A bag of dog food costs $8.99 and will feed the dog for 15 days. A bag of cat food costs $10.89 and will feed the cat for 17 days. Which animal is more costly to feed?

45. The local movie theater charges $8 for adults and $5.50 for children. How much is the cost for 82 adults and 153 children to watch a movie?

46. Darius drove 496 mi on one tank of gas. His car gets 32 mi/gal. About how many gallons of gas does the car's tank hold?

47. One supermarket sold 72 cases of *Zing* soap, but only one-third as many cases of *Essence* soap. If each case holds 156 bars of soap, how many bars were sold in all?

48. The Allen family is planning a 1200-mi trip. Assuming they travel at exactly 55 mi/h for 6 h/day, how many days will it take them to reach their destination?

49. A display in a store is between 5 ft and 6 ft tall. The display contains stacks of boxes and stacks of cans. Each box is 17 in. tall. Each can is 4 in. tall. The stacks of boxes and cans are the same height. How tall are the stacks?

50. Jan has 3-in. by 5-in. cards. She placed them next to each other to form a square (same length and width). What is the least number of cards Jan could use?

51. Tashi wants to buy one can of lemon-lime soda for every three cans of cola. If she buys 36 cans of cola, how many cans of lemon-lime soda should she buy?

52. A bakery offers 6 varieties of cookies and bakes 5 dozen of each kind every day, Monday through Friday. How many cookies are baked in four weeks.

The Language of Mathematics

Goals
- Use mathematical symbols to describe sets.
- Describe relationships among sets and elements of sets.

Applications   Finance, Retail, Communication

Work in groups of two students each.

1. Make a list of the different ways you can use language to tell who you are. You might say: "I am the son of Dennis Williams. My name is Jacob Ryan Williams. My nickname is Jake. I am a member of the family at 36 W. Main. I am Jill's brother."

2. Compare lists with your partner.

◤ BUILD UNDERSTANDING

You can find many different kinds of **sets** in your life, such as chess sets and salt-and-pepper sets. You will even find some salt-and-pepper chess sets! In mathematics, you will study many different kinds of sets.

A *well-defined* set makes it possible to determine whether an item is a member of that set. A set whose elements cannot be counted or listed is called an **infinite set**. If all the elements of a set can be counted or listed, it is called a **finite set**.

You can identify or describe yourself in different ways. You can also use different mathematical symbols to describe sets as well as relationships between sets and/or elements of sets.

You can use *description notation* such as "the set of all natural numbers," *roster notation* such as $S = \{1, 2, 3, \ldots\}$, or *set-builder notation* such as $\{x \mid x$ is a natural number$\}$. The symbol $\in$ can be used to show that an element is a member of a set. If every element of set A is also an element of set B, then A is called a **subset** *of* B, written as $A \subseteq B$.

If $A = \{e, f, g, h, i, j\}$ and $B = \{e, i\}$, you can write $f \in A$ and $f \notin B$, which means that f is a member of set A and f is not a member of set B. You can also write $B \subseteq A$, which means that set B is a subset of set A. Since every set is a subset of itself, you can also write $A \subseteq A$.

Consider the set D of days containing 26 hours. Since no day contains 26 hours, D has no elements. The set having no elements is called the **empty set** or the **null set**. To indicate that D is an empty set, write either $D = \{\ \}$ or $D = \varnothing$. The null set is a subset of every set.

Example 1

Determine all the possible subsets of the set {1, 5}.

Solution

There are two single-element subsets of {1, 5}: {1} and {5}. The set itself is a subset. The null set is also a subset. So, {1, 5} has four subsets: {1}, {5}, {1, 5}, ∅. Sets are useful in the study of mathematical sentences.

You can use **sentences** or **equations** of different types to describe the truth of a statement. An **equation** is a statement that two numbers or expressions are equal. It can be true, false, or neither true nor false.

Any sentence that contains one or more variables is called an **open sentence**. The set of all possible values for the variable in an open sentence is called the **replacement set**. An open sentence can be true or false, depending on what values are substituted for the variables.

The value of the variable that makes an equation true is called the **solution** of the equation. The **solution set** of an open sentence is the set of all elements in the replacement set that makes the sentence true.

Example 2

Which of the values -2 and 4 is a solution of the equation $4x + 3 = 19$?

Solution

Substitute -2 for x in the equation.

$$4x + 3 = 19$$
$$4(-2) + 3 \stackrel{?}{=} 19 \quad \text{Remember the order of operations: multiply first, then add.}$$
$$-8 + 3 \stackrel{?}{=} 19$$
$$-5 \neq 19$$

So, -2 is not a solution.

Substitute 4 for x in the equation.

$$4x + 3 = 19$$
$$4(4) + 3 \stackrel{?}{=} 19$$
$$16 + 3 \stackrel{?}{=} 19$$
$$19 = 19$$

So 4 is a solution.

Example 3

Use mental math to solve the equation $x - 8 = 11$.

Solution

Think: Eight subtracted from what number equals 11? You know that

$19 - 8 = 11$, so $x = 19$.

Example 4

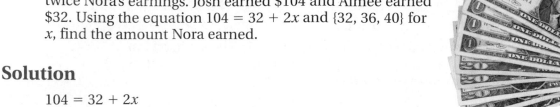

FINANCE Josh's earnings equaled the sum of Aimee's and twice Nora's earnings. Josh earned $104 and Aimee earned $32. Using the equation $104 = 32 + 2x$ and $\{32, 36, 40\}$ for x, find the amount Nora earned.

Solution

$$104 = 32 + 2x$$

Substitute:

32 for x in the equation.	36 for x in the equation.	40 for x in the equation.
$104 = 32 + 2(32)$	$104 = 32 + 2(36)$	$104 = 32 + 2(40)$
$104 = 32 + 64$	$104 = 32 + 72$	$104 = 32 + 80$
$104 \neq 96$ $\neq$ means "is not equal to."	$104 = 104$ 36 is a solution.	$104 \neq 112$

So, Nora earned $36.

◣ TRY THESE EXERCISES

1. Use set notation to write the following: 6 is not an element of $\{1, 3, 5, 7, 9\}$.

2. Determine all the possible subsets of $\{M, A, N\}$.

3. Which of the given values is a solution of the equation $2x + 1 = -1$?
$-2, -1, 0, 1, 2$

Use mental math to solve the following equations.

4. $x + 15 = 13$

5. $z - 2 = 9$

6. Ping scored 22 more points than Terri scored. Ping scored 81 points. Use the equation $81 = T + 22$ and the replacement set $\{51, 57, 59\}$ for T. Find T, the number of points Terri scored.

7. **YOU MAKE THE CALL** Ray says that any set containing three members has exactly three subsets. Is he correct? Explain.

◣ PRACTICE EXERCISES • For Extra Practice, see page 662.

Name each set using roster notation.

8. even natural numbers greater than 9

9. days having 27 hours

Determine if each statement is *true* or *false*.

10. $0 \in \{x \mid x \text{ is a natural number}\}$

11. $9 \in \{-3, 0, 3, 6, \ldots\}$

12. $\{a, b, c\} \subset \{a, e, i, o, u\}$

Write all the subsets of each set.

13. $\{a\}$

14. $\{b, c\}$

15. $\{t, e, n\}$

16. **WRITING MATH** For Exercise 15, you could write $\{a\} \not\subset \{t, e, n\}$. Write a definition of the symbol $\not\subset$.

Which of the given values is a solution of the equation?

17. $a - 9 = -5; 4, 5$

18. $c + 3 = -1; -4, 4$

19. $5n + 3 = 8; -1, 0, 1$

20. $\dfrac{3r}{2} = -3; -2, -4, 2$

Use mental math to solve each equation.

21. $y + 8 = 2$

22. $m - 5 = 2$

23. $6p = -18$

Determine whether the statement is *true* or *false*.

24. $\varnothing$ is not a subset of itself.

25. $\{2, 3\}$ is a subset of $\varnothing$.

Rewrite each statement so that it is correct.

26. $\{o, r\} \in \{r\}$

27. $\{1, 2\} \subseteq \varnothing$

Name each set using roster notation.

28. $\{x | x \text{ is a negative integer and } x > -3\}$

29. $\{x | x \text{ is a whole number and } 4x + 6 = 2\}$

Which of the given values is a solution of the equation?

30. $3b = -2\left(1 - \dfrac{1}{2}\right); \dfrac{1}{2}, -\dfrac{1}{3}, -1$

31. $-\dfrac{1}{3} + m = -\dfrac{1}{4}; \dfrac{1}{12}, -\dfrac{1}{12}, \dfrac{1}{4}$

32. A stack of quarters is worth $2.25. Use the equation $0.25q = 2.25$ and the replacement set $\{7, 8, 9\}$. Find q, the number of quarters in the stack.

33. **RETAIL** The sum of the number of John and Sarah's sales equals $\frac{1}{2}$ the sum of the number of Cora and Dan's sales. Sarah made 15 sales, Cora made 20, and Dan made 22. Use the equation $J + 15 = \frac{1}{2}(20 + 22)$ and the replacement set $\{17, 11, 6\}$ to find J, the number of John's sales.

34. **COMMUNICATION** A newspaper charges $3 per word for its classified advertisements. Don paid $54 to put an ad in the newspaper. Use the equation $3w = 54$ and the replacement set $\{16, 17, 18\}$ to find w, the number of words in the ad.

◼ EXTENDED PRACTICE EXERCISES

Determine the number of subsets for each set.

35. $\{a, b, c\}$

36. $\{a, b, c, d\}$

37. $\{a, b, c, d, e\}$

38. How many subsets do you think a set of 6 elements has? A set of 7 elements?

Use mental math to find two solutions of these equations.

39. $x^2 = 36$

40. $r^2 - 4 = 12$

◼ MIXED REVIEW EXERCISES

Use set notation to write each of the following. (Basic Math Skills)

41. The set P has no members.

42. The set G is a subset of the set K.

43. The number 8 is not a member of the set R.

44. The letter m is a member of the set W.

1-2 Real Numbers

Goals ■ Identify and graph real numbers.

Applications Temperature, Geography, Earth Science

Work in groups of four.

1. Make a set of cards and a number line as shown.

2. To play, each member draws a card, uses a calculator to find the answer, and places a marker on the number line to show the location of the answer.

3. The player whose marker is farthest to the right scores one point. Remove the markers and play again. The winner is the first person to score 5 points.

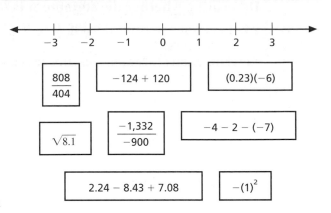

$\dfrac{808}{404}$ $-124 + 120$ $(0.23)(-6)$

$\sqrt{8.1}$ $\dfrac{-1{,}332}{-900}$ $-4 - 2 - (-7)$

$2.24 - 8.43 + 7.08$ $-(1)^2$

◥ BUILD UNDERSTANDING

The set of **natural** or counting numbers is {1, 2, 3, . . .}. The set of **whole** numbers is the set of the natural numbers and zero {0, 1, 2, 3, . . .}. The set of **integers** consists of the whole numbers and their **opposites** {. . . , −3, −2, −1, 0, 1, 2, 3, . . .}.

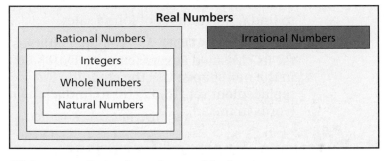

Real Numbers

Rational Numbers

Irrational Numbers

Integers

Whole Numbers

Natural Numbers

Numbers other than integers can also be shown on a number line. Numbers such as $\frac{1}{4}$, $-\frac{1}{2}$, 1.75, and −6.3 can be represented as locations that lie between points represented by integers. This type of number along with the integers represents another set of numbers, the **rational** numbers.

A **rational number** is a number that can be expressed as a ratio of two integers a and b, where b is not equal to zero. This is usually written $\frac{a}{b}$, $b \neq 0$. The symbol $\neq$ means "is not equal to."

Any rational number can be written as a fraction, and any fraction can be written as either a *terminating* decimal or a *repeating* decimal.

Some numbers, such as π (pi), the square root of 2, and 8.112111211112 . . . , are nonterminating and nonrepeating decimals.

Such numbers are called **irrational** numbers.

An **irrational number** is one that *cannot* be written as $\frac{a}{b}$, where a and b are integers and $b \neq 0$, but can still be designated by a point on the number line.

Reading Math

Positive integers are greater than zero. **Negative integers** are less than zero. Zero is *neither* positive nor negative.

Together, the set of rational numbers and the set of irrational numbers make up the set of **real numbers**.

All real numbers can be graphed on a number line. The number that corresponds to a point on a number line is called the **coordinate of the point**. Each real number corresponds to exactly one point on a number line. The point that corresponds to a number is called the **graph of the number**, and is indicated by a solid dot.

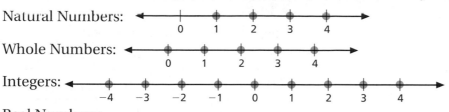

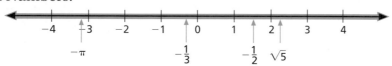

Check Understanding

Name the set or sets of numbers for each of the following:

1. $\sqrt{36}$

2. $-\sqrt{3}$

3. 8.63

4. 2.15115 . . .

Example 1

Graph this set of numbers on a number line.

$$\left\{ -1.75, \quad 3, \quad 2\frac{1}{4}, \quad -2\frac{2}{3}, \quad \sqrt{3}, \quad \frac{\pi}{4} \right\}$$

Solution

Draw a number line. Use a solid dot to graph each number.

Numbers on a number line *increase* as you move from left to right. A number to the left of another number on a number line is *less*. Likewise, a number to the right of another number is *greater*.

The mathematical symbols $<, >, \geq$, and $\leq$ are used to express these relationships. A mathematical sentence that contains one of the symbols $<, >, \geq$, or $\leq$ is called an **inequality**. The inequality symbols and the equal sign are used to compare numbers.

$<$ means *is less than* $\leq$ means *is less than or equal to*
$>$ means *is greater than* $\geq$ means *is greater than or equal to*
$=$ means *is equal to*

Math: Who Where, When

A sixteenth century English mathematician, Thomas Harriot, was the first to use the signs $>$ and $<$. He was considered one of the founders of algebra as we know it today. He was sent to survey and map what was called Virginia. That region is now North Carolina.

Example 2

Graph each set of numbers on a number line.

a. the set of integers from -2 to 3, inclusive

b. the set of real numbers from -2 to 3, inclusive

c. {all real numbers less than or equal to 2}

d. {all real numbers greater than -2}

Solution

a. The set consists of $-2, -1, 0, 1, 2, 3$.

b. The set consists of -2 and 3 and all the real numbers between.

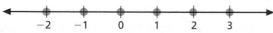

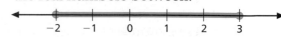

c. The set consists of 2 and all real numbers less than 2.

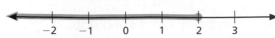

d. The set consists of all real numbers greater than −2.

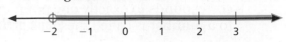

The **absolute value** of a number is the distance that number is from zero on the number line. Since opposite numbers are the same distance from zero, opposite numbers have the same absolute value. The absolute value of a number a is written as $|a|$.

> **Problem Solving Tip**
>
> When working with real numbers, the **opposite of the opposite** property can be useful. For every real number n, $-(-n) = n$.

Example 3

Evaluate each expression when $m = -5$.

a. $-m$ **b.** $-(-m)$ **c.** $|m|$ **d.** $|-m|$ **e.** $-|-m|$

Solution

a. Since $m = -5$, $-m = -(-5) = 5$.

b. Since $-m = 5$, then $-(-m) = -5$.

c. Since $m = -5$, then $|m| = |-5| = 5$.

d. Since $-m = 5$, then $|-m| = |5| = 5$.

e. Since $|-m| = 5$, then $-|-m| = -5$.

◤ TRY THESE EXERCISES

Determine if each statement is *true* or *false*.

1. $\sqrt{6}$ is a rational number.

2. $-\sqrt{25}$ is an integer.

3. $\dfrac{13}{16}$ is an irrational number.

4. 0.10 is a whole number.

Graph the given sets of numbers on a number line.

5. $\left\{-\sqrt{2}, -\dfrac{1}{3}, 0.75, 1\dfrac{1}{2}\right\}$

6. $\left\{1.\overline{6}, 2\dfrac{1}{2}, -\sqrt{5}, -0.50\right\}$

Graph each set of numbers on a number line.

7. the set of integers from −2 to 5, inclusive

8. {all real numbers less than −1}

Evaluate each expression.

9. $|t|$, when $t = -15$

10. $-|-a|$, when $a = -5$

◤ PRACTICE EXERCISES • For Extra Practice, see page 662.

Determine if each statement is *true* or *false*.

11. −1.010010001 . . . is an irrational number.

12. 2.17 is a rational number.

13. 0 is a natural number.

14. $\sqrt{-\dfrac{1}{8}}$ is not a real number.

Graph each set of numbers on a number line.

15. $\left\{-3.25, -\sqrt{3}, 0, 1\dfrac{1}{3}, \sqrt{4}\right\}$

16. whole numbers less than 6

17. {all real numbers greater than −3}

18. {all real numbers less than or equal to 2}

Evaluate each expression.

19. $|r|$, when $r = -9$

20. $-|-(-n)|$, when $n = 12$

Replace each ● with <, >, or =.

21. $-(-3)$ ● $|-3|$

22. $|-\sqrt{16}|$ ● $|-\sqrt{25}|$

23. $|-(-1.5)|$ ● -1.5

Name the set of numbers that are graphed on each number line.

24.

25.

26.

Use the information in each statement to write an inequality.

27. TEMPERATURE Water freezes at 32°F and boils at 212°F.

28. GEOGRAPHY The depth of the Red Sea is -1764 ft and the depth of the Black Sea is -3906 ft.

29. LANGUAGE It is estimated that there are 284 million people who speak Russian and 126 million who speak Japanese.

Red Sea

30. WRITING MATH When comparing two real numbers, will the number with the greater absolute value always be the greater number? Explain your thinking.

31. DATA FILE Refer to the data on principal rivers of the world on page 646 to find the lengths of the Volga and Mackenzie rivers. Compare to find which river is longer.

32. CHAPTER INVESTIGATION Make a number line from -12 to 13. Assign a letter of the alphabet to each position on the number line. Write a message, 3 to 4 sentences in length, and encode the message by substituting numbers for letters. The message will look like a series of math problems.

■ EXTENDED PRACTICE EXERCISES

33. Graph the values that make $|x| < 2$ true.

34. CRITICAL THINKING Determine if the statement is *true* or *false*. For any real number n, where $n < 0$, $|-n| = n$.

35. Are there numbers that are *not* real numbers? Research the term *imaginary numbers*.

■ MIXED REVIEW EXERCISES

Which of the given values is a solution of the equation? (Lesson 1-1)

36. $m - 3 = 8$; 5, 11

37. $g + 6 = -4$; -10, 10

38. $p - 6 = -2$; $-8, -4, 4$

39. $a + 3 = 1$; $-4, -2, 4$

40. $3b + 5 = -1$; $-2, \frac{1}{5}, 2$

41. $\frac{2x}{3} = 2$; $-3, 1, 3$

42. $4w - 2 = -10$; $-3, -2, 2$

43. $5f + 6 = 6$; $-\frac{1}{5}, 0, \frac{1}{5}$

Review and Practice Your Skills

1. List all the subsets of the set {10, 15}.

2. For "the set of positive even numbers less than 8":

   a. Name this set using roster notation.

   b. Name this set using set-builder notation.

   c. List all the subsets of this set.

Use mental math to solve each equation. Name all solutions.

3. $x + 9 = 4$

4. $21 - x = 7$

5. $x^2 = 64$

6. $\frac{1}{2}x = 5$

7. $x^2 = 6^2$

8. $2x + 2 = 10$

9. A stack of nickels is worth $1.65. Use the equation $0.05n = 1.65$ and the replacement set {31, 33, 35}. Find n, the number of nickels in the stack.

10. Define **null set**. Explain why the null set is a subset of all sets.

Which of the given values is a solution of the equation?

11. $n - 17 = -9$; 26, 8, −8

12. $\frac{3}{2}d = -12$; −9, −18, −8

13. $-2p = 11$; −5.5, 22

Use set notation to write the following.

14. 5 is not an element of {2, 4, 6, 8}

15. The null set is a subset of {b, u, g}

Determine if each statement is *true* or *false*.

16. $3\frac{1}{3}$ is a rational number.

17. $-\sqrt{49}$ is an integer.

18. −3 is a natural number

19. π is a rational number.

20. π is a real number.

21. $\sqrt{12}$ is an irrational number

22. 3.51 is a rational number

23. 0 is a natural number

24. $\sqrt{16}$ is an irrational number

Evaluate each expression when $b = -8$.

25. $|b|$

26. $-|-b|$

27. $-(-(-b))$

Graph each set of numbers on a number line.

28. {all real numbers greater than −3}

29. the set of natural numbers less than or equal to 6

30. the set of integers from −1 to 4, inclusive

31. $\{\sqrt{3}, \frac{1}{3}, -3, -6\frac{2}{3}, 3^2, 6\}$

Use <, >, or = to make a true sentence.

32. $-\sqrt{9}$ __?__ $-\pi$

33. $|5| - |-5|$ __?__ 10

34. $-\sqrt{9}$ __?__ $-\sqrt{16}$

35. $-|3.75|$ __?__ $|-3.75|$

36. 6 __?__ $-(-(-(-6)))$

37. $\frac{1}{3}$ __?__ 0.3

38. Graph the values that make $|x| \le 3$ true.

39. How many different letters are in the word *maximize*? Which letters are used more than once? (Lesson 1-1)

For Exercises 40–48, use $A = \{m, n, p, q, r\}$, $B = \{n, q, r\}$, $C = \{x, y, z\}$. Classify each statement as *true* or *false*. (Lesson 1-1)

40. $m \in A$

41. $m \in B$

42. $\varnothing \subseteq B$

43. $\{x, r\} \subseteq C$

44. B has 6 subsets.

45. $B \subseteq A$

46. C contains 3 elements.

47. A contains no vowels.

48. A has 32 subsets.

List all sets of numbers to which each number belongs. (Lesson 1-2)

49. $\sqrt{10}$

50. -17

51. 0

52. $4.3232323232\ldots$

53. $8.71765392\ldots$

54. $-\dfrac{3}{5}$

MathWorks Career – Cryptographer
Workplace Knowhow

Cryptography is a branch of mathematics that combines communication with math. Cryptographers create and try to break codes using step-by-step mathematical processes called algorithms. In the past, encryption was used mainly by the military. Today, businesses also employ cryptographers to protect their electronic messages and financial data.

Cryptographers encrypt text using a key. The key works like a recipe with steps for encoding and decoding messages. The key is the only way to decode the message.

Use this encoding key for Exercises 1–3.

Key: Each letter of the alphabet is assigned an odd positive integer beginning with 1 for Z and proceeding backwards to 51 for A. Then each odd positive integer is added to the even positive integer one number higher. Finally, subtract the sum from 105. Assign the result of this algorithm to each letter. Example: T corresponds to 13. Add: $13 + 14 = 27$. Subtract: $105 - 27 = 78$. T is represented by the number 78.

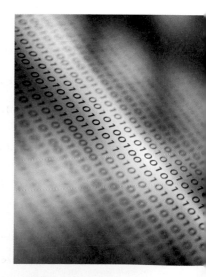

1. What number represents the letter L?

2. Encrypt the following message: Your bank account is overdrawn.

3. Use the key to decipher this message.

10 58 14 18 74 2 70 18 22 82 54
98 58 82 6 70 58 42 18 50 98 6 18 74 78 10 58 14 18

1-3 Union and Intersection of Sets

Goals
■ Identify unions and intersections of sets.
■ Use Venn diagrams to solve problems.

Applications Advertising, Language, Logic

Suppose that in your homeroom 8 students are on the basketball team and 11 students are on the soccer team. Five of the students are on both teams. Another 12 students are not on either team. How many students are on at least one team? Copy and complete the Venn diagram at the right to solve the problem.

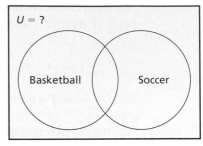

BUILD UNDERSTANDING

Two or more sets can be used to form new sets. Two of these new sets are the union and intersection of the sets. The **union** of any two sets A and B is symbolized as $A \cup B$ (read as "A union B").

The set $A \cup B$ contains all the elements that are in A, in B, or in both.

$$A \cup B = \{x \mid x \in A \text{ or } x \in B\}$$

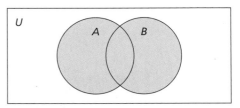

At the right is a Venn diagram of $A \cup B$. A **Venn diagram** uses circles inside a rectangle to represent the union and intersection of sets. The rectangle, labeled U, represents the **universe** or the **universal set**.

The **intersection** of two sets A and B is symbolized by $A \cap B$, read as "A intersect B." The set $A \cap B$ contains the elements that are common to both A and B.

$$A \cap B = \{x \mid x \in A \text{ and } x \in B\}$$

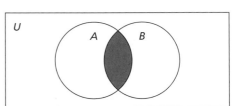

The subset of all elements of U that are *not* elements of A is called the **complement** of A and is symbolized by A'. $A' = \{x \mid x \in U \text{ and } x \notin A\}$.

If two sets have no elements in common, their intersection is the empty set, $\varnothing$. Two sets whose intersection is the empty set are called **disjoint sets**.

Example 1

Refer to the diagram to find each.
Let $A = \{1, 3, 5\}$, $B = \{3, 6\}$, and $C = \{2, 4\}$.

a. C'

b. $A \cup B$

c. $A \cap B$

d. $A \cap C$

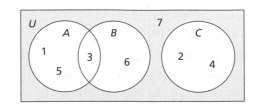

Solution

a. C' is the set of those elements of U that are *not* in C. From the diagram, $U = \{1, 2, 3, 4, 5, 6, 7\}$. So, $C' = \{1, 3, 5, 6, 7\}$.

b. $A \cup B$ is the set of those elements that are in A, in B, or in both.
$$\{1, 3, 5\} \cup \{3, 6\} = \{1, 3, 5, 6\}$$
$$A \cup B = \{1, 3, 5, 6\}$$

If an element is in both, it is still listed *only once* in a union.

c. $A \cap B$ is the set of elements common to A and B. The only element common to both A and B is 3.
$$\{1, 3, 5\} \cap \{3, 6\} = \{3\}$$
$$\text{So, } A \cap B = \{3\}$$

d. A and C have *no* elements in common. So, $A \cap C = \varnothing$.

An inequality that combines two inequalities is called a **compound inequality**. The solution set for a compound inequality containing the word *and* can be found from the *intersection* of the graphs of the individual inequality. The solution set for a compound inequality containing the word *or* can be found from the *union* of the graphs of the individual inequalities.

Example 2

Use the set of real numbers as the replacement set to find the solution set for $x \geq -1$ and $x < 4$.

Let $A = \{x \mid x \geq -1\}$
$\quad B = \{x \mid x < 4\}$
Graph $A \cap B$ on a number line.

Solution

Graph of A: $\quad x \geq -1$

Graph of B: $\quad x < 4$

Graph of $A \cap B$: $x \geq -1$
$\quad\quad\quad\quad\quad$ and $x < 4$

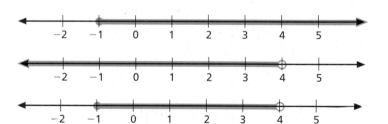

The graph of $x \geq -1$ and $x < 4$ is the *intersection* of the graphs of the two inequalities, $A \cap B$.

This can be written as $-1 \leq x < 4$.

The solution set is $\{x \mid -1 \leq x < 4\}$.

Example 3

Using the set of real numbers as the replacement set, find the solution set for $x \geq 4$ or $x < -1$.

Let $A = \{x \mid x \geq 4\}$
$\quad B = \{x \mid x < -1\}$

Graph $A \cup B$ on a number line.

Solution

Graph of A: $x \geq 4$

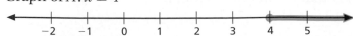

Graph of B: $x < -1$

Graph of $A \cup B$: $x \geq 4$ or $x < -1$

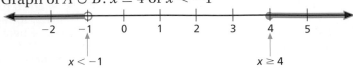

The graph of $x \geq 4$ or $x < -1$ is the *union* of the graphs of the two inequalities, $A \cup B$. The solution set is $\{x \mid x < -1 \text{ or } x \geq 4\}$.

If you change *or* to *and* in Example 3, you would need to find the *intersection* of the two graphs, $A \cap B$. The graphs of the sets have *no* values in common. They *do not* intersect. The solution set would be $\{ \ \}$ or $\varnothing$.

◤ TRY THESE EXERCISES

Refer to the diagram. Find the sets named by listing the members.

1. T'

2. $R \cap S$

3. $R \cup S$

4. $R \cap T$

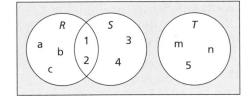

Using the set of real numbers as the replacement set, graph the solution set for each compound inequality.

5. $x \geq 1$ and $x < 5$

6. $x > 5$ or $x \leq -1$

◤ PRACTICE EXERCISES • For Extra Practice, see page 663.

Refer to the diagram. Find the set named by listing the members.

7. $A \cup B$

8. $A \cap B$

9. $A \cup C$

10. $A \cap C$

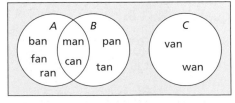

Let $U = \{c, h, a, r, t\}$, $M = \{h, a, r, t\}$, and $N = \{a, r, t\}$. Find each union or intersection.

11. M' **12.** $M' \cup N'$ **13.** $(M \cup N)'$ **14.** $M \cap N$ **15.** $(M \cap N)'$

16. Let $A = \{f, o, r, g, e\}$ and $B = \{m, a, j, o, r\}$. Find $A \cap B$.

17. Let $P = \{3, 6, 9\}$ and $Q = \{4, 8\}$. Find $P \cap Q$.

Use the set of real numbers as the replacement set to find the solution set for each compound inequality.

18. $x \le 3$ and $x \ge 3$

19. $x \le 1$ or $x \ge 2$

For Exercises 20–31 let $U = \{0, 1, 2, 3, \ldots, 9\}$, $A = \{x \mid x$ is a whole number such that $x > 2$ and $x < 8\}$, $B = \{1, 3, 5, 7, 9\}$, $C = \{0, 1, 2, 3, 4, 5\}$, and $D = \{0, 3, 6, 9\}$. Find each set.

20. A' **21.** D' **22.** $A \cap B$ **23.** $C \cup D$

24. $B' \cup C$ **25.** $A' \cap B$ **26.** $B \cap (C \cap D)$ **27.** $B \cup (C \cap D)$

28. $A \cap (B \cup C)$ **29.** $(A \cap B) \cap C$ **30.** $(A \cap B) \cap (C \cap D)$ **31.** $(A \cap B)' \cap (B \cap C)$

32. WRITING MATH Let $A = \{x \mid x > -3\}$
$B = \{x \mid x \le 1\}$
Find $A \cap B$ and write a brief description of the graph of $A \cap B$.

Draw Venn diagrams to solve these problems.

33. LANGUAGE Twenty-four students are on a tour of the United Nations building. Twelve of these students speak Russian. Six speak German, and fifteen speak Spanish. Only one student speaks all three languages. Two speak Russian and German, and two different students speak German and Spanish. Each student speaks at least one of these languages. How many students speak Russian and Spanish, but not German?

34. ADVERTISING A company runs commercials on radio and television for the 18 products it makes. Sixteen of the products are advertised on television and 14 are advertised on the radio. Four of the products are advertised on television only. How many products are advertised only on the radio?

■ EXTEND PRACTICE EXERCISES

Determine whether each statement is *true* or *false*. If the statement is false, provide the correct answer.

$U = \{0, 1, 2, 3, 4, 5, 6, \ldots\}$

$W = \{$whole numbers$\}$ $N = \{$natural numbers$\}$

$P = \{$positive integers$\}$ $M = \{1, 3, 5, 7\}$

35. $P \cup N' = W$ **36.** $P \cap N = \varnothing$ **37.** $N \cup M = M$

38. $N \cap W = N$ **39.** $P \cup N = P$ **40.** $P \cap M = M$

41. Graph $A = \{x \mid x \in W$ or $x \in N\}$ **42.** Graph $B = \{x \mid x \in P$ and $x \in W\}$

■ MIXED REVIEW EXERCISES

Graph each set of numbers on a number line. (Lesson 1-2)

43. the set of integers from -1 to 5, inclusive

44. the set of real numbers from -4 to 3, inclusive.

45. the set of all real numbers greater than or equal to -3

46. the set of all real numbers less than -2

1-4 Addition, Subtraction, and Estimation

Goals
- Add and subtract rational numbers.
- Estimate answers to addition and subtraction problems.

Applications   Recreation, Finance, Construction

You can use a number line to add rational numbers. To add a positive number, move right. To add a negative number, move left.

For example, to show $1\frac{1}{2} + \left(-2\frac{3}{4}\right)$ on a number line, start at 0. Move right $1\frac{1}{2}$ units, then move left $2\frac{3}{4}$ unit.

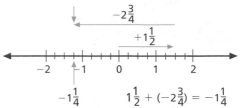

Add the following pairs of rational numbers. Locate each answer and its corresponding letter on the number line below. Then complete these generalizations using your answers.

$$RO \quad KJ \quad V \quad L \quad \quad I \quad E \quad W \quad S \quad Z \quad N \quad H \quad TG \quad F \quad \quad P \quad A \quad CB$$

(number line from −4 to 4)

1. −3.1, 3.1

2. 1.5, 0.5

3. $-2, -\frac{1}{2}$

4. $-1\frac{1}{2}, 2\frac{1}{2}$

5. −4.75, 6.50

6. $-1.25, -2\frac{3}{4}$

7. $-\frac{7}{8}, 4\frac{1}{4}$

8. $3\frac{1}{8}, -4\frac{1}{2}$

9. −0.4, −0.6

10. $4\frac{3}{4}, -8.5$

11. $1\frac{1}{3}, 1\frac{2}{3}$

12. The sum of two positive numbers is $\underline{}\ \underline{}\ \underline{}\ \underline{}\ \underline{}\ \underline{}\ \underline{}\ \underline{}$.
$$\quad\quad\quad\quad\quad\quad\quad\quad\quad\quad\quad\quad\quad\quad\quad\quad 11\ \ 10\ \ 1\ \ 8\ \ 5\ \ 8\ \ 3\ \ 9$$

13. The sum of two negative numbers is $\underline{}\ \underline{}\ \underline{}\ \underline{}\ \underline{}\ \underline{}\ \underline{}\ \underline{}$.
$$\quad\quad\quad\quad\quad\quad\quad\quad\quad\quad\quad\quad\quad\quad\quad\quad 4\ \ 9\ \ 2\ \ 7\ \ 5\ \ 8\ \ 3\ \ 9$$

14. The sign of the sum of a positive and a negative number is the same sign as the number with the $\underline{}\ \underline{}\ \underline{}\ \underline{}\ \underline{}\ \underline{}\ \underline{}$ absolute value.
$$\quad\quad\quad\quad\quad\quad\quad\quad\quad\quad\quad 2\ \ 6\ \ 9\ \ 7\ \ 5\ \ 9\ \ 6$$

◢ BUILD UNDERSTANDING

Using a number line to add and subtract numbers can become quite cumbersome. Rules that apply to all numbers make it easier.

If the numbers have the *same* sign, add the absolute values and use the sign of the addends.

If the numbers have *different* signs, subtract the absolute values and use the sign of the addend with the greater absolute value.

All the properties of addition of integers also apply to rational numbers. For every real number a, $a + (-a) = 0$. In the table, a, b, and c represent rational numbers.

Additive inverses are also used when you subtract integers.

$$-22 + 9 = -13$$
$$-22 - (-9) = -13$$
same result

Subtracting an integer is the same as adding its inverse.

Addition Properties

Property	In Symbols	Examples
Closure	$a + b$ is a rational number.	$2 - 4 = -2$ $\frac{1}{2} + \frac{7}{8} = 1\frac{3}{8}$
Commutative	$a + b = b + a$	$14 + 16 = 16 + 14$
Associative	$(a + b) + c$ $= a + (b + c)$	$(\frac{1}{8} + \frac{1}{4}) + \frac{1}{2}$ $= \frac{1}{8} + (\frac{1}{4} + \frac{1}{2})$
Identity	$a + 0 =$ $0 + a = a$	$-7.3 + 0 =$ $0 + (-7.3) = -7.3$
Inverse	$a + (-a) = 0$	$13.6 + (-13.6) = 0$ $\frac{1}{2} + (-\frac{1}{2}) = 0$

Example 1

Add or subtract.

a. $-3.6 - 4.9$ **b.** $3.5 + (-0.875)$

Solution

a. $-3.6 - 4.9 = -3.6 + (-4.9)$ Add the additive inverse of 4.9.
$= -8.5$ Add the numbers.

b. $3.5 + (-0.875) = 3.5 - 0.875$ Signs are different, so subtract absolute values.
$= 2.625$ Use the sign of the addend with the larger absolute value.

Evaluating expressions may involve adding or subtracting integers.

Example 2

Evaluate each expression when $a = -32$ and $b = 19$.

a. $a + b$ **b.** $b - a$ **c.** $a - b$

Solution

a. $a + b$ **b.** $b - a$ **c.** $a - b$
$-32 + 19$ $19 - (-32)$ $-32 - 19$
-13 $19 + 32$ $-32 + (-19)$
   51 -51

Reading Math

Suppose you are to find the sum $-5 + 5$. The addition involves a number and its opposite. The numbers are called **additive inverses**. Additive inverses are used to state the **addition property of opposites**.

Estimating the answer *before* you start your calculations can give you a "ballpark" figure which you can use to check the reasonableness of your answer.

Example 3

FINANCE Jeremy had $295.48 in his bank account. He deposited a check for $196.68 and withdrew $65.00. How much is in his account now?

Solution

Before solving the problem, estimate the answer. Begin by rounding the amounts.

$295.48 → $300.00 $196.68 → $200.00 $65.00 → $70.00

Then use the numbers to get an estimate of the total. $300 + $200 - $70 = $430

After completing the transactions, Jeremy will have about $430.00 in his account.

Add or subtract.

1. $-12 + (-15)$

2. $-3.7 - (-2.4)$

3. $1\frac{1}{2} - \left(-9\frac{3}{4}\right)$

Evaluate each expression when $m = -2.1$ **and** $n = 1.5$.

4. $m + n$

5. $m - n$

6. $-m - n$

7. $n - m$

8. FINANCE The balance in Dawn's checking account is $251.92. She wrote a check for $129.63. What's her new balance?

9. ERROR ALERT A submarine dove to a depth of 195 ft below the surface of the ocean. It then dove another 186 feet before climbing 229 ft. How many feet below the surface is the submarine? Clay solved the problem this way: $195 + (-186) + 229 = 238$. Is he correct?

10. RECREATION Carla is playing a game in which she needs 250 points to win. She started the game with no points and lost 180 points on the first round. She gained 165 points on the second round. How many points does she need to gain on the third round to win the game?

11. WRITING MATH The cost of sending four packages is $4.86, $7.65, $2.12, and $8.61. You have $25. Do you have enough money to cover the total cost of sending the packages. How do you know?

Add or subtract.

12. $-9 + (-45)$

13. $-23.6 + 19.7$

14. $-1\frac{3}{4} - 3\frac{1}{3}$

15. $-28.2 - (-3.8)$

16. $-10\frac{2}{3} + \left(-5\frac{1}{6}\right) + \left(-2\frac{1}{3}\right)$

17. $-7.6 + (-12.5) + 2.4 + 9.7$

18. $-12 + 18 + 29 + (-16)$

Evaluate each expression when $a = 27$ **and** $b = -16$.

19. $a - b$

20. $a + b$

21. $-a - b$

22. FINANCE The previous balance of Kendra's savings account is $816.85. She makes the following transactions to her savings account: deposit, $135; withdrawal, $185; deposit, $239.58; withdrawal, $395.50. What is her new balance?

Replace each ■ **with <, >, or =.**

23. $-4\frac{3}{4} + \left(-6\frac{1}{2}\right)$ ■ $-18.0 - 6.75$

24. $-23.6 + (-44.7)$ ■ $14.8 + 53.5$

25. $-89 + (-77)$ ■ $-143 - (-35)$

26. $131.0 - (-112.6)$ ■ $272.0 - 33.7$

27. **CONSTRUCTION** Tom wants to construct a garden next to his house with 75 ft of fencing, using the house as one side of the fence. The lengths of the sides of the garden using the fencing are 25.6 ft, 18.9 ft, and 24.8 ft. How much fencing will he have left?

28. **CHAPTER INVESTIGATION** Code makers often use layers of encryption to make the code harder to break. Work with the message and number line code you created in Lesson 1-2. Make the code more difficult by substituting operations that will lead from one letter to the next.

Example: If the word CAT was first encoded as +5 −3 +12, it becomes 5 − 8 + 15. The operation − 8 shows the change from +5 to −3. The operation + 15 shows the change from −3 to +12.

Exchange encoded messages and number line keys with a partner. Reveal the first letter in the passage. Decode your partner's message.

◢ EXTENDED PRACTICE EXERCISES

29. **CRITICAL THINKING** Evaluate $|-a - (a - b(a - b(a - b)))|$ when $a = -3$ and $b = -2$.

30. **TEMPERATURE** The temperature in Cheyenne, Wyoming was 3°F at 4:00 A.M. During the next five hours, the temperature changed as follows:

| 5 A.M., + 8°F; | 6 A.M., + 6°F; | 7 A.M., + 13°F; |
| 8 A.M., −10°F; | 9 A.M., −2°F. | |

What was the temperature at the end of the five hours?

◢ MIXED REVIEW EXERCISES

Refer to the diagram at the right. Let $A = \{0, 4, 5, 7\}$, $B = \{0, 3, 4, 6\}$, $C = \{1, 8\}$ and $U = \{0, 1, 2, 3, ..., 9\}$ (Lesson 1-3)

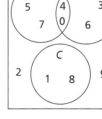

Find each of the following.

31. B' 32. C'

33. $A \cup B$ 34. $A \cap B$

35. $A' \cap C$ 36. $B \cap C$

Evaluate each expression when $x = -3$. (Lesson 1-2)

37. $|x|$ 38. $|-x|$ 39. $-|x|$ 40. $-(x)$

41. $-|-x|$ 42. $-(-x)$ 43. $-|-(-x)|$ 44. $-(-|x|)$

Simplify each numerical expression. (Basic Math Skills)

45. $4 \times 3 + 8$ 46. $16 \div 2 - 2 \times 3$ 47. $5 \times 6 + 12 - 4$ 48. $25 \div 5 + (7 \times 3)$

49. $(4^2 + 3) \times 2$ 50. $6^2 - 24 \div (5 + 3)$ 51. $5 + 9 \div 2 - 3$ 52. $4^2 \div 2^2 + (2 \times 8)$

Evaluate each expression when $x = -12$. (Lesson 1-2)

53. $0 - x$ 54. $-|x|$ 55. $x + |x|$

56. $\dfrac{x}{|-x|}$ 57. $12 + x - |2x|$ 58. $x^2 + |x|^2 + x^2$

Review and Practice Your Skills

PRACTICE ◤ LESSON 1-3

Let $U = \{c, o, m, p, u, t, e, r\}$, $A = \{c, o, m\}$, $B = \{o, m, p, u, t\}$, and $C = \{c, e, r\}$.
Find each set.

1. A'
2. $A \cup B'$
3. $A' \cap C$
4. $A \cap A'$
5. $B \cap C$
6. $(A' \cap B') \cup A$
7. $A' \cap (B \cup C)$
8. $(A \cup C) \cup (B \cap C)$
9. $(B' \cap C') \cup (B' \cup C')$

Use the set of real numbers as the replacement set. Graph the solution of each inequality.

10. $x \le 7$ and $x \ge -4$
11. $x \le -3$ or $x \ge 13$
12. $-11 < x \le -4$
13. $x > 5$ or $x \le 0$
14. $x \ge 6$ or $x \le 6$
15. $x \le -1$ and $x \ge -1$
16. $x \le 3$ and $x > 9$
17. $x > 10$ or $x < 10$
18. $x < -3$ and $x > -3$

Draw a Venn diagram to solve this problem.

19. Eighty-nine students are registered for beginning art electives. Forty-five of these students are enrolled in Drawing I. Thirty-four are enrolled in Painting I, and forty are enrolled in Photography. Two students are enrolled in all three courses. Five students are enrolled in both Drawing I and Painting I. Six students are enrolled in both Painting I and Photography. How many students are enrolled in Drawing I and Photography, but not Painting I?

PRACTICE ◤ LESSON 1-4

Add or subtract.

20. $-3.6 + 4.2 + (-8.7) - 2.5 + 11.3$
21. $133 - (-77) + (-110) + 100$
22. $3\frac{3}{4} + 6\frac{7}{8} - \left(-2\frac{1}{4}\right) + 8\frac{1}{8} - 10$
23. $10 - 9 + 8 - 7 + 6 - 5 + 4 - 3 + 2 - 1$

Evaluate each expression when $a = -48$ and $b = 19$.

24. $-a - b$
25. $-b + a$
26. $|a| - |-b|$
27. $a - b(b + a) + a$
28. $a(a - b) + a(b - a)$
29. $a + b - b - (-a) + (-b)$

30. The previous balance of Sam's savings account was $1345.67. He makes the following transactions: deposits, $228.48, $74.36; withdrawals, $435, $110. What is his new balance?

For Exercises 31–33, estimate to find the answer.

31. Arlene wants to install a fence around a play area for her dog. She will use her house as one side of the fence. She has 110 ft of fencing available. The lengths of the sides of the play area using the fencing are 33.1 ft, 27.8 ft, and 27.8 ft. How much fencing will she have left?

32. You have $20 to spend at the store. Do you have enough money to purchase five items that cost $4.47, $6.12, $2.76, $3.97, and $3.35?

33. The cost of sending four packages is $7.82, $2.92, $5.29, and $11.13. Is $25 enough to cover the total cost of sending the packages?

For Exercises 34–39, determine if each statement is *true* or *false*. (Lesson 1-1)

34. $\{r, c, e\} \subseteq \{l, o, c, k, e, r\}$ **35.** Every set is a finite set.   **36.** $\varnothing \subseteq \{4, 8, 12, 16\}$

37. $\{0, 1, 2, 3, 4\}$ has 16 subsets **38.** $\{n, o\} \subseteq \{o,n\}$ **39.** The null set has one element.

For Exercises 40–42, determine if each statement is *true* or *false*. (Lesson 1-2)

40. $0 \in \{x \mid x$ is a natural number$\}$ **41.** 13 is a rational number   **42.** $\left|-(-13.5)\right| = 21 - 7.5$

43. Evaluate $(y - 7) - \left|-y\right| + (-2 - y)$ when $y = 6$. (Lesson 1-2)

Find the following. $U = \{a, c, e, g, j, l, n, p, r, v, y, z\}, X = \{c, l, e, a, n\}$ and $Y = \{y, a, r, n\}$.
(Lesson 1-3)

44. $X \cap Y$

45. $X \cup Y$

46. $X' \cap Y'$

47. $(X \cup Y)'$

48. $(X \cap Y) \cup (X' \cap Y')$

49. U'

50. Graph $\{x \mid x$ is a real number less than 25 and greater than $-30\}$. (Lesson 1-3)

Add or subtract. (Lesson 1-4)

51. $3.5 + 3.05 + 3.055$

52. $12\frac{1}{2} + \left(-6\frac{5}{8}\right)$

53. $48 + (-48) - 48 - (-48) + 48$

54. $\left|-17\right| + 15 - \left|15\right| + 17$

55. $-6.9 + 9.5 - (-4.8)$

56. $0 - \left(3\frac{1}{4} - 5\frac{1}{3}\right)$

Mid-Chapter Quiz

1. List all the subsets of $\{3, 5, 9\}$. (Lesson 1-1)

2. Write the solution set of $\{x \mid x$ is a negative integer and $x > -5\}$ in roster notation. (Lesson 1-1)

Graph each set of numbers on a number line. (Lesson 1-2)

3. all real numbers greater than or equal to -5

4. all natural numbers less than 10 but greater than 2

5. all numbers with an absolute value greater than 3 but less than 7

Let $U = \{w, o, m, e, n\}, M = \{n, o, w\}$, and $T = \{w, o, n\}$. Find each set. (Lesson 1-3)

6. $M \cup T$ **7.** $(M \cup T)'$ **8.** $M \cap T$ **9.** $(M \cap T)'$

Simplify. (Lesson 1-4)

10. $-22 - (-39)$

11. $46.8 + (-21.5) + 18.6$

12. $-2\frac{1}{4} + 7\frac{3}{8} + \left(-6\frac{3}{4}\right)$

13. $-0.76 + 2.5 - (-0.82)$

Evaluate each expression when $a = 14$ and $b = -22$. (Lesson 1-4)

14. $a + b$ **15.** $a - b$ **16.** $-a - b$ **17.** $-a + (-b)$

18. The previous balance of Joel's account was \$317.73. He made the following transactions to his account: deposits, \$123.25, \$55.62; withdrawal, \$228.98. What is his new balance?

1-5 Multiplication and Division

Goals ■ Multiply and divide rational numbers.

Applications Science, Food service, Health, Fitness

1. Use the diagram to represent the product or quotient of two numbers. Begin at zero. Let the horizontal values represent the first number and the vertical values represent the second number.

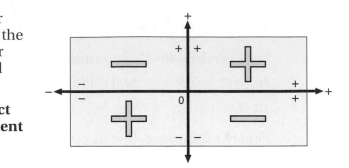

Square	First Number	Second Number	Product or Quotient
1	+	+	☐
2	−	+	☐
3	−	−	☐
4	+	−	☐

2. What pattern did you use to complete the sentences?

   a. If two positive numbers are multiplied, what is the sign of the product?

   b. If two negative numbers are multiplied, what is the sign of the product?

   c. If a positive number and a negative number are multiplied, what is the sign of the product?

◤ BUILD UNDERSTANDING

When you multiply and divide rational numbers, you follow the same rules you use when you multiply and divide integers.

The product or quotient of two numbers that have the *same* signs is positive.

The product or quotient of two numbers that have *different* signs is negative.

All the properties of multiplication of integers also apply to rational numbers. Multiplication and division are inverse or opposite operations.

The associative and commutative properties of multiplication may be used to arrange factors in the order easiest for multiplying.

In the table, a, b, and c represent rational numbers.

Two numbers whose product is 1 are **reciprocals** of each other. Another name for reciprocal is **multiplicative inverse**. Zero does *not* have a reciprocal.

Multiplication Properties

Property	In Symbols	Examples
Closure	ab is a rational number	$(-5) \cdot (4) = -20$ is a rational number
Commutative	$ab = ba$	$25 \cdot 5 = 5 \cdot 25$
Associative	$(ab)c = a(bc)$	$\left(\frac{3}{8} \cdot \frac{2}{5}\right) \cdot \frac{10}{9}$ $= \frac{3}{8} \cdot \left(\frac{2}{5} \cdot \frac{10}{9}\right)$
Identity	$a \cdot 1 = 1 \cdot a = a$	$26 \cdot 1 = 1 \cdot 26 = 26$
Property of Zero	$a \cdot 0 = 0 \cdot a = a$	$83.2 \cdot 0 = 0 \cdot 83.2 = 0$
Inverse	$a \cdot \frac{1}{a} = \frac{1}{a} \cdot a = 1,$ $a \neq 0$	$5 \cdot \frac{1}{5} = \frac{1}{5} \cdot 5 = 1$

Example 1

Find each product or quotient.

a. $\left(\dfrac{1}{7}\right)\left(-\dfrac{5}{6}\right)\left(-\dfrac{7}{25}\right)$

b. $(-5)(-0.9)(0.4)(-20)$

c. $-8.4 \div 7$

d. $-\dfrac{32}{9} \div \left(-\dfrac{8}{9}\right)$

Solution

a. $\left(\dfrac{1}{7}\right)\left(-\dfrac{5}{6}\right)\left(-\dfrac{7}{25}\right) = \left[\left(\dfrac{1}{7}\right)\left(-\dfrac{7}{25}\right)\right]\left(-\dfrac{5}{6}\right)$ Multiply $\dfrac{1}{7}$ and $-\dfrac{7}{25}$.

$$= \left(-\dfrac{1}{25}\right)\left(-\dfrac{5}{6}\right)$$

$$= \dfrac{1}{30}$$

b. $(-5)(-0.9)(0.4)(-20) = [(-0.9)(0.4)][(-5)(-20)]$
$$= (-0.36)(100)$$
$$= -36$$

c. $-8.4 \div 7 = -1.2$ Since the signs are different, the quotient is negative.

d. $-\dfrac{32}{9} \div \left(-\dfrac{8}{9}\right) = -\dfrac{32}{9} \times -\dfrac{9}{8}$
$$= 4$$

Since the signs are the same, the quotient is positive.

Remember to use the order of operations when simplifying expressions.

Example 2

Evaluate each expression when $q = 4$, $r = -0.5$, and $s = -8$.

a. rs

b. $-qs + r$

c. $r\left(\dfrac{s}{q}\right)$

Solution

a. rs

$(-0.5)(-8)$

$= 4$

b. $-qs + r$

$-(4)(-8) + (-0.5)$

$= 32 + (-0.5)$

$= 31.5$

c. $r\left(\dfrac{s}{q}\right)$

$= -0.5\left(\dfrac{-8}{4}\right)$

$= -0.5(-2)$

$= 1$

Example 3

Simplify.

a. $-0.8 + 3.9 \div 3$

b. $(-4.5 + 7) \div 5$

c. $5 + 25 \div 5 \div 5 \cdot 2 - 16 \div 8$

Solution

a. $-0.8 + 3.9 \div 3$
$= -0.8 + 1.3$
$= 0.5$

b. $(-4.5 + 7) \div 5$
$= 2.5 \div 5$
$= 0.5$

c. $5 + 25 \div 5 \div 5 \cdot 2 - 16 \div 8$
$= 5 + 5 \div 5 \cdot 2 - 16 \div 8$
$= 5 + 1 \cdot 2 - 2$
$= 5 + 2 - 2$
$= 5$

You will need to use the properties of multiplication along with the rules for multiplying and dividing integers to evaluate expressions.

◥ TRY THESE EXERCISES

Find each product or each quotient.

1. $(-1.7)(2.9)$

2. $-4\frac{1}{4} \div \frac{3}{4}$

3. $3\frac{1}{8}\left(-\frac{8}{25}\right)$

4. $-98 \div (-14)$

5. $(3.2)(-5)(2)$

6. $225 \div (-15)$

Simplify.

7. $6 \div (-3) + 7.1$ **8.** $-4(-9) - (-30.5)$ **9.** $(7.6 - 1.6)(-0.2)$ **10.** $2[-7.7 \div (-99.5 + 98.5)]$

Evaluate each expression when $a = 8$, $b = 0.5$, and $c = \frac{1}{4}$.

11. ac

12. $(a + b)c$

13. $a + bc$

14. $ab \div c$

15. TEMPERATURE From 1 A.M. to 5 A.M., the temperature dropped 16°F. What was the average change in temperature per hour?

 16. WRITING MATH When is the product of three rational numbers negative?

◥ PRACTICE EXERCISES • For Extra Practice, see page 664.

Perform the indicated operations.

17. $8.4(-2.6)$

18. $2.7(5)(-2)$

19. $-48 \div (-1.2)$

20. $2\frac{1}{4} \div \left(-\frac{3}{8}\right)$

21. $(-125) \div (0.25)$

22. $2\frac{1}{5} + (4)(-6)$

23. $5.4 \div 1.8 - 2.6$

24. $-25 \div (-10) + \left(-3\frac{1}{2}\right)$

25. $8\frac{1}{3} \div (-5) + 6$

Evaluate each expression when $m = -5$, $n = -1.6$, and $p = \frac{5}{8}$

26. mn

27. $m + np$

28. $m(n + p)$

29. $(m + n) \div p$

30. SCIENCE At the beginning of an experiment, the temperature of a solution was 68°C. During a 12-hour period, the temperature dropped 1.5 degrees each hour. Write an expression to show the temperature change during the entire time period.

31. FOOD SERVICE Sharon is planning a party for 25 people. If she caters the food, the cost will be $8.75 per person (tax and tip are included). Sharon has $200 to spend. Does she have enough money for 25 guests? If not, how many guests can she afford to invite?

CALCULATOR Use a calculator to perform the indicated operation. Express your answer as a decimal rounded to the nearest tenth.

32. $-1\frac{3}{16} \cdot 5.16$

33. $\frac{4}{9} \cdot \frac{3}{16}$

34. $\frac{3}{8}\left(-\frac{5}{8}\right)$

35. $-10\frac{7}{8} \div (-1.3)$

36. $4\frac{9}{32} - (-3.8502)$

37. $\left(5\frac{8}{9}\right)^2$

38. COMMUNICATION Martin is writing a magazine article. He will be paid $0.12 per word or $100, whichever is greater. How many words long will the article need to be in order for Martin to earn more than $100?

DATA FILE For Exercises 39–40, use the data on calories used by people of different weights on page 651.

39. How many calories does a 154-lb person use playing racquetball for 15 min?

40. Denise consumed 470 calories at lunch. If she weighs 110 lb, how long must she walk at 3 mi/h to use up those calories?

EXTENDED PRACTICE EXERCISES

Replace each ■ with +, −, ×, or ÷ to make a true sentence.

41. $(-9 \blacksquare 5) \blacksquare (-4) = 1$ **42.** $-6 \blacksquare 12 \blacksquare 8 = -4$

43. $21 \blacksquare (-7) \blacksquare 3 + 5 \blacksquare -5 = -1$

44. $(64 \blacksquare (-8)) \blacksquare 9 \blacksquare (-8) = 0$

45. Write 0.016 as a quotient of two integers.

MIXED REVIEW EXERCISES

Evaluate each expression when $a = 12$ and $b = -9$. (Lesson 1-4)

46. $a + b$

47. $b - a$

48. $2a - 3b$

49. $3a \div b$

50. $(b + 6) \div a$

51. $4a + 8b$

52. $2b - 6a$

53. $4b \div 2a$

54. $2(a + 4) - 3b$

55. $3(b - 1) + a$

56. $2(b + 1) - 4a$

57. $9a + 12b$

Let $U = \{0, 1, 2, 3, ..., 9\}$. Let $A = \{x \mid x \geq 2 \text{ and } x \leq 5\}$, $B = \{0, 2, 4, 6, 8\}$, $C = \{1, 3, 5, 7, 9\}$, and $D = \{0, 3, 5, 8\}$. Find each of the following. (Lesson 1-3)

58. D'

59. $C' \cup A$

60. $A \cap B$

61. $D \cup A$

62. $(A \cap B) \cup D$

63. $(B' \cap A) \cup C$

64. $(C \cap D)' \cup A$

65. $(B \cap D) \cup (A' \cap C)$

66. DATA FILE Use the data on sleep patterns on page 651. Approximately, what percent of the day does a 1-year-old sleep? (Prerequisite Skill)

67. DATA FILE Use the data on page 652 on Olympic record times for the 400-m freestyle swimming event. Compare the record men's time from 1924 to the record men's time for 2000. What is the percent decrease? (Prerequiste Skill)

1-6 Problem Solving Skills: Use Technology

Calculators and computers are useful tools to solve real-world problems. Graphing calculators can be used for a wide range of operations. They can also be used to store data, formulas and equations.

A computer **spreadsheet** is an application that stores data in **cells** formed by vertical columns and horizontal rows. The cells are identified by a row and a column. The columns are named by letters and the rows by numbers. By entering a **formula** in a cell, a spreadsheet will do calculations automatically.

Both calculators and computers follow the order of operations. Always think through the order of operations carefully when writing formulas.

Problem Solving Strategies

Guess and check

Look for a pattern

Solve a simpler problem

Make a table, chart or list

Use a picture, diagram or model

Act it out

Work backwards

Eliminate possibilities

✓ Use an equation or formula

PROBLEM

A printer is printing the bylaws of a civic organization. The organization wants 3 sizes, each of which has a width that is half the length. If the lengths they have chosen are 1 ft, 2 ft, and 3 ft, find each width, perimeter, and area.

Solve the Problem

To find the values of the perimeter and area for a rectangle with length equal to whole numbers 1, 2, and 3 and widths equal to half the length, a spreadsheet can be used. The rule or formula for generating each value is shown in each cell. Note that Row 1 and Column A are used for headings. The other cells are used to store data and formulas.

	A	B	C	D	E
1		Length	Width	Perimeter	Area
2	R1	1	$B2 * 0.5$	$2 * B2 + 2 * C2$	$B2 * C2$
3	R2	$B2 + 1$	$B3 * 0.5$	$2 * B3 + 2 * C3$	$B3 * C3$
4	R3	$B3 + 1$	$B4 * 0.5$	$2 * B4 + 2 * C4$	$B4 * C4$
5	R4	$B4 + 1$	$B5 * 0.5$	$2 * B5 + 2 * C5$	$B5 * C5$
6	R5	$B5 + 1$	$B6 * 0.5$	$2 * B6 + 2 * C6$	$B6 * C6$

After entering the formulas and data in the spreadsheet, your results should look like this.

	A	B	C	D	E
1		Length	Width	Perimeter	Area
2	R1	1	0.5	3	0.5
3	R2	2	1	6	2
4	R3	3	1.5	9	4.5
5	R4	4	2	12	8
6	R5	5	2.5	15	12.5

TRY THESE EXERCISES

Find the value in each cell named.

1. A2, B2, C2, D2

2. A3, B3, C3, D3

3. A4, B4, C4, D4

	A	B	C	D
1	1	2	5	10
2	2 * A1	2 * A2	B2 + 3	2 * C2
3	2 * A2	2 * A3	B3 + 3	2 * C3
4	2 * A3	2 * A4	B4 + 3	2 * C4

4. Write a formula to find the difference between twice a number (x) and the sum of 5 and that number. What is the result if 8 is the number? If 15 is the number?

5. **WRITING MATH** The civic organization decides to print their bylaws in a larger size with a 4 ft length. Explain how you would use a calculator to find the perimeter and area of the bylaws.

PRACTICE EXERCISES

SEWING You are making school pennants in the shape of isosceles triangles. The sizes will vary so that the height of each triangle is three times the length of the base. You also know that an additional 10% of the material will be used for hems or waste.

6. Write a formula for a calculator that can be used to find the material needed for any base (B).

Make a spreadsheet to find the material needed (including 10% extra) for each of the following bases.

7. 9 in.
8. 1 ft
9. 1 yd
10. 5 ft

DATA FILE For Exercises 11–13, use the data on the highest and lowest continental altitudes on page 647.

11. Write the feet below sea level of the lowest point in Asia as a negative integer.

12. What is the difference in elevation between the highest point in Asia and the lowest point in Asia?

13. What is the difference in elevation between the highest point in Australia and the lowest point in Australia?

14. **CHAPTER INVESTIGATION** Using the original message you wrote in Lesson 1-2, make a spreadsheet that will automatically calculate the percent for the letters E, T, A, O, N, R, I, S and H. You will need to know the total number of letters in the passage and the number of times each letter appears. Compare your percents to the table on page 5.

MIXED REVIEW EXERCISES

Perform the indicated operations. Round your answers to the nearest hundredth if necessary. (Lesson 1-5)

15. $4.3\,(-2.8)$
16. $9.60 \div 3.2$
17. -7.65×4.04
18. $-8.54 \div 2.91$

19. $-3.68(-6.14)$
20. $-8.5 \div 2.47$
21. $0.52 \times (-5.73)$
22. $38 \div 1.7$

Review and Practice Your Skills

For Exercises 1–9, perform the indicated operations.

1. $(-5.6)(3.9)$

2. $\left(4\frac{2}{3}\right)\left(3\frac{4}{5}\right)$

3. $\frac{4}{15} \div \frac{8}{3}$

4. $8\frac{2}{7} + (3)\left(-7\frac{1}{2}\right)$

5. $18 \div (-4) + 3\frac{1}{2}$

6. $\frac{1}{2} \div \frac{3}{4} \cdot \left(-\frac{5}{9}\right)$

7. $(-9)(0.4)(-0.8)(-3)$

8. $5[16 - (-5)]$

9. $\left(-\frac{3}{8}\right)^2$

Evaluate each expression when $a = -6\frac{1}{2}$, $b = -\frac{1}{3}$, and $c = 9$.

10. $c(a - b)$

11. $-bc + ac$

12. $a \div b$

13. $b^2 - c$

14. $bc + |(-bc)|$

15. $abc \div ac$

16. Steve earns \$8.50/h in his 40-h work week. For each hour over 40 h, he earns $1\frac{1}{2}$ times his hourly pay. How much does he earn by working 47 h in one week?

17. Write 0.625 as a quotient of two integers.

18. You are organizing a banquet. Expenses include: food, \$355; decorations, \$35; trophies, \$85; sound system rental, \$42.50. You expect 90 people to attend. What should you charge each person to cover expenses?

19. Ben earns \$8.82 for the first 40 h he works each week and 1.5 times his hourly pay for any hours over 40. Write a formula that can calculate the total weekly pay for Ben when he works more than 40 h.

Use your calculator to find the total weekly pay for Ben if he works the following number of hours.

20. 42

21. 36

22. 45

23. 31

Find the value in each cell.

		A	B	C
24.	1	0.5	A1 + 5	A1 − 2*B1
25.	2	2*A1	A2 + 5	A2 − 2*B2
26.	3	3*A1	A3 + 5	A3 − 2*B3
27.	4	4*A1	A4 + 5	A4 − 2*B4

28. The volume of a box is found using the formula $V = l \cdot w \cdot h$. The formula to find the surface area of a box is $SA = 2lw + 2wh + 2lh$. Make a spreadsheet which uses these formulas to calculate the volume and surface area of boxes with the following dimensions:

   a. $l = 1$, $w = 2$, $h = 3$

   b. $l = 2$, $w = 3$, $h = 4$

   c. $l = 3$, $w = 4$, $h = 5$

   d. $l = 4$, $w = 5$, $h = 6$

Write all subsets of the following sets. (Lesson 1-1)

29. $\left\{\dfrac{1}{2}, \dfrac{3}{5}\right\}$ 　　　　**30.** $\{c, a, n, e\}$ 　　　　**31.** $\{0\}$

32. Graph the set $\left\{-2, -\dfrac{1}{2}, |-2|, 2\dfrac{1}{2}, 2^2\right\}$ on a number line. (Lesson 1-2)

Let $U = \{x \mid x \text{ is an integer}\}$, $A = \{x \mid x \text{ is a multiple of 4}\}$, $B = \{x \mid x \text{ is a natural number}\}$, $C = \{0\}$. **For Exercises 33–38, find each set.** (Lesson 1-3)

33. $A \cap B$ 　　**34.** B' 　　**35.** $A' \cup B$ 　　**36.** $A \cap B'$ 　　**37.** $A \cap C$ 　　**38.** $B' \cup C$

39. Jean goes to the grocery store to purchase five items that cost $3.95, $5.12, $0.78, $7.05, and $14.88. She has $35 in her pocket. Use estimation to determine whether or not she has enough money to buy all five items. (Lesson 1-4)

Simplify. (Lessons 1-4, 1-5)

40. $-35.7 - (-26.3)$ 　　**41.** $(-35.7)(-26.3)$ 　　**42.** $(2)(-3.5) + \left(-8\dfrac{1}{2}\right)(6) - (-72 \div 12)$

43. $\dfrac{10}{3} - \dfrac{1}{3}[12 + (-3)]$ 　　**44.** $\dfrac{3}{4} + 2\left(5 - 1\dfrac{1}{2}\right)$ 　　**45.** $\dfrac{2}{5} + 5[9 - (-2)]$

Math*Works* Career – Cashier
Workplace Knowhow

Cashiers are employed by stores to add up your purchases, take your money or credit card, and give you change when you pay with cash. Communication and math skills are two large components of a cashier's job no matter the kind of business.

In addition to working with money, cashiers work with weights, length, area, and volume. Estimation and mental math skills enable cashiers to spot errors and check the reasonableness of cash register calculations.

1. Suppose you are a cashier at a pet store. A young boy has $20. He wants to buy as many cans of cat food as he can. Brand A is $0.49 per can; however, for every four cans you buy, the fifth one is free. Brand B is $0.39 per can. Which brand should the boy purchase?

2. In a clothing store, a customer asks you whether she has enough money for two blouses. She has $30. The first blouse is $17.00 with a 40% discount. The second is $20.00 with a 25% discount. Find the total cost of the blouses.

3. You are a cashier in a state in which the sales tax is 8.25%. The subtotal of a customer's purchases is $112.78. The register then adds a sales tax of $0.93 for a total of $113.71. Does the result seem reasonable? Explain your thinking.

Distributive Property and Properties of Exponents

Goals
- Use the distributive property to evaluate and simplify expressions.
- Use properties of exponents to evaluate and simplify expressions.

Applications Science, Number Sense, Sales

Suppose you are typing a term paper, and your typing rate is 4 pages per hour. On Monday night you type for 3.5 hours, and on Tuesday night you complete the paper by typing 2.5 hours. How long is the term paper? Show two ways to determine the total number of pages.   24 pages; 4(3.5) + 4(2.5); 4(3.5 + 2.5)

◥ BUILD UNDERSTANDING

As you can see from this question, you may have more than one way to solve a problem. Examine the expressions:

$$4(3.5 + 2.5) \qquad 4(3.5 + 2.5)$$
$$= 4 \cdot 3.5 + 4 \cdot 2.5 \qquad = 4(6)$$
$$= 14 + 10 \qquad = 24$$
$$= 24$$

Both ways have the same answer. This is an example of the distributive property.

The **distributive property** relates addition and multiplication. It states that a factor outside parentheses can be used to multiply each term within the parentheses.

Distributive Property	For any real numbers, a, b, and c, $\quad a(b + c) = ab + ac$

Other mathematical properties, including the distributive, are summarized in the box. Each applies to all real numbers a, b, and c.

These mathematical properties are useful in simplifying mathematical expressions. Some expressions will also involve exponents. There are special properties that apply to exponents. Exponents are used as a short way to indicate repeated multiplication of a number, or factor.

A number written in **exponential form** has a base and an exponent.

The **base** tells what factor is being multiplied. The **exponent** tells how many equal factors there are. The expression a^4 is read as "a to the fourth power."

Other Mathematical Properties

Property	In Symbols	Examples
Distributive	$a(b + c) = ab + ac$ or $a(b - c) = ab - ac$	$3(5 + 4) = 3 \cdot 5 + 3 \cdot 4$ $3(5 - 4) = 3 \cdot 5 - 3 \cdot 4$
Transitive	If $a = b$ and $b = c$, then $a = c$.	If $3 + 5 = 4 + 4$ and $4 + 4 = 2 + 6$, then $3 + 5 = 2 + 6$.
Reflexive	$a = a$ Any number is equal to itself.	$8 = 8$
Substitution	If $a = b$, then we can substitute b for a or substitiute a for b in any statement.	If $x = 4$, then $x + 2 = 4 + 2$.
Symmetric	If $a = b$, then $b = a$.	If $4 + 6 = 5 + 5$, then $5 + 5 = 4 + 6$.

When the base is a negative number, enclose it in parentheses. The expressions $(-3)^2$ and -3^2 represent different numbers.

$$(-3)^2 = (-3)(-3) = 9 \quad \text{and} \quad -3^2 = -(3)(3) = -9$$

The exponents 0 and 1 are special. Any number raised to the first power is the number itself.

$$n^1 = n, \quad \text{so } 13^1 = 13$$

Any number, except 0, raised to the zero power is equal to 1. For every nonzero real number a, $a^0 = 1$.

$$2^0 = 1, \quad 4^0 = 1$$

Mental Math Tip

Use the distributive property to multiply numbers mentally.

$15 \cdot 8 + 15 \cdot 12$

Think: $15 \cdot (8 + 12)$

$= 15(20)$

$= 300$

Example 1

Evaluate each expression. Let $a = 3$ and $b = -2$.

a. a^2 **b.** b^3 **c.** a^2b

Solution

a. a^2
$= 3^2$
$= 9$

b. b^3
$= (-2)^3$
$= (-2)(-2)(-2)$
$= -8$

c. a^2b
$= (3)^2(-2)$
$= (9)(-2)$
$= -18$

Several rules for exponents make simplifying expressions easier.

Properties of Exponents for Multiplication	For all real numbers a and b, if m and n are integers, then $a^m \cdot a^n = a^{m+n} \quad (a^m)^n = a^{mn} \quad (ab)^m = a^m b^m$

Example 2

Simplify.

a. $x^2 \cdot x^7$ **b.** $(a^3)^5$ **c.** $(5^2 \cdot n)^2$

Solution

a. $x^2 \cdot x^7 = x^{2+7}$
$= x^9$

b. $(a^3)^5 = a^{3 \cdot 5}$
$= a^{15}$

c. $(5^2 \cdot n)^2 = (5^{2 \cdot 2})(n)^2$
$= 5^4 n^2$

As with multiplication, some rules make simplifying division expressions easier.

Properties of Exponents for Division	For all real numbers a and b, if m and n are integers, then $\dfrac{a^m}{a^n} = a^{m-n}, \text{ if } a \neq 0 \quad \left(\dfrac{a}{b}\right)^m = \dfrac{a^m}{b^m}, \text{ if } b \neq 0$

To divide numbers with the same base, subtract the exponents.

$$a^m \div a^n = a^{m-n}$$

$$2^5 \div 2^3 = \frac{2^5}{2^3} = \frac{2 \cdot 2 \cdot 2 \cdot 2 \cdot 2}{2 \cdot 2 \cdot 2} = 2^{5-3} = 2^2$$

To find the power of a quotient, find the power of each number and divide.

$$\left(\frac{a}{b}\right)^m = \frac{a^m}{b^m} \qquad \left(\frac{3}{4}\right)^4 = \frac{3}{4} \cdot \frac{3}{4} \cdot \frac{3}{4} \cdot \frac{3}{4} = \frac{3^4}{4^4}$$

Example 3

Simplify.

a. $\dfrac{a^7}{a^3}$ **b.** $\left(\dfrac{t}{3}\right)^4$ **c.** $\left(\dfrac{x^2}{5}\right)^4$ **d.** $\left(\dfrac{c^5}{c^2}\right)^3$

Solution

a. $\dfrac{a^7}{a^3} = a^{7-3}$ **b.** $\left(\dfrac{t}{3}\right)^4 = \dfrac{t^4}{3^4}$ **c.** $\left(\dfrac{x^2}{5}\right)^4 = \dfrac{(x^2)^4}{5^4}$ **d.** $\left(\dfrac{c^5}{c^2}\right)^3 = \dfrac{(c^5)^3}{(c^2)^3}$

$= a^4$ $= \dfrac{t^4}{81}$ $= \dfrac{x^8}{625}$ $= \dfrac{c^{15}}{c^6} = c^{15-6}$ or c^9

◥ TRY THESE EXERCISES

Use the distributive property to find each product.

1. $16 \cdot 15 - 16 \cdot 5$

2. $18 \cdot 92$

3. $28\left(10\dfrac{1}{7}\right)$

Evaluate each expression when $x = -4$ and $y = 3$.

4. y^2 **5.** x^3 **6.** xy^2 **7.** x^3y^2

Simplify.

8. $(-2)^3(-2)^2$ **9.** $\dfrac{c^9}{c^5}, c \neq 0$ **10.** $(4^2 \cdot 3)^5$ **11.** $c^{10} \cdot c^5$

12. $(d^6)^3$ **13.** $(a^2b^3)^4$ **14.** 3^0 **15.** $x^9 \cdot x^2$

◥ PRACTICE EXERCISES • For Extra Practice, see page 664.

Use the distributive property to find each product.

16. $8.5 \cdot 92 + 8.5 \cdot 8$

17. $82 \cdot 53$

18. $45\left(2\dfrac{4}{9}\right)$

Evaluate each expression when $a = -3$ and $b = 4$.

19. a^2 **20.** $a^2 - b^2$ **21.** b^3 **22.** ab^2

23. $-2ab^2$ **24.** $(-2ab)^2$ **25.** $(2 - b)^3$ **26.** $(a^2 - 7)^3$

Simplify.

27. $2^{10} \cdot 2^{15}$ **28.** $\dfrac{a^9}{a^6}, a \neq 0$ **29.** $r^5 \cdot r^4$ **30.** $(a^9)^6$

31. $\left(\dfrac{1}{d}\right)^9$ **32.** $m^{15} \cdot m^3$ **33.** $(a^5)^5$ **34.** $(n^4)(n^5)(n^6)$

Evaluate mentally each sum or product when $a = 3.5$, $b = 2$, and $c = 0$. Use the properties of mathematics as needed.

35. $5ab$

36. $165a^2bc$

37. $(6.5 + a)(1.4b) + c$

Simplify. You may need to use more than one of the properties of exponents.

38. $m^4(m^5m^6)^2$

39. $\dfrac{(c^6 \cdot c^4)^2}{c^2}$

40. $\left(\dfrac{18g^6}{6g^3}\right)^2$

41. NUMBER SENSE Without multiplying, tell which of these numbers is the greatest. 100^5 1000^4 $10,000^3$ $100,000^2$ Explain your reasoning. (*Hint:* $100 = 10^2$; $1000 = 10^3$; $10,000 = 10^4$; $100,000 = 10^5$.)

42. SALES John and Brian bought 125 pieces of candy for $43.75. They sold each piece for $0.50. How much profit did they make on their total sale of candy?

43. The volume of a cube is the product of its length, width, and height. What is the volume of a cube with side length 3 in.? 5 cm? e cm? g in.?

44. WRITING MATH Does $3^2 = 2^3$? Explain your answer.

◼ EXTENDED PRACTICE EXERCISES

CRITICAL THINKING Write *true* or *false*. Give examples to support your answer.

45. If a and b are negative integers and $a < b$, then $a^2 > b^2$.

46. If a and b are negative integers and $a < b$, then $a^3 > b^3$.

47. SCIENCE A certain type of bacteria doubles in number every hour. At 10 A.M., there were 250 bacteria. How many bacteria will be present at 1 P.M.?

48. WRITING MATH Decide whether the distributive property works with the intersection and union of sets. For example, does $A \cap (B \cup C) = (A \cap B) \cup (A \cap C)$? Give examples to support your answer.

◼ MIXED REVIEW EXERCISES

Solve. (Lesson 1-6)

49. Adult passes to the amusement park cost $27.95 each, while a child's pass costs $19.95. What would the total cost be for a family of four adults and six children to buy passes for the park?

50. Kordell drove his car 396 miles on 15 gal of gas. Gas costs $1.32 per gal. How much did Kordell pay for gas per mile driven?

51. By buying a washing machine on sale, Natasha saved 30% off the regular price. The washer regularly sold for $418. How much did Natasha pay for the washer?

52. The paint crew has 2000 ft^2 of wall to paint in the office building. They painted 384 ft^2 on Monday, and 377 ft^2 on Tuesday. How much will they have to paint each day to finish the job on Friday?

1-8 Exponents and Scientific Notation

Goals
- Evaluate variable expressions with negative exponents.
- Write numbers in scientific notation.

Applications Chemistry, Astronomy, Business

Explore the meaning of exponents using a number line.

1. Draw a number line from 0 to 16.

2. Graph 2^4, 2^3, 2^2, 2^1 and 2^0 on the number line.

3. Describe the pattern shown by the graphed points.

◣ BUILD UNDERSTANDING

Exponents can be negative numbers. The quotient property of exponents can help you understand the meaning of **negative exponents**.

$$\frac{2^3}{2^4} = \frac{2 \cdot 2 \cdot 2}{2 \cdot 2 \cdot 2 \cdot 2} = \frac{1}{2}$$

If you use the quotient property, you obtain $\frac{2^3}{2^4} = 2^{3-4} = 2^{-1}$.

So, $2^{-1} = \frac{1}{2}$.

For every nonzero real number a, if n is an integer, than $a^{-n} = \frac{1}{a^n}$.

So, for any nonzero real number a, a^{-n} is the reciprocal, or multiplicative inverse, of a^n.

Check Understanding

Write each expression as a fraction.

1. 4^{-3} **2.** $(-3)^{-3}$

3. $(-3)^{-x}$ **4.** 2^{-5}

Example 1

Simplify each expression, using the properties of exponents.

a. $a^9 \div a^{-5}$ **b.** $x^4 \cdot x^{-3}$ **c.** $(c^2)^{-5}$

Solution

a. $a^9 \div a^{-5} = a^{9-(-5)}$
$= a^{14}$

b. $x^4 \cdot x^{-3} = x^{4+(-3)}$
$= x$

c. $(c^2)^{-5} = c^{2(-5)}$
$= c^{-10}$

You can evaluate variable expressions with negative exponents.

Example 2

Evaluate a^{-5} when $a = 2$.

Solution

$$a^{-5} = 2^{-5} = \frac{1}{2^5} = \frac{1}{32}$$

Exponential expressions having a base of 10 serve a very useful purpose in science.

Scientists often deal with very large or very small numbers. To save time and space and be able to write and compute with such numbers more easily, the system of **scientific notation** was developed. Scientific notation uses powers of 10 to write large and small numbers in a more concise manner.

A number written in scientific notation has two factors. The first factor is greater than or equal to 1 and less than 10. The second is a power of 10.

Standard form
$$496,000,000 = 4.96 \cdot 10^8$$
$$0.00059 = 5.9 \cdot 10^{-4}$$
Scientific notation

Mental Math Tip

To multiply 10^n when n is a positive integer, move the decimal point n places to the right.

$$8.302 \times 10^5 = 830,200$$

To multiply by 10^n when n is a negative integer, move the decimal point n places to the left.

$$6.03 \times 10^{-3} = 0.00603$$

Example 3

CHEMISTRY The mass of an oxygen atom is $2.66 \cdot 10^{-23}$ gram (g). What is the approximate mass of 1 billion oxygen atoms?

Solution

The mass of an oxygen atom = $2.66 \cdot 10^{-23}$ g.
Recall that 1 billion = $1 \cdot 10^9$.
To find the mass of 1 billion oxygen atoms, multiply 1 billion by the mass of 1 oxygen atom.

1 billion · the mass of 1 oxygen atom
$$= (10^9)(2.66 \cdot 10^{-23}) \text{ g}$$
$$= (2.66)(10^{9-23}) \text{ g}$$
$$= 2.66 \cdot 10^{-14} \text{ g}$$

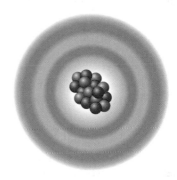

 If your calculator has an EE, EXP, or 10^x key you can use it as a quick way to enter a number in scientific notation. For example, to enter 2.66×10^{-23}, use the key sequence:

2.66 [2nd] [EE] [(−)] 23

Now try the multiplication from Example 3:

1 [2nd] [EE] 9 [X] 2.66 [2nd] [EE] [(−)] 23 [ENTER]

This equals 2.66E −14, which is the same as $2.66 \cdot 10^{-14}$.

The ability to measure quantities such as length, area, volume, mass, and elapsed time is of vital importance. The measurement system used most often in scientific work is the metric system (also called the SI system).

The metric system is based on the decimal system. It uses prefixes to indicate multiples of 10. You can use the prefixes shown in the table to change from one unit to another.

Prefix	Multiple	Scientific notation
mega	1,000,000	10^6
kilo	1,000	10^3
hecto	100	10^2
deka	10	10^1
base	1	10^0
deci	0.1	10^{-1}
centi	0.01	10^{-2}
milli	0.001	10^{-3}
micro	0.0000001	10^{-6}
nano	0.000000001	10^{-9}
pico		10^{-12}
femto		10^{-15}

Ratios of equal quantities can be used to change from one unit to another.

Example 4

MEASUREMENT Find the number of milligrams in 1 kilogram.

Solution

$$1 \text{ kg} \cdot \frac{10^3 \text{ g}}{1 \text{ kg}} \cdot \frac{1 \text{ mg}}{10^{-3} \text{ g}} = 10^6 \text{ mg}$$

◣ TRY THESE EXERCISES

Simplify.

1. $(-2)^{-4}$ **2.** $(-1)^{-5}$ **3.** $r^8 \div r^{12}$ **4.** $x^{-9} \cdot x^9, x \neq 0$

5. $(m^7)^{-3}$ **6.** $w^5 \cdot w^{-13}$ **7.** $d^6 \div d^{-11}$ **8.** $(x^2y)^{-5}$

Evaluate each expression when $x = -3$ and $y = 4$.

9. x^{-2} **10.** y^{-4} **11.** $x^0y^{-2}y^0$ **12.** $x^3x^{-6}y$

Write each number in scientific notation.

13. 59,300,000,000   **14.** 0.000059   **15.** 0.00006052

Write each number in standard form.

16. $3.6 \cdot 10^5$ **17.** $4.3 \cdot 10^{-4}$ **18.** $2.09 \cdot 10^{-7}$

19. ASTRONOMY The distance from the planet Mercury to the Sun is about 36,000,000 miles. Write the distance in scientific notation.

◣ PRACTICE EXERCISES • For Extra Practice, see page 665.

Simplify.

20. $(-1)^{-3} + (-1)^{-4}$ **21.** $d^{-15} \div d^{-3}$ **22.** $y^{-5} \cdot y^{-4}$

Evaluate each expression when $p = -2$ and $q = 2$.

23. p^{-6} **24.** q^{-3} **25.** $(pq)^{-3}$ **26.** $q^3q^{-2}p^{-3}$

Write each number in scientific notation.

27. 9,300   **28.** 0.00356   **29.** 0.00000215

CALCULATOR Your calculator displays these answers. Write the answers in scientific notation and standard form.

30. 2.7 E6   **31.** 3.9 E−5

32. LANGUAGE It is estimated that there are 750,000 words in the English language. Write the number in scientific notation.

Solve. Write your answer in scientific notation.

33. The speed of light is $3.00 \cdot 10^{10}$ meters per second. How far does light travel in 1 day?

34. WRITING MATH Which is greater $4.12 \cdot 10^3$ or $1.5 \cdot 10^4$? Write a rule for comparing numbers written in scientific notation.

Simplify. You may need to use more than one of the properties of exponents.

35. $\left(\dfrac{y^4}{y^3}\right)^{-2}$, $y \neq 0$ **36.** $\left(\dfrac{m^{-9}}{m}\right)^{-4}$, $m \neq 0$ **37.** $\dfrac{r^{-9}(r^2)^3}{r^{-6}r^{-5}}$, $r \neq 0$

Solve. Write your answer in scientific notation.

38. Simplify $0.000136 \times 18.05 \div 0.001$.

39. BUSINESS Greater Automotive employs $3.2 \cdot 10^2$ people. Each hour, the company pays $\$2.88 \cdot 10^3$ in wages. What is the average wage per hour for each employee?

40. TIME Find the number of nanoseconds in 5 milliseconds.

41. MEASUREMENT Find the number of centimeters (cm) in 1 foot (ft). Use the equivalent: 1 meter equals 3.2808 ft.

■ EXTENDED PRACTICE EXERCISES

42. CRITICAL THINKING You know that $3^2 = 9$ and $3^{-2} = \frac{1}{9}$. What do you think is the value of $\frac{1}{3^{-2}}$? If a is a nonzero real number and n is an integer, what does $\frac{1}{a^{-n}}$ represent?

Rewrite each with positive exponents.

43. $\dfrac{5^{-2}}{r^{-5}}$

44. $\dfrac{a^{-3}b}{b^{-2}a^{-1}}$

45. $\left(\dfrac{m^5 n^{-3} n}{n^{-2} m^{-2}}\right)^{-1}$

46. $\left[\dfrac{a^{-6}(b^2)^{-4}}{b^3 a^0}\right]^{-2}$, $a, b \neq 0$

■ MIXED REVIEW EXERCISES

Evaluate each expression when $a = 5$ and $b = -3$. (Lesson 1-7)

47. a^2 **48.** $a^2 - b^2$ **49.** $b^2 - a^2$ **50.** ab^2

51. $(ab)^2$ **52.** $3a^2b$ **53.** $-3ab^2$ **54.** $(b^2 - 4)^3$

55. $a^2 + b^2$ **56.** $-a^3$ **57.** $-(b^3)$ **58.** $-(a^2 - b^2)$

Simplify. (Lesson 1-7)

59. $m^3 \cdot m^7$ **60.** $(m^4)^3$ **61.** $\dfrac{m^8}{m^5}$, $m \neq 0$ **62.** $\left(\dfrac{1}{m}\right)^7$, $m \neq 0$

63. $m^2(m^4 \cdot m^3)^5$ **64.** $\dfrac{m^{10}}{m^2}$, $m \neq 0$ **65.** $\left(\dfrac{m^3}{m^2}\right)^8$, $m \neq 0$ **66.** $m^5 \cdot m^8 \cdot m^{10}$

67. $m^4(m^3)^4$ **68.** $m^4(((m^2)^3)^4)^2$ **69.** $m^5 \cdot m^6 \cdot m^3$ **70.** $m^2(m^6)^5$

Chapter 1 Review

VOCABULARY ◣

Choose the word from the list that completes each statement.

1. A(n) __?__ number can be expressed as either a terminating or a repeating decimal.

2. __?__ uses powers of 10 to write numbers in a more concise manner.

3. The __?__ property relates addition and multiplication.

4. The __?__ of a number is the distance that number is from zero on a number line.

5. The intersection of two disjoint sets is the __?__.

6. The subset of all elements in the universal set that are *not* elements of a given set is the __?__ of the set.

7. All elements in a replacement set that make an open sentence true form the __?__.

8. An open sentence with $<$, $>$, $\leq$, or $\geq$ is called a(n)__?__.

9. In the expression 5^x, the variable is the __?__.

10. Two numbers whose product is one are called __?__.

a. absolute value
b. complement
c. distributive
d. empty set
e. exponent
f. inequality
g. multiplicative inverses
h. natural
i. opposites
j. rational
k. scientific notation
l. solution set

LESSON 1-1 ◣ The Language of Mathematics, p. 6

▶ A **set** is a well-defined collection of items.

▶ Any sentence that contains one or more variables is called an **open sentence**.

▶ The set of all possible values for the variable in an open sentence is called the **replacement set**.

Determine if each statement is *true* or *false*.

11. $\{4, 5, 6\} \subseteq \{4, 5\}$ 12. $\varnothing \subseteq \{1, 2, 3\}$ 13. $\{a, b, c\}$ is a subset of $\{a, b, c\}$.

Which of the given values is a solution of the equation?

14. $3x + 2 = 5; -1, 0, 1$ 15. $2a + 7 = 1; -3, 0, 3$ 16. $\frac{5b}{3} = 10; 4, 5, 6$

LESSON 1-2 ◣ Real Numbers, p. 10

▶ A **rational** number can be represented as a ratio of two integers, $\frac{a}{b}$, $b \neq 0$. An **irrational** number cannot.

▶ The set of rational numbers and the set of irrational numbers make up the set of **real numbers**.

17. Graph the set of all whole numbers less than 5 on a number line.

18. Graph the set of all real numbers greater than -4 on a number line.

Replace each ● with $<$, $>$, or $=$.

19. $-4 ● |-4|$ 20. $-(-|-7|) ● -7$ 21. $|-2| + |-5| ● |7|$

LESSON 1-3 ◣ Union and Intersection of Sets, p. 16

▶ The **union** of two sets A and B, $A \cup B$, contains the elements in A, in B, or in both. The **intersection** of two sets A and B, $A \cap B$, contains elements common to A and B.

Refer to the diagram. Find the set named by listing the members.

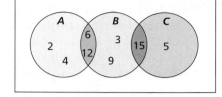

22. $A \cup B$

23. $A \cap B$

24. $A \cup C$

25. $A \cap C$

26. Let $A = \{1, 3, 5\}$ and $B = \{2, 3, 4\}$. Find $A \cup B$.

27. Given that $A = \{x \mid x \text{ is a real number and } x \geq 0\}$
$B = \{x \mid x \text{ is a real number and } x < 5\}$, graph $A \cup B$.

LESSON 1-4 ◣ Addition, Subtraction, and Estimation, p. 20

▶ To add and subtract rational numbers, use the same rules as for integers. For the following a, b, and c are rational numbers.

Closure	Commutative	Associative	Identity	Inverse
$a + b$ is rational.	$a + b = b + a$	$(a + b) + c =$ $a + (b + c)$	$a + 0 =$ $0 + a = a$	$a + (-a) = 0$

Add or subtract.

28. $-5.4 + (-9.8)$

29. $-2\frac{1}{2} + 3\frac{1}{8}$

30. $-39.6 - (-23.9)$

Evaluate each expression when $r = -\frac{3}{4}$ and $s = \frac{5}{6}$.

31. $r + s$

32. $r - s$

33. $s - r$

LESSON 1-5 ◣ Multiplication and Division, p. 26

▶ The product or quotient of two rational numbers with the same sign is positive.

▶ The product or quotient of two rational numbers with different signs is negative.

Closure	Commutative	Associative	Identity	Inverse	Property of Zero
ab is rational	$ab = ba$	$(ab)c = a(bc)$	$a \cdot 1 = 1 \cdot a = a$	$a \cdot \left(\frac{1}{a}\right) =$ $\left(\frac{1}{a}\right) \cdot a = 1$	$0 \cdot a = 0$

Perform the indicated operations.

34. $-4\frac{2}{3} \div 3\frac{1}{2}$

35. $-4\frac{1}{2}\left(-2\frac{2}{3}\right)$

36. $(-1.5)(3.2)(10)$

37. $20(-6.4)(-2.5 \div 0.5)$

38. $3.4 \div (-17) - 2.3$

39. $\frac{2}{5}\left(-\frac{1}{6}\right)\left(-\frac{5}{2}\right) + \frac{5}{6}$

Evaluate each expression when $x = -\frac{1}{2}$, $y = \frac{3}{5}$, and $z = \frac{5}{9}$.

40. xyz

41. $xy \div z$

42. $x + yz$

LESSON 1-6 ◾ Problem Solving Skills: Use Technology, p. 30

▶ Decide which computational method is best for solving a given problem—estimation, mental math, pencil and paper, or calculator. Then use your method to solve the problem.

43. The jazz band is selling water bottles and license plate holders to pay for a trip. The band earns a profit of $2.00 for each water bottle sold and $1.50 for each license plate holder sold. Each band member must earn $150. If Maria sells 50 license plate holders, and the remainder of the money comes from selling water bottles, how many water bottles does she need to sell?

44. The cost of a taxi is $3 plus $1.25 for each mile traveled. Write a formula to determine the cost of a taxi ride that is m miles long.

LESSON 1-7 ◾ Distributive Property and Properties of Exponents, p. 34

Distributive property: $a(b + c) = ab + ac$

Properties of one and zero: $a^1 = a$, $a^0 = 1$, $a \neq 0$

Multiplication: $a^m \cdot a^n = a^{m+n}$,
$(a^m)^n = a^{mn}$,
$(ab)^m = a^m b^m$

Division: $\dfrac{a^m}{a^n} = a^{m-n}$, $a \neq 0$,
$\left(\dfrac{a}{b}\right)^m = \dfrac{a^m}{b^m}$, $b \neq 0$

Simplify.

45. $x^3 \cdot x^8$

46. $\dfrac{z^6}{z^9}$, $z \neq 0$

47. $\left(\dfrac{a^2 b^3}{c}\right)^2$, $c \neq 0$

48. $(t^3)^5$

49. $\left(\dfrac{1}{2}\right)^4$

50. $\left(\dfrac{m^5 n^2}{m^3}\right)^4$

LESSON 1-8 ◾ Exponents and Scientific Notation, p. 38

▶ For every nonzero real number a, $a^{-n} = \dfrac{1}{a^n}$.

▶ A number written in scientific notation has two factors. The first factor is greater than or equal to 1 and less than 10. The second factor is a power of 10.

Simplify.

51. $x^5 \cdot x^{-6}$

52. $a^{-2} \div a^{-5}$

53. $(-2)^{-3}$

Write each number in scientific notation.

54. 457,000

55. 0.000392

56. 0.00000021

Write each number in standard form.

57. $4.6 \cdot 10^4$

58. $8.763 \cdot 10^2$

59. $2.78 \cdot 10^{-5}$

CHAPTER INVESTIGATION

EXTENSION Write a report about how you encoded your message and how you decoded your partner's message. Describe how your code differed from your partner's code.

Chapter 1 Assessment

1. Write the set of positive integers greater than 2. Use roster notation.

2. List all the subsets of $\{s, e, t\}$.

3. Which of the given values is a solution of the equation?

$$4x + 5 = -11; \quad -4, 0, 2, 4$$

4. Use mental math to solve $a - 9 = -2$.

5. *True* or *false*. 0 is a whole number.

6. Graph the set of all natural numbers greater than -1.

7. Graph the set of all real numbers less than or equal to 2.

8. Evaluate $-|r|$, when $r = -2.6$

Let $U = \{d, r, a, k, e\}$, $A = \{r, a, k, e\}$, $B = \{r, e, d\}$, and $C = \{e, d\}$. Find each union or intersection.

9. $A \cap B$

10. $B \cup C$

11. $A \cap C$

12. $A' \cap B'$

Let $A = \{x | x$ is a negative integer and $x > -4\}$ and $B = \{x | x$ is a real number and $x \leq 0\}$. Graph the following.

13. $A \cap B$

14. $A \cup B$

Add or subtract.

15. $-18.3 + (-17.8)$

16. $-85.6 - (-79.7)$

17. $-21.4 + (-35.8)$

18. $-71.02 - (-62.93)$

Multiply or divide.

19. $(-9.6) \div (-1.6)$

20. $\left(-8\frac{1}{3}\right)\left(\frac{12}{5}\right)(-5)(2)$

21. $(-24.8) \div (-12.5)$

22. $(4)(-3)\left(-\frac{3}{5}\right)\left(\frac{15}{21}\right)$

Simplify.

23. $z^{-9} \cdot z^{12}, z \neq 0$

24. $\dfrac{x^5}{x^6}, x \neq 0$

25. $\dfrac{(a^2 b^{-1})^2}{a^0 b^{-4}}, a \neq 0, b \neq 0$

26. $\left(\dfrac{x^4}{x^2}\right)^3$

27. $\dfrac{(a^3 b^4)}{a^5 b^{-4}}$

28. $\dfrac{(x^5 y^{-3} z^2)}{x^8 y^2 z^{-3}}$

29. Evaluate $-m - n$ when $m = -8\frac{1}{2}$ and $n = -5.7$.

30. Evaluate $m^{-2} n^5$ when $m = -3$ and $n = -2$.

31. Write 0.0000035 in scientific notation.

32. Write $2.79 \cdot 10^7$ in standard notation.

33. Sharie, Cal, and Nick each wrote a report for literature class. Sharie's report was twice as long as Nick's. Cal's report was 4 pages shorter than Nick's. If Cal's report was 8 pages, how long was Sharie's?

Standardized Test Practice

Record your answers on the answer sheet provided by your teacher or on a sheet of paper.

1. Which is *not* a subset of $\{D, O, G\}$? (Lesson 1-1)

(A) $\varnothing$

(B) $\{D, O, G\}$

(C) $\{O, G\}$

(D) $\{D, E\}$

2. Which is a solution of $3x - 5 = 16$? (Lesson 1-1)

(A) -7

(B) $3\frac{2}{3}$

(C) 7

(D) $10\frac{1}{3}$

3. What is the correct roster notation of $\{x \mid x \text{ is a negative integer and } x < -3\}$? (Lesson 1-1)

(A) $\{-3, -2, -1\}$

(B) $\{-2, -1, 0\}$

(C) $\{-2, -1\}$

(D) $\{-4, -5, -6, \ldots\}$

4. Which is *not* a rational number? (Lesson 1-2)

(A) -3.14

(B) $-\frac{3}{8}$

(C) $\sqrt{25}$

(D) $\sqrt{28}$

5. Evaluate $-|m|$ when $m = -5$. (Lesson 1-2)

(A) -5

(B) 0

(C) 5

(D) -151

6. Which graph represents the set of real numbers less than 5? (Lesson 1-2)

(A) ![number line from -3 to 6, closed circle at 5, shaded left]

(B) ![number line from -3 to 6, open circle at 5, shaded left]

(C) ![number line from -3 to 6, dots at 0,1,2,3,4,5]

(D) ![number line from -3 to 6, shaded left arrow]

7. Given the Venn diagram, find $R \cup S$. (Lesson 1-3)

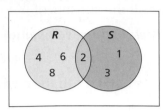

(A) $\{2\}$

(B) $\{1, 2, 3, 4, 6, 8\}$

(C) $\{1, 3, 4, 6, 8\}$

(D) $\{2, 4, 6, 8\}$

8. What is the value of $-1\frac{3}{5} - \frac{3}{10}$? (Lesson 1-4)

(A) $-1\frac{9}{10}$

(B) $-1\frac{1}{3}$

(C) $1\frac{3}{10}$

(D) $1\frac{9}{10}$

9. What is the value of $\frac{5}{16}\left(-\frac{4}{15}\right)$? (Lesson 1-5)

(A) $-\frac{1}{7}$

(B) $-\frac{1}{12}$

(C) $\frac{1}{12}$

(D) $\frac{1}{7}$

10. Simplify $x^4 \cdot x^8$. (Lesson 1-7)

(A) x^{32}

(B) x^{12}

(C) x^4

(D) x^2

11. Neptune is about 2,790,000,000 mi from the sun. How can this distance be represented in scientific notation? (Lesson 1-8)

(A) $279 \cdot 10^{-7}$

(B) $279 \cdot 10^7$

(C) $2.79 \cdot 10^{-9}$

(D) $2.79 \cdot 10^9$

12. A nitrogen atom is approximately $2.33 \cdot 10^{-23}$ g. How many nitrogen atoms are there in 1 mg of nitrogen atoms? (Lesson 1-8)

(A) $2.33 \cdot 10^{-26}$

(B) $2.33 \cdot 10^{-20}$

(C) $4.29 \cdot 10^{19}$

(D) $4.29 \cdot 10^{23}$

Test-Taking Tip

Question 8
Be careful when a question involves adding, subtracting, multiplying, or dividing negative integers. Check to be sure that you have chosen an answer with the correct sign.

Part 2 Short Response/Grid In

Record your answers on the answer sheet provided by your teacher or on a sheet of paper.

13. The following graph represents the values of *a*. Write an inequality for the graph. (Lesson 1-2)

14. Refer to the Venn diagram. How many elements are in $M \cap N$? (Lesson 1-3)

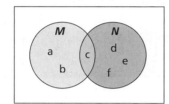

15. Leah's baby brother weighed $7\frac{1}{2}$ lb at birth. He weighed $8\frac{3}{4}$ lb at his one-month checkup. How much did the baby gain? (Lesson 1-4)

16. What is the value of $74.3 - (-45.9)$? (Lesson 1-4)

17. What is the perimeter of the figure? (Lesson 1-4)

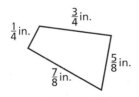

18. Evaluate the expression $2A - B$ when $A = 12$ and $B = -18$. (Lesson 1-4)

19. Alvin wants to buy a new video game. The game he wants costs $48. The sales tax is calculated by multiplying the price by 0.065. What is the total cost of the game including tax? (Lesson 1-5)

20. What is the value of $-2\frac{1}{3} \div \frac{1}{2}$? (Lesson 1-5)

21. Simplify $\frac{m^{15}}{m^{10}}$, where $m \neq 0$. (Lesson 1-7)

22. The Giganotosaurus weighed 14,000 lb. Write this number in scientific notation. (Lesson 1-8)

23. An organism called the *Mycoplasma laidlawii* has a diameter of $4 \cdot 10^{-6}$ in. Write this number in standard form. (Lesson 1-8)

24. Evaluate $(xy)^{-2}$ when $x = -4$ and $y = 3$. (Lesson 1-8)

25. Simplify $(-3)^2 + \left(\frac{1}{4}\right)^{-2}$. (Lesson 1-8)

Part 3 Extended Response

Record your answers on a sheet of paper. Show your work.

26. Describe how to use a Venn diagram to solve the following problem. Then give the solution. (Lesson 1-3)

    Shari has a flower garden that contains 24 flowers. Half of the flowers are grown from bulbs. Eight flowers are over 3 ft tall. One third of the flowers are her favorite color, pink. Only 1 pink flower comes from a bulb and it is over 3 ft tall. Five flowers from bulbs are over 3 ft tall. How many flowers over 3 ft tall are not from bulbs?

27. Each variable *v*, *w*, *x*, and *y* represents a different number. If $v = -\frac{2}{3}$ and $y \neq 1$, find each of the following values. Explain your reasoning using addition and multiplication properties. (Lessons 1-4 and 1-5).

    $$v \times w = v$$
    $$y \times x = x$$
    $$v + y = x$$

    a. *w*   **b.** *x*   **c.** *z*

Essential Algebra and Statistics

THEME: News Media

Newspapers and print magazines communicate information in words, symbols, and pictures. Many feature colorful graphs and charts to make data easier to understand.

Television is a primary source of news information for many people. Reporters research the facts of a story and explain the facts to the viewing audience. Pictures, lists, and colorful graphics help the viewers make sense of the story.

- **Environmental journalists** (page 61) research what is being done to harm and save our planet. They must be able to interpret data, understand graphs, and notice trends in numbers to present the facts without inserting opinions.

- **Transcriptionists or prompter operators** (page 81) prepare the copy that television newscasters read as they report the news. They must be able to work quickly and accurately.

Where Can You Find the News?
News Coverage by Story Type

Story type	Newspapers	Print news magazines	TV news	TV news magazines
Consumer News	7.8%	14.2%	12.3%	12.2%
Health, Medicine, and Lifestyle	9.3%	9.7%	14.8%	24.3%
Celebrity, Entertainment and Personal Profiles	3.6%	20.0%	5.6%	18.6%
Government and Economics	32.8%	21.3%	16.9%	0.6%
Foreign Affairs and Military	17.8%	7.7%	16.1%	2.5%
Crime and Justice	13.3%	9.1%	15.1%	23.1%
Accidents and Disasters	1.8%	1.3%	4.9%	4.5%
Science	2.7%	9.7%	9.4%	5.2%
Other	10.9%	7.0%	4.9%	9.0%

Data Activity: Where Can You Find the News?

Use the table for Questions 1–3.

1. Which type of news media gives the most coverage to the daily happenings in the U.S. Congress?

2. A TV news magazine show is about how many more times as likely as a newspaper to give coverage to problems in a rock star's marriage?

3. A print news magazine devotes an average of 4 pages per magazine to its coverage of foreign affairs and the military. About how many pages would you expect to find in the section entitled *Consumer News*?

CHAPTER INVESTIGATION

The news media use many different ways to present data to the public. Newspapers and magazines use charts, tables, and graphs. Television uses animated diagrams and graphs to make data appealing. Some types of presentations are more effective than others. Some graphs and charts may even be biased or misleading.

Working Together

Think of something that would improve your school or community. Discuss what kind of data would be needed to convince voters and community leaders to make the change. Develop a media plan that would include using television, newspapers, and magazines to encourage support for your plan. Use the Chapter Investigation icons to guide your group.

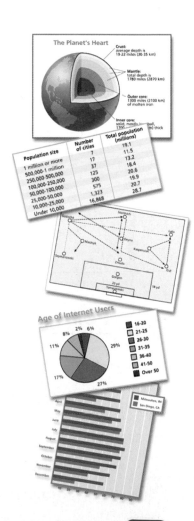

2 Are You Ready?

Refresh Your Math Skills for Chapter 2

The skills on these two pages are ones you have already learned. Use the examples to refresh your memory and complete the exercises. For additional practice on these and more prerequisite skills, see pages 654–661.

GRAPHING INEQUALITIES

Graphing an inequality on a number line can give you an easily understandable visual record of all the solutions for the inequality.

Example Graph the inequality $3x \geq -9$ on a number line.

Use the properties of inequalities.

$3x \geq -9$

$\dfrac{3x}{3} \geq -\dfrac{9}{3}$

$x \geq -3$

Graph the solution.

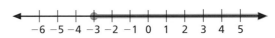

Use a closed dot if that number is included in the solution set. Use an open dot if it is not included.

Graph each inequality on a number line.

1. $2a \geq 6$

2. $2(m + 3) < 8$

3. $\dfrac{k}{2} > -1$

4. $4b - 3 \leq 13$

5. $\left(\dfrac{1}{2}\right)x > -4$

6. $3z + 5 \leq 14$

7. $3(b - 3) \leq -3$

8. $8c - 3 < 5c$

9. $\dfrac{m}{4} - 7 \geq -8$

10. $5(w - 2) > -15$

11. $2y + 7 < -13$

12. $9d + 2 \geq -4$

POINTS ON A COORDINATE PLANE

You will need to know how to plot points on the coordinate plane in order to graph linear equations and inequalities.

Plot each point on a coordinate plane. Label each point with its letter.

13. $A(-3, 2)$

14. $B(1, 8)$

15. $C(8, -1)$

16. $D(2, -6)$

17. $E(3, 4)$

18. $F(-7, 5)$

19. $G(-5, -5)$

20. $H(6, 2)$

21. $I(0, 8)$

22. $J(-7, 0)$

23. $K(7, -4)$

24. $L(-4, -3)$

MEASURES OF CENTRAL TENDENCY

Example Find the mean, median, mode, and range of this data:

8 10 9 9 8 7 5 12 8 6 6

The *mean* is the sum of the data divided by the number of data.

$(8 + 10 + 9 + 9 + 8 + 7 + 5 + 12 + 8 + 6 + 6) \div 11 = 8$

The *median* is the middle value when the data is arranged in numerical order.

5 6 6 7 8 **8** 8 9 9 10 12

If the number of data items is even, the median is the average of the two middle numbers.

The *mode* is the number that occurs most often in the set of data.

5 6 6 7 **8 8 8** 9 9 10 12

The *range* is the difference between the greatest and least values in the set of data.

5 6 6 7 8 8 8 9 9 10 **12**

$12 - 5 = 7$

Find the mean, median, mode, and range of each set of data.

25. 2 3 8 7 8 1
2 5 10 8 12

26. 5 8 11 13 5
9 2 4 6

27. 22 31 16 19 15 24
27 27 14 31 32 30

28. 42 40 38 46 51 28 37 44
30 29 45 36 27 43 34

LANGUAGE OF MATHEMATICS

Write each phrase as an algebraic expression. Use *n* for "a number."

29. seven less than five times a number

30. the product of two and the sum of a number and eight

31. the quotient of three times a number and five

32. the sum of thirteen and eight decreased by a number

33. fourteen decreased by five times a number

34. five times the difference of a number and ten

35. three increased by the quotient of a number and two

Patterns and Iterations

Goals ■ Identify the next terms in a sequence and the rule in an iterative process.

Applications Media, Finance, Business

1. Use blocks or tiles to make the next two figures in this pattern.

2. Make a chart like this to show the number of pieces used in each figure.

3. How many pieces will it take to make the 7th figure? How do you know your answer is correct?

Figure	1	2	3	5
Number of pieces				

◤ BUILD UNDERSTANDING

Mathematicians have always been interested in number patterns. Many such patterns exist naturally, both in mathematics and in nature. A **pattern**, or **sequence**, is an arrangement of numbers in a particular order. The numbers are called **terms**, and the pattern is formed by applying a rule. If a pattern exists, a prediction can be made about the terms in the pattern. For example, the numbers 2, 7, 12, 17, 22, . . . are arranged in a pattern. Each number is 5 more than the preceding number. The rule is "add 5."

Example 1

Identify the pattern for the sequence 1, 2, 4, 7, ____, ____, ____, ... Find the next three terms.

Solution

In this pattern, the number being added is 1 more than the number that was added to the previous term.

The next three terms are 11, 16, and 22.

In some patterns, a term can be calculated by applying a rule to the term's position number.

Example 2

In the sequence 1, 4, 9, 16, . . . , identify the rule relating each term to its position number. Then find the 10th term and the 15th term.

Solution

Each term in the sequence 1, 4, 9, 16, . . . , is found by taking the square of its position number. The 10th term is 10^2, or 100. Likewise, the 15th term is $15^2 = 225$.

Position number	1	2	3	4
Term	$1^2 = 1$	$2^2 = 4$	$3^2 = 9$	$4^2 = 16$

In mathematics, the term **iteration** is used to describe a process that is repeated over and over. You have already seen how iterations can be used to create number patterns. For example, the sequence 1, 2, 4, 8, . . . , is generated by using the iterative process of multiplying by 2. An iteration diagram can also be used to model the sequence.

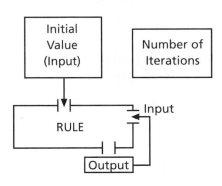

Example 3

The sequence 1, 3, 9, 27, . . . , can be modeled using an iteration diagram. Draw the diagram and calculate the output for 7 iterations.

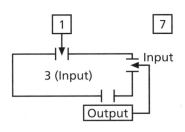

Solution

Initial value (input): 1   Number of iterations: 7

Rule: Multiply input by 3   Output: 3, 9, 27, 81, 243, 729, 2187

Many occupations require the use of iterative processes. Most assets a business owns, such as a car or a piece of business equipment, become less valuable over the time they are used. This is called **depreciation**. For example, if a new car was purchased for $13,000 and sold three years later for $5000, the value of the car has depreciated $8000 in value.

There are different methods to calculate depreciation. Many such methods are iterative, such as the one called the *declining-balance method*.

Example 4

NEWSPAPER A printing machine has an expected life of five years, a beginning book value (cost when bought) of $50,000, and a depreciation rate of 30% per year. Find the ending book value after five years.

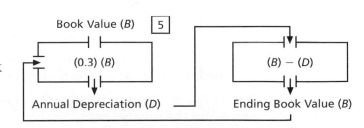

Solution

Calculate the output for 5 iterations.

Year	Beginning book value	Depreciation rate	Annual depreciation	Ending book value
1	$50,000.00	0.3	$15,000.00	$35,000.00
2	35,000.00	0.3	10,500.00	24,500.00
3	24,500.00	0.3	7350.00	17,150.00
4	17,150.00	0.3	5145.00	12,005.00
5	12,005.00	0.3	3601.50	8403.50

After five years the ending book value is $8403.50.

◣ TRY THESE EXERCISES

Identify the pattern for each sequence. Find the next three terms in each sequence.

1. $-15, -11, -7, -3,$ _____, _____, _____, . . .

2. $1, 7, 13, 19,$ _____, _____, _____, . . .

3. $5, 2, -1, -4,$ _____, _____, _____, . . .

4. $-1, 3, -9, 27,$ _____, _____, _____, . . .

5. $2, 6, 18, 54,$ _____, _____, _____, . . .

6. Draw the iteration diagram for this sequence:

$16, 4, 8, 2,$ _____, _____, _____, . . .

Identify each part of the diagram. Then calculate the output for the first 7 iterations.

◣ PRACTICE EXERCISES • For Extra Practice, see page 665.

Determine the next three terms in each sequence.

7. $2, 3, 5, 8,$ _____, _____, _____, . . .

8. $1, 2, 5, 10,$ _____, _____, _____, . . .

9. $1, 3, 7, 13,$ _____, _____, _____, . . .

10. $20, 8, -4, -16,$ _____, _____, _____, . . .

11. $1, 8, 27, 64,$ _____, _____, _____, . . .

12. $-1, \dfrac{1}{2}, -\dfrac{1}{4}, \dfrac{1}{8},$ _____, _____, _____, . . .

13. ERROR ALERT Ryan determines that the next term in the sequence 1.7, 6.9, 12.1, 17.3, . . . is 23.5. Explain what Ryan did wrong.

14. Draw the iteration diagram for the sequence 1, 2, 4, 8, 16, ... Calculate the output for the first 7 iterations.

15. Determine the output for the given iteration. Round each answer to the nearest cent.

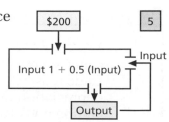

Draw the next three figures in the pattern.

16.

17.

TELEVISION Use the declining-balance method to find the book value at the end of the expected life for each of these assets.

18. sound system for $18,000, expected life of five years, depreciation rate of 30% per year

19. video equipment for $12,000, expected life of three years, depreciation rate of 50% per year

20. NEWSPAPER A local newspaper hopes to increase its circulation at the rate of 15% per year for the next five years. If its current circulation is 15,640, what will its circulation be at the end of five years? Round your answer to the nearest whole number.

EXTENDED PRACTICE EXERCISES

21. FINANCE John deposits $500 in his savings account which earns interest at a 7.5% rate. At the end of each year, he also adds $100 to his account. Determine the iteration diagram and the output for 5 iterations. Round each answer to the nearest cent.

22. WRITING MATH Examine the following sequence.

$$1, 1, 2, 3, 5, 8, 13, 21, \ldots$$

Identify the rule being applied to the pattern, and find the next three terms.

MIXED REVIEW

Graph each set of numbers on a number line. (Lesson 1-1)

23. $\left\{-6, -2.75, 0, \sqrt{5}, 4\frac{1}{2}\right\}$

24. whole numbers less than 3

25. real numbers greater than or equal to -4

26. real numbers less than -1

27. real numbers greater than 2

28. real numbers less than or equal to -3

Graph the intersection of each pair of sets. (Lesson 1-3)

29. $A = \{x \mid x \geq -3\}$ **30.** $A = \{x \mid x > -1\}$ **31.** $A = \{x \mid x > 0\}$

     $B = \{x \mid x < 5\}$ $B = \{x \mid x \geq 4\}$ $B = \{x \mid x \leq 5\}$

32. $A = \{x \mid x < 2\}$ **33.** $A = \{x \mid x > -5\}$ **34.** $A = \{x \mid x \leq 0\}$

     $B = \{x \mid x > 1\}$ $B = \{x \mid x \leq 2\}$ $B = \{x \mid x \geq -4\}$

2-2 The Coordinate Plane, Relations, and Functions

Goals
- Identify relations and their domains and ranges.
- Identify and evaluate functions.

Applications Engineering, Biology, Television

Determine the final text size by completing the table.

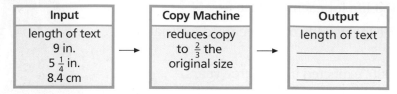

Input	Copy Machine	Output
length of text 9 in. $5\frac{1}{4}$ in. 8.4 cm	reduces copy to $\frac{2}{3}$ the original size	length of text _____ _____ _____

◥ BUILD UNDERSTANDING

You know that real numbers can be graphed on a number line. You can graph pairs of numbers on a grid system called a **coordinate plane**. A coordinate plane consists of two perpendicular number lines, dividing the plane into four regions called **quadrants**. The horizontal number line is the **x-axis**, and the vertical number line is the **y-axis**. The point where the axes cross is the **origin**. Points on the axes are not part of the quadrants.

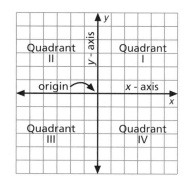

There are infinitely many points in the plane. Each point is unique and is assigned an **ordered pair** of real numbers, consisting of one **x-coordinate** and one **y-coordinate**. For example, the point $(2, -5)$ has an x-coordinate of 2, and a y-coordinate of -5. The x-coordinate, or *abscissa*, determines the horizontal location of the point, while the y-coordinate, or *ordinate*, determines the vertical location. The order of the numbers is important. $(2, -5)$ and $(-5, 2)$ refer to two different points.

> ### Reading Math
>
> The x-coordinate is *always* the first coordinate in ordered pairs. The y-coordinate is the second coordinate.
>
> (x, y)

A set of ordered pairs is defined as a **relation**. You can represent a relation by a table of values or a graph.

The **domain** is the set of all the x-coordinates of ordered pairs in the relation. The **range** is the set of all the y-coordinates of ordered pairs in the relation. A **mapping** is the relationship between the elements of the domain and the range.

The mapping at the right shows the relationship between the x-coordinates (domain) and the y-coordinates (range) for the set of ordered pairs $\{(0, 1), (2, -1), (3, 2)\}$.

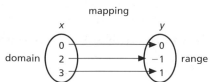

A special kind of relation that is important in mathematics is called a *function*. A **function** is a set of ordered pairs in which each element of the domain is paired with *exactly one* element in the range.

56 | Chapter 2 **Essential Algebra and Statistics**

Example 1

Determine whether each relation is a function. State the domain and range of each.

a.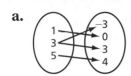

b.

x	−4	−2	0	2
y	2	0	2	4

c. $\{(0, 3), (1, 4), (−3, 0)\}$

Check Understanding

Write the set of ordered pairs as a table and a mapping; then graph each point.

Solution

a. No; the element 3 in the domain is paired with two elements in the range, −3 and 3. Domain: {1, 3, 5}, Range: {−3, 0, 3, 4}

b. Yes; each element of the domain is paired with exactly one element of the range. However, you will note that one element of the *range* can be paired with more than one element of the domain.
Domain: {−4, −2, 0, 2}, Range {0, 2, 4}

c. Yes. Domain: {−3, 0, 1}, Range: {0, 3, 4}

Below is another method to determine if a relation is a function.

Vertical Line Test: When a vertical line is drawn through the graph of a relation, the relation is *not a function* if the vertical line intersects the graph in more than one point.

Example 2

Determine whether $\{(2, 2), (4, −4), (2, −2), (4, 4)\}$ is a function by using the Vertical Line Test.

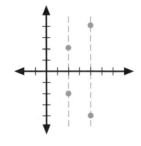

Solution

The relation is *not* a function. A vertical line passes through more than one point.

Examples of functions in everyday life are the relationship between the numbers of hamburgers sold and the price of each, or the weight of a package and the cost of postage. In mathematics, functions are usually given as *rules* that show the relationship of elements of the domain (input values) to elements of the range (output values). The variable whose values make up the domain is the **independent variable**, x. The variable that depends on x is the **dependent variable**.

Function notation can represent the rule that associates the input value with the output value. The most commonly used function notation is called the "f of x" notation. If f is the function that assigns to each real number x the value $x + 1$, then $f(x) = x + 1$.

Rule represented by	Example	Is read
equation in two variables	$y = x + 1$	"y is a function of x equal to $x + 1$" or "y equals $x + 1$"
f of x notation	$f(x) = x + 1$	"f of x equals $x + 1$"

Example 3

Evaluate each function.

a. $f(x) = 3x + 2; f(6)$ **b.** $g(x) = x^2 - 1; g(-1)$

Solution

a. $f(6) = 3(6) + 2$ **b.** $g(-1) = (-1)^2 - 1$

$= 18 + 2$ $= 1 - 1$

$= 20$ $= 0$

Because functions define the mathematical relationship between two variables, they are often used to model real-world problems.

Example 4

ENGINEERING The air conditioner in a car should produce air that is 26 degrees below the temperature outside the car. The formula for this function is $T(x) = x - 26$, where x is the outside air temperature. What is the temperature inside the car when the outside temperature is 92°F?

Solution

$T(92) = 92 - 26 = 66$ The temperature is 66°F inside the car.

TRY THESE EXERCISES

Graph each point on a coordinate plane.

1. $A(-1, 0)$ **2.** $B(2, -1)$ **3.** $C(0, -4)$ **4.** $D(-3, -2)$

Given $f(x) = x - 5$, evaluate each of the following.

5. $f(3)$ **6.** $f(0)$ **7.** $f(-2)$ **8.** $f(11)$

Determine if each relation is a function. Give the domain and range.

9.

x	-2	-3	1	0	1
y	1	0	-1	1	-3

10.
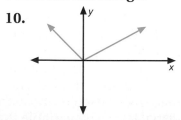

PRACTICE EXERCISES • For Extra Practice, see page 666.

Graph each point on a coordinate plane. Name the quadrant in which each point is located.

11. $A(3, 5)$ **12.** $B(-2, -3)$ **13.** $C(-4, 3)$ **14.** $D(1, -5)$

Given $f(x) = 4x - 1$, evaluate each of the following.

15. $f(-4)$ **16.** $f(0)$ **17.** $f(2)$

Write each relation as a set of ordered pairs. Give the domain and range.

18.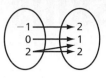

19.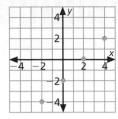

20.

x	0	1	2
y	−1	1	3

BIOLOGY Biologists have determined that the number of chirps made by a cricket in one minute is a function of the temperature (t) measured in degrees Fahrenheit. This relationship is modeled by the function $c(t) = \left(\frac{1}{4}\right)t + 37$. Calculate the number of chirps per minute for the given temperatures.

21. 48°F

22. 92°F

Given $f(x) = 5x + 2$, $g(x) = -2x + 1$, and $h(x) = 3x^2$, find each value.

23. $f(3)$

24. $g(5)$

25. $h(4)$

26. $f(1) + g(1)$

■ EXTENDED PRACTICE EXERCISES

Given $f(x) = ax + b$, $g(x) = cx^2$. Find each value if $a \neq 0$ and $c \neq 0$.

27. $f\left(\frac{1}{a}\right)$

28. $f\left(\frac{-b}{a}\right)$

29. $g(c)$

30. $g\left(\frac{-1}{c}\right)$

Use the Vertical Line Test to determine whether each graph represents a function.

31.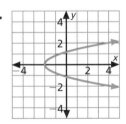

32.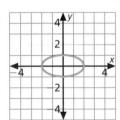

33.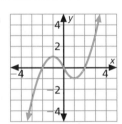

34. WRITING MATH Explain *why* the Vertical Line Test works.

35. TELEVISION A video technician charges $80 for the first hour. Each additional half-hour or part of a half-hour costs $30. What is the total charge for a $3\frac{1}{4}$-hour session?

■ MIXED REVIEW EXERCISES

Let $U = \{0, 3, 5, 6, 8, 10, 13, 14, 18, 20\}$, $A = \{0, 3, 6, 10, 14, 20\}$, $B = \{3, 5, 10, 13, 18\}$, and $C = \{0, 3, 6, 8, 14, 20\}$. Find each set. (Lesson 1-3)

36. $A' \cup B'$

37. $B' \cap C$

38. $(A \cup C) \cap B'$

39. $(B \cap C') \cup A$

40. $(B' \cap C') \cup A$

41. $(A \cup B') \cap C$

42. $(A \cup B) \cap C'$

43. $(B' \cup C) \cap (A \cup B')$

44. $(A' \cap C) \cup (C' \cap B)$

45. DATA FILE Use the data on page 644 on tall buildings of the world. What is the mean height in feet of the buildings? What is the median height of the buildings listed? (Prerequisite Skill)

Review and Practice Your Skills

Find the next three terms in each sequence.

1. 5, −10, 20, −40, _____, _____, _____, . . .

2. 2187, 729, 243, 81, _____, _____, _____, . . .

3. 3, 5, 9, 15, _____, _____, _____, . . .

4. 1, 8, 5, 12, _____, _____, _____, . . .

5. 5, 2, −1, −4, _____, _____, _____, . . .

6. $\frac{1}{2}, \frac{1}{3}, \frac{1}{4}, \frac{1}{5}$, _____, _____, _____, . . .

7. $\frac{1}{7}, \frac{2}{7}, \frac{4}{7}, \frac{8}{7}$, _____, _____, _____, . . .

8. 1, 4, 9, 16, _____, _____, _____, . . .

9. 2, 1, 0.5, 0.25, _____, _____, _____, . . .

10. 38, 65, 92, 119, _____, _____, _____, . . .

Draw the iteration diagram for each sequence. Calculate the output for the first 7 iterations.

11. −15, −10, −5, 0, _____, _____, _____, . . .

12. $\frac{1}{3}$, 1, 3, 9, _____, _____, _____, . . .

13. 1024, 256, 64, 16, _____, _____, _____, . . .

14. 47, 40, 33, 26, _____, _____, _____, . . .

15. 15, −15, 15, −15, _____, _____, _____, . . .

16. −0.5, 2, −8, 32, _____, _____, _____, . . .

17. Glenda's money market account has a starting balance of $10,000. Annual interest rate is 10%. At the end of each year, Glenda also deposits $500 into her account. Draw an iteration diagram and determine the output after 5 iterations. Round to the nearest cent.

Graph each point on a coordinate plane.

18. $J(3, -5)$

19. $K(5, -3)$

20. $L(0, 6)$

21. $M(-6, 0)$

22. $A(-1, -2)$

23. $B(-2, 7)$

24. $C(-5, 3)$

25. $D(0, -8)$

Given $f(x) = 11 - 7x$, evaluate each function.

26. $f(2)$

27. $f(-2)$

28. $f(0)$

29. $f(5)$

Given $f(x) = x^2 + 5x + 6$, evaluate each function.

30. $f(-2)$

31. $f(-3)$

32. $f(0)$

33. $f(10)$

Determine if each relation is a function. Give the domain and range.

34.

x	y
1	5
−1	6
−3	5
−5	6

35.

x	y
0	0
3	6
3	12
4	18

Write each as a set of ordered pairs. Given the domain and range. (Lesson 2-2)

36.

x	2	4	6	8
y	11	8	5	2

37.

x	0	−1	1	−2	2
y	3	4	4	7	7

Determine the next three terms or figures in each pattern. (Lesson 2-1)

38. 243, 260, 277, 294, _____, _____, _____

39. −748, −595, −442, −289, _____, _____, _____

40. 32, 48, 72, 108, _____, _____, _____

41. Z, ZY, ZYX, ZYXW, _____, _____, _____

42. *, **, ***, ****, _____, _____, _____

43. 10, −10², 10³, −10⁴, _____, _____, _____

Given $f(x) = 3x^2$ **and** $g(x) = -3x + 1\frac{1}{2}$, **find each value.** (Lesson 2-2)

44. $f(2)$ **45.** $f(-2)$ **46.** $g\left(\frac{1}{2}\right)$ **47.** $f(1) + g(1)$

Name the quadrant in which each point is located. (Lesson 2-2)

48. $W(9, -1.5)$ **49.** $X(-8, 17)$ **50.** $Z(-6.5, -6.5)$

MathWorks
Workplace Knowhow

Career – Environmental Journalism

Environmental journalists inform the public about what is being done to harm and to save the planet. They work with a great deal of mathematical information such as, how much pollution a smokestack emits or how much garbage is added to a landfill each year.

Environmental journalists study data gathered by scientists and the government. They must be able to analyze raw data, understand and create graphs, spot trends in numbers, and draw conclusions about what is happening to the environment.

1. The United States has 17 commercial incinerators that burn hazardous waste and release pollutants into the air. The hazardous waste size has increased from 1.3 billion to 2.4 billion to 4.6 billion pounds of waste. Project the growth for the next five years. (Hint: Plot the ordered pairs as (0, 0) for the starting point, the difference of 2.4 billion and 1.3 billion for the increase at end of year 1, and so on.)

2. Create a graph to go with your story. Place years on the x-axis and billions of pounds of waste (in 0.5 billion increments) on the y-axis.

3. A television reporter claims that fewer people in your town are recycling. But is the report true? Last year, 30% of the people in your town recycled regularly. This year recycling is down to 27%. Last year, your town had a population of 2500, but this year the population is 3100. Is the reporter's claim true? Explain your thinking.

2-3 Linear Functions

Goals
- Graph linear functions
- Evaluate absolute value functions.

Applications   Retail sales, Photography

Use Algeblocks to represent and simplify expressions.

The green blocks in a set of Algeblocks can be used to represent integers. Blocks on the top half of the mat represent positive integers, and blocks on the bottom half represent negative integers. An equal number of positive and negative blocks add to zero and may be removed from the mat. These pairs are called **zero pairs**. For example, $+3 + (-7)$ is simplified to -4.

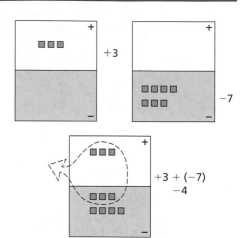

Use the units pieces on a mat to show how these expressions can be simplified.

a. $4 + (-1)$

b. $-6 + 2$

c. $7 - 4$ (rewrite as addition)

d. $6 - 8$

◣ BUILD UNDERSTANDING

In Lesson 2-2 you learned about number pairs produced by functions. Such pairs can be plotted on a coordinate plane and used to construct a graphical representation of the function. An equation that can be written in the form $Ax + By = C$, where A and B are not both zero, is called a **linear equation**. Graphs of such equations are *straight* lines. A function with ordered pairs that satisfy a linear equation is called a **linear function**.

Example 1

Graph $y = 2x + 3$.

Solution

x	$2x + 3$	y
-2	$2(-2) + 3$	-1
0	$2(0) + 3$	3
1	$2(1) + 3$	5

Choose at least three values for x, calculate the corresponding y-values, and make a table to show the ordered pairs. Then plot the points and draw the line containing them. The domain is the set of all real numbers. The range is the set of all real numbers.

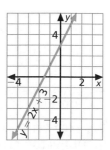

A **constant** function is a linear function with a domain of all real numbers and a range of only one value. The graph of a constant function is a horizontal line.

Example 2

Graph each equation. Determine if the relation is a function. Then determine the domain and range.

a. $x = 2$ b. $y = -1$

Solution

a. Any value of y results in an x value equal to 2.

$x = 2$ is not a function.

Domain: $x = 2$ Range: set of all real numbers.

b. Any value of x results in a y value equal to (-1).

$y = -1$ is a linear constant function.

Domain: set of all real numbers Range: $y = -1$

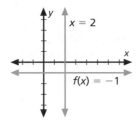

Example 3

TEMPERATURE The relationship between the scales used to measure temperature in degrees Fahrenheit (F) and degrees Celsius (C) can be represented by the linear equation $F = \frac{9}{5}C + 32$. Graph this function and determine the Fahrenheit temperature that is equivalent to 35°C.

Solution

Select three values for C. Calculate the corresponding F-values. Then plot the points and draw the line.

Find the point that has an x-coordinate of 35. The second coordinate of that point, 95, is the equivalent temperature measured in degrees Fahrenheit.

°C	°F
−10	14
0	32
20	68

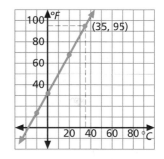

The **absolute value function** is defined as:

$$g(x) = |x| = \begin{cases} x \text{ if } x \geq 0 \\ -x \text{ if } x < 0 \end{cases}$$

For example, $g(-3) = |-3| = -(-3)$ or 3, because $-3 < 0$.

Example 4

Given $h(x) = |x + 2|$, find each value.

a. $h(-3)$ b. $h(0)$

Solution

a. $h(-3) = |-3 + 2|$

$= |-1|$

$= 1$, because $-1 < 0$

b. $h(0) = |0 + 2|$

$= |2|$

$= 2$, because $2 \geq 0$

A graphing calculator is useful for solving real-world problems. Functions and equations can be graphed more quickly than on paper. Follow these rules when using a graphing calculator.

Step 1 Use the viewing window to select the minimum and maximum values for the x- and y-axes. These values will be determined by the problem.

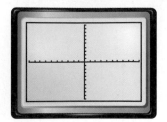

The minimum values (min) for x and y refer to the extreme low values on the x- and y-axes. Likewise, the maximum values (max) refer to the extreme high values on the x- and y-axes. The scale refers to what each tick-mark on the axes represents.

Step 2 Enter the equation into the calculator and graph.

Example 5

GRAPHING Use a graphing calculator to graph $y = x + 3$.

Solution

Calculator input: Y= | X,T,θ,*n* | + | 3 | Graph

Settings for Viewing Window:

 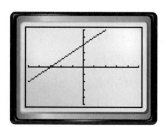

MODELING Use Algeblocks to represent each equation. Simplify where possible. Sketch your answer.

1. $3 - 2x + x = y - y + 3$ **2.** $3x + x + 1 = -4 - 2$

Given $g(x) = |2x - 1|$, find each value.

3. $g(1)$ **4.** $g(-2)$

5. RETAIL Rich's weekly salary is based on the number of pairs of shoes he sells. He is paid a base salary of $25, plus $5 for every pair of shoes he sells. The relationship between his pay (p) and pairs of shoes (s) sold can be represented by the linear equation $p = 25 + 5s$. Graph this function, and determine Rich's pay for a week in which he sold 7 pairs of shoes.

6. WRITING MATH Why should you use three values for x when graphing a linear equation or function?

Graph each function.

7. $y = x + 4$　　　　　　**8.** $f(x) = -5$　　　　**9.** $y = -2x + 3$

Given $h(x) = |-2x + 3|$, find each value.

10. $h(-6)$　　　　　　　　**11.** $h(2)$

GRAPHING Use a graphing calculator to graph each function.

12. $y + 1 = 2x - 3$　　　**13.** $2x + 4 = 4y$　　**14.** $y - 2 = 3x$

Given $F(x) = 2|2x| - 3|x + 1|$, find each value.

15. $F(-4)$　　　　　　　　**16.** $F(0)$

PHOTOGRAPHY A photographer charges a sitting fee of $15, and charges $4 for each 5-in. by 7-in. photograph the customer orders. The linear function $c = 15 + 4n$ can be used to calculate the customer's cost (c) based on the number of photographs (n) purchased.

17. Graph the function.

18. Use your graph to determine the total cost of 6 photographs.

19. How many photographs can be purchased if you cannot spend more than $50.00?

20. CHAPTER INVESTIGATION Working together, think of a positive change that you would like to make at your school or in your community. What kind of data would encourage others to adopt your proposal? Make a list.

◤ EXTENDED PRACTICE EXERCISES

Graph each function.

21. $y = |x + 2|$　　　　　**22.** $y = -|x + 2|$　　　　**23.** $y = |x| + 2$

24. Graph $y = \begin{cases} x & \text{for } x < 0 \\ 2x + 1 & \text{for } x \geq 0 \end{cases}$

◤ MIXED REVIEW EXERCISES

Add or subtract. (Lesson 1-4)

25. $-6 + (-3) - 12 - (-5)$　　**26.** $(-3) + (-2) + (-6)$　　**27.** $-(-4) - (-8)$

28. $5 + (-4) - 16 + (-2)$　　**29.** $(-8) + (-(-3)) + 2$　　**30.** $6 - 12 + (-7) - 4$

Estimate each sum or difference. (Lesson 1-4)

31.　5382
　　　$+7649$

32.　9764
　　　-3478

33.　5894
　　　$+9763$

34.　8043
　　　-5612

35.　$78.64
　　　$+ 85.06$

36.　$83.98
　　　$- 36.52$

37.　$94.76
　　　$+75.15$

38.　$52.25
　　　$- 18.96$

2-4 Solve One-Step Equations

Goals ■ Use the addition or multiplication properties of equality to solve one-step equations.

Applications Business, Finance, News media

In Lesson 2–3, you used Algeblocks to model equations. Model $x + 5 = 3$. Then add -5 to both sides and simplify each side. What is the result on both sides? Sketch your answer and complete the equations.

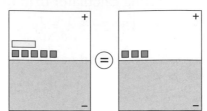

$$x + 5 = 3$$
$$x + 5 + \underline{\quad} = 3 + \underline{\quad}$$
$$\underline{\quad} = \underline{\quad}$$

◥ BUILD UNDERSTANDING

In the activity above, you used the opposite of a number to simplify and solve an equation. In the same way, you can use opposite, or **inverse operations** to get a variable alone on one side of an equation.

Example 1

Use Algeblocks to solve $x + 3 = 10$.

Solution

Represent the equation. Adding -3 to each mat will result in zeros and leave the x-piece alone on one mat.

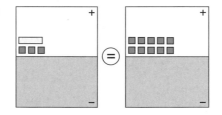

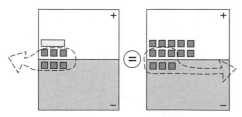 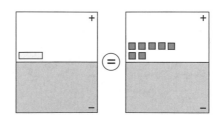

Read the answer, $x = 7$.

> **Reading Math**
>
> Mathematical notation can be used to show the steps in solving each equation in Example 1.
>
> $$x + 3 = 10$$
> $$x + 3 - 3 = 10 - 3$$
> $$x = 7$$
>
> Add the opposite. x is alone.

When equations involve the inverse operations of addition and subtraction, you can use opposites and the addition property of equality to solve them. This property states that adding the same number to both sides of an equation results in an **equivalent equation**.

Addition Property of Equality	For all real numbers a, b, and c, if $a = b$, then $a + c = b + c$ and $c + a = c + b$.

Example 2

Use mathematical notation to show the steps in solving the equation.

$$x - 3.7 = -0.1$$

Solution

$$x - 3.7 = -0.1$$
$$x - 3.7 + 3.7 = -0.1 + 3.7$$
$$x = 3.6$$

Check Understanding

Find the opposite and reciprocal of each number:

$2, -5, \dfrac{2}{3}, 0.5$

In a similar manner, reciprocals and the multiplication property of equality are used in solving equations involving multiplication. The multiplication property of equality states that multiplying both sides of an equation by the same number still maintains the equality.

Multiplication Property of Equality	For all real numbers a, b, and c, if $a = b$, then $ac = bc$ and $ca = cb$.

Example 3

Solve each equation.

a. $\left(\dfrac{2}{5}\right)y = 12$　　　　**b.** $-3q = 45$

Solution

a.
$$\left(\dfrac{2}{5}\right)y = 12$$
$$\left(\dfrac{5}{2}\right)\left(\dfrac{2}{5}\right)y = \left(\dfrac{5}{2}\right)12$$
$$1y = 30$$
$$y = 30$$

b.
$$-3q = 45$$
$$\left(-\dfrac{1}{3}\right)(-3q) = \left(-\dfrac{1}{3}\right)45$$
$$1q = -15$$
$$q = -15$$

You may need to simplify one or both sides of an equation before applying the properties of equality.

Reading Math

In Example 3:

a. Multiply both sides of the equation by the reciprocal of

$\dfrac{2}{5}$.

b. Multiply both sides of the equation by the reciprocal of -3.

In part b, dividing by -3 is the same as multiplying by the reciprocal of -3.

Example 4

Solve: $w - 9 + 23 = |-3 - 5|$

Solution

$$w - 9 + 23 = |-3 - 5|$$
$$w + 14 = 8$$
$$w + 14 + (-14) = 8 + (-14)$$
$$w = -6$$

The solution is -6.

Solving problems often involves translating a verbal problem into an algebraic equation. These equations can then be solved using the properties of equality.

Example 5

Translate the sentence into an equation using n to represent the unknown number. Then solve the equation for n.

When a number is decreased by 31, the result is the square of 3.

Solution

When a number is decreased by 31, the result is the square of 3.
The equation is:

$$n - 31 = 3^2$$
$$n - 31 = 9$$
$$n - 31 + 31 = 9 + 31$$
$$n = 40$$

The number is 40.

◤ TRY THESE EXERCISES

Solve each equation.

1. $q + 18 = 32$

2. $r - 5 = -2$

3. $-4z = 36$

4. $-16 = 21 + h$

5. $-7 = \dfrac{v}{8}$

6. $\left(\dfrac{5}{3}\right)k = 30$

7. $e + \dfrac{3}{4} = 1$

8. $\dfrac{2}{5} = m - \dfrac{1}{2}$

9. $w - 1.7 = -4.2$

WRITING MATH Translate each sentence into an equation using n to represent the unknown number. Do not solve.

10. The product of -8 and a number is the same as the square of -4.

11. Increasing a number by 15 yields the same result as taking half of 72.

12. The quotient of a number and 5 is 0.2.

13. The difference between a number and 26 is -9.

14. **DATA FILE** Use the data on page 646 on average daily temperatures. On November 16, the temperature in San Diego climbed 9° higher than the average daily temperature in that city for November and then dropped 12°. What was the temperature on November 16?

◤ PRACTICE EXERCISES • For Extra Practice, see page 667.

Solve each equation.

15. $f + 19 = 41$

16. $7m = -35$

17. $21 + a = -4$

18. $-10 = p - 1$

19. $25n = -10$

20. $0.9u = 0.63$

21. $12 = \left(\dfrac{-4}{3}\right)y$

22. $\left(\dfrac{3}{8}\right)x = 6$

23. $5.74 = j - 3.6$

24. YOU MAKE THE CALL Anthony says that $(-4)^2$ and -4^2 are equal. Do you agree with him? Explain.

Translate each sentence into an equation using n to represent the unknown number. Then solve the equation for n.

25. FINANCE When an account balance is increased by $25, the result is $-\$15$.

26. The difference between a number and 26 is the square of -3.

27. The quotient of a number and 8 is 0.7.

28. One-third of -81 is the same as the product of 3 and some number.

Solve each equation.

29. $(-2)(-3)(-4) = 12c$

30. $|13 - 19| = \left(\dfrac{-4}{5}\right)y$

31. $(2.5)(5) = m - 17 + 4$

32. $w + 3^4 = 4^3$

33. $a - 7 + 25 = 2^3$

34. $0.01k = (1 + 2 + 3 + 4)^2$

Find all solutions in each equation.

35. $|x| + 5 = 11$

36. $-48 = -4|z|$

37. $|w| - 3 = -3$

38. NEWS MEDIA A television news magazine has 48 minutes of airtime to fill. The producer decides to run an 8-minute health segment and a 9-minute science segment. At the last minute, a 12-minute celebrity feature is canceled. The producer decides to add a 20-minute segment. What length segment is needed to complete the broadcast? Write an equation to model the situation and solve.

EXTENDED PRACTICE EXERCISES

Replace each __?__ so that the equation will have the given solution.

39. $x + \underline{} = -4$; The solution is -15.

40. $\underline{}\, x = \dfrac{1}{2}$; The solution is $\dfrac{1}{12}$.

41. $-24 = x - \underline{}$; The solution is 16.

42. $-0.27 = \underline{}\, x$; The solution is 0.9.

43. Write an equation that has no solution.

44. Write an equation that has infinitely many solutions.

45. BUSINESS The cost of making a camera is 60% of its selling price (p). The rest is profit. If the camera cost $101.25 to make, how much is its selling price? Write an equation and solve.

MIXED REVIEW EXERCISES

Find each product or each quotient. (Lesson 1-5)

46. $(16)(-3.9)$

47. $-5\dfrac{1}{3} \div \dfrac{2}{3}$

48. $(-7)(-0.5)(-2)$

49. $345 \div (-15)$

50. $\left(-\dfrac{3}{4}\right)\left(\dfrac{1}{2}\right)\left(-\dfrac{5}{8}\right)$

51. $-63 \div (-0.7)$

52. $(-3)\left(-\dfrac{7}{8}\right)(-9)$

53. $54.6 \div (-4.2)$

Evaluate each expression when $a = 6$ and $b = -4$. (Lesson 1-4)

54. $a + (-b)$

55. $4a + 3b$

56. $6b - (-2a)$

Review and Practice Your Skills

Graph each function.

1. $y = 4x + 3$

2. $f(x) = -2x + 5$

3. $y = 7$

4. $f(x) = -\frac{1}{2}x + 6$

5. $y = 8 - x$

6. $y = 3x$

7. $x + y = 10$

8. $y - 2 = 2x + 6$

9. $y = x$

Given $g(x) = |-3x - 2|$, find each value.

10. $g(0)$

11. $g(-5)$

12. $g(3)$

Given $F(x) = 3|x| - 2|2x - 5|$, find each value.

13. $F(0)$

14. $F(-3)$

15. $F(3)$

A car rental agency charges a flat fee of \$30 to rent a car, and \$21 for each day the car is rented. The linear function $c = 30 + 21d$ can be used to calculate the customer's cost (c) based on the number of days (d) the car is rented.

16. Graph the function.

17. Determine the cost for a 7-day rental.

18. What is the maximum number of days Lakesha can rent a car if she has only \$140 to spend?

Solve each equation.

19. $4x = -12$

20. $x - 5 = -4$

21. $l + 8 = 11$

22. $y + 5 = 7.2$

23. $\frac{1}{3}p = -2$

24. $-5b = 65$

25. $\frac{2}{3} = -\frac{7}{12} + m$

26. $0.8t = 9.6$

27. $2u = \frac{1}{2}$

28. $-11 = n - 4$

29. $45 + m = 71$

30. $-1\frac{1}{8}x = 1$

31. $62.4 + k = -39.9$

32. $x - 4 = -4$

33. $\frac{y}{-3} = 27$

34. $38 = -43 + x$

35. $-8.4 = 0.12x$

36. $d - (-13) = 25$

37. $\frac{6}{13}a = 52$

38. $2.18 = r + 3.59$

39. $b - 5 = -41$

40. $\frac{e}{4} = -7$

41. $p - \frac{4}{5} = 1\frac{1}{5}$

42. $-6 = y - 6$

43. $-12n = 3$

44. $-12 + n = 3$

45. $\frac{n}{-12} = 3$

46. $3n = -12$

47. $\frac{m}{-3} = \frac{4}{9}$

48. $\frac{7}{3}x = 21$

49. $-4.7 = n - 2.5$

50. $0 = d + 11$

51. $-x = 16$

Determine the next three terms in each sequence. (Lesson 2-1)

52. 1600, 400, 100, 25, _____, _____, _____   **53.** 47, 36, 25, 14, _____, _____, _____

54. a, c, e, g, _____, _____, _____   **55.** $-174, -148, -122, -96,$ _____, _____, _____

Graph each function. (Lesson 2-3)

56. $f(x) = -x + 1$ **57.** $y = |x|$ **58.** $y = 9$

Translate each sentence into an equation using n to represent the unknown number. Then solve the equation for n. (Lesson 2-4)

59. When n is increased by 13, the result is -29.

60. The product of a number and 8 is the same as the square of -7.

61. The quotient of a number and -4 is 11.

62. The difference between n and 17 is 25.

Solve each equation. (Lesson 2-4)

63. $t + 1\frac{3}{8} = 3\frac{3}{4}$ **64.** $|x| + 4 = 10$ **65.** $|21 - 27| = -\frac{3}{2}y$

Mid-Chapter Quiz

Determine the next three terms in each sequence. (Lesson 2-1)

1. 4, 13, 22, 31, . . .   **2.** $2, -10, 50, -250, \ldots$ **3.** $-1, \frac{1}{3}, -\frac{1}{9}, \frac{1}{27}, \ldots$

4. Determine the output for the first five iterations: The initial input is 6; the rule is "add -4."

Use the relation, $\{(-1, 2), (0, 3), (2, 5)\}$ for Exercises 5–7. (Lesson 2-2)

5. What is the domain?   **6.** What is the range?   **7.** Is it a function?

Find each value. (Lesson 2-3)

8. $f(2)$ if $f(x) = -3x + 1$.   **9.** $f(-4)$ if $f(x) = 2$.   **10.** $f(-3)$ if $f(x) = |-x|$.

Solve each equation. (Lesson 2-4)

11. $9 = -3 + j$ **12.** $-4m = 1$ **13.** $\frac{2n}{3} = 7$

Translate the sentence into an equation using n to represent the unknown number. Then solve the equation for n. (Lesson 2-4)

14. The difference between a number and 6 is the product of 3 and 8.

15. The product of -5 and -4 is the product of 8 and a number.

Solve Multi-Step Equations

Goals ■ Solve equations with more than one step.

Applications Advertising, Finance, Recreation

Algeblocks can be used to solve two-step equations. Complete the equation to show algebraically the steps taken to solve $3x - 2 = 4$.

Algeblocks

a. Represent the equation.

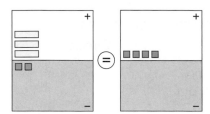

Algebraically

a. Write the equation.

b. Add the opposite of -2 to both sides and simplify.

c. Divide each side into three groups.

d. Read the solution.

a. $3x - 2 = 4$

b. $3x - 2 + \boxed{} = 4 + \boxed{}$
$\boxed{} = \boxed{}$

c. $\dfrac{3x}{\boxed{}} = \dfrac{\boxed{}}{\boxed{}}$

d. $x = \boxed{}$

◥ BUILD UNDERSTANDING

To solve some equations, it may take two or more steps to get the variable alone on one side of the equation. When solving these equations, use the addition property of equality first. Then use the multiplication property of equality.

Example 1

MODELING Solve $2x - 7 = -1$. Along with using Algeblocks, explain and represent each step algebraically.

Solution

Use Algeblocks to represent the equation.

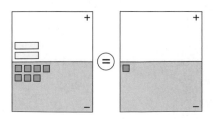

Add $+ 7$ to each side of the equation. Simplify.

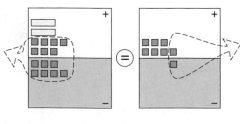

Separate into 2 groups.

The solution is $x = 3$.

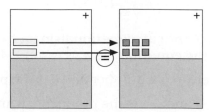

Some equations contain variables on both sides. For these equations, simplify the equation by using the addition property of equality to move like terms to the same side of the equation. Terms that have exactly the same variables are called **like** or **similar terms**.

Example 2

Solve $x + 5 = 2x - 3$. Check the solution.

Solution

$$x + 5 = 2x - 3$$

$$x + 5 + (-2x) = 2x + (-2x) - 3 \quad \text{Add } -2x \text{ to each side.}$$

$$-x + 5 = -3$$

$$-x + 5 + (-5) = -3 + (-5) \quad \text{Add } -5 \text{ to each side.}$$

$$-x = -8$$

$$-x(-1) = -8(-1) \quad \text{Multiply each side by } -1.$$

$$x = 8$$

The solution is 8.

Check

$$x + 5 = 2x - 3$$

$$8 + 5 = 2(8) - 3$$

$$13 = 16 - 3$$

$$13 = 13 \quad \checkmark$$

Sometimes you will need to simplify each side of an equation before applying the properties of equality.

Example 3

Solve $6(2x - 1) = -36 + 6$. Check the solution.

Solution

$$6(2x - 1) = -36 + 6 \quad \text{Apply the distributive property.}$$

$$12x - 6 = -36 + 6$$

$$12x - 6 = -30$$

$$12x - 6 + 6 = -30 + 6 \quad \text{Add 6 to each side.}$$

$$12x = -24$$

$$\left(\frac{1}{12}\right)(12x) = \left(\frac{1}{12}\right)(-24) \quad \text{Multiply each side by } \frac{1}{12}.$$

$$x = -2$$

The solution is -2.

Check Be sure to follow the order of operations.

$$6(2x - 1) = -36 + 6$$

$$6(2(-2) - 1) = -30$$

$$6(-4 - 1) = -30$$

$$6(-5) = -30$$

$$-30 = -30 \quad \checkmark$$

Problem Solving Tip

The equation in Example 3 can also be solved by dividing both sides by 6 first.

$$\frac{6(2x - 1)}{6} = \frac{-30}{6}$$

$$2x - 1 = -5$$

Then solve the equation.

$$2x - 1 + 1 = -5 + 1$$

$$2x = -4$$

$$x = -2$$

Example 4

ADVERTISING A local newspaper sells all classified ads for the same price. Larger boxed ads cost $24.50. Eun Ah bought three classified ads and one boxed ad. If the total cost for the ads was $79.25, what was the price of each classified ad?

Solution

Let a represent the price of each classified ad.

$$3a + 24.50 = 79.25$$
$$3a + 24.50 + (-24.50) = 79.25 + (-24.50)$$
$$3a = 54.75$$
$$\left(\frac{1}{3}\right)(3a) = \left(\frac{1}{3}\right)(54.75)$$
$$a = 18.25$$

Check   3 classified ads = 3($18.25): $54.75
1 larger ad: $24.50
Total: $79.25 ✓

Each classified ad costs $18.25.

◥ TRY THESE EXERCISES

1. MODELING Use Algeblocks to solve $2x - 5 = 7$. Show each step algebraically.

Solve each equation and check the solution.

2. $3a - 5 = 7$

3. $-4x + 1 = 25$

4. $52 = 4(2j + 5)$

5. $4u - 5 = 2u - 13$

6. $-6b + 9 = 4b - 41$

7. $2n + 14 = -8$

8. YOU MAKE THE CALL Maggie says that multiplying by the reciprocal of a number is the same as dividing by the number. Is Maggie correct?

9. RECREATION A carnival pass costs $15, and buys unlimited access to 10 rides. This pass costs $2.50 less than paying the individual price for each of the 10 rides. What is the individual price of each ride?

10. WRITING MATH Write a multi-step equation that has -4 as a solution.

◥ PRACTICE EXERCISES • For Extra Practice, see page 667.

Solve each equation and check the solution.

11. $4n + 3 = 15$

12. $-2d - 16 = 4$

13. $-28 = 3r - 7$

14. $-14 = 18 - 8e$

15. $2(5z - 3) = 34$

16. $-3(2h - 1) = 3$

17. $-5p - 1 = 3p + 15$

18. $4 - 7a = -1 - 2a$

19. $6v + 3 - 2v = 1 + 5v$

20. $9 - 4c + 15 = 0$

21. $\left(\frac{1}{2}\right)(12f + 30) = 9$

22. $8(1.25 - q) = 6$

Translate each sentence into an equation. Then solve.

23. Four more than 3 times a number is 31. Find the number.

24. When 12 is decreased by twice a number, the result is -14. Find the number.

Solve each equation and check the solution.

25. $-3(d - 5) = 2(4d - 9)$

26. $\left(\dfrac{1}{3}\right)(15z - 21) = \left(\dfrac{2}{5}\right)(10z - 35)$

27. $4(5 - 3m) - 9 = 3m - 4$

28. $-2(4k + 1) + k = 8 - 5k$

29. $9(a + 4) - 2a = 19 - 3(a + 6)$

30. $15x - 4(4 + 3x) = -5(2x - 5) + 11$

Translate each sentence into an equation. Then solve.

31. Fifteen more than twice a number is the same as 7 less than four times the number. Find the number.

32. When the sum of twice a number and 3 is multiplied by 5, the result is the same as decreasing the product of 6 and the number by 1. Find the number.

33. FINANCE The stereo system Doug wants to buy can be purchased by paying a $50.00 down payment, and paying the rest in equal monthly installments over the next 6 months. If the total cost of the stereo system is $228.50, what will be the amount of each monthly payment?

34. CHAPTER INVESTIGATION Think about how you will use the media to convince the public to support your message. Write a public relations plan and a budget. Estimate the cost of any advertisements or commercials you will need to run in local newspapers or on television.

▰ EXTENDED PRACTICE EXERCISES

35. Solve $2(3x + 2) + x = 3x + 4 + 4x$. Explain your solution.

36. Solve $5 + 4(2x - 1) = 3(x + 1) + 5x$. Explain your solution.

37. WRITING MATH Summarize, in writing, the steps used to solve equations.

▰ MIXED REVIEW EXERCISES

Evaluate each expression when $a = 2$ and $b = -3$. (Lesson 1-7)

38. ab^2

39. a^3b^2

40. $a^3 - b^3$

41. $a^2 + b^2$

42. $(a^2 - b)^2$

43. $-4ab^3$

44. $(a^2)(b^2)$

45. $(a^3 - 5)^3$

46. $-(b^2)(a^3)$

47. $-4a^3b$

48. $(b + 8)(a^2)$

49. $(a^3 + 2)^2$

Write each number in scientific notation. (Lesson 1-8)

50. 8,640,000,000,000

51. 0.000000045

52. 0.0000017

53. 0.000000000039

54. 128,000,000,000,000

55. 0.0000000026

56. DATA FILE Use the data on foreign trade on page 648. Write the dollar amount of United States imports to Mexico in scientific notation. (Prerequisite Skill)

Solve Linear Inequalities

Goals
- Solve an inequality in one or two variables.
- Graph the solution to an inequality in one or two variables.

Applications   News media, Sales

Consider the graph of the equation $x + y = 5$. The points that lie on this line have coordinates whose sum is 5. For example, $(0, 5)$, $(1, 4)$, and $(-2, 7)$ are points that lie on the line $x + y = 5$.

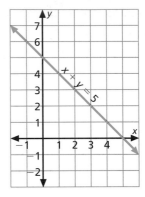

a. Are there any points not on the line that have coordinates whose sum is 5?

b. Select any three points below the line and find the sum of the coordinates. How do these sums compare with 5?

c. Select any three points above the line and find the sum of the coordinates. How do these sums compare with 5?

BUILD UNDERSTANDING

A mathematical sentence that contains one of the symbols $<$, $>$, $\leq$, or $\geq$ is an inequality. Inequalities are used to indicate the *order* of a comparison between two quantities.

A linear inequality in *one variable* is an inequality only in x or y. The techniques used to solve an inequality are similar to those used to solve equations. The addition property of inequality states that adding the same number to both sides of an inequality maintains the order of the inequality.

Addition Property of Inequality	For all real numbers a, b, and c: If $a < b$, then $a + c < b + c$. If $a > b$, then $a + c > b + c$.

The multiplication property of inequality states that multiplying both sides of an inequality by the same positive number still maintains the order of the inequality. However, if the number you are multiplying by is a negative number, you must reverse the order of the inequality.

Multiplication Property of Inequality	For all real numbers a, b, and c: $c > 0$: If $a < b$ then $ac < bc$. $\quad\quad\,$ If $a > b$ then $ac > bc$. $c < 0$: If $a < b$ then $ac > bc$. $\quad\quad\,$ If $a > b$ then $ac < bc$.

The transitive property of inequality relates two inequalities to produce a third.

Transitive Property of Inequality	For all real numbers a, b, and c: If $a < b$ and $b < c$, then $a < c$. If $a > b$ and $b > c$, then $a > c$.

For example, if $x + 3 < y$ and $y < 7$, then $x + 3 < 7$. This inequality can then be solved for x.

Example 1

Solve each inequality and graph the solutions on a number line.

a. $3x + 10 < 4$

b. $23 \geq 8 - 5y$

Solution

a.
$$3x + 10 < 4$$
$$3x + 10 + (-10) < 4 + (-10)$$
$$3x < -6$$
$$\left(\frac{1}{3}\right)3x < \left(\frac{1}{3}\right)(-6)$$
$$x < -2$$

The open circle indicates that -2 is not a solution.

b.
$$23 \geq 8 - 5y$$
$$23 + (-8) \geq 8 + (-8) - 5y$$
$$15 \geq -5y$$
$$\left(-\frac{1}{5}\right)15 \leq \left(-\frac{1}{5}\right)(-5y)$$
$$-3 \leq y$$
$$y \geq -3$$

The closed circle indicates that -3 is a solution.

A solution of a linear inequality in two variables, such as $2x + y < 4$, is an ordered pair that makes the inequality true. The graph of all such solutions is a region called a **half-plane**.

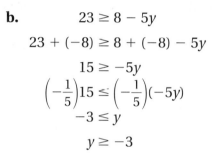

The edge of the half-plane is called the **boundary**. If the inequality is a strict inequality ($<$ or $>$), then the region is an **open half-plane**, and the boundary is not part of the solution set. If the inequality is inclusive ($\leq$ or $\geq$), then the region is a **closed half-plane**, and the boundary is part of the solution set.

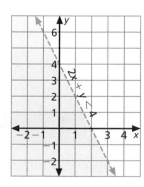

To graph an inequality in two variables, first graph the related equation. This line will serve as the boundary. If the solution will be a closed half-plane, draw the boundary as a solid line. Otherwise, draw it with a dashed line. Then shade the half-plane that contains the solutions of the inequality.

Example 2

Graph $y \leq 4x$.

x	y
−1	−4
0	0
1	4

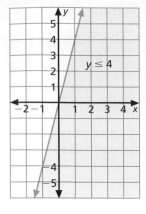

Solution

The related equation is $y = 4x$. Make a table of values that can be used to graph the boundary.

Note that the boundary is part of the solution set, and is drawn as a solid line. To decide which half-plane to shade, use a test-point not on the boundary. If it *is* a solution, then all points of that half-plane will also be solutions; so, shade that side. If the point is not a solution, shade the half-plane that does not contain the test point.

Test Point: $(−1, 1)$ $y \leq 4x$

$$1 \leq 4(−1)$$

$$1 \leq −4 \text{ (false)}$$

Because 1 *is not* less than or equal to −4, shade the half-plane that does not contain $(−1, 1)$.

Example 3

Graph $y > \dfrac{3}{2}x - 4$.

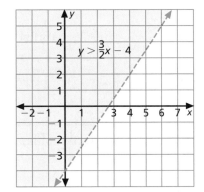

Solution

The related equation is $y = \dfrac{3}{2}x - 4$.

Make a table of values that can be used to graph the boundary. Note that the boundary is not included in the solution set, and is drawn as a dashed line.

$$y = \frac{3}{2}x - 4$$
Test Point: $(0, 0)$ $y > \dfrac{3}{2}x - 4$

$$(0) > \frac{3}{2}(0) - 4$$

$$0 > −4$$

x	y
0	−4
2	−1
4	2

Because 0 *is* greater than −4, shade the half-plane that contains $(0, 0)$.

◤ TRY THESE EXERCISES

Solve each inequality and graph the solution on a number line.

1. $4m + 5 > 25$

2. $−2k + 9 \geq 1$

3. $\left(\dfrac{3}{4}\right)c - 4 > −16$

Graph each inequality on the coordinate plane.

4. $y \leq 3x - 1$

5. $y \geq −x + 2$

6. $2x - y < 3$

7. $x + 2y < 10$

8. $4x + 3y \geq −6$

9. $2x - 5y \geq 15$

Solve each inequality and graph the solution on a number line.

10. $5b + 4 \le -11$

11. $\left(\dfrac{1}{2}\right)p - 10 > -7$

12. $9 - 4r > 5$

13. $13 < 3a - 8$

14. $26 \le -9n - 1$

15. $31 > 14 - 15z$

16. $\left(\dfrac{3}{2}\right)h + 12 \ge 6$

17. $3(4e + 3) < -9$

18. $8k - 7 > 6k - 9$

Graph each inequality on the coordinate plane.

19. $y > 2x - 5$

20. $y \le -3x + 4$

21. $x + y \ge -3$

22. $x - y > 2$

23. $2x - 3y < 6$

24. $6 > 2x - \left(\dfrac{2}{3}\right)y$

25. NEWS MEDIA A reporter estimates that $\dfrac{2}{3}$ of the hours (h) spent on a story increased by 15 h is less than 27 h. What values are possible for h?

26. SALES A jacket sells for \$55. Decreasing the price of the jacket by a discount amount (d) yields a result greater than one-half the sum of the discount and \$10. What are the possible values for the discount?

Solve each inequality and graph the solution on a number line.

27. $0.5c - 7.4 \ge 0.35 + 1.75c$

28. $6 - (5m + 7) < 3(2m + 1) - 10m$

29. $4(3d + 1) - 5d \le 8 - 2(5d + 2)$

30. $-10 \le \left(\dfrac{2}{5}\right)(10 - 5q)$

Write the inequality represented by each graph.

31.
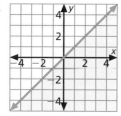

32.
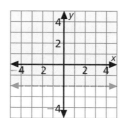

33.
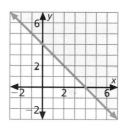

■ EXTENDED PRACTICE EXERCISES

34. Graph the solution to the inequality $|x| > 2$ on a number line.

35. WRITING MATH Write a paragraph in which you discuss two different ways to interpret and graph the inequality $y < 5$. How are these two interpretations and their graphs related?

36. Name three points that are solutions of both the inequality $x + y < 1$ and the equation $-3x + 2y = 10$.

■ MIXED REVIEW EXERCISES

Identify the pattern for each sequence. Determine the next three terms in each sequence. (Lesson 2-1)

37. $1, 4, 16, 64, 256, \ldots$

38. $100, 50, 25, 12.5, 6.25, \ldots$

39. $1, 4, 7, 10, 13, \ldots$

40. $200, 193, 186, 179, 172, \ldots$

41. $1, -3, 9, -27, 81, \ldots$

42. $50,000, -10,000, 2000, -400, 80, \ldots$

Review and Practice Your Skills

Solve each multi-step equation.

1. $4r - 1 = 35$
2. $5g + 1 = -29$
3. $-4q - 5 = 7$
4. $2(x - 3) = 14$
5. $6x - 13 = -13$
6. $\frac{1}{2}x + 5 = 16$
7. $0.4x - 3.8 = 4.2$
8. $-7(m - 3) = 2(4m + 3)$
9. $0.2(1.8 + z) = 0.3z$
10. $10b - 6b - 3 = 9$
11. $\frac{z}{3} + 70 = 98$
12. $12 - 3m = -15$
13. $7 - x = 23$
14. $-4x + 23 = 75$
15. $15(2 + x) - 3x = 114$
16. $14 = 8r - 58$
17. $\frac{2}{5}x - 7 = 11$
18. $7(x - 2) = -14$
19. $12 - 3(4 - 7x) = 9(3x + 2) + x$
20. $\frac{1}{3}(18y - 6) = -\frac{5}{6}(12 - 6y)$
21. $-m + 3(m + 1) = 11$
22. $-8 - 2w = 11$
23. $112 = 12 + 8y$
24. $\frac{4}{7}x = -(21 + 14)$

Translate each sentence into an equation. Then solve.

25. Six more than 5 times a number is -29. Find the number.
26. When 47 is decreased by twice a number, the result is -75. Find the number.
27. The sum of one-third of a number and $\frac{1}{2}$ is $3\frac{1}{2}$. Find the number.
28. 7 less than twice a number is 14. Find the number.
29. Three times the sum of a number and 2 is -27. Find the number.

Solve each inequality and graph the solutions on a number line.

30. $3 + x < 2$
31. $-g - 9 > 3$
32. $2x - 0.3 \leq 0.5$
33. $n + 5 < 2$
34. $-4x + 6 \geq 17$
35. $5d - 8 < -8$
36. $\frac{r}{4} \leq -2$
37. $c + \frac{2}{3} > 1\frac{1}{3}$
38. $2 - (3 - s) \leq 4$
39. $5 > \frac{1}{3}k + 14$
40. $-6(k + 2) > 48$
41. $72 \geq -3h + 4 - 5h$
42. $5(7 + r) \geq 12r$
43. $2x - 5(x + 3) \leq -20$
44. $\frac{5}{8}e - 3 - \frac{3}{8}e > -5$
45. $m - 19 > -15$
46. $\frac{1}{3}x - 2 \geq 11$
47. $-3x - 2 \geq 4.5$

Graph each inequality on the coordinate plane.

48. $y \leq x - 2$
49. $y > \frac{4}{5}x - 4$
50. $y < -2x$
51. $y \geq 5 - 3x$
52. $5x + 10y < -30$
53. $x - y > 4$
54. $y \geq -3$
55. $x < 8$
56. $y \geq -3x + 2$
57. $y < x$
58. $-2y \leq x$
59. $4x + 3y \leq 12$
60. $-5x + 4y > 20$
61. $0.2x - 0.8y \leq 3.2$
62. $-\frac{1}{4}x - \frac{1}{2}y > 2\frac{1}{2}$

Find the next three terms in each sequence. (Lesson 2-1)

63. 1.2, 1.5, 1.8, 2.1, _____, _____, _____

64. 1.2, 2.4, 4.8, 9.6, _____, _____, _____

65. 1.2, −1.2, −3.6, −6.0, _____, _____, _____

66. 1.2, −6, 30, −150, _____, _____, _____

67. 1.2, 1.21, 1.212, 1.2121, _____, _____, _____

68. 1.2, 1.44, 1.728, 2.0736, _____, _____, _____

Write each as a set of ordered pairs. Graph on a coordinate plane. List the domain and range. Determine if each is a function. (Lessons 2-2 and 2-3)

69.

x	−3	−4	−5	−6
y	1	6	1	6

70.

x	10	13	10	7
y	−5	−6	−7	−8

Solve each equation. (Lesson 2-4)

71. $x - 3 = 0$

72. $9 = (-4) + f$

73. $c + \dfrac{3}{5} = -1\dfrac{4}{5}$

74. $-3.2d = 48$

75. $\dfrac{w}{7} = -14$

76. $\dfrac{2}{3}x = 24$

Solve each equation. (Lesson 2-5)

77. $-8 - x = 14$

78. $-4x + 5 = 33$

79. $\dfrac{t}{-3} - 9 = -1$

MathWorks Career – Transcriptionist
Workplace Knowhow

TV newscasters can't possibly remember everything they need to say in a newscast. Instead, they read from a device known as a teleprompter. A teleprompter is a television screen that scrolls slowly through the script for the show. Transcriptionists make sure the copy is typed accurately and on time for the program.

1. You hire a transcription assistant at the rate of $4/page of typed copy. You also pay her a base salary of $25 per day. Her total earnings is represented by $e = 4p + 25$ when e is the total earnings and p is the number of pages. If you can afford to pay her up to $150 for one day, how many pages of copy can you ask her to type?

2. Samatha has six hours to get four tasks done. She spends 45 min talking to the producer, 1 h 45 min talking to a repair technician, and 2 h 15 min proofreading the copy of a speech. She still has to type 1,800 words of copy for the evening newscast. How many words per minute must she type?

3. Victor knows that he can type 30 words/min. He has 60 pages of copy to type, and there are about 100 words/page. How long will it take him to finish the job?

2-7 Data and Measures of Central Tendency

Goals
- Construct frequency tables for data.
- Determine mean, median, and mode for a set of data.

Applications Sports, Education, Marketing, News media

Consumers are constantly bombarded with facts and figures from advertisers who use statistics to entice people to buy their products. Newspaper and television advertisements are full of these statistics. For example, a television commercial makes the following claim:

"In a national taste-test, 7 out of 10 teenagers preferred our brand of cola to our competitor's."

a. How is this advertiser trying to influence consumers?

b. Is it possible that this statistic is not truly representative of the nation's preference? Explain.

◣ BUILD UNDERSTANDING

Statistics is a branch of mathematics that involves the study of **data**. Statisticians study methods of collecting, organizing, and interpreting data. The purpose of a statistical study is to reach a conclusion or make a decision about an entire group called a **population**. Often, it is not possible to survey or poll an entire population. In these cases, a **sample**, or representative part, of the population is used.

Once the sample is selected and the data are collected, the data must be organized so it can be analyzed. One common way to organize data is a **frequency table** or tally system. When data sets contain a wide range of items, it is sometimes useful to group the data into intervals.

Example 1

SPORTS In preparing a sports report for the newspaper, Juan recorded the batting averages of 2 baseball players systematically sampled from each of the ten teams in the league. Construct a frequency table for this data.

.243	.281	.255	.296	.278	.248	.267
.303	.254	.292	.304	.269	.253	.241
.249	.281	.277	.295	.244	.294	.266
.251	.270	.268	.261	.302	.276	.265

Solution

The lowest batting average is .241, and the highest is .304. Group the data into intervals. Then mark a tally for each data item in the appropriate interval, and record the total for each interval.

Batting Average	Tally	Frequency
.240−.249	⊪⊩	5
.250−.259	⊪⊩⊩	4
.260−.269	⊪⊩	6
.270−.279	⊪⊩⊩	4
.280−.289	⊪⊩	2
.290−.299	⊪⊩⊩	4
.300−.309	⊪⊩	3

Think About It

There are four common methods of sampling.

Random sampling: each member of the population has an equal chance of being selected.

Cluster sampling: members of the population are randomly selected from particular parts of the population and surveyed in groups.

Convenience sampling: members of a population are selected because they are readily available.

Systematic sampling: members of a population that has been ordered in some way are selected according to a pattern.

Once data has been organized, it can then be analyzed statistically. Three **measures of central tendency** that can be calculated are the mean, median, and mode.

The **mean**, or arithmetic average, is the sum of the data divided by the number of data. The mean is the most representative measure of central tendency for data sets that do not contain extreme values.

The **median** is the middle value of the data when arranged in numerical order. If the number of data items is even, the median is the average of the two middle numbers. The median is the most representative measure of central tendency for data sets that contain extreme values.

The **mode** is the number (or numbers) that occurs most often in the set of data. A set of data may contain one mode, more than one mode, or no mode. The mode is used to describe the most characteristic value of a set of data.

Example 2

TEST TAKING The SAT mathematics scores for 8 high school students are listed below.

539   541   576   505   548   576   565   558

a. Find the mean of the data.

b. Find the median of the data.

c. Find the mode of the data.

d. Which measure of central tendency is the best indicator of the typical SAT mathematics score for these students?

Solution

a. To find the mean, add the data and divide by the number of data.

$$\frac{539 + 541 + 576 + 505 + 548 + 576 + 565 + 558}{8}$$

$$= \frac{4408}{8} = 551$$

The mean is 551.

b. To find the median, first rewrite the data in numerical order.

505   539   541   548   558   565   576   576

Because there is an even number of data, the median is the average of the two middle numbers.

$$\frac{548 + 558}{2} = \frac{1106}{2} = 553$$

The median is 553.

c. The mode is the number that occurs most often. So the mode is 576.

d. The best indicator of the typical SAT mathematics score for the students is the median, 553, which is not affected by the extreme value (505).

Example 3

CALCULATOR A photographer sold photos to a magazine for the following: $150, $225, $175, $350, $635, $120, and $550. Find the mean and median of the amounts.

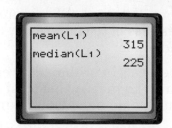

mean(L₁)	
	315
median(L₁)	
	225

Solution

Use the list feature to create and store a new list (L1). After entering the data, choose MATH from the LIST menu to find the mean and median of the new list.

The mean of the data is $315 and the median is $225.

◥ TRY THESE EXERCISES

EDUCATION A random sample of 20 student records was used to determine the average number of absences per student during the school year. The number of absences on each record is listed below.

3	2	1	4	3	1	2	1	0	2
2	3	5	1	4	8	9	0	4	1

1. Construct a frequency table for these data.

2. Find the mean, median, and mode of the data.

3. Which measure of central tendency is the best indicator of the average number of absences per student for this school year?

NEWS MEDIA The manager of the school newspaper researched and recorded the number of issues of each edition of the newspaper that were sold.

362	398	409	377	421	351	399	358	406	388
379	412	423	361	414	420	409	387	361	425
366	401	392	387	390	371	405	417	399	358

4. Construct a frequency table for these data. Group the data into intervals of 10.

5. Determine the interval that contains the median of the data.

◥ PRACTICE EXERCISES • For Extra Practice, see page 668.

MARKETING Thirty families were randomly sampled and surveyed as to the number of magazines to which they subscribe. The results are listed below.

3	1	0	0	2	3
1	4	5	1	0	2
2	0	1	1	1	4
3	2	1	3	4	4
1	0	2	3	2	1

6. Construct a frequency table for these data.

7. Find the mean, median, and mode of the data.

JOURNALISM Amanda systematically sampled every tenth student on the cafeteria lunch line to record the amount of money spent on lunch that day. The results of her survey are listed below.

$2.95	$3.10	$2.85	$2.95	$3.35	$3.15	$3.15	$2.80
$2.60	$2.85	$3.15	$2.70	$3.25	$3.00	$2.95	$3.20
$2.85	$2.95	$2.90	$3.00	$2.95	$2.65	$3.05	$2.75

8. Construct a frequency table for these data. Group the data into intervals.

9. Which interval contains the median of the data?

SPORTS The heights, in inches, of the members of the Hills High School Boys' Basketball Team are listed below.

| 75 | 74 | 66 | 76 | 71 | 74 |
| 78 | 77 | 67 | 76 | 77 | 74 |

10. Use a calculator to find the mean and median of the data.

11. Find the mode of the data.

12. Which measure of central tendency is the best indicator of the typical height of a member of the basketball team?

WEATHER For a television documentary on desert environments, a meteorologist recorded the highest temperature for each day of June in Death Valley, California. The data are displayed in the frequency table.

13. Find the interval that contains the median.

14. To the nearest percent, on what percent of the days was the recorded temperature at least 100°F?

Temperature	Tally	Frequency
80–89	III	3
90–99	ﷻ III	8
100–109	ﷻ IIII	9
110–119	ﷻ II	7
120–129	III	3

▪ EXTENDED PRACTICE EXERCISES

15. Refer to Exercises 13–14. Do you think the mean of the daily high temperatures for Death Valley is greater than or less than 100°F? Explain.

16. **WRITING MATH** Suppose you were interested in determining the *average* temperature during June (as opposed to the average daily high temperature). What sampling method would you use, and how would you collect the data?

Death Valley, California

▪ MIXED REVIEW EXERCISES

Evaluate each function.

17. $f(x) = 2x - 1; f(3)$

18. $f(x) = \frac{1}{2}x + 3; f(4)$

19. $f(x) = 3x + 5; f(7)$

20. $f(x) = \frac{x}{3} + 4; f(-9)$

21. $f(x) = 2x + 6; f(-3)$

22. $f(x) = x + 4; f(2)$

23. $f(x) = 3x - 2; f(6)$

24. $f(x) = -3x + 2; f(-4)$

25. $f(x) = 5x + 8; f(-2)$

26. Clarks Plumbing and Heating purchased a new computer for $6000. The depreciation rate for this computer is 30%. Use the declining-balance method to find the ending book value after the fourth year. (Lesson 2-2)

Display Data

Goals
- Construct stem-and-leaf plots.
- Construct histograms for a data set.

Applications Health, News, Entertainment

Work in small groups.

From a newspaper or magazine, find a table of information. Study the data presented in the table. Discuss whether a different type of display might have been more effective. If so, sketch your idea.

◥ BUILD UNDERSTANDING

Graphs and plots are often used to present a picture of the data. These types of displays provide visual representations of the distribution of the data. They also display characteristics about the data that are sometimes difficult to identify from charts and tables.

One type of visual data display is the **stem-and-leaf plot**. To construct a stem-and-leaf plot, first divide each piece of data into two parts: a stem and a leaf. The last digit of each number is referred to as its **leaf**; the remaining digits comprise the **stem**. The data is then organized by grouping together data items that have common stems.

Example 1

HEALTH For an article she was preparing for a women's health magazine, Sharon recorded the cholesterol levels of the twenty women on the magazine staff.

185	234	208	197
259	177	192	188
208	200	215	199
209	234	208	146
216	201	232	186

Construct a stem-and-leaf plot to display the data. Interpret the data using your plot.

Cholesterol Levels of Female Staff Members

Stems	Leaves
14	6
15	
16	
17	7
18	5 6 8
19	2 7 9
20	0 1 8 8 8 9
21	5 6
22	
23	2 4 4
24	
25	9

14 | 6 represents a cholesterol level of 146 mg/dL.

Solution

For these data, the digits in the hundreds and tens places form the stem, and the units digit is the leaf. Sort the data according to stems, and arrange the leaves in numerical order.

Be sure to provide a title and a key for your plot.

Because 146 is much less than the other data, and 259 is much greater, these data items are considered to be **outliers**. There is a large data **cluster** for cholesterol levels between 177 and 216, and a smaller cluster for levels in the lower 230s. **Gaps** exist between these two clusters, and between the outliers and the rest of the data. The mode is 208. The median cholesterol level is 204.5 (the average of the tenth and eleventh data pieces in the plot).

A **histogram** is a type of bar graph used to display data. The height of the bars of the graph are used to measure frequency. Histograms are frequently used to display data that have been grouped into equal intervals.

Example 2

The Town Gazette surveyed 40 families, and asked them to record the number of hours per week their television was in use. The results are shown in this frequency table. Construct a histogram to display these data.

Television Use

Number of Hours	Frequency
0–9	1
10–19	6
20–29	15
30–39	12
40–49	2
50–59	4

Solution

Let the horizontal axis represent the number of hours, and the vertical axis represent the frequency. Draw each bar so that its height corresponds to the frequency of the interval it represents.

A spreadsheet program can quickly create a variety of charts and graphs from a set of data.

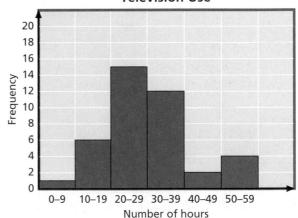

Example 3

SPREADSHEET A class earned the following scores on a science quiz: 89, 88, 72, 66, 89, 90, 94, 78, 95, 82, 84. Make a frequency table and a histogram of the data.

Solution

Create the frequency table on a spreadsheet. Use the intervals 61–70, 71–80, 81–90, and 91–100.

Highlight the cells and select **CHART** from the **INSERT** menu. From the list of types of charts and graphs, choose **column**. Format the width of the bars so there are no gaps. Add titles and your histogram is complete.

	A	B
1	Scores	Frequency
2	61–70	1
3	71–80	2
4	81–90	5
5	91–100	3

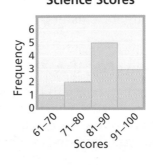

On an aptitude test measuring reasoning ability on a scale of 0 to 100, a class of 30 students received the following scores.

38	75	28	34	56	32	61	28	71	27
62	50	66	40	38	71	60	52	33	59
74	69	86	57	65	16	60	56	55	38

1. Construct a stem-and-leaf plot to display the data.

2. Identify any outliers, clusters, and gaps in the data.

3. Find the mode of the data.

4. Find the median of the data.

5. Find the mean of the data.

6. **NEWS MEDIA** A newspaper report on the price of gasoline contained this frequency table showing the amount of money spent weekly at the gas pump by 25 people surveyed. Construct a histogram to represent these data.

Money Spent on Gasoline

Amount of Money	Frequency
0–14.99	1
15.00–19.99	2
20.00–24.99	10
25.00–29.99	8
30.00–34.99	4

REPORTING For an article he was writing for the school newspaper, Norman surveyed 30 students about the average amount of time (in minutes) each student spent on homework on a weeknight. He recorded the following data.

30	43	58	50	41	98	75	30	72	45
38	75	81	45	17	43	55	52	78	47
31	45	46	55	77	53	58	46	43	35

7. Construct a stem-and-leaf plot to display the data.

8. Identify any outliers, clusters, and gaps in the data.

9. Find the mode of the data.

10. Find the median of the data.

11. Find the mean of the data to the nearest tenth.

12. **WRITING MATH** What conclusions could Norman draw from the data? Write the lead paragraph of his newspaper article.

13. ENTERTAINMENT An entertainment magazine surveyed a sample of its readers about the average number of movies they see in a year. The data are recorded in this frequency table. Construct a histogram to display the data.

Number of Movies Seen in One Year

Number of movies	Frequency
0–4	6
5–9	15
10–14	30
15–19	25
20–24	32
25–29	12

14. DATA FILE Use the data on page 652 on the All-American Girls Professional Baseball League Batting Champions. Make a stem-and-leaf plot of the at-bats for the champion each year.

Refer to the histogram for Exercises 15–18.

15. How many students earn less than $60.00 per week?

16. What percent of the students earn between $60.00 and $99.99 per week?

17. Which interval contains the median amount of earnings?

18. Is it possible to identify the mode of the data? Explain.

■ EXTENDED PRACTICE EXERCISES

19. WRITING MATH Write a paragraph comparing stem-and-leaf plots and histograms. Include in your comparison a discussion of the kinds of data for which each type of display is best suited, and describe the statistical conclusions that can be deduced from analyzing each type of display.

20. CHAPTER INVESTIGATION Develop a survey question to ask your classmates that could provide you with data to support your proposal. Collect the data, construct a frequency table, and draw the related histogram. Write a paragraph analyzing your results.

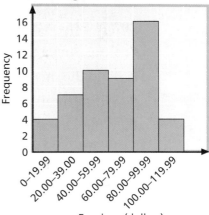

Weekly Earnings of Fifty High School Seniors

■ MIXED REVIEW EXERCISES

Graph each function. (Lesson 2-3)

21. $y = 3x - 2$

22. $y = \frac{1}{2}x + 3$

23. $y = 2x + 1$

24. $y = x - 4$

25. $y = -2x + 1$

26. $y = x - 3$

Solve each equation. (Lesson 2-4)

27. $m - 13 = 28$

28. $6n = -42$

29. $8 + g = -2$

30. $25 = p + 47$

31. $0.9x = 7.2$

32. $-7h = 28$

33. $\frac{a}{4} = 1.2$

34. $\left(\frac{2}{3}\right)c = 12$

35. $0.4w = -6$

Review and Practice Your Skills

Find the mean, median, and mode of each set of data.

1. 98, 77, 89, 93, 75, 81, 77, 88, 78

2. 4237, 4516, 4444, 4379, 4516, 4869

3. 280, 295, 235, 210, 230, 235, 195, 210, 270

4. 3.8, 2.6, 4.1, 4.8, 5.9, 2.7, 6.9, 4.1

5.
Number of students	Grade
3	90
7	80
10	70

6.
Number of students	Height
3	150 cm
5	155 cm
2	160 cm

Althea measures the following volumes of water in milliliters in beakers in a chemistry lab:

210	215	235	208	210	218	218	215	208
230	210	218	218	205	202	206	224	225
207	215	210	228	230	238	234	201	210

7. Construct a frequency table for these data. Group the data into intervals of 10.

8. Find the mean, median, and mode of the data.

9. Which interval contains the median? Which interval contains the mean?

10. Which measure of central tendency best describes the most commonly measured volume of water?

For the data given above for Exercises 7–9 (Althea's chemistry lab):

11. Construct a stem-and-leaf plot to display the data.

12. Identify any outliers, clusters, and gaps in the data.

13. Use your frequency table from Exercise 7 to construct a histogram.

The school newspaper surveyed 50 seniors about the average amount of time (in hours) per week that each student spent talking on the phone with friends.

Time (h)	1–3	4–6	7–9	10–12	13–15	16–18	19–21	22–24
Frequency	2	6	3	12	5	11	8	4

14. Construct a histogram to display the data.

15. How many students talk on the phone less than 13 h/wk?

16. What percent of the students talk on the phone between 7 and 15 h/wk?

17. Which interval contains the median of the data?

18. Is it possible to identify the mean of the data? Explain.

19. If each student increases his or her phone time by 3 h/wk, which measures of central tendency would it affect? How?

Find the next three terms in each sequence. (Lesson 2-1)

20. 4000, 800, 160, 32, _____, _____, _____

21. $-45, -17, 11, 39,$ _____, _____, _____

22. 4, 8, 7, 14, 13, 26, _____, _____, _____

23. A, E, I, M, _____, _____, _____

Graph each point on a coordinate plane. Name the quadrant in which each point is located. (Lesson 2-2)

24. $M(0, 5)$

25. $N(-6, 3)$

26. $P(-4, -1)$

27. $Q(-7, 0)$

28. $R(1.5, -1.5)$

29. $S\left(\frac{1}{2}, 4\right)$

30. $T\left(0, -8\frac{1}{3}\right)$

31. $U(-8, 5)$

Graph each function. (Lesson 2-3)

32. $y = 3x$

33. $f(x) = \frac{1}{2}x + 3$

34. $x + y = -8$

35. $f(x) = -2x + 3$

36. $y = |3x|$

37. $y = 4(x - 2)$

Solve each equation. (Lesson 2-4 and Lesson 2-5)

38. $x + 17 = 31$

39. $23 - a = 0$

40. $b - (-16) = -23$

41. $-4c = 18$

42. $13d = -78$

43. $\frac{x}{5} = -30$

44. $23 - m = -17$

45. $9e - 1 = 23$

46. $2(x - 5) = -24$

47. $\frac{y - 3}{2} = -11$

48. $17 = -41 + p + 16$

49. $3(x - 4) = -6 + 4x$

50. $2.8x - 11.7 = 24.42$

51. $8\frac{1}{3} - 2\frac{2}{3}n = 27$

52. $\frac{z}{3}$ $14 =$ 72

Solve each inequality and graph the solutions on a number line. (Lesson 2-6)

53. $x - \left(-1\frac{1}{4}\right) > 7$

54. $-3h + 38 \leq 56$

55. $\frac{10}{3}g - 2 \geq 18$

Graph each linear inequality on a coordinate plane. (Lesson 2-6)

56. $y < -x + 7$

57. $x - y \leq -7$

58. $24 > 6x + 4y$

Find the mean, median, and mode of each set of data. (Lesson 2-7)

59. 15, 25, 30, 22, 45, 35, 38, 22, 37, 20, 31

60. $\frac{1}{2}, \frac{1}{3}, \frac{1}{4}, \frac{2}{3}, \frac{3}{4}, \frac{1}{2}, \frac{1}{8}, \frac{7}{8}, \frac{1}{2}, \frac{-1}{3}, \frac{1}{4}, \frac{4}{3}, \frac{3}{4}, \frac{1}{2}$

The table gives salaries and number of workers at a manufacturer of auto parts. (Lesson 2-8)

Job	President	Group Manager	Line Manager	Machinist	Clerk
Number	1	2	5	25	4
Salary Range	$81–95,000	$66–80,000	$51–65,000	$35–50,000	$20–34,000

61. In a labor dispute, which measures of central tendency might the president use to show that the worker's wages were already high enough? Why?

62. What percent of the workers earn $51,000 per year or more?

Problem Solving Skills: Misleading Graphs

When examining or reading statistics, think critically. Although informative and useful, statistics can be misleading. Graphs can mislead when scales or dimensions are changed. Measures of central tendency can be misleading if they do not accurately represent the data. Advertising claims can be misleading if they are vague, omit information, or hint at something that may not be true.

Problem Solving Strategies

Guess and check

Look for a pattern

Solve a simpler problem

✔ Make a table, chart or list

Use a picture, diagram or model

Act it out

Work backwards

Eliminate possibilities

Use an equation or formula

PROBLEM

A sales representative for a new soft drink is trying to convince a chain of food stores to order a much larger quantity. The salesperson uses the graph at the right in the sales pitch.

1. What can you say about sales of the drink?

2. What is deceptive about the graph?

3. How would you change the graph so that it is not misleading?

Solve the Problem

1. Read the vertical scale. Sales are increasing. Sales in March are 1.5 times as great as the sales in January.

2. Dimensions other than height have been changed. Sales for February and March appear to be greater than they actually are.

3. Diameters should be the same, so only the heights are compared.

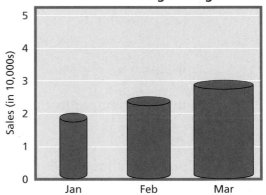

Sales of Mango Mango

TRY THESE EXERCISES

1. **SPORTS** "Just look at the graph," the player's agent said. "Phil's hits have doubled since last year. We are looking for a large raise and a long contract."

   Look at the graph. Is the agent's claim misleading? Why?

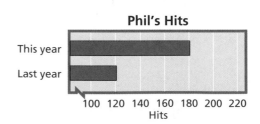

Phil's Hits

2. **ADVERTISING** Tell why you think each of the following advertisements is misleading.

   a. Eighty percent of all dentists surveyed agree: Zilch toothpaste tastes best.

   b. In the past 5 years, 25,000 cola drinkers have switched to Koala Kola.

   c. Thousands of teenagers wake in the morning to a glass of Zest. It has 20% real fruit juice and 100% bounce.

PRACTICE EXERCISE

Five-step
Plan

1 Read
2 Plan
3 Solve
4 Answer
5 Check

These data show sales of the *Earn A Million-A-Day At Home* video. Read the table, and then examine the graphs below.

Week	1	2	3	4	5
Sales	252	246	265	276	280

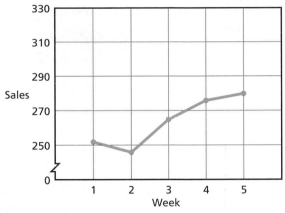

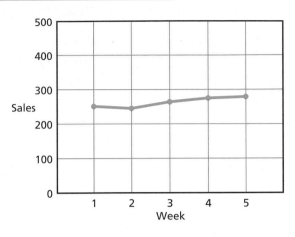

3. WRITING MATH The graphs show the same data. Why do they have different appearances?

4. Suppose you were a sales representative for the video and wanted to convince stores to stock more copies. Which graph would you use? Why?

5. BUSINESS Employee annual wages at a plant rose steadily, but very gradually, from one year to the next during one 5-year period. Make two graphs to show these changes, one from the perspective of the factory owner who wants to show that workers' wages are rising rapidly, and one from the perspective of an employee representative who wants to show that wages are rising minimally.

6. WRITING MATH Create your own advertisement that contains misleading statistics, or find one in a newspaper or magazine. See if a classmate can tell what is misleading about the ad.

■ MIXED REVIEW EXERCISES

Solve each equation. (Lesson 2-5)

7. $4b - 3 = 17$

8. $2(m + 5) = -9$

9. $-4d + 6 = 2d - 8$

10. $\left(\frac{1}{2}\right)(8x - 6) = -7$

11. $-12 + 3a + 14 = 4$

12. $(0.75)(8w - 12) = 9$

13. $-5k - 3 = -38$

14. $3(2m - 4) = 2m - 22$

15. $3(3x + 6.5) = -3$

Evaluate $g(x) = |3x - 4|$ for the given values of x. (Lesson 2-3)

16. $g(2)$

17. $g(-4)$

18. $g(3)$

19. $g(-2)$

20. $g(-1)$

21. $g(7)$

22. $g(0.5)$

23. $g(1.3)$

Chapter 2 Review

VOCABULARY ◣

Choose the word from the list at the right that completes each statement.

1. A type of bar graph used to display data is called a(n) __?__.

2. A(n) __?__ is a relation in which each element of one set is paired with *exactly one* element of another set.

3. The middle value in a set of data that are arranged in numerical order is called the __?__.

4. To calculate the __?__, divide the sum of the data by the number of data items.

5. An arrangement of terms in a particular order is called a(n) __?__.

6. Perpendicular lines divide a coordinate plane into __?__.

7. Values that are much less or much greater than the other data are called __?__.

8. A representative part of a population is called a(n) __?__.

9. The set of all input values in a relation is called the __?__.

10. The point where the *x*-axis crosses the *y*-axis is called the __?__.

a. domain
b. function
c. histogram
d. mean
e. median
f. mode
g. origin
h. outliers
i. quadrants
j. range
k. sample
l. sequence

LESSON 2-1 ◣ Patterns and Iterations, p. 52

▶ An arrangement of numbers in a pattern is a **sequence**.

▶ An **iteration** is a process that is repeated over and over again.

Find the next three terms in each sequence. Identify the rule.

11. 64, 16, 4, . . .

12. $1, \frac{3}{2}, \frac{9}{4}, \ldots$

13. 1, −2, 4, −8, . . .

14. The price of a new car is $20,000. If it has a depreciation rate of 18% per year, what is its value in 3 yr?

LESSON 2-2 ◣ The Coordinate Plane, Relations and Functions, p. 56

▶ A set of ordered pairs is defined as a **relation**. The **domain** of a relation is the set of all *x*-coordinates. The **range** of a relation is the set of all *y*-coordinates.

▶ A **function** is a set of ordered pairs in which each element of the domain is paired with exactly one element in the range.

Determine whether each relation is a function. Give the domain and range.

15. {(2, 3), (−1, 4), (0, 2), (1, 2)}

16. {(2, −1), (1, 0), (0, 1)}

Given $f(x) = 5x - 9$, find each value.

17. $f(1)$

18. $f(0)$

19. $f(-1)$

20. $f(-2)$

LESSON 2-3 ◾ Linear Functions, p. 62

▶ An equation that can be written in the form $Ax + By = C$ where A and B are not both zero is called a **linear equation**. A linear equation whose graph is not a vertical line represents a **linear function**. Graphs of such equations are *straight* lines.

An **absolute value** function is $g(x) = |x| = \begin{cases} x \text{ if } x \geq 0 \\ -x \text{ if } x < 0 \end{cases}$

Graph each function.

21. $y = 2x - 3$ **22.** $y = -x + 2$ **23.** $y = 2x - 7$

Evaluate $f(x) = |1 - 3x|$ for the given values of x.

24. $f(4)$ **25.** $f(-2)$ **26.** $f(3)$

LESSON 2-4 ◾ Solve One-Step Equations, p. 66

▶ You can add (subtract) the same number to (from) each side of an equation and/or multiply (divide) each side by the same number. Remember to perform the same operations on each side.

Solve each equation.

27. $a + 17 = 43$ **28.** $-15 = t - 55$ **29.** $1.7p = -1.87$

30. $\left(\frac{3}{5}\right)w = 33$ **31.** $\frac{2}{5} = \frac{3}{4} + t$ **32.** $\frac{d}{17} = -5$

LESSON 2-5 ◾ Solve Multi-Step Equations, p. 72

▶ Some equations contain variables on both sides. To solve these equations, use the addition property of equality to move terms with the variable to one side of the equal sign and the constants to the other side. Then, solve the equation using the multiplication property of equality.

Solve each equation and check the solution.

33. $11 + 9y = 119$ **34.** $341 = 71 + 2w$ **35.** $7 = 4 - \frac{n}{3}$

36. $10h = 8h + 6$ **37.** $6.2s + 7 = 3s - 1$ **38.** $2.4k + 1 = 7(2k - 4)$

39. $3 + \frac{5}{8}x = 2(x - 4)$ **40.** $4(5 - a) - 2.1(6a) = -4.9$ **41.** $5(2g - 3) + 2(g + 4) = 17$

LESSON 2-6 ◾ Solve Linear Inequalities, p. 76

▶ The **graph of an inequality** is the graph of the set of all ordered pairs that make the inequality true.

Solve each inequality and graph the solution set on a number line.

42. $-3x + 4 \geq 13$ **43.** $5x + 1 > -4$ **44.** $-6x + 2 < 12 - x$

Graph each inequality on the coordinate plane.

45. $y > 2x + 1$ **46.** $y \leq -x + 3$ **47.** $y - x > 2$

LESSON 2-7 ◣ Data and Measures of Central Tendency, p. 82

▶ The **mean** of a set of data is the sum of the items divided by the number of items.

▶ The **median** is the middle value (or the mean of the two middle values) of a set of data arranged in numerical order.

▶ A **mode** is a number that occurs most often in a set of data.

The results of the last math test are:

81 78 90 85 62 59 86 94 93 92 85 82 90 80 86 85

48. Construct a frequency table of interval width 10.

49. Find the mean, median, and mode.

LESSON 2-8 ◣ Display Data, p. 86

▶ Individual items can be displayed in **stem-and-leaf plots**. In these, the digit farthest to the right in a number is the leaf. The other digits make up the stem.

▶ The frequency of data can be displayed in a type of bar graph called a **histogram**.

50. Construct a stem-and-leaf plot to display the data in Exercises 48–49.

51. Identify any outliers, clusters, and gaps in the data.

LESSON 2-9 ◣ Problem Solving Skills: Misleading Graphs, p. 92

▶ Statistics can be helpful when trying to make a decision. However, they can be misleading.

52. A television advertisement says, "Over 100 dentists can't be wrong. XYZ toothpaste is the one you should use for a healthier smile." Tell why this advertisement might be misleading.

The two graphs below show the results of a taste test of Bill's cookies.

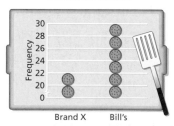

53. Do the graphs represent the same information? Explain.

54. Is one of the graphs misleading? Explain.

CHAPTER INVESTIGATION

 EXTENSION Create an advertisement for your school or community newspaper to promote your proposal. Include your survey question, the histogram of the data you collected, and a paragraph that will convince people to support your proposal.

Chapter 2 Assessment

1. Find the next three terms in the pattern 1, 2, 5, 10, ...

2. Determine whether {(−1, −2), (1, 2), (2, 2)} is a function. Give the domain and range.

3. Given $f(x) = -3x + 5$, find $f(-1)$.

4. Graph $y = 2x - 3$.

5. Evaluate $g(x) = -|2x - 1| + |-x|$ for $g(-1)$.

Solve each equation.

6. $7x - 5 = -4$

7. $\frac{2}{5}x + 1 = 3$

Translate each sentence into an equation. Then solve.

8. When three times a number is increased by 2, the result is 17. Find the number.

9. When three-sevenths of a number is decreased by 1, the result is 5. Find the number.

Solve and graph each inequality.

10. $8x + 5 > 12$

11. $-3x + 8 \geq 11$

Graph each inequality in the coordinate plane.

12. $y \leq 3x - 4$

13. $y > -x + 3$

14. Solve $ax = -a + ba + c$ for a

Teenagers polled about the number of evening meals they ate at home in one week reported the following number of meals.

5	7	1
3	5	4
5	4	6
2	6	4

15. Construct a frequency table for the data.

16. Find the mean of the data.

17. Find the median of the data.

18. Find the mode of the data.

Another group of teenagers polled about the number of evening meals they ate at home in one four-week period reported the following number of meals.

20	35	31
17	3	28
18	28	30
30	26	28

19. Construct a stem-and-leaf plot to display the data.

20. Identify the outliers, clusters, and gaps in the data.

Standardized Test Practice

Record your answers on the answer sheet provided by your teacher or on a sheet of paper.

1. Which of the following does *not* represent an integer? (Lesson 1-2)

   (A) $\sqrt{36}$ (B) $\dfrac{75}{5}$

   (C) -44.5 (D) $-|-4|$

2. If $M = \{1, 2, 3\}$ and $N = \{4, 5\}$, find $M \cap N$. (Lesson 1-3)

   (A) $\{1, 2, 3, 4, 5\}$ (B) $\{1, 2, 3\}$

   (C) $\{4, 5\}$ (D) $\varnothing$

3. Simplify $(a^4)^5$. (Lesson 1-7)

   (A) a^{20} (B) a^9

   (C) a^5 (D) a

4. Evaluate d^{-4} when $d = 3$. (Lesson 1-8)

   (A) -81 (B) -12

   (C) $\dfrac{1}{81}$ (D) 12

5. Which of the following graphs is a function? (Lesson 2-2)

   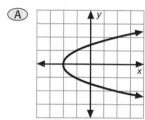
   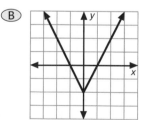

   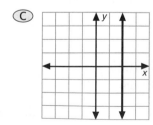
   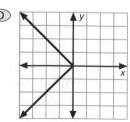

6. Which equation has the greatest solution? (Lesson 2-4)

   (A) $\dfrac{2}{3}x = 8$ (B) $x + 7 = 21$

   (C) $x - 5 = 8$ (D) $5x = 75$

7. Which would be the first step to solve $6 = 4t - 2$? (Lesson 2-5)

   (A) Add 2 to each side.

   (B) Add $-4t$ to each side.

   (C) Multiply each side by $\dfrac{1}{4}$.

   (D) Multiply each side by 4.

8. Which graph represents the solution of $3x - 5 < 1$? (Lesson 2-6)

   (A)
   ```
   ←———+——+——+——+——+——+——+——⊕——+——→
      −4 −3 −2 −1  0  1  2  3  4
   ```

   (B)
   ```
   ←———+——+——+——+——+——⊕━━━━━━→
      −4 −3 −2 −1  0  1  2  3  4
   ```

   (C)
   ```
   ←━━━━━⊕——+——+——+——+——+——+——→
      −4 −3 −2 −1  0  1  2  3  4
   ```

   (D)
   ```
   ←———+——⊕━━━━━━━━━━━━━━━→
      −4 −3 −2 −1  0  1  2  3  4
   ```

9. Salaries of the 12 employees at the XYZ Company are $28,600, $32,000, $29,400, $31,200, $28,600, $38,500, $20,100, $85,000, $36,000, $25,350, $26,500, and $19,850. What is the mean salary? (Lesson 2-7)

   (A) $28,500   (B) $29,050

   (C) $29,500   (D) $33,425

10. The stem-and-leaf plot records Molly's test scores. Which of the following is true? (Lesson 2-8)

Stems	Leaves
7	5 6 8
8	0 2 5 9
9	0 0 3 7

    7|5 represents a test score of 75.

    (A) 75 is an outlier.

    (B) The median equals the mean.

    (C) There is no mode.

    (D) The mode is less than the median.

Test-Taking Tip

Question 10
Always read every answer choice, particularly in questions that ask, "Which of the following is true?"

Preparing for Standardized Tests
For test-taking strategies and more
practice, see pages 709-724.

Part 2 Short Response/Grid In

**Record your answers on the answer sheet
provided by your teacher or on a sheet of paper.**

11. Chapa had $432.28 in her checking account.
She wrote two checks for $25.50 and $43.23.
Then she deposited $50. How much is in her
account now? (Lesson 1-4)

12. The Cabinet Shop made a desktop by gluing a
sheet of oak veneer to a sheet of $\frac{3}{4}$-in.
plywood. The total thickness of the desktop is
$\frac{13}{16}$ in. What is the thickness of the oak veneer?
(Lesson 1-4)

13. What is the area of the rectangle?
(Lesson 1-5)

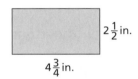

$2\frac{1}{2}$ in.

$4\frac{3}{4}$ in.

14. The United States purchased Alaska for
$7,200,000. Write this dollar amount in
scientific notation. (Lesson 1-8)

15. What is the next term in the sequence?
(Lesson 2-1)

$$192, -96, 48, -24, 12, \dots$$

16. A house costs $200,000. If it increases in value
4% each year, what is its value in 3 yr?
(Lesson 2-1)

17. The function $f(a) = (220 - a) \times 0.8 \div 4$ gives
the target 15-sec heart rate for an athlete
during a workout. In the function, a
represents the athletc's age. Find the target
15-sec heart rate for a 20-yr-old athlete during
a workout. (Lesson 2-2)

18. If $-2m + 3 = m - 12$, what is the value of
$10m$? (Lesson 2-5)

19. A company plans to open a new fitness
center. It conducts a survey of the number of
hours people exercise weekly. The results for
twelve people chosen at random are 7, 2, 4, 8,
3, 0, 3, 1, 0, 5, 3, and 4 h. What is the mode of
the data? (Lesson 2-7)

20. For their science fiction book reports,
students must choose one book from a list of
eight books. The numbers of pages in the
books are 272, 188, 164, 380, 442, 216, 360,
and 262. What is the median number of
pages? (Lesson 2-7)

Part 3 Extended Response

**Record your answers on a sheet of paper. Show
your work.**

21. The table gives prices two different bowling
alleys charge. You plan to rent shoes and play
some games.

Bowling Alley	Shoe rental	Cost per game
Bob's Bowling Alley	$2.50	$4.00
Midtown Bowling Alley	$3.50	$3.75

a. Write an equation to find the number of
games g for which the total cost to bowl at
each alley would be equal. Solve the
equation showing each step. (Lesson 2-5)

b. For how many games will Bob's Bowling
Alley be cheaper? Write an equation to
show the number of games where
Midtown Bowling Alley will be cheaper.
(Lesson 2-6)

22. Construct a stem-and-leaf plot to display the
following data. Then interpret the data.
(Lesson 2-8)

90	87	111
92	105	121
86	94	113
122	85	88
100	97	150
131	101	91

Geometry and Reasoning

THEME: Geography

Have you ever wondered how the early explorers found their way on their journeys? Navigators used their understanding of planetary objects and angle measurements to calculate how many degrees north or south their location was from the equator.

To locate a point on the Earth's surface, navigators lay an imaginary grid on the Earth's surface. The latitude of a point is given as the number of degrees (°) north or south of the equator. The longitude of a point is given as the number of degrees east or west of the prime meridian. For more accurate measurements, degrees are divided into minutes (′) and seconds (″).

- **Cross-country bus drivers** (page 113) read maps, plan routes, and make calculations to conserve fuel and stay on schedule.

- **Cartographers** (page 133) make maps. They must be able to interpret data from satellites and computers and translate actual distance to map distances. They also use a knowledge of geometry to draw three-dimensional surfaces on a flat plane.

Math Online

mathmatters3.com/chapter_theme

Latitude and Longitude of World Cities

City	Latitude	Longitude
Washington, D.C.	38° 54′ 18″ N	77° 00′ 58″ W
Sydney, Australia	33° 55′ 00″ S	151° 17′ 00″ E
Rio de Janeiro, Brazil	22° 27′ 00″ S	42° 43′ 01″ W
Greenwich, England	51° 28′ 00″ N	0° 00′ 00″
Athens, Greece	38° 01′ 36″ N	23° 44′ 00″ E
Honolulu, Hawaii	21° 19′ 02″ N	157° 48′ 15″ W
Johannesburg, South Africa	26° 08′ 00″ S	27° 54′ 00″ E
Beijing, China	39° 55′ 00″ N	116° 23′ 00″ E
Salt Lake City, Utah	40° 46′ 38″ N	111° 55′ 48″ W
Moscow, Russia	55° 45′ 00″ N	37° 37′ 00″ E
Tokyo, Japan	35° 41′ 00″ N	139° 44′ 00″ E
Panama City, Panama	9° 04′ 01″ N	79° 22′ 59″ W

Data Activity: Latitude and Longitude of World Cities

Use the table for Questions 1–6.

1. The prime meridian passes through the poles and Greenwich, England. Name one city that lies east of Greenwich.

2. Which city from the table lies south of the equator and west of the prime meridian?

3. What is the latitude of a point that lies on the equator?

4. Which is farther north: Washington, D.C. or Athens, Greece? How much farther? (*Hint:* There are 60 seconds in a minute and 60 minutes in a degree.)

5. *True or False*: All the "lines" of longitude are circles of the same size, and they intersect in just two points.

6. *True or False*: All the "lines" of latitude are circles of the same size parallel to the equator.

CHAPTER INVESTIGATION

Navigators use grid lines and angles to plot a course from one location to another. By drawing a ray due north from a destination, navigators can determine the bearing, or angle, they need to travel to arrive at the new destination.

Working Together

Draw a map of your neighborhood showing at least ten specific points of interest. Draw and label latitude and longitude lines on the map. Then plot a course from one location to another. Create a navigator's log in which you specify the direction and bearing for each leg of the journey.

3 Are You Ready?

Refresh Your Math Skills for Chapter 3

The skills on these two pages are ones you have already learned. Use the examples to jog your memory and complete the exercises. For additional practice on these and more prerequisite skills, see pages 654–661.

As you look forward to learning more about geometry, it might be helpful to refresh your memory about lines and angles.

IDENTIFYING ANGLES

Examples   Perpendicular lines are two lines that intersect to form four right angles.

Vertical angles are two angles whose sides form two pairs of opposite rays. Vertical angles are congruent.

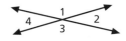

∠1 and ∠3 are vertical angles.

∠2 and ∠4 are vertical angles.

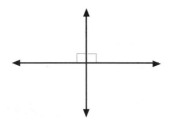

Use the figure at the right to name the following.

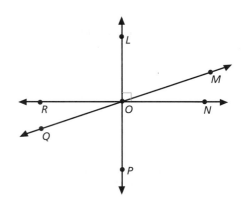

1. all pairs of perpendicular lines

2. all right angles

3. all pairs of vertical angles that are not right angles

Identify each angle as right, acute or obtuse.

4.

5.

6.

7.
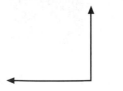

8.

9.
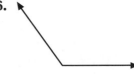

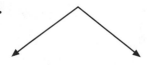
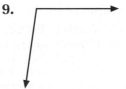

MEASURING AND DRAWING ANGLES

Use a protractor to measure each angle.

10.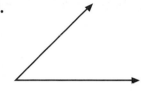

11.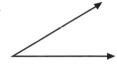

12.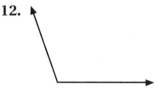

13.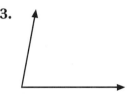

14.

15.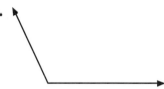

Use a protractor to draw an angle with the given measure.

16. $135°$ **17.** $15°$ **18.** $63°$ **19.** $122°$

LOGICAL REASONING

For each set of premises, write the conclusion that follows by deductive reasoning from the premises. If no conclusion is possible, write *none*.

20. If an animal is a horse, it has a tail.
This animal is a horse.

21. A person who wears white gloves to tea has impeccable manners.
Miss Jones wears white gloves to tea.

22. All parakeets sing with a sweet voice.
This bird has a lousy voice.

23. Flowers that are red attract hummingbirds.
Daffodils are yellow.

24. A red sports car is faster than a green sports car.
My car is blue.

25. Michael is taller than Scotty.
Kareem is shorter than Scotty.

26. Every day at 5 P.M., my dogs expect their evening meal.
It is 3:47 P.M. right now.

3-1 Points, Lines, and Planes

Goals ■ Define basic terms.
          ■ Apply postulates about points, lines, and planes.

Applications Geography and Carpentry

Work in groups of 2 or 3 students.

You will need three small, flat objects that will not roll when dropped.

1. Drop two of the objects onto a flat surface. Determine whether you can place a rod on the surface so that it touches both objects. Repeat the experiment ten times and record the result.

2. Perform the experiment using three objects. Repeat ten times.

3. Compare the results from the two experiments. What generalization can you make from your results?

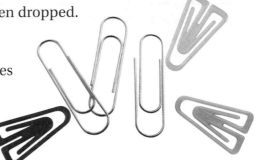

◥ BUILD UNDERSTANDING

In geometry, *point, line,* and *plane* are intuitive ideas. It is impossible to give a precise definition of these words, so they are *undefined terms.*

A **point** indicates a specific location. Although you use a dot to picture a point, it actually has no dimensions.

A **line** is a set of points that extends without end in two opposite directions. When you picture a line, it appears to have some thickness, but actually it has none.

A **plane** is a set of points that extends without end in all directions along a flat surface. Like a line, it has no thickness. Although a plane has no edges, you usually picture a plane as a four-sided shape.

Using these undefined terms as a base, it is now possible to develop *definitions* of other geometric terms. For instance, a figure is defined as any set of points. The set of all points is called **space.**

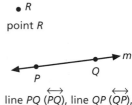

• R
point R

line PQ ($\overleftrightarrow{PQ}$), line QP ($\overleftrightarrow{QP}$),
or line m

plane W

Collinear points are points that lie on the same line. Points that do not lie on the same line are called *noncollinear* points.

Coplanar points are points that lie in the same plane. Points that do not lie in the same plane are called *noncoplanar* points.

R, S, and T are collinear.
R, S, and U are noncollinear.

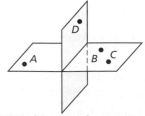

A, B, and C are coplanar.
A, B, C, and D are noncoplanar.

The **intersection** of two geometric figures is the set of all points common to both figures.

Just as the meanings of some words are accepted without definition, **postulates** are accepted as true without proof.

Postulate 1	**The Unique Line Postulate** Through any two points, there is exactly one line. (This is sometimes stated: *Two points determine a line.*)
Postulate 2	**The Unique Plane Postulate** Through any three noncollinear points, there is exactly one plane. (*Three noncollinear points determine a plane.*)
Postulate 3	If two points lie in a plane, then the line joining them lies in that plane.
Postulate 4	If two planes intersect, then their intersection is a line.

Example 1

Which postulate justifies the answer to each question?

a. Name three points that determine plane $\mathcal{J}$.

b. Name the intersection of planes $\mathcal{J}$ and $\mathcal{H}$.

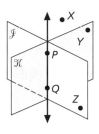

Solution

a. According to Postulate 2, three noncollinear points determine a plane. Three points that determine plane $\mathcal{J}$ are points P, Q, and Z.

b. According to Postulate 4, the intersection of two planes is a line. Planes $\mathcal{J}$ and $\mathcal{H}$ intersect in line PQ ($\overleftrightarrow{PQ}$).

A **ray** is part of a line that begins at one point, called the **endpoint**, and extends indefinitely in one direction.

A **line segment**, more simply called a **segment**, is part of a line that begins at one endpoint and ends at another.

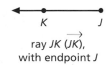

ray JK ($\overrightarrow{JK}$),
with endpoint J

segment FG ($\overline{FG}$), or segment GF ($\overline{GF}$),
with endpoints F and G

The *ruler postulate* is a basic assumption about segments.

Postulate 5	**The Ruler Postulate** The points on any line can be paired with the real numbers in such a way that any point can be paired with 0 and any other point can be paired with 1. The real number paired with each point is the **coordinate** of that point. The **distance** between any two points on the line is equal to the absolute value of the difference of their coordinates.

Example 2

Using the number line, find the length of $\overline{QS}$.

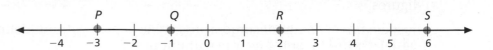

Solution

The coordinate of point Q is -1. The coordinate of point S is 6.

$$|-1 - 6| = |-7| = 7 \quad \text{or} \quad |6 - (-1)| = |7| = 7$$

So, the distance between points Q and S is 7. This means that the length of $\overline{QS}$ is 7, that is, $QS = 7$.

For collinear points A, B, and C, point B is **between** points A and C if and only if the coordinate of B is between the coordinates of A and C. This leads to the *segment addition postulate*.

Postulate 6	**The Segment Addition Postulate** If point B is between points A and C, then $AB + BC = AC$.

Example 3

In the figure at the right, $AC = 47$. Find AB.

Solution

From the figure, $AB = n - 5$ and $BC = n + 8$. You are given $AC = 47$. Use the segment addition postulate to write and solve an equation.

$$AB + BC = AC$$

$n - 5 + n + 8 = 47$ Combine like terms.

$2n + 3 = 47$ Add -3 to each side.

$2n = 44$ Multiply each side by $\frac{1}{2}$.

$n = 22$

The value of n is 22. To find AB, replace n with 22 in $n - 5$.

$$AB = n - 5 = 22 - 5 = 17$$

◥ TRY THESE EXERCISES

1. Refer to the figure at the right. Name a line that lies in plane $\mathcal{G}$. Which postulate justifies your answer?

2. Refer to the number line in Example 2 above. Find PR.

3. In the figure at the right, $JL = 88$. Find KL.

4. **WRITING MATH** The length of $\overline{JL}$ is 88, or $JL = 88$. What is the difference in meaning between the notations $\overline{JL}$ and JL?

Technology Note

Geometry software can help you draw geometric figures quickly and easily. Most software gives you a quick reading of the length of a segment. To learn how to use this feature on your software, try this activity.

1. Draw a segment. Label its endpoints A and B.

2. Find AB, the length of the segment. (The length should appear on the screen.)

3. Locate a point C between A and B.

4. Find AC and CB.

5. Calculate $AC + CB$.

6. Change the location of point C and repeat Steps 4 and 5.

Use the figure at the right for Exercises 5–8.
Which postulate justifies your answer?

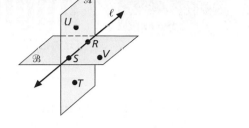

5. Name two points that determine line ℓ.

6. Name three points that determine plane $\mathcal{A}$.

7. Name the intersection of planes $\mathcal{A}$ and $\mathcal{B}$.

8. Name three lines that lie in plane $\mathcal{B}$.

Use the number line at the right for
Exercises 9–12. Find each length.

9. CF 10. GE 11. HD 12. GH

13. In the figure below, $MP = 104$. Find NP.

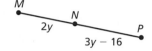

14. In the figure below, $XZ = 61$. Find YX.

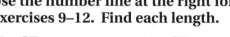

15. On a number line, the coordinate of point S is -8. The length of ST is 17. Give two possible coordinates for point T.

16. On a number line, the distance between points V and W is 39. The coordinate of point W is -3.25. Give two possible coordinates for point V.

17. **GEOGRAPHY** On a map, three cities, represented by points N, P, and Q, lie on a straight line, and N lies between P and Q. The distance from N to P is twice the distance from N to Q. The actual distance between P and Q is 51 mi. Find PN and NQ.

18. Point Z is between points J and K. The length of $\overline{KZ}$ is two less than three times the length of $\overline{JZ}$, and $JK = 18$. Find JZ and KZ.

◤ **EXTENDED PRACTICE EXERCISES**

19. **CARPENTRY** A four-legged table will wobble if its legs are of different lengths. However, a three-legged table will never wobble, even when the legs are of different lengths. Give a reason for this difference.

20. **YOU MAKE THE CALL** Eunsook says that six lines are determined by four noncollinear points. Is she correct?

◤ **MIXED REVIEW EXERCISES**

Graph each function on the coordinate plane. (Lesson 2-3)

21. $y = 3x - 2$ 22. $y - 2x + 1$ 23. $y = 2(x - 2)$ 24. $y = x - 5$

25. $y = 2x - 7$ 26. $y = 3x + 4$ 27. $y = \left(\frac{1}{2}\right)x + 3$ 28. $y = \left(\frac{1}{2}\right)x - 3$

Solve each equation. (Lesson 2-5)

29. $4m - 6 = -14$ 30. $3(d + 1) = 2d - 2$ 31. $-3a + 4 = 16$

32. $-2(4p + 2) = 8$ 33. $-16 = 5x + 4$ 34. $3 - 2n = 5n + 4$

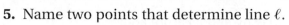

3-2 Types of Angles

Goals ■ Classify and measure angles.

Applications Tiling and Architecture

Draw and label a representation of each figure.

1. $\overline{JK}$

2. $\overrightarrow{JK}$

3. $\overleftrightarrow{JK}$

4. $\overrightarrow{KJ}$

5. $\overrightarrow{VX}$ and $\overrightarrow{VY}$, so that V is between X and Y

6. $\overrightarrow{VX}$ and $\overrightarrow{VY}$, so that V, X, and Y are noncollinear

◥ BUILD UNDERSTANDING

An **angle** is the union of two rays with a common endpoint. The endpoint is the **vertex** of the angle, and each ray is a **side** of the angle.

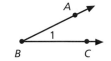

angle ABC (∠ABC),
angle CBA (∠CBA),
angle B (∠B),
or angle 1 (∠1)

The *protractor postulate* is a basic assumption about angles.

Postulate 7	**The Protractor Postulate** Let O be a point on $\overleftrightarrow{AB}$ such that O is between A and B. Consider $\overrightarrow{OA}$, $\overrightarrow{OB}$, and all the rays that can be drawn from O on one side of $\overleftrightarrow{AB}$. These rays can be paired with the real numbers from 0 to 180 in such a way that: 1. $\overrightarrow{OA}$ is paired with 0 and $\overrightarrow{OB}$ is paired with 180. 2. If $\overrightarrow{OP}$ is paired with x and $\overrightarrow{OQ}$ is paired with y, then the number paired with $m\angle POQ$ is $\lvert x - y \rvert$. This number is called the **measure**, or the **degree measure**, of $\angle POQ$.

Example 1

Find the measure of $\angle POQ$ ($m\angle POQ$), in the figure at the right.

Solution

Notice that the protractor has two scales.

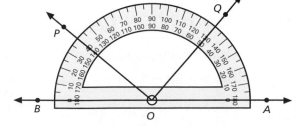

Using the inner scale, $\overrightarrow{OP}$ is paired with 140 and $\overrightarrow{OQ}$ is paired with 50.

$\lvert 140 - 5 \rvert = \lvert 90 \rvert = 90$

Using the outer scale, $\overrightarrow{OP}$ is paired with 40 and $\overrightarrow{OQ}$ is paired with 130.

$\lvert 40 - 130 \rvert = \lvert -90 \rvert = 90$

In either case, the measure of $\angle POQ$ is 90° or $m\angle POQ = 90°$.

GEOMETRY SOFTWARE You can draw an angle and find its measure using Cabri Jr. First, draw an angle. Then select and measure the angle. The measure of the angle will appear on the screen.

From your work in previous courses, recall that angles are classified according to their measures. Using the figure in Example 1, the chart below summarizes the types of angles and gives an example of each.

Type of angle	Measure	Example
acute	greater than 0° and less than 90°	$\angle QOA$
right	equal to 90°	$\angle POQ$
obtuse	greater than 90° and less than 180°	$\angle POA$
straight	equal to 180°	$\angle BOA$

An angle with a measure less than 180° divides a plane into three sets of points: the angle itself, the points in the *interior* of the angle, and the points in the *exterior* of the angle.

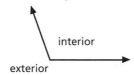

Sometimes, pairs of angles are classified by the *sum* of their measures.

Two angles are **complementary** if the sum of their measures is 90°. Each angle is called a *complement* of the other.

Two angles are **supplementary** if the sum of their measures is 180°. Each angle is called a *supplement* of the other.

$m\angle J + m\angle K = 39° + 51° = 90°$
$\angle J$ and $\angle K$ are complementary.

$m\angle G + m\angle H = 133° + 47° = 180°$
$\angle G$ and $\angle H$ are supplementary.

Another basic assumption about angles is the *angle addition postulate*.

Postulate 8	**The Angle Addition Postulate** If point *B* lies in the interior of $\angle AOC$, then: $m\angle AOB + m\angle BOC = m\angle AOC$ If $\angle AOC$ is a straight angle, and *B* is any point not on $\overleftrightarrow{AC}$, then: $m\angle AOB + m\angle BOC = 180°$

In each of the two figures that illustrate the angle addition postulate, $\angle AOB$ and $\angle BOC$ form a pair of *adjacent angles*. **Adjacent angles** are two angles in the same plane that share a common side and a common vertex, but have no interior points in common. The sides of the two adjacent angles that are not common to the adjacent angles are called **exterior sides**.

When complementary angles are adjacent, the exterior sides form a right angle.

∠RQT is a right angle.

When supplementary angles are adjacent, the exterior sides form a straight angle.

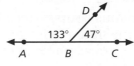

∠ABC is a straight angle.

Example 2

TILING To make a decorative tile border, two tiles must be joined as shown in the figure. The square bracket (⌐) indicates that ∠UYW is a right angle. Find $m\angle UYV$.

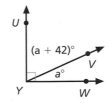

Solution

Since ∠UYW is a right angle, the adjacent angles ∠UYV and ∠VYW are complementary. This means that the sum of their measures is 90°. Use this fact to write and solve an equation.

$$m\angle UYV + m\angle VYW = 90°$$

$a + 42 + a = 90$ Combine like terms.

$2a + 42 = 90$ Add −42 to each side.

$2a = 48$ Multiply each side by $\frac{1}{2}$.

$a = 24$

So, the value of a is 24. From the figure, $m\angle UYV = (a + 42)°$. Substituting 24 for a, $m\angle UYV = (24 + 42)° = 66°$.

◥ TRY THESE EXERCISES

Find the measure of each angle. Then classify each angle as *acute*, *obtuse*, or *right*.

1. ∠HXL   **2.** ∠JXF   **3.** ∠KXG

4. ∠GXJ   **5.** ∠FXL   **6.** ∠HXK

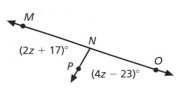

Find the measure of each angle in the figure at the right.

7. ∠MNP   **8.** ∠PNO

9. WRITING MATH In Lesson 3-1, you studied the ruler postulate. How are the ruler postulate and the protractor postulate alike?

10. CHAPTER INVESTIGATION Draw a map of your community showing at least ten specific points of interest. Make distances as accurate as possible. Overlay your map with a grid of latitude (east-west) and longitude (north-south) lines.

Exercises 11–14 refer to the protractor at the right.

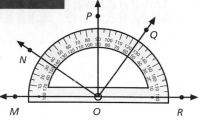

11. Name the straight angle. 12. Name the three right angles.

13. Name all the obtuse angles. Give the measure of each. 14. Name all the acute angles. Give the measure of each.

15. In the figure below, $m\angle FJG$ is $y°$ and $m\angle GJH$ is $4y°$. Find $m\angle GJH$.

16. In the figure below, $m\angle AZC$ is $(5x + 8)°$ and $m\angle CZD$ is $(2x - 17)°$. Find $m\angle BZC$.

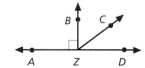

SPATIAL SENSE Without measuring, draw an angle that you think has the given measure. Then use a protractor to determine the actual measure of your angle.

17. 90°   18. 45°   19. 60°   20. 135°   21. 120°

22. The measure of $\angle JKL$ is 34° more than the measure of its complement. Find $m\angle JKL$.

23. The measure of $\angle XYZ$ is 15° less than twice the measure of its supplement. Find $m\angle XYZ$.

24. **WRITING MATH** The measure of an acute angle is represented by $x°$. Write expressions that represent the measures of its complement and its supplement.

25. **ARCHITECTURE** On the plans for a building, the sum of the measure of the complement of an angle and the measure of its supplement is 136°. Find the measure of the angle.

■ **EXTENDED PRACTICE EXERCISES**

Determine whether each statement is *always*, *sometimes*, or *never* true.

26. The supplement of an obtuse angle is an acute angle.

27. The complement of an acute angle is an obtuse angle.

28. Complementary angles are also adjacent angles.

■ **MIXED REVIEW EXERCISES**

Evaluate each expression when $a = -5$ and $b = 4$. (Lesson 1-7)

29. $a^2 + b^2$ 30. $a^3 - b$ 31. $(a + b)^2$ 32. $\dfrac{a^5}{a^2} - b^3$

33. $-2(a^2b^3)$ 34. $3(ab)^2$ 35. $(a^3 - 2) \cdot b$ 36. $\dfrac{b^5}{b} + a^2$

37. $-4(a^2b)^2$ 38. $a - b^3$ 39. $(a^2 - b^3)^2$ 40. $\dfrac{a^4}{a^2} \cdot b^2$

Write each number in scientific notation. (Lesson 1-8)

41. 24,000,000,000,000   42. 0.0000000000301

Review and Practice Your Skills

Use the number line at the right for Exercises 1–4. Find each length.

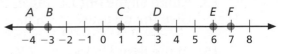

1. AC
2. BE
3. FC
4. DA

5. In the figure below, $QS = 78$. Find QR.

6. In the figure below, $LN = 131$. Find MN.

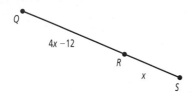

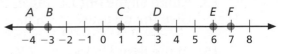

7. On a number line, the coordinate of point G is -5. The length of $\overline{GH}$ is 13. Give two possible coordinates for point H.

8. On a number line, the distance from R to S is twice the distance from S to T. The coordinate of point T is 15 and the coordinate of point S is -3. Give two possible coordinates for point R.

9. *True* or *false*: A plane has four edges.

10. *True* or *false*: Two points may be collinear and also noncoplanar.

11. *True* or *false*: Two planes intersect in a line.

12. *True* or *false*: The distance between two points on a number line can be a negative number.

PRACTICE ◥ LESSON 3-2

Find the measure of a complement of each angle.

13. $78°$
14. $27°$
15. $6°$
16. $51°$
17. $89°$
18. $35°$
19. $42°$
20. $104°$

Find the measure of a supplement of each angle.

21. $83°$
22. $55°$
23. $126°$
24. $3°$
25. $30.5°$
26. $77°$
27. $18°$
28. $180°$

Find the measure of each angle in the given figure.

29.

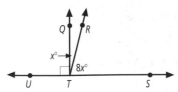

30.

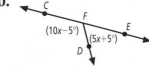

31.

32. The measure of $\angle ABC$ is $28°$ more than the measure of a complementary angle. Find $m\angle ABC$.

33. The measure of $\angle DEF$ is $48°$ less than twice the measure of a supplementary angle. Find $m\angle DEF$.

Use the figure at the right for Exercises 34–37. Which postulate justifies your answer? (Lesson 3-1)

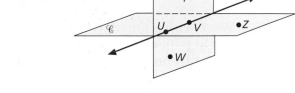

34. Name two points that determine line ℓ.

35. Name three points that determine plane $\mathscr{C}$.

36. Name the intersection of planes $\mathscr{C}$ and $\mathscr{D}$.

37. Name three lines that lie in plane $\mathscr{D}$.

38. In the figure below, $PR = 36$. Find PQ. (Lesson 3-1)

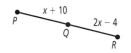

39. In the figure below, $GJ = 354$. Find GH. (Lesson 3-1)

40. The measure of a supplement of $\angle JKL$ is 2.5 times the measure of a complement of $\angle JKL$. Find $m\angle JKL$. (Lesson 3-2)

41. The measure of $\angle GHI$ is 5 times the measure of a complement of $\angle GHI$. Find $m\angle GHI$. (Lesson 3-2)

42. *True* or *false*: The measure of an angle is always less than the measure of a supplement of the angle. (Lesson 3-2)

Math*Works*
Workplace Knowhow

Career – Cross-Country Bus Driver

B us drivers do not only drive school buses and cross-town buses. Some drive great distances to towns and cities all over the country. These bus drivers are responsible for the safety and timely arrival of their passengers to the correct location.

Bus drivers need to be able to read road maps, follow directions, and use a compass. They use math to calculate the quickest route to their destination and to know how far they can drive before they need to refuel.

1. A bus driver drives from New York to Chicago on a route that goes through Pittsburgh and Cleveland. The entire route is 885 mi. The line segment to the right represents the driver's route. Using the segment addition postulate, what is the driving distance between Pittsburgh and Cleveland?

2. A driver needs to drive from point B to point A (see the diagram). There are two alternate routes: route 1 goes directly from B to A, and route 2 passes through point P. Calculate which route will take less time. Use the formula: rate × time = distance.

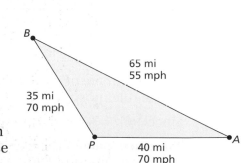

3-3

Segments and Angles

Goals
- Identify bisectors of angles and segments.
- Apply theorems about midpoints, angle bisectors, and vertical angles.

Applications Engineering, Art, Map Making

Work with a partner.

1. On paper, draw any two points R and S. Then draw $\overline{RS}$. Fold the paper so point R falls on top of point S. Unfold. Label the point where the fold intersects $\overline{RS}$ as point Y. With a ruler, find RS, RY, and SY. What do you notice?

2. On a sheet of paper, draw $\angle JKL$. Fold the paper so $\overrightarrow{KJ}$ falls on top of $\overrightarrow{KL}$. Unfold the paper. Choose a point along the fold that falls in the interior of $\angle JKL$, and label it Z. Draw $\overrightarrow{KZ}$. With a protractor, find $m\angle JKL$, $m\angle JKZ$, and $m\angle LKZ$. What do you notice?

◤ BUILD UNDERSTANDING

The **midpoint** of a segment is the point that divides it into two segments of equal length. A **bisector of a segment** is any line, segment, ray, or plane that intersects the segment at its midpoint.

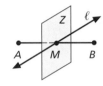

M is the midpoint of $\overline{AB}$.

$AM = MB$

Line ℓ and plane Z
bisect $\overline{AB}$.

In Lessons 3-1 and 3-2, you studied several fundamental postulates of geometry. In this lesson, you will look at other statements that are called *theorems*. Whereas a postulate is an assumption that is accepted as true without proof, a **theorem** is a statement that can be proved true.

The Midpoint Theorem	If point M is the midpoint of $\overline{AB}$, then $AM = \frac{1}{2}AB$ and $MB = \frac{1}{2}AB$.

E x a m p l e   1

Use the figure below. Find the midpoint of $\overline{SY}$.

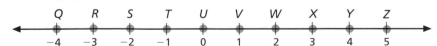

Solution

The coordinate of point S is -2. The coordinate of point Y is 4. By the ruler postulate, $SY = |-2 - 4| = |-6| = 6$.

Since $\frac{1}{2}(6) = 3$, the midpoint of $\overline{SY}$ is 3 units to the right of point S.

The coordinate of this point is $-2 + 3 = 1$.
So, the midpoint of $\overline{SY}$ is point V.

Check Understanding

Explain the difference between the *midpoint* of a segment and a *bisector* of a segment.

The **bisector of an angle** is the ray that divides the angle into two adjacent angles that are equal in measure. This definition leads to the following theorem.

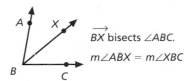

$\overrightarrow{BX}$ bisects $\angle ABC$.

$m\angle ABX = m\angle XBC$

The Angle Bisector Theorem	If $\overrightarrow{BX}$ is the bisector of $\angle ABC$, then: $$m\angle ABX = \tfrac{1}{2}m\angle ABC \text{ and } m\angle XBC = \tfrac{1}{2}m\angle ABC$$

Example 2

ENGINEERING The diagram shown represents a portion of the plans for adding angular supports to a bridge. In the figure, $\angle JKL$ is a straight angle, and $\overrightarrow{KH}$ bisects $\angle GKL$. Find $m\angle HKL$.

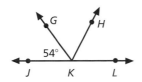

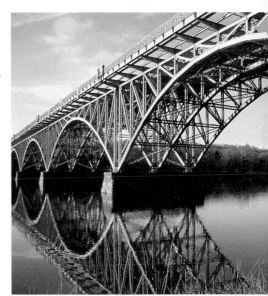

Solution

Since $\angle JKL$ is a straight angle, $m\angle JKL = 180°$.
By the angle addition postulate, $m\angle JKG + m\angle GKL = 180°$.

From the figure, $m\angle JKG = 54°$.
So, $54° + m\angle GKL = 180°$, and $m\angle GKL = 180° - 54° = 126°$.

Since $\overrightarrow{KH}$ bisects $\angle GKL$, $m\angle HKL = \tfrac{1}{2} m\angle GKL = \tfrac{1}{2}(126°) = 63°$.

Two rays $\overrightarrow{BA}$ and $\overrightarrow{BC}$ are called **opposite rays** if point B is on $\overleftrightarrow{AC}$ and is between points A and C.

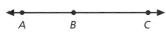

opposite rays $\overrightarrow{BA}$ and $\overrightarrow{BC}$

Two angles whose sides form two pairs of opposite rays are called **vertical angles**. The following theorem is an important statement about vertical angles.

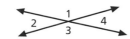

$\angle 1$ and $\angle 3$ are vertical angles.

$\angle 2$ and $\angle 4$ are vertical angles.

The Vertical Angles Theorem	If two angles are vertical angles, then they are equal in measure.

Technology Note

Use geometric software to explore vertical angles. Try this activity.

1. Draw two intersecting lines. Label points A, B, C, D, and E as shown.

2. Find $m\angle AEC$, $m\angle CEB$, $m\angle BED$, and $m\angle DEA$.

3. Change the position of each line and repeat Step 2.

Example 3

In the figure at the right, $\overleftrightarrow{AC}$ and $\overleftrightarrow{DB}$ intersect at point X. Find $m\angle DXA$.

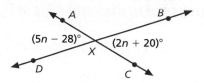

Solution

Since $\angle DXA$ and $\angle BXC$ are a pair of vertical angles, they are equal in measure. Use this fact to write and solve an equation.

$$m\angle DXA = m\angle BXC$$

$5n - 28 = 2n + 20$ Add $-2n$ to each side.

$3n - 28 = 20$ Add 28 to each side.

$3n = 48$ Multiply each side by $\frac{1}{3}$.

$n = 16$

So, the value of n is 16. From the figure, $m\angle DXA = (5n - 28)°$.
Substituting 16 for n, $m\angle DXA = (5 \cdot 16 - 28)° = (80 - 28)° = 52°$.

Check Understanding

In Example 3, how can you use the expression $2n + 20$ to check that the answer is correct?

◣ TRY THESE EXERCISES

Refer to the figure below. Find the midpoint of each segment.

1. $\overline{FM}$ **2.** $\overline{JN}$ **3.** $\overline{HM}$

4. In the figure below, $\overrightarrow{RT}$ bisects $\angle SRU$. Find $m\angle URV$.

5. In the figure below, find $m\angle QZW$ and $m\angle QZY$.

◣ PRACTICE EXERCISES • For Extra Practice, see page 670.

Exercises 6–9 refer to the figure below.

6. Name the midpoint of $\overline{PT}$.

7. Name the segment whose midpoint is point W.

8. Name all the segments whose midpoint is point T.

9. Assume that point Y is the midpoint of $\overline{PV}$. What is its coordinate?

ART A company logo has five line segments that seem to radiate from a single point. An artist is recreating the logo on a computer using the figure on the right. In the figure, $\overleftrightarrow{AD}$ and $\overleftrightarrow{BE}$ intersect at point F, and $\overrightarrow{FC}$ bisects $\angle BFD$. Find the measure of each angle.

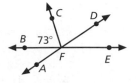

10. $\angle BFD$ **11.** $\angle AFE$ **12.** $\angle BFA$

In the figure at the right, $\overleftrightarrow{MN}$, $\overleftrightarrow{PQ}$, and $\overleftrightarrow{RS}$ intersect at point T, and $\overrightarrow{TP}$ bisects $\angle MTR$. Find the measure of each angle.

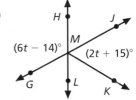

13. $\angle MTS$ **14.** $\angle MTR$ **15.** $\angle NTQ$

MAP MAKING The figure at the right represents the distances between four cities along a bus route. City O is located at the midpoint between cities P and Q. Find the distance between these cities.

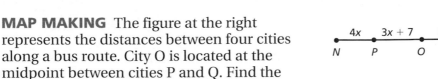

16. O and Q   **17.** N and P   **18.** N and O   **19.** N and Q

In the figure at the right, $\overleftrightarrow{GJ}$ and $\overleftrightarrow{HL}$ intersect at point M, and $\overrightarrow{MK}$ bisects $\angle JML$. Find the measure of each angle.

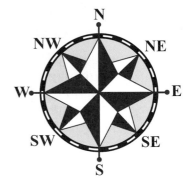

20. $\angle KML$ **21.** $\angle JML$ **22.** $\angle HMJ$

23. $\angle GML$ **24.** $\angle HMK$ **25.** $\angle KMG$

For Exercises 26–29, use the compass at the right to find information about the directions of the compass.

26. Which direction is directly opposite NE?

27. What is the measure of the angle formed by NW and W?

28. What is the measure of the angle formed by SE and SW?

29. Which two directions form a 135° angle with NW?

■ EXTENDED PRACTICE EXERCISE

WRITING MATH Exercises 30–35 refer to the figure at the right. Tell whether each statement is true or false, then write a brief explanation of your reasoning.

30. $\overrightarrow{XZ}$ and $\overrightarrow{XV}$ are opposite rays.

31. $\overrightarrow{XY}$ and $\overrightarrow{XU}$ are opposite rays.

32. $\overrightarrow{XV}$ bisects $\angle UXW$.   **33.** $\overrightarrow{XU}$ bisects $\angle TXV$.

34. $\angle TXU$ and $\angle ZXY$ are a pair of vertical angles.   **35.** Point X is the midpoint of ZV.

■ MIXED REVIEW EXERCISES

Solve each equation. (Prerequisite Skills)

36. $x - 6 = 2$ **37.** $2x + 3 = 8$ **38.** $-2x - 3 = 7$ **39.** $\left(\frac{1}{2}\right)x + 4 = -3$

Identify the pattern for each sequence. Find the next three terms in each sequence. (Lesson 2-1)

40. 1, 2, 4, 8, 16, 32, …   **41.** 1, −3, 9, −27, 81, …

42. 1, 13, 25, 37, 49, …   **43.** 100, 92.5, 85, 77.5, 70, …

44. 1000, −500, 250, −125, 62.5, …   **45.** 50, 54, 27, 31, 15.5, …

3-4 Constructions and Lines

Goals
- Use constructions of segment bisectors and angle bisectors to solve problems.
- Identify parallel, perpendicular and skew lines.
- Identify congruent angles formed by parallel lines and a transversal.

Applications Architecture, Surveying, Navigation

Work with a partner.

Draw a line segment on a piece of paper, and find its midpoint. Can you find the midpoint without measuring the segment?

◤ BUILD UNDERSTANDING

A geometric construction is a precise drawing of a geometric figure made with the aid of only two tools: a compass and a straightedge. One of the most basic geometric constructions is the segment bisector construction described below.

Example 1

Divide a line segment into two segments of equal length.

Solution

Step 1: Start with a line segment, $\overline{AB}$.

Step 2: Open the compass to a radius that is more than half AB. With the compass point at A, draw one arc above and one arc below $\overline{AB}$.

Step 3: Place the compass point at B. Using the same radius as in Step 2, draw arcs above and below $\overline{AB}$.

Step 4: Label the points where the two pairs of arcs intersect X and Y. Using a straightedge, draw $\overleftrightarrow{XY}$. Label the point where $\overleftrightarrow{XY}$ intersects $\overline{AB}$ as point M. This is the midpoint of $\overline{AB}$.

> **Reading Math**
>
> A **compass** is used to draw circles and arcs. The **straightedge** is used to draw lines, rays, and segments. You may use a ruler as a straightedge but you must ignore the markings.

Step 1   Step 2   Step 3   Step 4

Another important geometric construction is the angle bisector construction.

Example 2

ARCHITECTURE On the plans for a building, the angle of the roof is bisected by a length of decorative molding. Draw the bisector using a compass and a straightedge.

Step 1: Start with an angle, $\angle ABC$.

Step 2: With the compass point at B, draw an arc that intersects $\overrightarrow{BA}$ and $\overrightarrow{BC}$. Label the points of intersection P and Q.

Step 3: Use a radius that is more than half PQ. Place the compass tip at P. Draw an arc in the interior of $\angle ABC$. With the compass set to the same radius, place the compass tip at Q and draw an arc that intersects the first arc. Label the point of intersection X.

Step 4: Using a straightedge, draw $\overrightarrow{BX}$. This is the bisector of $\angle ABC$.

 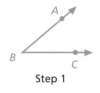 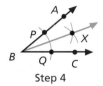

Step 1   Step 2   Step 3   Step 4

Technology Note

Another way to bisect segments and angles is to use geometry software. Try this activity.

1. Draw $\angle JKL$ with a measure of 60°.

2. Select $\angle JKL$ and use the construction menu to draw the angle bisector. Label point M on the bisector.

3. Measure the new angles: $\angle JKM$ and $\angle KML$. Each should measure 30°.

Two lines that intersect to form right angles are called **perpendicular lines**. Because a ray and a segment are each part of a line, the word *perpendicular* (shown by $\perp$) also can refer to rays and segments.

Line r is perpendicular to line s.

$r \perp s$

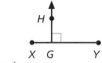

$\overrightarrow{GH}$ is perpendicular to $\overline{XY}$.

$\overrightarrow{GH} \perp \overline{XY}$

Example 3

In the figure, $\overleftrightarrow{AD} \perp \overleftrightarrow{BE}$. Find $m\angle AZF$.

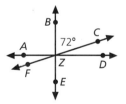

Solution

Since $\overleftrightarrow{AD} \perp \overleftrightarrow{BE}$, $\angle BZD$ is a right angle. The exterior sides of $\angle BZC$ and $\angle CZD$ form a right angle, so the angles are complementary. So, $72° + m\angle CZD = 90°$, and $m\angle CZD = 90° - 72° = 18°$. Because $\angle AZF$ and $\angle CZD$ are vertical angles, $m\angle AZF = m\angle CZD$. So, $m\angle AZF = 18°$.

Not all lines intersect.

Parallel lines ($\parallel$) are *coplanar* lines that do not intersect.

Line ℓ is parallel to line m.

$\ell \parallel m$

Noncoplanar lines are called **skew lines**.

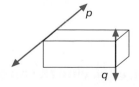

Lines p and q are skew lines.

As with the word perpendicular, *parallel* also is used in reference to rays and segments.

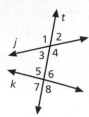

The figure at the right shows lines j and k intersected by a third line, t, called a *transversal*. A **transversal** is a line that intersects two or more coplanar lines in different points. The four angles between the lines—$\angle 3$, $\angle 4$, $\angle 5$, and $\angle 6$—are called **interior angles**. The other four angles are called **exterior angles**.

Alternate interior angles are nonadjacent interior angles on opposite sides of the transversal. There are two pairs of alternate interior angles shown above.

$\qquad$ $\angle 3$ and $\angle 6$ $\qquad$ $\angle 4$ and $\angle 5$

Alternate exterior angles are nonadjacent exterior angles on opposite sides of the transversal. There are two pairs of alternate exterior angles.

$\qquad$ $\angle 1$ and $\angle 8$ $\qquad$ $\angle 2$ and $\angle 7$

Corresponding angles are in the same position relative to the transversal and the lines. The figure above has four pairs of corresponding angles.

$\qquad$ $\angle 1$ and $\angle 5$ $\qquad$ $\angle 2$ and $\angle 6$ $\qquad$ $\angle 3$ and $\angle 7$ $\qquad$ $\angle 4$ and $\angle 8$

Postulate 9	**The Parallel Lines Postulate** If two parallel lines are cut by a transversal, then corresponding angles are equal in measure.

Example 4

In the figure, $\overleftrightarrow{GH} \parallel \overleftrightarrow{JK}$. Find $m\angle HQR$.

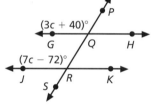

Solution

Since $\angle PQG$ and $\angle QRJ$ are corresponding angles, they are equal in measure.

$m\angle PQG = m\angle QRJ$

$3c + 40 = 7c - 72$ $\quad$ Add $-3c$ to each side.

$\qquad 40 = 4c - 72$ $\quad$ Add 72 to each side.

$\qquad 112 = 4c$ $\qquad$ Multiply each side by $\frac{1}{4}$.

$\qquad 28 = c$

Substituting 28 for c, $m\angle PQG = (3 \cdot 28 + 40)° = (84 + 40)° = 124°$.

Since $\angle PQG$ and $\angle HQR$ are vertical angles, $m\angle HQR = 124°$.

Check Understanding

In the figure below $b \parallel c$, which line is a transversal?

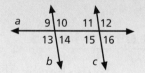

Which pairs of angles are alternate interior angles? alternate exterior angles? corresponding angles?

◤ TRY THESE EXERCISES

In the figure at the right, $m\angle TZV = m\angle UZS$, and $\overleftrightarrow{RS} \perp \overleftrightarrow{VW}$. Find the measure of each angle.

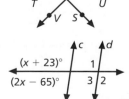

1. $\angle SZV$ $\qquad\qquad$ **2.** $\angle RZT$ $\qquad\qquad$ **3.** $\angle TZV$

In the figure at the right, $c \parallel d$. Find the measure of each angle.

4. $\angle 1$ $\qquad\qquad$ **5.** $\angle 2$ $\qquad\qquad$ **6.** $\angle 3$

7. **GEOMETRY SOFTWARE** Divide a line segment, $\overline{AB}$, into four segments of equal length using a compass and straightedge. Then try the same construction using geometry software.

8. **WRITING MATH** Can the segment bisector construction be used to divide a segment into three segments of equal length? Explain your thinking.

SURVEYING City streets form the angles shown at the right. In the figure, $\overrightarrow{CF} \parallel \overrightarrow{BG}$ and $\overrightarrow{AE} \perp \overrightarrow{DH}$. Find the measure of each angle.

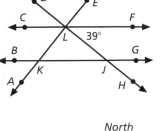

9. $\angle ELF$ 10. $\angle LKJ$

11. $\angle DLC$ 12. $\angle LJK$

13. $\angle KJH$ 14. $\angle CLJ$

NAVIGATION The **bearing** of a ship is the measure of the angle formed by a ray pointing due north from the harbor, $\overrightarrow{HN}$, and the ray that represents the ship's course out of the harbor, $\overrightarrow{HS}$, measured in a clockwise direction. A bearing may be any measure from 0° up to 360°, but may not include 360°. In the figure, the ship's bearing is $180° + 60° = 240°$

Draw a point R near the middle of a sheet of paper. Draw $\overrightarrow{RN}$ to show due north from R. Using a protractor, draw a ray to show each bearing.

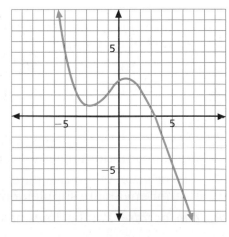

15. $\overrightarrow{RA}$, 40°   16. $\overrightarrow{RB}$, 190°

17. $\overrightarrow{RC}$, 330°   18. $\overrightarrow{RD}$, 100°

■ EXTENDED PRACTICE EXERCISES

19. **CHAPTER INVESTIGATION** Using the map you made, sketch a course to travel between two points. Write directions in which you specify the bearing of each leg of the journey.

20. **YOU MAKE THE CALL** Janeira says that three lines that intersect in one point are always coplanar. Do you agree?

■ MIXED REVIEW EXERCISES

Use the Vertical Line Test to determine if each graph is a function. (Lesson 2-2)

21.   22.

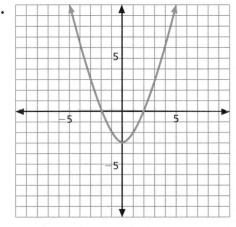

Review and Practice Your Skills

Exercises 1–4 refer to the figure at the right.

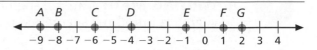

1. Name the midpoint of $\overline{AF}$.

2. Name the segment whose midpoint is point E.

3. Name all the segments whose midpoint is point C.

4. Assume the point Z is the midpoint of $\overline{AG}$. Find its coordinate.

In the figure at the right, $\overrightarrow{GH}$ bisects $\angle TUV$. Find the measure of each angle.

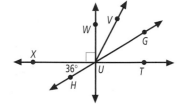

5. $\angle VUX$

6. $\angle GUW$

7. $\angle TUH$

8. $\angle WUV$

In the figure at the right, point B is the midpoint of $\overline{AC}$. Find the length of each segment.

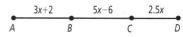

9. $\overline{BC}$

10. $\overline{CD}$

11. $\overline{AC}$

12. $\overline{AD}$

13. *True* or *false*: Only a line segment can be a bisector of a segment.

14. *True* or *false*: Vertical angles are never supplements of each other.

15. Draw a line segment $\overline{AB}$ that is 2 in. long. Using a compass and straightedge, construct a perpendicular bisector of this segment.

16. Trace $\overline{MN}$ onto a sheet of paper. Using a compass and straightedge, divide it into eight segments of equal length.

17. Using a compass and straightedge, construct a 22.5° angle.

In the figure at the right, $\overleftrightarrow{DE} \parallel \overleftrightarrow{RS}$ and $\overleftrightarrow{DE} \perp \overleftrightarrow{GH}$. Find the measure of each angle.

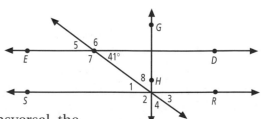

18. $\angle 1$ 19. $\angle 2$ 20. $\angle 3$ 21. $\angle 4$

22. $\angle 5$ 23. $\angle 6$ 24. $\angle 7$ 25. $\angle 8$

26. *True* or *false*: When two parallel lines are cut by a transversal, the alternate interior angles are supplements of each other.

27. *True* or *false*: When two parallel lines are cut by a transversal, the alternate exterior angles are supplements of each other.

28. *True* or *false*: Parallel lines have the same slope.

29. In the figure below, $PR = 200$. Find PQ. (Lesson 3-1)

30. In the figure below, $GJ = 35$. Find GH. (Lesson 3-1)

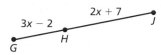

31. In the figure below, find $m\angle CDF$. (Lesson 3-2)

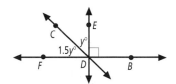

32. In the figure below, find $m\angle TPN$. (Lesson 3-2)

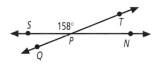

For Exercises 33–35, refer to the rectangular prism at the right. (Lesson 3-4)

33. Name three segments parallel to $\overline{AB}$.

34. Name four segments perpendicular to $\overline{EF}$.

35. Name four segments skew to $\overline{EJ}$.

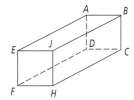

Mid-Chapter Quiz

Identify each of the following. (Lesson 3-1)

1. part of a line with one endpoint that extends without end in one direction

2. a statement that is always assumed to be true

3. points on the same plane

Find the measure of a complement of each angle. (Lesson 3-2)

4. $56°$ **5.** $x°$ **6.** $3x°$

Find the measure of a supplement of each angle. (Lesson 3-2)

7. $72°$ **8.** $b°$ **9.** $(2x + 1)°$

Find the midpoint of each segment. (Lesson 3-3)

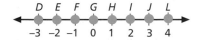

10. $\overline{IL}$

11. $\overline{DJ}$

$\overrightarrow{QV}$ **is the bisector of** $\angle UQW$. **Find the measure of each angle.** (Lesson 3-3)

12. $m\angle TQX$ **13.** $m\angle UQV$ **14.** $m\angle VQX$

15. Draw two parallel lines and a transversal intersecting both. (Lesson 3-4)

Inductive Reasoning in Mathematics

Goals ■ Use inductive reasoning to complete patterns.

Applications City Planning, Scheduling, Art

List the next three terms in each sequence. What do you think the twentieth term will be?

a. 1, 9, 17, 25, 33, 41, . . . **b.** 1, 2, 4, 7, 11, 16, . . .
c. 1, 2, 4, 8, 16, 32, . . . **d.** 1, 1, 2, 3, 5, 8, . . .

◥ BUILD UNDERSTANDING

In Lesson 2-1, you learned how to see a pattern in a sequence of numbers and use your observations to extend the pattern. In doing this, you were using *inductive reasoning*. **Inductive reasoning** is the process of observing data and making a generalization based on your observations.

The generalization is called a **conjecture**. In this lesson, you will learn how to apply your inductive reasoning skills to make conjectures about geometric figures.

Example 1

a. Draw the next figure in this pattern.

b. Describe the twentieth figure in the pattern.

Solution

Each figure in the pattern consists of a number of segments that intersect only at their endpoints. It is important to note that, in each figure, no three of the endpoints are collinear.

The 1st figure consists of 3 segments.

The 2nd figure consists of 4 segments.

The 3rd figure consists of 5 segments.

The 4th figure consists of 6 segments.

a. The 5th figure consists of 7 segments, as shown.

b. From the discussion above, you see that the *n*th figure consists of (*n* + 2) segments.

So, the 20th figure will consist of (20 + 2) segments, or 22 segments. The 22 segments intersect only at their endpoints, and no three of the endpoints are collinear.

Many problems in geometry involve not only a geometric pattern, but also a number pattern.

Example 2

CITY PLANNING A traffic engineer is using line segments determined by ten collinear points to represent the stops on a bus route. How many different segments of the route are there?

Solution

Solving the problem directly would involve identifying and counting all the segments in a figure such as this.

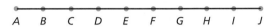

A   B   C   D   E   F   G   H   I   J

The process would be time-consuming, and it would be very easy to make a mistake. As an alternative, count the number of segments formed in a sequence of simpler cases and try to find a pattern in the results. Then use inductive reasoning to make a conjecture about the given problem. Organize the data into a table like this.

Number of Collinear Points		Number of Segments Formed	
A B	2	$\overline{AB}$	1
A B C	3	$\overline{AB}, \overline{AC}, \overline{BC}$	3
A B C D	4	$\overline{AB}, \overline{AC}, \overline{AD}, \overline{BC}, \overline{BD}, \overline{CD}$	6
A B C D E	5	$\overline{AB}, \overline{AC}, \overline{AD}, \overline{AE}, \overline{BC}, \overline{BD},$ $\overline{BE}, \overline{CD}, \overline{CE}, \overline{DE}$	10
A B C D E F	6	$\overline{AB}, \overline{AC}, \overline{AD}, \overline{AE}, \overline{AF}, \overline{BC}, \overline{BD},$ $\overline{BE}, \overline{BF}, \overline{CD}, \overline{CE}, \overline{CF}, \overline{DE}, \overline{DF}, \overline{EF}$	15

The numbers in the right-hand column of this table form the pattern of *add 2, add 3, add 4, add 5*, and so on. So, extend the sequence until the number of points is ten.

Number of collinear points	2	3	4	5	6	7	8	9	10
Number of segments formed	1	3	6	10	15	21	28	36	45

+2 +3 +4 +5 +6 +7 +8 +9

Using this table, you can now make the following conjecture: *The number of segments determined by ten collinear points is 45.*

> ## Check Understanding
>
> In Example 2, the number of segments formed depends on the number of collinear points. So, you can say that the number of segments formed *is a function of* the number of collinear points. The table at the bottom of this page is a table for this function.
>
> If *n* represents the number of collinear points, which of the following do you think is a rule for this function?
>
> **a.** $f(n) = n + 2$
>
> **b.** $f(n) = \dfrac{n}{2}$
>
> **c.** $f(n) = (n + 1)(n - 1)$
>
> **d.** $f(n) = \dfrac{n(n - 1)}{2}$

1. Draw the next figure in this pattern of points. Then describe the sixteenth figure in the pattern.

SCHEDULING Employees work in pairs to accomplish certain tasks. To find the number of possible pairings for any number of workers, the supervisor draws segments determined by noncollinear points. Each segment represents a pairing of two employees. The figures below show the segments determined by three, four, five and six noncollinear points.

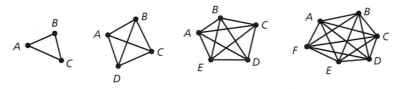

2. How many segments are in each of the figures shown?

3. How many pairings are possible for 12 employees? *Hint:* Find the number of segments determined by 12 noncollinear points.

Draw the next figure in each pattern. Then describe the fourteenth figure in the pattern.

4.

5.

The figures below show one, two, three, and four lines passing through the center of a circle.

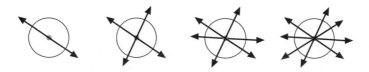

6. In each figure, the lines divide the interior of the circle into regions. How many individual regions are formed in each of the figures shown?

7. Find the number of regions that would be formed when eleven lines pass through the center of a circle.

The figures show two, three, four, and five rays sharing a common endpoint.

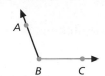 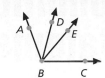 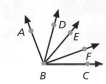

8. How many angles are formed in each of the figures shown?

9. Find the number of angles that would be formed when eighteen rays share a common endpoint.

10. ART A block structure similar to those shown below will be at the center of a fountain. How many blocks will the artist need to build the tenth figure in the pattern?

 11. WRITING MATH Describe the pattern below. Then write a function rule to represent the number of points in the nth figure in the pattern.

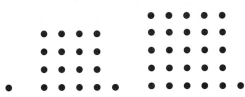

EXTENDED PRACTICE EXERCISES

Create a geometric pattern for each number pattern.

12. 1, 3, 9, 27, . . .   **13.** 1, 4, 9, 16, . . .   **14.** 1, 8, 27, 64, . . .

MIXED REVIEW EXERCISES

Graph each inequality on the number line. (Lesson 2-6)

15. $3a + 2 \leq -8$ **16.** $7 - 2b > 3$ **17.** $\left(\frac{1}{2}\right)c + 4 < -2$

18. $4d - 6 \geq -2$ **19.** $4x + 3 < 4$ **20.** $2m - \frac{3}{2} > -3$

Construct a frequency table for each set of data, using the given interval.
(Lesson 2-7)

21. interval = 10
62 71 59 94 86 67 82
75 74 81 79 40 93 72
79 80 78 99 85 65 75
83

22. interval = 5
124 117 114 122 103 102 100
107 123 112 129 106 103 108
107 100 104 117 116 102 108
112 101 113 108 103

23. interval = 3
85 72 87 77 70 79 82
85 83 86 73 85 87 84
77 80 85 82 79 81 83
86 80 83 87

24. interval = 8
80 59 50 92 77 79 82 97 84
84 75 68 70 59 63 68 75 85
77 89 79 72 76 84 79 78 63

3-6 Conditional Statements

Goals
- Identify and evaluate conditional statements.
- Identify converses and biconditionals.

Applications Drafting, Sports, Geography

Do you think each statement is true or false?
Explain your reasoning.

a. Denver is the capital of Colorado.
b. For all x, $x^2 > x$.
c. Lines m and n are parallel.

Denver, CO

◣ BUILD UNDERSTANDING

Many of the statements in this chapter are written
in *if–then* form. Statements like these are called
conditional statements, or simply **conditionals**. The clause following "if" is called the
hypothesis of the conditional. The clause following "then" is called the **conclusion**.
For example, the *parallel lines postulate* was presented as a conditional.

If ⎡ two parallel lines are cut by a transversal ⎤ ,   hypothesis

then ⎡ corresponding angles are equal in measure ⎤ .   conclusion

A conditional is either true or false. When a conditional is true, you can justify it
in a variety of ways. For instance, you may be able to show that the conditional is
true because it follows directly from a definition. When a conditional is a
postulate, such as the parallel lines postulate, it is *assumed* to be true. Still other
conditionals are theorems, and these must be *proved* true.

To demonstrate that a conditional is false, you need to find only one example for
which the hypothesis is true but the conclusion is false. An example like this is
called a **counterexample**.

Example 1

Tell whether each conditional is true or false.

a. If two lines are parallel, then they are coplanar.
b. If two lines do not intersect, then they are parallel.

Solution

a. Parallel lines are defined as *coplanar* lines that do not intersect. So, the
conditional is true.

b. Consider skew lines k and ℓ shown. By the definition of skew lines, k and ℓ
do not intersect, and so the hypothesis is true. However, also by the
definition of skew lines, k and ℓ are noncoplanar. Lines k and ℓ cannot be
parallel, and so the conclusion is false. Therefore, lines k and ℓ are a
counterexample, and the conditional is false.

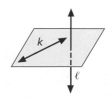

The **converse** of a conditional is formed by interchanging the hypothesis and the conclusion. The fact that a conditional is true is no guarantee that its converse is true.

Example 2

DRAFTING People who draw plans must apply this true statement: If two lines are parallel, then they do not intersect. Write the converse of the statement. Is it also true?

Solution

Interchange the hypothesis and the conclusion of the given statement.

Statement: If | two lines are parallel | , then | they do not intersect | .
        hypothesis   conclusion

Converse: If | two lines do not intersect | , then | they are parallel | .
        hypothesis   conclusion

By definition, parallel lines do not intersect, and so the given statement is *true*. Part b of Example 1 demonstrated that lines that do not intersect are not necessarily parallel, and so the converse is *false*.

The converse of the parallel lines postulate also is assumed to be true. It is stated as the *corresponding angles postulate* in the following manner.

Postulate 10	**The Corresponding Angles Postulate** If two lines are cut by a transversal so that a pair of corresponding angles are equal in measure, then the lines are parallel.

When a statement and its converse are both true, they can be combined into an "if and only if" statement. This type of statement is called a **biconditional statement**, or simply a **biconditional**. Every definition can be written as a biconditional.

Example 3

Write this definition as two conditionals and as a single biconditional.

A right angle is an angle whose measure is 90°.

Solution

The definition leads to two true conditionals.

 If an angle is a right angle, then its measure is 90°.

 If the measure of an angle is 90°, then it is a right angle.

These can be combined into a single biconditional as follows.

 An angle is a right angle if and only if its measure is 90°.

Reading Math

Many statements about everyday situations can be expressed as conditionals. For instance, the following is a true conditional.
 If it is raining, then it is cloudy.

Its converse is false.
 If it is cloudy, then it is raining.

Problem Solving Tip

When writing the converse of a conditional, you may need to change the wording of the hypotheses and the conclusion to make the converse read clearly.

1. **TALK ABOUT IT** Decide whether this conditional is true or false.

   If two lines are each perpendicular to a third line, then they are parallel to each other.

   Discuss your reasoning with a classmate.

2. Write the converse of this statement.

   If two angles are vertical angles, then they are equal in measure.

   Are the given statement and its converse true or false?

3. Write this definition as two conditionals and as a single biconditional.

   The bisector of an angle is the ray that divides the angle into two adjacent angles that are equal in measure.

4. **NUMBER SENSE** Tell whether this conditional is true or false.

   If a number is less than 1, the number is a proper fraction.

   Write the converse of the statement. Is the converse true or false?

5. **SPORTS** If a shortstop makes a bad throw to first base, the error is charged to the shortstop. This statement is true. Write the converse of the statement. Is it true or false?

6. **GEOGRAPHY** If a point is located north of the equator, it has a northern latitude. This statement is true. Write the converse of the statement. Is it true or false?

PRACTICE EXERCISES • For Extra Practice, see page 672.

Sketch a counterexample that shows why each conditional is false.

7. If line *t* intersects lines *g* and *h*, then line *t* is a transversal.

8. If $PQ = QR$, then point Q is the midpoint of $\overline{PR}$.

9. If points *A*, *B*, and *C* are collinear, then $\overrightarrow{BA}$ and $\overrightarrow{BC}$ are opposite rays.

10. If two angles share a common side and a common vertex, then they are adjacent angles.

Write the converse of each statement. Then tell whether the given statement and its converse are true or false.

11. If points *J*, *K*, and *L* are coplanar, then they are collinear.

12. If point *Y* is the midpoint of $\overline{XZ}$, then $XY + YZ = XZ$.

13. If the sum of the measures of two angles is 90°, then the angles are complementary.

14. If two angles are supplementary, then the sum of their measures is greater than 90°.

15. If two lines are perpendicular, then they do not intersect.

16. If $m\angle QRS = m\angle SRT$, then $\overrightarrow{RS}$ bisects $\angle QRT$.

Write each definition as two conditionals and as a single biconditional.

17. The midpoint of a segment is the point that divides it into two segments of equal length.

18. Perpendicular lines are two lines that intersect to form right angles.

19. A transversal is a line that intersects two or more coplanar lines in different points.

20. Vertical angles are two angles whose sides form two pairs of opposite rays.

GEOMETRIC CONSTRUCTION The corresponding angles postulate provides a method for constructing parallel lines.

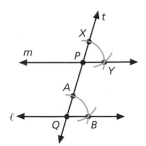

In the figure at the right, you see the finished construction of a line parallel to line ℓ through point P. Trace line ℓ and point P onto a sheet of paper and repeat the construction. Then complete the statements below that outline the steps of the construction.

21. *Step 1:* Using a straightedge, draw any line __?__ through point P intersecting line ℓ. Label the intersection point __?__.

22. *Step 2:* With the compass point at __?__, draw an arc intersecting lines t and ℓ. Label the intersection points __?__ and __?__.

23. *Step 3:* Using the same radius as in Step 2, place the compass point at point __?__ and draw an arc intersecting line t. Label the intersection point __?__.

24. *Step 4:* Place the compass point at point __?__ and the pencil at point __?__. Using this radius, draw an arc that intersects line ℓ.

25. *Step 5:* Using the same radius as in Step 4, place the compass point at point __?__ and draw an arc that intersects the arc you drew in Step 3. Label the intersection point __?__.

26. *Step 6:* Draw line __?__ through points P and Y. $m\angle$__?__ $= m\angle$__?__, and so __?__ $\parallel$ __?__.

▸ EXTENDED PRACTICE EXERCISES

WRITING MATH Explain why each of the following is *not* a good definition.

27. Vertical angles are two angles whose sides form opposite rays.

28. A line segment is part of a line.

29. Complementary angles are adjacent angles whose exterior sides form a right angle.

30. Skew lines are noncoplanar lines that do not intersect.

▸ MIXED REVIEW EXERCISES

Find each length. (Lesson 3-1)

31. In the figure below, $AC = 130$. Find BC.

32. In the figure below, $LM = 94$. Find LN.

33. In the figure below, $JL = 88$. Find KL.

34. In the figure below, $QS = 41$. Find QR.

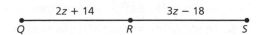

Review and Practice Your Skills

Draw the next figure in each pattern. Then describe the tenth figure in the pattern.

1.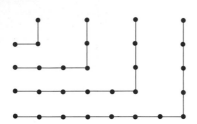

2.

3.

4.

5.

6.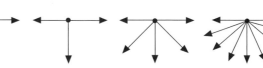

7.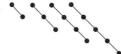

8.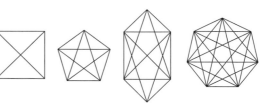

Sketch a counterexample to show why each conditional is false.

9. If $\angle ABC$ and $\angle DEF$ are supplements, then $m\angle ABC > m\angle DEF$.

10. If three points are coplanar, then they are collinear.

11. If two lines are skew, then they intersect.

Write the converse of each statement. Then tell whether the given statement and its converse are true or false.

12. If two lines intersect, then they are perpendicular.

13. If C is the midpoint of $\overline{AB}$, then $AB = 2(AC)$.

14. If two angles are vertical angles, then their supplements are equal.

Write each definition as two conditionals and as a single biconditional.

15. Perpendicular lines are lines that intersect to form right angles.

16. Skew lines are noncoplanar lines.

17. On a number line, the coordinate of point G is -6. The length of $\overline{GH}$ is 13. Give two possible coordinates for point H. (Lesson 3-1)

18. Point Z is between points X and Y. The length of $\overline{XZ}$ is three times the length of $\overline{ZY}$, and $XY = 48$. Find XZ and ZY. (Lesson 3-1)

19. In the figure below, find $m\angle PYV$ and $m\angle PYX$. (Lesson 3-3)

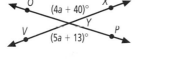

20. In the figure below, $\overrightarrow{QS}$ bisects $\angle RQT$. Find $m\angle TQU$. (Lesson 3-3)

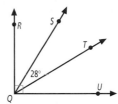

In the figure at the right, $\overrightarrow{NP} \parallel \overrightarrow{MQ}$. Find the measure of each angle. (Lesson 3-4)

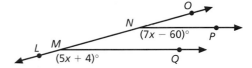

21. $\angle LMQ$

22. $\angle NMQ$

23. $\angle MNP$

24. $\angle ONP$

MathWorks
Workplace Knowhow
Career – Cartographer

A cartographer is a mapmaker. Before there were ways to see the Earth from above, cartographers made maps by visiting the places they were mapping. Lewis and Clark mapped their route all the way to the Pacific Ocean by drawing the features of the landscape as they traveled.

Today, a cartographer has many tools to make maps more accurate. Satellite photos give cartographers a birds-eye view of the Earth and its features. Mathematics is an essential tool for cartographers. Mapmakers must be able to convert distances in miles to distances in inches or centimeters. They must understand the principles of geometry to represent three-dimensional objects on a flat plane and check the accuracy of angle measurements.

1. A cartographer is making a road map. On the map, Beverly Avenue is transversed by two parallel roads, 3rd Street and 6th Street. On the map, if $\angle A$ measures 104°, what is the measure of $\angle D$?

2. Surveyors determine that a lake is 8.4 km wide. On a map, a cartographer is using a scale of 1 cm = 3 km. What will the width of the lake be on the map?

3. On a map of Oregon, Cottonwood Mountain is about $2\frac{1}{2}$ in. north of Cedar Mountain. The map scale shows that $\frac{7}{8}$ in. = 20 mi. To the nearest mile, how many miles apart are the mountains?

3-7 Deductive Reasoning and Proof

Goals ■ Write geometric proofs in two-column format.

Applications Surveying and Carpentry

Study this geometric pattern.

2 points
2 regions

3 points
4 regions

4 points
8 regions

5 points
16 regions

Use inductive reasoning to make a conjecture about the number of regions formed by six points on a circle.

◤ BUILD UNDERSTANDING

A theorem is a statement that can be proved true. To do this, you start with the hypothesis of the theorem and make a series of logical statements that demonstrates why the conclusion is true. Each statement must be accompanied by a reason—a definition, postulate, theorem, or algebraic property that justifies the statement.

The set of statements and reasons is called a **proof of the theorem**. When you prove a theorem by this process, you are using **deductive reasoning**.

Example 1 shows a two-column proof of the vertical angles theorem.

Theorem	If two angles are vertical angles, then they are equal in measure.

Example 1

Given $\angle 1$ and $\angle 3$ are vertical angles.

Prove $m\angle 1 = m\angle 3$

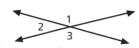

Solution

Statements	Reasons
1. $\angle 1$ and $\angle 3$ are vertical angles.	1. given
2. $m\angle 1 + m\angle 2 = 180°$ $m\angle 3 + m\angle 2 = 180°$	2. angle addition postulate
3. $m\angle 1 + m\angle 2 = m\angle 3 + m\angle 2$	3. transitive property of equality
4. $m\angle 1 = m\angle 3$	4. addition property of equality

> **Reading Math**
>
> When a proof is written in the two-column format, the hypothesis is usually labeled *Given*.

134 | Chapter 3 **Geometry and Reasoning**

Once a theorem has been proved true, you can use it to justify statements in subsequent proofs. For instance, Example 2 shows how the vertical angles theorem is used to prove the next theorem about parallel lines.

Theorem	If two parallel lines are cut by a transversal, then alternate interior angles are equal in measure.

Example 2

SURVEYING Two parallel property lines are cut by a third line. How can the surveyor know that $m\angle 1 = m\angle 3$?

Given $p \parallel q$

Prove $m\angle 1 = m\angle 3$

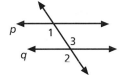

Solution

Statements	Reasons
1. $p \parallel q$	1. given
2. $m\angle 1 = m\angle 2$	2. parallel lines postulate
3. $m\angle 3 = m\angle 2$	3. vertical angles theorem
4. $m\angle 1 = m\angle 3$	4. transitive property of equality

Example 3 shows how the vertical angles theorem also can be used to prove the converse of the theorem above.

Theorem	If two lines are cut by a transversal so that alternate interior angles are equal in measure, then the lines are parallel.

Example 3

CARPENTRY A cross-brace is nailed to two wooden beams so that the alternate interior angles formed are equal in measure. How can the carpenter know that the beams are parallel?

Given $m\angle 1 = m\angle 3$

Prove $p \parallel q$

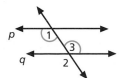

Solution

Statements	Reasons
1. $m\angle 1 = m\angle 3$	1. given
2. $m\angle 2 = m\angle 3$	2. vertical angles theorem
3. $m\angle 1 = m\angle 2$	3. transitive property of equality
4. $p \parallel q$	4. corresponding angles postulate

Reading Math

How does deductive reasoning differ from inductive reasoning?

Inductive reasoning often is described as "particular to general": you observe several particular cases and arrive at a general conclusion. The conclusion is *probably* true, but, as illustrated in the Explore activity, it is not *necessarily* true. In geometry, inductive reasoning is used primarily to explore ideas and to make conjectures.

In contrast, deductive reasoning is described as "general to particular": you use several general statements to arrive at a particular conclusion. If all the statements in a deductive argument are true, the conclusion *must* be true. For this reason, most geometric proofs are based in deductive reasoning.

You will be asked to prove two related theorems in *Try These Exercises*.

◣ TRY THESE EXERCISES

Copy and complete each proof.

| **Theorem** | If two parallel lines are cut by a transversal, then alternate exterior angles are equal in measure. |

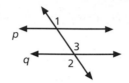

1. Given $p \parallel q$

   Prove $m\angle 1 = m\angle 2$

Statements	Reasons
1. $p \parallel q$	1. ___?___
2. $m\angle 1 = m\angle 3$	2. ___?___
3. ___?___	3. vertical angles theorem
4. $m\angle 1 = m\angle 2$	4. ___?___

| **Theorem** | If two lines are cut by a transversal so that alternate exterior angles are equal in measure, then the lines are parallel. |

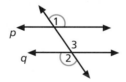

2. Given $m\angle 1 = m\angle 2$

   Prove $p \parallel q$

Statements	Reasons
1. ___?___	1. ___?___
2. ___?___	2. vertical angles theorem
3. $m\angle 1 = m\angle 3$	3. ___?___
4. ___?___	4. ___?___

◣ PRACTICE EXERCISES • For Extra Practice, see page 672.

3. WRITING MATH Describe a real-life example of an occasion when you used inductive reasoning to solve a problem. Then describe a situation when you used deductive reasoning.

4. YOU MAKE THE CALL Given: $\angle DEF$ and $\angle FEG$ are adjacent angles. Based on the given information, Dana makes the statement that $m\angle DEF + m\angle FEG = m\angle DEG$. She justifies the statement using the definition of complementary angles. Is her reasoning correct? Explain.

5. Copy and complete the following proof of this theorem.

If two angles are complementary to the same angle, then they are equal in measure to each other.

Given ∠1 is complementary to ∠2.

∠3 is complementary to ∠2.

Prove ___?___

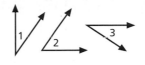

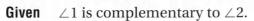

Problem Solving Tip

For Exercise 5, you probably will find it helpful to refer to the proof of the vertical angles theorem in Example 1 on page 134.

Statements	Reasons
1. ___?___	1. ___?___
2. $m\angle 1 + m\angle 2 = 90°$; $m\angle 3 + m\angle 2 = 90°$	2. ___?___
3. $m\angle 3 + m\angle 2 = 90°$	3. ___?___
4. ___?___	4. addition property of equality

6. Write a proof of the following theorem.

Theorem	If two angles are supplementary to the same angle, then they are equal in measure to each other.

(*Hint:* Use the proof in Exercise 5 as a model.)

◼ EXTENDED PRACTICE EXERCISES

Exercises 7–9 state three important theorems related to perpendicular and parallel lines. Write a proof of each theorem.

7. If two coplanar lines are each parallel to a third line, then the lines are parallel to each other.

8. If two coplanar lines are each perpendicular to a third line, then the lines are parallel to each other.

9. If a transversal is perpendicular to one of two parallel lines, then it is perpendicular to the other.

10. CHAPTER INVESTIGATION List at least two geometric theorems or postulates from Exercises 7–9 that are evident in your neighborhood map.

◼ MIXED REVIEW EXERCISES

Write each number in standard notation. (Lesson 1-8)

11. 1.46×10^{-8} **12.** 3.7×10^5 **13.** 7.02×10^7 **14.** 6.82×10^{-11}

15. 5.9×10^{-9} **16.** 8.13×10^8 **17.** 2.1×10^4 **18.** 4×10^{-7}

19. 3.97×10^{-4} **20.** 6.48×10^7 **21.** 5.12×10^9 **22.** 7.3×10^{-10}

Find the complement and supplement of each angle. (Lesson 3-2)

23. $m\angle ABC = 57°$ **24.** $m\angle CDE = 27°$ **25.** $m\angle EFG = 75°$ **26.** $m\angle GHI = 40°$

27. $m\angle JKL = 83°$ **28.** $m\angle LMN = 34°$ **29.** $m\angle NOP = 61°$ **30.** $m\angle PRS = 18°$

Problem Solving Skills: Use Logical Reasoning

To solve some problems, you can use logical reasoning without using any numbers. Making a table and using the *process of elimination* is one way to solve some problems.

Problem

TRAVEL Four friends talked about their trips to different states. Andrea and Don discussed the differences between the states they visited, even though each state has the word *New* in its name.

Natalie told Edward that she liked South Dakota a lot. Edward visited New Mexico or California; Don visited the other. One person talked about sightseeing in New York. Where had each person gone on vacation?

Solve the Problem

Make a table listing each person and each state.

Andrea and Don went to states with *New* in their names. Put an **X** in the boxes to show what states they could *not* have visited.

	Andrea	Don	Natalie	Edward
California	x	x		
New Mexico				
New York				
South Dakota	x	x		

Natalie visited South Dakota. Use an **0** to show that information. Notice that the rest of each row and column is filled with **X**s.

	Andrea	Don	Natalie	Edward
California	x	x	x	
New Mexico			x	
New York			x	
South Dakota	x	x	o	x

Notice, then, that Ed is the only one who could have visited California.

Finally, Don visited either New Mexico or California. Because *New* was part of the name, it must have been New Mexico. So, by process of elimination, Andrea must have visited New York.

	Andrea	Don	Natalie	Edward
California	x	x	x	o
New Mexico		x		x
New York		x		x
South Dakota	x	x	o	x

Five-step Plan

1 Read
2 Plan
3 Solve
4 Answer
5 Check

1. Make a table to help solve this problem. Cory, Srey, Molly, and Mao each visited a different city: Santa Clara, Seattle, Des Moines, and Pittsburgh. Cory and Molly visited cities with two words in their names. Mao and Cory visited cities whose names start with an **S**. Who visited each city?

2. Draw a Venn diagram to solve this problem. There are 160 students in the junior class. Of these students, 120 are taking English, 60 are taking Spanish, and 60 are in band. Only 10 students are taking all 3 classes. There are 15 students taking only English and Spanish and 20 taking only English and band. How many students are taking Spanish and band, but not English?

PRACTICE EXERCISE

3. Pedro, Carina, Ned, and Eva each visited different zoos during their vacations. The zoos were the San Diego Zoo, the San Francisco Zoo, the Miami Metrozoo, and the Dallas Zoo.

   ▶ Eva, Carina, and Pedro saw outstanding gorilla exhibits.

   ▶ Both Carina and Pedro went to zoos with fewer than 400 species.

   ▶ The zoo that Pedro visited has a greater budget than the zoo that Carina visited.

   Which zoo did each person visit?

Major U.S. Public Zoological Parks

Zoo	Budget (millions)	Number of species	Major attraction
Dallas	7.4	387	Gorilla Exhibit
Miami Metrozoo	7.0	250	African Plains
San Diego	72.4	800	Gorilla Tropics
San Francisco	12.0	270	Gorilla World

4. Use logical reasoning to solve this problem. Xenia corresponded on a regular basis with three pen pals. She wrote to all three on one Friday in 2003. She knew she would next write to Daksha in India on Thursday two weeks after that, then on Wednesday in the fourth week. She would write to Leonel in the Dominican Republic every Friday. Finally, she would write to Catherine every fourth day, since they had been good friends before Catherine moved to England. On what date in 2003 did Xenia write for the second time to all three pen pals on the same day?

MIXED REVIEW EXERCISES

Graph each point on the coordinate plane. Label each point. (Lesson 2-2)

5. $A(3, 6)$

6. $B(7, 3)$

7. $C(2, -2)$

8. $D(-6, 6)$

9. $E(-2, 3)$

10. $F(-7, -3)$

11. $G(8, -5)$

12. $H(-2, -7)$

Solve each equation. (Lesson 2-5)

13. $3(4x - 3) + 8 = 15x - 2(x + 1)$

14. $7(x + 4) - 8 = -3x + 8\left(x + \frac{1}{2}\right)$

Chapter 3 Review

VOCABULARY ▨

Write the letter of the word at the right that matches each description.

1. A line that intersects two coplanar lines in different points
2. An angle whose measure is greater than 90° and less than 180°
3. An example to show that a conditional statement is false
4. Two angles in the same plane that share a common side and a common vertex, but have no interior points in common
5. The type of reasoning in which a theorem is proved by a series of logical statements and reasons
6. The set of all points that two geometric figures have in common
7. The type of reasoning in which generalizations are made based on observations
8. Two angles whose sum of their measures is 180°
9. A part of a line with one endpoint
10. A line or plane that divides a segment into two segments of equal length

a. adjacent angles
b. bisector
c. complementary angles
d. corresponding angles
e. counterexample
f. deductive
g. inductive
h. intersection
i. obtuse angle
j. ray
k. supplementary angles
l. transversal

LESSON 3-1 ◼ Points, Lines, and Planes, p. 104

▶ In geometry, **point, line,** and **plane** are undefined terms. Other geometric definitions are developed by using these undefined terms.

▶ Statements that are accepted as true without proof are called **postulates.**

▶ The **segment addition** postulate states that if point B is between A and C, then $AB + BC = AC$.

Use the figure below for Exercises 11–13.

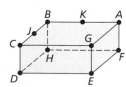

11. How many planes are represented in the figure?
12. Name three points that are collinear.
13. Are points A, C, D, and J coplanar?
14. In the figure, $RS = 75$. Find RN.

15. In the figure, $AB = 26$. Find AD.

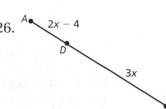

LESSON 3-2 ◼ Types of Angles, p. 108

▶ If point *B* lies in the interior of ∠*AOC*, then *m*∠*AOB* + *m*∠*BOC* = *m*∠*AOC*.

16. In the figure, *m*∠*ABC* = (2*x*)°, *m*∠*CBD* = (5*x* – 6)°, and *m*∠*ABD* = 120°. Find *m*∠*CBD*.

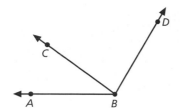

17. In the figure, *m*∠*RNX* = (3*x* – 2)° and *m*∠*XNW* = (2*x* + 7)°. Find *m*∠*TNX*.

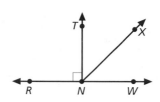

LESSON 3-3 ◼ Segments and Angles, p. 114

▶ Two angles whose sides form two pairs of opposite rays are called **vertical angles**. Vertical angles are equal in measure.

In the figure at the right, $\overleftrightarrow{AX}$ and $\overleftrightarrow{BZ}$ intersect at *Y*. Find the measure of each angle.

18. ∠*XYZ*

19. ∠*XYB*

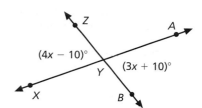

In the figure at the right, $\overrightarrow{NM}$ bisects ∠*LNR* and $\overrightarrow{NS}$ bisects ∠*MNR*. If *m*∠*RNS* = 25 – 2*x* and *m*∠*SNM* = 3*x* + 5, find the measure of each angle.

20. ∠*MNL*

21. ∠*LNR*

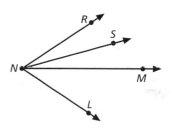

LESSON 3-4 ◼ Constructions and Lines, p. 118

▶ A precise drawing of a geometric figure made by using only a compass and an unmarked straightedge is called a **construction**.

22. Trace ∠*EFG* on another sheet. Using a compass and a straightedge, divide ∠*EFG* into four angles of equal measure.

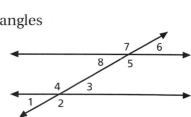

Two or more parallel lines cut by a transversal form several pairs of angles that are equal in measure.

∠1 and ∠8 are **corresponding angles**.

∠3 and ∠8 are **alternate interior angles**.

∠2 and ∠7 are **alternate exterior angles**.

23. Name the other pairs of corresponding angles in the figure.

24. Name the other pairs of alternate interior angles in the figure.

25. Name the other pairs of alternate exterior angles in the figure.

26. If *m*∠7 = 120°, find *m*∠4.

27. If *m*∠6 = 40°, find *m*∠2.

LESSON 3-5 ◾ Inductive Reasoning in Mathematics, p. 124

▶ **Inductive reasoning** is the process of observing data and making a generalization based on your observations.

Draw the next figure in each pattern. Then describe the tenth figure.

28.

29.

LESSON 3-6 ◾ Conditional Statements, p. 128

▶ The **converse** of a conditional statement is formed by interchanging the hypothesis and conclusion.

Write the converse of each statement. Then tell whether the given statement and its converse are *true* or *false*.

30. If the sum of the measures of two angles is 180°, the angles are supplementary.

31. If $m\angle ABC = m\angle DBE$, then $\angle ABC$ and $\angle DBE$ are vertical angles.

LESSON 3-7 ◾ Deductive Reasoning and Proof, p. 134

▶ Proofs of theorems are often written in a two-column format, with statements in one column and reasons in the other.

32. Complete the following proof.

Given E is between L and T.   **Prove** $ET = LT - LE$

LESSON 3-8 ◾ Problem Solving Skills: Use Logical Reasoning, p. 138

▶ Many problems are easier to solve if you use a table.

33. Marco, Sue, and Stephanie went on vacations last summer. They visited California, Florida, and Wisconsin, but not necessarily in that order. Marco did not visit a state bordering the ocean. Sue did not visit Florida. Which state did each visit?

34. Four friends have birthdays in January, July, September, and December. Amelia was not born in the winter. Brandon celebrates his birthday near Independence Day. Lianna's birthday is the month after Jason's birthday. In what month was each one born?

CHAPTER INVESTIGATION

EXTENSION Make a presentation to the class of your map. Show that the navigator's log you created accurately specifies the direction and bearing for each leg of the course you plotted.

Chapter 3 Assessment

Use the number line to find each length.

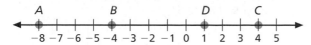

1. AB 2. AC 3. BD

4. In the figure at the right, $RS = 110$. Find RP.

Refer to the protractor at the right.

5. Name the right angles.

6. Name the obtuse angles and give the measure of each.

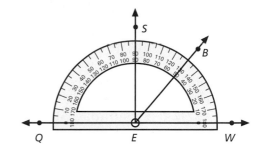

7. In the figure below, $m\angle RSY$ is $(4x + 6)°$ and $m\angle YSX$ is $(2x - 12)°$. Find $m\angle TSY$.

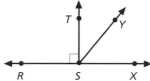

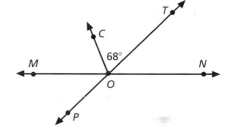

8. In the figure at the right, $\overleftrightarrow{MN}$ and $\overleftrightarrow{PT}$ intersect at point O. $\overrightarrow{OC}$ bisects $\angle MOT$. Find the measure of $\angle MOP$.

9. Trace $\overline{FP}$ onto a sheet of paper. Using a compass and a straightedge, divide it into four segments of equal length.

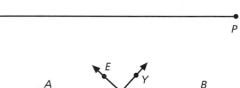

In the figure at the right $\overleftrightarrow{AB} \parallel \overleftrightarrow{CD}$ and $\overleftrightarrow{XY} \perp \overleftrightarrow{EF}$ at G. Find the measure of each angle.

10. $\angle AGE$ 11. $\angle XKH$ 12. $\angle DHF$

13. Write the converse of the following statement. Then tell if the given statement and the converse are true or false.

    If $AB + BC = AC$, then B is the midpoint of $\overline{AC}$.

14. In a two-column proof, the hypothesis is labeled ___?___ and the conclusion is labeled ___?___.

15. Draw the next figure in the pattern of dots.

16. Three students named Alf, Beth, and Chan play three different sports. One plays baseball, one plays tennis, and one plays hockey. Chan plays a sport that does not require a ball. Beth does not play baseball. What sport does each play?

Standardized Test Practice

Record your answers on the answer sheet provided by your teacher or on a sheet of paper.

1. What is the element with the least value in $\{x \mid x$ is a negative integer and $x > -3\}$? (Lesson 1-2)
   - (A) −4
   - (B) −3
   - (C) −2
   - (D) −1

2. Rosalinda worked three days last week. She earned $19.85 on Monday, $17.75 on Thursday, and $21.30 on Saturday. About how much did she earn in all? (Lesson 1-4)
   - (A) $50
   - (B) $55
   - (C) $60
   - (D) $65

3. What is the solution of $m - 1.2 = 3.4$? (Lesson 2-4)
   - (A) 2.2
   - (B) 2.8
   - (C) 3.6
   - (D) 4.6

4. What is the least y-value that will satisfy the inequality when $x = 6$? (Lesson 2-6)

   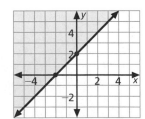

   - (A) 6
   - (B) 7
   - (C) 8
   - (D) 9

5. If $\angle ACD$ is a right angle, what is the relationship between $\angle ACF$ and $\angle DCF$? (Lesson 3-2)
   - (A) complementary angles
   - (B) corresponding angles
   - (C) supplementary angles
   - (D) vertical angles

6. Suppose $\overrightarrow{BR}$ bisects $\angle ABC$. If $m\angle CBR = 6x + 5$ and $m\angle ABR = 7x - 2$, find $m\angle ABC$. (Lesson 3-3)
   - (A) 7
   - (B) 47
   - (C) 63
   - (D) 94

7. Which of the following is an example of inductive reasoning? (Lesson 3-5)
   - (A) Carlos learns that the measures of all acute angles are less than 90°. He conjectures that if he sees an acute angle, its measure will be less than 90°.
   - (B) Carlos reads in his textbook that the measure of all right angles is 90°. He conjectures that the measure of each right angle in a square equals 90°.
   - (C) Carlos measures the angles of several triangles and finds that the sum of their measures is 180°. He conjectures that the sum of the measures of the angles of any triangle is always 180°.
   - (D) Carlos knows the sum of the measures of the angles in a square is always 360°. He conjectures that if he draws a square, the sum of the measures of the angles will be 360°.

8. Which of the following is the converse of the statement *If T is the midpoint of OB, then OT + TB = OB*? (Lesson 3-6)
   - (A) If $OT + TB = OB$, then T is the midpoint of OB.
   - (B) If T is *not* the midpoint of OB, then $OT + TB \neq OB$.
   - (C) If $OT + TB \neq OB$, then T is *not* the midpoint of OB.
   - (D) If T is the midpoint of OB, then $OT + TB \neq OB$.

Test-Taking Tip

Question 7
Review any terms that you have learned before you take a test. Remember that *deductive reasoning* involves drawing a conclusion from a general rule, and that *inductive reasoning* involves making a general rule after observing several examples.

Preparing for Standardized Tests
For test-taking strategies and more
practice, see pages 709-724.

Part 2 | Short Response/Grid In

**Record your answers on the answer sheet
provided by your teacher or on a sheet of paper.**

9. By law, the length of an official United States flag must be $1\frac{1}{10}$ times its width. If the width of a flag is $3\frac{1}{2}$ ft, what is the length? (Lesson 1-5)

10. Evaluate x^3y^2 when $x = 3$ and $y = -1$. (Lesson 1-7)

11. The speed of light is 3×10^8 m/sec. Write this speed in standard form. (Lesson 1-8)

12. What is the next term in the following sequence? (Lesson 2-1)

$$3, 6, 10, 15, 21, \dots$$

13. In general, a female's height is a function of the length of her tibia or shin bone t. The function is $h(t) = 72.6 + 2.5t$. How tall would you expect a woman to be with a 30.5 cm tibia? (Lesson 2-2)

14. Solve $3(a - 2) = 2(a + 1)$. (Lesson 2-5)

The stem-and-leaf plot records the number of CDs each of fifteen friends owns. Use the data for Questions 15–17. (Lesson 2-7)

Stems	Leaves
2	6 9 9
3	0 4 5 7 9
4	2 3 3 3 6 6 8

2|6 represents 26 CDs.

15. Find the mode of the data.

16. Find the median of the data.

17. Find the mean of the data.

18. In the figure, $AC = 33$. Find AB. (Lesson 3-1)

19. In the figure, $\overrightarrow{AB}$ and $\overrightarrow{CD}$ intersect at point E, and $\overrightarrow{EK}$ bisects $\angle AED$. Find $m\angle AEC$. (Lesson 3-3)

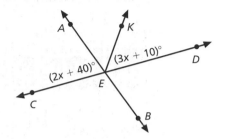

20. How many dots are in the next figure in the pattern? (Lesson 3-5)

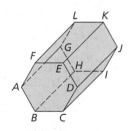

Part 3 | Extended Response

Record your answers on a sheet of paper. Show your work.

21. Describe each segment in the figure as either parallel, perpendicular, or skew to segment AB. If a segment could be extended to intersect $\overrightarrow{AB}$, describe the segment as intersecting. (Lesson 3-4)

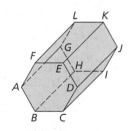

22. Write a proof for the following. (Lesson 3-7)

Given $\angle 1$ and $\angle 3$ are vertical angles.
$m\angle 1 = (3x + 5)°$, $m\angle 3 = (2x + 8)°$

Prove $m\angle 1 = 14°$

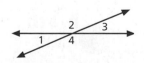

Triangles, Quadrilaterals, and Other Polygons

THEME: Art and Design

Many of the designs found on ancient murals and pottery derive their beauty from complex patterns of geometric shapes. Modern sculptures, buildings, and bridges also rely on geometric characteristics for beauty and durability. Today, computers give graphic designers, architects, and engineers a means for experimenting with design elements.

- **Jewelers** (page 159) design jewelry, cut gems, make repairs, and use their understanding of geometry to appraise gems. Jewelers need skills in art, math, mechanical drawing, and chemistry to practice their trade.

- **Animators** (page 177) create pictures that are filmed frame by frame to create motion. Many animators use computers to create three-dimensional backgrounds and characters. Animators need an understanding of perspective to create realistic drawings.

Math
nline

mathmatters3.com/chapter_theme

Suspension Bridges of New York

Name	Year opened	Length of main span	Height of towers	Clearance above water	Cost of original structure
Brooklyn Bridge	1883	1595.5 ft	276.5 ft	135 ft	$15,100,000
Williamsburg Bridge	1903	1600 ft	310 ft	135 ft	$30,000,000
Manhattan Bridge	1909	1470 ft	322 ft	135 ft	$25,000,000
George Washington Bridge	1931 1962*	3500 ft	604 ft	213 ft	$59,000,000
Verrazano Narrows Bridge	1964 1969*	4260 ft	693 ft	228 ft	$320,126,000

* lower deck added

Data Activity: Suspension Bridges of New York

Use the table for Questions 1–4.

1. For which bridge is the ratio of tower height to length of main span closest to 1 : 5?

2. The width of the George Washington Bridge is 119 ft. What is the area of the main span in square feet? (*Hint:* Use the formula $A = \ell w$, where ℓ = length and w = width.)

3. Suppose you want to make a scale model of the Verrazano Narrows Bridge. The entire model must be no more than 4 ft in length. Choose a scale and then find the model's length of main span, clearance above water, and height of towers.

4. By what percent did the cost of building a bridge increase from 1931 to 1964? Round to the nearest whole percent. Disregard differences in bridge size.

CHAPTER INVESTIGATION

A truss bridge covers the span between two supports using a system of angular braces to support weight. Triangles are often used in the building of truss bridges because the triangle is the strongest supporting polygon. The railroads often built truss bridges to support the weight of heavy locomotives.

Manhattan Bridge

Working Together

Design a system for a truss bridge similar to the examples shown throughout this chapter. Carefully label the measurements and angles in your design. Build a model of the truss using straws, toothpicks or other suitable materials. Use the Chapter Investigation icons to assist your group in designing the structure.

Are You Ready?

Refresh Your Math Skills for Chapter 4

The skills on these two pages are ones you have already learned. Stretch your memory and complete the exercises. For additional practice on these and more prerequisite skills, see pages 654–661.

You will be learning more about geometric shapes and their properties. Now is a good time to review what you already have learned about polygons and triangles.

POLYGONS

A **polygon** is a closed plane figure formed by joining 3 or more line segments at their endpoints. Polygons are named for the number of their sides.

Tell whether each figure is a polygon. If not, give a reason.

1. 　　2. 　　3. 　　4.

Give the best name for each polygon.

5. 　　6. 　　7. 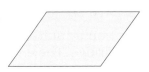　　8.

9. 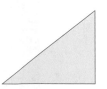　　10. 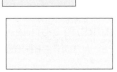　　11. 　　12.

CONGRUENT TRIANGLES

Triangles can be determined to be congruent, or having the same size and shape, by three tests:

Triangles with the same measure of two angles and the included side are congruent.

Triangles with the same measure of two sides and the included angle are congruent.

Triangles with the same measures of three sides are congruent.

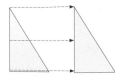

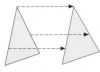

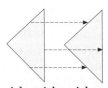

angle-side-angle
ASA

side-angle-side
SAS

side-side-side
SSS

Tell whether each pair of triangles is congruent. If they are congruent, identify how you determined congruency.

13.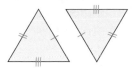

14.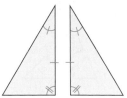

15.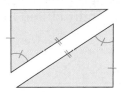

16.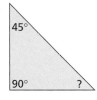

17.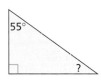

18.

ANGLES OF TRIANGLES

Example   The sum of the measures of the interior angles of any triangle is 180°. Find the unknown measure.

$$53° + 73° + x° = 180°$$
$$126° + x° = 180°$$
$$x° = 54°$$

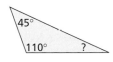

Find the unknown measures in each triangle.

19.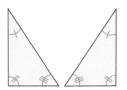

20.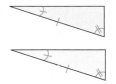

21.

22.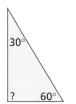

23.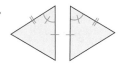

24.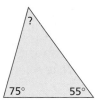

25.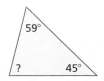

26.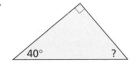

27.

4-1 Triangles and Triangle Theorems

Goals
- Solve equations to find measure of angles.
- Classify triangles according to their sides or angles.

Applications Technical Art, Construction, Engineering

Work with a partner.

1. Using a pencil and a straightedge, draw and label a triangle as shown below. Carefully cut on the straight lines. Then tear off the four labeled angles.

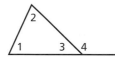 

a. What relationships can you discover among the four angles?

b. Using these relationships, make at least two conjectures that you think apply to all triangles.

BUILD UNDERSTANDING

A **triangle** is the figure formed by the segments that join three noncollinear points. Each segment is called a **side** of the triangle. Each point is called a **vertex** (plural: *vertices*). The angles determined by the sides are called the **interior angles** of the triangle.

triangle *ABC*
(△*ABC*)

sides: $\overline{AB}$, $\overline{BC}$, and $\overline{AC}$
vertices: points *A*, *B*, and *C*
interior angles: ∠*A*, ∠*B*, and ∠*C*

Often a triangle is classified by relationships among its sides.

| **Equilateral triangle** three sides of equal length | **Isosceles triangle** at least two sides of equal length | **Scalene triangle** no sides of equal length |

A triangle also can be classified by its angles.

Acute triangle three acute angles   **Obtuse triangle** one obtuse angle   **Right triangle** one right angle   **Equiangular triangle** three angles equal in measure

You probably remember a special property of the measures of the angles of a triangle. Since the fact is a theorem, it can be proved true.

| **The Triangle-Sum Theorem** | The sum of the measures of the angles of a triangle is 180°. |

As you read the proof of the theorem, notice that it makes use of a line that intersects one of the vertices of the triangle and is parallel to the opposite side. This line, which has been added to the diagram to help in the proof, is called an **auxiliary line.**

Given $\triangle ABC; \overleftrightarrow{DB} \parallel \overline{AC}$
Prove $m\angle 1 + m\angle 2 + m\angle 3 = 180°$

Statements	Reasons
1. $\triangle ABC; \overleftrightarrow{DB} \parallel \overline{AC}$	1. given
2. $m\angle 4 + m\angle 2 = m\angle DBC$ $m\angle DBC + m\angle 5 = 180°$	2. angle addition postulate
3. $m\angle 4 + m\angle 2 + m\angle 5 = 180°$	3. substitution property
4. $m\angle 4 = m\angle 1$ $m\angle 5 = m\angle 3$	4. If two parallel lines are cut by a transversal, then alternate interior angles are equal in measure.
5. $m\angle 1 + m\angle 2 + m\angle 3 = 180°$	5. substitution property

Check Understanding

Explain how the substitution property was used in both Step 3 and Step 5 of the proof.

The triangle-sum theorem is useful in art and design.

Example 1

TECHNICAL ART An artist is using the figure at the right to create a diagram for a publication. Using the triangle-sum theorem, find $m\angle Q$.

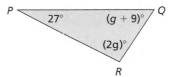

Solution

From the triangle-sum theorem, you know that the sum of the measures of the angles of a triangle is 180°. Use this fact to write and solve an equation.

$$m\angle P + m\angle Q + m\angle R = 180$$

$27 + (g + 9) + 2g = 180$ Combine like terms.

$3g + 36 = 180$ Add -36 to each side.

$3g = 144$ Multiply each side by $\frac{1}{3}$.

$g = 48$

So, the value of g is 48. From the figure $m\angle Q = (g + 9)°$. Substituting 48 for g, $m\angle Q = (48 + 9)° = 57°$.

An **exterior angle** of a triangle is an angle that is both adjacent to and supplementary to an interior angle, as shown at the right. The following is an important theorem concerning exterior angles.

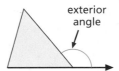

exterior angle

The Exterior Angle Theorem	The measure of an exterior angle of a triangle is equal to the sum of the measures of the two nonadjacent (remote) interior angles.

You will have an opportunity to prove the exterior angle theorem in Exercise 12 on the following page. Example 2 shows one way the theorem can be applied.

Example 2

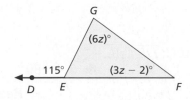

In the figure at the right, find $m\angle EFG$.

Solution

Notice that $\angle DEG$ is an exterior angle, while $\angle EGF$ and $\angle EFG$ are nonadjacent interior angles. Use the exterior angle theorem to write and solve an equation.

$m\angle DEG = m\angle EGF + m\angle EFG$

$115 = 6z + (3z - 2)$ Combine like terms.

$115 = 9z - 2$ Add 2 to each side.

$117 = 9z$ Multiply each side by $\frac{1}{9}$.

$13 = z$

So, the value of z is 13. From the figure, $m\angle EFG = (3z - 2)°$. Substituting 13 for z, $m\angle EFG = (3 \cdot 13 - 2)° = (39 - 2)° = 37°$.

◥ TRY THESE EXERCISES

Refer to $\triangle RST$ below. Find the measure of each angle.

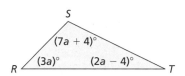

Refer to $\triangle XYZ$ below. Find the measure of each angle.

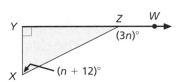

1. $\angle R$ **2.** $\angle S$ **3.** $\angle T$ **4.** $\angle YXZ$ **5.** $\angle XZW$ **6.** $\angle XZY$

◥ PRACTICE EXERCISES • For Extra Practice, see page 673.

Find the value of x in each figure.

7.

8.

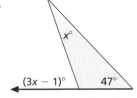

9.

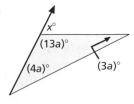

10. WRITING MATH How many exterior angles does a triangle have? Draw a triangle and label all its exterior angles.

11. The measure of the largest angle of a triangle is twice the measure of the smallest angle. The measure of the third angle is 10° less than the measure of the largest angle. Find all three measures.

12. In the figure below, $\overline{XZ} \perp \overline{YZ}$. Find $m\angle XYZ$.

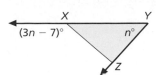

13. BRIDGE BUILDING The plans for a bridge call for the addition of triangular bracing to increase the amount of weight the bridge can hold. On the plans, $\triangle FGH$ is drawn so that the $m\angle F$ is 14° less than three times the $m\angle G$, and $\angle H$ is a right angle. Find the measure of each angle.

 GEOMETRY SOFTWARE On a coordinate plane, draw the triangle with the given vertices. Measure all sides and angles. Then classify the triangle, first by its sides, then by its angles.

14. $A(-5, 0); B(1, 2); C(1, -2)$

15. $J(-1, -3); K(6, 2); L(-7, 1)$

16. $R(1, -5); S(-3, -1); T(6, 0)$

17. $D(3, -5); E(-4, -3); F(-2, 4)$

18. Copy and complete this proof of the exterior angle theorem.

Given $\triangle ABC$, with exterior $\angle 4$

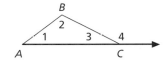

Prove ___?___

Statements	Reasons
1. ___?___	1. ___?___
2. $m\angle 1 + m\angle 2 + m\angle 3 = 180°$	2. ___?___
3. $m\angle 4 + m\angle 3 = 180°$	3. ___?___
4. $m\angle 1 + m\angle 2 + m\angle 3 = m\angle 4 + m\angle 3$	4. ___?___
5. ___?___	5. ___?___

■ EXTENDED PRACTICE EXERCISES

Determine whether each statement is *always*, *sometimes*, or *never* true.

19. Two interior angles of a triangle are obtuse angles.

20. Two interior angles of a triangle are acute angles.

21. An exterior angle of a triangle is an obtuse angle.

22. The measure of an exterior angle of a triangle is greater than the measure of each nonadjacent interior angle.

■ MIXED REVIEW EXERCISES

Find each length. (Lesson 3-1)

23. In the figure below, $AC = 75$. Find BC.

$$A \bullet \overset{2x - 3}{\underset{B}{\rule{2cm}{0.4pt}}} \overset{3x + 13}{\rule{2cm}{0.4pt}} \bullet C$$

24. In the figure below, $PR = 138$. Find PQ.

$$P \bullet \overset{2x + 5}{\rule{2cm}{0.4pt}} \overset{x - 17}{\underset{Q}{\rule{2cm}{0.4pt}}} \bullet R$$

Find the mean, median and mode of each set of data. (Lesson 2-7)

25. 4 8 7 10 8 8 3
      7 9 14 3 5

26. 8 7 3 9 9 5 7 9
      1 3 2 6 9 1 4

Congruent Triangles

Goals ■ Prove triangles are congruent.

Applications Animation, Construction, Engineering

For the following activity, use a protractor and a metric ruler, or use geometry software. Give lengths to the nearest tenth of a centimeter, and give angle measures to the nearest degree.

a. Draw $\triangle ABC$, with $m\angle A = 40°$, $AB = 7$ cm, and $m\angle B = 60°$. What is the measure of $\angle C$? What is the length of $\overline{AC}$? of $\overline{BC}$?

b. Draw $\triangle DEF$, with $DF = 5$ cm, $m\angle D = 120°$, and $DE = 6$ cm. What is the measure of $\angle E$? of $\angle F$? What is the length of $\overline{EF}$?

c. Draw $\triangle GHJ$, with $m\angle G = 35°$, $m\angle H = 45°$, and $m\angle J = 100°$. What is the length of $\overline{GH}$? of $\overline{GJ}$? of $\overline{HJ}$?

d. Draw $\triangle KLM$, with $KM = 3$ cm, $KL = 6$ cm, and $LM = 4$ cm. What is the measure of $\angle K$? of $\angle L$? of $\angle M$?

Check Understanding

Name all the pairs of congruent parts in the triangles below. Then state the congruence between the triangles.

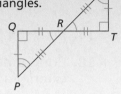

Is there a different way to state the congruence?

◤ BUILD UNDERSTANDING

When two geometric figures have the same size and shape, they are said to be **congruent**. The symbol for *congruence* is ≅.

It is fairly easy to recognize when segments and angles are congruent. **Congruent segments** are segments with the same length. **Congruent angles** are angles with the same measure.

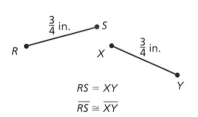

$$RS = XY$$
$$\overline{RS} \cong \overline{XY}$$

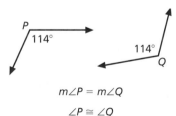

$$m\angle P = m\angle Q$$
$$\angle P \cong \angle Q$$

Congruent triangles are two triangles whose vertices can be paired in such a way that each angle and side of one triangle is congruent to a *corresponding angle* or *corresponding side* of the other. For instance, the markings in the triangles at the right indicate these six congruences.

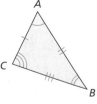

$$\angle A \cong \angle Z \qquad \overline{AB} \cong \overline{ZX}$$
$$\angle B \cong \angle X \qquad \overline{BC} \cong \overline{XY}$$
$$\angle C \cong \angle Y \qquad \overline{AC} \cong \overline{ZY}$$

So, the triangles are congruent, and you can pair the vertices in the following correspondence.

$$A \leftrightarrow Z \qquad\qquad B \leftrightarrow X \qquad\qquad C \leftrightarrow Y$$

$$\triangle ABC \cong \triangle ZXY$$

To state the congruence between the triangles, you list the vertices of each triangle in the *same order* as this correspondence.

You can use given information to prove that two triangles are congruent. One way to do this is to show the triangles are congruent *by definition*. That is, you prove that all six parts of one triangle are congruent to six corresponding parts of the other. However, this can be quite cumbersome.

Fortunately, triangles have special properties that allow you to prove triangles congruent by identifying only *three* sets of corresponding parts. The first way to do this is summarized in the *SSS postulate*.

Postulate 11	**The SSS Postulate** If three sides of one triangle are congruent to three sides of another triangle, then the triangles are congruent.

Example 1

ANIMATION The figure shown is part of a perspective drawing for a background scene of a city. How can the artist be sure that the two triangles are congruent?

Given $JK \cong JM$; $KL \cong ML$

Prove $\triangle JKL \cong \triangle JML$

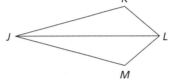

Reading Math

It logically follows from the statement $\overline{AB} \cong \overline{CD}$ that $AB = CD$. The same is true of the statements $\angle A \cong \angle B$ and $m\angle A = m\angle B$.

Solution

Statements	Reasons
1. $\overline{JK} \cong \overline{JM}$; $\overline{KL} \cong \overline{ML}$	1. given
2. $\overline{JL} \cong \overline{JL}$	2. reflexive property
3. $\triangle JKL \cong \triangle JML$	3. SSS postulate

GEOMETRY SOFTWARE Use geometry software to explore the postulate. Draw two triangles so that the three sides of one triangle are congruent to the three corresponding sides of the other triangle. Measure the interior angles of both triangles. They are also congruent.

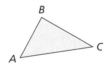

$\angle A$ is included between $\overline{AB}$ and $\overline{AC}$. $\overline{AB}$ is included between $\angle A$ and $\angle B$.

Sometimes it is helpful to describe the parts of a triangle in terms of their relative positions.

Each angle of a triangle is formed by two sides of the triangle. In relation to the two sides, this angle is called the **included angle**. Each side of a triangle is included in two angles of the triangle. In relation to the two angles, this side is called the **included side**.

Using these terms, it is now possible to describe two additional ways of showing that two triangles are congruent.

Postulate 12	**The SAS Postulate** If two sides and the included angle of one triangle are congruent to two corresponding sides and the included angle of another triangle, then the triangles are congruent.
Postulate 13	**The ASA Postulate** If two angles and the included side of one triangle are congruent to two corresponding angles and the included side of another triangle, then the triangles are congruent.

Example 2

Given $\overline{VW} \cong \overline{ZY}$; $\angle V \cong \angle Z$
$\overline{VW} \perp \overline{WY}$; $\overline{ZY} \perp \overline{WY}$

Prove $\triangle VWX \cong \triangle ZYX$

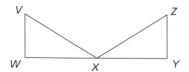

Problem Solving Tip

Before you start to write a proof, it is a good idea to develop a *plan* for the proof. When the proof involves congruent triangles, many students find it helpful to first copy the figure and mark as many congruences as they can. For instance, after reviewing the given information, the figure for Example 2 would be copied and marked as follows.

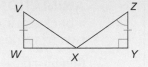

Once the figure is marked, it becomes clear that the plan for proof will involve the ASA Postulate.

Solution

Statements	Reasons
1. $\overline{VW} \cong \overline{ZY}$; $\angle V \cong \angle Z$ $\overline{VW} \perp \overline{WY}$; $\overline{ZY} \perp \overline{WY}$	1. given
2. $\angle W$ and $\angle Y$ are right angles.	2. definition of perpendicular lines
3. $m\angle W = 90°$; $m\angle Y = 90°$	3. definition of right angle
4. $m\angle W = m\angle Y$, or $\angle W \cong \angle Y$	4. transitive property of equality
5. $\triangle VWX \cong \triangle ZYK$	5. ASA postulate

◤ TRY THESE EXERCISES

1. Copy and complete this proof.

Given $\overline{RQ} \cong \overline{RS}$; $\overline{RT}$ bisects $\angle QRS$.

Prove $\triangle QRT \cong \triangle SRT$

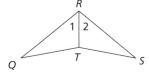

Statements	Reasons
1. ___?___	1. ___?___
2. $m\angle 1 = m\angle 2$, or $\angle 1 \cong \angle 2$	2. definition of ___?___
3. ___?___	3. ___?___ property
4. $\triangle QRT \cong \triangle SRT$	4. ___?___

2. CONSTRUCTION Plans call for triangular bracing to be added to a horizontal beam. Prove the triangles are congruent by writing a two-column proof.

Given $\overline{GL} \cong \overline{JK}$; $\overline{HL} \cong \overline{HK}$
Point H is the midpoint of $\overline{GJ}$.

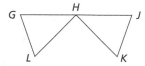

Prove $\triangle GHL \cong \triangle JHK$

◤ PRACTICE EXERCISES • For Extra Practice, see page 673.

3. Write a two-column proof.

Given $\overline{AB} \cong \overline{CB}$; $\overline{EB} \cong \overline{DB}$
$\overline{AD}$ and $\overline{CE}$ intersect at point B.

Prove $\triangle ABE \cong \triangle CBD$

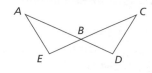

ENGINEERING The figures in Exercises 4–7 are taken from truss bridge designs. Each figure contains a pair of congruent overlapping triangles.

Use the given information to complete the congruence statement. Then name the postulate that would be used to prove the congruence. (You do *not* need to write the proof.)

4.

Given $\overline{PQ} \cong \overline{SR}$; $\overline{QS} \cong \overline{RP}$
$\triangle PQS \cong$ ___?___

5.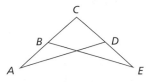

Given $\overline{AC} \cong \overline{EC}$; $\angle A \cong \angle E$
$\triangle ACD \cong$ ___?___

6.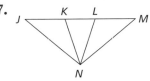

Given $\overline{EH} \cong \overline{FG}$; $\overline{EH} \perp \overline{EF}$; $\overline{FG} \perp \overline{EF}$
$\triangle HEF \cong$ ___?___

7.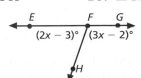

Given $\overline{JL} \cong \overline{MK}$; $\angle J \cong \angle M$; $\overline{JN} \cong \overline{MN}$
$\triangle JNL \cong$ ___?___

 GEOMETRY SOFTWARE Use geometry software or paper and pencil to draw the figures in Exercises 8–9 on a coordinate plane.

8. Draw $\triangle MNP$ with vertices $M(-5, 5)$, $N(3, 5)$, and $P(3, -6)$. Then draw $\triangle QRS$ with vertices $Q(-4, 2)$, $R(7, -6)$ and $S(-4, -6)$. Explain how you know that the triangles are congruent. Then state the congruence.

9. Draw $\triangle ABC$ with vertices $A(-3, 5)$, $B(6, 5)$, and $C(6, -8)$. Then graph points $X(2, 2)$ and $Y(-7, 2)$. Find two possible coordinates of a point Z so that $\triangle ABC \cong \triangle XYZ$.

 10. CHAPTER INVESTIGATION Design a 20-foot side section of a truss bridge. Draw your design to the scale 1 in. = 2 ft.

EXTENDED PRACTICE EXERCISES

11. Write a proof of the following statement.

If two legs of one right triangle are congruent to two legs of another right triangle, then the triangles are congruent.

12. WRITING MATH Write a convincing argument to explain why there is no SSA postulate for congruence of triangles.

MIXED REVIEW EXERCISES

Find the measure of each angle. (Lesson 3-2)

13. $\angle ABD$ **14.** $\angle CBD$ **15.** $\angle EFH$ **16.** $\angle GFH$

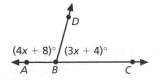

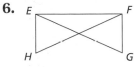

Review and Practice Your Skills

PRACTICE ◼ LESSON 4-1

Find the value of *x* in each figure.

1.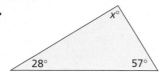

2.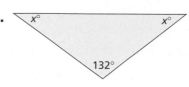

3.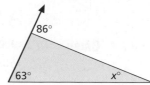

4.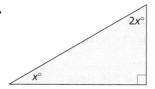

5.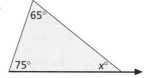

6.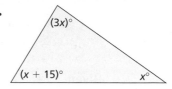

Determine whether each statement is *true* or *false*.

7. If two angles in a triangle are acute, then the third angle is always obtuse.

8. If one angle in a triangle is obtuse, then the other two angles are always acute.

9. If one exterior angle of a triangle is obtuse, then all three interior angles are acute.

10. If two angles in a triangle are congruent, then the triangle is equiangular.

On a coordinate plane, sketch the triangle with the given vertices. Then classify the triangle, first by its sides, then by its angles.

11. $A(2, 2)$; $B(-3, -3)$; $C(-3, 2)$ 12. $X(6, -2)$; $Y(-4, -2)$; $Z(1, 0)$ 13. $M(-1, 4)$; $N(1, 0)$; $P(-4, 0)$

PRACTICE ◼ LESSON 4-2

14. Copy and complete this proof.

    Given $\overline{AB} \parallel \overline{DE}$; C is the midpoint of $\overline{BD}$.
    Prove $\triangle ABC \cong \triangle EDC$

    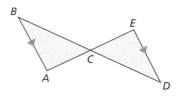

Statements	Reasons
1. $\overline{AB} \parallel \overline{DE}$	1. __?__
2. $\angle B \cong \angle D$	2. __?__
3. $\angle BCA \cong \angle ECD$	3. __?__
4. __?__	4. given
5. $\overline{BC} \cong \overline{CD}$	5. __?__
6. __?__	6. ASA postulate

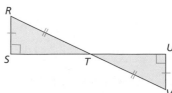

15. Name all the pairs of congruent parts in these triangles. Then state the congruence between the triangles.

16. *True* or *false*: If three angles of one triangle are congruent to three angles of another triangle, then the triangles are congruent.

Determine whether each statement is *always*, *sometimes*, or *never* true.
(Lesson 4-1)

17. There are two exterior angles at each vertex of a triangle.

18. An exterior angle of a triangle is an acute angle.

19. Two of the three angles in a triangle are complementary angles.

20. The sum of the measures of the angles in a triangle is 90°.

Use the given information to complete each congruence statement. Then name the postulate that would be used to prove the congruence. (You do *not* need to write the proof.) (Lesson 4-2)

21. Given $\overline{PQ} \cong \overline{NO}$; $\overline{QR} \cong \overline{MO}$; $\angle Q \cong \angle O$
Prove $\triangle PQR \cong$ ___?___

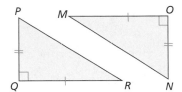

22. Given $\overline{EY} \cong \overline{EF}$; $\overline{DY} \cong \overline{LF}$; $\overline{DE} \cong \overline{LE}$
Prove $\triangle DYE \cong$ ___?___

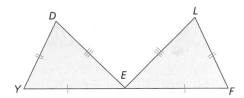

MathWorks — Career – Jeweler
Workplace Knowhow

A jeweler designs and repairs jewelry, cuts gems and appraises the value of gemstones and jewelry. Most jewelers go through an apprenticeship program where they work under an experienced jeweler to hone their skills and learn new techniques. A background in art, math, mechanical drawing and chemistry are all useful when working with gems and precious metals. Math skills help a jeweler in the areas of design and gem cutting. Jewelers use computer-aided design (CAD) programs to design jewelry to meet a customer's expectations. A symmetrically cut gem is a valuable gem. A poorly cut gem becomes a wasted investment for the jeweler.

In the gem cut shown to the right, all triangles shown can be classified as isosceles triangles.

1. What additional classifications can be given to triangle *ABC*?

2. What is the measure of $\angle BCE$?

3. Sides *CE* and *DE* are congruent and $\angle BCE$ and $\angle EDF$ are congruent. Angle *DEF* measures 38°. Are triangles *BCE* and *FDE* congruent? If so, what postulate could be used to prove the congruence?

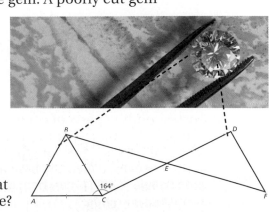

Congruent Triangles and Proofs

Goals
- Establish congruence between two triangles to show that corresponding parts are congruent.
- Find angle and side measures of triangles.

Applications Design, Architecture, Construction, Engineering

Fold a piece of paper and draw a segment on it as shown. Now cut both thicknesses of paper along the segment. Unfold and label the triangle.

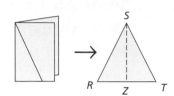

1. Are there any perpendicular segments on the triangle?

2. Does any segment lie on an angle bisector of the triangle?

3. List as many congruences as you can among the segments, angles, and triangles that you see on the folded triangle.

◤ BUILD UNDERSTANDING

The SSS, SAS, and ASA postulates help you determine a congruence between two triangles by identifying just three pairs of corresponding parts. Once you establish a congruence, you may conclude that *all* pairs of corresponding parts are congruent. Example 1 shows how this fact can be used to show that two angles are congruent.

Example 1

Given $\overline{AB} \cong \overline{CB}$; $\overline{AD} \cong \overline{CD}$

Prove $\angle A \cong \angle C$

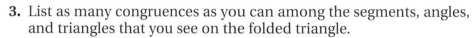

Solution

Statements	Reasons
1. $\overline{AB} \cong \overline{CB}$; $\overline{AD} \cong \overline{CD}$	1. given
2. $\overline{BD} \cong \overline{BD}$	2. reflexive property
3. $\triangle ABD \cong \triangle CBD$	3. SSS postulate
4. $\angle A \cong \angle C$	4. Corresponding parts of congruent triangles are congruent.

> **Reading Math**
>
> The final reason of the proof in Example 1 is *Corresponding parts of congruent triangles are congruent*. This fact is used so often that it is commonly abbreviated *CPCTC*.

Corresponding parts of congruent triangles are used in the proofs of many theorems. For example, an isosceles triangle is a triangle with two **legs** of equal length. The third side is the **base**. The angles at the base are called the **base angles**, and the third angle is the **vertex angle**. *CPCTC* can be used to prove the following theorem about base angles.

legs: $\overline{AB}, \overline{CB}$
base: $\overline{AC}$
base angles: $\angle A, \angle C$
vertex angle: $\angle B$

The Isosceles Triangle Theorem	If two sides of a triangle are congruent, then the angles opposite those sides are congruent. This is sometimes stated: *Base angles of an isosceles triangle are congruent.*

Given $\triangle ABC$ is isosceles, with base $\overline{AC}$.
$\overline{BX}$ bisects $\angle ABC$.

Prove $\angle A \cong \angle C$

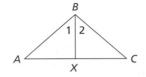

Statements	Reasons
1. $\triangle ABC$ is isosceles, with base $\overline{AC}$. $\overline{BX}$ bisects $\angle ABC$.	1. given
2. $\overline{AB} \cong \overline{CB}$	2. definition of isosceles $\triangle$
3. $\angle 1 \cong \angle 2$	3. definition of $\angle$ bisector
4. $\overline{BX} \cong \overline{BX}$	4. reflexive property
5. $\triangle AXB \cong \triangle CXB$	5. SAS postulate
6. $\angle A \cong \angle C$	6. CPCTC

Example 2

DESIGN An artist is positioning the design elements for a new company logo. At the center of the logo is the triangle shown in the figure. Find $m\angle P$.

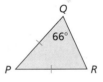

Solution

Since $PQ = PR$, $\triangle PQR$ is isosceles with base $\overline{QR}$. By the isosceles triangle theorem, $m\angle R = m\angle Q = 66°$. By the triangle-sum theorem, $m\angle P + 66° + 66° = 180°$, or $m\angle P = 48°$.

Check Understanding

How would the solution of Example 2 be different if the measure of $\angle Q$ were 54°?

A statement that follows directly from a theorem is called a **corollary**. The following are corollaries to the isosceles triangle theorem.

Corollary 1	If a triangle is equilateral, then it is equiangular.
Corollary 2	The measure of each angle of an equilateral triangle is 60°.

The converse of the isosceles triangle theorem is the *base angles theorem*.

The Base Angles Theorem	If two angles of a triangle are congruent, then the sides opposite those angles are congruent.
Corollary	If a triangle is equiangular, then it is equilateral.

1. Copy and complete this proof.

   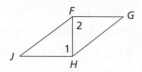

   Given $\overline{FG} \cong \overline{HJ}$; $\overline{FG} \perp \overline{FH}$; $\overline{JH} \perp \overline{FH}$

   Prove $\angle J \cong \angle G$

Statements	Reasons
1. ___?___	1. ___?___
2. $\angle 1$ and $\angle 2$ are right angles.	2. definition of ___?___
3. $m\angle 1 = 90°$; $m\angle 2 = 90°$	3. definition of ___?___
4. ___?___	4. transitive property of equality
5. ___?___	5. reflexive property
6. $\triangle JFH \cong \triangle GHF$	6. ___?___
7. ___?___	7. ___?___

Find the value of *n* in each figure.

2.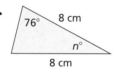

3.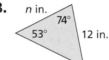

4.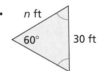

Find the value of *x* in each figure.

5.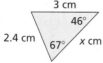

6.

7.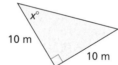

8. **ARCHITECTURE** An architect sees the figure at the right on a set of building plans. The architect wants to be certain that $\angle T \cong \angle R$. Copy and complete this proof.

   Given $\overline{PS} \cong \overline{QS}$; $\overline{PT} \cong \overline{QR}$

   Point *S* is the midpoint of $\overline{TR}$.

   Prove $\angle T \cong \angle R$

Statements	Reasons
1. ___?___	1. ___?___
2. ___?___	2. definition of ___?___
3. ___?___	3. SSS postulate
4. $\angle T \cong \angle R$	4. ___?___

9. **YOU MAKE THE CALL** A base angle of an isosceles triangle measures 70°. Cina says the two remaining angles must each measure 55°. What mistake has Cina made?

Name all the pairs of congruent angles in each figure.

10.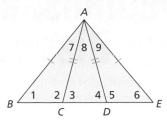

11.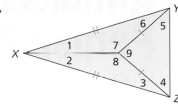

12. BRIDGE BUILDING On a truss bridge, steel cables cross as shown in the figure below. The inspector needs to be certain that $\overline{GL}$ and $\overline{JK}$ are parallel. Copy and complete the proof.

Given Point H is the midpoint of $\overline{GK}$.
Point H is the midpoint of $\overline{LJ}$.

Prove $\overline{GL} \parallel \overline{JK}$

Statements	Reasons
1. ___?___	1. ___?___
2. $\overline{GH} \cong \overline{KH}$; ___?___	2. ___?___
3. $\angle 1$ and $\angle 2$ are vertical angles.	3. definition of ___?___
4. ___?___	4. ___?___ theorem
5. ___?___	5. SAS postulate
6. $\angle G \cong \angle K$, or $m\angle G = m\angle K$	6. ___?___
7. $\overline{GL} \parallel \overline{JK}$	7. If ___?___, then ___?___.

DATA FILE For Exercises 13–16, use the data on the types of structural supports used in architecture on page 644. For each type of support, find the measure of each angle in the diagram using a protractor.

13. king-post   **14.** queen-post   **15.** scissors   **16.** Fink

EXTENDED PRACTICE EXERCISES

17. Suppose that you join the midpoints of the sides of an isosceles triangle to form a triangle. What type of triangle do you think is formed?

18. WRITING MATH Write a proof of the second corollary to the isosceles triangle theorem: *The measure of each angle of an equilateral triangle is 60°.*

MIXED REVIEW EXERCISES

Exercises 19–22 refer to the protractor at the right.
(Lesson 3-2)

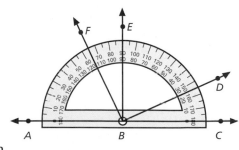

19. Name the straight angle.

20. Name the three right angles.

21. Name all the obtuse angles and give the measure of each.

22. Name all the acute angles and give the measure of each.

Altitudes, Medians, and Perpendicular Bisectors

Goals ■ Identify and sketch altitudes and medians of a triangle and perpendicular bisectors of sides of a triangle.

Applications Architecture, Physics, Service

Working with a partner, draw a large acute triangle *ABC*, as shown at the right.

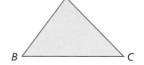

1. With compass point at point *A*, draw two arcs of equal radii that intersect $\overline{BC}$. Label the points of intersection *X* and *Y*.

2. Choose a suitable radius of the compass. With compass point first at point *X*, then at point *Y*, draw two arcs that intersect at *Z*.

3. Using a straightedge, draw $\overleftrightarrow{AZ}$.

4. Label point *D* where $\overleftrightarrow{AZ}$ intersects $\overline{BC}$.

5. Repeat steps **1** through **4**, but this time place the compass point at point *B* and construct a line that intersects $\overline{AC}$ at point *E*.

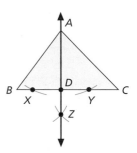

6. Repeat steps **1** through **4** again, but now place the compass point at point *C* and construct a line that intersects $\overline{AB}$ at point *F*.

7. What observations do you make about the lines you constructed?

▨ BUILD UNDERSTANDING

There are several special segments related to triangles. An **altitude** is a perpendicular segment from a vertex to the line containing the opposite side. A **median** is a segment with endpoints that are a vertex of the triangle and the midpoint of the opposite side.

Example 1

Sketch all the altitudes and medians of △*ABC*.

Solution

There are three altitudes, shown below in red.

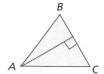 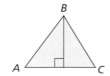

Similarly, there are three medians, shown below in blue.

 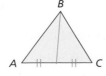 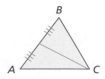

Any line, ray, or segment that is perpendicular to a segment at its midpoint is called a **perpendicular bisector** of the segment. In a given plane, however, there is exactly one line perpendicular to a segment at its midpoint. That line usually is called the perpendicular bisector of the segment. The following is an important theorem concerning perpendicular bisectors.

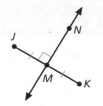

$\overleftrightarrow{MN}$ is the perpendicular bisector of $\overline{JK}$.

The Perpendicular Bisector Theorem	If a point lies on the perpendicular bisector of a segment, then the point is equidistant from the endpoints of the segment.

You will have a chance to prove this theorem in Exercise 22 on page 167.

Example 2

ARCHITECTURE A triangular construction is shown on a set of plans. The architect has determined that $\overline{DF}$ is a perpendicular bisector of $\overline{GE}$. She needs to know whether the following statements are true or false.

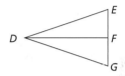

a. $\overline{EF} \cong \overline{GF}$ **b.** $\overline{DE} \cong \overline{DG}$

Solution

a. By the definition of perpendicular bisector, F is the midpoint of $\overline{GE}$. By the definition of midpoint, this means that $EF = GF$, or $\overline{EF} \cong \overline{GF}$. The given statement is *true*.

b. Point D is a point on the perpendicular bisector of $\overline{GE}$. By the perpendicular bisector theorem, this means that point D is equidistant from points G and E. That is, $DE = DG$, or $\overline{DE} \cong \overline{DG}$. The given statement is *true*.

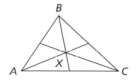

Two or more lines that intersect at one point are called **concurrent lines**. You can explore concurrence among the special segments in a triangle by using geomety software or making constructions with a compass and straightedge.

Example 3

GEOMETRY SOFTWARE Draw a scalene, acute triangle ABC. Locate the midpoints of $\overline{AB}$, $\overline{BC}$, and $\overline{AC}$. Draw the three medians of the triangle. What do you notice?

Solution

The medians are concurrent. Label the point of concurrence X.

Trace △RST onto a sheet of paper.

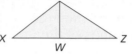

1. Sketch all the altitudes.

2. Sketch all the medians.

In △XYZ, $\overline{YW}$ is an altitude. Tell whether each statement is true, false, or cannot be determined.

3. $\overline{YW} \perp \overline{XZ}$ 4. $\overline{XW} \cong \overline{ZW}$

5. $\overline{XY} \cong \overline{ZY}$ 6. $\angle XWY \cong \angle ZWY$

7. **GEOMETRY SOFTWARE** Draw a scalene, acute triangle QPR. Construct its three altitudes. What do you observe?

8. **TALK ABOUT IT** Ezra says that an altitude and a median of a triangle could possibly be the same segment. Do you think Ezra's thinking is correct? Discuss the idea with a partner.

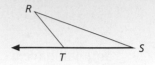

◤ **PRACTICE EXERCISES** • For Extra Practice, see page 674.

Trace △JKL, at the right, onto a sheet of paper.

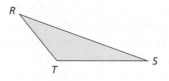

9. Sketch all the altitudes.

10. Sketch all the medians.

Exercises 11–17 refer to △EFG, at the right. Tell whether each statement is true, false, or cannot be determined.

11. $\overline{EG} \cong \overline{EF}$

13. $\overline{GH} \cong \overline{FH}$

15. $\overline{EH}$ is median of △EFG.

17. $\overline{EH}$ is an altitude of △EFG.

12. $\angle EHG \cong \angle EHF$

14. $\angle GEH \cong \angle FEH$

16. △EGH ≅ △EFH

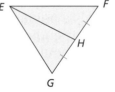

GEOMETRIC CONSTRUCTIONS Draw two copies of a scalene, acute triangle.

18. Label vertices *A*, *B*, and *C*. Bisect ∠A, ∠B, and ∠C. Label the point of concurrence *Z*. Now measure the perpendicular distance from point *Z* to each side of the triangle. What do you observe?

19. Draw the perpendicular bisectors of $\overline{AB}$, $\overline{BC}$, and $\overline{AC}$. Label the point of concurrence *W*. Measure the distance from point *W* to each vertex of the triangle. What do you observe?

20. **PHYSICS** The **center of gravity** of an object is the point at which the weight of the object is in perfect balance. Which point of concurrence do you think is the center of gravity of a triangle? Cut a large triangle out of cardboard. Using compass and straightedge, draw medians, altitudes, angle bisectors, and perpendicular bisectors. Place the eraser of a pencil at each point of concurrence. When the triangle balances, you have located the center of gravity.

21. WRITING MATH Can a side of a triangle also be an altitude or a median of the triangle? Explain your reasoning.

22. Copy and complete this proof of the perpendicular bisector theorem.

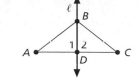

Given Line ℓ is the perpendicular bisector of $\overline{AC}$.
Point B lies on ℓ.

Prove $AB = BC$

Statements	Reasons
1. ___?___	1. ___?___
2. Point D is the midpoint of $\overline{AC}$.	2. definition of ___?___
3. $AD = CD$, or $\overline{AD} \cong \overline{CD}$	3. definition of ___?___
4. $\ell \perp \overline{AC}$	4. definition of ___?___
5. $\angle 1$ and $\angle 2$ are right angles.	5. definition of ___?___
6. $m\angle 1 = 90°$; $m\angle 2 = 90°$	6. definition of ___?___
7. $m\angle 1 = m\angle 2$, or $\angle 1 \cong \angle 2$	7. ___?___
8. $BD = BD$, or $\overline{BD} \cong \overline{BD}$	8. ___?___
9. ___?___	9. SAS postulate
10. $\overline{AB} \cong \overline{BC}$, or $AB = BC$	10. ___?___

◾ EXTENDED PRACTICE EXERCISES

23. WRITING MATH Make a list of at least eight true statements concerning $\triangle PQR$, shown at the right.

24. CHAPTER INVESTIGATION Imagine your bridge design is to be given to a construction crew. Provide information that will help the crew build the truss accurately. Indicate which line segments are parallel, label angle measures, mark right angles, and classify triangles.

◾ MIXED REVIEW EXERCISES

Exercises 25–28 refer to the figure below. (Lesson 4-4)

25. Name the midpoint of $\overline{AG}$.

26. Name the segment whose midpoint is J.

27. Name all the segments whose midpoint is E.

28. Assume that L is the midpoint of BI. What is its coordinate?

Given $f(x) = 3(x - 2)$, find each value. (Lesson 3-3)

29. $f(-3)$ **30.** $f(2)$ **31.** $f(-5)$ **32.** $f(8)$

Given $f(x) = -2(x - 3)$, find each value. (Lesson 2-2)

33. $f(3)$ **34.** $f(-4)$ **35.** $f(-2)$ **36.** $f(9)$

Review and Practice Your Skills

PRACTICE ◼ LESSON 4-3

Find the value of *n* in each figure.

1.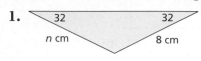
32   32
n cm   8 cm

2.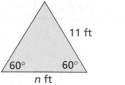
11 ft
60°   60°
n ft

3.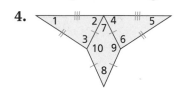
21 in.   67°
n°
21 in.

Name all pairs of congruent angles in each figure.

4.

5.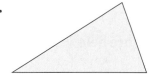

Determine whether each statement is *true* or *false*.

6. All equiangular triangles are equilateral.

7. If two sides and one angle in a triangle are congruent to two sides and one angle in another triangle, then the triangles are congruent.

8. The symmetric property applies to both congruent sides and congruent angles.

9. If two triangles share a common side, then they are congruent.

10. Given $\triangle ABC \cong \triangle DEF$, it can be shown that $\angle B \cong \angle F$ and $\overline{AB} \cong \overline{DE}$.

PRACTICE ◼ LESSON 4-4

Trace each triangle onto a sheet of paper. Sketch all the altitudes and all the medians.

11.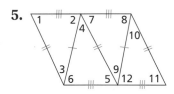

12.

13.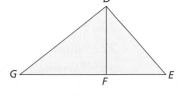

For Exercises 14–19, refer to $\triangle DEG$ at the right with altitude $\overline{DF}$. Tell whether each statement is *true* or *false*.

14. $\angle E \cong \angle G$ **15.** $\angle DFE \cong \angle GFD$ **16.** $m\angle EFD = 90°$

17. $\overline{GF} \cong \overline{EF}$ **18.** $\angle FDG \cong \angle FDE$ **19.** $\overline{DF} \perp \overline{GE}$

20. Copy and complete this proof.

Given $\overline{JK} \cong \overline{ML}$; $\overline{KM} \cong \overline{LJ}$

Prove ____?____ $\cong \triangle LMJ$

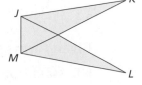

Statements	Reasons
1. ___?___	1. Given
2. $\overline{JM} \cong \overline{JM}$	2. ___?___
3. ___?___ $\cong \triangle LMJ$	3. SSS Postulate

Find the value of *x* in each figure. (Lesson 4-1)

21.

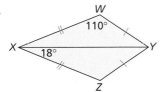

22.

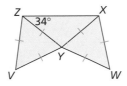

23.

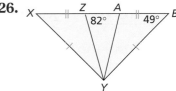

Find *m∠XYZ* in each figure. (Lesson 4-2)

24.

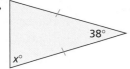

25.

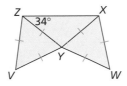

26.

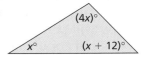

Mid-Chapter Quiz

Find the unknown measures of the interior angles of each triangle. (Lesson 4-1)

1. an isosceles triangle with an interior angle of 100°

2. a triangle with interior angles of $x°$, $(2x + 10)°$, and $(2x - 5)°$

Sketch each pair of triangles and state either that they are congruent or that no conclusion is possible. If they are congruent, name the postulate that could be used to prove the congruence. (Lesson 4-2)

3. Triangles *ABD* and *CBD* share side $\overline{BD}$. Side $\overline{BD}$ is the perpendicular bisector of $\overline{AC}$.

4. Triangles *EFH* and *GFH* share side $\overline{FH}$. Sides $\overline{EF}$ and $\overline{GF}$ are congruent. Angles *E* and *G* are congruent.

5. Triangles *RSU* and *TSU* share side $\overline{SU}$. Side $\overline{SU}$ bisects angles *RST* and *RUT*.

Determine whether each statement is *true* or *false*. (Lesson 4-3)

6. Two isosceles triangles with congruent vertex angles always have congruent base angles.

7. If two sides of a triangle are congruent, then the base angles and the vertex angle must be congruent.

Determine whether each statement is *always*, *sometimes*, or *never* true. (Lesson 4-4)

8. An altitude of a triangle divides the corresponding side of the triangle into two congruent parts.

9. The median of a side of a triangle is perpendicular to that side of the triangle.

10. An altitude of a triangle is a segment that is inside the triangle.

Problem Solving Skills: Write an Indirect Proof

The proofs that you have studied so far in this book have been *direct proofs.* That is, the proofs proceeded logically from a hypothesis and known facts to show that a desired conclusion is true. In this lesson, you will study *indirect proof.* In an **indirect proof**, you begin with the desired conclusion and assume that it is *not* true. You then reason logically until you reach a contradiction of the hypothesis or of a known fact.

Problem Solving Strategies

✔ Guess and check

  Look for a pattern

  Solve a simpler problem

  Make a table, chart or list

  Use a picture, diagram or model

  Act it out

  Work backwards

  Eliminate possibilities

  Use an equation or formula

Problem

Prove the following statement.

If a figure is a triangle, then it cannot have two right angles.

Solve the Problem

Begin by drawing a representative triangle, such as $\triangle ABC$ at right.

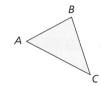

Step 1: Assume that the conclusion is false. That is, assume that a triangle *can* have two right angles. In particular, in $\triangle ABC$, assume that $\angle A$ and $\angle B$ are right angles.

Step 2: Reason logically from the assumption, as follows. By the definition of a right angle, $m\angle A = 90°$ and $m\angle B = 90°$. By the addition property of equality, $m\angle A + m\angle B = 90° + 90° = 180°$. By the protractor postulate, $m\angle C = n°$, where n is a positive number less than or equal to 180. By the addition property of equality, $m\angle A + m\angle B + m\angle C = 180° + n°$. By the triangle-sum theorem, $m\angle A + m\angle B + m\angle C = 180°$.

Step 3: Note that the last two statements in Step 2 are contradictory. Therefore, the assumption that a triangle can have two right angles is false. The given statement must be true.

The solution of the problem above illustrates the following general method for writing an indirect proof.

Writing an Indirect Proof	
	Step 1 Assume temporarily that the conclusion is false.
	Step 2 Reason logically until you arrive at a contradiction of the hypothesis or a contradiction of a known fact (a definition, a postulate, or a previously proved theorem).
	Step 3 State that the temporary assumption must be false, and that the given statement must be true.

Five-step Plan

1 Read
2 Plan
3 Solve
4 Answer
5 Check

Suppose you are asked to write an indirect proof of each statement. Write Step 1 of the proof.

1. **ART** If the triangle in a sculpture is a right triangle, then it cannot be an obtuse triangle.

2. **ARCHITECTURE** If the triangle in a building design is equilateral, then it is an isosceles triangle.

PRACTICE EXERCISE

Copy and complete the indirect proof of each theorem.

3. Theorem: *If two lines intersect, then they intersect in one point.*

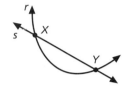

Step 1: Assume that two lines can intersect in __?__ points. In particular, in the figure at the right, assume that there are lines __?__ and __?__ that intersect at points __?__ and __?__.

Step 2: By the unique line postulate (postulate 1), there is exactly __?__ line through points *X* and *Y*.

Step 3: The statements in Step 1 and Step 2 are __?__. Therefore, the assumption that two lines can intersect in two points is __?__. The given statement must be __?__.

4. Theorem: *Through a point not on a line, there is exactly one line parallel to the given line.*

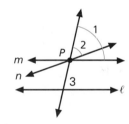

Step 1: Assume that there are __?__ lines parallel to the given line. In particular, in the figure at the right, assume that, through point *P*, __?__ ∥ ℓ and __?__ ∥ ℓ.

Step 2: By the parallel lines postulate, *m*∠ __?__ = *m*∠3 and *m*∠ __?__ = *m*∠3. By the transitive property of equality, *m*∠ __?__ = *m*∠ __?__.

However, because *m* and *n* are different lines, *m*∠1 ≠ *m*∠2.

Step 3: The last two statements in Step 2 are __?__. Therefore, the assumption that there can be two lines parallel to a given line through a point outside the line is __?__. The given statement must be __?__.

5. Write an indirect proof of the following theorem. *Through a point outside a line, there is exactly one line perpendicular to the given line.* (*Hint:* Use the proof in Exercise 4 above as a model.)

6. **WRITING MATH** Write what you would do to prove indirectly that a triangle cannot have two obtuse angles.

MIXED REVIEW EXERCISES

Simplify each expression. (Lesson 1-4)

7. $-3 + 4 - (-6) + (-2)$

8. $-9 + (-4) - (-3) + 8$

9. $4 - (-6) + 2 + (-3)$

10. $(4) + 9 - 3 - (-2)$

11. $8 - (-3) + 2 - 3 + 9$

12. $2 - (-8) - (-(-12))$

4-6 Inequalities in Triangles

Goals ■ Understand relationships among sides and angles of a triangle.

Applications Construction, Art, Architecture

Work in groups of two or three students.

The figure at the right shows four paths that ants took from point *A* to point *B*.

1. Using a centimeter ruler, find the length of each path. (You will need to use some ingenuity to measure path ②!)

2. Trace points *A* and *B* onto a sheet of paper. Can you draw a path from point *A* to point *B* that is *longer* than any of the given paths? Use the ruler to find the length of your path.

3. Can you draw a path from point *A* to point *B* that is *shorter* than any of the given paths? Use the ruler to find the length of your path.

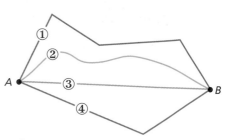

◣ BUILD UNDERSTANDING

In the activity above, you had an opportunity to investigate yet another fundamental postulate of geometry.

Postulate 14	**The Shortest Path Postulate** The length of the segment that connects two points is shorter than the length of any other path that connects the points.

The shortest path postulate leads to some important conclusions about triangles. As an example, consider the following proof.

Given $\triangle ABC$

Prove $AB + BC > AC$

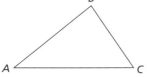

Proof

Assume that $AB + BC \not> AC$. Then one of these two statements must be true:
$AB + BC = AC$ or $AB + BC < AC$.

If $AB + BC = AC$, then there is a path other than AC that connects points *A* and *C* that is equal to AC; this contradicts the shortest path postulate.

Similarly, if $AB + BC < AC$, then there is a path connecting points *A* and *C* that is shorter than AC; this also contradicts the shortest path postulate.

Therefore, the assumption $AB + BC \not> AC$ must be false. It follows that the desired conclusion, $AB + BC > AC$, is true.

> ### Reading Math
>
> Just as the symbol $\neq$ means *is not equal to*, the symbol $\not>$ means *is not greater than*. What do you think the symbol $\not<$ means?

In Exercises 23 and 24 on page 174, you will prove that $AB + AC > BC$ and $AC + BC > AB$ are true statements also. So, you will have completed the proof of the following theorem.

The Triangle Inequality Theorem	The sum of the lengths of any two sides of a triangle is greater than the length of the third side.

Example 1

CONSTRUCTION A frame must be built to pour a triangular cement slab to complete a walkway. The lengths of two sides of the triangle are 5 ft and 9 ft. Find the range of possible lengths for the third side.

Solution

Use the variable n to represent the length in feet of the third side. By the triangle inequality theorem, these three inequalities must be true.

$$\textbf{I.}\ 5 + 9 > n \qquad \textbf{II.}\ 5 + n > 9 \qquad \textbf{III.}\ 9 + n > 5$$
$$14 > n \qquad\qquad n > 4 \qquad\qquad n > -4$$

Inequality **III** is not useful, since a length must be a positive number.

From inequalities **I** and **II**, you obtain the combined inequality $14 > n > 4$.

So, the length of the third side must be less than 14 ft and greater than 4 ft.

The following two theorems also involve inequalities in triangles. In this book, we will accept these theorems as true without proof.

The Unequal Sides Theorem	If two sides of a triangle are unequal in length, then the angles opposite those sides are unequal in measure, in the same order.
The Unequal Angles Theorem	If two angles of a triangle are unequal in measure, then the sides opposite those angles are unequal in length, in the same order.

Example 2

In $\triangle KLM$, KL = 8 in., LM = 10 in., and KM = 7 in. List the angles of the triangle *in order* from largest to smallest.

Solution

Draw and label $\triangle KLM$, as shown at the right.

The angle opposite $\overline{LM}$ is $\angle K$.
The angle opposite $\overline{KL}$ is $\angle M$.

Since $10 > 8$, $LM > KL$.

So, by the unequal sides theorem, $m\angle K > m\angle M$.
By similar logic, $m\angle M > m\angle L$.
So, from largest to smallest, the angles are $\angle K$, $\angle M$, and $\angle L$.

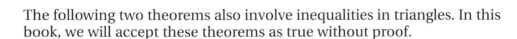

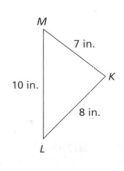

ART The design for a sculpture has three triangular platforms. The lengths of two sides of each platform are given. Find the range of possible lengths for the third side.

1. 6 ft, 9 ft

2. 7 ft, 7 ft

3. 2 ft, 7 ft

List the angles of each triangle *in order* from largest to smallest.

4.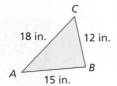

5.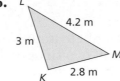

6.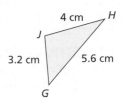

7. In $\triangle XYZ$, $m\angle X = 56°$ and $m\angle Z = 19°$. List the sides of the triangle *in order* from shortest to longest.

8. ARCHITECTURE The base for an indoor fountain has a triangular shape. On the plans, the base is shown as $\triangle RST$. If $m\angle S > m\angle R$ and $m\angle R > m\angle T$, which is the shortest side of the triangle?

◤ **PRACTICE EXERCISES** • **For Extra Practice, see page 675.**

Determine if the given measures can be lengths of the sides of a triangle?

9. 7 cm, 2 cm, 6 cm

10. 7.3 m, 15 m, 7.3 m

11. $9\frac{1}{4}$ ft, $3\frac{1}{2}$ ft, $5\frac{3}{4}$ ft

12. 24 in., 5 ft, 54 in.

13. 34 yd, 34 yd, 34 yd

14. 3 mm, 5 cm, 7 mm

Which is the longest side of each triangle? the shortest?

15.

16.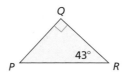

17.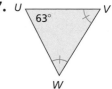

In each figure, give a range of possible values for *x*.

18.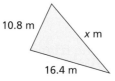

19.

20.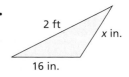

21. In $\triangle CDE$, $CD < DE$ and $CE < CD$. Which is the largest angle of the triangle?

22. GEOMETRY SOFTWARE Use the following information to draw $\triangle QRS$: $QS = 17$, $RS = 23$, and $QR = 20.5$. List the angles of the triangle *in order* from largest to smallest.

For Exercises 23 and 24, refer to the proof on page 172.

23. Given $\triangle ABC$
  Prove $AB + AC > BC$

24. Given $\triangle ABC$
  Prove $AC + BC > AB$

List all the segments in each figure *in order* from longest to shortest.

25.
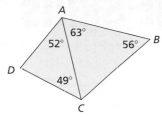

26.
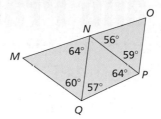

Give a range of possible values for *z*.

27.
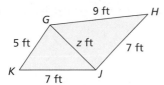

28.
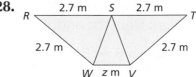

EXTENDED PRACTICE EXERCISES

29. WRITING MATH In a right triangle, the side opposite the right angle is called the *hypotenuse*. Explain why the hypotenuse must be the longest side.

30. ERROR ALERT A blueprint calls for the construction of a right triangle with sides measuring 5 ft, 6 ft, and 11 ft. How do you know the measurements are incorrect?

CONSTRUCTION Manuella is building an A-frame dog house with the front in the shape of an isosceles triangle. Two sides of the front will each be 4 ft long.

31. Under what conditions will the base of the front of the dog house be exactly 4 ft?

32. Under what conditions will the base of the front of the dog house be greater than 4 ft?

33. Under what conditions will the base of the front of the dog house be shorter than 4 ft?

34. CHAPTER INVESTIGATION Using your design for a truss bridge, build a section of the truss using straws or toothpicks. Use a ruler and protractor to make sure your construction matches the plans.

MIXED REVIEW EXERCISES

Write a function rule to represent the number of points in the *n*th figure in the patterns below. (Lesson 3-5)

35.

36.

Write each number in scientific notation. (Lesson 1-8)

37. 371,000,000,000

38. 0.000000074

39. 256,000,000,000

40. 0.00000942

41. 8,900,000,000,000

42. 0.00000007

Review and Practice Your Skills

Write Step 1 of an indirect proof of each statement.

1. If a triangle is not isosceles, then it is not equilateral.

2. If a point lies on the perpendicular bisector of a segment, then the point is equidistant from the endpoints of the segment.

3. If two angles are vertical angles, then they are equal in measure.

4. If two parallel lines are cut by a transversal, then alternate interior angles are equal in measure.

5. If two lines are perpendicular, then they intersect.

6. The sum of the measures of the angles of a triangle is 180°.

Write an indirect proof of each statement.

7. If a triangle is a right triangle, then it cannot be an obtuse triangle.

8. If a triangle is equilateral, then it is isosceles.

9. If two angles are vertical angles, then they are equal in measure.

10. If two sides of a triangle are not congruent, then the angles opposite those sides are not congruent.

Can the given measures be the lengths of the sides of a triangle?

11. 5.5 ft, 8.2 ft, 12.9 ft   12. 14 cm, 35 cm, 21 cm   13. 21 m, 13.2 m, 7 m

In each figure, give a range of possible values for *x*.

14.
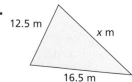
12.5 m, *x* m, 16.5 m

15.
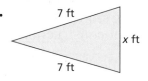
7 ft, *x* ft, 7 ft

16.
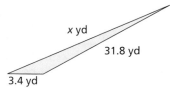
x yd, 31.8 yd, 3.4 yd

List all the segments in each figure in order from shortest to longest.

17.
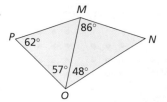
M 86°, P 62°, N, 57° 48°, O

18.
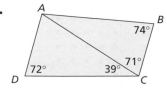
A, B 74°, 72°, 39° 71°, D, C

19.
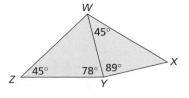
W 45°, Z 45°, 78° 89° X, Y

Determine whether each statement is *true* or *false*.

20. In a scalene triangle, no two angles are equal in measure. (Lesson 4-6)

21. A triangle can have sides of length 178 cm, 259 cm, and 440 cm. (Lesson 4-6)

Determine whether each statement is *true* or *false*.

22. All equilateral triangles are also isosceles triangles. (Lesson 4-1)

23. If $\triangle ABC \cong \triangle DEF$, then $\triangle BAC \cong \triangle EDF$. (Lesson 4-2)

24. If two angles of a triangle are congruent, then the triangle is isosceles.

25. All altitudes of a triangle lie in the interior of the triangle. (Lesson 4-4)

26. In an indirect proof, one starts by assuming that the conclusion is false. (Lesson 4-5)

Find the value of *x* in each figure. (Find the range of possible values for *x* in Exercise 32.)

27. (Lesson 4-1)

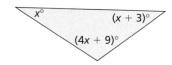

28. (Lesson 4-1)

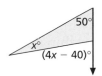

29. (Lesson 4-2)

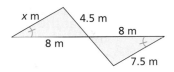

30. (Lesson 4-3)

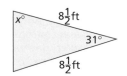

31. (Lesson 4-4)

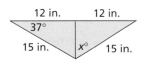

32. (Lesson 4-6)

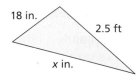

MathWorks Career – Animator
Workplace Knowhow

Traditional animation involves making many hand-drawn pictures with slight differences and filming them frame by frame to create the illusion of motion. The newest form of animation is computer-assisted animation. Knowledge of coordinates, area of curved surfaces, conics and polygons are all important pieces of an animator's tool kit for drawing great pictures.

To give objects depth, animators use perspective drawing. For instance, to make a house look three-dimensional, it must be drawn so that the house's front walls look larger than those in the rear of the house.

1. The front wall of the house in the drawing has a perimeter of $6\frac{1}{4}$ in. Find the measure of *x*.

2. The roof panels and side wall shown are drawn as parallelograms. Find the measures of *a*, *b*, *c*, and *d*.

3. The altitude of the triangle formed by the roof is 0.5 in. Find the length of the sides of the triangle to the nearest hundredth inch.

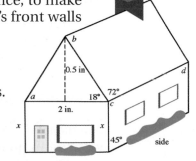

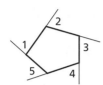

4-7 Polygons and Angles

Goals
- Find the measures of interior angles of polygons.
- Find the measures of exterior angles of polygons.

Applications Surveying, Sign making, Games, Sports

Work with a partner.

Draw and label a pentagon as shown at the left below. Then cut out the five exterior angles and arrange them as shown at the right.

 $\Rightarrow$

1. What is the relationship among the five exterior angles?

2. Repeat the experiment, this time drawing a hexagon and labeling six exterior angles. What is the relationship among the exterior angles?

◥ BUILD UNDERSTANDING

A **polygon** is a closed plane figure that is formed by joining three or more coplanar segments at their endpoints. Each segment is called a **side** of the polygon. Each side intersects exactly two other sides, one at each endpoint. The point at which two sides meet is called a **vertex** of the polygon.

polygons   not polygons

A polygon is **convex** if each line containing a side contains no points in the interior of the polygon. A polygon that is not convex is called **concave**. In this book, when the word polygon is used, assume the polygon is *convex*. The angles determined by the sides are called the **angles**, or the **interior angles**, of the polygon.

Two sides of a polygon that intersect are called **consecutive sides**. The endpoints of any side of a polygon are **consecutive vertices**. When naming a polygon, you list consecutive vertices in order. For example, two names for the pentagon at the right are "pentagon *ABCDE*" and "pentagon *BCDEA*." It is *not* correct to call the figure "pentagon *ABCED*."

A **diagonal** of a polygon is a segment that joins two *nonconsecutive* vertices. In pentagon *ABCDE*, the diagonals are shown in red.

convex   concave

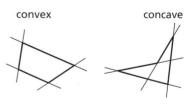

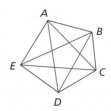

If you draw all the diagonals from just one vertex of a polygon, you divide the interior of the polygon into nonoverlapping triangular regions. The sum of the measures of the angles of the polygon is the product of the number of triangular regions formed and 180°.

4 sides
2 triangular regions
2 × 180° = 360°

5 sides
3 triangular regions
3 × 180° = 540°

6 sides
4 triangular regions
4 × 180° = 720°

In each case, the number of triangular regions formed is two fewer than the number of sides of the polygon. This leads to the following theorem.

The Polygon-Sum Theorem	The sum of the measures of the angles of a convex polygon with n sides is $(n - 2)180°$.

Check Understanding

Refer to pentagon *ABCDE*, on page 178. Name the following.

▶ all the sides

▶ all the angles

▶ all the vertices

▶ all the diagonals

Give at least two names for the pentagon other than those names given in the text.

Example 1

SURVEYING A playground has the shape shown in the figure to the right. A surveyor measures six of the angles of the playground. Find the unknown measure.

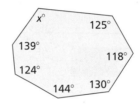

Solution

The polygon has 7 sides. Use the polygon-sum theorem to find the sum of the angle measures.

$$(n - 2)180° = (7 - 2)180° = (5)180° = 900°$$

Add the *known* angle measures.

$$139° + 124° + 144° + 130° + 118° + 125° = 780°$$

Subtract this sum from 900°: $900° - 780° = 120°$

The unknown angle measure is 120°.

An **exterior angle** of any polygon is an angle both adjacent to and supplementary to an interior angle. Since the sum of the *interior* angles of a polygon depends on the number of sides of the polygon, you might expect that the same would be true for the exterior angles. So, the following theorem about exterior angles may come as a surprise to you.

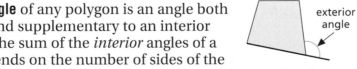

exterior angle

The Polygon Exterior Angle Theorem	The sum of the measures of the exterior angles of a convex polygon, one angle at each vertex, is 360°.

Technology Note

Explore the theorem using geometric software.

1. Draw four rays to form a polygon. Mark and label a point on each ray outside the polygon.

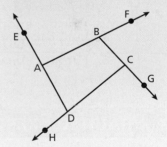

2. Use the software to measure each exterior angle.

3. Calculate the total of the angles.

4. Change the positions of the rays to change the measures of the angles. What happens to the sum?

A polygon with all sides of equal length is called an **equilateral polygon**. A polygon with all angles of equal measure is an **equiangular polygon**. A **regular polygon** is a polygon that is *both* equilateral and equiangular.

Hexagon

Equilateral
hexagon

Equiangular
hexagon

Regular
hexagon

Example 2

a. Find the measure of each interior angle of a regular octagon.

b. Find the measure of each exterior angle of a regular octagon.

Solution

a. Using the polygon-sum theorem, the sum of the measures of the interior angles is $(n - 2)180° = (8 - 2)180° = (6)180° = 1080°$.

Because the octagon is regular, the interior angles are equal in measure.

So, the measure of one interior angle is $1080° \div 8 = 135°$.

b. By the polygon exterior angle theorem, the sum of the measures of the exterior angles is $360°$.

So, the measure of one exterior angle is $360° \div 8 = 45°$.

◣ TRY THESE EXERCISES

Find the unknown angle measure or measures in each figure.

1.

2.

3.

4. Find the measure of each interior angle of a regular polygon with 15 sides.

5. Find the measure of each exterior angle of a regular decagon.

◣ PRACTICE EXERCISES • For Extra Practice, see page 675.

Find the unknown angle measure or measures in each figure.

6.
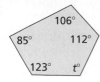

7.
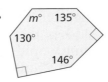

8.
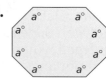

9.
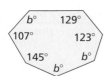

10.
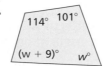

11.
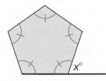

12. A road sign is in the shape of a regular hexagon. Find the measure of each interior angle.

13. **RECREATION** A game board is in the shape of a regular polygon with 18 sides. Find the sum of the measures of the interior angles.

14. Find the sum of the measures of the exterior angles of a regular nonagon.

15. Find the measure of each exterior angle of a regular polygon with 24 sides.

Each figure is a regular polygon. Find the values of *x*, *y*, and *z*.

16.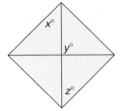

17.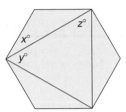

18.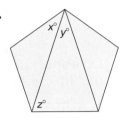

Find the number of sides of each regular polygon.

19. The measure of each exterior angle is 9°.

20. The sum of the measures of the interior angles is 1980°.

21. The measure of each interior angle is 162°.

For Exercises 22–23, use the Reading Math feature on page 221 to locate information about convex regular polyhedrons.

22. A **polyhedron** is a closed three-dimensional figure in which each surface is a polygon. Why do you think these are called *regular* polyhedrons?

23. **SPORTS** At the right is a soccer ball. It is shaped like a polyhedron with faces that are all regular polygons. However, this shape is not pictured with the convex regular polyhedrons on page 221. Explain.

EXTENDED PRACTICE EXERCISES

For Exercises 24 and 25, consider a regular polygon with *n* sides. Write an expression to represent each quantity.

24. the measure in degrees of one exterior angle

25. the measure in degrees of one interior angle

 WRITING MATH For Exercises 26–28, consider what happens as the number of sides of a regular polygon becomes larger and larger.

26. What happens to the measure of each exterior angle?

27. What happens to the measure of each interior angle?

28. What happens to the overall appearance of the polygon?

MIXED REVIEW EXERCISES

Classify each triangle by its sides and angles.

29.

30.

31.

32.

4-8 Special Quadrilaterals: Parallelograms

Goals ■ Apply properties of parallelograms to find missing lengths and angle measures.

Applications Art, Construction, Engineering, Architecture

On a sheet of paper, draw line ℓ. Mark and label a point A on line ℓ as shown at the right.

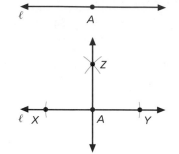

1. With compass point at point A, draw two arcs of equal radii that intersect ℓ. Label the points of intersection X and Y.

2. With compass point first at point X, then at point Y, draw two arcs that intersect at Z.

3. Using a straightedge, draw $\overleftrightarrow{AZ}$. What do you observe about the line you constructed?

4. Use this method to construct a rectangle. Using a straightedge, draw the diagonals of your rectangle. What observations do you make about the diagonals?

▧ BUILD UNDERSTANDING

Opposite sides of a quadrilateral are two sides that do not share a common endpoint. **Opposite angles** are two angles that do not share a common side.

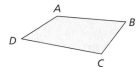

opposite sides	opposite angles
$\overline{AB}$ and $\overline{CD}$	∠A and ∠C
$\overline{BC}$ and $\overline{DA}$	∠B and ∠D

A **parallelogram** is a quadrilateral with both pairs of opposite sides parallel.

parallelogram PQRS

▱PQRS

$\overline{PQ} \parallel \overline{SR}$

$\overline{PS} \parallel \overline{QR}$

The following theorems identify some properties of all parallelograms.

The Parallelogram-Side Theorem	If a quadrilateral is a parallelogram, then its opposite sides are equal in length.
The Parallelogram-Angle Theorem	If a quadrilateral is a parallelogram, then its opposite angles are equal in measure.
The Parallelogram-Diagonal Theorem	If a quadrilateral is a parallelogram, then its diagonals bisect each other.

The proofs of these theorems are based on properties of parallel lines and congruent triangles. You will have a chance to prove them in Exercises 22–25 on page 185.

Mental Math Tip

Most students find it helpful to memorize the fact that the sum of the interior angles of a quadrilateral is 360°. If you remember this, problems such as Example 1 involve far less work.

Example 1

Find $m\angle J$ in $\square$ JKLM.

Solution

Since $\angle K$ and $\angle M$ are opposite angles, by the parallelogram-angle theorem, $m\angle K = m\angle M = 48°$.

Use the polygon-sum theorem to find the sum of the measures of the interior angles.

$$(n - 2)180° = (4 - 2)180° = 2(180°) = 360°$$

Notice that $m\angle M + m\angle K = 48° + 48° = 96°$. It follows that $m\angle J + m\angle L = 360° - 96° = 264°$.

Since $\angle J$ and $\angle L$ are opposite angles, by the parallelogram-angle theorem, $m\angle J = 264° \div 2 = 132°$.

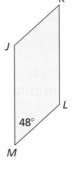

Other special quadrilaterals are *rectangles*, *rhombuses*, and *squares*.

A **rectangle** is a quadrilateral with four right angles.

A **rhombus** is a quadrilateral with four sides of equal length.

A **square** is a quadrilateral with four right angles *and* four sides of equal length.

Rectangle

Rhombus

Square

Rectangles, rhombuses, and squares are parallelograms, and so they have all the properties of parallelograms. In addition, however, they have the special properties summarized in the following theorems. In this book, these theorems will be accepted as true without proof.

Math: Who, Where, When

In 1981, when she was a 21-year-old senior at Yale University, Maya Ying Lin won a nationwide competition to design the proposed Vietnam Veterans Memorial in Washington, D.C. At the time, her design was criticized as being too simple—two large walls of polished black granite, joined at a 130° angle and engraved with the names of all those killed or missing in the conflict. However, since its official dedication on November 11, 1982, the memorial has become the most visited monument in the nation's capital. Lin has since graduated from Yale and become a highly respected architect and sculptor.

The Rectangle-Diagonal Theorem	If a quadrilateral is a rectangle, then its diagonals are equal in length.
The Rhombus-Diagonal Theorem	If a quadrilateral is a rhombus, then its diagonals are perpendicular and bisect each pair of opposite angles.

Example 2

ART A rectangular mural is reinforced from the back using wire diagonals. The diagram at the right shows how the wires are attached. If $ZO = 8$ ft, find WY.

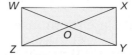

Solution

A rectangle is a parallelogram. By the parallelogram-diagonal theorem, the diagonals bisect each other. So, $XZ = 2(ZO) = 2(8\text{ ft}) = 16$ ft.

Then, by the rectangle-diagonal theorem, you know that the diagonals are equal in length. So, $WY = XZ = 16$ ft.

■ TRY THESE EXERCISES

In Exercises 1–2, each figure is a parallelogram. Find the values of x and z.

1.

2.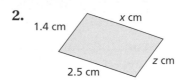

BRIDGE BUILDING A portion of a truss bridge forms quadrilateral $XYZW$, shown at the right. Given that $XYZW$ is a rhombus and $m\angle YXZ = 32°$, find the measure of each angle.

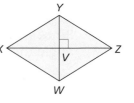

3. $\angle YXW$ **4.** $\angle XYW$ **5.** $\angle XVY$

6. $\angle YZW$ **7.** $\angle YVZ$ **8.** $\angle XWZ$

■ PRACTICE EXERCISES • For Extra Practice, see page 676.

ARCHITECTURE The parallelograms in Exercises 9–12 are from building plans. Find the values of a, b, c, and d.

9.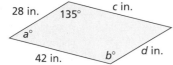

10.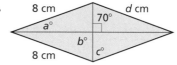

11. $MJ = 1\frac{1}{2}$ yd, $MK = 3\frac{1}{4}$ yd

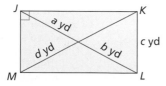

12. $RS = 4.9$ mm, $RQ = 5.6$ mm, $RP = 9.7$ mm

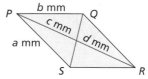

ERROR ALERT Dillon made the following statements about quadrilaterals. Decide whether each statement is *true* or *false*.

13. A rectangle is a parallelogram.

14. No rhombus is a square.

15. Every quadrilateral is a parallelogram.

16. Some rectangles are rhombuses.

17. The diagonals of a square are not equal in length.

18. Consecutive angles of a parallelogram are supplementary.

WRITING MATH Do you think that the given figure is a parallelogram? Write *yes* or *no*. Then explain your reasoning.

19.
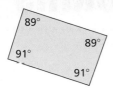
89°
89°
91°
91°

20.

6   6
5   5

21.
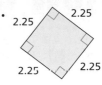
2.25   2.25
2.25
2.25   2.25

22. Copy and complete this proof.
Given *ABCD* is a parallelogram.
Prove $m\angle A = m\angle C$

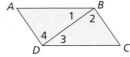

Statements	Reasons
1. __?__	1. __?__
2. $\overline{AB} \parallel \overline{DC}; \overline{AD} \parallel \overline{BC}$	2. definition of __?__
3. $m\angle 1 = m\angle 3$, or $\angle 1 \cong \angle 3$ $m\angle 2 = m\angle 4$, or $\angle 2 \cong \angle 4$	3. If __?__, then __?__
4. __?__	4. reflexive property
5. __?__	5. ASA postulate
6. $m\angle A = m\angle C$	6. __?__

23. The proof in Exercise 22 is the beginning of a proof of the parallelogram-angle theorem. Using this proof as a model, write the second part of the proof. That is, prove $m\angle B = m\angle D$.

24. Write a proof of the parallelogram-side theorem.

◼ EXTENDED PRACTICE EXERCISES

 25. WRITING MATH Suppose that you are asked to prove the parallelogram-diagonal theorem. Write a paragraph that explains how you would proceed. (Do not write the two-column proof.)

26. DESIGN Suppose you need to describe the figure at the right to a graphics designer. State as many facts as you can about the figure.

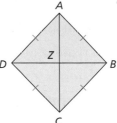

◼ MIXED REVIEW EXERCISES

Refer to the figure at the right for Exercises 27–29.

27. Name all the alternate exterior angles.

28. Name all the corresponding angles.

29. Name all the alternate interior angles.

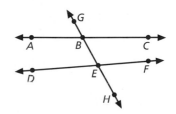

Determine if each relation is a function. Give the domain and range.

30.
a	0	1	2	2	3
b	−1	−3	−4	−5	−6

31.
x	2	3	4	5	6
y	4.5	6.5	8.5	10.5	12.5

32.
m	−1	0	1	0	−1
n	−4	−1	0	2	5

Review and Practice Your Skills

Find the unknown angle measure or measures in each figure.

1.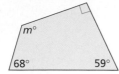

2.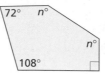

3.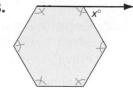

4. Find the measure of each interior angle of a regular polygon with 13 sides.

5. Find the measure of each exterior angle of a regular polygon with 20 sides.

6. Find the sum of the measures of the interior angles of a regular heptagon.

7. Find the sum of the measures of the exterior angles of a regular heptagon.

8. Using diagonals from one vertex, into how many nonoverlapping triangular regions can you divide a nonagon? a polygon with 21 sides?

Find the number of sides of each regular polygon.

9. The measure of each exterior angle is 40°.

10. The sum of the measures of interior angles is 2160°.

11. The measure of each interior angle is 165°.

Determine whether each statement is *true* or *false*.

12. The diagonals of a rhombus are equal in length.

13. Every square is a rhombus.

14. Quadrilaterals include squares, parallelograms, pentagons, and rectangles.

15. A square is a regular polygon.

16. In all quadrilaterals, the opposite sides are equal in length.

For the following parallelograms, find the values of *a*, *b*, *c*, and *d*.

17.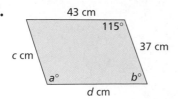

18.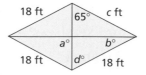

19.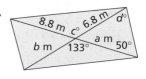

Is the given figure a parallelogram? Write *yes* or *no*. Then explain your reasoning.

20.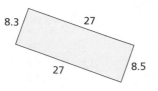

21.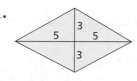

22.

Find the value of *x* in each figure. (Lesson 4-1)

23.

24.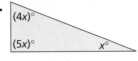

25.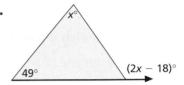

26. Copy and complete this proof. (Lesson 4-2)

Given $\overline{AE} \cong \overline{EC}$; $\overline{DE} \cong \overline{EB}$
Prove $\triangle DAE \cong$ ___?___

Statements	Reasons
1. ___?___	1. Given
2. ___?___	2. Vertical Angles Theorem
3. $\triangle DAE \cong$ ___?___	3. ___?___

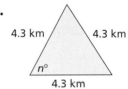

Find the value of *n* in each figure. (Lesson 4-3)

27.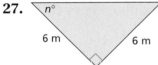

28.

29.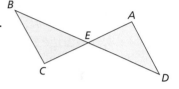

Give the range of possible values for *x* in each figure. (Lesson 4-6)

30.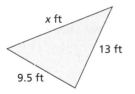

31.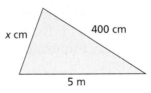

32.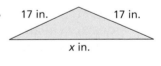

Find the unknown angle measure or measures in each figure. (Lesson 4-7)

33.

34.

35.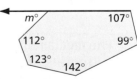

36.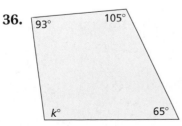

37.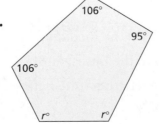

38.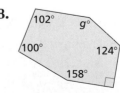

4-9 Special Quadrilaterals: Trapezoids

Goals ■ Apply properties of trapezoids to find missing lengths and angle measures.

Applications Stage Design, Construction, Art

Work with a partner.

The *tangram* is an ancient Chinese puzzle consisting of the seven pieces shown at the right. Use a manufactured set of tangram pieces or trace the figure onto a sheet of paper and then cut out the pieces along the lines.

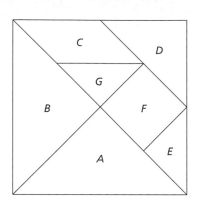

1. Arrange pieces *E*, *F*, and *G* to form a rectangle.

2. Arrange *E*, *F*, and *G* to form a parallelogram that is not a rectangle.

3. Arrange *E*, *F*, and *G* to form a quadrilateral that is not a parallelogram.

4. Arrange pieces *A*, *C*, *E*, and *G* to form a square.

5. Arrange all seven tangram pieces to form a quadrilateral that is not a parallelogram.

6. Form as many different rectangles that are not squares as possible. (For each rectangle, use as many tangram pieces as needed.)

◥ BUILD UNDERSTANDING

A **trapezoid** is a quadrilateral with exactly one pair of parallel sides. The parallel sides are called the **bases** of the trapezoid. Two consecutive angles that share a base form a pair of **base angles**; every trapezoid has two pairs of base angles. The nonparallel sides are called the **legs**.

$\overline{AB} \parallel \overline{DC}$

ABCD is a trapezoid.

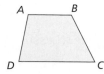

bases: $\overline{AB}$ and $\overline{DC}$

legs: $\overline{AD}$ and $\overline{BC}$

base angles: $\angle A$ and $\angle B$; $\angle D$ and $\angle C$

The **median** of a trapezoid is the segment that joins the midpoints of the legs. Two important properties of the median are stated in the following theorem, which will be accepted as true without proof.

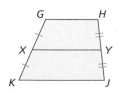

$\overline{GH} \parallel \overline{KJ}$

$\overline{XY}$ is the median of trapezoid *GHJK*.

The Trapezoid-Median Theorem	If a segment is the median of a trapezoid, then it is:
	1. parallel to the bases; and
	2. equal in length to one half the sum of the lengths of the bases.

Example 1

STAGE DESIGN The plans for two panels of a stage setting are shown in the figure at the right. In the figure, $\overline{QT} \parallel \overline{RS}$. Find AB.

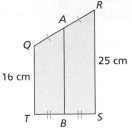

Solution

Quadrilateral $QRST$ is a trapezoid. $\overline{QT}$ and $\overline{RS}$ are the bases, and $\overline{AB}$ is the median. To find AB, apply the trapezoid-median theorem.

$$AB = \frac{1}{2}(QT + RS)$$

$$AB = \frac{1}{2}(16 + 25)$$

$$AB = \frac{1}{2}(41)$$

$$AB = 20.5$$

So, the length of $\overline{AB}$ is 20.5 cm.

A trapezoid with legs of equal length is called an **isosceles trapezoid**.

$\overline{PQ} \parallel \overline{SR}$

$PS = QR$

$PQRS$ is an isosceles trapezoid.

Technology Note

A **kite** is a quadrilateral that has exactly two pairs of consecutive sides of the same length.

Draw a kite using geometric drawing software. Use the figure to explore the following questions.

1. What relationship exists among the angles of a kite?

2. What relationships exist between the diagonals of a kite?

3. Connect the midpoints of the sides of the kite. What type of figure do you obtain?

The following theorem states an important fact about isosceles trapezoids. This theorem also will be accepted as true without proof.

The Isosceles Trapezoid Theorem	If a quadrilateral is an isosceles trapezoid, then its base angles are equal in measure.

Example 2

In the figure at the right, $\overline{ST} \parallel \overline{WV}$. Find $m\angle V$.

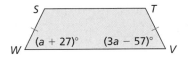

Solution

Quadrilateral $STVW$ is an isosceles trapezoid, with bases $\overline{ST}$ and $\overline{WV}$. So, $\angle W$ and $\angle V$ are a pair of base angles, and they are equal in measure. Use this fact to write and solve an equation.

$a + 27 = 3a - 57$	Add $-a$ to each side.
$a + 27 + (-a) = 3a - 57 + (-a)$	Combine like terms.
$27 = 2a - 57$	Add 57 to each side.
$27 + 57 = 2a - 57 + 57$	
$84 = 2a$	Multiply each side by $\frac{1}{2}$.
$42 = a$	

Check Understanding

In Example 2, what is the measure of $\angle S$? $\angle T$?

So, the value of a is 42. From the figure, $m\angle V = (3a - 57)°$.
Substituting 42 for a, $m\angle V = (3 \cdot 42 - 57)° = (126 - 57)° = 69°$.

A trapezoid and its median are shown. Find the value of *x*.

1.

14 ft

x ft

18 ft

2.

9 cm   6.5 cm

x cm

3.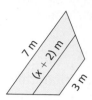

7 m

(*x* + 2) m

3 m

CONSTRUCTION The given figures are part of a design for a wrought-iron railing. Find all unknown angle measures.

4.

P

S

126°

R   Q

5.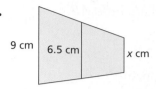

C   (4*n* + 27)°   (3*n* + 39)°   D

F   E

A trapezoid and its median are shown. Find the value of *z*.

6.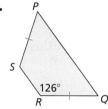

38 in.

z in.

27 in.

7.

4.9 m

z m

2.3 m

8.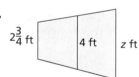

$2\frac{3}{4}$ ft   4 ft   *z* ft

9.

19 yd

14 yd

z yd

10.

17 in.

(*z* − 4) in.

25 in.

11.

z mm

14 mm

3*z* mm

The given figure is a trapezoid. Find all the unknown angle measures.

12.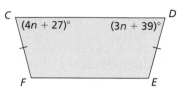

G   H

61°

K   J

13.

T   U

(6*a* − 31)°

(4*a* + 7)°   V

W

 14. CHAPTER INVESTIGATION Make a list of the quadrilaterals that you can see in your truss bridge design. Compare your design with those of your classmates. Which design do you think will support the most weight? Why?

 15. WRITING MATH Compare the median of a trapezoid to the median of a triangle. How are they alike? How are they different?

In Exercises 16–21, give as many names as are appropriate for the given figure. Choose from *quadrilateral*, *parallelogram*, *rhombus*, *rectangle*, *square*, *trapezoid*, and *isosceles trapezoid*. Then underline the *best* name for the figure.

16.

17.

18.

19.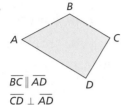

$\overline{BC} \parallel \overline{AD}$
$\overline{CD} \perp \overline{AD}$

20.

$\overline{XY} \parallel \overline{WZ}$
$\overline{XW} \parallel \overline{YZ}$

21.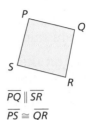

$\overline{PQ} \parallel \overline{SR}$
$\overline{PS} \cong \overline{QR}$

Copy and complete the following table that summarizes what you have learned about quadrilaterals. For each entry, write *yes* or *no*.

	Property	Quadrilateral	Parallelogram	Rectangle	Rhombus	Square	Trapezoid	Isos. Trap.
22.	sum of interior angles 360°	■	■	■	■	■	■	■
23.	all opposite sides equal in length	■	■	■	■	■	■	■
24.	all opposite angles equal in measure	■	■	■	■	■	■	■
25.	diagonals bisect each other	■	■	■	■	■	■	■
26.	diagonals are perpendicular	■	■	■	■	■	■	■
27.	diagonals equal in length	■	■	■	■	■	■	■
28.	diagonals bisect vertex angles	■	■	■	■	■	■	■

■ EXTENDED PRACTICE EXERCISES

29. ART The side view of the marble base of a statue is a trapezoid with bases $\overline{AB}$ and $\overline{DC}$, shown at the right. Prove that $\angle A$ and $\angle D$ are supplementary. (*Hint:* Extend $\overline{AD}$ to show $\overleftrightarrow{AD}$.)

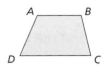

30. What type of figure do you obtain if you join the midpoints of all the sides of an isosceles trapezoid?

■ MIXED REVIEW EXERCISES

Use the number line below for Exercises 31–36. Find each length. (Lesson 3-1)

31. $\overline{AF}$ **32.** $\overline{BE}$ **33.** $\overline{DG}$

34. $\overline{AH}$ **35.** $\overline{CH}$ **36.** $\overline{DF}$

CHAPTER 4 REVIEW

VOCABULARY

Choose the word from the list that best completes each statement.

1. When two geometric figures have the same size and shape, they are said to be __?__.

2. If a point lies on the __?__ of a segment, then the point is equidistant from the endpoints of the segment.

3. A __?__ is a quadrilateral with both pairs of opposite sides parallel.

4. A __?__ of a polygon is a segment that joins two nonconsecutive vertices.

5. A __?__ is a quadrilateral with exactly one pair of parallel sides.

6. A __?__ follows directly from a theorem.

7. A __?__ of a triangle is a segment where one endpoint is a vertex and the other endpoint is the midpoint of the opposite side.

8. In a __?__, all angles have the same measure and all sides have the same length.

9. Two or more lines that intersect at one point are called __?__.

10. A polygon where the lines containing the side have no points in the interior of the polygon is called __?__.

a.	concave
b.	concurrent
c.	congruent
d.	convex
e.	corollary
f.	diagonal
g.	median
h.	midpoint
i.	parallelogram
j.	regular polygon
k.	rhombus
l.	trapezoid

LESSON 4-1 ■ Triangles and Triangle Theorems, p. 150

▶ The sum of the measures of the angles of a triangle is 180°.

▶ The measure of an exterior angle of a triangle is equal to the sum of the measures of the two nonadjacent (remote) interior angles.

Find the value of *x* in each figure.

11.

12.
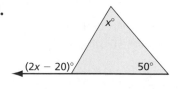

13.
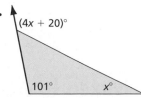

14.
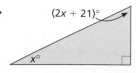

15.

16.
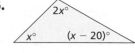

LESSON 4-2 ◼ Congruent Triangles, p. 154

▶ Three postulates for proving that two triangles are congruent are the SSS (Side-Side-Side) Postulate, the SAS (Side-Angle-Side) Postulate, and the ASA (Angle-Side-Angle) Postulate.

In each case, name a pair of congruent triangles. Then name the postulate you could use to prove the triangles congruent. You do not need to write a proof.

17.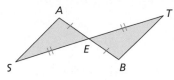

18.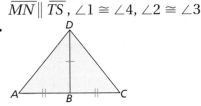

$\overline{MN} \parallel \overline{TS}$, $\angle 1 \cong \angle 4$, $\angle 2 \cong \angle 3$

19.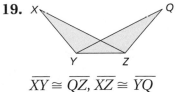

$\overline{XY} \cong \overline{QZ}$, $\overline{XZ} \cong \overline{YQ}$

20.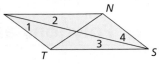

LESSON 4-3 ◼ Congruent Triangles and Proofs, p. 160

▶ Base angles of an isosceles triangle are congruent.

Find the value of x in each figure.

21.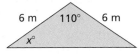

22.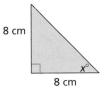

23.

LESSON 4-4 ◼ Altitudes, Medians, and Perpendicular Bisectors, p. 164

▶ An altitude of a triangle is the perpendicular segment from a vertex to the line containing the opposite side. A median of a triangle is a segment whose endpoints are on a vertex of the triangle and the midpoint of the opposite side.

For Exercises 24–25, use the figure at the right.

24. Name a median of $\triangle ABC$.

25. Name an altitude of $\triangle ABC$.

26. Draw an obtuse triangle. Sketch all the altitudes and the medians.

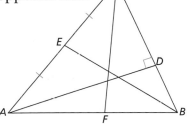

LESSON 4-5 ◼ Problem Solving Skills: Write an Indirect Proof, p. 170

▶ To write an indirect proof, the first step is to assume temporarily that the conclusion is false.

Write Step 1 of an indirect proof of each statement.

27. If a triangle is obtuse, then it cannot have a right angle.

28. If two parallel lines are cut by a transversal, then alternate exterior angles are congruent.

29. The angle bisector of the vertex angle of an isosceles triangle is also an altitude of the triangle.

LESSON 4-6 ◼ Inequalities in Triangles, p. 172

▶ The sum of the lengths of two sides in a triangle is greater than the length of the third side.

Determine if the given measures can be lengths of the sides of a triangle.

30. 19 cm, 10 cm, 8 cm   **31.** 6 ft, 7 ft, 13 ft   **32.** 8 m, 8m, 15 m

LESSON 4-7 ◼ Polygons and Angles, p. 178

▶ The sum of the measures of the angles of a convex polygon with n sides is $(n - 2)180°$.

Find the unknown angle measure or measures in each figure.

33.

34.

35.

36. Find the sum of the measures of the angles of a polygon with 11 sides.

LESSON 4-8 ◼ Special Quadrilaterals: Parallelograms, p. 182

▶ If a figure is a parallelogram, the opposite sides are equal in length, the opposite angles are equal in measure, and the diagonals bisect each other.

In the figure $OE = 19$ and $EU = 12$. Find each measure.

37. LE **38.** OJ **39.** $m\angle OJL$

40. OU **41.** OL **42.** JU

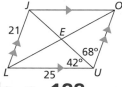

LESSON 4-9 ◼ Special Quadrilaterals: Trapezoids, p. 188

▶ The length of the median of a trapezoid equals half the sum of the lengths of the bases.

A trapezoid and its median are shown. Find the value of a.

43.
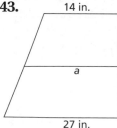

44.
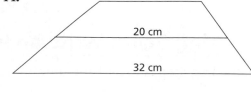

45.
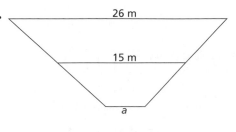

CHAPTER INVESTIGATION

EXTENSION Write a report about the design and model of your truss bridge. Include an explanation as to why an actual bridge constructed from your model would support the necessary weight.

Chapter 4 Assessment

Find the value of *x* in each figure.

1.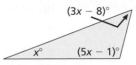

2.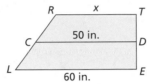

3.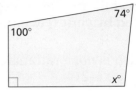

4.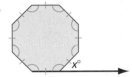

5.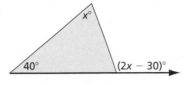

LETR is a trapezoid.
$\overline{CD}$ is a median.

6.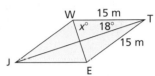

JETW is a parallelogram.

Complete the congruence statement. Name the postulate you could use to prove the triangles congruent. (You do not need to write a proof.)

7.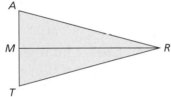

Given $\overline{AR} \cong \overline{TR}$, $\angle ARM \cong \angle TRM$
$\triangle ARM \cong$ _____

8.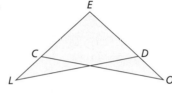

Given $\angle L \cong \angle O$, $\overline{LE} \cong \overline{OE}$
$\triangle LED \cong$ _____

9. Write a two-column proof.
Given $\overline{TR} \cong \overline{RE}$, $\overline{WT} \cong \overline{WE}$
Prove: $\triangle TRW \cong \triangle ERW$

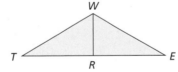

10. Draw an acute triangle and sketch the perpendicular bisectors of all the sides.

11. Suppose you are asked to write an indirect proof of the following statement: *If a triangle is equilateral, then it cannot have two sides of unequal lengths.* Write Step 1 of the indirect proof.

12. Can a triangle have sides that measure 45 mm, 19 mm, and 23 mm? Explain.

13. In the figure at the right, give a range of possible values for *x*.

14. Find the sum of the measures of the angles of a polygon with 13 sides.

15. A figure and the result of the first two iterations are shown. Show the result of the third iteration.

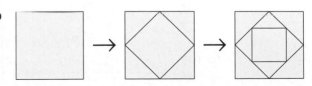

Standardized Test Practice

Part 1 Multiple Choice

Record your answers on the answer sheet provided by your teacher or on a sheet of paper.

1. Which is *not* a rational number? (Lesson 1-2)
   - (A) $-\sqrt{144}$
   - (B) 0
   - (C) 0.3
   - (D) $\sqrt{50}$

2. Evaluate t^{-3} when $t = -4$. (Lesson 1-8)
   - (A) -64
   - (B) $-\dfrac{1}{64}$
   - (C) $\dfrac{1}{64}$
   - (D) 64

3. Given $f(x) = 4x - 1$ and $g(x) = 2x^2$, evaluate $f(10) + g(10)$. (Lesson 2-2)
   - (A) 39
   - (B) 139
   - (C) 200
   - (D) 239

4. Which inequality is represented by the graph? (Lesson 2-6)
   - (A) $y < x$
   - (B) $y \le x$
   - (C) $y > x$
   - (D) $y \ge x$

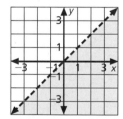

5. What is $m\angle POM$ if $m\angle MOD = (5x)°$ and $m\angle POM = (x - 12)°$? (Lesson 3-2)
   - (A) $5°$
   - (B) $17°$
   - (C) $13.5°$
   - (D) $85°$

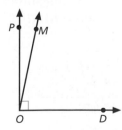

6. Which segment is an altitude of $\triangle RST$? (Lesson 4-4)
   - (A) $\overline{AS}$
   - (B) $\overline{AT}$
   - (C) $\overline{BR}$
   - (D) $\overline{CS}$

7. If the following statement is to be proved using indirect proof, what assumption should you make at the beginning of the proof? (Lesson 4-5)

   If two sides of a triangle are not congruent, then the angles opposite those sides are not congruent.

   - (A) If two sides of a triangle are congruent, then the angles opposite those sides are congruent.
   - (B) If two sides of a triangle are congruent, then the angles opposite those sides are not congruent.
   - (C) If two sides of a triangle are not congruent, then the angles opposite those sides are congruent.
   - (D) If two angles of a triangle are congruent, then the sides opposite those angles are congruent.

8. Determine which set of numbers can be lengths of the sides of a triangle. (Lesson 4-6)
   - (A) 5 m, 10 m, 20 m
   - (B) 9 in., 10 in., 14 in.
   - (C) 1 km, 2 km, 3 km
   - (D) 8 ft, 15 ft, 29 ft

9. The figure below is a parallelogram with diagonals. Which statement is *not* true? (Lesson 4-8)

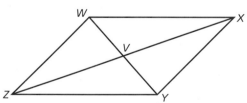

   - (A) $VZ = VX$
   - (B) $WX = ZY$
   - (C) $\overline{WZ} \parallel \overline{XY}$
   - (D) $WY = XZ$

Test-Taking Tip

Question 5
Read the question carefully to check that you answered the question that was asked. In Question 5, you are asked to find the measure of $\angle POM$, not the value of x or the measure of $\angle MOD$.

Preparing for Standardized Tests
For test-taking strategies and more
practice, see pages 709–724.

Part 2 Short Response/Grid In

**Record your answers on the answer sheet
provided by your teacher or on a sheet of paper.**

10. Refer to the diagram below to find the
    number of elements in C'. (Lesson 1-3)

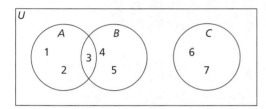

11. On Mercury, the temperatures range from
    805°F during the day to −275°F at night. Find
    the difference between these temperatures.
    (Lesson 1-4)

12. The bee hummingbird of Cuba is $\frac{1}{4}$ the
    length of the giant hummingbird. If the
    length of the giant hummingbird is $8\frac{1}{4}$ in.,
    find the length of the bee hummingbird.
    (Lesson 1-5)

13. At the beginning
    of each week,
    Lina increases
    the time of her
    daily jog. If she
    continues her

Week	Time Jogging
1	8 min
2	16 min
3	24 min
4	32 min

    pattern, how many minutes will she spend
    jogging each day during her fifth week of
    jogging? (Lesson 2-1)

14. What is the ordered pair for the point in the
    graph below? (Lesson 2-2)

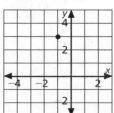

15. The following are Tom's test scores. What is
    the mean of the data? (Lesson 2-7)

    81, 87, 92, 97, 83

16. In the figure, $AB = 45$. Find AQ.
    (Lesson 3-1)

17. In the figure, $\overline{RS} \parallel \overline{PY}$. Find $m\angle RPY$.
    (Lesson 3-4)

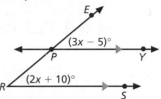

18. The measure of one acute angle of a right
    triangle is 63°. What is the measure of the
    other acute angle? (Lesson 4-1)

19. Find the value of x in the figure.
    (Lesson 4-3)

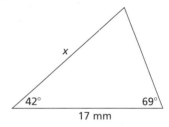

20. If a convex polygon has 8 sides, find the sum
    of the interior angles. (Lesson 4-7)

Part 3 Extended Response

**Record your answers on a sheet of paper. Show
your work.**

21. Write a two column proof. (Lesson 4-2)

    Given $\overline{FB}$ is a
    perpendicular
    bisector of $\overline{AC}$.

    Prove $\triangle AFB \cong$
    $\triangle CFB$

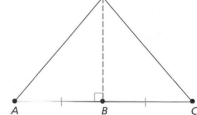

22. Two segments with lengths 3 ft and 5 ft form
    two sides of a triangle. Draw a number line
    that shows possible lengths of the third side.
    Explain your reasoning. (Lesson 4-6)

Measurement

THEME: Lost Cities of Ancient Worlds

What happens to a city once the people are gone? Often, it lies buried in the earth waiting for discovery. Once it is excavated, its roads, buildings, sewers, and drains provide clues to the habits and ingenuity of the people who lived there. Fragments of pottery, sculptures, paintings, and toys offer glimpses into the society's values, beliefs, and daily lives.

Archaeologists discover and decipher these clues from the past. Through painstaking digging and sifting through the remains of ancient cities, archaeology has given us a remarkable portrait of the world as it once was and the wonders of the past.

- **Heavy Equipment Operators** (page 211) clear the land, dig, and move dirt, debris, rock, and water to uncover archaeological ruins.

- **Archaeologists** (page 229) trace the histories of ancient civilizations by studying maps, artifacts, and the writings of ancient people. Archaeologists must be able to catalog items and draw conclusions from the clues to the past that they find.

The Seven Wonders of the Ancient World

Name	Location	Date built	Size
The Great Pyramid of Khufu	Giza	c. 2700–2500 B.C.	height: 480 ft width of square base: 756 ft
The Hanging Gardens of Babylon	Baghdad	c. 600 B.C.	height: 75 ft width of square base: 400 ft
The Statue of Zeus at Olympia	Greece	c. 457 B.C.	height: 40 ft
The Colossus of Rhodes	Greece	c. 290 B.C.	height: 120 ft
The Temple of Artemis	Turkey	c. 550 B.C.	length: 370 ft; width: 170 ft
The Mausoleum at Halicarnassus	Turkey	c. 353 B.C.	length: 126 ft; width: 105 ft; height: 140 ft
Lighthouse of Alexandria	Egypt	c. 270 B.C.	height: 400 ft

(c. stands for *circa*, meaning *about* or *approximately*)

Data Activity: The Seven Wonders of the Ancient World

Use the table for Questions 1–3.

1. Find the volume in cubic feet of the Great Pyramid of Khufu. Use the formula for volume of a pyramid: $V = \frac{1}{3}Bh$, where B = area of the base and h = height.

2. The Lighthouse of Alexandria was toppled by an earthquake in the fourteenth century A.D. Approximately how long did it stand?

3. The largest pyramid ever built is not the Pyramid of Khufu. It is the Quetzalcoatl, located in the ancient city of Cholula in modern-day Central America. This monument, about 177 ft tall, has a volume estimated at about 116.5 million ft^3. About how long would it take to walk around it? Explain how you figured it out.

CHAPTER INVESTIGATION

Archaeologists often build three-dimensional models as an aid to understanding how an artifact, monument, or building may have looked at the time it was built. Scientists use ancient writings and their knowledge of the customs of the people to make educated guesses about the features and functions of structures that no longer exist.

Working Together

Choose an ancient structure for further research. Gather data about the measurements and known features of the structure. Then make a scale model or drawing of the structure. Using your data, estimate the exterior surface area and volume of the structure. Use the Chapter Investigation icons to guide your group.

5

Are You Ready?

Refresh Your Math Skills for Chapter 5

The skills on these two pages are ones you have already learned. Stretch your memory and complete the exercises. For additional practice on these and more prerequisite skills, see pages 654–661.

In this chapter you will solve problems involving perimeter, circumference, and area. It may be helpful to review a few of the basic formulas.

PERIMETER, CIRCUMFERENCE, AND AREA

You can use these formulas to find the perimeter and area of a rectangle, the circumference and area of a circle, and the perimeter and area of a triangle.

Examples

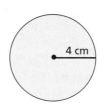

4 cm

6 cm   10 cm

8 cm

$P = 2b + 2h$

$\quad = 2(7) + 2(5)$

$\quad = 14 + 10$

$\quad = 24$ cm

$A = b \times h$

$\quad = 7 \times 5$

$\quad = 35$ cm^2

$C = 2\pi r$

$\quad \approx 2 \times 3.14 \times 4$

$\quad \approx 25.12$ cm

$A = \pi r^2$

$\quad \approx 3.14 \times 4^2$

$\quad \approx 3.14 \times 16$

$\quad \approx 50.24$ cm^2

$P = a + b + c$

$\quad = 6 + 8 + 10$

$\quad = 24$ cm

$A = \frac{1}{2}(b \times h)$

$\quad = \frac{1}{2}(8 \times 6)$

$\quad = \frac{1}{2}(48)$

$\quad = 24$ cm^2

Find the perimeter or circumference and the area of each figure. Round answers to the nearest hundredth if necessary. Use 3.14 for π.

1.

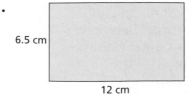

6.5 cm

12 cm

2.

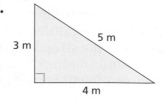

3 m   5 m

4 m

3.

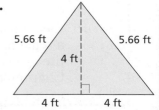

5.66 ft   5.66 ft

4 ft

4 ft   4 ft

4.

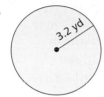

3.2 yd

5.

9 cm

9 cm

6.

12 in.

7.

6 in.

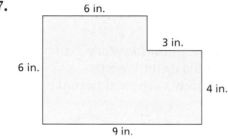

3 in.

6 in.

4 in.

9 in.

8.

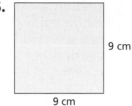

8.4 cm

9.

11.46 in.

5.6 in.

6.88 in.

10 in.

4 in.

PROBABILITY

The probability of an event can be expressed as a ratio:

$$P(\text{any event}) = \frac{\text{number of ways an event can occur}}{\text{total number of possible outcomes}}$$

Example   You toss a number cube. What is the probability that it will show an even number?

Number of ways to show an even number: $\{2, 4, 6\} = 3$

Total number of possible numbers: $\{1, 2, 3, 4, 5, 6\} = 6$

$$P(\text{even number}) = \frac{3}{6} = \frac{1}{2}$$

Find the probability of each event.

10. tossing a "heads" on a coin

11. tossing a 3 on a number cube

12. tossing a number less than 5 on a number cube

13. picking a "diamond" from a standard deck of cards

14. picking a "5" from a standard deck of cards

15. picking a "jack of clubs" from a standard deck of cards

16. picking a red marble from a bag of 8 blue marbles and 7 red marbles

17. picking a brown sock from a drawer of 12 black socks and 3 brown socks

5-1 Ratios and Units of Measure

Goals ■ Use ratios and rates to solve problems.

Applications Engineering, Number Sense, Archaeology, History

Imagine you are an archaeologist of the 25th century, and you have discovered your room, or another room in your home, looking exactly as it does today! List what you would find there. Describe how you would measure the contents. Then sketch the room as it would look when seen from above. Use a ruler and graph paper. Show all furniture and any rugs or other features you would see. Make your drawing as accurate as you can.

◣ BUILD UNDERSTANDING

Measurement is a process we use to find size, quantities, or amounts. When you make measurements, you can use either **customary** or **metric** units. We can measure to varying degrees of accuracy. Different instruments are used to make different measurements.

The **compass** is used for drawing curved lines and circles. The **protractor** is an instrument for measuring and drawing angles. **Steel scales**, or rules, measure length.

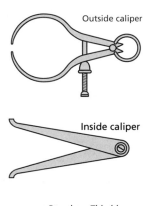
Outside caliper

Inside caliper

Tool-and-die makers use **calipers** and **micrometers** to make precise measurements. Outside calipers are used to transfer the measurement of an object to a scale or drawing. Inside calipers are often used to measure diameters of objects. Micrometers are used to measure length and/or thickness.

The **precision** of a measurement is related to the unit of measure used. The smaller the unit of measure, the more precise the measurement. The **greatest possible error** (GPE) of any measurement is $\frac{1}{2}$ the smallest unit used to make the measurement.

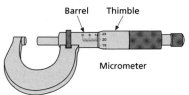
Barrel   Thimble

Micrometer

Example 1

ENGINEERING An engineer is using a steel scale. The smallest markings on the scale are $\frac{1}{64}''$. What is the GPE of any measurement made with the scale?

Solution

Find half of $\frac{1}{64}$: $\frac{1}{64} \times \frac{1}{2} = \frac{1}{128}$. The GPE is $\frac{1}{128}$ in.

Measurements are often made in order to compare quantities. To compare quantities in the same unit, such as the length of a table in a drawing with the length of the actual table, you are using a ratio. A **ratio** is a quotient of two numbers that compares one number with the other.

There are three different ways to write a ratio. The order in which the terms appear is important. Each form below can be read "six to eleven."

analogy form	*fraction form*	*word form*
6:11	$\dfrac{6}{11}$	6 to 11

For both customary and metric measures, you divide to change from a smaller unit of measure to an equivalent larger unit.

Example 2

Change 17 ft to yards.

Solution

3 ft = 1 yd

Divide 17 by 3 to find how many yards are in 17 ft.

$$17 \div 3 = 5\frac{2}{3}$$

So, 17 ft = $5\frac{2}{3}$ yd.

Use multiplication to change from a larger unit of measure to an equivalent measure in a smaller unit.

Example 3

Change 3.426 kg to grams.

Solution

1 kg = 1000 g

Multiply 3.426 by 1000 to find how many grams are in 3.426 kg.

$$(3.426)(1000) = 3426$$

So, 3.426 kg = 3426 g.

When you write ratios involving measurements, it is sometimes necessary to rename measurements using **like units**.

> ## Mental Math Tip
>
> To multiply by a multiple of 10, move the decimal point to the right one place for each zero in the multiplier. To divide by a multiple of 10, move the decimal point to the left one place for each zero in the divisor.

Example 4

Write the ratio of measurements 3 in. to 20 ft in lowest terms.

Solution

$$\frac{3 \text{ in.}}{20 \text{ ft}}$$ Write the ratio as a fraction.

$$\frac{3 \text{ in.}}{20 \text{ ft}} = \frac{3 \text{ in.}}{240 \text{ in.}}$$ Rename the measurements using inches.

$$\frac{3 \div 3}{240 \div 3} = \frac{1}{80}$$ Divide to write the fraction in lowest terms.

The ratio of measurements is 1 to 80.

A ratio that compares two *different* quantities is called a **rate**. When you compare a quantity to *one unit* of another quantity, you are finding the **unit rate**. The **unit price** of an item is its cost per unit. Consumers can determine which of two items is the *better buy* by comparing the unit prices.

Example 5

COST ANALYSIS A 10-oz box of Cat Cravings costs $1.80. A 16-oz box of Kitty Yummies sells for $2.56. Which box is the better buy?

Solution

Write a ratio of price per weight for each product to find each unit price. Then compare prices.

$$\frac{1.80}{10} = \frac{0.18}{1} = 0.18 \quad \text{Cat Cravings}$$

$$\frac{2.56}{16} = \frac{0.16}{1} = 0.16 \quad \text{Kitty Yummies}$$

$$\uparrow \qquad \uparrow$$
$$\text{unit} \quad \text{unit}$$
$$\text{rate} \quad \text{price}$$

Kitty Yummies costs $0.16 per oz. Cat Cravings costs $0.18 per oz. Kitty Yummies is the better buy.

◥ TRY THESE EXERCISES

Change each unit of measure as indicated.

1. 2 gal to cups

2. 162 in. to yards

3. 3.5 L to milliliters

4. 6.25 km to meters

Write each ratio in lowest terms.

5. 12 m to 30 m

6. 9 yd : 4 ft

7. $\dfrac{135 \text{ g}}{15 \text{ g}}$

8. 6 m : 25 cm

9. ARCHAEOLOGY It took a team of archaeological volunteers 24 days of steady work to excavate a 30-ft wall. At that rate, how much did the volunteers excavate each day?

10. WRITING MATH Which measurement is more precise, 3 in. or $3\frac{1}{4}$ in.? Explain.

◥ PRACTICE EXERCISES • For Extra Practice, see page 677.

Complete.

11. 8 qt = ___?___ c

12. 444 in. = ___?___ yd ___?___ ft

13. 2 gal = ___?___ fl oz

14. 3.2 T = ___?___ oz

15. 0.4 cm = ___?___ m

16. 300 mg = ___?___ g

17. 0.006 kg = ___?___ g

18. 8.7 mL = ___?___ L

Name the best customary unit for expressing the measure of each.

19. weight of a TV

20. height of a room

Name the best metric unit for expressing the measure of each.

21. capacity of a reservoir

22. mass of a shovel

Write each ratio in lowest terms.

23. 14 kg : 35 kg

24. 80 m to 400 cm

25. $\dfrac{16\,\text{h}}{2\,\text{days}}$

Find each unit rate.

26. 135 mi in 3 h

27. $15 for 250 copies

28. HISTORY An Egyptian merchant ship from 1500 B.C. was about 90 ft long and a Roman galley was about 235 ft long. Write a ratio in lowest terms to express the relationship between lengths of the two ships.

29. Which holds more liquid, a 4-L vase or a 3500-mL vase?

30. Use a centimeter ruler. Measure the width of your desktop to the nearest centimeter. What is the measurement? What is the GPE of the measurement?

31. DATA FILE Use the data on rectangular structures on page 645. What is the GPE for the measurements of the length and width of the Wat Kukut Temple? of the Parthenon?

32. NUMBER SENSE The capacity of a cup is either 0.25 L, 2.5 L, or 25 L. Which measurement makes the most sense?

33. Loch Ness has a capacity of about 2000 billion gal of water. If you were to drain the lake to look for the "monster," how many quarts of water would you have to remove?

34. The ratio of boys to girls at O'Neal High is 5:6. If there are 308 students, how many are boys and how many are girls?

◣ EXTENDED PRACTICE EXERCISE

35. The 1.5-mi walking trail through the main ruins area at Bandelier National Monument in New Mexico passes by caves, cliff ruins, petroglyphs, and rock carvings from this ancient village. Most people walk the trail in 45 min. What is their walking speed in miles per hour?

36. According to the early Greeks, if the ratio of the length to the width of a rectangle is 1.6:1, it is a *Golden Rectangle.* Why do you think rectangles with this shape are "golden"? What objects in the classroom or in daily life have nearly this same shape?

◣ MIXED REVIEW EXERCISE

Find the measure of each angle. (Lesson 4-1)

37. ∠ABC

∠BCA

∠CAB

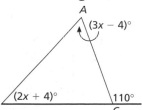

38. ∠DEF

∠EFD

∠FDE

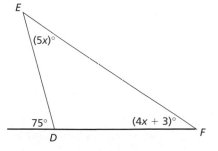

Perimeter, Circumference, and Area

Goals ■ Apply perimeter, circumference, and area formulas.

Applications Archaeology, Stage Design, Sports, and Recreation

Work with a partner.

What if the ancient Romans had invented basketball or football? Imagine spectators cheering slam dunks and touchdowns instead of gladiator fights!

The arena within the Roman Colosseum was oval-shaped and had an area of about 40,000 ft². Could a basketball court fit within the arena? Was the arena large enough for a football field? Explain your thinking.

Roman Colosseum, Italy

◥ BUILD UNDERSTANDING

When solving a problem involving measuring a plane figure, you may need to decide whether the problem requires finding the distance around the figure, or the amount of surface the figure covers, or both. When you know which measurement you want, apply the correct formula.

Recall that the distance around a polygon is its **perimeter**, the distance around a circle is its **circumference**, and the amount of surface a figure covers is its **area**.

Example 1

ARCHAEOLOGY What is the width of the fence around an archaeological dig if the region enclosed is a rectangle with a perimeter of 68 m and a length of 24.4 m?

Solution

The situation involves perimeter. Use the formula $P = 2l + 2w$.

$$P = 2l + 2w$$

$68 = 2(24.4) + 2w$ Substitute.

$68 = 48.8 + 2w$ Subtract 48.8 from each side.

$19.2 = 2w$ Multiply each side by $\frac{1}{2}$.

$9.6 = w$

The fence is 9.6 m wide.

Reading Math

The perimeter of a figure means "the measure all around it." The term comes from two Greek words—*peri*, meaning "all around," and *metron*, "measure." What other words can you think of that are derived from *peri* and/or *metron*? Check your choices with a dictionary.

Example 2

The largest pizza ever made measured 122 ft 8 in. in diameter. If your classmates were to share this pizza equally, about how many square inches of pizza would each get? Use 3.14 for π.

Solution

First, find the area of the pizza. Use the formula $A = \pi r^2$.

$$A = \pi r^2$$
$$\approx 3.14 \times 736^2 \quad \text{Use a calculator.}$$
$$\approx 1,700,925.44 \quad \text{Round your answer.}$$

The pizza had an area of about 1,700,925 in.^{2}. Divide by the number of students in your class to find out how much pizza each gets.

Example 3

Find the area of this figure.

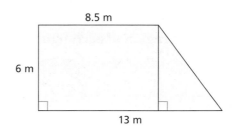

Solution

The figure can be divided into a rectangle and a triangle.

rectangle	triangle
$A = lw$	$A = \dfrac{1}{2}bh$
$= (8.5)(6)$	$= (0.5)(4.5)(6)$
$= 51$	$= 13.5$

The area of the figure is the sum of the areas of the rectangle and triangle. The area is 51 m^2 + 13.5 m^2, or 64.5 m^2.

Example 4

What is the area of the shaded region of this figure? Use 3.14 for π.

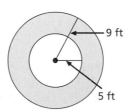

Solution

The shaded area is the difference between the areas of the circles.

$A = \pi r^2$	$A = \pi r^2$
$\approx 3.14(9^2)$	$\approx 3.14(5^2)$
≈ 254.34 ft^2	≈ 78.5 ft^2

Subtract: $254.34 - 78.5 = 175.84$

The area of the shaded region is about 175.84 ft^2.

Find the perimeter or circumference of each. Then find the area of each. If necessary, round answers to the nearest whole number.

1.
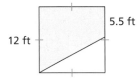
2.6 m

2.
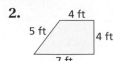
4 ft
5 ft
4 ft
7 ft

3.

3.5 cm

4.

├─11 ft─┤

5. What is the perimeter of a regular octagon with 5-cm sides?

6. What is the circumference of a circle with a radius of 6.6 m?

7. Find the height of a triangle if area $= 24$ cm^2 and base $= 10$ cm.

Find the area of the shaded region of each figure.

8.
12 ft
5.5 ft

9.
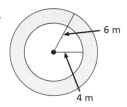
6 m
4 m

10.
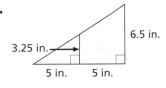
6.5 in.
3.25 in.
5 in.   5 in.

11. If you triple the length of the radius of a circle, how does the circumference change?

12. **DATA FILE** Use the data on page 645 to find the information needed to answer this question. Which has the greater area, the base of the Ziggurat of Ur or the Parthenon in Athens? Use mental math.

13. **ARCHAEOLOGY** A 1000-year-old Anasazi *kiva* is in the shape of a circle. If the area of the *kiva* is about 1661 ft^2, what is the distance around it?

14. **STAGE DESIGN** A stage from an ancient amphitheatre is shaped like a trapezoid. The front of the stage is 30 ft across, the back is 40 ft across, and the distance from front to back is 25 ft. If a circular region of the stage, 6 ft across, is designated as a pond, what is the area of the space left for actors to walk in?

15. **SPORTS** The distance from one base to the next in a standard baseball diamond is 90 ft. If the ratio of that length to the length of a basepath in a Little League diamond is 3:2, what is the area of a Little League diamond?

16. **WRITING MATH** Since π is an irrational number, many calculations involving π are found using the approximations 3.14 or $\frac{22}{7}$. When might it be easier to use $\frac{22}{7}$ rather than 3.14 to estimate area or circumference?

17. **TALK ABOUT IT** Irene and Luis both used calculators to find the area of a circle with a radius measuring 2.5 cm. Irene got 19.63495408 cm^2 and Luis got 19.625 cm^2. How can you account for the difference in their answers?

18. The side of a square is equal in length to the diameter of a circle. Which figure will have the greater area, the square or the circle?

RECREATION A community has set aside a rectangular area 250 ft by 200 ft to use for a swim center. The board of directors wants the park to have four pools:

- Pool A—a circular wading pool for small children
- Pool B—a large rectangular lap pool
- Pool C—a smaller L-shaped pool
- Pool D—a pool with an appealing irregular shape that has semi-circular as well as rectangular regions

The park should also have a small circular fountain and a shower area. It may also have picnic tables, chairs, and a food concession.

19. Imagine that you are on the planning board. Design a park layout. Submit a detailed sketch showing the size and location of each pool and feature. (You may want to use graph paper.)

20. Find the area of each pool.

21. Suppose you decide to make the pools safer by placing a border of 1-ft² tiles around each. If each tile costs $5, what will be the total cost of the number of tiles you need?

22. CHAPTER INVESTIGATION Choose a manmade structure, either ancient or modern. Begin research to learn the structure's length, width, and height. Identify any special characteristics. Then make a rough sketch of the structure and label any known measurements.

▨ EXTENDED PRACTICE EXERCISES

23. Suppose you were going to paint all the walls of your classroom. What must you know to find how much it will cost and how long it will take?

24. If all sides of the figure at the right are either parallel or perpendicular, then what is the perimeter of the unshaded portion? Does it matter what size the shaded region is?

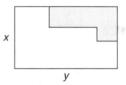

25. A farmer has 100 ft of fence. Can the farmer enclose more pasture for grazing with a square or a circular enclosure? Is this true for any length of fence?

▨ MIXED REVIEW EXERCISES

Can the given measures be the lengths of the sides of a triangle? (Lesson 4-6)

26. 5 cm, 8 cm, 10 cm   **27.** 6 in., 9 in., 5 in.

28. 8 ft, 9 ft, 17 ft   **29.** 3 m, 9 m, 11 m

30. 14 yd, 18 yd, 36 yd   **31.** 7 km, 3 km, 5 km

32. 9 dm, 9 dm, 16 dm   **33.** 4 mi, 10 mi, 10 mi   **34.** 8.6 m, 5.8 m, 15.3 m

Use the number line at the right for Exercises 35–40. Find each length. (Lesson 3-1)

35. $\overline{MP}$ **36.** $\overline{QS}$ **37.** $\overline{PS}$

38. $\overline{MR}$ **39.** $\overline{NR}$ **40.** $\overline{MS}$

Review and Practice Your Skills

PRACTICE ◣ LESSON 5-1

Complete.

1. 48 c = _____ qt

2. 512 fl oz = _____ gal

3. 11,600 oz = _____ T

4. 7000 mm = _____ cm

5. 7.03 L = _____ mL

6. 48 g = _____ kg

Write each ratio in lowest terms.

7. 120 m : 2700 cm

8. 3 yd : 48 in.

9. 7 days to 120 hours

10. 5 kg to 5,000,000 mg

11. 1200 min : 1 day

12. $1 \text{ yd}^2 : 3 \text{ ft}^2$

Choose the best estimate for each.

13. length of basketball court   **a.** 90 in.   **b.** 90 ft   **c.** 90 yd

14. weight of an infant   **a.** 9 kg   **b.** 9 g   **c.** 9 mg

15. length of a drinking straw   **a.** 20 m   **b.** 20 mm   **c.** 20 cm

Solve.

16. What is the unit rate for a train which travels 805 mi in 7 h?

17. Which is the better buy, 1 gal of milk for $1.79, or 1 pt of milk for $0.25?

18. The ratio of girls to boys at South High School is 3:4. If there are 980 students, how many are boys?

PRACTICE ◣ LESSON 5-2

Find the perimeter or circumference of each.

19.
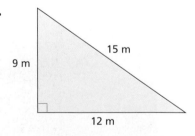

20.
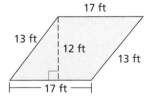

21.

22–24. Find the area of each figure in Exercises 19–21.

Find the area of the shaded region in each figure.

25.
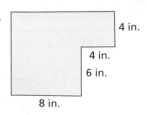

26.
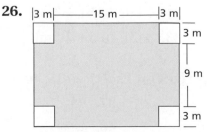

27.
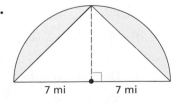

28. If you triple the length of the radius of a circle, how does the area change? (Lesson 5-2)

Complete. (Lesson 5-1)

29. 6.5 gal = _____ fl oz   **30.** $6\frac{5}{8}$ kg = _____ g   **31.** 816 in. = _____ yd _____ ft

32. 2,500,000 mL = _____ kL   **33.** 0.04 cm = _____ m   **34.** 56 c = _____ gal _____ qt

Write each ratio in lowest terms. (Lesson 5-1)

35. 1000 m : 10 km   **36.** 8.5 T to 34,000 lb   **37.** 75 cL : 25 mL

38. 10 yd to 540 in.   **39.** 9.2 mL : 9.2 L   **40.** 1 gal : 1 fl oz

41. Which holds more liquid, a 200 L barrel or a 22,500 mL barrel? (Lesson 5-1)

Find the perimeter or circumference of each figure. Then find the area of each figure. (Lesson 5-2)

42.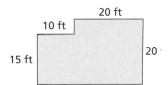

43.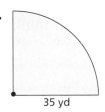

44.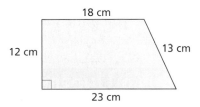

MathWorks Career – Equipment Operators
Workplace Knowhow

Ancient city ruins often lay under many tons of dirt, debris, rock, or even water. Some sites are overgrown with vegetation. To get to these ruins, archaeologists must employ workers who can operate earth-moving machines such as bulldozers, conveyors, trench excavators, hoists, winches, backhoes, and cranes.

To remove trees from the ground area above ancient ruins, the project site has been divided into three sections—a rectangle, a triangle and a half circle.

1. Find the area in square feet of each section.

To clear 10 ft^2 of vegetation requires 3 workers paid $12.50/h and 2 machine operators paid $16/h. Working together, these workers can clear 10 ft^2 in 2 h.

2. How long will it take the workers to clear each section?

3. What is the cost to clear the triangular section of vegetation?

5-3 Probability and Area

Goals ■ Determine probabilities using areas.

Applications Games, Archaeology, Weather

Play this game in groups of 3–4 students. You will need the bottom of a large 16-in. pizza box, a 4-in. square paper, and a small coin.

1. One person plays against the other members of the group. This person puts the box on a table and tapes the paper square anywhere on the bottom of the box.

2. The remaining members of the group sit on the floor a few feet away (so that the placement of the square cannot be seen). These players take five turns each tossing the counter into the box.

3. A player on the floor wins if the coin comes to rest completely within the paper square once. The person placing the square wins if a counter never lands on the paper square.

4. Is there a way to determine the chance of landing within the paper square? Explain. How can you find the chance of landing elsewhere within the box? Do you think the game is fair? Explain your thinking.

◤ BUILD UNDERSTANDING

You can use what you know about **probability** to solve problems like the one above, in which you need to determine the likelihood that an event will occur. Recall that the probability of an event can be expressed as a ratio:

$$P(\text{any event}) = \frac{\text{number of favorable outcomes}}{\text{number of possible outcomes}}$$

Check Understanding

The probability of an event is a number between 0 and 1. What is the probability of an event that will always occur? What is the probability of an impossible event? Give an example of each.

Example 1

What is the probability that a point chosen at random from within _M_ is also in _N_?

Solution

Find the probability.

$P = \dfrac{\text{area of } N}{\text{area of } M}$ area of N = favorable outcome
area of M = possible outcome

$= \dfrac{16}{256}$

$= \dfrac{1}{16}$ or 0.0625

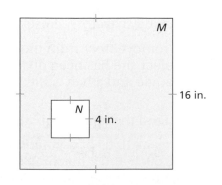

So the probability that a point chosen at random is in N is $\frac{1}{16}$, or 0.0625.

Example 2

A dart is dropped onto a foam board shown at the right. What is the probability that the dart lands in the blue region? In the green region?

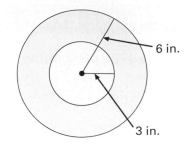

6 in.

3 in.

Solution

$$P(\text{blue}) = \frac{\text{area of blue circle}}{\text{area of larger circle}}$$

$$= \frac{(\pi)(3^2)}{(\pi)(6^2)} \quad \begin{array}{l}\text{Since } \pi \text{ is a common factor,} \\ \text{you do not have to calculate each area.}\end{array}$$

$$= \frac{9}{36} = \frac{1}{4}$$

Since $P(\text{green or blue}) = 1$, then $P(\text{green}) = 1 - P(\text{blue})$. $P(\text{green}) = \frac{3}{4}$.

Example 3

A treasure chest was buried long ago beneath what is now school property. No one knows where the chest lies. If the school property is a rectangle measuring 600 ft by 540 ft, what is the probability that the chest could be found by excavating the baseball diamond, a square with sides of 90 ft each?

Solution

$$P(\text{chest in diamond}) = \frac{\text{area of diamond}}{\text{area of property}}$$

area of diamond	**area of property**
$A = s^2$	$A = lw$
$A = 90^2 = 8100$	$A = (600)(540) = 324{,}000$

$$P(\text{chest in diamond}) = \frac{8100}{324{,}000} = \frac{1}{40} = 0.025$$

The probability is $\frac{1}{40}$ or 0.025.

Example 4

GAMES Twenty-five darts are randomly thrown at a circular dartboard and all of the darts land within the dartboard. Four hit the bull's-eye. If the diameter of the bull's-eye is 24 cm, what is the approximate area of the dartboard?

Solution

Since 4 of 25 darts landed in the bull's-eye, the probability of a single dart hitting the bull's-eye is $\frac{4}{25}$.

$$P(\text{dart landing in bull's-eye}) = \frac{\text{area of bull's-eye}}{\text{area of dartboard}} = \frac{4}{25}$$

Area of bull's-eye $= \pi r^2 \approx 3.14 \cdot 12^2 \approx 452.16 \text{ cm}^2$

Let x = area of dartboard. $\frac{4}{25} \approx \frac{452.16}{x}$; $x \approx 2826$

The area of the dartboard is about 2826 cm².

> ## Problem Solving Tip
>
> In Example 4, you know that the ratio of the area of the bull's-eye to the area of the whole target is 4:25. How many times greater is the area of the target? How can you use this information to solve the problem?

Find the probability that a point selected at random is in the shaded region.

1.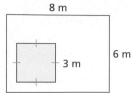

2.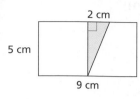

3.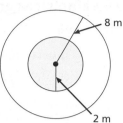

4. GAMES Refer to the game in the opening situation. What is the probability of landing the coin on the small square if the box measures 12 in. on each side? Disregard the area of the coin.

GAMES A standard deck of playing cards has 52 cards. A card is drawn at random from a shuffled deck. Find each probability.

5. P(queen) **6.** P(red card) **7.** P(black face card)

Find the probability that a point selected at random is in the shaded region.

8.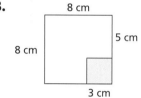

9.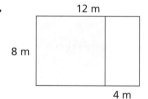

10.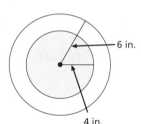

11.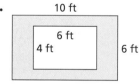

12.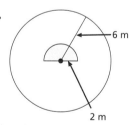

13.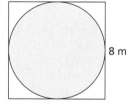

14. Suppose a cordless phone has been left somewhere within a 2000 ft² house. What is the probability it is in the 20-ft by 15-ft living room?

15. The total area of the state of Oklahoma is 69,919 mi². The area of its capital, Oklahoma City, is 604 mi². If a meteor were to land somewhere in the state, estimate the probability that it would land with in the city limits of the capital.

16. ARCHAEOLOGY The rectangular foundation for an ancient building measures 150 ft by 80 ft. The foundation is made from stone cubes, with sides measuring 2 ft each. A scroll is hidden within one of the blocks. What is the probability of finding the scroll if a block is chosen at random?

Oklahoma City

17. **YOU MAKE THE CALL** A square (side = 2 in.) is placed inside a larger square (side = 6 in.). Evan says that the probability of selecting a point at random within the smaller square is $\frac{1}{3}$, since the ratio of the smaller square's side to the larger square's side is 2:6, or 1:3. Do you agree with Evan's thinking? Explain your reasoning.

Tell whether each event is *certain*, *likely*, *unlikely*, or *impossible*.

18. **WEATHER** It will snow in July where you live.

19. You will roll a sum of 5 or greater using two number cubes.

20. You left a pencil in one of 5 classrooms, but you don't know which one. You find it in the first room you search.

21. There is life on other planets.

Find the probability that a point selected at random in each figure is in the shaded region.

22.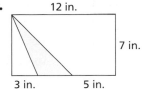

23.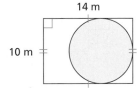

24.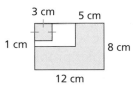

25. Suppose a leak occurs from above the room shown at the right. What is the probability the leak will be over the carpet?

26. Draw a figure containing a shaded region, so that the probability is 1 out of 6 that a point selected at random will be in the shaded region.

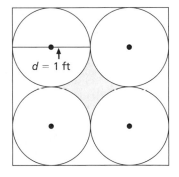

EXTENDED PRACTICE EXERCISES

27. **WRITING MATH** In a scale drawing of the ruins of Pompeii, 1 in. = 400 ft. A student is erasing a pencil mark accidentally made somewhere on the drawing. What additional information do you need to know to find the probability that the mark was made on the Palaestra?

28. Suppose a square target looks like the one at the right. What is the probability of hitting the shaded region?

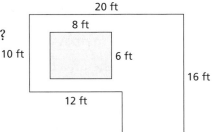

MIXED REVIEW EXERCISES

On a coordinate plane, sketch the triangle with the given coordinates. Then classify the triangle by both its angles and its sides. (Lesson 4-1)

29. $A(5, 5)$, $B(-4, -4)$, $C(-4, 5)$

30. $L(3, 1)$, $M(9, -1)$, $N(-2, -4)$

31. $R(6, 1)$, $S(-3, -2)$, $T(6, -5)$

32. $X(-2, -2)$, $Y(-1, 2)$, $Z(6, -3)$

Given $f(x) = 5x - 8$ and $g(x) = 1.3x - 4.7$, find each value. (Lesson 2-2)

33. $f(-5)$

34. $f(13)$

35. $f(-8)$

36. $f(9)$

37. $g(12)$

38. $g(-17)$

39. $g(34)$

40. $g(-27)$

5-4 Problem Solving Skills: Irregular Shapes

Sometimes you can find the answer to a difficult problem by breaking it into smaller problems you already know how to solve.

Problem

An architect's sketch of the plan for one floor of a new archaeological museum is shown. What is the area of the floor?

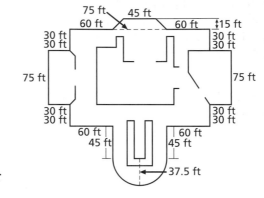

Solution

Solve a simpler problem. Copy or trace the *outline* of the floor plan on graph paper. Ignore all the inner walls. Divide the floor into 6 figures: a trapezoid, 4 rectangles, and a half-circle. Label them A–F.

Find the area of each figure. Use 3.14 for π.

A. $\left(\dfrac{75 + 45}{2}\right)(15) = 900$ ft^2

B. $30(195) = 5850$ ft^2

C. $75(255) = 19{,}125$ ft^2

D. $30(195) = 5850$ ft^2

E. $45(75) = 3375$ ft^2

F. $\pi \cdot 37.5^2 \cdot 0.5 \approx 2208$ ft^2

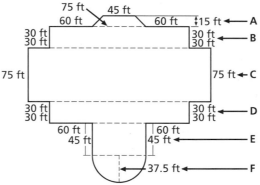

Add to find the total area of all the regions. The area of the first floor of the museum is 37,308 ft^2.

Check your answer by tracing the outline again and dividing it into a different arrangement of plane figures.

Problem Solving Strategies

- Guess and check
- Look for a pattern
- ✔ Solve a simpler problem
- Make a table, chart or list
- Use a picture, diagram or model
- Act it out
- Work backwards
- Eliminate possibilities
- Use an equation or formula

Check Understanding

Why do you multiply by 0.5 when finding the area of region F?

◤ TRY THESE EXERCISES

1. **CARPETING** This layout of a wing at a natural history museum shows the African Peoples, Asian Peoples, and Birds of the World galleries. How much carpeting is needed for this wing?

2. An archaeological team has 30 ft of fencing with which to enclose a rectangular region. If the length and width are whole numbers, what different areas, in square feet, are possible?

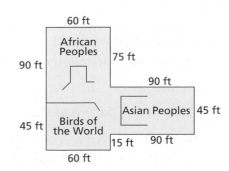

3. A display case planned for showing fragments of 3500-year-old frescoes will have the shape at right. If there will be 3 glass shelves, how much glass, in square feet, is needed?

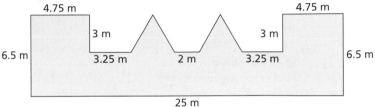

23.5 ft
8 ft   4 ft   4 ft   8 ft
4.75 ft   4.75 ft

◥ PRACTICE EXERCISES

4. How many small squares are in a checkerboard? How many squares of all sizes are there?

5. ART Painters are to cover the entire side of this wall from an abandoned factory with a mural. What is the area of the region to be painted?

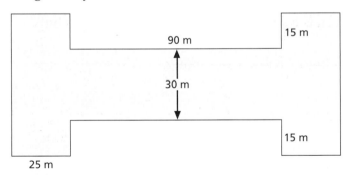

4.75 m   4.75 m
6.5 m   3 m   3 m   6.5 m
3.25 m   2 m   3.25 m
25 m

6. WRITING MATH Write your own problem involving the area of irregular figures that can be solved by first solving a simpler problem or problems.

7. RECREATION The Mayans played a game resembling our game of volleyball on courts that looked like the one shown below. Imagine for a moment that you are the coordinator of a tournament for which you will need a rectangular field containing three side-by-side courts like these. You decide that the courts must be at least 20 m apart and that there must be at least 5 m between an edge of a court and the perimeter of the field. How long must your field be? How wide?

90 m
15 m
30 m
25 m
15 m

◥ MIXED REVIEW EXERCISES

Use the polygon-sum theorem to find the sum of the measures of convex polygons with the given number of sides. Then find the measure of each interior angle, assuming each polygon is regular. Round to the nearest hundredth if necessary. (Lesson 4-7)

8. 23   **9.** 45   **10.** 38   **11.** 28

12. 18   **13.** 34   **14.** 50   **15.** 42

16. DATA FILE Use the data on page 645 on housing units. What percentage of mobile homes or trailers are vacant? (Prerequisite Skill)

Review and Practice Your Skills

PRACTICE ◣ LESSON 5-3

Find the probability that a point selected at random is in the shaded region.

1.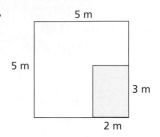
5 m
5 m
3 m
2 m

2.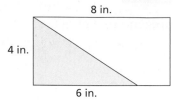
8 in.
4 in.
6 in.

3.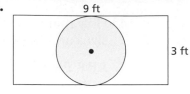
9 ft
3 ft

4.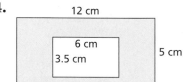
12 cm
6 cm
3.5 cm
5 cm

5.
3 m
3 m
3 m

6.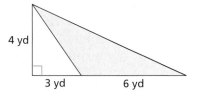
4 yd
3 yd
6 yd

7. Suppose a book has been left somewhere within a 2700 ft² house. What is the probability that the book is either in the 15-ft by 10-ft den or a 12-ft by 10-ft bedroom?

Tell whether the event is *certain, likely, unlikely,* or *impossible*.

8. The temperature will hit 90 degrees Farenheit in December.

9. You will draw a red face card from a standard, shuffled deck of cards.

10. The area of a rectangular region is calculated by multiplying its length by its width.

11. You will leave school grounds before 6 P.M. this evening.

PRACTICE ◣ LESSON 5-4

Find the area of each figure.

12.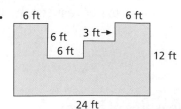
6 ft   6 ft
6 ft   3 ft→
6 ft
12 ft
24 ft

13.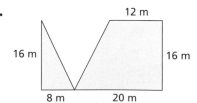
12 m
16 m
16 m
8 m   20 m

14.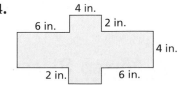
4 in.
6 in.   2 in.
4 in.
2 in.   6 in.

15. A counter top for a kitchen will have the shape shown. How much laminate material, in square inches, is needed to make this counter top?

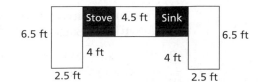
Stove   4.5 ft   Sink
6.5 ft   6.5 ft
4 ft   4 ft
2.5 ft   2.5 ft

16. How many rectangles of all sizes are in the figure shown?

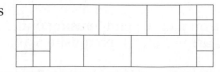

Write each ratio in lowest terms. (Lesson 5-1)

17. 98 m:49 km

18. 13 c to 4 fl oz

19. 1760 yd:10,560 ft

Find the perimeter or circumference of each figure. Then find the area of each figure. Use 3.14 for π. (Lesson 5-2)

20.

7 in.

21.

10 m

22.

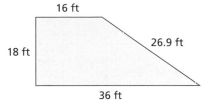

16 ft

18 ft   26.9 ft

36 ft

23. Suppose a parachutist will be landing in the region at the right. What is the probability that she will land in the shaded part of the region? (Lesson 5-3)

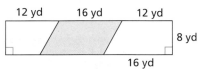

12 yd   16 yd   12 yd

8 yd

16 yd

24. If everyone in this classroom shakes hands with everybody else exactly once, how many handshakes will occur? (Lesson 5-4)

Mid-Chapter Quiz

Write each ratio in lowest terms. (Lesson 5-1)

1. 25 in. to 5 ft

2. 48 c : 320 oz

3. Which is the better buy, a 19-oz box of cereal for $2.66 or the 12-oz box of the same cereal for $1.92?

4. The diagonal of a rectangle is 25 cm. If the long side of the rectangle is 24 cm, what is the ratio of the short side to the diagonal?

Find the perimeter and area of each figure. Round answers to the nearest tenth. (Lesson 5-2)

5.

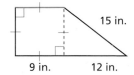

15 in.

9 in.   12 in.

6.

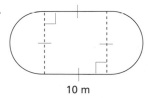

10 m

Draw the described figures. Find the probability that a randomly chosen point inside the larger figure lies outside the shaded area. (Lesson 5-3)

7. Draw a circle with a radius of 6 cm. Draw a smaller circle inside the first with a radius of 3 cm. Shade the smaller circle.

8. Draw a square with sides measuring 10 cm each. Find the midpoint of one side. Connect the point to one of the endpoints of the opposite side. Shade the right triangle.

Three-dimensional Figures and Loci

Goals ■ Analyze space figures.

Applications Archaeology, Architecture, Art

Trace and cut out each pattern below. Try to fold each to form a three-dimensional figure. What do you notice?

◤ BUILD UNDERSTANDING

A **polyhedron** (plural: *polyhedra*) is a closed, three-dimensional figure in which each surface is a polygon. The surfaces are called **faces**. Two faces intersect at an **edge**. A **vertex** is a point where three or more edges intersect.

A polyhedron with two identical parallel faces is called a **prism**. Each of these faces is called a **base**. Every other face is a parallelogram. A **pyramid** is a polyhedron with only one base. The other faces are triangles that meet at a vertex. A prism is named by the shape of its bases and a pyramid by the shape of its base. The **lateral faces** are those that are not bases. The edges of these faces are called **lateral edges** and can be parallel, intersecting, or skew.

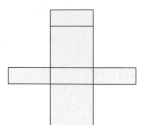

Right rectangular prism

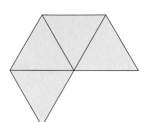

Triangular pyramid

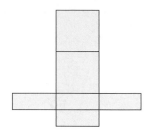

Hexagonal prism

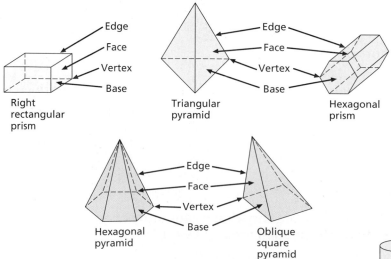

Hexagonal pyramid

Oblique square pyramid

Some three-dimensional figures have flat *and* curved surfaces. A **cylinder** has a curved region and two parallel congruent circular bases. Its **axis** joins the centers of the two bases.

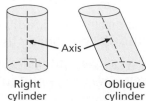

Right cylinder

Oblique cylinder

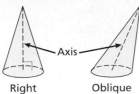

Right cone   Oblique cone

A **cone** is a three-dimensional figure with a curved surface and one circular base. Its axis is a segment from the vertex to the center of the base.

A **sphere** is the set of points in space that are the same distance from a given point called the **center** of the sphere.

Center

Sphere

Example 1

Identify the figure.

a.

b.

Solution

a. Square pyramid—it has one square base and triangular faces.

b. Cone—it has a curved surface and one circular base.

Reading Math

A polyhedron is a **regular polyhedron** if all its faces are congruent regular polygons. The Greek scholar, Plato, studied these figures, also known as the five **Platonic solids**.

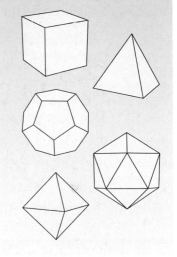

Example 2

For the pentagonal prism at the right, identify the bases, a pair of intersecting faces and the edge at which they intersect, and a pair of skew edges.

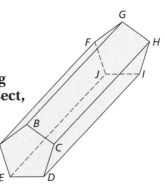

Solution

Some answers may vary.

Bases: *ABCDE* and *FGHIJ*

Pair of intersecting faces: *AFGB* and *BGHC*

Edge where these two faces intersect: $\overline{BG}$

Pair of skew edges: $\overline{CH}$ and $\overline{ED}$

Example 3

ARCHAEOLOGY An archaeologist says that a Greek artifact is in the shape of a right hexagonal prism. Draw the prism.

Solution

Step 1: Draw two congruent hexagons on graph paper.

Step 2: Use a straightedge to connect the corresponding vertices. Use dotted lines to show the unseen lateral edges.

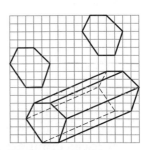

Sometimes you will be asked to describe or identify a set of points that meets particular requirements. The mathematical term for specifying points is **locus**, the set of all points that satisfy a given set of conditions. The word *locus* comes from Latin and means *place*; its plural is *loci*.

Example 4

Describe the locus of points 6 cm from a given point, *P*. All points lie within the same plane.

Solution

Draw point *P* on a sheet of paper. Locate and mark several points 6 cm from it. If you continue to add points to the drawing, what figure is formed? A circle with a radius of 6 cm.

TRY THESE EXERCISES

Identify each figure. Then identify the base(s), a pair of parallel edges, intersecting faces, and intersecting edges.

1.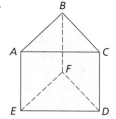

2.

3.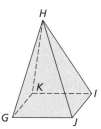

4. Draw a cone.

5. Describe the locus of points equidistant from two parallel lines.

PRACTICE EXERCISES • For Extra Practice, see page 678.

Name the polyhedra shown below. Then state the number of faces, vertices, and edges each has.

6.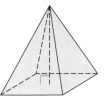

7.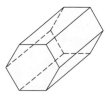

8.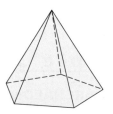

Draw the figure.

9. triangular prism

10. cylinder

11. **WRITING MATH** Examine your answers for Exercises 6–8. For each polyhedron, what can you say about how the sum of its faces and vertices compares with the number of edges? Write a rule to describe the relationship among the faces, vertices and edges of a polyhedron.

12. Describe and draw the locus of points in a plane that are 4 m from a given line in the plane.

13. Draw a picture to show the locus of points equidistant from the two sides of ∠*ABC* that are in the interior of ∠*ABC*.

14. Describe the locus of points *in space* 3 ft from point *O*. (*Hint:* A locus of points in space may form a three-dimensional object.)

15. Describe the locus of points in space that are a given positive distance from a given line.

16. **ARCHITECTURE** A 6-story building is 72 ft high. All stories are the same height. Describe the locus of points that are within the building 24 ft from the floor of the fourth floor of the building.

17. **ART** A sculpture is formed by placing an oblique square pyramid on top of a right rectangular prism. The rectangular prism has a square base and its height is twice the length of an edge of the base. The base of the pyramid is the same size as the base of the prism. Draw the sculpture.

18. **CHAPTER INVESTIGATION** Build a three-dimensional model of the structure you have chosen. Break the structure into smaller three-dimensional figures or sections. Then assemble them to make the final product.

■ EXTENDED PRACTICE EXERCISES

19. A **cross section** is the two-dimensional figure formed when you cut a three-dimensional shape with a plane. If you cut a cross section of a square pyramid parallel to the base, what polygon will be formed?

20. What would a triangular prism look like if seen from the side? What would it look like from above? Assume that the prism is resting on its base.

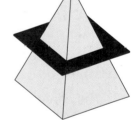

■ MIXED REVIEW EXERCISES

Find the value of *x* in each figure. (Lesson 4-3)

21.

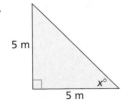
5 m
5 m
x°

22.

1.7 cm   1.7 cm
x°
1.7 cm

23.

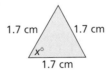
2.9 in.   3.8 in.
82°   49°
x in.

Solve each equation. (Lesson 2-5)

24. $3(2x + 1) - 8 = 5x + 2(x + 1)$

25. $-3x + 4(x - 1) = 6 - 4(x + 2)$

26. $-2(x - 3) + 5 = -5x + 9$

27. $3(x - 3) + 2x = 7x - 3(x + 1)$

28. $5 - 4(x - 8) = -5(4x - 1)$

29. $-4(x - 2) + 3 = x + 2(x - 5)$

30. $2 - 3(2x - 6) + x = 3x - 5$

31. $2(x + 4) - 3x + 8 = -3(x - 4) + 5x$

32. $4(x + 2) - 3x + 11 = 2(3x - 2) + 3(x + 4)$

33. $-5(2x + 1) + 2(x - 3) = 2x + 6(x + 3)$

Surface Area of Three-dimensional Figures

Goals ■ Find surface areas of three-dimensional figures.

Applications Packaging, Manufacturing, Sports, Astronomy

Work with a partner.

Construct a square pyramid out of construction paper, using only the following: straightedge, compass, scissors, and tape. Write a description of how you did it.

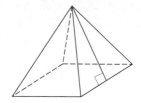

◥ BUILD UNDERSTANDING

When you are asked to find the **surface area** of a three-dimensional figure, think about whether the figure is a prism, pyramid, cylinder, cone, or sphere, or whether it is a combination of figures. To help you identify the shape of each surface, think about what the figure would look like if it were cut apart. Notice whether any surfaces are congruent.

Example 1

PACKAGING A box of Teen Chow cereal is 11.5 in. high, 7.5 in. wide, and 2.5 in. deep. What is the surface area of the box?

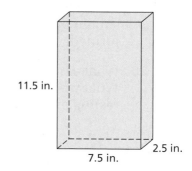

11.5 in.

2.5 in.
7.5 in.

Solution

The cereal box is a rectangular prism, so it has 3 pairs of congruent rectangular faces. To find its surface area, find the area of each face. Use the formula $A = lw$.

SA means "surface area."

$SA = 2(\text{area of front}) + 2(\text{area of side}) + 2(\text{area of top})$

$= 2(11.5 \times 7.5) + 2(11.5 \times 2.5) + 2(7.5 \times 2.5)$

$= 172.5 + 57.5 + 37.5$

$= 267.5$

The surface area is 267.5 in.^{2}.

Example 2

MANUFACTURING At Farrow's Ceramic Factory ceramic replicas of the Great Pyramid at Giza are made. Each model has a square base 10 cm in length, and triangular faces each with a height of 12 cm. Farrow's plans to paint the models. What is the surface area of each?

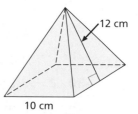

12 cm

10 cm

Solution

The model is a square pyramid. Each of the four triangular faces has the same area. To find its surface area, find the area of each face and of the base.

$SA = 4(\text{area of triangular face}) + \text{area of square base}$

area of triangular face

$$A = \frac{1}{2}bh$$

$$= \left(\frac{1}{2}\right)(10)(12)$$

$$= 60$$

area of square base

$$A = s^2$$

$$= 10^2$$

$$= 100$$

$SA = 4(60) + 100 = 340$

The surface area is 340 cm².

Example 3

8 cm

14 cm

A can of bread crumbs is 14 cm high and 8 cm across. What is the surface area of the can?

Solution

The can is a cylinder. To find its surface area, add the area of the curved surface to the area of the two bases.

area of the curved surface

$$A = 2\pi rh$$

$$= (2)(\pi)(4)(14)$$

$$\approx 351.68$$

area of each circular base

$$A = \pi r^2$$

$$= (\pi)(16)$$

$$\approx 50.24$$

The can has two congruent circular bases. $(2)(50.24) = 100.48$

$$SA \approx 351.68 + 100.48 \approx 452.16$$

The surface area of the can is approximately 452.16 cm².

> ### Check Understanding
>
> If the curved surface of the cylinder is laid flat, it forms a rectangle. What dimension of the cylinder equals the length of this rectangle? To what dimension is the width equal?

Example 4

A tent in the shape of a tepee is 4 m across with a slant height of 2.6 m. What is the surface area of the canvas, including the floor?

Solution

The tent approximates a cone. To find its surface area, add the area of the curved surface to the area of the base.

$$SA = \pi rs + \pi r^2 \qquad s = \text{slant height}$$

$$= \pi (2)(2.6) + \pi (2)^2$$

$$\approx 16.328 + 12.56$$

$$\approx 29$$

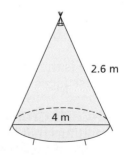

2.6 m

4 m

The surface area of the tent is approximately 29 m².

Example 5

SPORTS What is the surface area of a soccer ball with a diameter of about 9 in.?

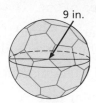
9 in.

Solution

The soccer ball is a sphere. To find its surface area, use the formula $SA = 4\pi r^2$.

$$SA \approx (4)(3.14)(4.5)^2$$

$$\approx 254$$

The soccer ball has a surface area of about 254 in.^{2}

◥ TRY THESE EXERCISES

Find the surface area of each figure. Assume that the pyramid is a regular pyramid. Use 3.14 for π.

1.
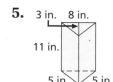
3 cm
8 cm
5 cm

2.
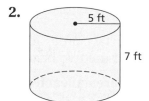
5 ft
7 ft

3.
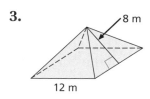
8 m
12 m

4. **DATA FILE** Use the data on the sizes and weights of various balls used in sports on page 653. Calculate the surface area of a volleyball.

◥ PRACTICE EXERCISES • For Extra Practice, see page 679.

Find the surface area of each figure. Assume that all pyramids are regular pyramids. Use 3.14 for π. Round answers to the nearest whole number.

5.
3 in.   8 in.
11 in.
5 in.   5 in.

6.

2.6 m
2 m

7.
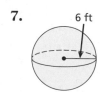
6 ft

8.
3 cm   6 cm
4.5 cm

9. What is the surface area of a square pyramid whose base length is 8 m and whose faces have heights of 6.4 m?

10. **ARCHITECTURE** The Marina Towers in Chicago are cylindrical shaped buildings that are 586 ft tall. There is a 35-ft diameter cylindrical core in the center of each tower. If the core extends 40 ft above the roof of the tower, find the exposed surface area of the core.

Find the surface area of each figure. Use 3.14 for π. Round answers to the nearest whole number.

11.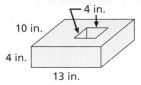
10 in.
4 in.
4 in.
13 in.

12.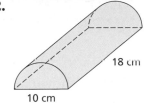
18 cm
10 cm

13.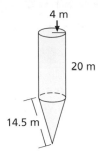
4 m
20 m
14.5 m

14. Use mental math. Which has the greater surface area, the can or the box?

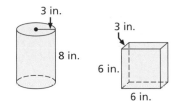
3 in.
8 in.
3 in.
6 in.
6 in.
6 in.

15. ART The base of a sculpture is a regular pentagonal prism with sides 10 cm high and 6 cm wide. What additional information do you need to find the surface area of the base of the sculpture?

16. WRITING MATH The surface area of a rectangular prism is 178 in.² What is the height of the figure if its length is 3 in. and its width is 4 in.? Explain how you got your answer.

17. SPORTS The "shots" that shot-putters toss are heavy spheres that range in diameter from 95 mm to 130 mm. What is the difference in surface area between the largest and smallest shot?

18. ASTRONOMY Jupiter, the largest planet, has an equator with a diameter of about 88,000 mi. To the nearest million miles, what is the surface area of Jupiter? Assume that it is a sphere.

▮ EXTENDED PRACTICE EXERCISES

19. What happens to the surface area of a cube if you (a) double the length of a side, or (b) divide the length of a side by 3?

20. One way to express the formula for finding the surface area of a cylinder is $SA = 2\pi rh + 2\pi r^2$. How else can this be expressed?

21. To paint the sides of a cube, 1 quart of paint is used. Suppose two such cubes are glued together to form a rectangular solid. How much paint will it take to paint the new rectangular solid?

22. CHAPTER INVESTIGATION Estimate the surface area of the ancient structure you have chosen. Explain how you got your answer.

▮ MIXED REVIEW EXERCISES

Find each length. (Lesson 3-1)

23. In the figure below, $\overline{AC}$ = 106. Find $\overline{AB}$.

A •———2x + 7———•———3x + 4———• C
       B

24. In the figure below, $\overline{RT}$ = 170. Find $\overline{ST}$.

R •———4(x + 3)———•———2(x − 2)———• T
        S

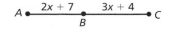

Review and Practice Your Skills

PRACTICE ◣ LESSON 5-5

Name each three-dimensional figure shown below. Then state the number of faces, vertices, and edges for each.

1.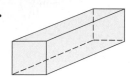

2.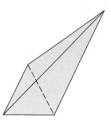

3.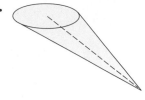

4–6. For the figures in Exercises 1–3, identify the following:
   a. base(s)
   b. a pair of parallel edges
   c. a pair of intersecting faces
   d. a pair of intersecting edges

Draw each figure.

7. oblique cone

8. pentagonal prism

9. oblique hexagonal pyramid

10. Describe and draw the locus of points that are inside or on a square and equidistant from two adjacent sides of the square.

11. Describe and draw the locus of points in a plane equidistant from a line and a point that is not on the line.

PRACTICE ◣ LESSON 5-6

Find the surface area of each figure. Assume the pyramid is a regular pyramid. Round answers to the nearest tenth.

12.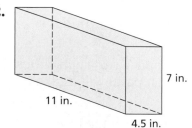
11 in. 7 in. 4.5 in.

13.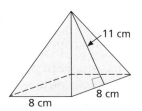
11 cm 8 cm 8 cm

14.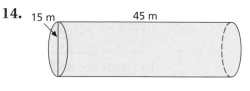
15 m 45 m

Find the surface area of each figure. Round answers to the nearest tenth.

15.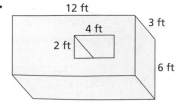
12 ft 4 ft 3 ft 2 ft 6 ft

16.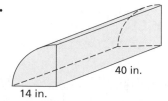
40 in. 14 in.

17.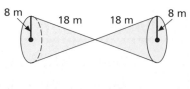
8 m 18 m 18 m 8 m

18. Find the surface area of a square pyramid with base length = 10 m and faces with heights of 8.2 m.

19. What happens to the surface area of a sphere if you triple the radius? (Lesson 5-6)

20. What happens to the surface area of a cone if you double the radius? (Lesson 5-6)

Find each unit rate. (Lesson 5-1)

21. 260 mi in 4 h

22. $42 for 1400 stamps

23. 51 gal in 1.5 min

Find the probability that a point selected at random in each figure is in the shaded region. (Lesson 5-3)

24.

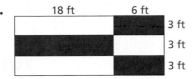

25.

26.

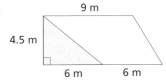

Find the surface area of each figure. Assume the pyramid is a regular pyramid. Round answers to the nearest whole number. (Lesson 5-6)

27.

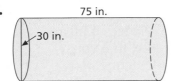

28.

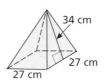

MathWorks Career – Archaeologist
Workplace Knowhow

Archaeologists trace the histories of ancient civilizations by studying ancient records and artifacts. Many hours are spent in the field searching for buried cities and artifacts. Archaeologists make detailed maps of each site they discover. They analyze the layout of buildings and rooms. Each item found is labeled and catalogued. Archaeologists note the exact location artifacts are found in order to determine their purpose.

Suppose you are excavating a building buried under several feet of ancient volcanic ash and silt. By reading inscriptions on stones, you expect to find a cylindrical pedestal, 3 ft in diameter, somewhere in the interior of the room.

1. If the building is circular with a diameter of 25 feet, what is the probability of finding the pedestal if a dig site is chosen randomly?

2. The site can be divided into two rectangular sections: the first measuring 15 ft by 13 ft, and the second, 20 ft by 20 ft. Find the total area of the site.

5-7 Volume of Three-dimensional Figures

Goals ■ Find the volume of three-dimensional figures.

Applications Manufacturing, Astronomy, Archaeology

Work in groups of three or four students.

Many environmental groups criticize manufacturers for over-packaging their products. On the other hand, over-packaging is one way to make customers think they are getting more for their money.

1. Choose a product that you think uses too much packaging.

2. Develop a new way to package the product that uses less packaging material. Remember, the packaging must keep the product from breaking, fit neatly in shipping cartons, and look appealing to the consumer.

3. Make a packaging sample for display.

◣ BUILD UNDERSTANDING

Recall that volume is a measure of the number of cubic units needed to fill a region of space. To find the volume of a three-dimensional figure, first you must determine whether the figure is a prism, pyramid, cylinder, cone, sphere, or a combination of shapes. Then apply the appropriate formula or formulas for volume.

Example 1

Find the volume of the figure at the right.

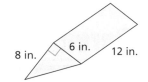

8 in.   6 in.   12 in.

Solution

The figure is a prism. To find the volume (V) of any prism, multiply the area of the base (B) by the height (h) of the prism. First find the area of the base, which is a right triangle.

$$B = \frac{1}{2}bh$$
$$= \left(\frac{1}{2}\right)(8)(6)$$
$$= 24$$

The area of the base is 24 in.^{2}. Then use the volume formula.

$$V = Bh$$
$$= (24)(12)$$
$$= 288$$

The volume is 288 in.^{3}.

A three-dimensional figure may be a combination of shapes. Mentally break the figure into smaller pieces. Then find the volume of each piece. Finally, use the information to solve the problem.

If either the area of the base or the height of a pyramid is evenly divisible by 3, you can use mental math to find the volume. For example, what is V, if $B = 18$ m^2 and $h = 11$ m?

Example 2

Find the volume of the shaded part of the figure shown.

Solution

To find the volume of the shaded part, find the difference between the volume of the small pyramid and the volume of the large pyramid.

To find the volume of any pyramid, multiply $\frac{1}{3}$ of the area of its base (B) by its height (h).

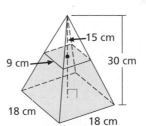

The base of each of these pyramids is a square.

Find the volume of the small pyramid.

$$V = \frac{1}{3}Bh$$
$$= \left(\frac{1}{3}\right)(9^2)(15)$$
$$= 405$$

Find the volume of the large pyramid.

$$V = \frac{1}{3}Bh$$
$$= \left(\frac{1}{3}\right)(18^2)(30)$$
$$= 3240$$

The volume of the large pyramid is 3240 cm^3. The volume of the small pyramid is 405 cm^3.

$$3240 - 405 = 2835$$

The volume of the shaded portion is 2835 cm^3.

Example 3

MANUFACTURING A candy company decides to sell its new Blast Off candy bars in a package shaped like a rocket. The body of the rocket is shown at the right. Find the volume of the figure.

Solution

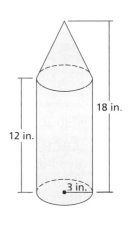

A cylinder and cone combine to form the figure shown. Add the volume of the cone to the volume of the cylinder.

Find the volume of the cylinder.

$$V = \pi r^2 h$$
$$= (\pi)(3^2)(12)$$
$$\approx 339$$

Find the volume of the cone.

$$V = \frac{1}{3}\pi r^2 h$$
$$= \left(\frac{1}{3}\right)(\pi)(3^2)(6)$$
$$\approx 57$$

The volume of the cylinder is about 339 in.^{3}.

The volume of the cone is about 57 in.^{3}.

$$339 + 57 = 396$$

The volume of the figure is about 396 in.^{3}.

Math Online mathmatters3.com/extra_examples Lesson 5-7 **Volume of Three-dimensional Figures** | **231**

Example 4

ASTRONOMY The only asteroid visible to the naked eye is 4 Vesta, discovered in 1807. Its diameter is 323 mi. What is its volume? Assume that 4 Vesta is a sphere.

Solution

To find the volume of a sphere, use the formula $V = \frac{4}{3}\pi r^3$.

$$V = \frac{4}{3}\pi r^3$$

$$\approx \left(\frac{4}{3}\right)(3.14)(161.5^3) \approx 17{,}635{,}426$$

The volume of 4 Vesta is approximately 17,635,426 mi^3.

◣ TRY THESE EXERCISES

Find the volume of each figure. Use 3.14 for π. Round answers to the nearest whole number.

1.
8.5 mm
11 mm
14 mm

2.
8 in.

3.
9 cm
14 cm

4.
3 m
5 m
6 m
12 m

5. A prism has a hexagonal base with an area of 24 cm². If the volume of the figure is 144 cm³, what is its height?

◣ PRACTICE EXERCISES • For Extra Practice, see page 679.

Find the volume to the nearest whole number. Use 3.14 for π.

6.
24 ft²
8.5 ft

7.
20 cm
15 cm

8.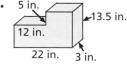
5 in.
13.5 in.
12 in.
22 in.
3 in.

9.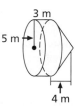
3 m
5 m
4 m

10. ARCHAEOLOGY One room in a 12th-century cliff dwelling is in the shape of a rectangular prism. The floor measures 10 ft by 12.5 ft, and the ceiling is 7 ft high. What is the volume of the room?

11. How many cubic meters of water can a water tank hold if the tank is a cylinder 9 m high and 6 m in diameter?

12. A pyramid with a volume of 384 cm³ has a base of area 64 cm². What is the height of the figure?

13. WRITING MATH A pentagonal prism and a pentagonal pyramid prism have congruent bases and the same height. Describe the relationship between the volumes of the two figures.

Cliff Palace, Colorado

Find the volume of each. Use 3.14 for π. Round answers to the nearest whole number.

14.
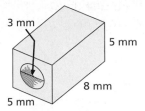
3 mm
5 mm
8 mm
5 mm

15.
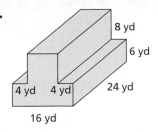
8 yd
6 yd
4 yd   4 yd   24 yd
16 yd

16.
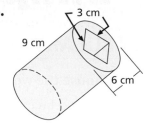
3 cm
9 cm
6 cm

17. ARCHAEOLOGY The height of the pyramid at Cheops is greater than the height of the Quetzalcoatl pyramid at Cholula, but its volume is less. How can that be?

18. The base of a rectangular pyramid has sides of 3.14 m and 12 m. A cone has the same height and volume. What is its radius? *Hint*: Write and solve an equation; use 3.14 for π.

19. What happens to the volume of a cone if its height is tripled?

20. If the radius of a cylinder is tripled, what happens to its volume?

21. Describe how you could find the volume of a crystal belonging to the tetragonal system.

Tetragonal system

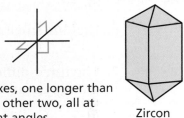

3 axes, one longer than the other two, all at right angles.

Zircon

22. CHAPTER INVESTIGATION Estimate the volume of your structure using your understanding of three-dimensional figures. Using an index card, make a fact list about your structure. Include the date the structure was built, the name of the architect, materials used, dimensions, surface area and volume. Display your model and fact list.

■ EXTENDED PRACTICE EXERCISES

23. COST ANALYSIS A can of Iguana Goodies that is 14 cm high with a radius of 6 cm sells for $1.70. Another can is the same height, but with a radius of 3 cm. It sells for $0.40. Which can of Iguana Goodies is the better buy?

24. A rectangular prism is 12 ft long and 3 ft wide. If its volume is 288 ft³, what is its surface area?

25. PACKAGING Suppose you have a square sheet of cardboard 24 cm on a side. You want to cut equal squares from the corners of the large square to fold the cardboard into an open box. If you cut corners of 1 cm, 2 cm, 3 cm and so on, which size cuts will give you the box with the greatest volume?

■ MIXED REVIEW EXERCISES

Find the perimeter or circumference and the area of each. Use 3.14 for π. Round answers to the nearest hundredth if necessary. (Lesson 5-2)

26.
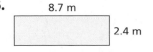
8.7 m
2.4 m

27.
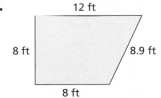
12 ft
8 ft
8.9 ft
8 ft

28.

5.8 cm

29.
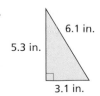
6.1 in.
5.3 in.
3.1 in.

Chapter 5 Review

VOCABULARY ▰

Choose the word from the list that best completes each statement.

1. Two faces of a polyhedron intersect along a(n) __?__.

2. A(n) __?__ is used for measuring distances and drawing curved lines and circles.

3. A ratio that compares two quantities that use different units is a(n) __?__.

4. A polyhedron with one base is called a(n) __?__.

5. A three-dimensional figure with a curved surface and one circular base is called a(n) __?__.

6. Half the smallest unit used to make a measurement is the __?__.

7. The three-dimensional figure where all the points on the surface are the same distance from the center is called a(n) __?__.

8. A(n) __?__ has two bases and lateral faces that are parallelograms.

9. The measure of the amount of space enclosed by a three-dimensional figure is its __?__.

10. The distance around a polygon is its __?__.

a. area

b. compass

c. cone

d. cylinder

e. edge

f. greatest possible error

g. perimeter

h. prism

i. pyramid

j. rate

k. sphere

l. volume

LESSON 5-1 ▰ Ratios and Units of Measure, p. 202

▶ A ratio compares two numbers by division and can be written in three ways. For example: $4 : 5$, $\frac{4}{5}$, or 4 to 5.

▶ For both customary and metric measures, multiply to change from a larger unit of measure to an equivalent measure in a smaller unit. Divide to convert a smaller unit of measure to an equivalent measure in a larger unit.

11. Write 8:20 in lowest terms in three ways.

12. How many grams are in 3 kg?

13. What is the greatest possible error (GPE) of a measurement of 2.46 m?

LESSON 5-2 ▰ Perimeter, Circumference and Area, p. 206

▶ To solve a problem involving the distance around a plane figure or the surface covered by it, you must choose the correct formula.

Find the perimeter or circumference of each. Then find the area of each. If necessary, round answers to the nearest whole number.

14.
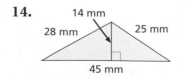

15.
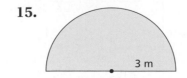

16.
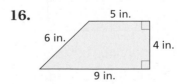

LESSON 5-3 ◢ Probability and Area, p. 212

▶ To determine the probability that a randomly chosen point in a figure will be within a given region inside the figure, use the ratio $\frac{\text{area of given region}}{\text{total area of figure}}$.

Find the probability that a point selected at random is in the shaded region.

17.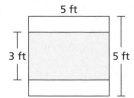

18.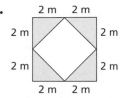

19.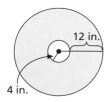

20. An ice skater skates in a rectangular rink 40 m by 125.6 m. If she were to fall at random inside the rink, what is the probability she will fall within a given circle of diameter 8 m?

LESSON 5-4 ◢ Problem Solving Skills: Irregular Shapes, p. 216

▶ To find the total area of an irregular figure, first break it down into smaller shapes whose areas you can add.

Find the area of each figure.

21.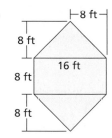

22.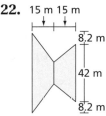

23.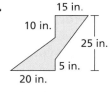

24. A wing of a new house has the shape shown at the right. What is the area of the wing?

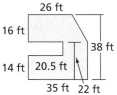

LESSON 5-5 ◢ Three-dimensional Figures and Loci, p. 220

▶ A polyhedron is a three-dimensional figure in which each face is a polygon.

Identify each figure. Then state the number of faces, vertices, and edges for each figure.

25.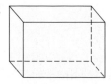

26.

27.

Draw each figure.

28. oblique cylinder

29. right cone

30. rectangular prism

LESSON 5-6 ◼ Surface Area of Three-dimensional Figures, p. 224

▶ To find the surface area of a three-dimensional figure, choose the correct formula for the area of each surface of the figure.

Find the surface area of each figure to the nearest tenth.

31.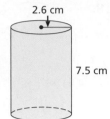
2.6 cm
7.5 cm

32.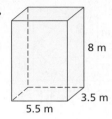
8 m
3.5 m
5.5 m

33.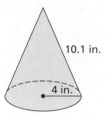
10.1 in.
4 in.

34. A cube has a surface area of 294 cm². What is the length of a side?

35. A can of corn is in the shape of a cylinder. The diameter of the base is 8 cm and the height of the can is 9 cm. If the curved surface of the can is to be covered with a label, what is the area of the label?

36. Each side of a cube is 5 m long. The height of a cylinder is 5 m and the base of the cylinder has a diameter of 5 m. Which figure has the greatest surface area?

LESSON 5-7 ◼ Volume of Three-dimensional Figures, p. 230

▶ To find the volume of a three-dimensional figure, use the correct formula or formulas.

Find the volume of each figure to the nearest tenth.

37.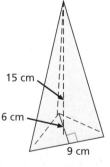
15 cm
6 cm
9 cm

38.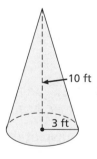
10 ft
3 ft

39.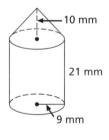
10 mm
21 mm
9 mm

40. How many cubic feet of concrete will be needed for a driveway that is 40 ft long, 18 ft wide, and 4 in. deep?

41. A rectangular prism and a rectangular pyramid have congruent bases. The height of the prism is 12 in. If both the prism and pyramid have the same volume, what is the height of the pyramid?

42. How many cubic centimeters are in a cubic meter?

CHAPTER INVESTIGATION

EXTENSION Write a report about the structure you chose. Be sure to include the measurements and known features of the structure. Explain why you chose the scale you did for your model. State your estimate of the exterior surface area and volume of your structure, and explain what these estimates tell you about the structure.

Chapter 5 Assessment

Write each ratio in lowest terms.

1. 4 yd to 16 ft

2. 400 m to 2 km

3. 6 qt to 9 gal

4. What is the height of a triangle with an area of 40 cm² and a base of 12 cm?

5. How many faces, vertices, and edges does a hexagonal prism have?

For Exercises 6–9, use 3.14 for π. Round your answers to the nearest tenth.

Find the area of the shaded region of each.   **Find the surface area of each.**

6.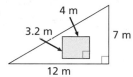
3.6 cm
9 cm

7.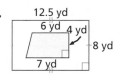
2.5 m
8 m
11 m

8.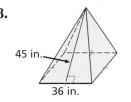
45 in.
36 in.

9.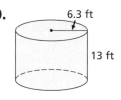
6.3 ft
13 ft

Find the probability that a point selected at random in each figure is in the shaded region.

10.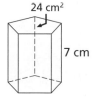
4 m
3.2 m
7 m
12 m

11.
12.5 yd
6 yd
4 yd
8 yd
7 yd

12.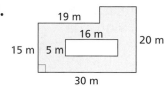
19 m
16 m
20 m
15 m   5 m
30 m

Find the volume of each figure to the nearest tenth. Use 3.14 for π.

13. 24 cm²
7 cm

14.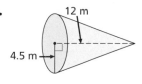
12 m
4.5 m

15.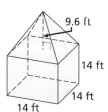
9.6 ft
14 ft
14 ft
14 ft

16. You want to paint the figure shown. Which formula will you use to find how much paint you will need?

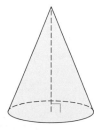

17. How many 1-ft² floor tiles are needed to cover the floor shown below?

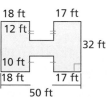
18 ft   17 ft
12 ft
32 ft
10 ft
18 ft   17 ft
50 ft

18. Which measurement is more precise, 4.003 L or 4.99 L?

19. What is the GPE of the measurement 100.9 cm?

20. A micrometer setting shows the measurement 24.46 mm. What are the upper and lower limits for the measurement if the tolerance is ±0.03 mm?

21. A square board with each side measuring 30 cm has a circle in the center with a diameter of 12 cm. If a dart is thrown at the board, what is the probability that it will land in the center circle? Use 3.14 for π. Round to the nearest hundredth.

Standardized Test Practice

Part 1 Multiple Choice

Record your answers on the answer sheet provided by your teacher or on a sheet of paper.

1. Kwan-Yong bought $6\frac{1}{2}$ lb of ground beef for $11.70 and 2 dozen cookies for $3.78. What was the cost per pound of the ground beef? (Lesson 1-5)

    Ⓐ $1.80   Ⓑ $1.82
    Ⓒ $1.89   Ⓓ $2.38

2. Which expression has the least value? (Lesson 1-8)

    Ⓐ $(-6)^{-3}$ Ⓑ $(-6)^3$
    Ⓒ 6^{-3} Ⓓ 6^3

3. Which is *not* a term in the following sequence? (Lesson 2-1)

    $$16, -8, 4, -2, \ldots$$

    Ⓐ $-\dfrac{1}{2}$ Ⓑ 1
    Ⓒ 0 Ⓓ $\dfrac{1}{4}$

4. Which graph represents the solution of $4x + 1 < 9$? (Lesson 2-6)

    Ⓐ
    Ⓑ
    Ⓒ
    Ⓓ

5. Point E is between M and N. If $ME = 102$ and $MN = 120$, find EN. (Lesson 3-1)

    Ⓐ 18   Ⓑ 36
    Ⓒ 122   Ⓓ 222

6. What is the measure of $\angle TOZ$? (Lesson 3-2)

    Ⓐ 17.5°
    Ⓑ 21.25°
    Ⓒ 62.5°
    Ⓓ 117.5°

    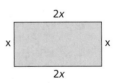
    $(7x - 5)°$ $(3x + 10)°$

7. In the figure below, $\overleftrightarrow{ZW}$ and $\overleftrightarrow{KN}$ are intersecting lines, $\overline{KR} \cong \overline{RN}$, and $\overline{ZR} \cong \overline{RW}$. Which postulate could you use to prove $\triangle KRZ$ is congruent to $\triangle NRW$? (Lesson 4-2)

    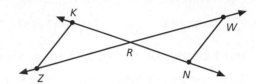

    Ⓐ Angle-Angle-Side Postulate
    Ⓑ Angle-Side-Angle Postulate
    Ⓒ Side-Angle-Side Postulate
    Ⓓ Side-Side-Side Postulate

8. Which quadrilateral has exactly one pair of parallel sides? (Lesson 4-9)

    Ⓐ parallelogram   Ⓑ rhombus
    Ⓒ square   Ⓓ trapezoid

9. Express the ratio 16 in. to 3 ft as a fraction in simplest form. (Lesson 5-1)

    Ⓐ $\dfrac{3}{16}$ Ⓑ $\dfrac{4}{9}$
    Ⓒ $\dfrac{8}{15}$ Ⓓ $\dfrac{16}{9}$

10. The perimeter of the rectangle is 42 ft. What is the area of the rectangle? (Lesson 5-2)

    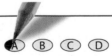

    Ⓐ 7 ft²   Ⓑ 14 ft²
    Ⓒ 98 ft²   Ⓓ 196 ft²

Test-Taking Tip
Ⓐ Ⓑ Ⓒ Ⓓ

Question 10
Sometimes more than one step is required to find the answer. In Question 10, you need to find the value of *x*. Then you can use the value of *x* to determine the length and width of the rectangle. Finally, you can multiply the length times the width to find the area.

Part 2 Short Response/Grid In

Record your answers on the answer sheet provided by your teacher or on a sheet of paper.

11. Let $A = \{1, 3, 5, 7, 9\}$ and $B = \{4, 5, 6, 7\}$. What is the least member in $A \cup B$? (Lesson 1-3)

12. Cole had $\frac{7}{8}$ of a tank of gas in the lawn mower. After mowing the grass, he had $\frac{1}{4}$ of a tank. What fraction of a tank did Cole use mowing the lawn? (Lesson 1-4)

13. The Mariana Trench is the deepest part of the Pacific Ocean with a depth of about 36,000 ft. Write this depth in scientific notation. (Lesson 1-8)

14. If $g(x) = |x - 10|$, find $g(7)$. (Lesson 2-3)

15. If $8.1d + 2.3 = 5.1d - 3.1$, what is the value of $d + 10$. (Lesson 2-5)

Ten friends recorded the number of minutes they needed to complete their math homework. Use their data below to answer Questions 16–17.

16, 23, 21, 17, 17, 20, 22, 17, 14, 23

16. What is the mean number of minutes the friends spent on their math homework? (Lesson 2-7)

17. What is the median number of minutes the friends spent on their math homework? (Lesson 2-7)

18. What is the measure of $\angle RME$? (Lesson 3-4)

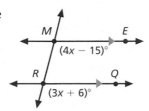

$(4x - 15)°$

$(3x + 6)°$

19. What is the value of x? (Lesson 4-1)

$(3x - 10)°$ $40°$ $2x°$

20. What is the probability that a point selected at random is in the shaded area? (Lesson 5-3)

4 m

5 m

8 m

21. The sides and bottom of a cylindrical swimming pool with a diameter of 25 ft and height of 4 ft are lined with vinyl. Find the area of the vinyl to the nearest square foot. (Lesson 5-6)

Part 3 Extended Response

Record your answers on a sheet of paper. Show your work.

22. A prism with a triangular base has 9 edges. A prism with a rectangular base has 12 edges.

9 edges   12 edges

Explain in words or symbols how to determine the number of edges for a prism with a 9-sided base. Be sure to include the number of edges in your explanation. (Lesson 5-5)

23. The diagrams show the design of the trashcans in the school cafeteria.

Front   Back

$\frac{3}{4}$ ft

3 ft

2 ft $1\frac{1}{2}$ ft

a. The top and sides of the cans need to be painted. Find the area that needs to be painted. Show your work. (Lesson 5-6)

b. Find the volume of the trash can. Show your work. (Lesson 5-7)

Linear Systems of Equations

THEME: Manufacturing Industry

Think back through your day and mentally make a list of all the products that you have used. Every one of those products was designed, tested, and built by a manufacturer.

Once a new product is designed and thoroughly tested, engineers break the assembly of the product into steps. Each step will be performed by workers on an assembly line. The engineers, technicians, and line workers use math to ensure that each product is manufactured according to the design specifications.

- **Precision Assemblers** (page 253) work in factories to produce complex goods and must know how to use specialized measuring instruments. They read engineering diagrams called *schematics*.

- **Engineering Technicians** (page 273) design and develop new products to meet all required safety standards. Engineering technicians test designs for product quality and look for ways to keep costs down to make new products affordable to the consumer.

Math Online

mathmatters3.com/chapter_theme

U.S. Goods–Imports and Exports, 2004 (millions of dollars)

Category	Exports	Imports	Category	Exports	Imports
Aircraft	7995	3245	TVs, VCRs	1164	9899
Apparel	1846	13,037	Synthetic fabrics	1833	1495
Jewelry	1329	3486	Vehicles	28,042	74,507
Vegetables	1072	1803	Toys and games	2121	7656
Glassware, Chinaware	128	753	Computers	2944	7311
Rugs	258	590	Printed Materials	1423	1219

Data Activity: U.S. Goods–Imports and Exports

Use the table for Questions 1–4.

1. What is the dollar value of the apparel exports?

2. Much of the toys and games sold in the United States is manufactured in other countries. How much greater is the value of imports than exports?

3. Of total exports and imports of computers, what percent is imported? Round to the nearest percent.

4. For which category are exports and imports most nearly balanced?

CHAPTER INVESTIGATION

Manufactured items go through an extensive design and development process. Many companies use focus groups made up of panels of consumers to find new ideas for product improvement. Before a new product is sold in stores, it is tested by selected consumers.

Working Together

Choose a simple product that you use frequently. Gather a focus group of four to five students and brainstorm ideas for product improvement. Finally, draw a model or prototype of the new product and determine how the improvements will add to the cost of the product. Use the Chapter Investigation icons to guide your group.

The skills on these two pages are ones you have already learned. Read the examples and complete the exercises. For additional practice on these and more prerequisite skills, see pages 654-661.

GRAPHING LINEAR EQUATIONS

In this chapter you will learn more about graphing equations on the coordinate plane. It will be helpful to review some of these basic skills.

Example   Graph the equation $3x - y = -5$.

Change the equation to the form $y = mx + b$.

$$3x - y = -5$$
$$-y = -3x - 5$$
$$y = 3x + 5$$

Make a table. Substitute each number for x to find the value of y.

x	y
1	8
0	5
−1	2

Plot the points and draw a line.

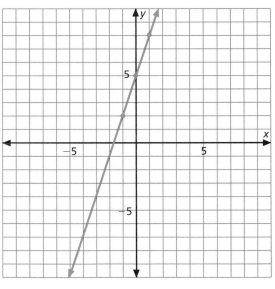

Graph each equation on the coordinate plane.

1. $y = 5x - 3$
2. $3x + 8 = 4y$
3. $-4x - 3y = 4$
4. $3y - x = 7$
5. $-3x + 2y = -7$
6. $y = -4x - 3$
7. $3y = -5x + 7$
8. $\frac{1}{2}(x - 3) = 2y$
9. $-2x = y + 3$

SOLVING EQUATIONS

In this chapter you will be working with various forms of equations. Remember to always apply the order of operations.

Solve each equation.

10. $5a - 3 = 12$
11. $-2a + 4 = 4a - 2$
12. $3(a - 2) = -3a + 12$
13. $-4(a - 2) = 16$
14. $12 - 3a = 2a + 2$
15. $-2(a + 4) = 16$
16. $\frac{1}{2}(a + 3) - 4 = -2$
17. $8 - 2a + 7 = 5$
18. $4a - 6 - a = 2$
19. $3(a + 3) - 2 = 4(a - 6) + a$
20. $2(a - 4) + 6 = 3(a + 2) - 4$
21. $a + 2(3a - 6) = 30 - 2(a + 3)$
22. $-4(a - 2) + 8 = -2a + 3(a - 2)$

Solving Compound Inequalities

Example Using the replacement set of the real numbers, graph the solution set for the compound inequality $x \leq 3$ and $x > -5$.

Graph each inequality: $x \leq 3$

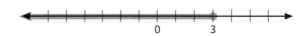

$x > -5$

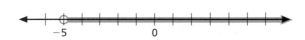

Graph the intersection:

Graph the solution set for each compound inequality on a number line. Use the replacement set of the real numbers.

23. $x > 3$ and
  $x < 9$

24. $x \leq -1$ and
  $x > -8$

25. $x > -2$ and
  $x \geq 1$

26. $x \leq 5$ and
  $x \geq -3$

27. $x < 4$ and
  $x \leq 2$

28. $x > 5$ and
  $x < -2$

29. $x \geq -4$ and
  $x < -3$

30. $x \geq -2$ and
  $x < 5$

31. $x < 4$ and
  $x > 1$

Ordering Rational Numbers

To compare rational numbers that are in mixed formats, you need to convert all values to the same format. Decimal form is often the preferred way to compare and order numbers.

Write the rational numbers in order from least to greatest value.

32. $1.156, 1\frac{1}{8}, \frac{7}{6}, \sqrt{2}, -0.2$

33. $0.7856, \frac{15}{16}, \frac{5}{7}, \sqrt{0.49}, -1, 1$

34. $-6.42, -6\frac{5}{8}, -0.6, \frac{32}{5}, -6.042$

35. $0.0025, 0.14500, 0.19, \frac{1}{7}, \sqrt{\frac{4}{5}}$

36. $0.03, \frac{2}{3}, \frac{3}{10}, \sqrt{6}, \frac{5}{6}$

37. $0.05, \frac{1}{24}, \frac{\sqrt{2}}{15}, 0.049, \frac{3}{26}$

38. $-\frac{3}{11}, \frac{1}{3}, -0.31, \frac{\sqrt{2}}{3}, -3.1$

39. $-2.9, -2.95, -2\frac{9}{11}, -2\frac{13}{14}$

6-1 Slope of a Line and Slope-intercept Form

Goals
- Find the slope of a line.
- Write the slope-intercept form of an equation and graph the equation.

Applications   Manufacturing, Finance, Recreation

Work with a partner.

Make identical line segments on geoboards. Lengthen or shorten one of the line segments. Make a right triangle on each geoboard using the line segment as the hypotenuse. Write the ratio of the length of the vertical side to the length of the horizontal side for each triangle. How do the ratios compare?

◣ BUILD UNDERSTANDING

Many examples of slope exist in everyday life. We measure the **slope**, or steepness, of roads, stairs and ramps. In mathematics, an important aspect of a line is its slope. We measure slope as a ratio of the vertical change (**rise** or fall) to the horizontal change (**run**).

$$\text{slope} = \frac{\text{rise}}{\text{run}} = \frac{\text{vertical change (change in } y\text{-coordinates)}}{\text{horizontal change (change in } x\text{-coordinates)}}$$

You can find the slope of a line on a coordinate plane by finding the units of change between the coordinates of any two points on the line. The vertical change, or change in y, is found by determining the difference of the y-coordinates. Likewise, the horizontal change, or change in x, is found by determining the difference in corresponding x-coordinates of the same two points.

The slope of a line segment containing the point $A(x_1, y_1)$ and $B(x_2, y_2)$ is

$$m = \frac{y_2 - y_1}{x_2 - x_1} \quad \text{or} \quad \frac{y_1 - y_2}{x_1 - x_2}.$$

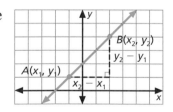

Example 1

Find the slope of $\overleftrightarrow{AB}$ containing points $A(-1, 2)$ and $B(3, -4)$.

Solution

$$m = \frac{y_2 - y_1}{x_2 - x_1} = \frac{-4 - 2}{3 - (-1)} = \frac{-6}{4} = \frac{-3}{2}$$

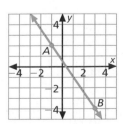

Example 1 shows that a line with a *negative* slope falls from left to right. As the value of x increases, the value of y decreases. A line with a *positive* slope rises from left to right. As the value of x increases, the value of y increases.

If you know one point on a line and the slope of the line, you can also graph the line.

Example 2

Graph the line that passes through $G(1, 1)$ and has a slope of $\frac{-3}{4}$.

Solution

First, plot the point $G(1, 1)$. The slope is $\frac{-3}{4}$, so from G, go *down* 3 units (because the rise is -3) and *right* 4 units (because the run is 4). The point is $(1 + 4, 1 - 3)$ or $(5, -2)$.

Draw a line through $(1, 1)$ and $(5, -2)$.

Check Understanding

A horizontal line has a slope of 0 because all y-values are the same. The slope of a vertical line is undefined because all x-values are the same.

Find the slope of a line containing these points:

1. $(-2, 3)$ and $(4, 3)$

2. $(1, 3)$ and $(1, 2)$

In a linear equation, you can find the **x-intercept**, or x-coordinate of the point at which the line crosses the x-axis, by substituting 0 for y and solving for x. You can find the **y-intercept**, or y-coordinate of the point at which the line crosses the y-axis, by substituting 0 for x and solving for y. You can then draw a line through these two points to graph the equation. Writing a linear equation in **slope-intercept form** is another way to easily find the slope and the y-intercept of the line. In the slope-intercept form, $y = mx + b$, m represents the slope of the line and b represents the y-intercept.

Example 3

Find the slope and y-intercept for the line with the equation $5x + 6y = 6$.

Solution

$$5x + 6y = 6$$
$$6y = -5x + 6 \quad \text{First, write the equation in slope-intercept form.}$$
$$y = \frac{-5x}{6} + 1$$
$$m = \frac{-5}{6} \text{ and } b = 1$$

The slope is $\frac{-5}{6}$, and the y-intercept is 1.

Technology Note

Write a program for your graphing calculator that can find the slope of a line when the coordinates of two points are known.

1. Prompt for the values of x_1, x_2, y_1 and y_2. You may want to call these values X, Y, A, and B.

2. Enter an expression that compares the ratio of the rise to the run.

$$(B - Y) / (A - X)$$

3. Run the program to find the slope of a line passing through points $(1, 4)$ and $(-2, 3)$. Did your calculator correctly display a slope of .333 or $\frac{1}{3}$?

Example 4

MANUFACTURING Production figures for an assembly plant are represented by a line with a slope of $\frac{1}{2}$ and a y-intercept of -1. Find the equation of the line. Then draw the graph of the line.

Solution

$$m = \frac{1}{2} \text{ and } b = -1$$
$$y = mx + b$$
$$y = \frac{1}{2}x - 1 \qquad \text{slope-intercept form}$$

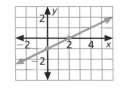

To draw the graph, start at point $(0, -1)$. Then using a slope of $\frac{1}{2}$, locate a point 1 unit up and 2 units right at $(2, 0)$. Draw a line through the points.

Find the slope of the line containing the given points.

1. $A(3, 5)$ and $B(0, 6)$ **2.** $P(-2, -6)$ and $Q(2, 1)$ **3.** $G(-2, 1)$ and $H(-2, 10)$

Graph the line that passes through the given point *P* and has the given slope.

4. $P(-1, 2)$, slope $= -1$ **5.** $P(4, -1)$, slope $= \dfrac{2}{3}$

Find the slope and *y*-intercept of each line.

6. $-5x + 13y = 15$ **7.** $y = -6x + 4$ **8.** $y = 5x - 1$ **9.** $y = -\dfrac{1}{10}x + \dfrac{1}{5}$

Find an equation of the line with the given slope and *y*-intercept.

10. slope $= -1$, *y*-intercept $= 3$ **11.** slope $= 4$, *y*-intercept $= -2$

Graph each equation.

12. $-2x + 3y = 3$ **13.** $2x - y = 4$ **14.** $y = 1$

▼ **PRACTICE EXERCISES** • **For Extra Practice, see page 679.**

Find the slope of each line.

15.   **16.** 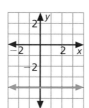   **17.**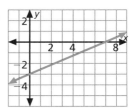

18. Find the slope of the line containing the points $A\left(\dfrac{1}{2}, \dfrac{-3}{4}\right)$ and $B\left(\dfrac{2}{3}, \dfrac{5}{8}\right)$.

GRAPHING Plot the point $P(1, 1)$. Then, on the same axes, graph the lines that pass through *P* and have the given slope.

19. $m = 0$ **20.** $m = 1$ **21.** $m = -1$

DATA FILE For Exercises 22–24, use the data on the desired weight range based on corresponding height for men and women on page 650.

22. Use *y* as the *lower male* weight (in pounds) and *x* as the height (in inches). Plot the corresponding weights for $x = 62, 63, 64$. Connect the points. Label them $P_1(62, w_1)$, $P_2(63, w_2)$, and $P_3(64, w_3)$.

23. Determine the slope for the line segment between P_1 and P_2 and between P_2 and P_3. Are the points collinear?

24. Repeat Exercise 23 using *y* as the *lower female* weight (in pounds) and *x* as the height (in inches). Plot $(62, w_1)$, $(63, w_2)$, $(64, w_3)$. Connect the points.

Find the slope and *y*-intercept of each line.

25. $y = 4x$ **26.** $-6x - y = 12$ **27.** $x = -9$

28. $-\dfrac{x}{2} + 2y = 0$ **29.** $x = 2$ **30.** $3y = -x + \dfrac{1}{2}$

Write an equation of the line with the given slope and y-intercept.

31. $m = -5$, $b = 4$

32. $m = 2$, $b = -\dfrac{3}{4}$

33. $m = 6$, $b = 0$

34. $m = 0$, $b = -1$

CALCULATOR Graph each equation.

35. $2x + 3y = -6$

36. $-x + y = 3$

37. $y = -3x + 2$

FINANCE Dave receives a salary of $200 a week plus a commission of 10% of his weekly sales. An equation $y = mx + b$ represents Dave's weekly earnings. The y-intercept is Dave's base salary. The slope of the line is his commission.

38. Write an equation representing Dave's weekly earnings.

39. Graph the equation.

40. If Dave sells $1500 of goods for one week, what is his salary for the week?

41. What value of goods does Dave need to sell in one week to have weekly earnings of $500?

▌ EXTENDED PRACTICE EXERCISES

42. RECREATION A ski resort is building a new ski slope. The desired slope of the new hill is 0.8. The horizontal distance from the crest of the hill to the bottom will be 400 ft. What should the height of the hill be?

Find the slope and y-intercept for each line.

43. $y = ax + r - m$

44. $\dfrac{b}{a}x + \dfrac{c}{b}y = 1$

45. The standard form of an equation is $Ax + By + C = 0$. Write the slope-intercept form of the equation.

46. WRITING MATH How is each set of equations different?

a. $y = 2x$ and $y = -2x$

b. $y = x - 3$ and $y = x + 3$

47. YOU MAKE THE CALL A line passes through $P(-5, 2)$ and $Q(-5, 5)$. Geoff says the slope of the line must be 0 because there is no change in x. Does Geoff's reasoning make sense?

▌ MIXED REVIEW EXERCISES

Find each unit rate or unit price. Round answers to the nearest hundredth if necessary. (Lesson 5-1)

48. $3.87 for 18 oz of soup

49. 496 mi in 8 h

50. 487.5 ft in 325 steps

51. $5.95 for 5 lb of nuts

52. 426.4 mi on 13 gal of gas

53. 9 mi in 2 h 15 min

54. 128 cookies in 8 equal sacks

55. $96.67 for 58 jars of jam

56. 528 fish in 6 tanks

Find all the solutions for each equation. (Lesson 2-4)

57. $|x| - 6 = 8$

58. $-24 = -3|x|$

59. $|x| + 4 = 8$

60. $3|x| = 21$

61. $\dfrac{1}{2}|x| = 13$

62. $36 = 4|x|$

63. $|x| - 7 = 8$

64. $-48 = -8|x|$

6-2 Parallel and Perpendicular Lines

Goals ■ Use slope to determine whether two lines are parallel or perpendicular.

Applications Electronics, Mapmaking, Manufacturing

GRAPHING Use the zoom feature of your graphing calculator to change the viewing window to a square. Then graph the following equations on the same screen.

$$y = 2x - 1 \qquad\qquad y = 2x \qquad\qquad y = 2x + 1$$

$$y = -\frac{1}{2}x + 1 \qquad\quad y = -\frac{1}{2}x + 2 \qquad\quad y = -\frac{1}{2}x - 2$$

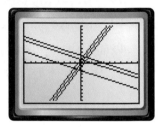

Which lines seem to be parallel? Which seem to be perpendicular? What do you notice about the slopes and y-intercepts of parallel and perpendicular lines?

◣ BUILD UNDERSTANDING

From the activity above, you can see that two lines with the same slope and different y-intercepts are parallel. Conversely, two lines are **parallel** if they have the same slope and different y-intercepts. Vertical lines are parallel and have slopes that are undefined.

Two nonvertical lines are **perpendicular** if the product of their slopes is -1. Conversely, if the slopes of two lines are negative reciprocals of each other, then the lines are perpendicular.

Example 1

Find the slope (m_1) of a line parallel to the given line and the slope (m_2) of a line perpendicular to the given line.

a. The line containing points $A(-2, 5)$ and $B(0, -1)$.

b. The line containing points $C(4, -1)$ and $D(-5, -1)$.

c. $x = 2$

> **Think Back**
>
> Two rational numbers that have a product of -1 are called negative reciprocals.

Solution

	m	m_1	m_2
a.	$\dfrac{-1-5}{0-(-2)} = \dfrac{-6}{2} = -3$	-3	$\dfrac{-1}{-3} = \dfrac{1}{3}$
b.	$\dfrac{-1-(-1)}{-5-4} = \dfrac{0}{-9} = 0$	0	undefined
c.	undefined	undefined	0

Example 2

Determine whether each pair of lines is *parallel, perpendicular,* or *neither*.

a. $7x + 2y = 14$

   $7y = 2x - 5$

b. $-5x + 3y = 2$

   $3x - 5y = 15$

c. $2x - 3y = 6$

   $\dfrac{8}{3}x - 4y = 4$

Solution

Rewrite each equation in slope-intercept form and find the slope of each line.

a. $7x + 2y = 14 \rightarrow y = -\dfrac{7}{2}x + 7 \qquad m_1 = -\dfrac{7}{2}$

   $7y = 2x - 5 \ \rightarrow y = \dfrac{2}{7}x - \dfrac{5}{7} \qquad m_2 = \dfrac{2}{7}$

   $m_1 \neq m_2$

   $m_1 \cdot m_2 = -\dfrac{7}{2} \cdot \dfrac{2}{7} = -1$

   Because $m_1 \cdot m_2 = -1$, the lines are perpendicular.

b. $-5x + 3y = 2 \rightarrow y = \dfrac{5}{3}x + \dfrac{2}{3} \qquad m_1 = \dfrac{5}{3}$

   $3x - 5y = 15 \ \rightarrow y = \dfrac{3}{5}x - 3 \qquad m_2 = \dfrac{3}{5}$

   $m_1 \neq m_2$

   $m_1 \cdot m_2 = \dfrac{5}{3} \cdot \dfrac{3}{5} = 1 \neq -1$

   Because $m_1 \neq m_2$ and $m_1 \cdot m_2 \neq -1$, the lines are neither parallel nor perpendicular.

c. $2x - 3y = 6 \rightarrow y = \dfrac{2}{3}x - 2 \qquad m_1 = \dfrac{2}{3}$

   $\dfrac{8}{3}x - 4y = 4 \ \rightarrow y = \dfrac{2}{3}x - 1 \qquad m_2 = \dfrac{2}{3}$

   $m_1 = m_2$

   Because $m_1 = m_2$ and the y-intercepts are different, the lines are parallel.

Example 3

ELECTRONICS A manufacturer of circuit boards uses a grid system to insert connecting pins. The design for a board requires pins at points $M(0, 1)$, $N(3, 0)$, $P(5, 6)$ and $Q(2, 7)$. When the pins are connected, will $MNPQ$ be a parallelogram?

Solution

Plot the points and draw the quadrilateral. Find the slope of $\overline{NP}$ and $\overline{MQ}$, and the slope of $\overline{MN}$ and $\overline{QP}$.

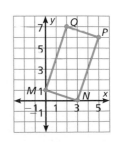

$$m_{NP} = \frac{6 - 0}{5 - 3} = \frac{6}{2} = 3; \qquad m_{MN} = \frac{0 - 1}{3 - 0} = \frac{-1}{3}$$

$$m_{MQ} = \frac{7 - 1}{2 - 0} = \frac{6}{2} = 3; \qquad m_{QP} = \frac{6 - 7}{5 - 2} = \frac{-1}{3}$$

Since the slopes of $\overline{NP}$ and $\overline{MQ}$ are the same, $\overline{NP} \parallel \overline{MQ}$. Because the slopes of $\overline{MN}$ and $\overline{QP}$ are the same, $\overline{MN} \parallel \overline{QP}$. Thus, opposite sides of the quadrilateral are parallel. Therefore, $MNPQ$ is a parallelogram.

TRY THESE EXERCISES

Find the slope of a line parallel to the given line and a line perpendicular to the given line.

1. The line containing points $M(8, 3)$ and $N(-1, 5)$

2. The line containing points $A(0, 5)$ and $B(-6, 5)$

3. The line containing points $S(2, -1)$ and $T(4, 7)$

4. The line containing points $P(3, -5)$ and $Q(3, 2)$

Determine whether each pair of lines is *parallel*, *perpendicular*, or *neither*.

5. The line containing points $A(9, 5)$ and $B(-1, 6)$
   The line containing points $C(-8, 2)$ and $D(12, 4)$

6. $-6x - 14y = 5$
   $7x = 3y + 21$

7. $-5x + 2y = 10$
   $3x - 5y = 15$

8. **CARTOGRAPHY** A road atlas is laid out on a grid. A mapmaker notes intersections at points $A(1, 1)$, $B(2, 0)$, $C(3, 2)$ and $D(2, 3)$. If the points are connected, will $ABCD$ form a parallelogram?

PRACTICE EXERCISES • For Extra Practice, see page 680.

Find the slope of a line parallel to the given line and a line perpendicular to the given line.

9. The line containing $(6, -3)$ and $(4, 5)$

10. The line containing $(-3, 2)$ and $(-7, 8)$

11. $10x + 12y = -6$ 12. $-3y + 2x = 7$ 13. $y = -7x + 6$

Determine whether each pair of lines is *parallel*, *perpendicular*, or *neither*.

14. The line containing points $P(9, -4)$ and $Q(-2, 7)$
    The line containing points $M(14, 8)$ and $N(19, 13)$

15. $-4x + 6y = 6$ 16. $8y = 12x - 20$
    $5x - 2y = 10$ $9x = 6y + 5$

17. **MANUFACTURING** A machine is used to apply a heat-sensitive transfer to a product. The transfer is placed by locating three points on a grid. The points are $A(0, 0)$, $B(5, 3)$ and $C(7, -3)$. Do the connected points form a right triangle?

Determine whether each pair of lines is *parallel, perpendicular,* or *neither*.

18. $1.5x - 4.5y = x + 15$
$2 - 3x = 13 + \frac{1}{3}y$

19. $\frac{2}{3}y - \frac{1}{4}x = 1$
$15 + 3x = 15 + 8y$

Determine the value of *x* so that the line containing the given points is parallel to another line whose slope is also given.

20. $A(x, 9)$ and $B(-5, 6)$

slope $= -2$

21. $M(-1, 7)$ and $N(x, -1)$

slope $= -\frac{1}{3}$

Determine the value of *y* so that the line containing the given points is perpendicular to another line whose slope is also given.

22. $A\left(\frac{-5}{6}, y\right)$ and $B\left(\frac{2}{3}, 1\right)$

slope $= \frac{-1}{3}$

23. $C\left(\frac{1}{3}, -\frac{2}{5}\right)$ and $D(1, y)$

slope is undefined

24. WRITING MATH Given the coordinates of four points which are the vertices of a polygon, how can you determine if the polygon is a rectangle?

25. CHAPTER INVESTIGATION Brainstorm a list of products that the members of your group have used during the past week. Select a product that all agree is in need of improvement. Prepare a list of questions that could be used to find out whether consumers share your concerns.

◤ EXTENDED PRACTICE EXERCISES

26. The vertices of a triangle are $A(a, 0)$, $B(a + b, c)$ and $C(c + a + b, c - b)$. Determine if $\triangle ABC$ is a right triangle.

Use your graph from Exercise 27 for Exercises 28–30.

27. Plot $L(3, 4)$, $I(9, 4)$, $N(7, 1)$, and $E(1, 1)$ on a coordinate plane.

28. Use what you know about the slopes of lines to show that quadrilateral *LINE* is a parallelogram.

29. Draw diagonals $\overline{LN}$ and $\overline{EI}$. Use the slopes of the diagonals to show whether they are perpendicular to each other.

30. On graph paper, draw two different shaped quadrilaterals in which the diagonals are perpendicular to each other. Name each type of quadrilateral. Give the slope of each side and each diagonal of each quadrilateral.

◤ MIXED REVIEW EXERCISES

Find the area of the shaded region of each figure. Use 3.14 for π. Round to the nearest hundredth if necessary. (Lesson 5-2)

31.

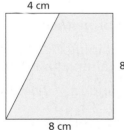

4 cm

8 cm

8 cm

32.

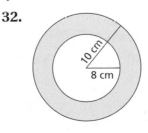

10 cm

8 cm

33.

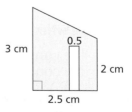

3 cm

0.5

2 cm

2.5 cm

Review and Practice Your Skills

Find the slope of the line containing the given points.

1. $A(2, 4)$ and $B(1, 3)$ **2.** $C(3, 2)$ and $D(5, 6)$ **3.** $E(0, 5)$ and $F(4, 0)$

4. $G(-3, 5)$ and $H(-5, 6)$ **5.** $Z(6, -2)$ and $Y(-3, 2)$ **6.** $X(-2, -2)$ and $W(-12, -8)$

7. $V(5, 7)$ and $U(-4, 7)$ **8.** $T(0, 0)$ and $S(-6, 5)$ **9.** $R(-2.4, 6)$ and $Q(-1, -2.4)$

Graph the line that passes through the given point P and has the given slope.

10. $P(0, 1)$, $m = 2$ **11.** $P(-4, 1)$, $m = -3$ **12.** $P(-2, -1)$, $m = \frac{4}{3}$

13. $P(-5, -3)$, $m = 0$ **14.** $P(-4, 2)$, $m = -\frac{2}{3}$ **15.** $P(0, 0)$, $m = -4$

Find the slope and y-intercept for each line.

16. $y = 3x + 2$ **17.** $y = -5x + 9$ **18.** $y = x$ **19.** $y = \frac{2}{3}x - 1$

20. $x = -5$ **21.** $2x + 3y = 12$ **22.** $5x - 2y = 20$ **23.** $\frac{4}{3}x + 4y = 1$

Write an equation of the line with the given slope and y-intercept.

24. $m = 4$, $b = -4$ **25.** $m = 0$, $b = 15$ **26.** $m = \frac{1}{3}$, $b = -2$

27. $m = -\frac{7}{2}$, $b = -3$ **28.** $m = 2$, $b = \frac{1}{2}$ **29.** $m = -20$, $b = 5$

Graph each equation.

30. $y = x$ **31.** $y = 2x$ **32.** $y = x + 2$

33. $5x - y = 10$ **34.** $x + 5y = 10$ **35.** $5x - y = 0$

36. Write an equation of the line containing the points $C(-5, -3)$ and $D(5, 9)$.

Find the slope of a line parallel to the given line and a line perpendicular to the given line.

37. The line containing $(4, -9)$ and $(-3, 5)$ **38.** The line containing $(8, 13)$ and $(15, -6)$

39. The line containing $(-3, -6)$ and $(5, 2)$ **40.** The line containing $(-5, 10)$ and $(-5, -17)$

41. The line containing $(2, -9)$ and $(15, -14)$ **42.** The line containing $(6, 6)$ and $(-7, 7)$

43. $y = -3x + 10$ **44.** $y = \frac{-3}{5}x - 7$ **45.** $-12x - 5y = -1$

Determine whether each pair of lines is *parallel*, *perpendicular*, or *neither*.

46. The line containing $(-3, -1)$ and $(4, -3)$
The line containing $(13, 9)$ and $(27, 5)$

47. The line containing $(7, -2)$ and $(-8, -2)$
The line containing $(0, 14)$ and $(0, -31)$

48. $y = \frac{1}{4}x - 6$ and $x + 4y = -8$

49. $y = \frac{7}{2}x + 17$ and $7x - 2y = 28$

50. $x + y = 10$ and $y = 4 - x$

51. $\frac{1}{5}x + \frac{2}{5}y = -2$ and $2y = x - 14$

Find the slope of the line containing the given points. (Lesson 6-1)

52. $A(0, 6)$ and $B(-9, 0)$

53. $C(-3, 14)$ and $D(-3, 8)$

54. $E(-4, -8)$ and $F(-8, -4)$

55. $G(-725, 630)$ and $H(435, -240)$

Find the slope and *y*-intercept for each line. (Lesson 6-1)

56. $y = x - 4.5$

57. $y = -\frac{1}{4}x + 7$

58. $y = 8 - 2x$

59. $9x - 3y = 12$

60. $\frac{1}{2}x + 2y = 0$

61. $-\frac{5}{6}x + \frac{10}{3}y = 30$

Find the slope of a line parallel to the given line and a line perpendicular to the given line. (Lesson 6-2)

62. The line containing $(9, -4)$ and $(-5, 3)$

63. The line containing $(7, 12)$ and $(16, -7)$

64. The line containing $(-5, -6)$ and $(1, 0)$

65. The line containing $(-8, 13)$ and $(-8, -22)$

66. The line containing $(4, -18)$ and $(30, -28)$

67. The line containing $(6.8, 6.5)$ and $(2.8, 7)$

MathWorks Career – Precision Assemblers
Workplace Knowhow

Precision assemblers work in factories to produce manufactured goods. These workers must be able to follow complex instructions and complete many steps with very little error.

Precision assemblers work on the assembly of aircraft, automobiles, computers and many other electrical and electronic devices. They work closely with engineers to build prototypes.

Precision assemblers must know how to read engineering schematics and how to use specialized and precise measuring instruments. An incorrect measurement may cause a product to work improperly or to endanger its users.

1. You must be certain that the slope of an airplane wing is exactly 0.13. If the wing is 5 ft long, how many inches should it rise from the fuselage to the wing tip?

2. You are mounting a Global Positioning System (GPS) onto the dashboard of a car. The slope of the GPS must be identical to the slope of the dashboard in order to work properly. The dashboard has a depth of 18 in. wide and rises 6 in. Find the slope of the dashboard.

3. The GPS unit has a depth of $4\frac{3}{8}$ in. How much should it rise in order to match the slope of the dashboard?

Write Equations for Lines

Goals ■ Write equations for lines in slope-intercept and point-slope forms.

Applications Product design, Fitness, Real Estate

Complete the table by matching the correct slope and *x*- and *y*-intercept with the linear equation.

	Equation	Slope	x-intercept	y-intercept
a.				
b.	$y = -\frac{1}{2}x + 2$			
c.		-2		-1
d.	$y = 2x - 1$			
e.			0	0

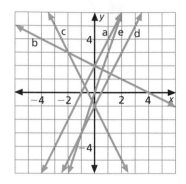

■ BUILD UNDERSTANDING

You can write an equation of a line if you know the information summarized in the following table.

You wrote equations using the slope-intercept form in Lesson 6–1. You can write an equation in **point-slope form** if you know the slope of the line and the coordinates of any point on the line.

If you know	You can write an equation in
1. The slope m and y-intercept b	1. Slope-intercept form: $y = mx + b$
2. A point (x_1, y_1) and the slope m	2. Point slope form: $y - y_1 = m(x - x_1)$
3. Two points (x_1, y_1) and (x_2, y_2)	3. Point slope form: $y - y_1 = m(x - x_1)$ or $y - y_2 = m(x - x_2)$ where $m = \dfrac{y_2 - y_1}{x_2 - x_1}$
4. The graph with points $A(x_1, y_1)$ and $B(x_2, y_2)$	4. Same as 3.

Example 1

Write an equation of the line with a slope of -2 and passes through the point $P(-1, 3)$.

Solution

$y - y_1 = m(x - x_1)$ point-slope form

$y - 3 = -2(x - (-1))$ Substitute for 2 for m, -1 for x_1, and 3 for y_1.

$y - 3 = -2(x + 1)$ Solve for y.

$y - 3 = -2x - 2$

$y = -2x + 1$ slope-intercept form

You can also write an equation in point-slope form, if you know the coordinates of two points on the line.

Example 2

Write an equation of the line containing the points $A(1, -3)$ and $B(3, 2)$.

Solution

Given: $x_1 = 1$, $y_1 = -3$, $x_2 = 3$, $y_2 = 2$

Find the slope: $m = \dfrac{y_2 - y_1}{x_2 - x_1} = \dfrac{2 - (-3)}{3 - 1} = \dfrac{5}{2}$

Find the equation using the point-slope form.

$$y - y_1 = m(x - x_1)$$

$$y - (-3) = \frac{5}{2}(x - 1) \quad \text{Solve for } y.$$

$$y + 3 = \frac{5}{2}x - \frac{5}{2}$$

$$y = \frac{5}{2}x - \frac{11}{2} \quad \text{Write equation in slope-intercept form.}$$

An equation of the line is $y = \dfrac{5}{2}x - \dfrac{11}{2}$.

Check Understanding

Find the slope of the line.

1. the line containing $P(0, 0)$ and $Q(-5, 2)$

2. $9y = 2x + 3$

3. $\frac{6}{5}x + \frac{2}{3}y = 1$

You can write an equation for a line if you are given the graph of the line and are able to read information from the graph.

Example 3

PRODUCT DESIGN A technician is using a coordinate grid to design a schematic for a circuit board. A connection aligns with the line shown at the right. Write an equation of the line.

Solution

y-intercepts: The line intersects the *y*-axis at the point $(0, 3)$. The *y*-intercept is 3.

slope: Use two points on the line whose coordinates are easily determined. Use $(x_1, y_1) = (0, 3)$ and $(x_2, y_2) = (2, -1)$.

$$m = \frac{-1 - 3}{2 - 0} = \frac{-4}{2} = -2$$

An equation of the line: $y = mx + b$
$\qquad\qquad\qquad\qquad\quad y = -2x + 3 \quad \text{slope-intercept form}$

GRAPHING When writing an equation from a graph, check your work using a graphing calculator. Using the graphing keys, enter the slope-intercept form of the equation and display the graph. You may need to adjust the size or shape of the viewing window.

Example 4

Write an equation of a line parallel to $y = -\frac{1}{3}x + 1$ and containing point $R(1, 1)$.

Solution

$y = -\frac{1}{3}x + 1 \quad m = -\frac{1}{3}$ Slope of line

Because parallel lines have equal slopes, $m = -\frac{1}{3}$.

$y - y_1 = m(x - x_1)$ Point-slope form.

$y - 1 = -\frac{1}{3}(x - 1)$ $x_1 = 1, y_1 = 1$

$y - 1 = -\frac{1}{3}x + \frac{1}{3}$

$y = -\frac{1}{3}x + \frac{4}{3}$

An equation of the line is $y = -\frac{1}{3}x + \frac{4}{3}$.

> ### Check Understanding
>
> Write an equation of the line whose slope and point on the line are given.
>
> **1.** $m = \frac{3}{4}, P(-1, 6)$
>
> **2.** $m = 0, R(2, -2)$

◤ TRY THESE EXERCISES

Write an equation of the line with the given slope and y-intercept.

1. $m = -3, b = -2$

2. $m = \frac{-3}{5}, b = 0$

Write an equation of the line that passes through the given point with the given slope.

3. $m = 7, Q(-1, -5)$

4. $m = \frac{3}{7}, S(5, 3)$

Write an equation of the line containing the given points.

5. $M(0, -1)$ and $N(-1, 4)$

6. $T(-2, -3)$ and $V(2, 2)$

7. $R(-6, 9)$ and $S(9, 9)$

8. $A(-5, 6)$ and $B(-5, 10)$

Write an equation for the lines graphed below.

9.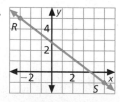

10.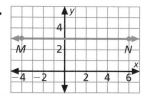

◤ PRACTICE EXERCISES • For Extra Practice, see page 681.

Write an equation of the line with the given slope and y-intercept.

11. $m = \frac{2}{3}, b = -3$

12. $m = 7, b = 2$

13. Write an equation of the line that is parallel to $3y + x = 6$ containing the point $P(1, -1)$.

14. TEMPERATURE The temperature of water at the freezing point is 0°C or 32°F. The temperature of water at the boiling point is 100°C or 212°F. Use these two data points to find an equation to convert the temperature from Celsius to Fahrenheit.

Use the given information to write an equation of each line.

15. $m = -\frac{1}{2}$, $B\left(0, -\frac{2}{5}\right)$

16. m is undefined, $D\left(\frac{1}{3}, \frac{1}{2}\right)$

17. $P(4, 0)$ and $Q(-9, 3)$

18. $M(2, -5)$ and $N(1, 3)$

19.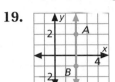

20.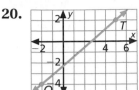

21. FITNESS Kim's average walking speed is 6 feet every 3 seconds. Complete the table. Then write an equation for the distance Kim walks in a given time.

Distance Kim walks (d) feet	0	2	6	12
Time (t) sec				

22. TALK ABOUT IT Marta is asked to write an equation of a line with a slope of 3 that contains point $D(2, 5)$. Marta writes $y - 2 = 3(x - 5)$, so $y = 3x - 13$. Where did Marta make her mistake?

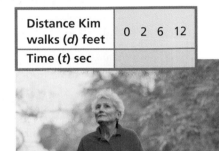

■ EXTENDED PRACTICE EXERCISES

The coordinates of the vertices of a parallelogram are $(2, 6)$, $(7, 2)$, $(3, 0)$, and $(-2, 4)$.

23. Write an equation for each diagonal of the parallelogram.

24. WRITING MATH Is the parallelogram a rhombus? Justify your answer.

25. REAL ESTATE A realtor is paid a fixed amount per week plus a commission on the total sales. If the fixed amount is F and the commission rate is $p\%$, find an equation to represent the amount of pay he receives each week.

■ MIXED REVIEW EXERCISES

Find the surface area of each figure. Assume that the pyramid is a regular pyramid. Use 3.14 for π. (Lesson 5-6)

26.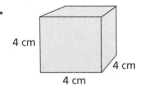

27.

28.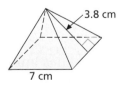

Refer to the diagram. Name each set using roster notation. (Lesson 1-3)

29. $A \cup B$

30. $B \cup C$

31. $A' \cap C$

32. $B' \cap C'$

33. $A' \cap B'$

34. $A' \cup C'$

6-4 Systems of Equations

Goals ■ Solve a system of equations by graphing.

Applications Income Tax, Manufacturing, Finance

Graph the following pairs of equations on the same coordinate plane. Then complete the table.

Equations	Point of Intersection (x, y)
1. $y = x$ and $y = 1$	(,)
2. $2y = x + 2$ and $y = x$	(,)
3. $y = 2x + 1$ and $y = x$	(,)
4. $y = x + 1$ and $y = x$	(,)

◥ BUILD UNDERSTANDING

Two linear equations with the same two variables form a **system of equations**. A *solution* of a system of equations is the ordered pair that makes both equations true. One way to solve a system of two equations is to graph both equations on the same coordinate plane. The *point of intersection* of the two lines is the **solution** of the system. If the graphs intersect in *one* point, the system is known as an **independent system**.

Example 1

Solve the system of equations by graphing. $y = -2x + 1$
$y = -3x + 4$

Solution

Graph each equation.

$y = -2x + 1$ $b = 1, m = -2$; use $(1, -1)$
$y = -3x + 4$ $b = 4, m = -3$; use $(1, 1)$

The lines intersect at the point $(3, -5)$.
The solution of the system of equations is $(3, -5)$.

Check: Substitute $(3, -5)$ in each equation.

$y = -2x + 1$ $y = -3x + 4$
$-5 = -2(3) + 1$ $-5 = -3(3) + 4$
$-5 = -6 + 1$ $-5 = -9 + 4$
$-5 = -5$ ✔ $-5 = -5$ ✔

If the graphs of the equations *do not intersect*, the system is known as an **inconsistent system**. The lines are parallel and have no common points. There is *no solution* to this system of equations.

Technology Note

Solve a system of equations using a graphing calculator.

1. Use the graphing features to enter all equations in slope-intercept form. Then press GRAPH.

2. Use TRACE to find the point of intersection.

3. Use ZOOM to read the x- and y-values of the intersection point.

4. Continue to zoom and trace until you reach the desired degree of accuracy.

Example 2

Solve the system of equations by graphing. $y = \frac{1}{2}x + 3$

$$2y = x - 2$$

Solution

Graph each equation, then read the solution from the graph.

$y = \frac{1}{2}x + 3$ $b = 3, m = \frac{1}{2}$; use $(-2, 2)$

$y = \frac{1}{2}x - 1$ Rewritten in slope-intercept form.

              $b = -1, m = \frac{1}{2}$; use $(2, 0)$

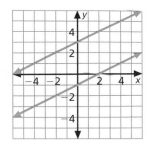

The lines are parallel and do not intersect. Therefore, there is no solution.

If the graphs of the equations are the same line—i.e., the lines coincide—the system is a **dependent system**. Any point on the line is a solution. There are *infinitely many solutions*.

Example 3

Solve the system of equations by graphing. $4x + 2y = 8$

                                $3y = -6x + 12$

> **Problem Solving Tip**
>
> If all the coefficients and constants in an equation have a common factor, dividing by that common factor results in an equation that is easier to work with.

Solution

Graph each equation. Then read the solution.

$4x + 2y = 8$ Subtract $4x$ from both sides.

    $y = -2x + 4$ Divide both sides by 2.

$3y = -6x + 12$

    $y = -2x + 4$ Divide both sides by 3.

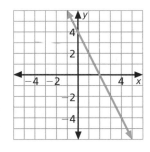

The equations are equivalent. The graphs are the same line. The system has an infinite number of solutions.

To summarize:

If the graph of two lines:	then the system is:	and has:
intersect in a point	independent	one solution
do not intersect	inconsistent	no solution
coincide	dependent	an infinite number of solutions

Equations can give you a better understanding of everyday life. For example, the rate of taxation is a function of the amount of taxable income earned. The table below shows the tax rate schedule for single individuals for 2003.

If you are single . . .

Line	If taxable income		The tax is	
	Is Over	But not over	This amount plus this %	of the excess over
1	$0	$7000	$0 + 10%	$0
2	$7000	$28,400	$700 + 15%	$7000
3	$28,400	$68,800	$3910 + 25%	$28,400

You can graph the equations defined in the table to gain an understanding of how the tax law works. Let T represent the amount of income tax owed and x represent the amount of taxable income. Using these variables, construct three linear equations that reflect the information contained in the table. Notice that each equation has a different slope.

Line 1: If $0 < x \le 7000$,
then $T = 0.1x$

Line 2: If $7000 < x \le 28{,}400$,
then $T = 700 + 0.15(x - 7000)$,
or $T = 0.15x - 350$

Line 3: If $28{,}400 < x \le 68{,}800$,
then $T = 3910 + 0.25(x - 28{,}400)$,
or $T = 0.25x - 3190$

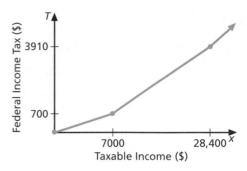

Example 4

INCOME TAX Rick's taxable income was \$18,000. Find the amount of income tax he will owe.

Solution

His income is less than \$28,400, so multiply by 0.15 and subtract 350.

$T = 0.15x - 350$

$\quad = 0.15(18{,}000) - 350$

$\quad = 2700 - 350$

$\quad = 2350$ Rick will owe \$2350 in income tax.

◣ TRY THESE EXERCISES

Determine whether the given ordered pair is a solution.

1. $(2, 1)$ $5y = 3x - 1$
            $2x - 3y = 1$

2. $(-1, 3)$ $7x + 2y = -1$
             $-4x + y = 15$

Solve each system of equations by graphing.

3. $x = -2y - 3$

    $x + 2y = -5$

4. $x = \dfrac{4}{3}y - 1$

    $4y = 3x + 3$

5. $-x + \dfrac{3y}{2} = 1$

    $4x = 2y + 4$

◣ PRACTICE EXERCISES • For Extra Practice, see page 681.

Determine the solution of each system of equations.

6.

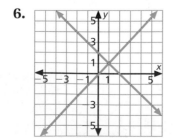

7.

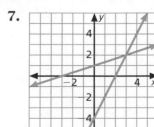

GRAPHING Solve each system of equations by graphing.

8. $2x - y = 3$
$x - \frac{1}{2}y = 1$

9. $4x - 2y = 4$
$3x + 6y = 3$

10. $4y = 3x - 5$
$x = 2y + 1$

11. $\frac{x}{3} + \frac{y}{5} = \frac{7}{15}$
$\frac{x}{4} + \frac{3y}{8} = \frac{1}{8}$

12. $\frac{c}{3} = \frac{d}{4} + \frac{1}{8}$
$\frac{3}{4}d = c + \frac{1}{2}$

13. MANUFACTURING The Food Division and Personal Care Division introduced a total of 10 new products this year. The Food Division introduced 4 more products than the Personal Care Division. How many products did each group introduce this year?

14. FINANCE Candice's monthly savings is twice the amount that she spends on transportation each month. The total of her monthly savings and transportation bill is $135. Find both amounts.

15. INCOME TAX Phil's taxable income was $48,000. Use the table on page 259 to find the amount of income tax he had to pay.

16. WRITING MATH Is it possible for a system of equations to have exactly two solutions? Explain your thinking.

17. CHAPTER INVESTIGATION Make a detailed drawing or build a prototype of your new product. Choose at least five selling points to emphasize in your marketing materials.

Determine the number of solutions for each system. Do not graph.

18. $4x + 5y = 3$
$3x - 2y = 8$

19. $\frac{a}{2} = \frac{b}{3} + \frac{1}{6}$
$\frac{2}{3}b = a + \frac{1}{2}$

20. $y - 2 = \frac{1}{2}x$
$\frac{x}{4} - \frac{1}{2}y = -1$

■ EXTENDED PRACTICE EXERCISES

21. The graphs of the equations $2x - y = 1$, $x + 2y = 8$, and $x = 4$ form a triangle. Graphically determine the location of each vertex.

22. Determine whether the ordered pair $\left(-a, \frac{b}{2}\right)$ is a solution to the system of equations.

$4ay = -3bx - ab$
$5bx - 2ay = -6ab$

■ MIXED REVIEW EXERCISES

Find the probability that a point selected at random in each figure is in the shaded region.

23.

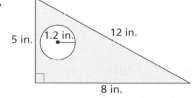

24.

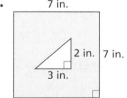

25.

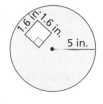

Write each number in standard form.

26. $3.84 \cdot 10^{-6}$

27. $1.9 \cdot 10^{8}$

28. $7 \cdot 10^{-9}$

29. $6.52 \cdot 10^{12}$

Review and Practice Your Skills

Write an equation of the line with the given slope and *y*-intercept.

1. $m = 4.5, b = -7.5$

2. $m = -3, b = 16$

3. $m = 0, b = -4$

4. $m = \dfrac{1}{6}, b = 6$

5. $m = \dfrac{5}{4}, b = 13$

6. $m = \dfrac{11}{8}, b = -44$

7. $m = 1, b = 0.05$

8. $m = -9.25, b = 0$

9. $m = 300, b = 530$

Write an equation of the line that passes through the given point with the given slope.

10. $m = \dfrac{-2}{3}, G(0, 0)$

11. $m = 2, H(-4, -1)$

12. $m = \dfrac{3}{5}, J(5, 0)$

13. $m = \dfrac{7}{2}, K(2, 9)$

14. $m = -\dfrac{1}{2}, L(-8, 1)$

15. $m = -4, N(3, -6)$

16. $m = 0, P(-8, 9)$

17. slope undefined, $Q(13, -12)$

18. $m = -1.25, R(12, -15)$

Write an equation of the line containing the given points.

19. $A(-8, 3)$ and $B(4, 9)$

20. $C(2, -9)$ and $D(-3, 11)$

21. $E(3, -5)$ and $F(3, 7)$

22. $G(-5, 3)$ and $H(10, -6)$

23. $J(2, 8)$ and $K(-3, -7)$

24. $L(4, -2)$ and $M(-12, -2)$

25. $P(-7, -8)$ and $Q(6, 5)$

26. $R(12, -7)$ and $S(-6, -4)$

27. $T(-4, -2)$ and $U(8, 7)$

Write an equation for the line with the given information.

28. parallel to $-2x + y = -14$ and passes through $Z(5, 3)$

29. perpendicular to $3y = -2x + 18$ and has *x*-intercept 8

30. parallel to the line through points $G(-4, 8)$ and $H(2, 5)$ and has same *y*-intercept as $y = 8x - 11$

Solve each system of equations by graphing.

31. $y = \dfrac{2}{3}x - 2$
$x = -3$

32. $y = -\dfrac{1}{2}x + 4$
$y = 3x - 3$

33. $y = 8$
$x - y = 0$

34. $y = 2x - 4$
$y = \dfrac{1}{7}x + 9$

35. $4y - 3x = 12$
$y = \dfrac{3}{4}x - 5$

36. $y = x + 2$
$y = -3x - 18$

37. $x - 3y = -3$
$3y = 2x - 6$

38. $y = 5 - x$
$y = 4x - 20$

39. $3x + 12y = -12$
$5x + 2y = 16$

Determine the number of solutions for each system. Do not graph.

40. $y = 3x - 8$
$3x - 4 = -5$

41. $y = -x + 4$
$-3x + 2y = 8$

42. $y = \dfrac{1}{4}x - 1$
$2x - 8y = 8$

43. The sum of two numbers is -3. Their difference is 13. Find the numbers.

Find the slope and y-intercept for each line. (Lesson 6-1)

44. $y = -\frac{1}{5}x + 4$

45. $y = x - 3\frac{1}{2}$

46. $y = 15 - \frac{2}{3}x$

47. $6x - 4y = 28$

48. $3x + 7y = -28$

49. $y = -12$

Determine whether each pair of lines is *parallel, perpendicular,* or *neither*. (Lesson 6-2)

50. $y = 4x - 6$ and $4x + y = 8$

51. $y = 7 - 2x$ and $4x + 2y = 16$

52. $y = -\frac{3}{8}x + 11$ and $24x - 9y = 1$

53. $y - 16$ and $x = -3$

Write an equation for the line with the given information. (Lesson 6-3)

54. $m = -\frac{1}{8}, B(0, -7)$

55. $M(-1, -7), N(2, -10)$

56. m undefined, $Z(-8, 6.5)$

57. $m = 0, Q(-2.7, 36)$

58. $m = 4, b = -11.4$

59. $m = 3,$ x-intercept -5

Solve each system of equations by graphing. (Lesson 6-4)

60. $y = -3x$
  $y = \frac{1}{2}x - 7$

61. $y = x - 9$
  $y = \frac{4}{5}x$

62. $x - 2y = 6$
  $y = -\frac{3}{2}x + 5$

Mid-Chapter Quiz

Find the slope of the line containing the given points. (Lesson 5-1)

1. $K(7, 1)$ and $L(4, 7)$

2. $I(2, 9)$ and $J(-2, 6)$

Write an equation of the line with the given slope and y-intercept. (Lesson 5-1)

3. $m = -5, b = 3$

4. $m = \frac{3}{4}, b = -2$

Graph each equation. (Lesson 5-1)

5. $x = -3$

6. $x + 2y = 4$

Determine whether each pair of lines is *parallel, perpendicular,* or *neither*. (Lesson 5-2)

7. $4x + y = 3$ and $x - 4y = -8$

8. $x + 2y = 4$ and $y = -2x + 2$

9. Plot and connect the points $J(-3, 6)$, $K(-5, 0)$, $L(1, -2)$ and $M(3, 4)$. Determine whether $JKLM$ is a rectangle.

Write an equation of each line. (Lesson 5-3)

10. containing the points $Y(-3, -2)$ and $Z(-1, 4)$

11. that is perpendicular to $x - 3y = 2$ and contains the point $(2, 4)$

12. that is parallel to $2x + 5y = 4$ and contains the point $(2, 1)$

Solve each system of equations by graphing. (Lesson 5-4)

13. $3m + n = -8$
  $m + 6n = 3$

14. $x + 2y = 5$
  $-6y = 3x - 15$

15. $p - 2q = 4$
  $2q = p + 2$

6-5 Solve Systems by Substitution

Goals ■ Solve systems of equations using substitution.

Applications Shipping, Packaging, Recreation

Work with a partner to practice using the distributive property.

Write one of the equations below on a piece of paper. Pass the paper back and forth, adding the next line to solve the equation.

1. $x + 2(3x - 6) = 2$

2. $-(4x - 2) = 2(x + 7)$

3. $5(2x + 4) - 10 = 70$

4. $3(2x + 9) = 81$

◥ BUILD UNDERSTANDING

You can use algebraic methods to solve a system of equations. One of these methods is **substitution**. This method is useful when one equation has already been solved for one of the variables. In any algebraic method, you need to eliminate one variable so you will have an equation in one variable to solve.

Example 1

Find the solution to the system of equations. $3x - y = 6$
$$x + 2y = 2$$

Solution

$3x - y = 6$
$\quad y = 3x - 6$ Solve the first equation for y in terms of x.

$\quad\quad x + 2y = 2$ Write the second equation.
$x + 2(3x - 6) = 2$ Substitute $(3x - 6)$ for y.
$\quad x + 6x - 12 = 2$ Solve for x.
$\quad\quad\quad 7x = 14$
$\quad\quad\quad\quad x = 2$

Choose one of the original equations.

$3x - y = 6$
$3(2) - y = 6$ Substitute 2 for x.
$\quad 6 - y = 6$ Solve for y.
$\quad\quad\quad y = 0$

Check $x = 2$, $y = 0$ in each original equation.

$\begin{aligned} 3x - y &= 6 \\ 3(2) - 0 &= 6 \\ 6 - 0 &= 6 \\ 6 &= 6 \checkmark \end{aligned}$ $\begin{aligned} x + 2y &= 2 \\ 2 + 2(0) &= 2 \\ 2 + 0 &= 2 \\ 2 &= 2 \checkmark \end{aligned}$

The solution is (2, 0).

Check Understanding

If you solve the second equation in Example 1 for x, does the solution change?

The substitution method is most useful when one of the coefficients is 1 or -1.

Example 2

Find the solution to the system of equations. $2x + 3y = 6$

$$4x + 6y = 6$$

Solution

$2x + 3y = 6$

$3y = -2x + 6$ Solve for y.

$y = -\dfrac{2}{3}x + 2$

$4x + 6y = 6$

$4x + 6\left(-\dfrac{2}{3}x + 2\right) = 6$ Substitute for y.

$4x - 4x + 12 = 6$

$12 = 6$

There is no solution. The lines are parallel.

Example 3

Solve the system of equations. $4x - 2y = 10$

$$-2x + y = -5$$

Solution

$-2x + y = -5$

$y = 2x - 5$ Solve for y.

$4x - 2y = 10$

$4x - 2(2x - 5) = 10$ Substitute for y.

$4x - 4x + 10 = 10$

$10 = 10$

The lines are the same. There are infinitely many ordered pairs that satisfy both of the equations $4x - 2y = 10$ and $-2x + y = -5$.

Problems in everyday life can lead to a system of equations that can be solved using the substitution method.

Example 4

SHIPPING An appliance store delivers large appliances using vans and trucks. When loaded, each van holds 4 appliances and each truck holds 6. If 42 appliances are delivered and 8 vehicles are full, how many vans and trucks are used?

Solution

Define each of the variables.

Write and solve a system of equations relating to the variables.

Let t = number of trucks used
v = number of vans used

There are 8 vehicles.

$v + t = 8$

$t = 8 - v$ Solve for t.

42 appliances are delivered; 6 in each truck, and 4 in each van.

$$4v + 6t = 42$$
$$4v + 6(8 - v) = 42 \quad \text{Substitute } 8 - v \text{ for } t.$$
$$4v + 48 - 6v = 42$$
$$-2v = -6 \quad \text{Solve for } v.$$
$$v = 3$$

$$v + t = 8$$
$$3 + t = 8 \quad \text{Substitute for } v.$$
$$t = 5 \quad \text{Solve for } t.$$

There are 3 vans and 5 trucks delivering appliances.

◤ TRY THESE EXERCISES

Solve and check each system of equations by the substitution method.

1. $2x + y = 0$
    $x - 5y = -11$

2. $x + 3y = -9$
    $-5x - 2y = -7$

3. $x = \frac{1}{2}y$
    $-x + 6y = -11$

4. $x - 5y = 6$
    $y = -\frac{1}{2}x - \frac{1}{2}$

5. PACKAGING The perimeter of a rectangular picture frame is 78 cm. If the width is $\frac{5}{8}$ of the length, find the dimensions of the frame.

◤ PRACTICE EXERCISES • For Extra Practice, see page 682.

Solve and check each system of equations by the substitution method.

6. $3x = y + 9$
    $2x - 4y = 16$

7. $-4x + 3y = -16$
    $-x + 2y = -4$

8. $\frac{x}{2} - y = \frac{5}{4}$
    $8x + 3y = 1$

9. $5x + 2y = 1$
    $3x + 4y = -5$

10. $6x - 3y = -9$
    $13x - 5y = -15$

11. $10x - 5y = 65$
    $10y - 5x = -55$

12. RECREATION Jake is going on a 20-day vacation to the beach and to the mountains. He wants the time spent at the beach to be $\frac{2}{3}$ the time spent in the mountains. How many days will he spend at the beach, and how many in the mountains?

13. WRITING MATH A friend asks you how to know which variable to solve for first when solving a system of equations by substitution. What advice would you give?

14. FINANCE Shari has 17 coins consisting of dimes and quarters worth $3.35. How many quarters and how many dimes does she have?

15. At the Hearty Hut, Chad bought 4 hamburgers and 5 fries and paid $8.71. Alisa bought 1 hamburger and 3 fries and paid $3.56. Find the cost of each hamburger and each order of fries.

Solve each system of equations by the substitution method. Check the solutions.

16. $4y + 5x - 5 = 2x + 5y$

    $9x + 2y = -4x - 8y$

17. $\dfrac{-3a + 5b}{2} = 1$

    $\dfrac{6a - 5b}{2} = 1$

DATA FILE For Exercises 18–20, use the data on the shrinking value of the dollar on page 649. Use a system of equations to find the year in which these prices existed.

18. 3 qt of milk and 5 lb of round steak cost $5.33;
    5 qt of milk and 1 lb of round steak cost $1.99

19. 20 lb of potatoes and 15 lb of flour cost $4.92;
    10 lb of potatoes and 25 lb of flour cost $5.89

20. 5 lb of flour and 4 qt of milk cost $0.79;
    15 lb of flour and 5 qt of milk cost $1.39

◼ EXTENDED PRACTICE EXERCISES

Solve each system of equations by the substitution method. Check the solutions.

21. $\dfrac{1}{x} + \dfrac{1}{y} = -1$

    $\dfrac{4}{x} - \dfrac{5}{y} = 23$

22. $\dfrac{2}{p} - \dfrac{1}{q} = -11$

    $\dfrac{-3}{p} - \dfrac{2}{q} = 13$

23. $-x + y - z = 4$
    $y = -x$
    $x - 4z = 7$

24. $-2a - 3b + c = 6$
    $c = a + 2b + 5$
    $a + 2b + 3c = -1$

25. **FINANCE** Coins consisting of nickels, dimes, and quarters total $2.40. The number of dimes is equal to one less than $\frac{2}{3}$ the number of nickels. Three times the number of quarters plus the number of dimes is 18. How many of each coin are there?

◼ MIXED REVIEW EXERCISES

26. Find the volume of the figure shown. (Lesson 5-7)

Evaluate each expression when $a = -4$ and $b = 2$. (Lesson 1-8)

27. $a^2 - b^2$

28. $a^2 b^2$

29. $(a - b)^2$

30. $(a^2 + b^2)^2$

31. $a^3 - b^3$

32. $4(ab)^3$

33. $-3(a^3 - ab)^2$

34. $2(ab^2 + a^2 b)$

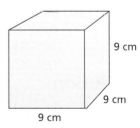

9 cm
9 cm
9 cm

35. **CIVICS** In the United States, a president is elected every four years. Members of the House of Representatives are elected every two years and senators are elected every six years. If a voter had the opportunity to vote for a president, a representative, and a senator in 2004, what will be the next year the voter has a chance to make a choice for president, a representative, and the same seat in the Senate? (Prerequisite Skill)

35. Justin rented three times as many DVDs as Cole last month. Cole rented four fewer than Maria, but four more than Paloma. Maria rented 10 DVDs. How many DVDs did each person rent? (Prerequisite Skill)

6-6 Solve Systems by Adding and Multiplying

Goals ■ Solve systems of equations by adding, subtracting, and multiplying.

Applications Entertainment, Community Service, Manufacturing

Work with a partner and use Algeblocks to simplify these expressions.

a. $(3x - y - 6) - (3x - 2y - 6)$

b. $(4x + y + 4) + (2x - y - 10)$

■ BUILD UNDERSTANDING

Another algebraic method for solving a system of equations is the **addition/subtraction method**. To eliminate one of the variables to get one equation in one variable, you can add or subtract the two equations. If the *coefficients* of one of the variables are *opposites* or the *same*, simply adding or subtracting the equations eliminates one of the variables.

Steps to follow when using the addition/subtraction method:

1. If the coefficients of one of the variables are opposites, add the equations to eliminate one of the variables.
   If the coefficients of one of the variables are the same, subtract the equations to eliminate one of the variables.

2. Solve the resulting equation for the remaining variable.

3. Substitute the value for the variable in one of the original equations and solve for the other variable.

4. Check the solution in both of the original equations.

> **Problem Solving Tip**
>
> The addition/subtraction method is best done with equations written in standard form:
> $Ax + By = C.$

Example 1

Solve: $2x + 7y = -5$
$-5x + 7y = -12$

Solution

The y-coefficients are the *same*, so *subtract* the equations.

$$\begin{array}{rl} 2x + 7y = -5 & \rightarrow \quad 2x + 7y = -5 \\ -(-5x + 7y = -12) & \quad \underline{5x - 7y = 12} \\ \text{Add.} \quad \rightarrow & \quad 7x = 7 \\ & \quad x = 1 \end{array}$$

Choose one of the original equations.

$$\begin{array}{rl} 2x + 7y = -5 & \\ 2(1) + 7y = -5 & \text{Substitute for } x. \\ 7y = -7 & \\ y = -1 & \text{The check is left to you. The solution is } (1, -1). \end{array}$$

Unless the coefficients of one variable are the same or are opposites, you will still have an equation with two variables when you add or subtract the equations. In that case, you first need to multiply one or both of the equations by a number to obtain an equivalent system of equations in which the coefficients of one of the variables are the same or opposite. Then add or subtract.

This method combines the multiplication property of equations with the addition/subtraction method, and is known as the **multiplication and addition method**.

Both the addition/subtraction method and the multiplication and addition method are best used when the equations are written in standard form.

Example 2

Solve: $3x - 4y = 10$
$3y = 2x - 7$

Solution

$3x - 4y = 10$
$-2x + 3y = -7$ Rewrite the second equation in standard form.

To eliminate x, multiply the first equation by 2 and the second by 3. Then add.

$3x - 4y = 10 \quad \rightarrow \quad 2(3x - 4y = 10) \quad \rightarrow \quad 6x - 8y = 20$
$-2x + 3y = -7 \quad \rightarrow \quad 3(-2x + 3y = -7) \quad \rightarrow \quad \underline{-6x + 9y = -21}$

Add. $\rightarrow \quad y = -1$

$3x - 4y = 10$ Choose one of the original equations.

$3x - 4(-1) = 10$ Substitute for y.

$3x + 4 = 10$

$3x = 6$

$x = 2$

The check is left to you. The solution is $(2, -1)$.

> ### Problem Solving Tip
>
> Look for the least common multiple of the coefficients of one of the variables. In Example 2, the LCM of 3 and 2 is 6. We can eliminate the x terms if we multiply the first equation by 2, and the second equation by 3.

Example 3

ENTERTAINMENT Tim sold 25 movie tickets for a total of $132. If each adult ticket sold for $6 and each children's ticket sold for $4, how many of each kind did he sell?

Solution

Make a chart for the number and values of the tickets.

	Adult	Child	Total
Number	A	C	$A + C$
Value	$6A$	$4C$	$6A + 4C$

Write and solve a system of equations to represent the problem.

The number of tickets sold is 25. $A + C = 25$
The value of the tickets is \$132. $6A + 4C = 132$

$$\begin{aligned} A + C = 25 &\rightarrow -4A - 4C = -100 \qquad \text{Multiply the first equation by } -4. \\ 6A + 4C = 132 &\rightarrow \underline{6A + 4C = 132} \\ \text{Add.} &\rightarrow 2A = 32 \\ & \ A = 16 \end{aligned}$$

$A + C = 25$ Choose one of the original equations.
$16 + C = 25$ Substitute for A.
$C = 9$

Tim sold 16 adult tickets and 9 children's tickets.

▀ TRY THESE EXERCISES

Solve each system of equations. Check the solutions.

1. $x - y = -1$
$x + y = 9$

2. $3x + y = -7$
$5x - y = -9$

3. $-3x - 5y = 4$
$2x + y = -5$

4. $4x = 2y - 12$
$3y = x + 13$

5. COMMUNITY SERVICE Adult tickets for a benefit breakfast cost \$2.50. Children's tickets cost \$1.50. If 56 tickets were sold for total sales of \$97, how many of each kind were sold?

6. FINANCE Afton invested \$5400 in two products, a new mouthwash and a new line of frozen dinners. She invested $\frac{1}{2}$ as much money in mouthwash as she did in frozen dinners. How much did she invest in each product?

7. MANUFACTURING It takes 18 months from the time a shoe design is approved for the shoes to arrive in stores for sale. The actual assembly of the shoe takes 3 months less than the time spent on development. How many months does the assembly take?

8. WRITING MATH Using the multiplication and addition method, how can you know when a system has an infinite number of solutions?

▀ PRACTICE EXERCISES • For Extra Practice, see page 682.

Solve each system of equations. Check the solutions.

9. $-x + 2y = 3$
$3x + 2y = -1$

10. $x = 5y + 7$
$5y = 9x + 15y - 8$

11. $-2x + 6y = 10$
$2x - 9y = -19$

12. $-10x + 6y = 25$
$9y = -2x + 12$

13. $4x - 9y = -1$
$9y = 8x + 1$

14. $7x + 4y = 6$
$5y + 7x = 15$

15. $8x + 3y = -4$
$6x + 5y = 8$

16. $12x - 3y = 18$
$8x = 2y + 12$

17. FARMING The Deckerts grow wheat and barley on their 1200-acre farm. The amount of wheat they plant is 200 acres more than 3 times the number of acres of barley. How many acres of wheat and how many acres of barley do they plant?

18. Jason has 15 coins in his pocket, consisting of nickels and dimes. The total value of the coins is \$1.15. How many of each coin does Jason have?

Solve each system of equations. Check the solutions.

19. $\dfrac{x}{2} - \dfrac{4y}{3} = -3$

$-3x + 4y = 6$

20. $2x - 5 = 3y - 7$

$6y + 4 = 5x + 6$

21. $\dfrac{-x}{3} + \dfrac{2}{9}y = 1$

$\dfrac{2}{3}x + \dfrac{y}{18} = \dfrac{5}{2}$

22. $\dfrac{1}{3}a - \dfrac{1}{2}b = -9$

$\dfrac{1}{5}b = \dfrac{1}{4}a + 5$

23. $\dfrac{5x + 3y}{10} = \dfrac{-1}{2}$

$\dfrac{9x}{16} + \dfrac{7y}{25} = \dfrac{-17}{20}$

24. $\dfrac{s}{2} - \dfrac{r+3}{4} = \dfrac{-1}{2}$

$\dfrac{2r-1}{3} - \dfrac{3s}{4} = \dfrac{-5}{12}$

25. A number divided by 3 plus another number divided by 9 have a sum of 7. If the first number is multiplied by 4 and divided by 5 and then subtracted *from* the second number divided by 2, the result is -3. What are the numbers?

◼ EXTENDED PRACTICE EXERCISES

Solve each system of equations. Check the solutions.

26. $\dfrac{9}{x} + \dfrac{4}{y} = -19$

$\dfrac{-6}{x} - \dfrac{2}{y} = 14$

27. $\dfrac{7}{s} = \dfrac{2}{r} + \dfrac{2}{15}$

$\dfrac{5}{r} = \dfrac{3}{s} + \dfrac{5}{14}$

28. $3x - 2y + z = 9$
$2x + y - z = 2$
$4x + 5y + 2z = 5$

29. $-a + 3b - 2c = 9$
$9a + 4b = 2c + 1$
$12a + 4b + 8c = -12$

30. $y = ax^2 + bx + c$ is an equation for a quadratic function. $(0, 0)$, $(1, -2)$ and $(2, 3)$ are solutions of the equation. Find a, b, and c.

◼ MIXED REVIEW EXERCISES

Complete the table. Use 3.14 for π. Assume each planet is a sphere. Round answers to the nearest million. (Lesson 5-6)

Planet	Diameter at equator	Radius	Surface area
31. Mercury	3031 mi	1515.5 mi	
32. Mars	4200 mi		
33. Saturn	71,000 mi		
34. Uranus	32,000 mi		
35. Neptune	30,600 mi		
36. Pluto	715 mi		

Solve each inequality. Graph the solution set on the number line. (Lesson 2-6)

37. $7 - x > -3$

38. $6 + 5x \le 11$

39. $x - 4 < -8$

40. $8 - x \ge 4$

41. $3x - 5 < 13$

42. $4(x + 3) \ge -4$

43. $\dfrac{1}{2}(x - 4) > 3$

44. $2(3 - x) \le -1$

SPACE For Exercises 45 and 46, use the following information.

Objects weigh six times more on Earth than they do on the moon because the force of gravity is greater. (Prerequisite Skill)

45. Write an expression for the weight of an object on Earth if its weight on the moon is x.

46. A scientific instrument weights 34 lb on the moon. How much does the instrument weigh on Earth?

Review and Practice Your Skills

PRACTICE ◼ LESSON 6-5

Solve and check each system of equations by the substitution method.

1. $x = 5y$
$x - 3y = 6$

2. $m = -2n - 2$
$3m - 2n = 10$

3. $x - y = 1$
$y = -x + 5$

4. $y + 6 = 3x$
$9x - 2y = 3$

5. $3x - 2y = -3$
$3x + y = 3$

6. $3p - 4q = 8$
$4p + q = 17$

7. $x - 2y = 16$
$4x + y = 1$

8. $s + 2t = 6$
$4s + 3t = 4$

9. $y = -4x + 5$
$2x - 3y = 13$

10. $5n - v = -23$
$3n - v = -15$

11. $-2x + y = 2$
$2x + 3y = 6$

12. $x + 5y = 11$
$4x - y = 2$

13. $2x + 3y = 11$
$3x + 3y = 18$

14. $5a + 3b = 4$
$4a - 2b = 1$

15. $5x + 7y = -3$
$2x + 14y = 2$

16. Leonard has 23 coins consisting of quarters and nickels worth $4.15. How many quarters and how many nickels does he have?

17. Lisa bought 7 bagels and 4 peaches and paid $5.25. Emily bought 1 bagel and 6 peaches and paid $2.65. Find the cost of each bagel and each peach.

PRACTICE ◼ LESSON 6-6

Solve each system of equations. Check the solutions.

18. $x - y = -5$
$x + y = 1$

19. $2r + s = -1$
$-2r + s = 3$

20. $x + y = 5$
$x + 2y = 8$

21. $3m + n = -6$
$m - n = -2$

22. $3x + 4y = 8$
$5x - 4y = 24$

23. $x + 2y = 0$
$x - y = -3$

24. $8x - 3y = 17$
$-7x + 6y = 2$

25. $7p - 10q = -1$
$3p + 2q = -13$

26. $4x - 3y = 15$
$8x + 2y = -10$

27. $2a + 8b = -1$
$-10a + 4b = 16$

28. $5x + 3y + 9 = 0$
$3x - 4y + 17 = 0$

29. $6g - 5 = 2g - 7h$
$2g = 5h - 6$

30. $6x + 7y = -11$
$4x - 7y = 21$

31. $3c + 4d = -15$
$3c + 9d = 10$

32. $3x - 5y = -1$
$6x + 2y = 10$

33. $-6x + 2y = -10$
$-5x + 7y = -35$

34. $3x + 3y + 4 = 0$
$9x - 5y = -20 - 6x$

35. $3x - y = 15$
$y = 3(x - 7)$

PRACTICE ◼ LESSON 6-1–LESSON 6-6

Graph the line that passes through the given point P and has the given slope.
(Lesson 6-1)

36. $P(-4, 5)$, $m = -\dfrac{1}{2}$

37. $P(0, 6)$, $m = \dfrac{2}{3}$

38. $P(-7, -1)$, $m = 0$

39. $P(4, -8)$, $m = 3$

40. $P(9, 7)$, $m = -5$

41. $P(1.5, 1.5)$, $m = 1.5$

Determine whether each pair of lines is *parallel, perpendicular,* or *neither.*
(Lesson 6-2)

42. The line containing $A(0, 5)$ and $B(-3, 7)$

The line containing $Y(10, 2)$ and $Z(4, 6)$

43. The line containing $C(-4, -9)$ and $D(5, 9)$

The line containing $M(-7, -1)$ and $N(1, -5)$

44. $y = 4.5x - 6.3$
$18x - 4y = 36$

45. $y = -13$
$y = -13x$

46. $3x + y = 0$
$3y = x$

Write an equation for the line with the given information. (Lesson 6-3)

47. $m = 3, Q(8, 5)$

48. $m = \dfrac{3}{4}, K(8, -1)$

49. $m = 1.5, x$-intercept -4

50. $G(10, 3)$ and $H(4, -3)$

51. $P(-5, 2)$ and $Q(10, -1)$

52. $T(-4, 7.5)$ and $U(-4, -12)$

Solve each system of equations by graphing. (Lesson 6-4)

53. $x + y = 2$
$y = x - 4$

54. $x + y = 1$
$y = x + 5$

55. $3x + 2y = -8$
$2x - 3y = -1$

Math*Works* Career – Engineering Technician
Workplace Knowhow

Engineering technicians help design and build new products. They use mathematics, engineering, and science to solve technical problems. They may be asked to create specifications for materials, establish quality testing procedures, and improve manufacturing efficiency. Engineering technicians work in laboratories, offices, industrial plants, and construction sites. They use math to make precise measurements and create schematics. They must be able to apply mathematical reasoning to find ways to cut costs and save time.

1. You are testing two products for your company. Your boss gives you a budget of $800 to spend on the testing. You ended up spending $\dfrac{1}{4}$ as much on product A as on product B. If you spent the whole amount, how much did you spend on each product?

2. You study efficiency in your plant by using the formula:
$$\text{efficiency rate} = \frac{\text{\# of products}}{\text{\# of hours}}$$

You discover that the efficiency rating for the second shift at the plant is 61.125 for $7\dfrac{1}{2}$ h of work. How many products are they producing each day?

3. After research, you discover that the efficiency rating of the second shift increases by 1.75 points when the workers are given three 15-min breaks instead of only two. Recalculate the number of products the shift produces based on the loss of 15 min from the shift but a gain of 1.75 in efficiency rating. Should the plant increase the number of breaks?

Problem Solving Skills: Determinants & Matrices

Another method of solving a system of equations is the method of determinants. A 2×2 **determinant** is a square array in the form

$\begin{vmatrix} a & b \\ c & d \end{vmatrix}$ consisting of two rows and 2 columns.

The **determinant of a system of equations**, det A, is formed using the coefficient of the variables when the equations are written in standard form.

$$\text{System of equations} \begin{cases} ax + by = e \\ cx + dy = f \end{cases} \quad \leftarrow \quad \det A = \begin{vmatrix} a & b \\ c & d \end{vmatrix} \begin{matrix} \text{Determinant} \\ \text{of coefficients} \end{matrix}$$

Problem Solving Strategies

Guess and check

Look for a pattern

Solve a simpler problem

Make a table, chart or list

Use a picture, diagram or model

Act it out

Work backwards

Eliminate possibilities

✔ Use an equation or formula

The value of the determinant is given by det $A = ad - bc$, which is the difference of the product of the diagonals.

You can find the solution to a system of equations using this determinant and another determinant formed by replacing the x or y column of the determinant with the constant column.

$$A_x = \begin{vmatrix} e & b \\ f & d \end{vmatrix} \begin{matrix} \text{Replace the } x\text{-column} \\ \text{with the constant column.} \end{matrix} \qquad A_y = \begin{vmatrix} a & e \\ c & f \end{vmatrix} \begin{matrix} \text{Replace the } y\text{-column} \\ \text{with the constant column.} \end{matrix}$$

To find x, divide A_x by determinant A.   To find y, divide A_y by determinant A.

$$x = \frac{A_x}{A} = \frac{\begin{vmatrix} e & b \\ f & d \end{vmatrix}}{\begin{vmatrix} a & b \\ c & d \end{vmatrix}} = \frac{ed - bf}{ad - bc}, A \neq 0 \qquad y = \frac{A_y}{A} = \frac{\begin{vmatrix} a & e \\ c & f \end{vmatrix}}{\begin{vmatrix} a & b \\ c & d \end{vmatrix}} = \frac{af - ec}{ad - bc}, A \neq 0$$

A **matrix** is an array of numbers. Each number in the array is called an **element**. A *column matrix* is an array of only one column. A *row matrix* is an array of only one row. A *square matrix* is an array with the same number of rows and columns. Examples are:

Column matrix: $\begin{bmatrix} 2 \\ 1 \end{bmatrix}$,   Row matrix: $[1 \ 2]$,   Square matrix: $\begin{bmatrix} 2 & 1 \\ -1 & 5 \end{bmatrix}$

A system of equations $2x - y = 4$ can be written in matrix form by using $-3x + 2y = 5$
a square matrix and 2 column matrices.

The matrix equation is $AX = B$. $\begin{bmatrix} 2 & -1 \\ -3 & 2 \end{bmatrix} \begin{bmatrix} x \\ y \end{bmatrix} = \begin{bmatrix} 4 \\ 5 \end{bmatrix}$

Problem

Solve by the method of determinants.

$$x + 3y = 4$$
$$-2x + y = -1$$

Solution

$$x = \frac{A_x}{A} = \frac{\begin{vmatrix} 4 & 3 \\ -1 & 1 \end{vmatrix}}{\begin{vmatrix} 1 & 3 \\ -2 & 1 \end{vmatrix}} = \frac{4(1) - (3)(-1)}{1(1) - 3(-2)} = \frac{4 + 3}{1 + 6} = \frac{7}{7} = 1$$

$$y = \frac{A_y}{A} = \frac{\begin{vmatrix} 1 & 4 \\ -2 & -1 \end{vmatrix}}{\begin{vmatrix} 1 & 3 \\ -2 & 1 \end{vmatrix}} = \frac{1(-1) - (4)(-2)}{1(1) - 3(-2)} = \frac{-1 + 8}{1 + 6} = \frac{7}{7} = 1$$

The solution is (1, 1). The check is left to you.

▼ TRY THESE EXERCISES

Write the matrix equations for Exercises 1–5. Then for each of the systems of equations, calculate det A, the determinant of the coefficients, and find the solution of the system using determinants. For Exercises 4 and 5, also define the system of equations.

1. $5x + y = 6$
$-3x + 4y = 2$

2. $3x - y = 2$
$x + 2y = -4$

3. $4x - 7y = 2$
$-2x + y = -4$

4. FINANCE Deanna has $2.15 in dimes and quarters. If the dimes were nickels and the quarters were dimes, she would have $1.25 less. How many of each coin does Deanna have?

5. BUSINESS Car Rental Company A charges $25 per day plus $0.35 per mile. Company B charges $35 per day, plus $0.25 per mile. Wayne Know-it-all determines the cost of a trip he will take will be $230 for Company A and $250 for Company B. How many miles and for how many days will Mr. Know-it-all's trip be?

6. WRITING MATH If det A = 0, what is the solution of a system of equations?

◣ MIXED REVIEW EXERCISES

Find the perimeter of each figure. (Lesson 5-2)

7.

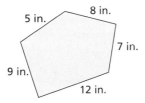
5 in. 8 in. 7 in. 9 in. 12 in.

8.

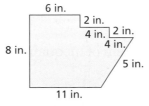
6 in. 2 in. 4 in. 2 in. 4 in. 8 in. 5 in. 11 in.

9.

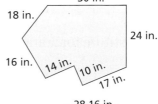
30 in. 18 in. 24 in. 16 in. 14 in. 10 in. 17 in.

10.

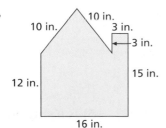
10 in. 10 in. 3 in. 3 in. 15 in. 12 in. 16 in.

11.

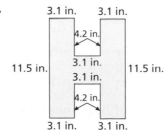
3.1 in. 3.1 in. 4.2 in. 3.1 in. 11.5 in. 11.5 in. 3.1 in. 4.2 in. 3.1 in. 3.1 in.

12.

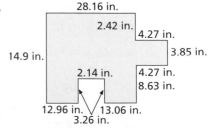
28.16 in. 2.42 in. 4.27 in. 3.85 in. 14.9 in. 2.14 in. 4.27 in. 8.63 in. 12.96 in. 13.06 in. 3.26 in.

6-8 Systems of Inequalities

Goals ■ Use graphing to solve systems of linear inequalities.

Applications Manufacturing, Health, Budgeting

Work with a partner.

1. On a coordinate plane, graph $y = x + 1$ and $y = x - 1$.

2. Plot each point listed in the table. For each point, replace □ with $<$, $=$, or $>$.

3. On the graph, shade the region where $y < x + 1$.

4. Shade the region of the graph where $y > x - 1$.

5. Find the section on the graph where the shading overlaps. What conclusion can you draw about this area?

Point	Column A	Column B
	$y \square x + 1$	$y \square x - 1$
(0, 0)		
(1, 1)		
(1, −1)		
(−1, 1)		

◤ BUILD UNDERSTANDING

The activity above shows a system of linear inequalities. A **system of linear inequalities** can be solved by graphing each related equation and determining the region where each inequality is true. The intersection of the graphs of the inequalities is the **solution set** of the system.

Reading Math

The graph of a linear equation separates the coordinate plane into two regions—one above the line, one below the line— and points on the line. The line is called the **boundary** of the region, and the regions are called **half-planes.**

Example 1

Determine whether the given ordered pair is a solution to the given system of inequalities.

a. (3, 1); $x + 2y < 5$
$2x - 3y \leq 1$

b. (2, −5); $4x - y \geq 5$
$8x + 5y \leq 3$

c. (1, 2); $x + y \geq 3$
$3x - y < 1$

Solution

Substitute for x and y in each system of inequalities.

a. $x = 3, y = 1$

$x + 2y < 5$
$3 + 2 < 5$ False

$2x - 3y \leq 1$
$6 - 3 \leq 1$ False

The ordered pair is not a solution for either inequality. Therefore, (3, 1) is not a solution of this system.

b. $x = 2, y = -5$

$4x - y \geq 5$
$8 + 5 \geq 5$
$13 \geq 5$ True

$8x + 5y \leq 3$
$16 - 25 \leq 3$
$-9 \leq 3$ True

The ordered pair is a solution for both inequalities. Therefore, (2, −5) is a solution of this system.

---

c. $x = 1, y = 2$

$x + y \geq 3$ $3x - y < 1$

$1 + 2 \geq 3$ $3 - 2 < 1$

$3 \geq 3$ True $1 < 1$ False

The ordered pair is a solution for only one of the inequalities. Therefore, $(1, 2)$ is not a solution of this system.

Example 2

Write a system of linear inequalities for the graph at the right.

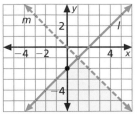

Problem Solving Tip

A solid line is used for $\leq$, and a dashed line for $<$. The dashed line shows the boundary without including it in the solution.

Solution

1. Determine the equation of each line	2. Determine shading	3. Determine inequality symbol	4. Inequality
line *l*: $b = -2, m = 1$ $y = x - 2$	below and including line	$\leq$	$y \leq x - 2$
line *m*: $b = 0, m = -1$ $y = -x$	below line	$<$	$y < -x$

The system of linear inequalities for the graph is
$y \leq x - 2$
$y < -x$

Example 3

MANUFACTURING A company writes a system of inequalities, shown below, to analyze how changes in plastic and paper affect a product's cost. Graph the solution set of the system.

$$2x - 3y \leq 6$$
$$x + 2y < 2$$

Solution

First graph the related equation for each inequality. Write each inequality in slope-intercept form. Then make a chart to use for graphing.

$2x - 3y \leq 6$ $x + 2y < 2$

$-3y \leq -2x + 6$ $2y < -x + 2$

$y \geq \dfrac{2}{3}x - 2$ $y < -\dfrac{1}{2}x + 1$

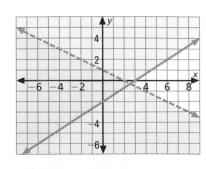

The solution set consists of all the points in the region that has been doubly shaded. The solution set includes points on the solid boundary line $y = \dfrac{2}{3}x - 2$, but not on the dashed boundary line $y = -\dfrac{1}{2}x + 1$.

Determine whether the given ordered pair is a solution of the given system of inequalities.

1. $(1, -3); 3x + 4y \le 12$
$-5x + y \le 5$

2. $(-2, 1); 2x + y < 4$
$2x - 2y \ge 3$

Technology Note

Most graphing calculators allow you to shade portions of a graph. Perform these steps on the $\boxed{Y=}$ screen.

1. Write the inequality in slope-intercept form and enter as an equation.

2. If the inequality contains $<$ or $>$ symbols, change the display to show a dashed line.

3. Determine which side of the line must be shaded and choose a shading option.

Write a system of linear inequalities for the given graph.

3.

4.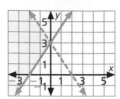

Graph the solution set of the system of linear inequalities.

5. $x \le 5$
$y \ge 1$

6. $x < y$
$x + y \ge 2$

7. $2x - y \le 2$
$x + y > -1$

8. $4 < 3x - y$
$y > 2x - 1$

■ PRACTICE EXERCISES • For Extra Practice, see page 682.

Determine whether the given ordered pair is a solution of the given system of inequalities.

9. $(3, 5); x - y \le -4$
$x - 2y \le 1$

10. $(-2, -1); x + 3y \le 6$
$4x - 2y \ge 4$

Write a system of linear inequalities for the given graph.

11.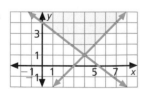

12.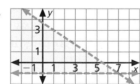

Graph the solution set of the system of linear inequalities.

13. $y > -2$
$x < 1$

14. $y > -x$
$x + y \le 2$

15. $y \ge 2x + 5$
$x - \dfrac{1}{3}y < 1$

16. $x - 2y < 6$
$3x \ge 2y - 6$

17. WRITING MATH How is the solution set of $y < 2x + 3$ different from the solution set of $y \le 2x + 3$?

Write a system of linear inequalities for the given graph.

18.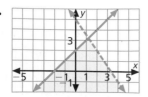

19.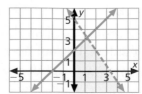

DATA FILE Use the data on height and weight for men and women on page 650. For Exercises 20–21, use y to represent weight (in pounds) and x to represent height (in inches).

20. Determine an equation for the lower male weights for heights up to 64 in. Determine an equation for the upper male weights for heights up to 64 in. Using the equations, write and graph a system of inequalities by shading the corresponding range of weights.

21. Determine equations for the lower and upper female weight for heights up to 60 in. Write a system of inequalities.

Graph the solution set of the system of linear inequalities.

22. $3 < x + y < 6$
$x \geq 0$
$y \geq 0$

23. $2 \leq 2x + y \leq 6$
$x > 1$
$y \geq 0$

24. **BUDGETING** Jasmine needs to earn at least $100 this week. She earns $6 per hour doing gardening and $8 per hour as part-time receptionist. She has only 18 h available to work during the week. Write and graph a system of linear inequalities that models the weekly number of hours Jasmine can work at each job and how much money she needs to earn.

◼ EXTENDED PRACTICE EXERCISES

25. **NUMBER THEORY** Find all numbers such that the ordered pairs (x, y) have the following conditions:
   1. x is greater than 1;
   2. y is greater than 0;
   3. the sum of the two numbers is less than 9, and the value of $x + 3y$ is at least 6.
   (*Hint:* Graph the solution set of the system of inequalities.)

26. Graph the system of inequalities and identify the figure.

   $y \geq -2x + 4$
   $y < -2x + 8$
   $-3 < x - y \leq 3$
   $1 < x \leq 9$

◼ MIXED REVIEW EXERCISES

Solve. (Lesson 5-1)

27. On one farm, the ratio of brown eggs to white eggs produced by the chickens is 1:3. If 312 eggs are produced, how many are brown?

28. At the amusement park, the ratio of children to adults is 5:2. If 63,000 people visit the park, how many are children?

29. At the mall, the ratio of people buying to people just looking is 8:7. If 9000 people come to the mall, how many will buy something?

30. In Seattle, WA, the ratio of rainy days to non-rainy days is approximately 2:3. In 365 days, how many days will it rain in Seattle?

Review and Practice Your Skills

Use determinants to solve each system of equations. Check your answers.

1. $4x - y = 9$
$x - 3y = 16$

2. $3x - 5y = -23$
$5x + 4y = 11$

3. $x + 4y = 13$
$5x - 7y = -16$

4. $-2x + 3y = 10$
$3x - 5y = 14$

5. $-x - y = -15$
$2x - y = 6$

6. $2x - 7y = -18$
$x - 2y = -6$

7. $3x + y = -1$
$2x - 3y = -8$

8. $2y + 3 = 9x$
$3x - y = 6$

9. $x - 7 = 3y$
$2(3y + 7) = 5x$

10. $3x - 7y = 2$
$6x - 4 = 13y$

11. $9x = 6y$
$3x - 4y = -18$

12. $x - 6y = 3$
$x + 2y = 5$

13. $4x + 6y = 16$
$x = 2y + 1.2$

14. $-2x + 3y = 15$
$2x - 3y = 6$

15. $4y = 10 - 5x$
$6x + 22 = 2y$

16. Nadine has $2.35 in dimes and quarters. She has 6 fewer quarters than dimes. How many of each type of coin does she have?

Determine whether the given ordered pair is a solution of the given system of inequalities.

17. $(-2, 3)$; $2x + y < 4$
$-2x + y \geq 2$

18. $(-2, 3)$; $x + 3y \leq 3$
$x + 3y > 3$

19. $(-2, 3)$; $x < 2$
$y \geq 3$

Graph the solution set of the system of linear inequalities.

20. $x + 2y \leq 3$
$2x - y \geq 1$

21. $x \geq -1$
$y > -2$

22. $y > 2x - 4$
$y \geq -x - 1$

23. $x + y > 2$
$x - y < -2$

24. $2x - 3y \leq 9$
$x + 2y < 6$

25. $y > 2x$
$y - x \leq 5$

26. $x + y < -1$
$3x - y > 4$

27. $5x + 2y \geq 12$
$2x + 3y \leq 10$

28. $y > -3$
$y < 2x + 4$

29. $y < -3x + 2$
$3y \geq x + 15$

30. $y > x$
$x \leq -3$
$-x - 3y > 3$

31. $2y - x \geq 0$
$x + 5y < 15$
$y \geq x + 1$

32. $0.1x + 0.4y \geq -0.8$
$4x - 8y \geq -24$

33. $y < -2x + 2$
$6x - 3y \leq -6$

34. $0.5x - y < 1$
$x - 2y \geq -6$

Find the slope and *y*-intercept for each line. (Lesson 6-1)

35. $y = \dfrac{5}{7}x - 11$

36. $5x - 7y = 77$

37. $154 + 10x = 14y$

38. $y = -3x$

39. $6x + 2y = 26$

40. $0.012x + 0.0004y = 0.096$

Determine whether each pair of lines is *parallel*, *perpendicular*, or *neither*. (Lesson 6-2)

41. $y = 3.2x - 64$
$16x - 5y = 30$

42. $y = -4x$
$y = 4x$

43. $-7x + 2y = 0$
$3.5y = -x$

Write an equation for the line with the given information. (Lesson 6-3)

44. $m = -\dfrac{1}{8}$, $Q(-8, 3)$

45. $m = \dfrac{2}{3}$, $K(12, -5)$

46. $m = 7.5$, $b = -2$

47. $G(10, 3)$, $b = 7$

48. $P(-6, 2)$ and $Q(6, -2)$

49. $T(-3, 3.5)$ and $U(2, 16)$

Solve each system of equations by graphing. (Lesson 6-4)

50. $x + y = -1$
$y = x + 5$

51. $x + y = 5$
$y = 2x - 10$

52. $3x + 2y = 13$
$2x - 3y = 26$

53. $x + 4y = -2$
$y = -0.25x - 0.5$

54. $4x - 2y = 18$
$y = 4x - 12$

55. $3x - 6y = 6$
$-x = -2y + 15$

Solve and check each system of equations by the substitution method. (Lesson 6-5)

56. $y = 2x$
$x + y = -9$

57. $n = 2m - 6$
$2m + n = 10$

58. $y = 2x + 5$
$x - 2y = 8$

59. $r + 3s = -5$
$3r + 2s = -15$

60. $4x = y + 3$
$9 = 12x - 3y$

61. $3a + 4b = -12$
$2b - a = 14$

Solve each system of equations. Check the solutions. (Lesson 6-6)

62. $2x + y = 8$
$-2x - 3y = -24$

63. $4x + 3y = 8$
$4x - 3y = 32$

64. $2x + y = 4$
$2x + 3y = 24$

65. $2x + y = -8$
$0.1x + 0.2y = -1.0$

66. $2x - y = 7$
$0.03x + 0.20y = 0.75$

67. $6x + 3y = 0$
$-4y = 2x + 12$

Use determinants to solve each system of equations. Check your answers. (Lesson 6-7)

68. $x + 3y = 11$
$2x - 3y = 13$

69. $5x - y = 16$
$5x + 2y = 13$

70. $x = 4y$
$4x + 2y = -36$

71. $y = 7x - 1$
$42x - 7y = 49$

72. $2y - 3x = -10$
$2x - y = 6$

73. $0.2x + 0.2y = 0.6$
$-9x + 3y = -3$

Graph the solution set of the system of linear inequalities. (Lesson 6-8)

74. $2x + y < 7$
$y \geq 2(1 - x)$

75. $x > -2$
$y \leq 3$

76. $y > -\dfrac{2}{3}x + 7$
$2x + 3y \leq 9$

6-9

Linear Programming

Goals ■ Write, minimize, and maximize an objective function.

Applications Business, Manufacturing, Farming

Work with a partner.

1. On graph paper, find the region defined by the following inequalities.

$$x \geq 0 \qquad y \geq 0 \qquad y \leq 5 \qquad y + x \leq 10$$

2. On the same coordinate axes, draw the line $y = -\frac{1}{2}x$.

3. Place a pencil on its side over the line $y = -\frac{1}{2}x$. Slowly slide the pencil over the polygonal region keeping it parallel to line $y = -\frac{1}{2}x$. Name the coordinates of the last point in the region that the pencil passes over.

4. Place the pencil on its side anywhere outside of the polygonal region but not parallel to any of its sides. Slowly slide the pencil over the region. What are the coordinates of the last point in the polygonal region that the pencil passes over?

5. What conclusion can you draw about the last point in the polygonal region that the pencil passes over?

◥ BUILD UNDERSTANDING

Linear programming is a method used by business and government to help manage resources and time. Limits to available resources are called **constraints**. In linear programming, such constraints are represented by inequalities. The intersection of the graphs of a system of constraints is known as a **feasible region**. The feasible region includes all the possible solutions to the system.

In the activity above, you determined that the last point in the polygonal region that the pencil passed over was located at a vertex. The line represented by the pencil is known as the **objective function**. The equation of this line can represent quantities such as revenue, profit or cost. In business, the objective function is used to determine how to make the maximum profit with minimum cost.

Example 1

MANUFACTURING High Tops Corporation makes two types of athletic shoes: running shoes and basketball shoes. The shoes are assembled by machine and then finished by hand. It takes 0.25 h for the machine assembly and 0.1 h by hand to make a running shoe. It takes 0.15 h on the machine and 0.2 h by hand to make the basketball shoe. At their manufacturing plant, the company can allocate no more than 900 machine hours and 500 hand hours per day. The profit is $10 on each type of running shoe and $15 on each basketball shoe. How many of each type of shoe should be made to maximize the profit?

Solution

If x represents the number of running shoes made and y represents the number of basketball shoes made, then the profit objective function (P) is $P = 10x + 15y$.

We can write inequalities to represent each constraint.

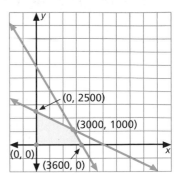

Machine hours: $0.25x + 0.15y \leq 900$

Hand hours: $0.1x + 0.2y \leq 500$

To make sure the feasible region is completely within the first quadrant of the coordinate plane, we include the constraints $x \geq 0$ and $y \geq 0$. Graph the system of inequalities.

The vertices of the feasible region are at (0, 0), (3600, 0), (0, 2500) and (3000, 1000). Evaluate the objective function at each of the vertices of the feasible region.

Vertex	10x	+ 15y	Profit P, dollars
(0, 0)	10(0)	+ 15(0)	0
(3600, 0)	10(3600)	+ 15(0)	36,000
(0, 2500)	10(0)	+ 15(2500)	37,500
(3000, 1000)	10(3000)	+ 15(1000)	45,000 (maximum)

Under the given daily constraints, the maximum daily profit the shoe company should expect to make is $45,000. To do this, they would have to produce and sell 3000 running shoes and 1000 basketball shoes per day.

Example 2

GRAPHING Using a graphing calculator, graph the solution set of the system of inequalities below to determine the maximum value of $P = 6x + 2y$.

$$x + y \leq 3 \qquad x \geq 0$$
$$y \geq -x + 1 \qquad y \geq 0$$

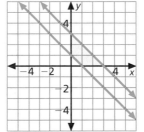

Solution

Graph the equation that corresponds to each inequality. Make a table of the constraints, related boundary equations, and shading.

Inequality	$x + y \leq 3$	$y \geq -x + 1$	$x \geq 0$	$y \geq 0$
Boundary equation	$x + y = 3$	$y = -x + 1$	$x = 0$	$y = 0$
Shading	below	above	right of y-axis	above x-axis
Line	solid	solid	solid	solid

Then locate the vertices of the feasible region using the trace, zoom, and intersect features to determine the coordinates. Make a table of the vertices and the value of P for each vertex.

Vertex	6x + 2y	P
(0, 1)	6(0) + 2(1)	2
(0, 3)	6(0) + 2(3)	6
(3, 0)	6(3) + 2(0)	18
(1, 0)	6(1) + 2(0)	6

The maximum value of P is 18 when $x = 3$ and $y = 0$.

◥ TRY THESE EXERCISES

Determine if each point is within the feasible region for $x \geq 0$, $y \geq 0$, and $5x + 2y \leq 30$.

1. $(10, 7)$ **2.** $(1, 6)$ **3.** $(2.75, 6.1)$ **4.** $(5, 2.5)$

Determine the maximum value of $P = 15x + 12y$ for each feasible region.

5. 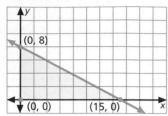   **6.**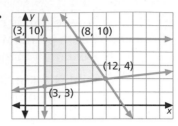

Find the feasible region for each system of constraints. Determine the maximum or minimum value of P as directed.

7. $-x + y \geq -2$ $x \geq 0$
      $y \leq 3$ $y \geq 0$

     maximum, $P = 5x + y$

8. $x + y \geq 2$ $x \geq 0$
      $y \leq 5$ $y \geq 0$
      $x \leq 3$

     minimum, $P = 4x + 3y$

◥ PRACTICE EXERCISES • For Extra Practice, see page 683.

For the feasible region whose vertices are given, find the minimum and maximum value of the objective function and identify the coordinates at which they occur.

9. $P = 10x + 6y$ $(0, 10)$ $(5, 15)$ $(8, 8)$ $(12, 0)$

10. $P = 4x + 5y$ $(2, 5)$ $(2, 9)$ $(6, 11)$ $(8, 5)$

11. $P = 1.25x + 0.75y$ $(0, 4)$ $(9, 15)$ $(20, 2)$

12. $P = 120x + 180y$ $(6, 6)$ $(6, 10)$ $(10, 12)$ $(13, 11)$ $(13, 6)$

13. WRITING MATH Why do you think each system of constraints in this lesson contains the inequalities $x \geq 0$ and $y \geq 0$? What do these constraints accomplish?

Identify the vertices of the feasible region defined by the constraints.

14. $x \geq 0$; $y \geq 0$; $7x + 9y \leq 63$

15. $x \geq 0$; $y \geq 0$; $y + 2x \geq 8$

16. $x \geq 0$; $y \geq 0$; $y + x \leq 10$

Determine the minimum value of the objective function $C = 3x + 2y$ for the graph of each feasible region.

17. 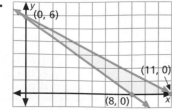   **18.**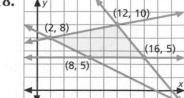

19. **MANUFACTURING** Glimmering Hobbies manufactures remote control cars and airplanes. The plant manufactures at least 50 items but not more than 75 items each week. If the profit is determined by $P = 25x + 35y$, where x is the number of cars and y is the number of planes manufactured, determine the number of cars and airplanes that should be manufactured to maximize the profit.

SMALL BUSINESS A group of students are making and selling custom-printed T-shirts and sweatshirts. Their costs are $3.00/T-shirt and $5.00/sweatshirt. A local store owner has agreed to sell their shirts but will only take up to a total of 50 T-shirts and sweatshirts combined. In addition, the store owner said they must sell at least 15 T-shirts and 10 sweatshirts to continue selling at the store. Let x equal the number of T-shirts and y equal the number of seatshirts.

20. The students need to minimize cost. Write the objective function for cost (C).

21. Write the inequalities that express the constraints.

22. Graph the inequalities and determine coordinates of the vertices of the feasible region.

23. How many of each type must they sell to minimize cost?

24. **CHAPTER INVESTIGATION** Write a magazine, newspaper, or radio advertisement for your product. Be sure to include a selling price. To establish a selling price, consider manufacturing costs such as materials and labor and the selling price of the other similar products on the market today.

■ EXTENDED PRACTICE EXERCISES

25. **AGRICULTURE** Valley Farms owns a 3600-acre field. The farmers want to plant Iceberg lettuce which yields $200 per acre and Romaine lettuce which yields $250 per acre. To prevent loss due to disease, the farmers should plant no more than 3000 acres of Iceberg and no more than 2500 acres of Romaine. How many acres of each crop should Valley Farms plant in order to maximize profits? What is the maximum profit?

26. **SMALL BUSINESS** Sasha owns and operates the Stand-In-Line Skate Shop. She makes a profit of $40 on each pair of adult skates and $20 on each pair of child skates sold. Sasha can stock at most 80 pairs of skates on her shelves. Sasha orders skates once every 6 weeks and can order up to 50 pairs of each type of skate. How many of each type of skate must Sasha stock and sell in a 6-week period in order to maximize her profits? What is the maximum profit?

■ MIXED REVIEW EXERCISES

Find each product or quotient. (Lesson 1-5)

27. $(-3.9)(-4.8)(-7.6)$

28. $-387 \div [4 \cdot (-3)]$

29. $125 \div (-10) + 38$

30. $(8.36)(9.74)(-3.85)$

31. $[-4 \cdot (-6)] \div (-2)$

32. $(-60) \div [-3 \cdot (-5)]$

Chapter 6 Review

VOCABULARY ◣

Choose the word from the list at the right that completes each statement.

1. The __?__ is the ratio of the vertical change to the horizontal change.

2. The graphs of a(n) __?__ system of linear equations do not intersect.

3. The graphs of a(n) __?__ system of two linear equations intersect in one point.

4. Two lines are __?__ if the product of their slopes is -1.

5. In the equation $y = 2x - 4$, -4 is the __?__.

6. Two lines are __?__ if they have the same slope and different y-intercepts.

7. The graphs of a(n) __?__ system of equations are the same line.

8. A(n) __?__ is a square array of numbers enclosed between two parallel lines.

9. In the __?__ method for solving a system of equations, a variable in one equation is replaced with an equivalent expression derived from the other equation.

10. The intersection of the graphs of a system of constraints is the __?__.

a. dependent

b. determinant

c. feasible region

d. inconsistent

e. independent

f. matrix

g. parallel

h. perpendicular

i. slope

j. substitution

k. x-intercept

l. y-intercept

LESSON 6-1 ◣ Slope of a Line and Slope-Intercept Form, p. 244

▶ slope $= \dfrac{\text{rise}}{\text{run}} = \dfrac{\text{vertical change (change in } y\text{-coordinates)}}{\text{horizontal change (change in } x\text{-coordinates)}}$

▶ A horizontal line has a 0 slope. The slope of a vertical line is undefined.

▶ The slope-intercept form of an equation of a line is written as $y = mx + b$, where m represents the slope of the line and b represents the y-intercept.

11. Find the slope of the line containing the points $A(3, -2)$ and $B(2, 9)$.

12. Graph the line that passes through the point $P(2, -1)$ and has a slope of $\dfrac{1}{2}$.

13. Graph the line that passes through point $P(8, -6)$ and has a slope of $-\dfrac{7}{8}$.

14. Find the slope and y-intercept for the line with the equation $4y = 2x - 8$. Then graph the equation.

15. Find an equation of the line with slope $= -2$ and y-intercept $= 1$.

LESSON 6-2 ◣ Parallel and Perpendicular Lines, p. 248

▶ Two lines with the same slope and different y-intercepts are parallel.

▶ Two lines are perpendicular if the product of their slopes is -1.

16. $\overrightarrow{MN}$ contains the points $M(4, 6)$ and $N(-1, 3)$. Find the slope of a line parallel to $\overrightarrow{MN}$ and the slope of a line perpendicular to $\overrightarrow{MN}$.

17. Are $7x - 3y = 21$ and $7y = 3x + 4$ parallel, perpendicular, or neither?

18. Determine whether the line containing the points $T(0, 3)$ and $U(3, 0)$ is perpendicular or parallel to the line containing the points $V(7, 1)$ and $W(1, 7)$.

19. Find the slope of the line parallel to the graph of $y = -\dfrac{2}{5}x + 7$.

20. Find the slope of the line perpendicular to the graph of $y = 6x - 4$.

LESSON 6-3 ◤ Write Equations for Lines, p. 254

▶ An equation of a line can be written in point-slope form if you know the slope of the line and the coordinates of any point on the line or the coordinates of two points on the line.

21. Write an equation of the line with slope of -1 containing point $A(2, 3)$.

22. Write an equation of the line containing points $C(3, 6)$ and $D(1, -2)$.

23. Write an equation of a line perpendicular to $y = \dfrac{1}{2}x - 5$ containing point $R(0, -4)$.

24. Write an equation of the line that is perpendicular to $4y - 3x = 12$ containing the point $Z(2, -1)$.

25. Write an equation of the line whose slope is $-\dfrac{1}{4}$ and contains point $T(8, 2)$.

LESSON 6-4 ◤ Systems of Equations, p. 258

▶ Two linear equations with the same two variables form a system of equations. A solution of the system is an ordered pair that makes both equations true. Graphing both equations can be used to solve the system. The point of intersection of the two lines is the solution.

▶ In an independent system, the graphs intersect in one point. In an inconsistent system, the graphs do not intersect. In a dependent system, the lines coincide.

Solve each system of equations by graphing.

26. $2x + y = 6$
$2y + 2 = 3x$

27. $-6x + 3y = 18$
$y = 6 - 2x$

28. $y = \dfrac{1}{2}x$
$2x + y = 10$

29. The perimeter of a rectangle is 40 m. The length is one less than twice its width. What are the dimensions of the rectangle?

LESSON 6-5 ◤ Solve Systems by Substitution, p. 264

▶ A system of equations can be solved algebraically. To solve a system using substitution, solve one equation for a variable and then substitute that expression into the second equation.

Solve each system of equations by the substitution method.

30. $y = 3x$
$x + 2y = -21$

31. $y = x + 7$
$x + y = 1$

32. $x - 3y = -9$
$5x - 2y = 7$

33. Angle A and $\angle B$ are supplementary. The measure of $\angle A$ is 24 degrees greater than the measure of $\angle B$. Find the measures of $\angle A$ and $\angle B$.

LESSON 6-6 ◢ Solve Systems by Adding and Multiplying, p. 268

▶ A system of equations can be solved by adding or subtracting two equations to eliminate one variable. Sometimes, one or both equations must be multiplied by a number or numbers before the equations are added or subtracted.

Solve each system of equations by the substitution method. Check the solutions.

34. $3x + 2y = -3$
$-4x + 2y = 4$

35. $2x + 7y = -1$
$3x = -2y + 7$

36. $3x - 5y = -1$
$-5x + 7y = -1$

37. The concession stand sells hot dogs and soda during Beck High School football games. John bought 6 hot dogs and 4 sodas and paid $10.50. Jessica bought 4 hot dogs and 3 sodas and paid $7.25. What is the cost of one hot dog?

LESSON 6-7 ◢ Problem Solving Skills: Determinants & Matrices, p. 274

▶ Determinants can be used to solve a system of equations.
$$\begin{cases} ax + by = e \\ cx + dx = f \end{cases} \rightarrow \det A = ad - bc; \quad x = \frac{ed - bf}{ad - bc}, \quad y = \frac{af - ec}{ad - bc}, \quad ad - bc \neq 0$$

Use determinants to solve each system of equations. Check your answers.

38. $5y + 3x = 4$
$2x - 3y = 5$

39. $2y = 4x - 10$
$3y + x = -1$

40. $4x + 3y = 19$
$3x - 4y = 8$

41. Danielle is 5 years less than twice Mario's age. In 15 years, Mario will be the same age as Danielle is now. Find the ages of Danielle and Mario in 5 years.

LESSON 6-8 ◢ Systems of Inequalities, p. 276

▶ A system of linear inequalities can be solved graphically. The intersection of the graphs of the inequalities is the solution set of the system.

Graph the solution set of the system of linear inequalities.

42. $2x + 5y \geq 10$
$3x - y \leq 3$

43. $2x - 3y \leq 6$
$x + y \geq 1$

44. $3x + y \geq 2$
$y \geq 2x - 1$

LESSON 6-9 ◢ Linear Programming, p 282

▶ Linear programming can be used to solve business-related linear inequalities.

Determine the maximum value of $P = 5x + 2y$ for each feasible region.

45.

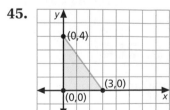

46.

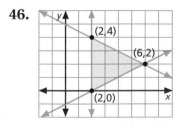

47.

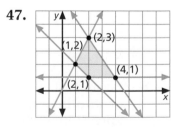

CHAPTER INVESTIGATION

EXTENSION Compare your improved product to the original product. Write a report about how your improvements will make the product better. Include an explanation as to how the improvements to the product will justify the increased cost of the product.

Chapter 6 Assessment

Find the slope and y-intercept for each line. Then graph the equation.

1. $-3x + 4y = 12$

2. $x + 5y = -5$

Find the slope of each line. Then give the slope of a line parallel to the given line and the slope of a line perpendicular to the given line.

3. The line containing the points $R(1, -1)$ and $S(-4, 1)$.

4. $4x - y = 8$

5. Graph the line that passes through $P(2, -1)$ and has a slope of $-\dfrac{1}{2}$.

Write an equation of the line with the given information.

6.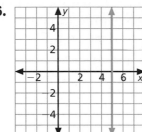

7. $m = -\dfrac{4}{5}, b = 2$

8. $m = \dfrac{1}{2}, A(-1, 2)$

9. $C(-4, 1)$ and $D(1, 4)$

Use a graph to solve each system.

10. $8x - 2y = 6$
    $3x = 4y - 1$

11. $2y = 3x$
    $3y - 2x = 10$

12. $x - 2y < 4$
    $3 \geq x + y$

13. $y - 3 < 0$
    $x + 2y > 2$
    $x \leq 2$

Solve.

14. $-4x - 5y = -2$
    $6y = -x + 10$

15. $x + 3y = -9$
    $7 - 2y = 5x$

16. $5x - 2y = 9$
    $x - 4y = 9$

17. $-9x - 3y = 9$
    $x = \dfrac{y}{3} + 1$

18. The sum of the digits of a two-digit number is 5. The units digit is one more than 3 times the tens digit. Find the original number.

19. Kari is 6 years older than Adam. In 9 years, $\dfrac{1}{2}$ of Adam's age will equal $\dfrac{1}{3}$ Kari's age. Find the ages of Kari and Adam in 2 years.

20. One solution contains a 40% acid solution. Another contains a 60% acid solution. Determine the number of liters of each solution needed to make 25 L of a 56% acid solution.

Standardized Test Practice

Part 1 Multiple Choice

Record your answers on the answer sheet provided by your teacher or on a sheet of paper.

1. If $A = \{x \mid x \leq 3\}$ and $B = \{x \mid x > -1\}$, what is the least integer in $A \cap B$? (Lesson 1-3)
   - (A) -1
   - (B) 0
   - (C) 1
   - (D) 2

2. Molly received grades of 79, 92, 68, 90, 72, and 92 on her history tests. What measure of central tendency would give her the highest grade for the term? (Lesson 2-7)
   - (A) mean
   - (B) median
   - (C) mode
   - (D) range

3. B is the midpoint of $\overline{AC}$. What is the value of x if $AC = 26$ and $AB = 3x - 5$? (Lesson 3-3)
   - (A) 6
   - (B) 7
   - (C) 13
   - (D) 16

4. In the figure, $\overline{BA} \cong \overline{BC}$ and $\angle A \cong \angle C$. Which postulate could you use to prove $\triangle ABF \cong \triangle CBD$? (Lesson 4-2)

   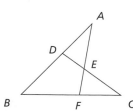

   - (A) Angle-Angle-Angle Postulate
   - (B) Angle-Side-Angle Postulate
   - (C) Side-Angle-Side Postulate
   - (D) Side-Side-Side Postulate

5. The lengths of two sides of a triangle are 7 in. and 10 in. Which length could *not* be the measure of the third side? (Lesson 4-6)
   - (A) 5 in.
   - (B) 7 in.
   - (C) 12 in.
   - (D) 18 in.

6. The area of $\triangle ABC$ is 15 cm². What is the value of x? (Lesson 5-2)

   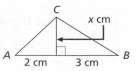

   - (A) 3 cm
   - (B) 5 cm
   - (C) 6 cm
   - (D) 10 cm

7. Which equation represents a graph that is perpendicular to the graph of $2x + 6y = 8$? (Lesson 6-2)
   - (A) $3x - 9y = 16$
   - (B) $12x + 4y = 4$
   - (C) $5y = 15x + 10$
   - (D) $21y = 3 - 7x$

8. When solving the following system of equations, which expression could be substituted for x? (Lesson 6-5)

   $$x + 4y = 1$$
   $$2x - 3y = -9$$

   - (A) $4y - 1$
   - (B) $1 - 4y$
   - (C) $3y - 9$
   - (D) $-9 - 3y$

9. Which graph is the solution of the following system of inequalities? (Lesson 6-8)

   $$y \geq 2x$$
   $$2y + x \leq 3$$

   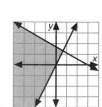

Test-Taking Tip

Question 9
If you are allowed to write in the test booklet, cross off each answer choice that you know is not the answer, so you will not consider it again.

Part 2 Short Response/Grid In

Record your answers on the answer sheet provided by your teacher or on a sheet of paper.

10. A developer is going to divide some land for single-family homes. If she buys 12 acres of land, how many $\frac{3}{4}$-acre lots can she sell? (Lesson 1-5)

11. Mt. Everest is the highest mountain in the world. It is about 2.9×10^4 ft above sea level. Write this height in standard form. (Lesson 1-8)

12. What is the next term in the following sequence? (Lesson 2-1)
$$1, -\frac{1}{2}, \frac{1}{4}, -\frac{1}{8}, \dots$$

13. If $g(x) = -3x^2 - |x| + 6$, what is the value of $g(2)$? (Lesson 2-3)

14. Solve $-\frac{5}{6}x + 6 = -9$. (Lesson 2-5)

15. If $\overrightarrow{AC} \perp \overrightarrow{DB}$, $m\angle EBC = (4x + 4)°$, $m\angle DBE = (3x + 9)°$, find $m\angle DBE$. (Lesson 3-2)

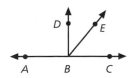

16. What is the value of n in the figure below? (Lesson 4-7)

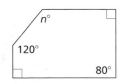

17. What is the surface area of the figure? (Lesson 5-6)

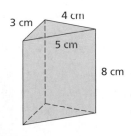

18. Find the height x if the volume of the rectangular box is 220 cm^3. (Lesson 5-7)

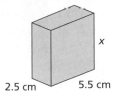

19. What is the slope of $\overleftrightarrow{MN}$ containing points $M(-1, 4)$ and $N(-5, -2)$? (Lesson 6-1)

20. What is the solution of the system of equations represented by the graph? (Lesson 6-4)

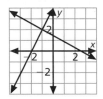

21. If $2x + 3y = 9$ and $8x - 5y = 19$, what is the value of $2x$? (Lesson 6-6)

Part 3 Extended Response

Record your answers on a sheet of paper. Show your work.

22. The vertices of a triangle are $A(-2, -1)$, $B(-1, 3)$, and $C(2, 1)$.

   a. Write the equations of the lines that contain the sides of the triangle. (Lesson 6-3)

   b. Write the inequalities that would represent the triangle and its interior. Then, draw the graph. (Lesson 6-9)

23. The manager of a movie theater found that Saturday's sales were $3675. He knew that a total of 650 tickets were sold Saturday. Adult tickets cost $7.50 each and child tickets cost $4.50 each. (Lessons 6-5, 6-6, and 6-7).

   a. Write a system of equations to represent the situation.

   b. What method would you use to solve the equations? Explain. Solve the equations.

Similar Triangles

THEME: Photography

Photography is a blend of science and art. A camera produces an image on film by allowing light from an object to pass through a lens in a dark box. The amount of light is controlled by the size of the opening and the amount of time that the shutter is open.

Today, photographers use highly sophisticated cameras and computers to manipulate images. Digital cameras store images electronically. Digital images can be easily edited for special effects. They can also be instantly transmitted using the Internet.

- **Police Photographers** (page 305) work with forensic scientists to record details at a crime scene. Police photographers must take pictures from all angles to record all possible clues.

- **Photographic Process Workers** (page 325) develop film, make prints or slides, and enlarge or retouch photographs. Photographic process workers use computers to enhance or alter photographs. They use their knowledge of ratio and proportion to make sure images look right.

Math
Online

mathmatters3.com/chapter_theme

Camera Settings for Outdoor Lighting Conditions (seconds)

Lighting	Very bright	Bright	Partly cloudy	Overcast
Shutter Speed	1/125	1/60	1/60	1/60
Aperture	f/16	f/16	f/8	f/5.6
Shutter Speed	1/250	1/125	1/125	1/125
Aperture	f/11	f/11	f/5.6	f/4

Image Sizes

Film format	Image size ratio
Disc	1 : 1
110	13 : 17
126	1 : 1
135	2 : 3
Panoramic	1 : 3

Data Activity: Camera Settings and Image Sizes

Use the tables for Questions 1–4.

1. Shutter speed is measured in fractions of a second. On a partly cloudy day, which combination of settings should a photographer use if a faster shutter speed is desired?

2. Suppose the aperture in a camera is stuck at f/5.6. What shutter speed should be used if the day is overcast?

3. An image from a roll of 135 film is enlarged so that the width is 5 in. If the image is not cropped, what is the length of the print?

4. A print from a roll of panoramic film must be reduced to fit in a magazine layout. The layout space is 4.5 in. in length. Find the width of the reduced image.

CHAPTER INVESTIGATION

Artists often work from photographs to paint realistic portraits and murals. The image shown in a photograph can be enlarged using a ratio or scale. If the artwork is to appear realistic, the larger work must be proportional to the photograph.

Working Together

Choose a photograph (either an actual photograph or a photograph published in a magazine). Make a proportional sketch of the subject of the photograph. Enlarge the photograph by a factor of 5. Use the Chapter Investigation icons to guide your group.

Are You Ready?

The skills on these two pages are ones you have already learned. Review the examples and complete the exercises. For additional practice on these and more prerequisite skills, see pages 654–661.

RATIOS

A ratio is used to compare two numbers. Ratios can be written three different ways, but always in lowest terms.

Example   400 miles traveled in 5 days

analogy form: $400 : 5 = 80 : 1$

fraction form: $\dfrac{400}{5} = \dfrac{80}{1}$

word form: 400 to 5 = 80 to 1

Write each as a ratio three different ways. Use lowest terms.

1. 6 tents for 24 campers

2. 72 horseshoes for 18 horses

3. 12 donuts for $4.56

4. 444 calories for 6 oz candy

5. 15 gardens hold 810 plants

6. 318 mi on 7 gal of gas

7. 96 min to make 8 pizzas

8. 4 elephants weight 14,000 lb

9. 279 bricks to cover 62 ft^2

10. 28 yd of fabric to make 8 curtains

CONGRUENT TRIANGLES

You can determine that triangles are congruent by three different methods.

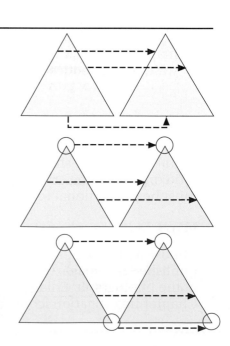

Examples   Triangles are congruent if:
. . . three sides of one triangle are congruent to three sides of another.

. . . two sides and the included angle of one triangle are congruent to two sides and the included angle of another.

. . . two angles and the included side of one triangle are congruent to two angles and the included side of another.

State whether each pair of triangles is congruent by SSS, SAS, or ASA. If the triangles are not congruent or you cannot determine congruency, write *not congruent*.

11.
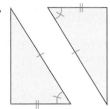

12.
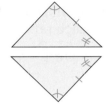

13.
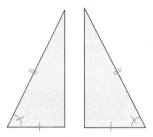

14.
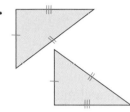

15.
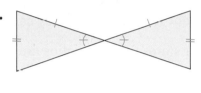

16.
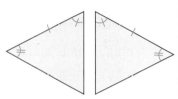

PARALLEL LINES

Coplanar lines that never intersect are parallel lines. If two parallel lines are cut by a transversal, corresponding angles are congruent. The converse is also true. If two lines are cut by a transversal such that corresponding angles are congruent, the lines are parallel.

Find the missing measures. If the lines are not parallel, write *not parallel*.

17.
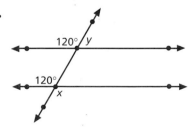

18.
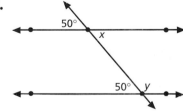

19.
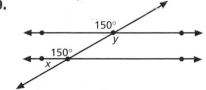

20.
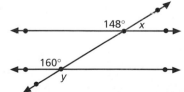

Ratios and Proportions

Goals
- Find equivalent ratios.
- Use ratios and proportions to solve problems.

Applications   Recreation, Real Estate, Retail, Business, Art

---

Work with a partner.

The photo lab at Glenisle has 315 technicians and 30 supervisors. The lab at Skatetown has 450 technicians and 36 supervisors.

1. Write the ratio of technicians to supervisors in one lab while your partner does the same for the other lab. Write each ratio in lowest terms.

2. Determine which lab has more technicians per supervisor.

BUILD UNDERSTANDING

Two ratios that can both be named by the same fraction are called **equivalent ratios**. 4:8 and 7:14 are equivalent ratios because they can each be written as $\frac{1}{2}$.

A **proportion** is an equation that states that two ratios are equivalent.

$$a{:}b = c{:}d \qquad\qquad \frac{a}{b} = \frac{c}{d}$$

The four numbers a, b, c, and d that are related in the proportion are called its **terms**. The first and last terms are called the **extremes**. The second and third are called the **means**.

$$
\overset{\textbf{extremes}}{\overbrace{a{:}b = c{:}d}}
\qquad\qquad
\overset{\textbf{extreme} \quad \textbf{mean}}{\frac{a}{b} = \frac{c}{d}}
$$
$$\underset{\textbf{means}}{\phantom{a{:}b=c{:}d}} \qquad\qquad \underset{\textbf{mean} \quad \textbf{extreme}}{\phantom{\frac{a}{b}=\frac{c}{d}}}$$

In a proportion, the product of the extremes equals the product of the means. This is the same as saying that the **cross products** are equal.

$$\frac{a}{b} \diagup\!\!\!\!\diagdown \frac{c}{d} \qquad ad = bc \qquad \frac{3}{4} \diagup\!\!\!\!\diagdown \frac{12}{16} \qquad \begin{array}{c} 3(16) = 4(12) \\ 48 = 48 \end{array}$$

Use cross products to find the missing term in a proportion.

Example 1

Solve the proportion $\dfrac{x}{18} = \dfrac{12}{27}$.

Solution

Use cross products to write another equation. Solve that equation for x.

$$\frac{x}{18} = \frac{12}{27}$$
$$27x = 18(12)$$
$$27x = 216$$
$$x = 8$$

Check your answer by substituting it in the original proportion.

$$\frac{8}{18} \stackrel{?}{=} \frac{12}{27}$$

$$\frac{8}{18} = \frac{8 \div 2}{18 \div 2} = \frac{4}{9} \qquad \frac{12}{27} = \frac{12 \div 3}{27 \div 3} = \frac{4}{9}$$

Because the ratios are equivalent when $x = 8$, the proportion is solved.

You can use proportions to solve a wide variety of problems.

Problem Solving Tip

When writing a proportion to solve a problem, it may help to write a word ratio first. In Example 2, the word ratio, enlargements:cost, makes it easier to write the terms in the correct place.

Example 2

PHOTO PROCESSING Fine Photo charges $3 for 2 enlargements. How much does the company charge for 5 enlargements?

Solution

Write a proportion. Let $x =$ the cost of 5 enlargements.

$$\frac{2}{3} = \frac{5}{x} \qquad \frac{\text{enlargements}}{\text{cost}}$$

$$2x = 15$$

$$x = 7.5$$

So, the company charges $7.50 for 5 enlargements.

Sometimes the information you are given in a problem is a ratio of two quantities.

Example 3

RECREATION The ratio of counselors to campers is 2:15. There are 102 people at a camp. How many are counselors?

Solution

Let $2x$ represent the number of counselors. Let $15x$ represent the number of campers.

The ratio of counselors to campers is $2x:15x$, which is the same as 2:15. Write an equation for the total number of people at camp.

$$2x + 15x = 102$$

$$17x = 102$$

$$x = 6$$

Because $2x$ represents the number of counselors, the answer is 12.

Is each pair of ratios equivalent? Write *yes* or *no*.

1. 28:49, 16:28

2. 39 to 13, 36 to 9

3. $\dfrac{4}{5} = \dfrac{6}{7.5}$

Solve each proportion.

4. $\dfrac{x}{21} = \dfrac{18}{27}$

5. $\dfrac{9}{y} = \dfrac{36}{8}$

6. $0.04{:}0.06 = m{:}0.24$

7. REAL ESTATE Two families decide to split the cost of renting a vacation house in a ratio of 3:4. The total cost is $2100. What will be each family's share of the cost?

8. RETAIL A used book store is selling 5 paperback books for $2. How much will 12 paperback books cost?

▧ **PRACTICE EXERCISES** • **For Extra Practice, see page 683.**

Is each pair of ratios equivalent? Write *yes* or *no*.

9. 3:6, 15:18

10. $\dfrac{3.5}{4.2}, \dfrac{10}{12}$

11. $\dfrac{4}{5} : \dfrac{2}{5}$, 4:8

Solve each proportion.

12. $\dfrac{3}{18} = \dfrac{10}{s}$

13. $\dfrac{n}{0.9} = \dfrac{0.7}{0.3}$

14. $\dfrac{27}{81} = \dfrac{k}{45}$

CALCULATOR **Use a calculator to solve these proportions.**

15. $\dfrac{119}{476} = \dfrac{247}{r}$

16. $\dfrac{245}{372} = \dfrac{t}{1488}$

17. $\dfrac{426}{z} = \dfrac{1491}{2205}$

18. A recipe for fruit punch calls for 3 parts pineapple juice to 5 parts orange juice. How much pineapple juice should be added to 16 L of orange juice?

19. BUSINESS The manager of Music World stocks audio cassettes and CDs in the ratio 2:7. This month, she is ordering 400 audio cassettes. How many CDs will she order?

20. INVESTING Two business partners purchased stock. The ratio of the money invested by one partner to the money invested by the other was 4:5. The stock earned $31,500. What is each partner's share?

21. Ricardo mixes dried fruit and nuts in a 3:5 ratio. He wants to make 12 lb of the mixture. How many pounds of nuts does he need?

Arrange the given terms to form a proportion. Supply the missing term.

22. 1.5, 5, 6

23. 100, 3, 30

24. 35, 36, 14

Solve each proportion.

25. $\dfrac{2x}{3} = \dfrac{x+5}{4}$

26. $\dfrac{4}{x+1} = \dfrac{5}{2x-1}$

27. $\dfrac{3x}{25} = \dfrac{8x+2}{70}$

28. **ART** Louisa wants to mix 1 part yellow paint to 3 parts blue to make a certain shade of green. How many pints of blue paint will she need if she wants 1 gal of green paint?

29. Eddie must read a biography by the end of the month. The book has 317 pages. He found that he could read 10 pages in 15 min. Estimate how many hours it will take him to read the whole book.

30. Otis and Steve bought an old car and fixed it up. Otis spent $2000 and Steve spent $1500. They were able to sell the car for $4900. How much should each receive from the profit made on the car?

31. An angle and its complement have measures in the ratio 2:3. What are the measures of the angles?

32. **WRITING MATH** If $\frac{a}{b} = \frac{c}{d}$, is it always true that $\frac{a}{c} = \frac{b}{d}$? Explain.

33. **ERROR ALERT** A survey showed that 3 out of 10 students have a regular physical fitness program. Wanitta knows that 48 students said that they exercised regularly. To find out how many students were surveyed, she wrote the proportion $\frac{3}{10} = \frac{x}{48}$ and solved for x. Wanitta suspects that her solution of 14.4 is not correct. What went wrong?

34. **CHAPTER INVESTIGATION** Prepare to make an enlargement of a photograph. Choose an actual photograph or a photograph published in a magazine. Measure the outer dimensions of the photograph and increase the dimensions by a factor of 5. Cut a piece of poster paper the size of the enlargement.

◾ EXTENDED PRACTICE EXERCISES

In the proportion $a:b = b:c$, b is called the **mean proportional** or the **geometric mean** between a and c where a, b, and c are greater than 0.

35. Find b if $a = 3$ and $c = 12$.

36. If b is a positive integer, what must be true about the product, ac?

◾ MIXED REVIEW EXERCISES

Find the slope of the line containing the given points. (Lesson 6-1)

37. $(-3, 2)$, $(4, 8)$ 38. $(1, 6)$, $(0, -2)$ 39. $(4, -3)$, $(1, 2)$ 40. $(3, -4)$, $(-2, -1)$

41. $(4, 6)$, $(-2, -3)$ 42. $(-4, 1)$, $(2, -3)$ 43. $(3, -2)$, $(3, 4)$ 44. $(5, 5)$, $(4, -2)$

45. $(7, 4)$, $(3, -3)$ 46. $(2, -1)$, $(-5, -1)$ 47. $(5, -2)$, $(-1, 3)$ 48. $(-3, 2)$, $(3, -2)$

The lengths of two sides of a triangle are given. Find the range of possible lengths for the third side. (Lesson 4-6)

49. 7 cm, 8 cm   50. 5 in., 12 in.   51. 14 cm, 13 cm   52. 7 m, 15 m

53. 3 dm, 9 dm   54. 23 ft, 18 ft   55. 3 ft, 15 in.   56. 32 in., 2 ft

Similar Polygons

Goals
- Identify similar polygons.
- Find missing measures of similar polygons.

Applications Photography, Architecture, Framing

Work with a partner. You will need a protractor and a centimeter ruler.

A copy machine is used to enlarge a company logo.

1. Measure the angles in the lower left and upper left corners of both drawings.

2. Measure the sides of both drawings.

3. Compare corresponding measurements. What do you notice?

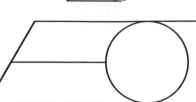

◤ Build Understanding

Two figures are **similar** if they have the same shape. The figures may not necessarily be the same size. Two polygons are similar if all corresponding angles are congruent and the measures of all corresponding sides form the same ratio (are in proportion).

The symbol for similarity is ~. Polygon *ABCDE* ~ polygon *KLMNO*.

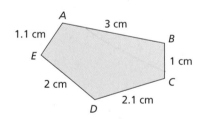

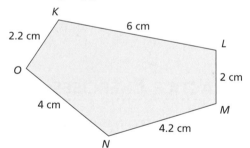

Corresponding angles are congruent.

$\angle A \cong \angle K, \angle B \cong \angle L, \angle C \cong \angle M, \angle D \cong \angle N, \angle E \cong \angle O$

Corresponding sides are in proportion.

$$\frac{AB}{KL} = \frac{BC}{LM} = \frac{CD}{MN} = \frac{DE}{NO} = \frac{EA}{OK} = \frac{1}{2}$$

Example 1

Is *WXYZ* similar to *EFGH*?

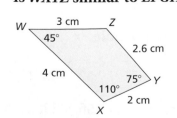

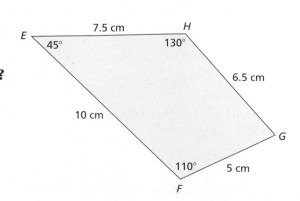

Solution

Find the missing angle measures.

$m\angle Z = 360° - (45° + 110° + 75°) = 130°$. So, $\angle Z \cong \angle H$.

$m\angle G = 360° - (45° + 110° + 130°) = 75°$. So, $\angle G \cong \angle Y$.

All four pairs of corresponding angles are congruent.

Find the ratios of all pairs of corresponding sides.

$$\frac{WX}{EF} = \frac{4}{10} = \frac{2}{5} \qquad \frac{XY}{FG} = \frac{2}{5}$$

$$\frac{YZ}{GH} = \frac{2.6}{6.5} = \frac{2}{5} \qquad \frac{WZ}{EH} = \frac{3}{7.5} = \frac{2}{5}$$

Each pair of corresponding sides has the same ratio. So, corresponding sides are in proportion. The two polygons are similar.

Example 2

PHOTOGRAPHY Two mats, shown at the right, are cut to display photographs. $PS = 40$ cm, $TW = 60$ cm and $QR = 50$ cm. If the mats are similar figures, what is the measure of $\overline{UV}$?

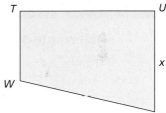

Solution

Because the figures are similar, their corresponding sides are in proportion. Write and solve a proportion to find x.

$$\frac{PS}{TW} = \frac{QR}{UV} \qquad \text{Ratios of corresponding sides are equivalent.}$$

$$\frac{40}{60} = \frac{50}{x}$$

$$40x = 3000$$

$$x = 75$$

So, $UV = 75$ cm.

> ### Check Understanding
>
> If $m\angle S = 115$, which angle of $TUVW$ has that measure?

Example 3

Name a pair of similar triangles.

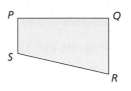

Solution

$\angle L \cong \angle P$, $\angle M \cong \angle N$, $\angle MOL \cong \angle NOP$.

There are 3 pairs of congruent angles.

$$\frac{LM}{PN} = \frac{3}{6} = \frac{1}{2} \qquad \frac{MO}{NO} = \frac{2}{4} = \frac{1}{2} \qquad \frac{OL}{OP} = \frac{2.5}{5} = \frac{1}{2}$$

Corresponding sides are proportional. To name the similar triangles, name corresponding vertices in the same order.

$\triangle LOM \sim \triangle PON$

Since congruent angles are important, you should know how to copy an angle by geometric construction. Follow these steps to copy ∠ABC.

Step 1: Draw an arc with the center at point *B* so that it intersects both rays of the angle. Label the points *P* and *Q*.

Step 2: Draw a ray *DE*.

Step 3: Place the compass point at *D* and use the same compass setting to draw an arc that intersects $\overline{DE}$ at *F*.

Step 4: Place the compass point at *F* and draw another arc that is the same measure as *PQ*. Label the point where the two arcs intersect point *G*. Draw ray *DG*. Now ∠ABC ≅ ∠GDF.

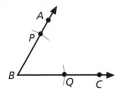 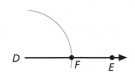 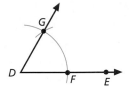

◤ TRY THESE EXERCISES

Determine if the polygons are similar. Write *yes* or *no*.

1.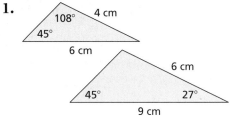

2.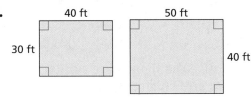

Find *x* in each pair of similar polygons.

3.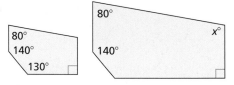

4.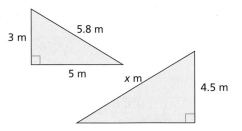

◤ PRACTICE EXERCISES • For Extra Practice, see page 684.

Determine if the polygons are similar. Write *yes* or *no*.

5.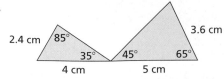

6.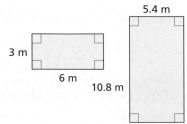

Find _x_ in each pair of similar figures.

7.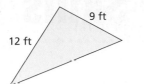
9 ft
5 ft
_x_
12 ft

8.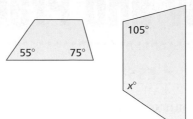
105°
55°   75°
_x_°

9. **WRITING MATH** When given two similar figures, how can you tell which angles are corresponding angles?

10. Draw an obtuse angle. Copy the angle using a straightedge and compass.

11. **PHOTOGRAPHY** A rectangular photograph that measures 3 in. by 4 in. is enlarged so that the 4-in. side measures 6 in. How long are the shorter sides of the enlargement?

12. _ABCD_ and _STUV_ are similar rectangles. If _AB_ = 3 cm, _BC_ = 8 cm, and _ST_ = 6.6 cm, what is the perimeter of _STUV_?

13. **ARCHITECTURE** On a blueprint, a diagonal brace forms similar triangles _RST_ and _WXY_. If _ST_ = 9 ft and _XY_ = 12 ft, what is the ratio of the perimeter of _RST_ to the perimeter of _WXY_?

14. _PQR_ and _STU_ are similar triangles. $\angle Q$ and $\angle T$ are right angles. If _PQ_ = 5 cm, _QR_ = 7 cm, and _ST_ = 17.5 cm, what is the area of $\triangle STU$?

15. **FRAMING** Odetta has 34 in. of beautiful oak molding she would like to use for a picture frame. The photo she wants to frame measures 8 in. by 10 in. Find the dimensions of a reduced photo that would have the same shape as the original and would have a perimeter of exactly 34 in.

■ EXTENDED PRACTICE EXERCISES

16. Draw two hexagons that have the same angle measures but are not similar.

17. Draw two quadrilaterals that have proportional corresponding sides but are not similar.

18. Point _D_ is said to divide $\overline{AB}$ externally. Two segments, $\overline{AD}$ and $\overline{BD}$, are formed. If _AD_ = 6 and $\frac{AD}{AB} = \frac{3}{2}$, find _AB_.

A   B   D

■ MIXED REVIEW EXERCISES

Find the _y_-intercept of the graph of each equation. (Lesson 6-1)

19. $y = 4x - 3$ 20. $y = \frac{2}{3}x + 2$ 21. $2x - 8y = 12$

22. $4y - 2x = 8$ 23. $4x - 3y = -2$ 24. $7 - 2y = 5x$

25. $12x = 24y + 48$ 26. $7x - 4 = 3y$ 27. $15 - 3y = -2x$

Review and Practice Your Skills

PRACTICE ◣ LESSON 7-1

Is each pair of ratios equivalent? Write *yes* or *no*.

1. $4 : 7, 12 : 14$

2. $3.5 : 12, 14 : 48$

3. $-3 : -2, 15 : 10$

4. $\dfrac{4.5}{13.5}, \dfrac{3}{1}$

5. $\dfrac{8}{20}, \dfrac{12}{30}$

6. $\dfrac{3}{7} : \dfrac{4}{7}, 1.8 : 2.4$

Solve each proportion.

7. $\dfrac{4}{7} = \dfrac{x}{42}$

8. $\dfrac{3}{-2} = \dfrac{10}{d}$

9. $\dfrac{8}{m} = \dfrac{20}{12.5}$

10. $\dfrac{p}{0.2} = \dfrac{0.5}{0.1}$

11. $\dfrac{-9}{400} = \dfrac{-2.25}{f}$

12. $\dfrac{1024}{x} = \dfrac{16}{3}$

13. $\dfrac{117}{36} = \dfrac{585}{g}$

14. $\dfrac{e}{47} = \dfrac{616}{5264}$

15. $\dfrac{2}{300} = \dfrac{n}{200}$

16. $\dfrac{-13}{-39} = \dfrac{x}{-663}$

17. $\dfrac{0.072}{z} = \dfrac{1.2}{0.6}$

18. $\dfrac{p}{136{,}974} = \dfrac{10}{-1110}$

19. The ratio of flour to sugar in a recipe is 5:2. How much flour must be added to $\frac{1}{2}$ c sugar?

20. The ratio of blue to red in a paint mixture is $8 : 15$. How many pints of red paint must be added to 72 pt of blue paint?

21. An angle and its complement have measures in the ratio $9 : 11$. What are the measures of the two angles?

22. The ratio of boys to girls in a class is $4 : 5$. There are 27 students in the class. How many are boys?

PRACTICE ◣ LESSON 7-2

Determine if the polygons are similar. Write *yes* or *no*.

23.

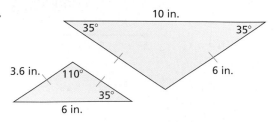

24.

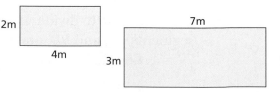

25.

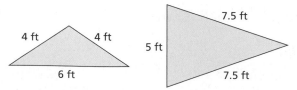

26.

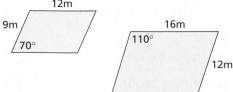

Find x in each pair of similar polygons.

27.

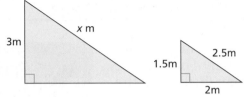

28.

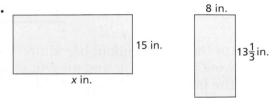

29.

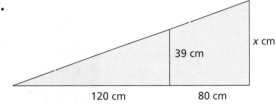

30.

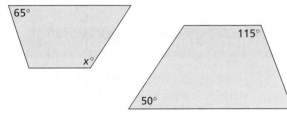

MathWorks Career – Police Photography
Workplace Knowhow

Police photographers record the details of a crime scene by taking photographs. Later, the photos are used by forensic scientists to search for clues. The solving of a crime may depend on a piece of information found in a crime photograph.

Police photographers must be prepared to work indoors or outdoors and in all types of weather and lighting conditions. They must choose the best camera and film for the situation. Sometimes, these workers use special filters to enhance details in their subject. Police photographers must be meticulous to make sure that every angle of a crime scene has been covered. Photos taken from odd angles will distort the perspective and slow down the investigation.

You have use of a surveillance video tape to produce several still photographs of a robbery suspect. The picture is taken at eye level and the suspect is standing in front of a counter with a known height.

1. The counter, which measures 3.5 ft in real life, measures 2.5 in. in the photograph. What is the scale of the photograph?

2. If the suspect's height is 4.2 in. in the photograph, how tall is the suspect in real life?

In another picture, there are 4 objects on the floor between a desk and a wall safe that was robbed. The safe is exactly $6\frac{1}{2}$ ft from the desk. The distance in the photograph is $3\frac{1}{4}$ in.

3. What is the scale of the photograph?

4. A glove in the photo is 1.8 in. from the wall. What is its actual distance from the wall?

7-3 Scale Drawings

Goals ■ Find actual or scale length using scale drawings.

Applications Architecture, Engineering, Photography

Suppose a new student has joined your class. Sketch a map of the school building. Try to represent distances accurately. For example, if the principal's office is further from the gym than the library, the distance from the gym on the map should be greater also.

◤ BUILD UNDERSTANDING

A **scale drawing** is a representation of a real object. All lengths on the drawing are proportional to actual lengths of the object. The **scale** of the drawing is the ratio of the size of the drawing to the actual size.

Work with a group of 2 or 3 students.

1. Think of two well-known locations within your community. For instance, you might choose your school and the public library.

2. Sketch a map showing how to get from one location to the other. Try to represent distances accurately on the map.

Example 1

ARCHITECTURE This is a scale drawing of a room in a house.

Use a ruler to find the actual distance along the wall between the window and the door.

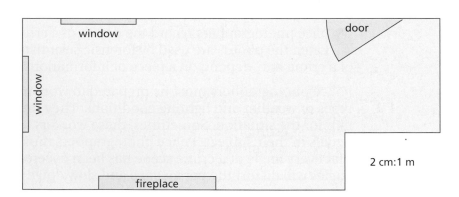

Solution

The ratio of the scale drawing to the actual size of the room is 2 cm:1 m. The first step is to measure the drawing. Because the scale is given in terms of centimeters, measure the distance to the nearest centimeter.

In the drawing, the distance between the window and the door is 5 cm. Write and solve a proportion.

Let x = the actual distance in meters.

$$\frac{2 \text{ cm}}{1 \text{ m}} = \frac{5 \text{ cm}}{x} \quad \substack{\text{scale distance} \\ \text{actual distance}} \qquad 2x = 5 \rightarrow x = 2.5$$

The actual distance between the window and the door is 2.5 m.

> ### Check Understanding
>
> How would the scale drawing of this room change if the scale were changed to 1 cm:1 m?

Scale drawings are used in engineering. The scale can be stated as a ratio without any reference to a particular unit of measure. For example, a model car often relates to the actual car by a scale of 1:24.

Example 2

ENGINEERING The distance between the front wheels of a model car is 4.5 centimeters. What is the actual distance on the car if the scale is 1:24?

Solution

Write and solve a proportion. Let $d =$ the actual distance in centimeters.

$$\frac{1}{24} = \frac{4.5}{d} \quad \begin{array}{l} \text{scale distance} \\ \text{actual distance} \end{array} \qquad d = 108$$

The actual distance between the front wheels is 108 cm or 1.08 m.

Satellite photographs are sometimes used to map terrain and roadways. The scale of a map can be determined by comparing distances on the map to known distances. Then the scale is given as a bar length.

Example 3

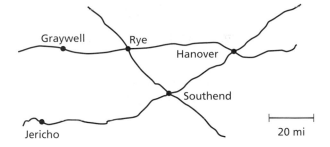

SATELLITE PHOTOGRAPHY The map to the right was made from a satellite photo. Using the scale bar, estimate the driving distance from Jericho to Hanover on the map.

Solution

Cut a piece of string that is as long as the route between the two points on the map. Then compare the string's length with the scale length. The string's length is equal to nearly 6 scale lengths.

Multiply the number of scale lengths it takes to cover the distance by the actual distance given for the scale length.

$$6 \times 20 = 120 \text{ mi}$$

Therefore, the actual distance between the two cities is about 120 mi.

Geometric iterations can produce a figure called a fractal. Fractal shapes also appear in nature. Mathematicians have discovered that coastlines are better described as fractals than as smooth curves.

Example 4

Kent and Baywater are two villages along the coast. From the map below, we can see that the distance along the coast highway between the towns is about 2 miles. Is the bikepath between Kent and Baywater also 2 miles long?

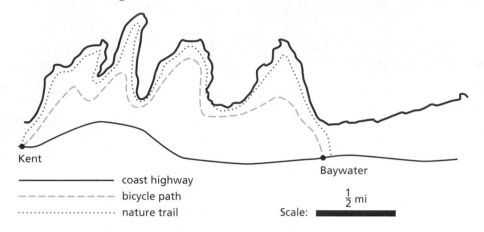

Kent

Baywater

————————— coast highway
- - - - - - - - - - bicycle path
⋯⋯⋯⋯⋯⋯⋯ nature trail

Scale: $\frac{1}{2}$ mi

Solution

No. The bike path is closer to the actual coast than the highway. Using a piece of string, we can estimate that the distance along the bike path is approximately 5 mi. ∎

TRY THESE EXERCISES

Find the actual length of each of the following.

1. scale length is 2 in., scale is $\frac{1}{2}$ in.:3 ft

2. scale length is 4 cm, scale is 1:200

Find the scale length for each of the following.

3. actual length is 5 m
   scale is 1 cm:4 m

4. actual distance is 175 mi
   scale is $\frac{1}{4}$ in.:25 mi

5. Use the map in Example 3. Estimate the actual distance between Rye and Hanover.

6. **PHOTOGRAPHY** A photo that measures 3.5 in. by 6 in. will be enlarged so that its width will be 8 in. Will the length of the enlargement be less than or greater than 15 in.?

7. **MODEL BUILDING** A model plane has the scale of 1:500. The wingspan on the model is 6.4 cm. How many meters is the wingspan of the plane itself?

8. **CONSTRUCTION** The blueprint for a garage indicates that a wooden beam measures 4.1 cm. The scale of the plan is 1:300. What is the actual length of the beam in meters?

Find the actual length of each of the following.

9. scale length is 3 cm, scale is 2 cm:5 m   10. scale distance is 2.1 cm, scale is 1:300

Find the scale length for each of the following.

11. actual length is 10 m, scale is 1:20   12. actual distance is 200 km, scale is 1.5 cm:25 km

Find the actual distances using the map at the right.

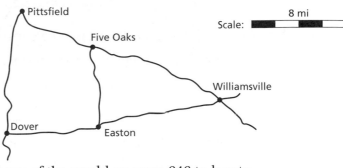

13. Easton to Williamsville

14. Pittsfield to Five Oaks

15. Dover to Williamsville

16. Five Oaks to Williamsville

 17. **DATA FILE** Use the data on principal rivers of the world on page 646 to locate information about the lengths of the St. Lawrence and Columbia rivers. On a map using a scale of 200 mi $= \frac{1}{4}$ in., what would be the lengths of the rivers on the map?

18. **WRITING MATH** Use the map in Example 4. Using a piece of string, estimate the distance between Kent and Baywater using the nature trail. What do you think would happen to the length of the trail between Kent and Baywater if you got even closer to the water and used an inch as the measuring unit?

 19. **CHAPTER INVESTIGATION** Select at least five prominent features in the photograph and plot their locations on your enlargement. You may want to draw a coordinate grid over the surface of the photograph and draw a corresponding grid on your enlargement. Once the main features of the photograph are placed correctly on the enlargement, sketch in the remaining details from the photograph.

◥ EXTENDED PRACTICE EXERCISES

20. Lisa wants to make a map of the school that will fit on a sheet of paper that measures $8\frac{1}{2}$ in. by 11 in. The longest length of the school is 600 ft and its longest width is 350 ft. What would be a good scale to use so that the map is as large as possible, but will fit on the paper?

21. A **hectare** (abbreviated ha) is a metric unit of land area equal to 10,000 m^2. On a map, a rectangular plot of land measures 5 cm by 12 cm. The scale of the map is 1:5000. How many hectares does the plot include?

◥ MIXED REVIEW EXERCISES

Find the slope of a line parallel to the given line and the slope of a line perpendicular to the given line. (Lesson 6-2)

22. the line containing points $A(2, 3)$ and $B(-1, -4)$

23. the line containing points $C(1, 3)$ and $D(-3, 8)$

24. the line containing points $G(1, 0)$ and $H(-2, 5)$

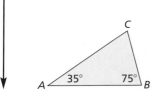

7-4 Postulates for Similar Triangles

Goals ■ Use the AA, SSS, and SAS similarity postulates to determine if two triangles are similar.

Applications Art, Surveying, Photography

Work with a partner. You will need a compass and a straightedge.

1. Using the straightedge, draw any triangle and label it *ABC*.

2. Draw a line segment on another sheet of paper that is longer than *AC*. Your partner should draw a line segment longer than *BC*.

3. Copy ∠*A* at one end of your line segment and ∠*C* at the other end. Have your partner copy ∠*B* and ∠*C* at the ends of his or her line segment.

4. Both you and your partner should now extend the outer rays of the angles you have drawn to form triangles.

5. Compare both triangles with the original triangle. What seems to be true?

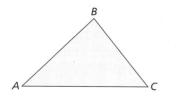

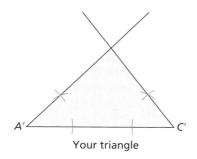

Your triangle

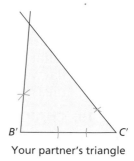

Your partner's triangle

▌ BUILD UNDERSTANDING

In Chapter 3, you learned that some statements in geometry are considered to be true without proof. Usually, these statements are based on direct observation of principles that always work.

Postulate 15 (The AA Similarity Postulate)	If two angles of a triangle are congruent to two angles of another triangle, the two triangles are similar.

Example 1

Is △*ABC* similar to △*DEF*?

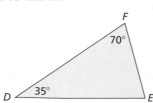

Solution

Find one of the missing angle measures in either triangle. To find $m\angle C$, subtract the sum of $m\angle A$ and $m\angle B$ from 180°.

$$180° - (35° + 75°) = 70°$$

Because $\angle A \cong \angle D$ and $\angle C \cong \angle F$, the two triangles are similar by the AA Similarity Postulate. $\triangle ABC \sim \triangle DEF$

There are other ways to determine whether or not two triangles are similar.

Postulate 16 (The SSS Similarity Postulate)	If the corresponding sides of two triangles are proportional, then the two triangles are similar.

Example 2

ART A wire sculpture is formed from triangles of the two sizes shown below. Is $\triangle PQR$ similar to $\triangle STU$?

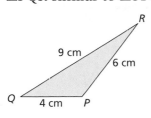 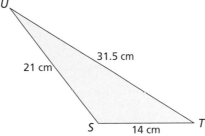

Solution

Find the ratio of each pair of corresponding sides.

$$\frac{PQ}{ST} = \frac{4}{14} \qquad \frac{QR}{TU} = \frac{9}{31.5} \qquad \frac{PR}{SU} = \frac{6}{21}$$

$$= \frac{2}{7} \qquad\qquad = \frac{2}{7} \qquad\qquad = \frac{2}{7}$$

Technology Note

In Example 2, you could use a calculator to find each ratio is equal to 0.285714.

Because all three pairs of corresponding sides are proportional, the triangles are similar.

Another way of proving that two triangles are similar involves two pairs of corresponding sides and the angle between those sides.

Postulate 17 (The SAS Similarity Postulate)	If an angle of one triangle is congruent to an angle in another triangle, and the two sides that include that angle are proportional to the corresponding sides in the other triangle, then the two triangles are similar.

Example 3

If $PS = 3ST$ and $XS = 3SY$, is $\triangle PSX$ similar to $\triangle TSY$?

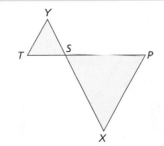

Check Understanding

If $m\angle P = 60$, which angle in $\triangle TSY$ has a measure of 60°?

Solution

Use the given information to show that two pairs of corresponding sides are proportional.

$$\frac{PS}{ST} = \frac{3ST}{ST} = \frac{3}{1} \qquad \frac{XS}{SY} = \frac{3SY}{SY} = \frac{3}{1}$$

$\angle PSX \cong \angle TSY$ because they are vertical angles. Therefore, $\triangle PSX \sim \triangle TSY$ by the SAS Similarity Postulate.

◥ TRY THESE EXERCISES

Determine whether each pair of triangles is similar. If the triangles are similar, give a reason: write AA, SSS, or SAS.

1.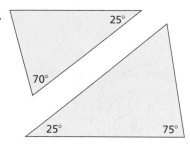

2.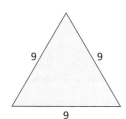

3.

4.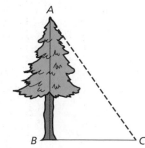

◥ PRACTICE EXERCISES • For Extra Practice, see page 685.

Determine whether each pair of triangles is similar. If the triangles are similar, give a reason: write AA, SSS, or SAS.

5.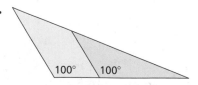

6.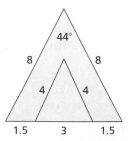

7. SURVEYING A surveyor measures the shadows cast by a tree and a pole at 4 P.M. and makes the drawing at the right. Explain why $\triangle ABC \sim \triangle DEF$.

8. PHOTOGRAPHY For an exhibit, Bruce crops a photo in the shape of $\triangle PQR$. He wants to create a montage of smaller photos in the shape of similar triangles. To find a similar triangle, he marks S, the midpoint of PQ, and T, the midpoint of PR. Then he connects the midpoints. Is $\triangle PQR \sim \triangle PST$? Explain.

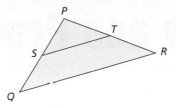

9. Suppose $\angle 1 \cong \angle 2$. Can you prove that $\triangle ABD \sim \triangle CED$? Explain.

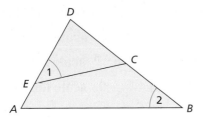

10. WRITING MATH Write a paragraph to prove that $\triangle DEF \sim \triangle FHG$?

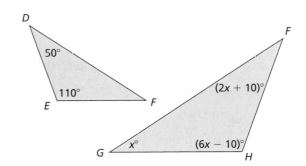

EXTENDED PRACTICE EXERCISES

Tell whether each statement is always true, sometimes true, or never true. Write *always*, *sometimes*, or *never*.

11. Two equilateral triangles are similar.

12. Two isosceles triangles are similar.

13. Two isosceles triangles that each have a 45° angle are similar.

14. An acute triangle and a right triangle are similar.

MIXED REVIEW EXERCISES

Write an equation of the line with the given information. (Lesson 6-3)

15. $m = \frac{1}{2}, b = -1$

16. $P(-3, 1), Q(4, -2)$

17. $m = 1, b = \frac{1}{2}$

18. $m = -3, b = -2$

19. $A(4, -5), B(-3, 4)$

20. $m = \frac{3}{4}, b = -3$

21. $m = 2, b = 4$

22. $R(1,4), S(-2, -3)$

23. $m = -2, b = -2$

Trapezoids and their medians are shown. Find the value of x. (Lesson 4-9)

24.

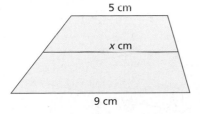

25.

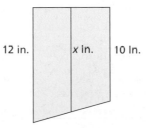

26.

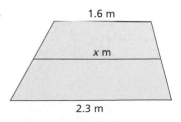

Review and Practice Your Skills

Find the actual length of each of the following.

1. scale length is 40 cm

   scale is 2.5 cm: 15 km

2. scale length is $4\frac{3}{4}$ in.

   scale is $\frac{1}{4}$ in.:3 mi

3. scale length is 14 ft

   scale is 700:1

4. scale length is 12.5 cm

   scale is 1:30

5. scale length is $\frac{7}{16}$ in.

   scale is $\frac{1}{8}$ in.:15 mi

6. scale length is 23 cm

   scale is 5 cm:28 m

7. scale length is 37 yd

   scale is 10 yd:$\frac{3}{2}$ yd

8. scale length is 1440 mm

   scale is 1 mm:0.001 m

9. scale length is 6.5 in.

   scale is 1.5 in.:237 mi

Find the scale length for each of the following.

10. actual length is 52 mi

    scale is 0.5 in.:4 mi

11. actual length is 75 yd

    scale is 3 in.:18 yd

12. actual length is 2450 mi

    scale is $\frac{3}{4}$ in.:5 mi

13. actual length is 256 km
    scale is 5 cm:32 km

14. actual length is 17,500 m
    scale is 2 cm:875 m

15. actual length is 0.003 mm
    scale is 5000:1

16. actual length is 817 mi
    scale is 4 in:19 mi

17. actual length is 26 ft
    scale is 1:6.5

18. actual length is 7500 mi
    scale is 10 in.:1.5 mi

19. The blueprints for a new house have a scale of $\frac{1}{2}$ in.:1.5 ft. The dimensions of one bedroom on the drawing are 6 in. by 4 in. What are the actual dimensions of the bedroom?

Determine whether each pair of triangles is similar. If the triangles are similar, give a reason: write AA, SSS, or SAS.

20.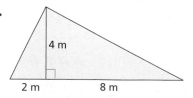

21.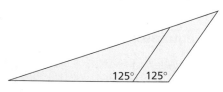

22.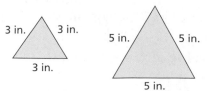

23.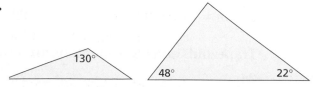

24.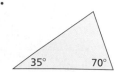

25.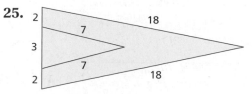

Solve each proportion. (Lesson 7-1)

26. $\dfrac{x}{4} = \dfrac{55}{22}$

27. $\dfrac{156}{a} = \dfrac{-8}{3}$

28. $-\dfrac{31}{112} = \dfrac{y}{-7}$

29. $\dfrac{0.6}{5} = \dfrac{0.72}{m}$

30. $\dfrac{x+1}{36} = \dfrac{1}{3}$

31. $\dfrac{x-1}{8} = \dfrac{65}{104}$

32. A photograph that measures 4 in. by 6 in. is enlarged so that the 4-in. side measures 15 in. How long does the 6 in. side become in the enlargement? (Lesson 7-2)

33. *ABCD* and *JKLM* are similar rectangles. If *BC* = 22 cm, *CD* = 42 cm, and *KL* = 165 cm, what is the perimeter of *JKLM*? (Lesson 7-2)

Find the actual distance using the map. (Lesson 7-3)

34. Clarktown to Pinckney

35. Pinckney to Grove City

36. Clarktown to Dwyer

37. Dwyer to Gurville

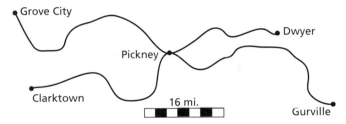

Mid-Chapter Quiz

Solve. (Lesson 7-1)

1. $\dfrac{0.45}{x} = \dfrac{0.9}{0.2}$

2. $\dfrac{42}{112} = \dfrac{x}{24}$

3. Tatiana and Simon bought art supplies. Tatiana spent $3.00 for every $2.00 Simon spent. If they spent $76.25 in all, how much did each one spend?

4. Triangles *ABC* and *DEF* are similar. If $m\angle B = 55°$ and $m\angle C = 98°$, what is $m\angle D$? (Lesson 7-2)

5. Rectangles *PQRS* and *ABCD* are similar. Find *QR* if *AB* = 72 cm, *BC* = 30 cm, and *RS* = 6 cm. (Lesson 7-2)

6. A map shows a distance of 7.3 cm. The map scale is 1:400. What is the actual distance in meters? (Lesson 7-3)

7. Find the scale length when the actual length is 3.75 km and the scale is 2 cm:15 km. (Lesson 7-3)

If you can determine from the given information that the triangles are similar, write *yes*, and give a reason. Otherwise, write *no*. (Lesson 7-4)

8. $\triangle ABC$ has an altitude *BD* such that point *D* is between points *A* and *C*. If *AD* = 3, *BD* = 4, and *DC* = $5\frac{1}{3}$, is $\triangle ADB \sim \triangle BDC$?

9. For $\triangle LMN$, $m\angle M = 95°$, *LM* = 12, *MN* = 8, and *LN* = 15. For $\triangle PQR$, $m\angle Q = 95°$, *PQ* = 36, *QR* = 24, and *PR* = 45. Are the triangles similar?

7-5 Triangles and Proportional Segments

Goals
- Prove theorems involving similar triangles.
- Find unknown lengths of sides of triangles.

Applications Scale models, Photography

You will need a compass and straightedge.

Draw any triangle XYZ. Construct the midpoint of $\overline{XY}$ and label it P. Construct the midpoint of $\overline{XZ}$ and label it Q. Then, use the straightedge to draw PQ.

How does $\triangle XYZ$ compare to $\triangle XPQ$? How does the measure of $\overline{YZ}$ compare with the measure of $\overline{PQ}$?

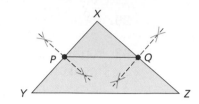

◤ BUILD UNDERSTANDING

Recall that a theorem is a statement that can be proven true. In Example 1, a proof is given for the following theorem.

Theorem	If a segment connects the midpoints of two sides of a triangle, then the length of the segment is equal to one-half the length of the third side.

Example 1

Given P is the midpoint of $\overline{XY}$.

           Q is the midpoint of $\overline{XZ}$.

Prove $PQ = \frac{1}{2}YZ$

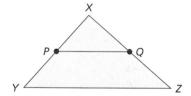

Solution

Statements	Reasons
1. P is the midpoint of XY. Q is the midpoint of XZ.	1. given
2. $XP = \frac{1}{2}XY$; $XQ = \frac{1}{2}XZ$	2. definition of midpoint
3. $\frac{XP}{XY} = \frac{1}{2}$; $\frac{XQ}{XZ} = \frac{1}{2}$	3. division property of equality
4. $m\angle X = m\angle X$	4. reflexive property of equality
5. $\angle X \cong \angle X$	5. definition of congruent angles
6. $\triangle XYZ \cong \triangle XPQ$	6. SAS similarity postulate
7. $\frac{PQ}{YZ} = \frac{1}{2}$	7. corresponding parts of similar triangles are proportional
8. $PQ = \frac{1}{2}YZ$	8. multiplication property of equality

---

Technology Note

Explore this theorem using geometry software. Try this activity.

1. Draw any triangle and label the vertices A, B, and C.

2. Construct midpoints of segments AB and BC and label them D and E, respectively. Connect the midpoints.

3. Measure $\overline{AC}$ and $\overline{DE}$.

4. Change the measure of the segments and angles by selecting and moving a vertex of $\triangle ABC$. What do you notice about the measures of $\overline{AC}$ and $\overline{DE}$?

There is also a theorem about altitudes of similar triangles.

> **Theorem**   If two triangles are similar, their altitudes are in the same proportion as the sides of the triangles.

Example 2

SCALE MODELS Jan is building a scale model of a tower. Steel bracing forms large and small similar triangles throughout the structure. Jan believes that she will not need to measure the altitudes of all the triangles in her model since the altitudes should be in the same proportion as the sides of the triangles. To be certain, she draws $\triangle ABC$ and $\triangle DEF$ to prove the theorem stated above.

Given $\triangle ABC \sim \triangle DEF$

$\overline{AX} \perp \overline{BC}, \overline{DY} \perp \overline{EF}$

Prove $\dfrac{AX}{DY} = \dfrac{AB}{DE}$

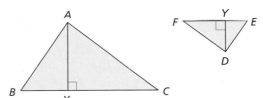

Solution

Statements	Reasons
1. $\triangle ABC \sim \triangle DEF$, $\overline{AX} \perp \overline{BC}$, $\overline{DY} \perp \overline{EF}$	1. given
2. $\angle AXB$ and $\angle DYE$ are right angles	2. definition of perpendicular lines
3. $m\angle AXB = 90°$, $m\angle DYE = 90°$	3. definition of right angles
4. $m\angle AXB = m\angle DYE$	4. substitution
5. $\angle AXB \cong \angle DYE$	5. definition of congruent angles
6. $\angle B \cong \angle E$	6. corresponding angles of similar triangles are congruent
7. $\triangle ABX \sim \triangle DEY$	7. AA similarity postulate
8. $\dfrac{AX}{DY} = \dfrac{AB}{DE}$	8. corresponding parts of similar triangles are in proportion

There is a similar theorem about the medians of similar triangles.

> **Theorem**   If two triangles are similar, their medians are in the same proportion as the sides of the triangles.

In $\triangle ABC$ and $\triangle PQR$, median $\overline{AD}$ and median $\overline{PS}$ have the same ratio as any corresponding sides of the triangles.

$$\frac{AD}{PS} = \frac{AC}{PR}$$

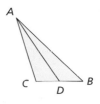
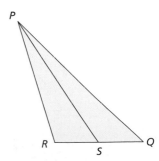

An altitude drawn to the hypotenuse of any right triangle always forms two similar triangles. The following theorem is used in Example 3 below to find the length of a missing segment.

Theorem	If the altitude to the hypotenuse of a right triangle is drawn, the altitude separates the original triangle into two triangles that are similar to the original triangle and to each other.

In right triangle ABC, $\overline{AD}$ is the altitude to the hypotenuse. Each of these pairs of triangles is similar.

$\triangle DAC \sim \triangle ABC$

$\triangle DBA \sim \triangle ABC$

$\triangle DAC \sim \triangle DBA$

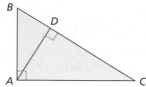

Example 3

Find x in right triangle RST if $\overline{SW}$ is the altitude to the hypotenuse.

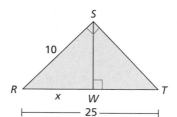

Solution

Identify which two of the three triangles include x. You may find it helpful to redraw the two triangles separately.

Because $\triangle RWS \sim \triangle RST$,

$$\frac{RW}{RS} = \frac{RS}{RT}$$

$$\frac{x}{10} = \frac{10}{25}$$

$$x = 4$$

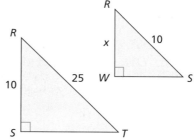

So, the value of x in this triangle is 4.

◤ TRY THESE EXERCISES

Find x in each pair of similar triangles to the nearest tenth.

1.

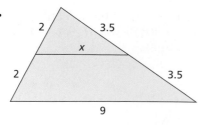

2.

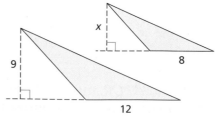

3.

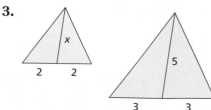

4.

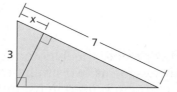

5. Copy and complete this proof.

Given $\triangle EFG \sim \triangle JKL$, $\overline{EH} \mid \overline{GF}$,

$\overline{JM} \perp \overline{LK}$

Prove $\dfrac{EH}{JM} = \dfrac{FG}{KL}$

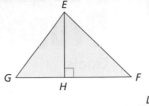

Statements	Reasons
1. $\triangle EFG \sim \triangle JKL$; $\overline{EH} \perp \overline{GF}$; $\overline{JM} \perp \overline{LK}$	1. __?__
2. $\overline{EH}$ is an altitude of $\triangle EFG$. $\overline{JM}$ is an altitude of $\triangle JKL$	2. __?__
3. $\dfrac{EH}{JM} = \dfrac{FG}{KL}$	3. __?__

Find x in each pair of similar triangles to the nearest tenth.

6.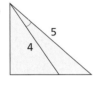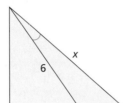

7.

8.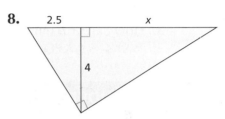

9. Prove the theorem about medians in similar triangles.
Given $\triangle WRY \sim \triangle KAL$, $\overline{WX}$ and $\overline{KB}$ are medians. **Prove** $\dfrac{WX}{KB} = \dfrac{WR}{KA}$

10. **WRITING MATH** If two rectangles are similar, do you think their diagonals are proportional to corresponding sides? Explain your thinking.

■ **EXTENDED PRACTICE EXERCISES**

11. $\triangle ABC$ has a base of x and a height of y. $\triangle DEF$ is similar to $\triangle ABC$ and $AB:DE = 2:7$. What is the area of $\triangle DEF$ in terms of x and y?

12. **PHOTOGRAPHY** A photographer wants to create a series of similar rectangular prisms to display her work in an exhibit. How could the artist determine that two rectangular prisms are similar?

13. Two square pyramids are similar. One has a base with side lengths of 3 cm and a height of 10 cm. The lengths of the other's base is 10 cm. Find its height.

■ **MIXED REVIEW EXERCISES**

Solve each system of equations by graphing. (Lesson 6-4)

14. $y = 3x - 2$
$y - x + 5$

15. $y = 2x + 2$
$y = -x - 3$

16. $2x - y = 4$
$3x + y = -1$

Solve each equation. (Lesson 2-5)

17. $4(x - 3) + 8 = 3x + 2$

18. $2(x - 5) - 4 = 3(x - 1)$

19. $x - 3(x + 2) = 5(x - 2) + 1$

20. $-2(x - 4) = 5(x - 2) + 8$

7-6 Parallel Lines and Proportional Segments

Goals
- Use theorems involving parallel lines and proportional segments to find unknown lengths.
- Divide a line segment into congruent parts.

Applications Model Building, Architecture, Real Estate

GEOMETRY SOFTWARE Use geometry software to construct similar triangles.

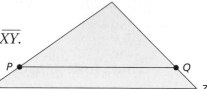

1. Draw any triangle XYZ. Mark any point P between X and Y on $\overline{XY}$.

2. Construct a line through P that is parallel to $\overline{YZ}$. Label the point where the parallel line meets $\overline{XZ}$ as Q.

3. How can you use the software features to prove $\triangle XYZ \sim \triangle XPQ$?

◤ BUILD UNDERSTANDING

When a line segment intersects two sides of a triangle and is parallel to the third side, two similar triangles are formed.

In $\triangle ABC$, $\overline{DE} \parallel \overline{AC}$. Notice that $\angle ABC \cong \angle DBE$ by the reflexive property, and $\angle BDE \cong \angle BAC$ because they are both corresponding angles formed by parallel lines and a transversal. Therefore, $\triangle ABC \sim \triangle DBE$. Because the triangles are similar, $BD : BA = BE : BC$.

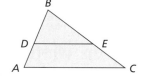

The same reasoning applies to a segment parallel to any other side of the triangle, so you can state the following theorem.

Theorem	If a line is parallel to one side of a triangle and intersects the other sides at any points except the vertex, then the line divides the sides proportionally.

So, by this theorem, $BD : DA = BE : EC$.

Example 1

Given $\triangle STR$, $\overline{WP} \parallel \overline{ST}$. Find the measure of x.

Solution

According to the theorem, because $\overline{WP} \parallel \overline{ST}$, P divides $\overline{TR}$ proportionally and W divides $\overline{SR}$ proportionally.

$$\frac{TP}{PR} = \frac{SW}{WR}$$

$$\frac{3}{x} = \frac{2}{7}$$

$$10.5 = x$$

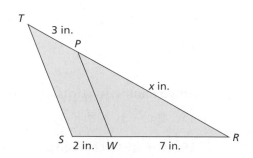

So, the measure of x is 10.5 in.

Recall that when a segment joins the midpoints of two sides of a triangle, that segment measures one-half the length of the third side of the triangle. Example 2 is a proof that such a segment is also parallel to the third side.

Example 2

MODEL BUILDING Len is building a miniature electrical tower for the filming of a movie. Working from a photograph, Len draws a diagram of the triangular tower. A catwalk, represented by $\overline{PQ}$ on the diagram shown below, seems to be parallel to the base of the tower.

Len measures to determine that P is the midpoint of $\overline{XY}$ and Q is the midpoint of $\overline{XZ}$. How can he prove that $\overline{PQ}$ is parallel to $\overline{YZ}$?

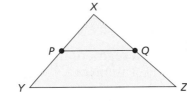

Solution

As in the proof in Example 1 of Lesson 7–5, $\triangle XYZ \sim \triangle XPQ$ by the SAS similarity postulate. Therefore, $\angle XPQ \cong \angle XYZ$, because they are corresponding angles of similar triangles. This fact also leads to the conclusion that $\overline{PQ} \parallel \overline{YZ}$, because corresponding angles formed by a transversal (XY) are congruent.

A similar theorem is true about medians in trapezoids. The **median** of a trapezoid is a segment that joins the midpoints of the legs.

Theorem	The median of a trapezoid is parallel to its bases, and its length is half the sum of the lengths of the bases.

Example 3

Given Trapezoid $WXYZ$, median $\overline{AB}$. Find AB.

Solution

The length of the median is half the sum of the lengths of the bases.

$$AB = \frac{WX + ZY}{2}$$

$$AB = \frac{5 + 6.6}{2}$$

$$AB = \frac{11.6}{2} = 5.8$$

The length of the median is 5.8 cm.

A compass and straightedge can be used to divide a given segment into congruent parts using parallel segments.

Example 4

Divide $\overline{AB}$ into three congruent parts.

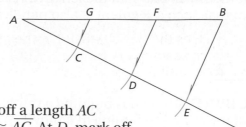

Solution

Step 1: Draw a ray with A as an endpoint.

Step 2: On the ray, use a compass to mark off a length AC that is shorter than $\overline{AB}$. At C, mark off $\overline{CD} \cong \overline{AC}$. At D, mark off $\overline{DE} \cong \overline{AC}$.

Step 3: Draw $\overline{BE}$.

Step 4: At D, construct $\overline{DF} \parallel \overline{BE}$. At C, construct $\overline{CG} \parallel \overline{BE}$.

Because $\triangle ACG \sim \triangle ADF \sim \triangle AEB$ and $AC = CD = DE$, it follows that $AG = GF = FB$.

This construction can be used to divide a segment into any given number of congruent parts.

◥ TRY THESE EXERCISES

In each figure, $\overline{AB} \parallel \overline{CD}$. Find x to the nearest tenth.

1.

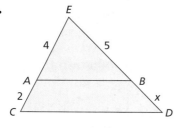

2.

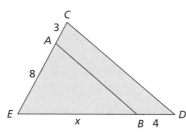

3.

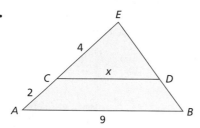

4.

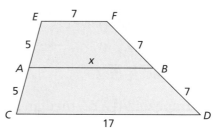

5. Draw any segment XY on a sheet of paper. Divide $\overline{XY}$ into five congruent parts.

◥ PRACTICE EXERCISES • For Extra Practice, see page 686.

In each figure, $\overline{AB} \parallel \overline{CD}$. Find the value of x to the nearest tenth.

6.

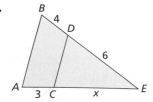

7.

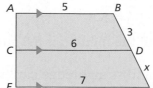

8. **WRITING MATH** Look at the trapezoids in Exercises 4 and 7. What is the relationship of the median of each trapezoid to its bases?

9. **CHAPTER INVESTIGATION** Complete your drawing by adding details and shading. Does the sketch resemble the original photograph? Are there any areas which seem out of proportion? Check measurements and make corrections until the sketch is an accurate enlargement.

10. **REAL ESTATE** The map shows a triangular lot bought by a real-estate developer. Copy the map and construct proportional subdivisions along River Alley.

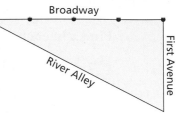

In each figure, $\overline{AB} \parallel \overline{CD}$. Find the value of x to the nearest tenth.

11.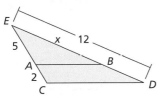

12.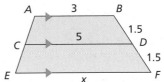

13. The angle bisector theorem for triangles states that any angle bisector in a triangle divides the opposite side into segments that are proportional to the other two sides of the triangle. Write a proportion based on this theorem for $\triangle XYZ$ if $\overline{XW}$ is an angle bisector.

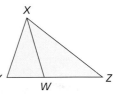

14. **ARCHITECTURE** A blueprint calls for an angle bisector to be added to a triangular structure as shown in the figure above. What is YW, if $XY = 10$ ft, $WZ = 9$ ft, and $XZ = 12$ ft?

◼ EXTENDED PRACTICE EXERCISES

To prove the angle bisector theorem, $\overline{EC}$ is drawn parallel to $\overline{BD}$. Given that $\overline{BD}$ is the bisector of $\angle B$ and $\overline{CE} \parallel \overline{BD}$, use the drawing to answer Exercises 15–18.

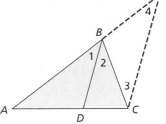

15. For $\triangle ACE$, complete this proportion: $\dfrac{AD}{DC} = \dfrac{AB}{?}$

16. Why is $\angle 3 \cong \angle 4$?

17. Why is $\overline{BC} \cong \overline{BE}$?

18. Why is $\dfrac{AD}{DC} = \dfrac{AB}{BC}$?

◼ MIXED REVIEW EXERCISES

Solve each system of equations by substitution. (Lesson 6-5)

19. $y = 3x - 2$
    $2x + y = 5$

20. $-3y + 2x = 8$
    $x - 2y = -7$

21. $9 = 2y - 4x$
    $3x + y = 4$

22. $x - 5y = -7$
    $3y + 4x = 2$

23. $-2x - 3y = -9$
    $4y + x = 3$

24. $y - 3x = 1$
    $2x - 5y = -18$

Find the mean, median and mode of each set of data. (Lesson 2-7)

25. 9   2   13   10   3   19   5   15   8
    2   20   11   4   3   10   9   22   23
    10   2

26. 26   23   28   22   25   24   20   25
    23   29   29   21   20   29   28

Review and Practice Your Skills

PRACTICE ▪ LESSON 7-5

Find x in each pair of similar triangles to the nearest tenth.

1.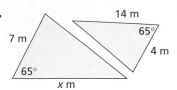
14 m
65°
7 m
4 m
65°
x m

2.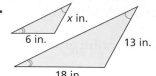
x in.
6 in.
13 in.
18 in.

3.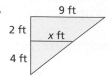
9 ft
2 ft
x ft
4 ft

4.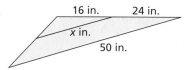
4 cm
6 cm
6 cm
x cm

5.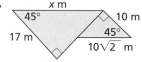
x m
45°
10 m
17 m
45°
10√2 m

6.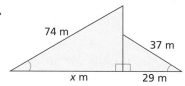
2.0 yd
121°
2.1 yd
121°
1.5 yd
2.7 yd

7.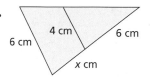
16 in. 24 in.
x in.
50 in.

8.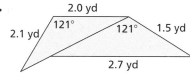
74 m
37 m
x m
29 m

9.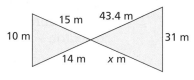
15 m 43.4 m
10 m
31 m
14 m x m

10. Two similar triangles have a 3:8 ratio of corresponding sides. What is the ratio of their areas?

PRACTICE ▪ LESSON 7-6

In each figure, $\overline{AB} \parallel \overline{CD}$. Find the value of x to the nearest tenth.

11.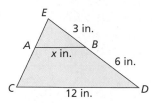
E
3 in.
A B
x in.
6 in.
C
12 in.
D

12.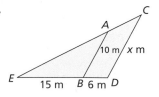
C
A
10 m x m
E
15 m B 6 m D

13.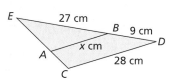
E
27 cm
B 9 cm
x cm
D
A
28 cm
C

14.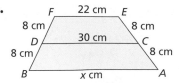
F 22 cm E
8 cm
8 cm
D
30 cm
C
8 cm
8 cm
B
x cm
A

15.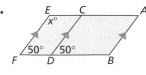
E C
A
x°
50° 50°
F D
B

16.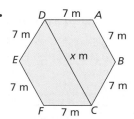
D 7 m A
7 m
7 m
E
x m
B
7 m
7 m
F 7 m C

17. *True* or *false*: The line that joins the midpoints of two sides of a triangle is parallel to the third side.

18. *True* or *false*: The line that joins the midpoints of two sides of a triangle creates a similar triangle whose perimeter is $\frac{1}{4}$ that of the original triangle.

Solve each proportion. (Lesson 7-1)

19. $\dfrac{b}{-9} = \dfrac{-4}{27}$

20. $\dfrac{13}{g} = \dfrac{78}{30}$

21. $\dfrac{1.6}{4.5} = \dfrac{k}{14.4}$

22. $\dfrac{-7}{26} = \dfrac{1.75}{m}$

23. $\dfrac{x-3}{20} = \dfrac{3}{2}$

24. $\dfrac{2x+1}{18} = \dfrac{26}{39}$

Find x in each pair of similar polygons. (Lesson 7-2)

25.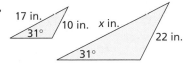

26.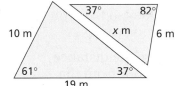

27.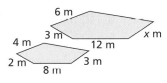

MathWorks
Workplace Knowhow

Career – Photographic Processors

Photographic process workers develop film, make prints or slides and enlarge or retouch photographs. They operate many special types of machines. Specialized workers handle delicate tasks, such as retouching negatives and prints. They restore damaged and faded photographs and may color or shade drawings to enhance images using an airbrush. Some photographic process workers use computers to enhance or alter images digitally. These workers may work for magazines to touch up portraits of models. They can also eliminate images from photographs or combine images from different photographs. To be successful in this field, workers must use ratio and proportion to make sure images look right.

1. A customer wants to crop and enlarge a portion of a 3 in. by 5 in. picture. The portion is 0.75 in. by 0.95 in. Find the ratio of the length to the width of the cropped portion.

Another customer brings in a group of old photographs in non-standard sizes for enlargement. You will convert the length to a standard photographic size. For each photo below, find the width of the enlarged photo when the length is converted as indicated.

Current photo dimensions	*New length*
2. $2\frac{1}{2}$ in. by $3\frac{1}{2}$ in.	5 in.
3. 4 in. by 7 in.	7 in.
4. 7 in. by 9 in.	10 in.
5. 6 in. by 9 in.	10 in.

Problem Solving Skills: Indirect Measurement

Properties you have learned about similar triangles can be used to measure heights and distances indirectly. For example, you can find the height of a tree by measuring its shadow on a sunny day.

Indirect measurement can be used even when the sun is not shining.

Work with a partner to measure the height of your classroom in meters. You will need a mirror and a centimeter ruler.

1. Place the mirror on the floor so that you can see the place where the ceiling meets a classroom wall.

2. Working together, measure the distance in centimeters along the floor from the place you are standing to the mirror (*a*). Measure the distance from the mirror to the wall (*b*). Measure the distance from the floor to your eye level (*c*). Draw a diagram. Record your measurements on the diagram.

3. The triangles formed are similar because of a property of light reflection. Solve the proportion $a:b = c:x$ for *x*, the height of the classroom in centimeters. Change the measure to meters.

Problem Solving Strategies
Guess and check
Find a pattern
Solve a simpler problem
Make a table, chart or list
✔ Use a picture, diagram or model
Act it out
Work backwards
Eliminate possibilities
Use an equation or formula

Problem

A tree casts a shadow 3.3 m long. A meterstick placed perpendicular to the ground at the same time of day casts a shadow that is 0.75 m long. How tall is the tree?

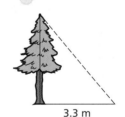

Solve the Problem

A sketch of the problem shows that the tree, the sun's rays, and the shadow form a right triangle similar to the triangle formed by the meterstick and its shadow.

Let *h* represent the height of the tree. Because the triangles are similar, l:*h* = 0.75:3.3. By cross multiplying, you get 0.75*h* = 3.3 and *h* = 4.4. Therefore, the tree is 4.4 m high.

◤ TRY THESE EXERCISES

1. Use the shadow method described above to find the height of a tree, flagpole, or streetlight near your home or school.

2. Use the mirror method described above to find the height of your school, home, or other structure.

3. **SURVEYING** The diagram below shows some measurements that a surveyor was able to take. Describe how she can find the width of the pond on her property by using similar triangles.

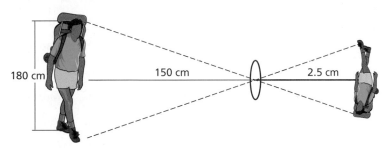

4. **PHOTOGRAPHY** A person is 150 cm from the camera lens. The film is 2.5 cm from the lens. If the person is 180 cm tall, how tall is his image on the film?

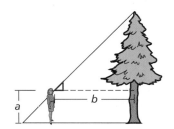

5. Ming took a square index card and folded it exactly in half, to form a 45° angle. She walked back from a tree until she could sight the tree at the very edge of the card she was holding at eye level. Ming stated that her distance from the tree (b) plus the distance from the ground to the card (a) is equal to the height of the tree. Was she correct? Explain.

Five-step Plan

1 Read
2 Plan
3 Solve
4 Answer
5 Check

▚ **MIXED REVIEW EXERCISES**

Solve each system of equations by adding and multiplying. (Lesson 6-6)

6. $y = 4 - 3x$
   $2y - x = -3$

7. $2y - 4x = -5$
   $3x - y = 8$

8. $-4x + 3y = -8$
   $3x - 5y = 2$

9. $3y = 4x + 2$
   $7x = y + 5$

10. $-2y - 4x = 7$
    $-3x = 7 - 2y$

11. $5y = 7 - 3x$
    $2x - 5 = 2y$

Find the probability that a point selected at random in each figure is in the shaded region. Round to the nearest hundredth if necessary. (Lesson 5-3)

12.

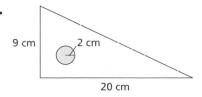

13.

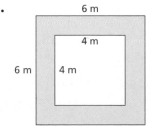

14.

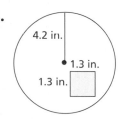

Chapter 7 Review

VOCABULARY ◥

Match the letter of the word in the right column with the description at the left.

1. type of measurement that uses similar triangles

2. two figures that have the same shape, but are *not* necessarily the same size

3. the first and last terms in a proportion

4. the ratio of the size of a drawing to the actual size

5. an equation that states two ratios are equivalent

6. a perpendicular segment from a triangle's vertex to the line containing the opposite side

7. two figures that have the same shape and are the same size

8. the middle two terms in a proportion

9. a segment that joins the midpoints of the legs of a trapezoid

10. the product of the extremes and the product of the means

a. altitude

b. congruent

c. corresponding

d. cross products

e. extremes

f. hypotenuse

g. indirect

h. means

i. median

j. proportion

k. scale

l. similar

LESSON 7-1 ◣ Ratios and Proportions, p. 296

▶ **Equivalent ratios** can be named by the same fraction.

▶ A **proportion** is an equation that states two ratios are equivalent. In a proportion, the product of the extremes equals the product of the means. If $\frac{a}{b} = \frac{c}{d}$, then $ad = bc$.

Is each pair of ratios equivalent? Write *yes* or *no*.

11. 8:12, 10:15

12. $\frac{10}{16}, \frac{25}{45}$

13. $\frac{0.6}{2.4}$, 0.4:0.16

Solve each proportion.

14. $\frac{5}{4} = \frac{y}{12}$

15. $\frac{7}{a} = \frac{2}{3}$

16. $\frac{t}{7} = \frac{5}{8}$

17. A recipe calls for 2 c of sugar to 5 c of flour. How much flour would be added to 5 c of sugar?

LESSON 7-2 ◣ Similar Polygons, p. 300

▶ Two figures are similar if they have the same shape. All corresponding angles are congruent. All corresponding sides are proportional.

Determine if the polygons are similar. Write *yes* or *no*.

18.

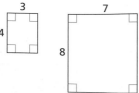

19.

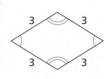
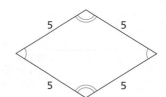

Find the measure of x in each pair of similar polygons.

20.
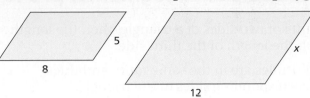

21.
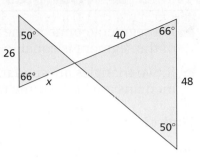

LESSON 7-3 ◣ Scale Drawings, p. 306

▶ A **scale drawing** is a representation of a real object. All lengths on the drawing are proportional to actual lengths in the objects. The **scale** of the drawing is the ratio of the size of the drawing to the actual size of the object.

Find the actual length of each of the following.

22. scale length is 4 cm
scale is 1 cm = 2.5 m

23. scale length is $\frac{1}{4}$ in.
scale is 2 in. = 420 mi

24. scale length is $1\frac{1}{2}$ ft
scale is $\frac{1}{4}$ ft = 15 yd

Find the scale length for each of the following.

25. actual length is 2 ft
scale is $\frac{1}{2}$ in. = 4 ft

26. actual length is 15 yd
scale is 0.5 in. = 3 yd

27. actual length is 350 km
scale is 2 cm = 70 km

LESSON 7-4 ◣ Postulates for Similar Triangles, p. 310

▶ Two triangles are similar if any of these conditions are true:

1. Two pairs of corresponding angles are congruent (AA Similarity Postulate).

2. All pairs of corresponding sides are proportional (SSS Similarity Postulate).

3. Two pairs of corresponding sides are proportional and the angles between those sides are congruent (SAS Similarity Postulate).

Determine whether each pair of triangles is similar. If the triangles are similar give the reason: write AA, SSS, or SAS.

28.

29.
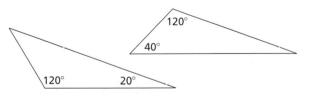

Find x to the nearest tenth for each pair of similar triangles.

30.
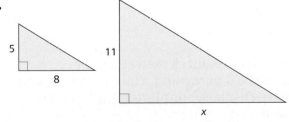

31.
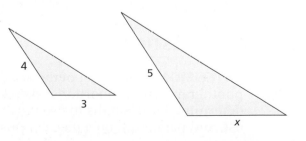

LESSON 7-5 ◣ Triangles and Proportional Segments, p. 316

▶ If a segment connects the midpoints of two sides of a triangle, then the length of the segment is equal to one-half the length of the third side.

▶ If two triangles are similar, their altitudes are in the same ratio, and their medians are in the same ratio as corresponding sides of the triangle.

Find *x* in each pair of similar triangles to the nearest tenth.

32.
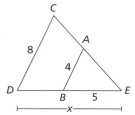

33.
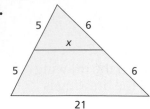
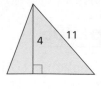

34.
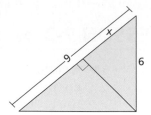

LESSON 7-6 ◣ Parallel Lines and Proportional Segments, p. 320

▶ If a line is parallel to one side of a triangle and intersects the other sides at any points except the vertex, then the line divides the sides proportionally.

▶ The median of a trapezoid is parallel to its base, and its length is half the sum of the lengths of the bases.

In each figure, $\overline{AB} \parallel \overline{CD}$. Find the value of *x* to the nearest tenth.

35.
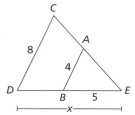

36.
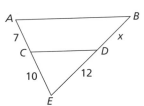

37.
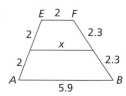

LESSON 7-7 ◣ Problem Solving Skills: Indirect Measurement, p. 326

▶ Properties of similar figures can be used to measure lengths and distances indirectly.

Use a diagram to solve.

38. A flagpole casts a shadow 16 ft long. At the same time, a yardstick casts a shadow 4 ft long. How tall is the flagpole?

39. To find the height of a tree, a forest ranger places a mirror on the ground 21 ft from the base of the tree. The ranger stands an additional 3 ft from the mirror so that she can see the top of the tree reflected in the mirror. If the ranger's eye level is 5 ft from the ground, what is the height of the tree?

CHAPTER INVESTIGATION

EXTENSION Make a list of the similarities and differences between your photograph and proportional sketch. Write a short paragraph explaining how you determined the lengths to use in your sketch. Display your photograph, sketch, list, and paragraph on a piece of posterboard.

Chapter 7 Assessment

Solve each proportion.

1. $\dfrac{x}{7} = \dfrac{3}{21}$

2. $\dfrac{4}{11} = \dfrac{6}{x}$

3. $\dfrac{9}{16} = \dfrac{x}{2}$

Determine if the polygons are similar. Write *yes* or *no*.

4.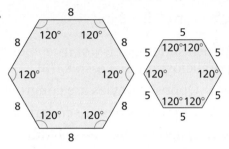

5.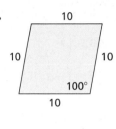

Find *x* in each pair of similar figures.

6.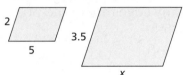

7.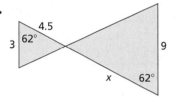

8.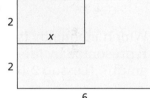

9. Find the actual length:
scale length: 5 cm, scale: 1 cm = 3.5 m

10. Find the scale length:
actual length: 2 mi, scale: $\dfrac{1}{4}$ in = 1 mi

Are the triangles similar? If so, give a reason: write AA, SSS, or SAS.

11.

12.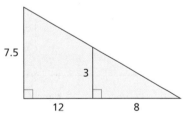

13.

Find *x* in each pair of similar figures.

14.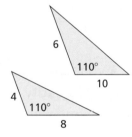

15.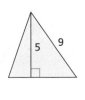

16.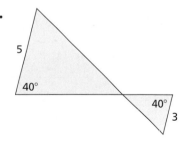

17. Luz placed a mirror on the ground and stood so that she could see the top of the tree. What is the height of the tree?

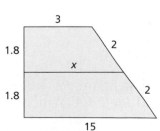

70 in.

80 in.   120 in.

Standardized Test Practice

Part 1 Multiple Choice

Record your answers on the answer sheet provided by your teacher or on a sheet of paper.

1. Which expression is *not* equivalent to a^{-6}? (Lesson 1-8)

   Ⓐ $\dfrac{1}{a^6}$
   Ⓑ $\left(-\dfrac{1}{a}\right)^6$
   Ⓒ $-a^6$
   Ⓓ $(-a)^6$

2. Given $f(x) = -3x + 1$, what is $f(-3)$? (Lesson 2-2)

   Ⓐ -10 Ⓑ -8
   Ⓒ 8 Ⓓ 10

3. Which inequality is represented by the graph? (Lesson 2-6)

   Ⓐ $y < x$
   Ⓑ $y > x$
   Ⓒ $y \leq x$
   Ⓓ $y \geq x$

   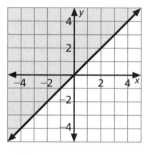

4. In the figure below $r \parallel s$. What is $m\angle 2$? (Lesson 3-4)

   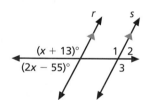

   Ⓐ 93°   Ⓑ 87°
   Ⓒ 85°   Ⓓ 74°

5. In the figure, $\overline{RL} \cong \overline{PA}$, $\overline{RL} \perp \overline{LA}$, and $\overline{PA} \perp \overline{LA}$. Which postulate could you use to prove $\triangle RLA \cong \triangle PAL$? (Lesson 4-2)

   Ⓐ Angle-Angle-Angle Postulate
   Ⓑ Angle-Side-Angle Postulate
   Ⓒ Side-Angle-Side Postulate
   Ⓓ Side-Side-Side Postulate

6. Which measures *cannot* be the lengths of the sides of a triangle? (Lesson 4-6)

   Ⓐ 16 m, 12 m, 20 m
   Ⓑ 12 cm, 8 cm, 21 cm
   Ⓒ 6 ft, 3 ft, 8 ft
   Ⓓ 7 in., 7 in., 8 in.

7. Which statement is *not* true about parallelograms? (Lesson 4-8)

   Ⓐ Opposite angles are congruent.
   Ⓑ Opposite sides are congruent.
   Ⓒ Diagonals bisect each other.
   Ⓓ Diagonals bisect the angles.

8. How many edges are in the prism? (Lesson 5-5)

   Ⓐ 2
   Ⓑ 7
   Ⓒ 10
   Ⓓ 15

   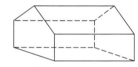

9. Determine whether the triangles are similar. If they are similar, state the reason. (Lesson 7-4)

   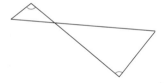

   Ⓐ no
   Ⓑ yes by the AA Similarity Postulate
   Ⓒ yes by the SAS Similarity Postulate
   Ⓓ yes by the SSS Similarity Postulate

10. Find x if $\overline{RM}$ is the altitude to the hypotenuse of right triangle RST. (Lesson 7-5)

    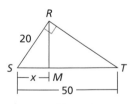

    Ⓐ 4   Ⓑ 8   Ⓒ 15   Ⓓ 20

Test-Taking Tip

Question 10
In similar triangles, corresponding angles are congruent and corresponding sides are proportional. When you set up a proportion, be sure that it compares corresponding sides.

Preparing for Standardized Tests
For test-taking strategies and more
practice, see pages 709-724.

Part 2 Short Response/Grid In

**Record your answers on the answer sheet
provided by your teacher or on a sheet of paper.**

11. Jaya works 21 h per week and earns $135.45.
How much money does Jaya earn per hour?
(Lesson 1-5)

12. Solve $5(x - 1) = 5 + 2x$. (Lesson 2-5)

13. The number of apples sold at the school
cafeteria each day for 2 wk is listed below.
What is the median number of apples sold?
(Lesson 2-7)

46, 48, 44, 40, 46, 48, 46, 49, 47, 50

14. Use the number line to find the length of
$\overline{CH}$. (Lesson 3-1)

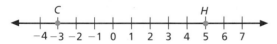

15. The measures of two complementary angles
are $x°$ and $(5x)°$. Find the measures of the
angles. (Lesson 3-2)

16. What is the value of
y in the figure?
(Lesson 4-1)

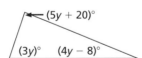

17. A 4-H group made a pair of jeans for a large
statue of Smokey the Bear. The distance
around the waist was 154 in. Change this
distance to feet. (Lesson 5-1)

18. What is the volume of the triangular prism?
(Lesson 5-7)

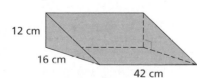

19. What is the slope of the line with the equation
$6x + 2y = 18$? (Lesson 6-1)

20. Solve the system of equations.
(Lessons 6-5, 6-6, and 6-7)

$$3x - 2y = 14$$
$$4x + y = 4$$

21. If a car can travel 536 mi on 16 gal of gasoline,
how far can it travel on 10 gal of gasoline?
(Lesson 7-1)

22. On a map the distance from Springfield to
Pleasantville is 6 in. The map scale is
$\frac{1}{2}$ in. = 20 mi. Find the actual distance
between Springfield and Pleasantville.
(Lesson 7-3)

23. What is the height of the telephone pole?
(Lesson 7-7)

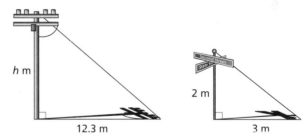

Part 3 Extended Response

**Record your answers on a sheet of paper. Show
your work.**

24. Triangle *RST* is similar to △*XYZ*. Find all the
missing measures. (Lesson 7-2)

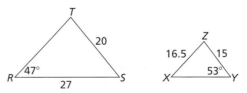

25. Given that $\overline{AB} \parallel \overline{CD}$, describe how you know
that △*ABE* is similar to △*CDE*. Then, find the
value of *x*. (Lesson 7-6)

Transformations

THEME: Amusement Parks

Suppose your family is planning a one-week vacation. If you are like millions of Americans, your plans would probably include a day or two at an amusement park.

The first amusement park in the United States was built in 1895 at Coney Island in New York City. Today, there are hundreds of amusement and theme parks throughout the country. More than 160 million people visit amusement parks across America each year. Many companies specialize in the design and construction of new rides and adventures.

- **Construction Supervisors** (page 347) oversee the construction of new attractions. These workers must pay particular attention to details to assure public safety. Construction supervisors must follow complicated plans, oversee large budgets, and supervise carpenters, electricians, artists, and many other workers.

- **Aerospace Engineers** (page 367) design roller coasters. They use their knowledge of aerodynamics, propulsion, stress tolerances, and gravitational forces to design roller coasters that are fast, exciting, and safe to enjoy.

Math Online

mathmatters3.com/chapter_theme

Classic Wooden Roller Coasters

Coaster names	Top speed	Height	Track length	Ride duration	Angle of descent of first hill	Vertical drop
The Rattler	55 mi/h	179.6 ft	5080 ft	2:15	61.4°	124 ft
Shivering Timbers	65 mi/h	125 ft	5384 ft	2:30	53.25°	120 ft
Texas Giant	65 mi/h	143 ft	4920 ft	2:30	53°	137 ft
Mean Streak	65 mi/h	161 ft	5427 ft	2:45	52°	155 ft
Georgia Cyclone	50 mi/h	95 ft	2970 ft	1:48	53°	78.5 ft
The Beast	65 mi/h	135 ft	7400 ft	3:40	45°	141 ft

Data Activity: Classic Wooden Roller Coasters

Use the table for Questions 1–4.

1. Find the average speed in feet per second of Mean Streak and The Beast. Which coaster has the fastest average speed? (*Hint:* Use the formula $d = rt$, where d = distance, r = rate, and t = time.)

2. Some say that the characteristic that most influences the top speed of a coaster is the angle of descent of the first hill. The method for measuring the angle of descent is shown in the diagram at the right. Do you agree with this thinking? Explain your reasoning.

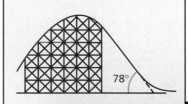

3. How many miles longer is The Beast than The Rattler?

4. If Texas Giant could maintain its top speed for the entire length of the track, what would be the duration of the ride?

CHAPTER INVESTIGATION

Amusement parks are constantly building new rides to attract new and returning customers. Designing new rides requires an understanding of geometry, physics, and construction techniques. Engineers are always looking for safe ways to provide greater thrills.

Working Together

Design an "out-and-back" roller coaster with eight hills. Make a scale drawing of the coaster indicating the height of each hill and the angle of descent. Estimate the track length and top speed of your coaster. Use the Chapter Investigation icons to guide your group's drawing.

The skills on these two pages are ones you have already learned. Stretch your memory and complete the exercises. For additional practice on these and more prerequisite skills, see pages 654-661.

In this chapter, you will be using matrices to solve equations. It is helpful to know how to find the determinant of a matrix.

BASIC OPERATIONS WITH INTEGERS

To perform basic operations with matrices, you must be able to do basic operations with integers. Recall that when adding integers whose signs are different, you actually subtract their absolute values and use the sign of the number with the greater absolute value for the answer. When multiplying and dividing integers whose signs are different, your answer is negative.

Perform the indicated operation.

1. $-6 + 10$
2. $-8(-3)$
3. $48 \div 2$
4. $-13 - 8$
5. -2×11
6. $-19(0)$
7. $4 - 21$
8. $-100 - 1$
9. $45 \div (-9)$
10. $37 + 99$
11. $-13(-1)$
12. $-1 - 2$
13. $-56 \div (-8)$
14. $-3 + (-14)$
15. $12 \times (-12)$

DETERMINANT OF A MATRIX

Example Find the determinant of this matrix: $\begin{bmatrix} 4 & -2 \\ 3 & -1 \end{bmatrix}$

To find the determinant, $\begin{vmatrix} a & c \\ b & d \end{vmatrix}$ use the formula $ad - bc$.

$$\begin{vmatrix} 4 & -2 \\ 3 & -1 \end{vmatrix} = 4(-1) - 3(-2) = -4 + 6 = 2$$

Find the determinant of each matrix.

16. $\begin{bmatrix} 5 & 3 \\ 2 & 4 \end{bmatrix}$
17. $\begin{bmatrix} 8 & -7 \\ 3 & 2 \end{bmatrix}$
18. $\begin{bmatrix} -5 & -1 \\ 1 & 4 \end{bmatrix}$

19. $\begin{bmatrix} 9 & 4 \\ -3 & -2 \end{bmatrix}$
20. $\begin{bmatrix} -1 & 6 \\ 8 & -7 \end{bmatrix}$
21. $\begin{bmatrix} 4 & -5 \\ 2 & 6 \end{bmatrix}$

22. $\begin{bmatrix} 8 & 2 \\ -5 & 6 \end{bmatrix}$
23. $\begin{bmatrix} -7 & -3 \\ -2 & -5 \end{bmatrix}$
24. $\begin{bmatrix} 3 & 0 \\ 4 & 2 \end{bmatrix}$

25. $\begin{bmatrix} 0 & 5 \\ -6 & 2 \end{bmatrix}$
26. $\begin{bmatrix} 1 & 1 \\ 1 & 1 \end{bmatrix}$
27. $\begin{bmatrix} 0 & 3 \\ 2 & 0 \end{bmatrix}$

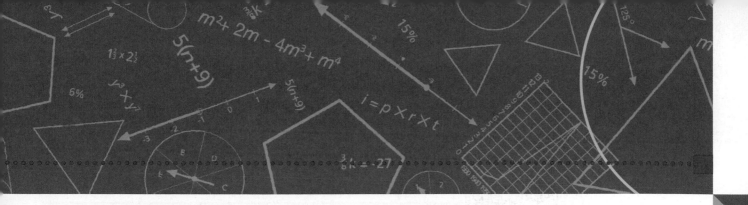

SYMMETRY

A line of symmetry is a line along which a figure can be folded so that the two parts match exactly.

Copy each figure on a sheet of paper. Sketch all the possible lines of symmetry for each figure.

28.

29.

30.

31.

32.

33.

MIDPOINT FORMULA

When you are working with figures in the coordinate plane, you may find the midpoint formula is often useful.

Example Find the midpoint of the $\overline{AB}$ when $A = (1, 6)$ and $B = (9, 2)$.

Remember to solve for both coordinates of the midpoint:

$$x(\text{midpoint}) = \frac{x_1 + x_2}{2} \qquad\qquad y(\text{midpoint}) = \frac{y_1 + y_2}{2}$$

$$x_m = \frac{1 + 9}{2} \qquad\qquad y_m = \frac{6 + 2}{2}$$

$$x_m = \frac{10}{2} \qquad\qquad y_m = \frac{8}{2}$$

$$x_m = 5 \qquad\qquad y_m = 4$$

The coordinates of the midpoint of $\overline{AB}$ are (5, 4).

Find the midpoint of each segment that has the given endpoints.

34. $C(0, 3)$ and $D(-4, 9)$

35. $E(2, 5)$ and $F(8, -1)$

36. $G(3, 3)$ and $H(7, 7)$

37. $I(1, 3)$ and $J(-5, 9)$

38. $K(4, 3)$ and $L(-6, -1)$

39. $M(8, 0)$ and $N(-4, 8)$

40. $O(4, -3)$ and $P(4, -9)$

41. $Q(6, -2)$ and $R(-2, 6)$

8-1 Translations and Reflections

Goals ■ Graph translation and reflection images on a coordinate plane.

Applications Recreation, Art, Amusement Park Design

Work with a partner. You will need a piece of graph paper, a ruler, and scissors.

1. Using the ruler, draw an isosceles triangle near one bottom corner of your graph paper. Make the base $\overline{AB}$ 6 units long and the height 4 units. Cut out the triangle and label the vertices *A*, *B*, and *C*.

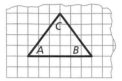

2. Draw a coordinate plane and label each axis from −10 to 10.

3. Place the triangle in the second quadrant so that $\overline{AB}$ is parallel to the *x*-axis. Situate each vertex at the intersection of a horizontal and a vertical line in such a way that *C* is above $\overline{AB}$. Trace the triangle and label each vertex to match the original triangle. Label this figure "Triangle 1."

4. Slide your triangle 9 units straight to the right. Trace the triangle in this new position and label each vertex. Label this figure "Triangle 2."

5. Turn your triangle so vertex *C* is below $\overline{AB}$. Now place the triangle in the fourth quadrant with $\overline{AB}$ parallel to the *x*-axis. Trace this triangle. Label this figure "Triangle 3."

6. Record the slopes of the sides of the three triangles in a table like the one shown.

7. Compare the slopes of the sides of Triangles 1 and 2, Triangles 2 and 3, and Triangles 1 and 3. What do you notice?

	Triangle 1	Triangle 2	Triangle 3
$\overline{AB}$			
$\overline{AC}$			
$\overline{BC}$			

◥ BUILD UNDERSTANDING

A **translation** is a *slide* of a figure. It produces a new figure exactly like the original. The new figure is the **image** of the original figure, and the original figure is the **preimage** of the new figure. A translation is an example of a **transformation** of a figure.

Another kind of transformation that yields a congruent figure is a **reflection**, or *flip*. Under a reflection, a figure is *reflected*, or *flipped*, over a **line of reflection**.

If a line can be drawn through a geometric figure so that the part of the figure on one side of the line is the reflection of the part of the figure on the opposite side, the figure is said to exhibit line symmetry and the line is a **line of symmetry**, or *axis of symmetry*. A reflection image and its preimage combined will always be a figure that has line symmetry.

> **Check Understanding**
>
> Write the formula for finding the slope when given two points and their coordinates.

Example 1

Graph the image of parallelogram *MNOP* with vertices *M*(2, 1), *N*(4, 7), *O*(7, 7), and *P*(5, 1) under each transformation from the original position.

a. 9 units down

b. reflected across the *y*-axis

Solution

a. To move the given figure 9 units down, subtract 9 from the y-coordinate of each vertex.

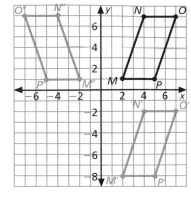

$M(2, 1) \rightarrow M'(2, 1 - 9) = M'(2, -8)$

$N(4, 7) \rightarrow N'(4, 7 - 9) = N'(4, -2)$

$O(7, 7) \rightarrow O'(7, 7 - 9) = O'(7, -2)$

$P(5, 1) \rightarrow P'(5, 1 - 9) = P'(5, -8)$

b. The reflection of the point (x, y) across the y-axis is the point $(-x, y)$.

$M(2, 1) \rightarrow M''(-2, 1)$ $N(4, 7) \rightarrow N''(-4, 7)$

$O(7, 7) \rightarrow O''(-7, 7)$ $P(5, 1) \rightarrow P''(-5, 1)$

Example 2

Compare the slopes of corresponding non-horizontal sides for the preimage and each transformation image in Example 1.

Solution

For the first transformation, compare the slopes of $\overline{MN}$ and $\overline{M'N'}$, as well as the slopes of $\overline{OP}$ and $\overline{O'P'}$. For the second transformation, compare the slopes of $\overline{MN}$ and $\overline{M''N''}$, as well as the slopes of $\overline{OP}$ and $\overline{O''P''}$.

Side	$\overline{MN}$	$\overline{M'N'}$	$\overline{OP}$	$\overline{O'P'}$	$\overline{M''N''}$	$\overline{O''P''}$
Slope	3	3	3	3	−3	−3

For the translation in part **a**, corresponding sides have equal slopes. For the reflection in part **b**, corresponding sides have opposite slopes.

Example 3

RECREATION At a miniature golf course, a hole is designed so that the ball must travel along a line of reflection between two congruent triangular blocks.

Graph the image of $\triangle ABC$ with vertices $A(3, 5)$, $B(5, 8)$, and $C(1, 7)$ under a reflection across the line whose equation is $y = x$. Compare the slopes of the corresponding sides of $\triangle ABC$ and $\triangle A'B'C'$.

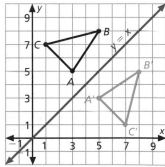

Solution

Graph the line $y = x$. Graph the reflection image as directed. Use the rule $(x, y) \rightarrow (y, x)$. Make a table to compare the slopes of the corresponding sides. The slopes of corresponding sides are reciprocals of each other.

Side	$\overline{AB}$	$\overline{BC}$	$\overline{CA}$	$\overline{A'B'}$	$\overline{B'C'}$	$\overline{C'A'}$
Slope	$\frac{3}{2}$	$\frac{1}{4}$	−1	$\frac{2}{3}$	4	−1

If an object is symmetrical with respect to a line and you only have half of it, you can draw the other half.

Example 4

ART The figure shown at the right is half of a figure that has line symmetry. Complete the figure.

Solution

Draw a reflection across the line of symmetry for the half of the figure shown.

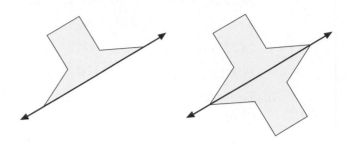

◥ **TRY THESE EXERCISES**

Copy parallelogram *DEFG* at the right on a coordinate plane. Then graph its image under each transformation from the original position.

1. 7 units down ($D'E'F'G'$)

2. reflected across the *y*-axis ($D''E''F''G''$)

3. Copy and complete the chart below.

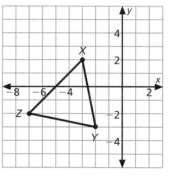

Side	$\overline{DE}$	$\overline{D'E'}$	$\overline{D''E''}$	$\overline{EF}$	$\overline{E'F'}$	$\overline{E''F''}$
Slope						

4. Copy △*XYZ* at the right on a coordinate plane and graph its image under a reflection across the line with equation $y = -x$. Compare the slopes of $\overline{XY}$, $\overline{X'Y'}$, $\overline{YZ}$, $\overline{Y'Z'}$, $\overline{XZ}$, and $\overline{X'Z'}$.

5. **WRITING MATH** How can you recognize a line of symmetry?

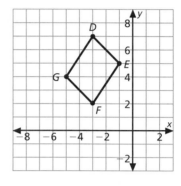

◥ **PRACTICE EXERCISES** • **For Extra Practice, see page 686.**

On a coordinate plane, graph parallelogram *HIJK* with vertices $H(1, 1)$, $I(5, 1)$, $J(8, 4)$, and $K(4, 4)$. Then graph its image under each transformation from the original position.

6. 10 units left

7. reflected across the *x*-axis

8. Compare the slopes of the non-horizontal sides of parallelogram *HIJK* in all three positions above.

9. Graph the image of △*ABC* with vertices $A(1, 4)$, $B(5, 6)$, and $C(2, 7)$ under a reflection across the line with equation $y = x$. Compare the slopes of the sides of △*ABC* and △*A'B'C'*.

10. **YOU MAKE THE CALL** Jenna says that the rule $(x, y) \rightarrow (4 - x, y)$ can be used to translate an image 4 units to the left. Do you agree with Jenna's thinking?

Copy each figure below on graph paper along with its line of symmetry. Then complete the figure.

11.

12.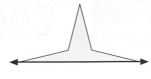

13.

14.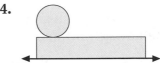

15. **AMUSEMENT PARK DESIGN** A new roller coaster has two entrances to the boarding platform. The eastern entrance is a reflection of the western entrance shown at the right. Copy the entrance on graph paper. Then draw its reflection.

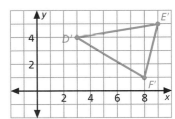

■ EXTENDED PRACTICE EXERCISES

16. Triangle $D'E'F'$ is the image of a figure that was translated under the rule $(x, y) \rightarrow (x + 3, y - 2)$. What are the vertices of the preimage of $\triangle D'E'F'$? What are the slopes of the sides $\overline{D'E'}$, $\overline{E'F'}$, and $\overline{D'F'}$? Are the slopes of the corresponding sides of the preimage the same?

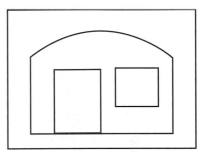

17. How do the slopes of a segment and a translation image compare?

18. **WRITING MATH** Consider a nonvertical, nonhorizontal segment and its reflection image across each of the lines. How do the slopes of the image and preimage compare? Support your answer with an example for each.
    a. x-axis   b. y-axis   c. $y = x$

■ MIXED REVIEW EXERCISES

Solve each proportion and find the value of x. Round to the nearest hundredth if necessary. (Lesson 7-1)

19. $\dfrac{2}{5} = \dfrac{6}{x}$

20. $\dfrac{3}{8} = \dfrac{x}{24}$

21. $\dfrac{75}{90} = \dfrac{15}{x}$

22. $\dfrac{23}{48} = \dfrac{x}{60}$

23. $\dfrac{2x}{10} = \dfrac{6}{15}$

24. $\dfrac{x - 4}{12} = \dfrac{x}{8}$

25. $\dfrac{3}{x + 2} = \dfrac{18}{36}$

26. $\dfrac{1}{3} = \dfrac{5x}{45}$

27. $\dfrac{72}{48} = \dfrac{x + 8}{4}$

28. $\dfrac{125}{3x} = \dfrac{5}{3}$

29. $\dfrac{169}{286} = \dfrac{13}{4x + 2}$

30. $\dfrac{x - 2}{54} = \dfrac{8}{48}$

Exercises 31–34 refer to the figure at the right. (Lesson 3-3)

31. Name the midpoint of $\overline{HJ}$.

32. Name the segment whose midpoint is B.

33. Name all the segments whose midpoint is E.

34. Assume Z is the midpoint of $\overline{BE}$. What is its coordinate?

A B C D E F G H I J K L
−6 −5 −4 −3 −2 −1 0 1 2 3 4 5

Rotations in the Coordinate Plane

Goals ■ Graph rotation images and identify centers, angles and directions of rotations.

Applications Ride Management, Computer Graphics, Art

Think about how gears mesh and turn one another. A gear is a mechanical device that transfers rotating motion and power from one part of a machine to another.

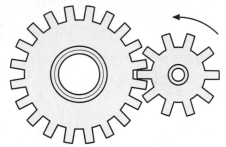

1. From the side at which you see the gears, would you say the larger gear turns clockwise or counterclockwise?

2. What fractional part of a turn does it take for a gear tooth to get from the top to a horizontal position?

3. How many degrees does a gear tooth travel from the top to the bottom of the gear?

◣ BUILD UNDERSTANDING

Another transformation that produces a figure congruent to the original is a **rotation**, or *turn*. A figure is rotated, or turned, about a point.

Rotation is described by three pieces of information:

- the point about which the figure is rotated, or the **center of rotation**.

- the amount of turn expressed as a fractional part of a whole turn, or as an **angle of rotation** in degrees.

- the rotation direction—*clockwise* or *counterclockwise*.

When you rotate a point 180° clockwise about the origin, both the *x*-coordinate and the *y*-coordinate are transformed into their opposites.

Check Understanding

How many degrees are in a one-quarter turn? How many degrees are in a one-half turn? How many degrees are in a three-quarter turn? How many degrees are in a full turn?

Example 1

Graph △*QRS* and its image after a 180° clockwise rotation about the origin. Then compare the slopes of $\overline{QR}$, $\overline{Q'R'}$, $\overline{QS}$, and $\overline{Q'S'}$.

Solution

Use the rule $(x, y) \rightarrow (-x, -y)$.

$Q(3, 4) \rightarrow Q'(-3, -4)$

$R(1, 1) \rightarrow R'(-1, -1)$

$S(5, 1) \rightarrow S'(-5, -1)$

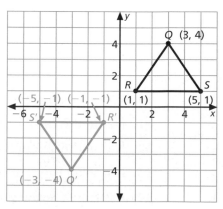

Side	$\overline{QR}$	$\overline{QS}$	$\overline{Q'R'}$	$\overline{Q'S'}$
Slope	$\frac{3}{2}$	$-\frac{3}{2}$	$\frac{3}{2}$	$-\frac{3}{2}$

The slopes of corresponding segments are the same.

When you rotate a figure 90° counterclockwise, the y-coordinate is multiplied by -1, and then the x-coordinate and y-coordinate are transposed. That is, $(x, y) \rightarrow (-y, x)$.

Example 2

RIDE MANAGEMENT Computers are used to signal ride operators when it is safe to begin a new ride cycle. The ride can start when the screen shows a raised flag. A lowered flag tells the operator to wait. On the computer screen, the raised flag contains the points $A(0, 0)$, $B(-2, 2)$, $C(-5, 5)$, and $D(-5, 2)$. Graph the flag and its image after a 90° counterclockwise rotation about the origin. Label the points of the image A', B', C', and D'. Then compare the slope of $\overline{BC}$ with the slope of $\overline{B'C'}$.

Solution

Use the rule $(x, y) \rightarrow (-y, x)$

$A(0, 0) \rightarrow A'(0, 0)$

$B(-2, 2) \rightarrow B'(-2, -2)$

$C(-5, 5) \rightarrow C'(-5, -5)$

$D(-5, 2) \rightarrow D'(-2, -5)$

slope of $\overline{BC} = \dfrac{5 - 2}{-5 - (-2)}$

$\qquad = \dfrac{3}{-3} = -1$

slope of $\overline{B'C'} = \dfrac{-5 - (-2)}{-5 - (-2)}$

$\qquad = \dfrac{-3}{-3} = 1$

The product of the slopes is -1. The lines are perpendicular.

Example 3

Triangle XYZ is rotated twice about the origin. Compare the slopes and determine the angle of rotation first for rotation 1 and then for rotation 2.

Original position		After rotation 1		After rotation 2	
Side	Slope	Side	Slope	Side	Slope
$\overline{XZ}$	2	$\overline{X'Z'}$	2	$\overline{X''Z''}$	$-\frac{1}{2}$
$\overline{YZ}$	$-\frac{3}{4}$	$\overline{Y'Z'}$	$-\frac{3}{4}$	$\overline{Y''Z''}$	$\frac{4}{3}$
$\overline{XY}$	$\frac{1}{6}$	$\overline{X'Y'}$	$\frac{1}{6}$	$\overline{X''Y''}$	$\frac{1}{6}$

Solution

The first rotation is 180° or 360°, because the slopes are equal to the slopes in the original position.

The second rotation is 90° or 270°, because the slopes are the negative reciprocals of the original position slopes. That is, the product of the slopes is -1.

1. Triangle *DEF* has vertices *D*(1, 1), *E*(5, 3), and *F*(2, 5). Graph △*DEF* and its image after a 180° counterclockwise rotation about the origin. Then compare the slopes of the corresponding sides of △*DEF* before and after the rotation.

2. **COMPUTER GRAPHICS** A figure contains the points *M*(0, 0), *N*(2, −4), *O*(4, −8), and *P*(5, −5). Graph the figure and its image after a 90° clockwise rotation about the origin. Use the rule $(x, y) \rightarrow (y, -x)$. Then compare the slopes of $\overline{NO}$ and $\overline{N'O'}$.

3. **ANIMATION** For a television commercial, a triangular logo is animated so that it rotates twice about the origin in a clockwise direction, as shown in the table below. Compare the slopes and determine how much of a rotation was done each time.

Original position		After rotation 1		After rotation 2	
Side	Slope	Side	Slope	Side	Slope
$\overline{QR}$	−1	$\overline{Q'R'}$	1	$\overline{Q''R''}$	1
$\overline{RS}$	$\frac{3}{5}$	$\overline{R'S'}$	$-\frac{5}{3}$	$\overline{R''S''}$	$-\frac{5}{3}$
$\overline{QS}$	3	$\overline{Q'S'}$	$-\frac{1}{3}$	$\overline{Q''S''}$	$-\frac{1}{3}$

4. **WRITING MATH** Describe how translations, reflections, and rotations could be used to create a pattern. Sketch an example.

For each figure, draw the image after the given rotation about the origin. Then calculate the slope of each side before and after the rotation.

5. Use the rule $(x, y) \rightarrow (y, -x)$ for a 90° clockwise rotation.

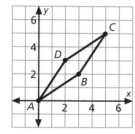

6. Use the rule $(x, y) \rightarrow (-x, -y)$ for a 180° clockwise rotation.

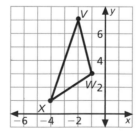

7. Use the rule $(x, y) \rightarrow (-y, x)$ for a 90° counter clockwise rotation.

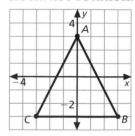

8. **ART** To create a pattern, △*DEF* is rotated twice about the origin in a clockwise direction, as shown. Compare the slopes to determine how much of a rotation was completed each time.

Original position		After rotation 1		After rotation 2	
Side	Slope	Side	Slope	Side	Slope
$\overline{DE}$	−1	$\overline{D'E'}$	−1	$\overline{D''E''}$	1
$\overline{EF}$	$\frac{1}{7}$	$\overline{E'F'}$	$\frac{1}{7}$	$\overline{E''F''}$	−7
$\overline{DF}$	1	$\overline{D'F'}$	1	$\overline{D''F''}$	−1

Use the figure at the right for Exercises 9–13.

9. Which triangle is the rotation image of Triangle 1 about the point $(-3, 0)$?

10. Which is the translation image of Triangle 1?

11. Which triangle is the reflection image of Triangle 1 across the x-axis?

12. Which triangle is the reflection image of Triangle 1 across the y-axis?

13. Which triangle is the rotation image of Triangle 1 180° clockwise about the origin?

14. **CHAPTER INVESTIGATION** An "out-and-back" coaster returns to its starting point when the ride ends. Design an "out-and-back" coaster with eight hills. You may join the hills with lengths of track that wind, dip and turn. Make a scale drawing of your coaster on graph paper.

◼ EXTENDED PRACTICE EXERCISES

15. **WRITING MATH** Make a generalization about corresponding slopes in each situation. Assume that no sides of the triangle are horizontal or vertical.

   a. A triangle is rotated 180° clockwise about the origin.

   b. A triangle is rotated 90° counterclockwise about the origin.

   c. A triangle is rotated 90° clockwise about the origin.

16. If you rotate a figure about its center and it fits back on top of itself in less than a 360° rotation, the figure is said to have **point symmetry**. Point symmetry is a type of rotation symmetry. For example, a square has point symmetry because, when it is rotated 90° about its center, each image vertex falls on top of one of the original vertices. Through how many degrees would you have to rotate a regular hexagon for this to happen? A regular pentagon?

◼ MIXED REVIEW EXERCISES

Find x in each pair of similar polygons. (Lesson 7-2)

17.

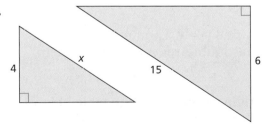

18.

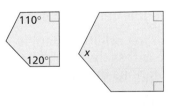

Determine the slope of the line containing the given pair of points. (Lesson 6-1)

19. $A(-2, -3)$, $B(-9, -4)$

20. $C(3, 6)$, $D(-2, 8)$

21. $E(-4, 5)$, $F(3, -1)$

22. $G(4, 8)$, $H(-2, -5)$

23. $I(5, -2)$, $J(0, -8)$

24. $K(-4, 2)$, $L(3, 2)$

25. $M(-1, 6)$, $N(5, 2)$

26. $O(4, -2)$, $P(-6, 3)$

27. $Q(-3, 1)$, $R(-3, -5)$

28. $S(-6, -2)$, $T(5, -3)$

29. $U(6, -7)$, $V(2, -4)$

30. $W(4, 3)$, $X(7, 12)$

Review and Practice Your Skills

Graph the image of parallelogram *PQRS* with vertices *P*(2, −1), *Q*(9, −1), *R*(8, −4), and *S*(1, −4). Then graph its image under each transformation from the original position.

1. 9 units up   **2.** reflected across *x*-axis   **3.** reflected across *y*-axis

4. Compare the slopes of the non-horizontal sides of parallelogram *PQRS* with the slopes of the sides of the image.

Graph the image of triangle *DEF* with vertices *D*(1, 7), *E*(3, 4), and *F*(5, 11). Then graph its image under each transformation from the original position.

5. 15 units down   **6.** reflected across *y*-axis   **7.** reflected across *y* = *x*

8. Compare the slopes of the sides of triangle *DEF* with the slopes of the corresponding sides of the three images.

Graph the image of trapezoid *MKLN* with vertices *M*(−2, 3), *K*(−4, 7), *L*(−3, 11), and *N*(0, 11). Then graph its image under each transformation from the original position.

9. reflected across *y*-axis   **10.** reflected across *y* = *x*   **11.** reflected across *y* = −*x*

12. Compare the slopes of the sides of trapezoid *MKLN* with the slopes of the corresponding sides of the three images.

Triangle *ABC* has vertices *A*(−3, 1), *B*(−5, 2), and *C*(−1, 4). Graph the triangle and its image after each of the following rotations about the origin.

13. 90° clockwise   **14.** 180° clockwise   **15.** 270° clockwise

16. Reflect the original triangle *ABC* across the *y*-axis. Then reflect this new image across the *x*-axis. To which of the rotations in Exercises 13–15 does this double-reflection correspond?

17. Reflect the original triangle *ABC* across the line *y* = *x*. Then reflect this new image across the line *y* = −*x*. How does this double-reflection compare to the rotations in Exercises 13–15?

18. Triangle *FGH* is rotated twice about the origin, as shown in the table below. Each rotation angle is between 0° and 360°. Compare the slopes and determine how much the triangle was rotated each time.

Original position		After rotation 1		After rotation 2	
side	slope	side	slope	side	slope
$\overline{FG}$	$-\frac{2}{3}$	$\overline{F'G'}$	$\frac{3}{2}$	$\overline{F''G''}$	$-\frac{2}{3}$
$\overline{GH}$	$\frac{1}{3}$	$\overline{G'H'}$	-3	$\overline{G''H''}$	$\frac{1}{3}$
$\overline{FH}$	$\frac{5}{6}$	$\overline{F'H'}$	$-\frac{6}{5}$	$\overline{F''H''}$	$\frac{5}{6}$

**Graph parallelogram *PQRS* with vertices *P*(−1, 2), *Q*(2, 3), *R*(4, 7), and *S*(1, 6).
Then graph its image under each transformation from the original position.**
(Lesson 8-1)

19. 12 units to the left   **20.** reflected across $y = x$ **21.** reflected across *y*-axis

22. Compare the slopes of the sides of parallelogram *PQRS* with the slopes of the
corresponding sides of the three images.

**Graph quadrilateral *ABCD* with vertices *A*(5, 4), *B*(8, −1), *C*(0, 0), and *D*(−2, 1).
Then graph its image under each transformation from the original position.**
(Lesson 8-1)

23. 5 units to the left, 4 units down   **24.** reflected across *x*-axis   **25.** reflected across *y*-axis

26. Compare the slopes of the sides of quadrilateral *ABCD* with the slopes of the
corresponding sides of the three images.

**Triangle *ABC* has vertices *A*(2, 0), *B*(8, −2), and *C*(7, 3). Graph the triangle and
its image after each of the following rotations about the origin.** (Lesson 8-2)

27. 90° clockwise   **28.** 180° clockwise   **29.** 270° counterclockwise

30. Compare the slopes of the sides of triangle *ABC* with the slopes of the
corresponding sides of the three images.

Math*Works* Career – Construction Supervisor
Workplace Knowhow

Amusement parks hire construction workers and supervisors to build new
attractions. Whether the attraction is a roller coaster or a carousel, the
construction supervisor must adhere to building and safety codes. The
supervisor oversees costs, materials, labor, and transportation related to the
project. Construction supervisors must be able to follow blueprints and design
specifications. For high-speed rides, the supervisor may work closely with an
engineer or specialist. When problems arise, the supervisor and the engineer
work together to find solutions.

An engineer has designed a coaster with six
hills. The final three hills are a reflection of the
first three hills. In the diagram below, the first
inclines have been straightened.

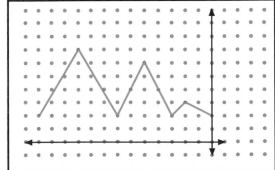

1. Identify the coordinates of apex of each of
   the three hills shown.

2. Draw the coaster's reflection over the *y*-axis.
   Write the coordinates of the reflections of
   the points you identified in Exercise 1.

Dilations in the Coordinate Plane

Goals ■ Draw dilation images on a coordinate plane.

Applications Graphic Design, Business, Art

Work with a partner to study the quilt pattern shown.

1. The pattern of a rotated square is used repeatedly in the design. How many different sizes appear?

2. How does the length of a side in the largest pattern compare to the length of a side in the smallest?

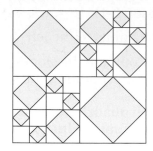

◥ BUILD UNDERSTANDING

A **dilation** is a transformation that produces an image in the same shape as the original figure, but usually of a different size. An image larger than the original figure is called an **enlargement**. An image smaller than the original figure is called a **reduction**. A figure and its dilation image are similar.

The lengths of the sides of a dilation image are obtained by multiplying the lengths of the sides of the original figure by a number called the **scale factor**. If the scale factor is greater than 1, you create an enlargement. If the scale factor is smaller than 1, you create a reduction. The description of a dilation includes the scale factor and the **center of dilation**. The distance from the center of dilation to each point on the image is equal to the distance from the center of dilation to each corresponding point of the original figure times the scale factor.

> **Check Understanding**
>
> Explain the meaning of: "A figure and its dilation are similar."

When the center of dilation is at the origin, you use the following rule to locate points on the dilation image. Let (x, y) represent a point on the original figure, and let k represent the scale factor.

$$(x, y) \rightarrow (kx, ky)$$

Example 1

Draw the dilation image of quadrilateral *MNPQ* below with vertices at *M*(3, 1), *N*(6, 1), *P*(6, 3), and *Q*(3, 3). The center of dilation is the origin, and the scale factor is 2.

Solution

Because the center of dilation is at the origin, use the rule $(x, y) \rightarrow (2x, 2y)$.

$M(3, 1) \rightarrow M'(6, 2)$

$N(6, 1) \rightarrow N'(12, 2)$

$P(6, 3) \rightarrow P'(12, 6)$

$Q(3, 3) \rightarrow Q'(6, 6)$

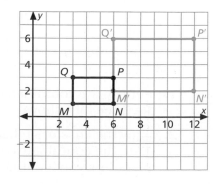

Sometimes the center of dilation is not at the origin. For example, the center of dilation might be a vertex of the original figure. In this case, the center of dilation and the corresponding vertex of the dilation image are the same.

Example 2

GRAPHIC DESIGN A triangular flag is designed to promote a new attraction at an amusement park. The flags, represented by $\triangle ABC$ on the grid shown to the right, will also be manufactured in a smaller size. Draw a dilation image of $\triangle ABC$ with center of dilation at A and a scale factor of $\frac{2}{3}$.

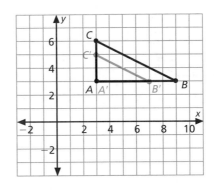

Solution

The distance from the center of dilation, A, to B is 6 units. So the distance from A to B' is $\frac{2}{3} \times 6$, or 4. Count over 4 units from A to locate B'. The distance from A to C' is $\frac{2}{3} \times 3$, or 2. Count up 2 units from A to locate C'. Points A and A' coincide because A is the center of dilation.

Technology Note

You can use geometry drawing software to transform figures. To draw a dilation of a figure, follow these steps:

1. Plot the vertices of the figure and draw segments.

2. Select the center of dilation. Many programs allow you to select the center of dilation by double-clicking on the desired point.

3. Enter the scale factor as a ratio of new distance to old distance.

4. Draw the dilation.

◥ TRY THESE EXERCISES

Refer to the figure at the right for Exercises 1–4.

1. What is the image of square $ABCD$?

2. What is the center of dilation?

3. How do the lengths of the sides in the image compare to lengths of the sides in square $ABCD$?

4. What is the scale factor?

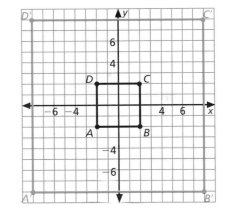

Copy each graph on graph paper. Then draw each dilation image.

5. The center of dilation is the origin and the scale factor is 4. Use the rule $(x, y) \rightarrow (4x, 4y)$.

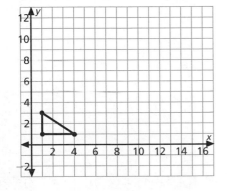

6. The center of dilation is the origin and the scale factor is $\frac{1}{2}$.

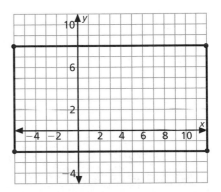

7. The center of dilation is the origin and the scale factor is 1.5.

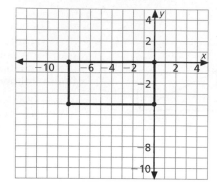

8. The center of dilation is point A and the scale factor is $\frac{1}{4}$.

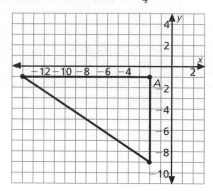

9. The center of dilation is point M and the scale factor is $\frac{1}{3}$.

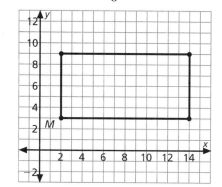

10. The center of dilation is point S and the scale factor is $\frac{1}{5}$.

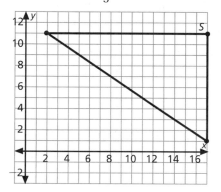

The following sets of points are the vertices of figures and their dilation images. For each two sets of points, give the scale factor.

11. $C(0, 0)$, $D(0, 4)$, $E(8, 0)$
$C'(0, 0)$, $D'(0, 5)$, $E'(10, 0)$

12. $S(0, 0)$, $T(0, -6)$, $U(-3, -9)$
$S'(0, 0)$, $T'(0, -4)$, $U'(-2, -6)$

13. $A(0, 2)$, $B(2, 2)$, $C(2, -1)$, $D(0, -1)$
$A'(0, 6)$, $B'(6, 6)$, $C'(6, -3)$, $D'(0, -3)$

14. BUSINESS Alice designed a logo for her business. She drew it on a grid, as shown in the diagram below. She wants to create an enlargement that is 5 times as big. Draw the logo on a separate sheet of graph paper. Then draw the enlargement by multiplying all the coordinates by 5.

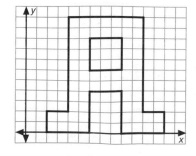

15. ART An artist needs to enlarge the drawing shown at the right so that the resulting figure is twice as large. She uses another method for making enlargements and reductions which does not require a grid. To use this method, pick any convenient point outside the figure. Label the point O. Draw line segments from O to various points on the figure. For example, draw a segment from O to point A. Then extend the segment from A to point A' so that $OA = AA'$. Continue until you have enough points to complete the drawing.

16. Trace pentagon $ABCDE$ on another sheet. Use the method described in Exercise 15 to make an enlargement of the pentagon 3 times as big. *Hint:* The distance OA will be $\frac{1}{2}$ of the distance from A to A'.

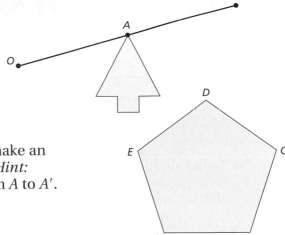

EXTENDED PRACTICE EXERCISES

17. GEOMETRY SOFTWARE Draw rectangle $TUVW$ with vertices $T(2, 2)$, $U(2, 6)$, $V(8, 6)$, and $W(8, 2)$ on a coordinate plane using geometric-drawing software. Draw the dilation images of rectangle $TUVW$ with the center of dilation at T and scale factors of 2 and $\frac{1}{2}$.

    a. Find the area of rectangle $TUVW$.

    b. Find the area of its enlargement.

    c. Find the area of its reduction.

    d. How does the area of the enlargement compare to the area of the original figure?

    c. How does the area of the reduction compare to the area of the original figure?

18. CRITICAL THINKING Describe the effect a scale factor of $-\frac{1}{2}$ would have on a dilation image of rectangle $TUVW$ from Exercise 17.

MIXED REVIEW EXERCISES

Find the actual length of each of the following. (Lesson 7-3)

19. scale length = 6 in.
scale is $\frac{1}{2}$ in. : 4 ft

20. scale length = 4.5 cm
scale is 1 cm : 8 mi

21. scale length = 5 cm
scale is $\frac{1}{2}$ cm : 6 km

22. scale length = 4 in.
scale is 1 in. : 12 mi

23. scale length = 9 cm
scale is $\frac{1}{2}$ cm : 3 m

24. scale length = 3.75 in.
scale is $\frac{1}{4}$ in. : 4 ft

25. scale length = 6.75 cm.
scale is 1 cm. : 12 m

26. scale length = $3\frac{3}{4}$ in.
scale is $\frac{1}{2}$ in. : 3.8 ft

27. scale length = 4.25 cm
scale is 1 cm : 24 km

Determine whether each relation is a function. Give the domain and range of each. (Lesson 2-2)

28. $\{(1, 0), (0, -1), (1, -1), (2, 3)\}$

29. $\{(-1, 0), (-2, 1), (1, -1), (0, -2)\}$

30. $\{(-4, 2), (-5, 3), (-6, 4), (-7, 5)\}$

31. $\{(2, 1), (3, 0), (-3, -1), (-2, -2)\}$

8-4 Multiple Transformations

Goals   ■ Identify and find composites of transformations.

Applications   Engineering, Art, Graphics Design

Work with a partner.

Locate the basic pattern outlined in blue in the upper left corner.

1. Could all the other images in the pattern be obtained by using only translations, rotations, or reflections?

2. Describe how you could obtain the pattern outlined in red by applying two successive transformations to the pattern outlined in blue.

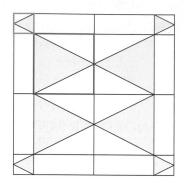

◥ BUILD UNDERSTANDING

Two or more successive transformations can be applied to a given figure. This is called a **composite of transformations**.

E x a m p l e 1

Begin with square $PQRS$ with vertices $P(-5, 2)$, $Q(-1, 2)$, $R(-1, 6)$, and $S(-5, 6)$. First, perform a reflection over the x-axis. Then using $P'Q'R'S'$, perform a dilation with center at the origin and a scale factor of 2 to obtain $P''Q''R''S''$.

Solution

Start with square $PQRS$. Use the rule $(x, y) \rightarrow (x, -y)$.

$P(-5, 2) \rightarrow P'(-5, -2)$

$Q(-1, 2) \rightarrow Q'(-1, -2)$

$R(-1, 6) \rightarrow R'(-1, -6)$

$S(-5, 6) \rightarrow S'(-5, -6)$

Then apply a dilation with center at the origin and a scale factor of 2 to square $P'Q'R'S'$. Use the rule $(x, y) \rightarrow (2x, 2y)$.

$P'(-5, -2) \rightarrow P''(-10, -4)$

$Q'(-1, -2) \rightarrow Q''(-2, -4)$

$R'(-1, -6) \rightarrow R''(-2, -12)$

$S'(-5, -6) \rightarrow S''(-10, -12)$

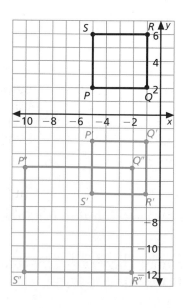

Example 2

ENGINEERING A ride designer is using a computer to map the movement of the car for a new amusement park ride. The two triangles at the right represent the car at the beginning and end of a short section of track after two transformations. Describe these transformations.

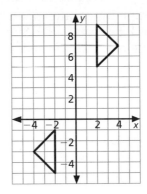

Solution

Think of the triangle in Quadrant III as the preimage and the triangle in Quadrant I as the image. Use what you know about transformations to identify the two transformations performed on the preimage.

The preimage has been reflected, or flipped, across the *y*-axis. The triangle was then moved up, or slid, in a vertical direction for 10 units.

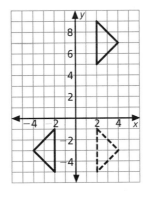

A composite of a reflection followed by a translation (or vice versa) is called a **glide reflection**.

▌ TRY THESE EXERCISES

1. Perform the following two transformations on rectangle *JKLM*: a translation five units to the left followed by a reflection over the *x*-axis.

2. **WRITING MATH** Begin with rectangle *CDEF* at $C(1, 2)$, $D(1, 4)$, $E(-3, 4)$, and $F(-3, 2)$. What two transformations could be used to create $C''D''E''F''$ at $C''(9, -2)$, $D''(9, -4)$, $E''(5, -4)$, and $F''(5, -2)$?

3. Does the order in which you perform the transformations for Exercise 2 affect the final image?

4. Describe two transformations that could be used to create the image in blue.

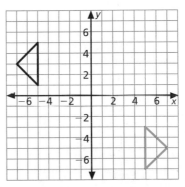

ART To produce a repeating pattern, an artist is asked to perform the following transformations. For Exercises 5–8, draw the result of the first transformation as a dashed figure and the result of the second transformation in red.

5. a translation 8 units to the left, followed by a translation 3 units down.

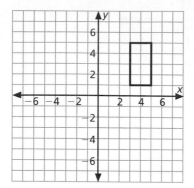

6. a translation 6 units up, followed by a reflection over the *y*-axis.

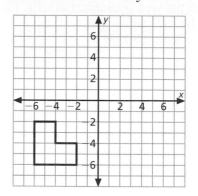

7. a reflection over the *y*-axis, followed by a dilation with center at the origin and a scale factor of 2.

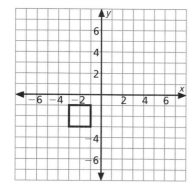

8. a clockwise rotation of 180° around the origin, followed by a clockwise rotation of 90° around the origin.

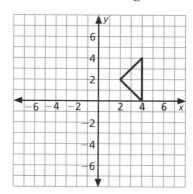

9. **GRAPHICS DESIGN** A designer has been asked to create a border by repeating a simple pattern that has both vertical and horizontal symmetry. Create the basic pattern using a triangle and multiple transformations. Write a paragraph describing the transformations.

In Exercises 10–11, describe two transformations that would create the image in blue. There may be more than one possible answer.

10.

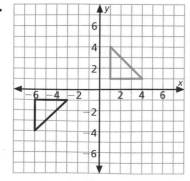

11.

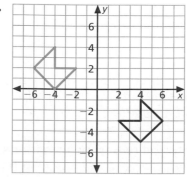

Tell whether the order in which you perform each pair of transformations affects the image produced. If it does affect the image, sketch an example.

12. a translation followed by another translation

13. a reflection followed by a rotation

14. a rotation followed by another rotation

15. a translation followed by a reflection

16. a reflection followed by another reflection

17. a translation followed by a rotation

◣ EXTENDED PRACTICE EXERCISES

18. In the figure at the right, $\triangle ABC$ is reflected over line p, and the image is then reflected over line q, where $p \parallel q$. What kind of single transformation would produce the same result as a composite of reflections over two parallel lines?

19. In the figure at the right, $\triangle DEF$ is reflected over line m, and then the image is reflected over line n, where lines m and n are intersecting lines. What kind of single transformation would produce the same result as a composite of reflections over two intersecting lines?

20. Trace the trapezoid shown at the right on another sheet of paper. Make three other copies of the trapezoid. Cut out the four trapezoids and rearrange them to form a larger trapezoid that is the same shape as the smaller trapezoid.

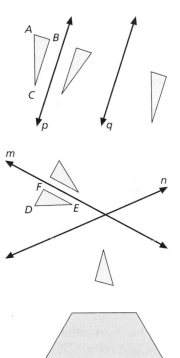

◣ MIXED REVIEW EXERCISES

Determine whether each pair of triangles is similar. If the triangles are similar, give a reason: write AA, SSS, or SAS. (Lesson 7-4)

21.

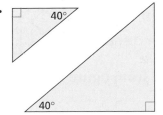

22.

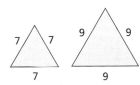

23.

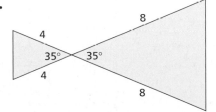

24.

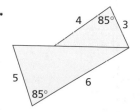

Review and Practice Your Skills

Using the figure on the right, draw each dilation image on a separate coordinate grid.

1. center of dilation is origin; scale factor is 3

2. center of dilation is origin; scale factor is $\frac{1}{2}$

3. center of dilation is A; scale factor is 2.5

4. center of dilation is B; scale factor is $\frac{3}{4}$

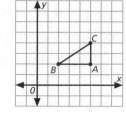

Using the figure on the right, draw each dilation image on a separate coordinate grid.

5. center of dilation is origin; scale factor is 1.5

6. center of dilation is origin; scale factor is $\frac{1}{4}$

7. center of dilation is F; scale factor is 2

8. center of dilation is G; scale factor is 3

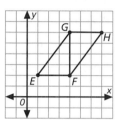

The following two sets of points are the vertices of triangles and their dilation images. Name the scale factor and the center of dilation for each.

9. $D(-5, -5), E(10, 0), F(-5, 0)$
   $D'(-5, -1), E'(-2, 0), F'(-5, 0)$

10. $S(2, 2), T(4, 8), U(6, 4)$
    $S'(6, 6), T'(12, 24), U'(18, 12)$

For each exercise, draw the result of the first transformation as a dashed figure and the result of the second transformation in red.

11. clockwise rotation of 90°; reflection across line $y = -x$

12. reflection across x-axis clockwise rotation of 270°

13. dilation—center at origin, scale factor of 2; translation 5 units up

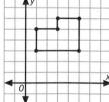

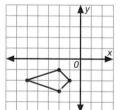

Determine the transformations necessary to create figure 2 from figure 1. There may be more than one possible answer.

14.

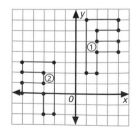

15.

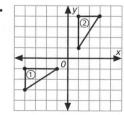

16.

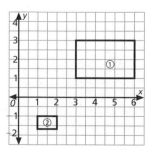

Graph the image of parallelogram *PQRS* with vertices *P*(5, 0), *Q*(−1, −1), *R*(1, 3), and *S*(7, 4). Then graph its image under each transformation from the original position. (Lesson 8-1)

17. 5 units down, 6 units left   **18.** reflected across $y = x$ **19.** reflected across $y = -2$

20. Compare the slopes of parallelogram *PQRS* in all four positions above.

Triangle *ABC* has vertices *A*(−2, 0), *B*(1, 3), and *C*(2, 7). Graph the triangle and its image after each of the following rotations about the origin. (Lesson 8-2)

21. 90° clockwise   **22.** 180° counterclockwise   **23.** 90° counterclockwise

24. Compare the slopes of triangle *ABC* in all four positions above.

The following sets of points are the vertices of triangles and their dilation images. Name the scale factor and the center of dilation for each. (Lesson 8-3)

25. $D(-5, -3), E(-5, 7), F(-1, 7)$
$D'(-15, -9), E'(-15, 21), F'(-3, 21)$

26. $D(-5, -3), E(-5, 7), F(-1, 7)$
$D'(-5, -18), E'(-5, 7), F'(5, 7)$

Mid-Chapter Quiz

On a coordinate grid, graph parallelogram *ABCD*, with vertices *A*(1, 6), *B*(2, 9), *C*(4, 6), and *D*(3, 3) and the following transformation images. (Lesson 8-1)

1. reflected across *x*-axis

2. translated two units right

3. $\triangle D'E'F'$ is the image after a rotation of 270° clockwise about the origin of $\triangle DEF$ with vertices $D(-5, -2)$, $E(-2, -1)$, and $F(-6, -7)$. Find the coordinates of D', E', and F' and the slopes of $\overline{DF}$ and $\overline{D'F'}$. (Lesson 8-2)

4. $\triangle ABC$ with vertices $A(1, -3)$, $B(6, -1)$, and $C(4, -5)$ has been rotated to $\triangle A'B'C'$ with vertices $A'(3, 1)$, $B'(1, 6)$, and $C'(5, 4)$. Name two rotations that could have been used. (Lesson 8-2)

5. Graph parallelogram *DEFG*, with vertices $D(-3, -1)$, $E(1, 3)$, $F(6, 2)$, and $G(2, -2)$, and the dilation image of parallelogram *DEFG* if the center of dilation is *D* and the scale factor is $\frac{1}{2}$. (Lesson 8-3)

6. Find the scale factor for $\triangle ABC$, with vertices $A(0, 0)$, $B(3, 6)$, and $C(9, 3)$, and dilation image $A'B'C'$, with vertices $A'(0, 0)$, $B'(4, 8)$, and $C'(12, 4)$. (Lesson 8-3)

7. $\triangle ABC$ has vertices $A(-2, -1)$, $B(-3, -4)$, and $C(-5, -3)$. Perform a reflection across the *y*-axis to obtain $\triangle A'B'C'$ followed by a reflection across the *x*-axis to obtain $\triangle A''B''C''$. (Lesson 8-4)

8. $\triangle D''E''F''$ with vertices $D''(1, 3)$, $E''(6, 3)$ and $F''(6, 1)$ is the image after two transformations of the $\triangle DEF$ with vertices $D(-4, -3)$, $E(1, -3)$, and $F(1, -1)$. Describe the two transformations used to create $\triangle D''E''F''$. (Lesson 8-4)

Addition and Multiplication with Matrices

Goals
- Identify matrices and elements within each matrix by rows and columns.
- Perform addition and scalar multiplication on matrices.

Applications Souvenir Sales, Cryptography, Manufacturing

Work with a partner to make a table of the following information.

Alsip and Bell work for Ski Park West. During the winter months, the company rents snowmobiles to park visitors. During December, January, and February, Alsip rented out 18, 12 and 15 snowmobiles, respectively, while Bell rented out 21, 15 and 8 during the same months.

Snowmobiles Sold

	Alsip	Bell
December		
January		
February		

1. Copy and complete the table shown at the right.

2. Make another table with the information, but reverse the rows and columns.

BUILD UNDERSTANDING

A **matrix** (plural: matrices) is a rectangular array of numbers arranged into rows and columns. Usually, square brackets enclose the numbers in a matrix. An array with m rows and n columns is called an $m \times n$ (read m by n) matrix.

The **dimensions** of the matrix are m and n. The numbers that make up the matrix are the **elements** of the matrix. The information in the snowmobile problem above can be shown by two different matrices, M_1 and M_2. The titles of rows and columns are not part of the matrices.

Technology Note

You can perform operations with matrices using a graphing calculator. To define a matrix, follow these steps:

1. Access the matrix menu.

2. Select a matrix name and press EDIT.

3. Enter dimensions (*row* $\times$ *column*). Then enter the matrix elements.

Matrix M_1

$$\begin{array}{c} \\ \text{Dec.} \\ \text{Jan.} \\ \text{Feb.} \end{array} \begin{array}{cc} \text{Alsip} & \text{Bell} \\ \begin{bmatrix} 18 & 21 \\ 12 & 15 \\ 15 & 8 \end{bmatrix} \end{array}$$

Matrix M_2

$$\begin{array}{c} \\ \text{Alsip} \\ \text{Bell} \end{array} \begin{array}{ccc} \text{Dec.} & \text{Jan.} & \text{Feb.} \\ \begin{bmatrix} 18 & 12 & 15 \\ 21 & 15 & 8 \end{bmatrix} \end{array}$$

Matrix M_1 has 3 rows and 2 columns. It is a 3×2 matrix. Matrix M_2 has 2 rows and 3 columns. So it is a 2×3 matrix. Although both matrices show the same information, they are not considered to be equal.

Two matrices are **equal matrices** if and only if they have the same dimensions and corresponding elements are equal.

Example 1

a. Find the dimensions of matrix C.

b. Identify the elements C_{32}, C_{21}, and C_{13}.

$$C = \begin{bmatrix} 17 & -2 & 3 \\ 0 & -1 & 5 \\ 6 & 1 & -7 \\ 5 & -4 & 2 \end{bmatrix}$$

Solution

a. *C* has 4 rows and 3 columns. So *C* is a 4 × 3 matrix.

b. C_{32} means the element in row 3, column 2.

$$C_{32} = 1, \ C_{21} = 0, \ C_{13} = 3$$

Math: Who, Where, When

The theory of matrices was first developed around 1858 by Arthur Cayley, a British mathematician.

SOUVENIR SALES A theme park sells red, white and blue sweatshirts in small, medium and large sizes. A manager at a souvenir stand receives two shipments of sweatshirts.

| | **Shipment 1** | | | | **Shipment 2** | | |
---	red	blue	white		red	blue	white
Small	18	12	15	Small	15	13	12
Med	15	33	14	Med	18	15	16
Large	16	15	18	Large	13	8	16

To find the combined inventory, she adds the corresponding elements in each matrix. This illustrates the concept of **matrix addition**.

If two matrices *M* and *N* have the same dimensions, their sum *M* + *N* is the matrix in which each element is the sum of the corresponding elements in *M* and *N*.

Example 2

Find the sum of the matrices shown for Shipment 1 and Shipment 2 above.

Solution

Call the matrices *A* and *B*.

Check Understanding

How is the addition of matrices just like adding two numbers? How is the scalar multiplication of matrices just like multiplying two numbers?

$$A = \begin{bmatrix} 18 & 12 & 15 \\ 15 & 33 & 14 \\ 16 & 15 & 18 \end{bmatrix} \quad B = \begin{bmatrix} 15 & 13 & 12 \\ 18 & 15 & 16 \\ 13 & 8 & 16 \end{bmatrix}$$

$$A + B = \begin{bmatrix} 18 + 15 & 12 + 13 & 15 + 12 \\ 15 + 18 & 33 + 15 & 14 + 16 \\ 16 + 13 & 15 + 8 & 18 + 16 \end{bmatrix} = \begin{bmatrix} 33 & 25 & 27 \\ 33 & 48 & 30 \\ 29 & 23 & 34 \end{bmatrix}$$

Suppose the shop owner wanted to double her total inventory for the holiday season. She could simply double each element of the matrix. This illustrates an operation on matrices called **scalar multiplication**.

A matrix can be multiplied by a constant *k* called a **scalar**. The product of a scalar *k* and matrix *A* is the matrix *kA* in which each element is *k* times the corresponding element in *A*.

Example 3

CRYPTOGRAPHY The matrix below represents numerical information that must be transmitted electronically. As the first step in encrypting the information, the matrix is multiplied by 5.

Find the product: $5 \begin{bmatrix} 8 & 12 & 10 \\ 3 & 4 & -3 \\ -2 & 0 & 6 \end{bmatrix}$

Solution

Every element in the matrix must be multiplied by 5.

$$5 \begin{bmatrix} 8 & 12 & 10 \\ 3 & 4 & -3 \\ -2 & 0 & 6 \end{bmatrix} = \begin{bmatrix} 5 \cdot 8 & 5 \cdot 12 & 5 \cdot 10 \\ 5 \cdot 3 & 5 \cdot 4 & 5 \cdot (-3) \\ 5 \cdot (-2) & 5 \cdot 0 & 5 \cdot 6 \end{bmatrix} = \begin{bmatrix} 40 & 60 & 50 \\ 15 & 20 & -15 \\ -10 & 0 & 30 \end{bmatrix}$$

◥ TRY THESE EXERCISES

1. Find the dimensions of M.

2. Identify the elements M_{34} and M_{12}.

$$M = \begin{bmatrix} 2 & -1 & 4 & 3 \\ 1 & 0 & -4 & 5 \\ 6 & -1 & 2 & 10 \end{bmatrix}$$

3. How many elements are in a 5×2 matrix?

4. Refer to matrix M above at the right. Find $6M$.

5. Find $C + D$ if $C = \begin{bmatrix} 3 & -5 \\ 4 & 2 \end{bmatrix}$ and $D = \begin{bmatrix} -3 & 2 \\ 1 & -2 \end{bmatrix}$.

6. Three schools have the following win–loss record: Appleton, 6 wins, 9 losses; Carrollton, 14 wins, 2 losses; Prestonsville, 12 wins, 5 losses. Show this information in a matrix.

◥ PRACTICE EXERCISES • For Extra Practice, see page 688.

Find the dimensions of each matrix.

7. $\begin{bmatrix} 2 & -5 & 3 \\ 1 & 4 & -6 \end{bmatrix}$

8. $\begin{bmatrix} 11 & -8 & 3 & 5 \\ \frac{1}{2} & 0 & 0 & 7 \\ -8 & 0 & 0 & 0 \end{bmatrix}$

9. $\begin{bmatrix} 0 & 3 \\ 1 & -1 \\ 5 & -5 \\ 2 & 7 \\ 8 & 1.9 \end{bmatrix}$

Use the following matrices in Exercises 10–18.

$$A = \begin{bmatrix} 2 & -5 & 1 \\ 3 & 1 & -7 \end{bmatrix} \quad B = \begin{bmatrix} -12 & 4 & 6 \\ 8 & -2 & -4 \end{bmatrix} \quad C = \begin{bmatrix} 6 & -2 & -3 \\ -4 & 1 & 2 \end{bmatrix}$$

Find each of the following.

10. $2A$

11. $-\frac{1}{2}B$

12. $\frac{3}{2}C$

13. $A + B$

14. $A + C$

15. $A + (-2B)$

16. $B + 2C$

17. $A + \frac{1}{2}B$

18. **MANUFACTURING** In June, a boat manufacturer produced 12 sailboats, 18 catamarans, and 8 yachts. In July, the company produced 10 sailboats, 9 catamarans, and 9 yachts. In August, it produced 9 sailboats, 8 catamarans, and 12 yachts. Write two different 3×3 matrices to show this information.

19. **DATA FILE** Use the data on the calorie count of foods on page 650. Create two different matrices showing the amount of the serving and the calorie count for white bread, whole milk, and spaghetti with meatballs.

20. **CHAPTER INVESTIGATION** Using the scale drawing of your roller coaster, estimate the track length of the entire coaster. Write a paragraph explaining how you arrived at your estimate.

21. **WRITING MATH** Describe a method for remembering the difference between rows and columns.

Matrix subtraction is defined by using the scalar -1.
If A and B are matrices with dimensions $m \times n$, then $A - B = A + (-1)B$.

Use the above definition in Exercises 23–26.

22. $\begin{bmatrix} 2 \\ 3 \end{bmatrix} - \begin{bmatrix} -2 \\ 7 \end{bmatrix}$

23. $\begin{bmatrix} 5 & -1 \\ 7 & 6 \end{bmatrix} - \begin{bmatrix} 1 & -5 \\ 4 & 3 \end{bmatrix}$

24. $\begin{bmatrix} 0 & 4 & -2 \\ -5 & 1 & 3 \end{bmatrix} - \begin{bmatrix} 6 & -7 & -1 \\ 0 & -4 & 8 \end{bmatrix}$

25. $[17 \quad 6 \quad 12] - [-3 \quad -5 \quad 9]$

Remember that two matrices are equal if and only if they have the same dimensions and all their corresponding elements are equal. Use this definition to solve for x and y.

26. $[x \quad 4y] = [y + 5 \quad 2x + 10]$

27. $\begin{bmatrix} x + y \\ x - y \end{bmatrix} = \begin{bmatrix} 9 \\ 4 \end{bmatrix}$

◼ EXTENDED PRACTICE EXERCISES

28. **YOU MAKE THE CALL** Dawnae says that any two matrices can be added together. Do you agree? If not, why not?

Refer to matrices R and S for Exercises 30 and 31.

$R = \begin{bmatrix} 4 & 6 \\ 3 & 6 \end{bmatrix} \quad S = \begin{bmatrix} 6 & 5 \\ 1 & -3 \end{bmatrix}$

29. Find $R + S$ and $S + R$. Does addition of matrices seem to be a commutative operation?

30. Find $R - S$ and $S - R$. Does subtraction of matrices seem to be a commutative operation?

31. **POPULATION** The matrices E, W, and N, shown at the right, give the enrollments by gender and grade at East, West, and North High Schools. In each matrix, Row 1 gives the number of boys and Row 2 the number of girls. Columns 1 to 4 give the number of students in grades 9 through 12, respectively. Calculate entries in matrix T that show the total enrollment by gender and grade in the three schools.

$$E = \begin{matrix} & 9 & 10 & 11 & 12 & \\ \begin{bmatrix} 180 & 220 & 265 & 250 \\ 205 & 231 & 255 & 260 \end{bmatrix} & \begin{matrix} \text{boys} \\ \text{girls} \end{matrix} \end{matrix}$$

$$W = \begin{bmatrix} 306 & 300 & 340 & 310 \\ 290 & 314 & 270 & 350 \end{bmatrix} \begin{matrix} \text{boys} \\ \text{girls} \end{matrix}$$

$$N = \begin{bmatrix} 408 & 410 & 406 & 389 \\ 380 & 420 & 444 & 370 \end{bmatrix} \begin{matrix} \text{boys} \\ \text{girls} \end{matrix}$$

◼ MIXED REVIEW EXERCISES

32. Find x to the nearest tenth in the pair of similar triangles. (Lesson 7-5)

Given $f(x) = -2x - 5$ **and** $g(x) = 3x^2$, **find each value.** (Lesson 2-2)

33. $f(-3)$

34. $f(2)$

35. $f(-8)$

36. $f(5)$

37. $g(-2)$

38. $g(3)$

39. $g(-5)$

40. $g(4)$

More Operations on Matrices

Goals
- Determine dimensions of product matrices in matrix multiplication.
- Perform row-by-column multiplication of matrices.

Applications   Ticket Sales, Encryption, Inventory

A theme park sells three kinds of tickets. Adults over age 18 pay $15, students from 13 through 18 pay $10, and children under 13 pay $5. On one day, the park sells 280 adult tickets, 420 student tickets, and 382 children's tickets. Notice how this information can be shown by two matrices.

number of tickets

$$[280 \quad 420 \quad 382]$$

adult   student child

cost per ticket

$$\begin{bmatrix} \$15 \\ \$10 \\ \$5 \end{bmatrix} \begin{matrix} \text{adult} \\ \text{student} \\ \text{child} \end{matrix}$$

Write an expression to show how to compute the total receipts for the day, and then find the total receipts.

◥ BUILD UNDERSTANDING

Matrix multiplication is done by using **row-by-column multiplication**. *You can only multiply two matrices when the number of columns in the first matrix is equal to the number of rows in the second matrix.*

To multiply a row by a column, multiply the first element in the row by the first element in the column, the second element in the row by the second element in the column, and so on. (Thus, the number of elements in a row must equal the number of elements in a column.) Finally, add the products.

The product of an $m \times n$ matrix by an $n \times p$ matrix is an $m \times p$ matrix.

$$\begin{matrix} A & \times & B & = & AB \\ m \times n & & n \times p & & m \times p \end{matrix}$$

Example 1

TICKET SALES Multiply the two matrices at the top of the page to find the total receipts from ticket sales for the theme park.

Solution

The first matrix is a 1×3 matrix and the second is a 3×1 matrix. So the product will be a 1×1 matrix. Use row-by-column multiplication.

$$[280 \quad 420 \quad 382] \cdot \begin{bmatrix} 15 \\ 10 \\ 5 \end{bmatrix} = 280 \cdot 15 + 420 \cdot 10 + 382 \cdot 5$$
$$= 4200 + 4200 + 1910$$
$$= 10{,}310$$

The total receipts equal $10,310.

Example 2

Let $M = \begin{bmatrix} 1 & 3 & 4 \\ 5 & -2 & 6 \end{bmatrix}$ and $N = \begin{bmatrix} 1 & 4 & 3 & 0 \\ 5 & -2 & 1 & 6 \\ -4 & 0 & 5 & 7 \end{bmatrix}$. **Find the dimensions of MN.**

Solution

Because M is a 2×3 matrix and N is a 3×4 matrix, MN is a 2×4 matrix.

$$\underset{2 \times 4}{\underbrace{\underset{2 \times 3}{\rule{0pt}{1em}} \qquad \underset{3 \times 4}{\rule{0pt}{1em}}}}$$

(Notice that you cannot find the product NM, because the number of columns in N is not the same as the number of rows in M.)

Example 3

ENCRYPTION A business uses a coding matrix to encrypt customer account numbers. Matrix A includes the last four digits of a customer's account number. Matrix B is the coding matrix.

Let $A = \begin{bmatrix} 1 & 3 \\ 5 & -2 \end{bmatrix}$ and $B = \begin{bmatrix} 8 & 7 & 0 \\ 4 & 2 & 6 \end{bmatrix}$. Find AB.

Solution

Because A is a 2×2 matrix and B is a 2×3 matrix, the product is a 2×3 matrix. The product of row 1 of A and column 1 of B is $1(8) + 3(4) = 20$. Write 20 in row 1 and column 1 of the product matrix.

$$\begin{bmatrix} \boxed{1 \quad 3} \\ 5 \quad -2 \end{bmatrix} \begin{bmatrix} \boxed{8} & 7 & 0 \\ \boxed{4} & 2 & 6 \end{bmatrix} = \begin{bmatrix} 20 & \rule{1em}{0.4pt} & \rule{1em}{0.4pt} \\ \rule{1em}{0.4pt} & \rule{1em}{0.4pt} & \rule{1em}{0.4pt} \end{bmatrix}$$

The product of row 1 of A and column 2 of B is $1(7) + 3(2) = 13$. Write 13 in row 1 and column 2 of the product.

$$\begin{bmatrix} \boxed{1 \quad 3} \\ 5 \quad -2 \end{bmatrix} \begin{bmatrix} 8 & \boxed{7} & 0 \\ 4 & \boxed{2} & 6 \end{bmatrix} = \begin{bmatrix} 20 & 13 & \rule{1em}{0.4pt} \\ \rule{1em}{0.4pt} & \rule{1em}{0.4pt} & \rule{1em}{0.4pt} \end{bmatrix}$$

The other elements in the product are formed by using this row $\boxed{}$ by column $\boxed{}$ pattern.

For instance, the element in the second row, third column of the product is found by multiplying the second row of A by the third column of B. This answer is shown below, along with the final result.

$$\begin{bmatrix} 1 & 3 \\ 5 & -2 \end{bmatrix} \begin{bmatrix} 8 & 7 & 0 \\ 4 & 2 & 6 \end{bmatrix} = \begin{bmatrix} 20 & 13 & 18 \\ 32 & 31 & -12 \end{bmatrix}$$

Refer to the matrices below. Find the dimensions of each product, if possible. *Do not multiply.* If it is not possible to multiply, write NP.

$$P = \begin{bmatrix} 1 \\ 3 \\ 4 \end{bmatrix} \quad Q = [1 \quad 6 \quad 7 \quad 9] \quad R = \begin{bmatrix} 2 & 3 \\ 1 & 4 \\ 6 & 2 \end{bmatrix} \quad S = \begin{bmatrix} 5 \\ 3 \end{bmatrix}$$

1. PQ
2. PR
3. SQ
4. RS
5. QP
6. SR
7. QS
8. SP

Find each product, if possible. If not possible, write NP.

$$A = [10 \quad 18 \quad 5] \quad B = \begin{bmatrix} 5 \\ 6 \\ 1 \end{bmatrix} \quad C = \begin{bmatrix} 6 & 1 \\ 5 & 0 \\ 3 & 3 \end{bmatrix}$$

9. AB
10. AC
11. CA

Refer to the matrices below. Find the dimensions of each product, if possible. *Do not multiply.* If it is not possible to multiply, write NP.

$$D = \begin{bmatrix} 1 & 3 & 5 \\ 0 & 6 & 4 \end{bmatrix} \quad E = \begin{bmatrix} 0 & 4 \\ 5 & 6 \\ 6 & 1 \end{bmatrix} \quad F = \begin{bmatrix} 3 \\ 4 \\ 5 \end{bmatrix} \quad G = [2 \quad 3 \quad 6]$$

12. DE
13. ED
14. FG
15. GF
16. EG
17. FD
18. DF
19. GE

MATRICES Find each product using a graphing calculator. If not possible, write NP.

20. $\begin{bmatrix} -1 & 0 \\ 0 & -1 \end{bmatrix} \begin{bmatrix} 4 \\ 2 \end{bmatrix}$

21. $[2 \quad 1 \quad 5] \begin{bmatrix} -2 \\ -3 \\ 7 \end{bmatrix}$

22. $\begin{bmatrix} 1 & 0 \\ 0 & 1 \end{bmatrix} \begin{bmatrix} 3 \\ 5 \end{bmatrix}$

23. $\begin{bmatrix} -2 & 4 & 0 \\ -3 & 0 & -8 \end{bmatrix} \begin{bmatrix} 1 & -1 \\ 2 & -1 \end{bmatrix}$

Find each product. If not possible, write NP.

24. $\begin{bmatrix} 1 & 2 \\ 3 & 4 \end{bmatrix} \begin{bmatrix} 1 & 2 \\ -3 & 4 \end{bmatrix}$

25. $\begin{bmatrix} 3 & 0 & 1 \\ 5 & 0 & 6 \end{bmatrix} [6 \quad 4 \quad 1]$

26. $\begin{bmatrix} -2 & 4 & 0 \\ -3 & 0 & -8 \end{bmatrix} \begin{bmatrix} -1 & -2 & -3 \\ 0 & 1 & 0 \\ 4 & 5 & 2 \end{bmatrix}$

> **Problem Solving Tip**
>
> To prevent inappropriate application of the row-by-column multiplication method, learn this rhyme:
>
> Row-Col is Pro
> Col-Row is No

27. INVENTORY Find the product JK, which gives the number of small, medium and large T-shirts in inventory at two souvenir stands.

$$J = \begin{bmatrix} 2 & 3 \\ 1 & 5 \end{bmatrix} \quad K = \begin{bmatrix} 3 & 5 & 1 \\ 2 & 1 & 3 \end{bmatrix}$$

28. Use the rule $A(kB) = (kA)B = k(AB)$ to compute: $\begin{bmatrix} 2 & 8 \\ 4 & 2 \end{bmatrix} \left(\frac{1}{2} \begin{bmatrix} 3 & 1 \\ -2 & 1 \end{bmatrix} \right)$.

29. **ENCRYPTION** The data in A must be encrypted by multiplying by B. Find AB and BA.

$$A = \begin{bmatrix} 2 & 1 \\ 4 & 3 \end{bmatrix} \qquad B = \begin{bmatrix} 5 & 1 \\ -3 & 2 \end{bmatrix}$$

30. **WRITING MATH** What can you conclude about the multiplication of matrices from the products in Exercise 29?

31. For $A = \begin{bmatrix} 3 & 2 \\ -1 & 5 \end{bmatrix}$ and $I = \begin{bmatrix} 1 & 0 \\ 0 & 1 \end{bmatrix}$, find AI and IA.

32. **WRITING MATH** What can you conjecture about matrix I in Exercise 31?

33. For $A = \begin{bmatrix} 1 & -1 \\ 0 & 3 \end{bmatrix}$, $B = \begin{bmatrix} 3 & 2 \\ 1 & 5 \end{bmatrix}$ and $C = \begin{bmatrix} 0 & 1 \\ -3 & 2 \end{bmatrix}$, show that $A(BC) = (AB)C$.

■ EXTENDED PRACTICE EXERCISES

34. If M^2 means $M \times M$, find the matrix M^2 if $M = \begin{bmatrix} 3 & 2 \\ 1 & 4 \end{bmatrix}$.

35. Find X^2, X^3, and X^4 if $X = \begin{bmatrix} 0 & -1 \\ -1 & 0 \end{bmatrix}$.

36. For $A = \begin{bmatrix} 1 & 0 & 0 \\ 0 & 1 & 0 \\ 0 & 0 & 1 \end{bmatrix}$, find A^3.

PARK ADMISSIONS A theme park offers a discount to members of a travel club. Table A shows the daily ticket sales for the park. Table B shows the average cost per person for park attractions.

A.

	Club	Non-Club
Adult	2000	1700
Children	5400	4200

B.

	Admission	Food	Souvenirs
Club	$23	$18	$30
Non-Club	$26	$20	$35

37. Write matrices for the A and B tables and find the product AB to find the amount spent by park visitors by category.

38. What do the rows and columns of the product matrix AB represent?

39. **CRITICAL THINKING** Solve for x and y: $\begin{bmatrix} 6 & 2 \\ 8 & 14 \end{bmatrix} \begin{bmatrix} x \\ y \end{bmatrix} = \begin{bmatrix} 4 \\ -6 \end{bmatrix}$.

■ MIXED REVIEW EXERCISES

Solve each system of equations by graphing. (Lesson 6-4)

40. $y = -3x - 4$
$y = -2x - 8$

41. $2y = 4x + 10$
$3y = -3x + 6$

42. $4x = 2y - 6$
$-3y = x + 5$

43. $5x = y + 7$
$-4y + x = -10$

44. $-9(x - 3) = 6y$
$-4y + 28 = 8x$

45. $3y = 2(x + 1.5)$
$5x = -2(-3y)$

Review and Practice Your Skills

Use matrices A, B, and C to find each of the following.

$$A = \begin{bmatrix} 4 & 0 & -5 & 7 \\ -3 & 8 & -2 & 1 \end{bmatrix} \qquad B = \begin{bmatrix} -11 & 3 & -1 & 6 \\ 5 & 2.5 & 0 & -6 \end{bmatrix} \qquad C = \begin{bmatrix} 0 & -8 & 13 & -5 \\ 1 & -4 & 7.5 & 9 \end{bmatrix}$$

1. $3A$

2. $A + B$

3. $C - A$

4. $2A + 2C$

5. $A - C$

6. $B + A$

7. $B + \frac{1}{2}A$

8. $A + B + C$

9. $C - B + A$

10. $5A - 2B$

11. $2(A + C)$

12. $\frac{2}{3}C$

13. $-\frac{2}{5}B$

14. $3A + 2B - 4C$

15. $(A + 2B) - 2(B + A)$

16. $-10B + 20C$

17. element A_{12}

18. element B_{24}

19. dimensions of C

20. $\frac{1}{2}$(element C_{21})

Solve for x and y.

21. $\begin{bmatrix} x + y \\ x - y \end{bmatrix} = \begin{bmatrix} -2 \\ -8 \end{bmatrix}$

22. $\begin{bmatrix} 3x + 2y \\ x - 5y \end{bmatrix} = \begin{bmatrix} -4 \\ 27 \end{bmatrix}$

23. $\begin{bmatrix} 15 - 6x \\ 8x + 3y \end{bmatrix} = \begin{bmatrix} 3y \\ 13 \end{bmatrix}$

24. $[3x \quad -2.5y] = [51 \quad 40]$

25. $[x + 2y \quad 3x - 5y] = [-1 \quad 2.5]$

26. $\begin{bmatrix} \dfrac{1}{x} & \dfrac{1}{y+3} \end{bmatrix} = \begin{bmatrix} -\dfrac{3}{4} & -\dfrac{13}{52} \end{bmatrix}$

Use the given matrices to find each of the following. If not possible, write NP.

$$M = \begin{bmatrix} 4 & -3 \\ 2 & 1 \end{bmatrix} \quad N = \begin{bmatrix} 0 & 2 \\ -5 & 4 \end{bmatrix} \quad P = \begin{bmatrix} 6 & 0 & 10 \\ -1 & 1 & 7 \end{bmatrix} \quad Q = \begin{bmatrix} -8 & 2 \\ 1 & 9 \\ -3 & 5 \end{bmatrix} \quad R = \begin{bmatrix} 6 \\ -10 \end{bmatrix}$$

27. MN

28. MP

29. MQ

30. MR

31. NM

32. PM

33. QM

34. RM

35. NP

36. NQ

37. NR

38. PN

39. QN

40. RN

41. PQ

42. PR

43. QP

44. RP

45. QR

46. RQ

47. $MN + PQ$

48. $MR - NR$

49. M^2

50. N^3

51. NMP

52. PQN

53. MPQ

54. $-QPQ$

55. QNR

56. $R(M + N)$

57. $(M + N)R$

58. $QMP + \frac{1}{2}QP$

Solve for x and y.

59. $\begin{bmatrix} 6 & -2 \\ -1 & 5 \end{bmatrix}\begin{bmatrix} x \\ y \end{bmatrix} = \begin{bmatrix} -8 \\ -22 \end{bmatrix}$

60. $\begin{bmatrix} 14 & 5 \\ -2 & 13 \end{bmatrix}\begin{bmatrix} x \\ y \end{bmatrix} = \begin{bmatrix} 1 \\ -7 \end{bmatrix}$

Trapezoid *ABCD* has vertices *A*(−5, 0), *B*(−5, 3), *C*(−2, 4), and *D*(1, 2). Graph the trapezoid and its image after each of the following rotations about the origin from the original position. (Lesson 8-2)

61. 90° counterclockwise   **62.** 180° clockwise   **63.** 45° clockwise

The following sets of points are the vertices of *PQRS* and its dilation image. Name the scale factor and the center of dilation. (Lesson 8-3)

64. *P*(−2, 2), *Q*(−2, 5), *R*(−3, 5), *S*(−6, 2)
  P′(−2, 2), *Q′*(−2, 20), *R′*(−8, 20) *S′*(−26, 2)

Tell whether the order in which you perform each pair of transformations affects the image produced. If it does affect the image, sketch an example. (Lesson 8-4)

65. rotation and translation

66. dilation (center at figure vertex) and translation

67. dilation (center at the origin) and reflection

68. rotation and reflection

MathWorks — Career – Aerospace Engineer
Workplace Knowhow

Aerospace engineers use their knowledge of structural design, aerodynamics, propulsion, thermodynamics, and acoustics to design roller coasters. They know how tight a turn can be without endangering passengers or damaging the coaster's structure. These workers use science to make a roller coaster fast, fun and safe.

Aerospace engineers are also employed to build aircraft and spacecraft and to develop military technology. These engineers apply technology learned in other industries to transportation on land, sea and air. Aerospace engineers must understand math and physics and how to use computers, calculators and other tools to test their ideas.

You have designed a roller coaster in the shape shown at the right. This coaster is for people over 48 in. tall. The amusement park now wants you to design a children's version of the coaster on a smaller scale with a less steep first hill.

1. Find the slope between points *A* and *B*.

2. Find the slope between points *B* and *C*.

3. Replot point *B* so that the slope of $\overline{AB}$ is 2 and slope of $\overline{BC}$ is −2. What are the new coordinates for *B*?

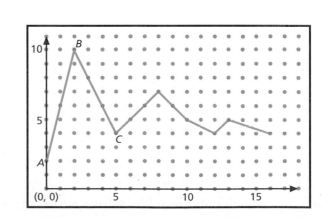

Transformations and Matrices

Goals
- Represent geometric figures on the coordinate plane by matrices.
- Identify and perform transformations with matrices.

Applications Graphic Art, Ride Design

Work with a partner. Use the point (2, 6) to answer each of the following.

1. What point is (2, 6) reflected over the *x*-axis?

2. What point is (2, 6) reflected over the *y*-axis?

3. What point is (2, 6) reflected over the line $y = x$?

4. What point is (2, 6) reflected over the line $y = -x$?

◥ BUILD UNDERSTANDING

A point can be represented by a matrix, as well as an ordered pair.

The ordered pair (2, 6) can be represented by the matrix $\begin{bmatrix} 2 \\ 6 \end{bmatrix}$.

The element in the first row is the *x*-coordinate, and the element in the second row is the *y*-coordinate.

In general, the ordered pair (*x*, *y*) is represented by the matrix $\begin{bmatrix} x \\ y \end{bmatrix}$.

In a similar way, a matrix can be used to denote a polygon. Because each vertex is a point, each point can be represented by a matrix. These four matrices for the vertices can be combined into a single matrix.

Example 1

First, represent each vertex of quadrilateral *ABCD* with a 2 × 1 matrix. Then combine these matrices into a single 2 × 4 matrix.

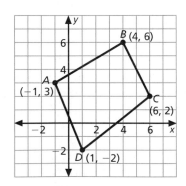

Solution

The vertices can be represented as follows.

$$\begin{bmatrix} -1 \\ 3 \end{bmatrix}, \begin{bmatrix} 4 \\ 6 \end{bmatrix}, \begin{bmatrix} 6 \\ 2 \end{bmatrix}, \begin{bmatrix} 1 \\ -2 \end{bmatrix}$$

Putting the column matrices into a single 2 × 4 matrix, you have the following.

$$\begin{bmatrix} -1 & 4 & 6 & 1 \\ 3 & 6 & 2 & -2 \end{bmatrix}$$

Each column refers to one vertex of the quadrilateral.

Check Understanding

What matrix represents △*ABC* if the vertices are *A*(1, 4), *B*(−2, 3), and *C*(5, −7)?

Just as points and polygons can be represented by matrices, you can represent transformations with matrices. Below is a table of matrices for reflections.

Reflection	Matrix	Reflection	Matrix
over the x-axis	$\begin{bmatrix} 1 & 0 \\ 0 & -1 \end{bmatrix}$	over the line $y = x$	$\begin{bmatrix} 0 & 1 \\ 1 & 0 \end{bmatrix}$
over the y-axis	$\begin{bmatrix} -1 & 0 \\ 0 & 1 \end{bmatrix}$	over the line $y = -x$	$\begin{bmatrix} 0 & -1 \\ -1 & 0 \end{bmatrix}$

E x a m p l e   2

Find the reflection image of $\triangle ABC$ with vertices at $A(1, -2)$, $B(6, -2)$, and $C(4, -5)$ when the triangle is reflected over the line $y = x$. Use matrices.

Solution

Triangle ABC can be represented by

$$\begin{bmatrix} 1 & 6 & 4 \\ -2 & -2 & -5 \end{bmatrix}.$$

The matrix representing a reflection over the line $y = x$ is $\begin{bmatrix} 0 & 1 \\ 1 & 0 \end{bmatrix}$.

Multiply the two matrices.

$$\begin{bmatrix} 0 & 1 \\ 1 & 0 \end{bmatrix} \begin{bmatrix} 1 & 6 & 4 \\ -2 & -2 & -5 \end{bmatrix}$$

So, the image of $\triangle ABC$ is $\begin{bmatrix} -2 & -2 & -5 \\ 1 & 6 & 4 \end{bmatrix}$.

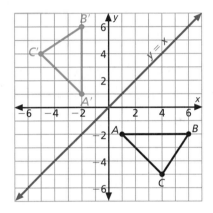

E x a m p l e   3

GRAPHIC ART An artist is creating a border using design software. To create the basic pattern, she enters the coordinates for $\triangle XYZ$ with vertices $X(0, 0)$, $Y(2, -3)$, and $Z(6, -3)$. She wants to draw the triangle reflected over the y-axis. Find the coordinates of the reflection image using matrices.

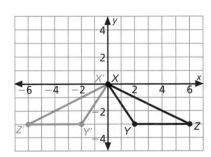

Solution

Let $\begin{bmatrix} 0 & 2 & 6 \\ 0 & -3 & -3 \end{bmatrix}$ represent the triangle.

Then multiply by $\begin{bmatrix} -1 & 0 \\ 0 & 1 \end{bmatrix}$, the matrix for a reflection over the y-axis.

$$\begin{bmatrix} -1 & 0 \\ 0 & 1 \end{bmatrix} \begin{bmatrix} 0 & 2 & 6 \\ 0 & -3 & -3 \end{bmatrix} = \begin{bmatrix} 0 & -2 & -6 \\ 0 & -3 & -3 \end{bmatrix}$$

So, the coordinates of the $X'Y'Z'$ are shown in the matrix $\begin{bmatrix} 0 & -2 & -6 \\ 0 & -3 & -3 \end{bmatrix}$.

Represent each geometric figure with a matrix.

1.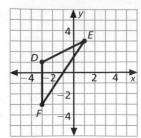

2.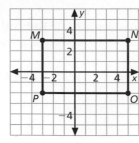

3.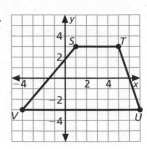

RIDE DESIGN Amusement park rides are tested using computer simulations. A triangle with vertices $\begin{bmatrix} -1 \\ 2 \end{bmatrix}$, $\begin{bmatrix} 3 \\ 7 \end{bmatrix}$, and $\begin{bmatrix} 7 \\ 3 \end{bmatrix}$ is used to represent a moving platform to which the ride's cars are attached. Find the reflection images of the triangle.

4. over the x-axis.　　**5.** over the line $y = x$.　　**6.** over the y-axis.　　**7.** over the line $y = -x$.

Represent each geometric figure with a matrix.

8.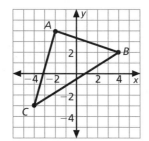

9.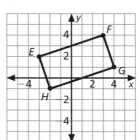

10.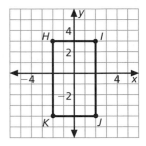

 CALCULATOR Multiply matrices using a graphing calculator to find the following reflection images of the quadrilateral represented by the matrix $\begin{bmatrix} -2 & 1 & 7 & 4 \\ 7 & 3 & 4 & 7 \end{bmatrix}$.

11. over the line $y = x$　　**12.** over the y-axis　　**13.** over the line $y = -x$　　**14.** over the x-axis

Interpret each equation as indicating:
The reflection image of point ___?___ over ___?___ is the point ___?___.

15. $\begin{bmatrix} 0 & -1 \\ -1 & 0 \end{bmatrix} \begin{bmatrix} 4 \\ 4 \end{bmatrix} = \begin{bmatrix} -4 \\ -4 \end{bmatrix}$

16. $\begin{bmatrix} -1 & 0 \\ 0 & 1 \end{bmatrix} \begin{bmatrix} 3 \\ 2 \end{bmatrix} = \begin{bmatrix} -3 \\ 2 \end{bmatrix}$

17. $\begin{bmatrix} 0 & 1 \\ 1 & 0 \end{bmatrix} \begin{bmatrix} -2 \\ 4 \end{bmatrix} = \begin{bmatrix} 4 \\ -2 \end{bmatrix}$

18. $\begin{bmatrix} 1 & 0 \\ 0 & -1 \end{bmatrix} \begin{bmatrix} 5 \\ -3 \end{bmatrix} = \begin{bmatrix} 5 \\ 3 \end{bmatrix}$

 19. TALK ABOUT IT Toni says that you can produce any dilation with center at the origin and a scale factor of k using the matrix $\begin{bmatrix} k & 0 \\ 0 & k \end{bmatrix}$. Does Toni's thinking make sense?

GAME DEVELOPMENT In a hand-held game, the player must click on falling stars within a time limit. To develop the game, the programmer specifies the coordinates for a star to appear. Two seconds later, the image of the star appears under a reflection. Exercises 20–23 specify image and preimage points used in the game. For each, name the reflecting line and verify your answer by matrix multiplication.

20. preimage $(5, -1)$, image $(-1, 5)$

21. preimage $(2, 0)$, image $(0, -2)$

22. preimage (b, a), image $(b, -a)$

23. preimage $(7, 3)$, image $(-7, 3)$

24. **GEOMETRY SOFTWARE** Find the image of rhombus $ABCD$ under the transformation associated with matrix M. Graph both the preimage and its image using geometric-drawing software.

$$M = \begin{bmatrix} 3 & 0 \\ 0 & 3 \end{bmatrix} \qquad ABCD = \begin{bmatrix} 3 & 5 & 3 & 1 \\ 2 & 0 & -2 & 0 \end{bmatrix}$$

■ EXTENDED PRACTICE EXERCISES

25. **WRITING MATH** What type of transformation is represented by the matrix $\begin{bmatrix} -2 & 0 \\ 0 & -2 \end{bmatrix}$?

26. Draw any triangle in the coordinate plane. Represent it with a 2×3 matrix. Then apply each transformation below. Draw each preimage and image on a coordinate grid.

a. $\begin{bmatrix} 0 & -1 \\ 1 & 0 \end{bmatrix}$ b. $\begin{bmatrix} -1 & 0 \\ 0 & -1 \end{bmatrix}$ c. $\begin{bmatrix} 0 & 1 \\ -1 & 0 \end{bmatrix}$

27. Refer to the graphs you drew in Exercise 26.

a. Which shows a clockwise rotation of 90°?

b. Which shows a clockwise rotation of 180°?

c. Which shows a counterclockwise rotation of 90°?

28. **CHAPTER INVESTIGATION** Measure the angle of descent for each hill in your roller coaster design. Estimate your coaster's top speed by comparing its features to the coasters shown in the table on page 337. Write a paragraph to justify your estimate.

■ MIXED REVIEW EXERCISES

Find the slope and y-intercept for each line. (Lesson 6-1)

29. $y = \frac{1}{2}x - 3$

30. $y = -3x + 4$

31. $2y - x = 6$

32. $3(x - 4) = 5y$

33. $3y - 4x - 7 = 0$

34. $2x = 4y + 2$

Problem Solving Skills: Use a Matrix

Some problems can be solved by translating directly to a matrix and then performing matrix operations. Consider using matrix operations to solve problems whenever information can be easily organized into tables with corresponding elements.

Problem Solving Strategies

Guess and check

Look for a pattern

Solve a simpler problem

✔ Make a table, chart or list

Use a picture, diagram or model

Act it out

Work backwards

Eliminate possibilities

Use an equation or formula

Problem

BUSINESS An orchard grows Delicious, Jonathan, and Granny Smith apples. The apples are sold in boxes to two different markets. The profit is $5.85 on a box of Delicious apples, $4.25 on Jonathans, and $3.75 on Granny Smiths. The table shows the number of boxes sold.

Find the amount of profit generated by sales to each market.

Apples	Markets	
	Bill's	Jan's
Delicious	250	225
Jonathan	320	295
Granny Smith	175	190

Solve the Problem

a. Represent the market data in a 3 × 2 matrix as shown.

$$A = \begin{bmatrix} 250 & 225 \\ 320 & 295 \\ 175 & 190 \end{bmatrix}$$

b. Write a matrix to represent the respective profits. Think about the dimensions necessary for matrix multiplication. Since matrix A has three rows, matrix B must have three columns. The dimensions of B must be 1 × 3.

$$B = [5.85 \quad 4.25 \quad 3.75]$$

c. The product BA will be a 1 × 2 matrix that determines the profit from each market.

$$BA = [5.85 \quad 4.25 \quad 3.75] \begin{bmatrix} 250 & 225 \\ 320 & 295 \\ 175 & 190 \end{bmatrix} = [3478.75 \quad 3282.50]$$

■ The profit from Bill's market is $3478.75. The profit from Jan's market is $3282.50.

Use matrices to solve each problem.

1. **FOOD DISTRIBUTION** A farm raises two crops, which are shipped to three distributors. The table shows the number of crates shipped to distributors.

	Distributor		
	A	B	C
Crop 1	350	275	550
Crop 2	200	310	260

The profit on crop 1 is $2.75 per crate. The profit on crop 2 is $3.20 per crate.

Find the amount of profit from each distributor.

2. **FOOD CONCESSIONS** A large amusement park owns four bakeries, each of which produces three types of bread: white, rye and whole wheat. The bread is used to supply food vendors throughout the park. The number of loaves produced daily at each bakery is shown in the table at the right.

	Bakery			
	A	B	C	D
White	190	215	240	112
Rye	65	80	110	60
Whole wheat	205	265	290	170

By baking its own bread, the park can reduce the amount of overhead and increase its profits. The profit on each loaf of bread is 75 cents for white, 50 cents for rye, and 60 cents for whole wheat. Find the amount of profit from each bakery.

3. **SALES** A sneaker manufacturer makes five kinds of sneakers: basketball, running, walking, cross-trainer, and tennis. The sneakers are shipped to three retail outlets. The number of pairs of sneakers shipped to each outlet is shown.

Profit on each pair of sneakers is as follows:

basketball	$4.50	cross-trainer	$5.25
running	$3.50	tennis	$5.00
walking	$3.75		

	Outlets		
	A	B	C
Basketball	30	40	35
Running	20	25	30
Walking	15	20	25
Cross-trainer	15	15	20
Tennis	25	30	25

Find the amount of profit for each outlet.

4. **SOUVENIRS** An amusement park sells hats, T-shirts and stuffed toys. The table on the left gives the number of each type of souvenir sold during a two-week period. The table on the right gives the price of each souvenir. Find the total amount spent on souvenirs each week.

	Hats	T-shirts	Toys
Week 1	128	240	58
Week 2	130	215	89

Item	Price
Hats	$14
T-shirts	$18
Toys	$24

Find the sum of the measures of the angles of a convex polygon with the given number of sides. (Lesson 4-7)

5. 37	**6.** 52	**7.** 29	**8.** 45
9. 62	**10.** 40	**11.** 58	**12.** 19

Chapter 8 Review

VOCABULARY ◣

Choose the word from the list that best completes each statement.

1. A translation, reflection, rotation, or dilation is known as a ___?___ of a figure.

2. A dilation image is obtained by multiplying the length of each side of a figure by a number called the ___?___.

3. Under a transformation, the new figure is called the image and the original figure is called the ___?___.

4. The number of rows and columns are the ___?___ of the matrix.

5. A ___?___ is a rectangular array of numbers arranged into rows and columns

6. Sliding a figure is called a ___?___.

7. When a matrix is multiplied by a number, the number is called a ___?___.

8. Turning a figure around a point is called a ___?___.

9. Flipping a figure over a line is called a ___?___.

10. Reducing or enlarging a figure is called a ___?___.

a.	composite
b.	dilation
c.	dimensions
d.	glide reflection
e.	matrix
f.	preimage
g.	reflection
h.	rotation
i.	scalar
j.	scale factor
k.	transformation
l.	translation

LESSON 8-1 ◣ Translations and Reflections, p. 338

▶ Under a **translation**, an image is produced by sliding every point of the original figure the same distance in the same direction.

▶ Under a reflection, a figure is flipped over a line of reflection.

Copy each graph on graph paper. Then draw the image of each figure under the given translation.

11. 4 units up

12. 3 units left

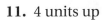
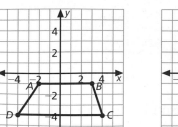

Copy each graph on graph paper. Then draw the image of each figure under the given reflection.

13. over the x-axis

14. over the y-axis

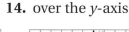
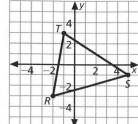
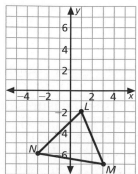
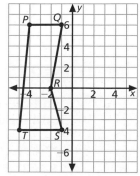

LESSON 8-2 ◣ Rotations in the Coordinate Plane, p. 342

▶ Under a **rotation**, a figure is turned about a point.

Copy each graph on graph paper. Then draw the image of each figure under the given rotation about the origin.

15. 90° counterclockwise

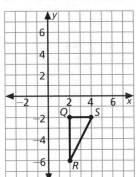

16. 180° clockwise

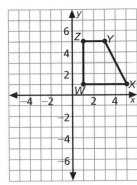

LESSON 8-3 ◣ Dilations in the Coordinate Plane, p. 348

▶ A **dilation** is a transformation that produces an image of the same shape, but a different size.

Copy each graph on graph paper. Then draw the image of each figure under the given dilation with the center at the origin.

17. scale factor of $\frac{1}{3}$

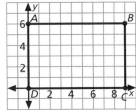

18. scale factor of 2

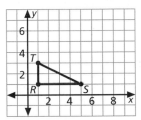

LESSON 8-4 ◣ Multiple Transformations, p. 352

▶ Two or more successive transformations can be applied to a given figure. This is called a **composite of transformations**.

Describe two transformations that would create the image in blue. There may be more than one possible answer.

19.

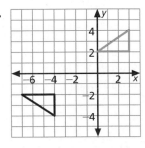

20.

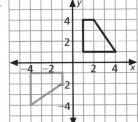

LESSON 8-5 ◣ Addition and Multiplication with Matrices, p. 358

▶ A **matrix** is a rectangular array of numbers arranged into rows and columns. The number of rows and columns are the **dimensions** of the matrix. The numbers that make up the matrix are the **elements** of the matrix.

21. Find the dimensions of D. $D = \begin{bmatrix} 1 & -2 & 3 & 2 \\ -4 & 5 & -6 & 4 \\ 7 & -8 & 9 & 6 \end{bmatrix}$

22. Identify the elements D_{32}, D_{21}, and D_{13}.

23. Find kD for $k = -2$.

Use matrices A–C to find each of the following.

$$A = \begin{bmatrix} -3 & 5 & 0 & 2 \\ 1 & -7 & 6 & -1 \end{bmatrix} \quad B = \begin{bmatrix} 0 & 1 & -3 & 11 \\ 13 & 2 & -4 & 1 \end{bmatrix} \quad C = \begin{bmatrix} 6 & 0 & 8 & -2 \\ 0 & 15 & -1 & 5 \end{bmatrix}$$

24. $A + B - C$

25. $C - B - A$

26. $3(A + C)$

LESSON 8-6 ◼ More Operations on Matrices, p. 362

▶ The product of an $m \times n$ matrix and an $n \times p$ matrix is an $m \times p$ matrix.

Find each product. If not possible, write NP.

27. $\begin{bmatrix} 3 & -2 & -4 \\ 2 & 1 & 3 \end{bmatrix} \begin{bmatrix} 0 & 1 \\ 2 & -2 \\ 6 & 5 \end{bmatrix}$

28. $\begin{bmatrix} 5 & -2 & -1 \\ 8 & 0 & 3 \end{bmatrix} \begin{bmatrix} -4 & 2 \\ 1 & 0 \end{bmatrix}$

LESSON 8-7 ◼ Transformations and Matrices, p. 368

▶ The point represented by (x, y) can also be represented by the matrix $\begin{bmatrix} x \\ y \end{bmatrix}$.

▶ Polygons and transformations can also be represented by matrices.

29. Triangle DEF is represented by $\begin{bmatrix} 4 & 6 & 5 \\ 2 & 1 & 4 \end{bmatrix}$. Use the matrix $\begin{bmatrix} -1 & 0 \\ 0 & 1 \end{bmatrix}$ to find the image when $\triangle DEF$ is reflected over the y-axis.

30. Quadrilateral $PQRS$ is represented by $\begin{bmatrix} 1 & 2 & 5 & 3 \\ 1 & 3 & 2 & -1 \end{bmatrix}$. Use the matrix $\begin{bmatrix} 0 & -1 \\ -1 & 0 \end{bmatrix}$ to find the image when the quadrilateral is reflected over the line $y = -x$.

LESSON 8-8 ◼ Problem Solving Skills: Use a Matrix, p. 372

▶ Some problems can be solved by translating to a matrix and using matrix operations.

31. The table at the right shows the number of students in a school's beginning and advanced orchestra classes. The students pay a fee for instruction books: $5 for brass, $5 for woodwinds, and $8 for strings. Find the amount each class will spend on books.

	Advanced	Beginning
Brass	12	28
Strings	22	25
Woodwinds	18	30

CHAPTER INVESTIGATION

EXTENSION Make a presentation to your class of your roller coaster. Explain why you designed the ride as you did. Display and use your marketing brochure during your presentation.

Chapter 8 Assessment

Use the figure at the right for Exercises 1–3.

1. Graph the image of $\triangle ABC$ under a translation 5 units to the right. Label the image points A', B', and C'.

2. Graph the image of $\triangle ABC$ under a reflection across the x-axis. Label the image points A'', B'', and C''.

3. Graph the image of $\triangle ABC$ under a 180° clockwise rotation about the origin. Label the image points A''', B''', and C'''.

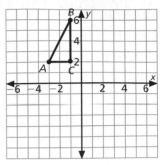

4. Draw the dilation image of rectangle *LIMB* with the center of dilation at the origin and a scale factor of $\frac{1}{3}$.

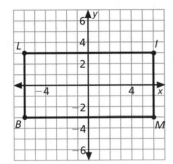

5. Describe two transformations that together could have been used to create the image shown in blue.

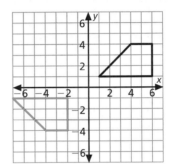

Use matrix *T* at the right for Exercises 6–8.

6. Find the dimensions of *T*.

7. Identify the elements T_{32}, T_{23}, and T_{11}.

8. Find kT for $k = -3$.

$$T = \begin{bmatrix} 4 & 2 & 0 & -2 \\ 6 & -1 & -4 & 7 \\ 3 & 8 & 5 & 9 \end{bmatrix}$$

9. Find the product *MN*: $M = \begin{bmatrix} 5 & 1 & -2 \\ 3 & 6 & 1 \end{bmatrix}$ $N = \begin{bmatrix} 4 & 3 \\ 8 & -1 \\ 0 & 7 \end{bmatrix}$

10. Find the image of the rectangle with vertices $\begin{bmatrix} 3 \\ 3 \end{bmatrix}$, $\begin{bmatrix} -5 \\ 3 \end{bmatrix}$, $\begin{bmatrix} -5 \\ -2 \end{bmatrix}$, and $\begin{bmatrix} 3 \\ -2 \end{bmatrix}$ under the transformation represented by $\begin{bmatrix} 1 & 0 \\ 0 & -1 \end{bmatrix}$.

11. A quilt maker has three retail outlets where quilts and pillows are sold. The table shows the number of quilts and pillows sold at each outlet. Find the profit from each outlet. Use matrices.

    The profit from each quilt is $90 and the profit from each pillow is $25.

Outlet

Item	A	B	C
quilts	21	20	28
pillows	30	19	27

Standardized Test Practice

Part 1 Multiple Choice

Record your answers on the answer sheet provided by your teacher or on a sheet of paper.

1. If $x > 1$, which value is the greatest? (Lesson 1-8)

   Ⓐ $(x^5)^{-2}$ 　　　　　Ⓑ $(-x)^5$

   Ⓒ $\dfrac{x^6}{x^3}$ 　　　　　Ⓓ $\dfrac{x^2}{x^7}$

2. Which graph represents the solution of $2x - 1 < 11$? (Lesson 2-6)

   Ⓐ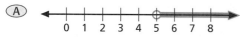

   Ⓑ

   Ⓒ

   Ⓓ

3. In the figure, $a \parallel b$. Which statement is true? (Lesson 3-4)

   Ⓐ $m\angle 1 - m\angle 2 = 0°$
   Ⓑ $m\angle 1 - m\angle 2 = 45°$
   Ⓒ $m\angle 1 + m\angle 2 = 90°$
   Ⓓ $m\angle 1 + m\angle 2 = 180°$

4. Which statement is *not* true about an isosceles trapezoid? (Lesson 4-9)

   Ⓐ The bases are parallel.
   Ⓑ The base angles are congruent.
   Ⓒ The diagonals bisect each other.
   Ⓓ The legs are congruent.

5. What is the best estimation of the area of the shaded region? (Lesson 5-2)

   Ⓐ 50 m²
   Ⓑ 75 m²
   Ⓒ 100 m²
   Ⓓ 150 m²

   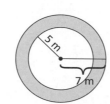

6. The graph of which equation has the greatest slope? (Lesson 6-1)

   Ⓐ $2x - 5y = 7$
   Ⓑ $4x - 5y = 10$
   Ⓒ $4y - 7x = 15$
   Ⓓ $3y - 4x = 6$

7. What is the value of y for the following system of equations? (Lessons 6-5, 6-6, and 6-7)

   $$7x - 2y = 22$$
   $$3x + y = 15$$

   Ⓐ 2 　　Ⓑ 3 　　Ⓒ 4 　　Ⓓ 5

8. In the figure below, $\overline{MN} \parallel \overline{YZ}$. Find the value of a. (Lesson 7-6)

   Ⓐ $2\dfrac{2}{3}$
   Ⓑ 6
   Ⓒ 10
   Ⓓ 13.5

   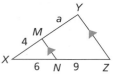

9. Which transformation is a reflection? (Lesson 8-1)

   Ⓐ 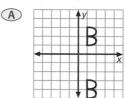 　　Ⓑ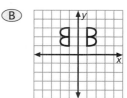

   Ⓒ 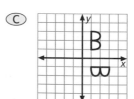 　　Ⓓ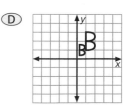

10. What are the dimensions of the following product? (Lesson 8-6)

    $$\begin{bmatrix} 10 & 2 & 0 \\ 1 & 3 & 5 \end{bmatrix} \begin{bmatrix} 6 & 4 \\ 0 & -1 \\ 3 & 2 \end{bmatrix}$$

    Ⓐ 2×2 　　　　Ⓑ 2×3
    Ⓒ 3×2 　　　　Ⓓ 3×3

Preparing for Standardized Tests
For test-taking strategies and more
practice, see pages 709-724.

Part 2 Short Response/Grid In

Record your answers on the answer sheet
provided by your teacher or on a sheet of paper.

11. Space shuttles encounter temperatures that
    range from −250°F while in orbit to 3000°F
    during the reentry of Earth's atmosphere. Find
    the range of temperatures that space shuttles
    encounter. (Lesson 1-4)

12. Given $f(x) = 6x - 3$, find $f(-1)$.
    (Lesson 2-2)

13. How many dots are in the tenth figure of the
    pattern below? (Lesson 3-5)

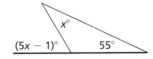

14. What is the value
    of x in the figure?
    (Lesson 4-1)

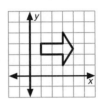

15. Find the volume of the figure below.
    (Lesson 5-7)

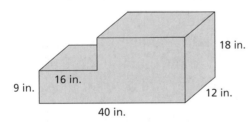

16. Find the slope of a line perpendicular to the
    line containing points $A(7, -2)$ and $B(-3, 4)$.
    (Lesson 6-2)

17. The area of 2500 ft² of grass produces enough
    oxygen for a family of 4. What is the area of
    grass needed to supply a family of 5 with
    oxygen? (Lesson 7-1)

18. The triangles below are similar. What is the
    value of x? (Lesson 7-2)

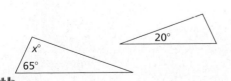

19. The distance on a map is 2.5 cm and the scale
    is 1 cm = 50 km. What is the actual distance?
    (Lesson 7-3)

Part 3 Extended Response

Record your answers on a sheet of paper.
Show your work.

20. Copy the graph on graph paper.

a. Graph the image of the figure under a
   translation 7 units down. (Lesson 8-1)

b. Graph the image of the original figure
   under a reflection across the y-axis.
   (Lesson 8-1)

c. Graph the image of the original figure
   under a rotation 180° clockwise about the
   origin. (Lesson 8-2)

d. Graph the image of the original figure
   under a dilation with the center at the
   origin and scale factor of 2. (Lesson 8-3)

e. Which images are arrows that point in the
   same direction as the original figure?

f. Which images are the same size as the
   original figure?

Test-Taking Tip   Ⓐ Ⓑ Ⓒ Ⓓ

Question 20
Always check your work for careless errors. To check your
answer to this question, remember the following rules. To
translate a figure 7 units down, add −7 to the y-coordinate.
To reflect a figure across the y-axis, change the sign of the
x-coordinate. To rotate a figure 180° clockwise, change the
sign of both coordinates. To dilate a figure, multiply each
coordinate by the scale factor.

Probability and Statistics

THEME: Sports

When the Cubs send their right-handed power hitter to the plate in the ninth inning, the Mets counter with a left-handed pitcher. Why? The manager of the Mets is simply "playing the odds."

Since its inception, baseball has kept meticulous records. By studying the data, managers, players, announcers, and fans use the concepts of probability and chance to make predictions. Today, players and managers use computers and calculators to record and analyze data. They know that the more effectively they use statistics and probability, the better they will do their jobs.

- **Team Dietitians** (page 391) plan meals and nutritional plans for athletes. They use their knowledge of nutrition to help team members maintain overall health, muscle health, and bone strength.

- **Physical Therapists** (page 411) determine exercises to strengthen muscles after injuries. Through specially designed exercise programs, they improve mobility, relieve pain, and reduce injuries.

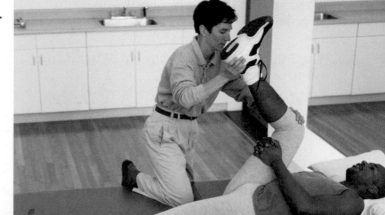

Math Online

mathmatters3.com/chapter_theme

Home Run Greats–Then and Now						
Player	Year	Home runs	Games played	At bats	Batting average	Runs batted in
Jim Thome	2003	47	159	578	.266	131
Barry Bonds	2001	73	153	476	.328	137
Alex Rodriguez	2002	57	162	624	.300	142
Roger Maris	1961	61	161	590	.269	142
Babe Ruth	1927	60	151	540	.356	164

Home Runs Month by Month								
Player	Year	Apr	May	Jun	Jul	Aug	Sept	Oct
Jim Thome	2003	4	8	9	6	10	10	0
Barry Bonds	2001	11	17	11	6	12	12	4
Alex Rodriguez	2002	9	8	7	12	12	9	0
Roger Maris	1961	1	11	15	13	11	9	1
Babe Ruth	1927	4	12	9	9	9	17	0

Data Activity: Home Run Greats

Use the tables for Questions 1–4.

1. For each player, divide the at bats by the total of home runs and round to the nearest hundredth. Compare the unit rates. Which player had the best unit rate?

2. What percent of Barry Bond's home runs were hit in August and September?

3. Find the average number of at bats per game for each player. Round to the nearest hundredth. Which player had the greatest average?

4. In 2001, Bonds hit 49 singles. For what percent of his at bats did he reach first base by hitting a single?

CHAPTER INVESTIGATION

Baseball has been called "America's Pastime." In recent years, the game's appeal has become international. Using statistics and probability, fans at home can predict what will happen when the bases are loaded in the bottom of the ninth inning.

Working Together

Gather baseball statistics and design your baseball simulation game using dice and percentile cards. Make a lineup and play a game. Decide whether the game's outcome matches your predictions. Use the Chapter Investigation icons to guide your group.

Are You Ready?

Refresh Your Math Skills for Chapter 9

The skills on these two pages are ones you have already learned. Stretch your memory and complete the exercises. For additional practice on these and more prerequisite skills, see pages 654-661.

PERCENTS, DECIMALS, AND FRACTIONS

Convert each fraction or decimal to a percent. Round to the nearest tenth if necessary.

1. $\frac{3}{4}$
2. 0.86
3. $\frac{7}{8}$
4. 0.93
5. $\frac{2}{3}$

6. 0.5
7. $\frac{3}{8}$
8. 0.42
9. $\frac{7}{16}$
10. 0.38

11. $\frac{1}{6}$
12. 0.64576
13. $\frac{6}{21}$
14. 0.19823
15. $\frac{27}{46}$

Convert each percent to a decimal. Round to the nearest thousandth if necessary.

16. 46%
17. 83%
18. 29%
19. 15%
20. 12%

21. 18.76%
22. 9.3825%
23. 78.6215%
24. 64.93%
25. 21.748%

Convert each percent to a fraction in lowest terms.

26. 80%
27. 50%
28. 68%
29. 92%
30. 75%

31. 64.2%
32. 39.8%
33. 20%
34. 19.6%
35. 51.9%

MEASURES OF CENTRAL TENDENCY

Find the mean, median, mode and range of each set of data. Round to the nearest tenth if necessary.

36. 74 75 79 76 77
    74 78 72 71

37. 30 32 34 36 38
    39 37 35 34 33

38. 40 48 52 47 56 49
    43 55 46 48 51

39. 17 12 13 16 22 21
    19 18 14 20 15 11

40. 88 87 81 92 86 87 89
    90 93 91 85 92 94

41. 62 63 67 68 65 69 64
    61 65 66 60 67 65 63

REDUCING FRACTIONS

Determine the greatest common factor of the numerator and denominator. Divide both the numerator and denominator by the factor to write the fraction in lowest terms.

42. $\frac{3}{24}$
43. $\frac{19}{57}$
44. $\frac{4}{52}$
45. $\frac{2}{110}$
46. $\frac{4}{144}$
47. $\frac{34}{17}$

48. $\frac{13}{52}$
49. $\frac{16}{48}$
50. $\frac{25}{35}$
51. $\frac{77}{121}$
52. $\frac{17}{85}$
53. $\frac{15}{75}$

HISTOGRAMS

Draw a histogram for each frequency chart.

54.

Interval	Tally	Frequency
21-24	‖	2
17-20	‖‖	4
13-16	‖‖‖ ‖‖‖	8
9-12	‖‖‖ ‖‖‖	10
5-8	‖‖‖	5
1-4	‖	1

55.

Interval	Tally	Frequency
96-100	‖	2
91-95	‖‖	3
86-90	‖‖‖ ‖‖	8
81-85	‖‖‖ ‖‖‖	10
76-80	‖‖‖ ‖‖	9
71-75	‖‖‖ ‖	7

56.

Interval	Tally	Frequency
82-85	‖‖‖ ‖‖‖ ‖	11
78-81	‖‖‖ ‖‖‖ ‖‖	14
74-77	‖‖‖ ‖‖‖ ‖‖‖ ‖‖	19
70-73	‖‖‖ ‖‖‖ ‖‖‖ ‖‖‖ ‖‖‖	25
66-69	‖‖‖ ‖‖‖ ‖‖‖ ‖‖‖ ‖	22
62-65	‖‖‖ ‖‖‖ ‖‖‖ ‖	18
58-61	‖‖‖ ‖‖‖ ‖‖‖ ‖	16
54-57	‖‖‖ ‖‖	8

57.

Interval	Tally	Frequency
71-80	‖‖‖ ‖‖	9
61-70	‖‖‖ ‖‖‖ ‖	11
51-60	‖‖‖ ‖‖‖ ‖‖‖	15
41-50	‖‖‖ ‖‖‖ ‖‖	13
31-40	‖‖‖ ‖‖‖	10
21-30	‖‖‖ ‖‖	8
11-20	‖‖‖	5
1-10	‖	1

58.

Interval	Tally	Frequency
900-999	‖‖‖ ‖‖‖	10
800-899	‖‖‖ ‖‖‖ ‖‖‖	15
700-799	‖‖‖ ‖‖‖ ‖‖‖ ‖	17
600-699	‖‖‖ ‖‖‖ ‖	12
500-599	‖‖‖ ‖‖	9
400-499	‖‖‖ ‖	7
300-399	‖‖‖ ‖	6

59.

Interval	Tally	Frequency
31-35	‖‖‖ ‖‖	8
26-30	‖‖‖ ‖‖‖ ‖‖	14
21-25	‖‖‖ ‖‖‖ ‖‖‖ ‖‖‖	20
16-20	‖‖‖ ‖‖‖ ‖‖‖ ‖‖‖ ‖‖	23
11-15	‖‖‖ ‖‖‖ ‖‖‖ ‖‖‖	19
6-10	‖‖‖ ‖‖‖ ‖	12
1-5	‖‖‖	5

AREA

Find the area of the shaded region of each figure. Use 3.14 for π. Round to the nearest hundredth if necessary.

60.

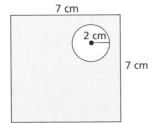

7 cm

2 cm

7 cm

61.

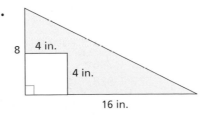

8

4 in.

4 in.

16 in.

62.

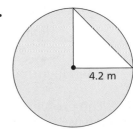

4.2 m

9-1 Review Percents and Probability

Goals ■ Find experimental and theoretical probabilities.

Applications Sports, Card Games, Test Taking

Work with a partner.

1. Discuss: How large must a group of people be for there to be a 50% chance that two members of the group will share a birthday? Make a guess.

2. Take a survey of the students in one of your classes to look for common birthdays. What did you learn?

◥ BUILD UNDERSTANDING

Recall that **probability** theory is the mathematics of chance. Probability is used to describe how likely it is that an event will occur. Probabilities are reported using fractions, decimals, and percents. The greater the probability of an event, the more likely the event is to occur.

One way to find the likelihood of an event occurring in the real world is to conduct an experiment. In an **experiment**, you either take a measurement or make an observation. A probability determined by observation or measurement is called **experimental probability**. An **outcome** is the result of each trial of an experiment.

The experimental probability of an event E is defined as follows.

$$P(E) = \frac{\text{number of favorable observations of } E}{\text{total number of observations}}$$

Example 1

RECREATION Lions fans attending a recent 3-game series were asked whether the team should have a mascot. The table shows how many fans thought it should.

According to these results, what is the probability that a Lions fan wants the team to have a mascot?

Game	Attendance	In favor of mascot
Friday	681	388
Saturday	527	428
Sunday	928	786

Solution

Use the experimental probability formula.

$$P(E) = \frac{\text{number of favorable observations of } E}{\text{total number of observations}}$$

$$P(\text{fan favoring mascot}) = \frac{388 + 428 + 786}{681 + 527 + 928} = \frac{1602}{2136} = 0.75$$

The probability that a fan interviewed at a Lions game will favor having a team mascot is 0.75.

The set of all possible outcomes of an experiment is called the **sample space**.

Problem Solving Tip

Probability is often expressed as a percent. A *percent* is a ratio of a number compared to 100. For example, 87% means

$87 : 100$ or $\frac{87}{100}$ or 0.87.

A decimal can be converted to a percent by moving the decimal two places to the right. So, 0.4 can also be written as 40%.

Example 2

SPORTS A baseball team has 8 pitchers and 3 catchers. The manager is choosing a pitcher-catcher combination. How many are possible?

Solution

One way to show all possible outcomes is to use ordered pairs. For example, use the numbers 1–8 for pitchers and the letters A–C for the catchers.

(1, A)	(2, A)	(3, A)	(4, A)	(5, A)	(6, A)	(7, A)	(8, A)
(1, B)	(2, B)	(3, B)	(4, B)	(5, B)	(6, B)	(7, B)	(8, B)
(1, C)	(2, C)	(3, C)	(4, C)	(5, C)	(6, C)	(7, C)	(8, C)

There are 24 possible pitcher-catcher combinations.

Another way to show the sample space is to use a **tree diagram**.

You can use probability to predict the number of times an event will occur.

Pitchers	Catchers	Outcomes
1	A	(1, A)
	B	(1, B)
	C	(1, C)
2	A	(2, A)
	B	(2, B)
	C	(2, C)
3	A	(3, A)
	B	(3, B)
	C	(3, C)
4	A	(4, A)
	B	(4, B)
	C	(4, C)
5	A	(5, A)
	B	(5, B)
	C	(5, C)
6	A	(6, A)
	B	(6, B)
	C	(6, C)
7	A	(7, A)
	B	(7, B)
	C	(7, C)
8	A	(8, A)
	B	(8, B)
	C	(8, C)

Example 3

SPORTS A softball player has had 24 hits in her first 60 times at bat. Predict her total number of hits in 330 at bats.

Solution

First, use the outcomes that have already occurred to find the experimental probability of the player getting a hit each time at bat.

$$P(\text{hit}) = \frac{24}{60} = 0.4$$

Then multiply that result by the total number of times at bat.

$$0.4(330) = 132$$

Based on the player's first 60 times at bat, you can predict that she will get 132 hits in 330 at bats.

As you increase the number of trials in a probability experiment, the experimental probability will probably get closer to the **theoretical probability**. For example, when tossing a fair coin, $P(\text{heads}) = \frac{1}{2}$. The more often you toss the coin, the closer you will come to tossing an equal number of heads and tails.

The theoretical probability of an event, $P(E)$, is the ratio of the number of favorable outcomes to the number of possible outcomes in the sample space.

$$P(E) = \frac{\text{number of favorable outcomes}}{\text{number of possible outcomes}}$$

Reading Math

The **odds** in favor of an event are expressed as the ratio of the number of favorable outcomes to the number of unfavorable outcomes. For example, when you roll a die, the odds of getting a 4 are 1:5—because there is 1 way the event can occur, and 5 ways it cannot.

Example 4

CARD GAMES A card is picked at random from a standard deck of 52 cards. Find P(face card).

Solution

There are 52 possible outcomes. There are 12 favorable outcomes— 4 jacks, 4 queens, and 4 kings.

$$P\text{(face card)} = \frac{12}{52} = \frac{3}{13}$$

TRY THESE EXERCISES

1. **SPORTS** Of the first 1500 fans to pass through the turnstiles at the stadium, 1050 had reserved seats. What is the probability that the next person through will have a reserved seat?

2. **WRITING MATH** A person flips a penny, a nickel, and a dime. Each coin can land with heads up (H) or tails up (T). Make a tree diagram to show what different outcomes are possible.

3. You roll a die 60 times. Predict the number of times you will roll an even number greater than 2.

4. **GAMES** A spinner for a game is divided into ten equal sections numbered 1 through 10. What is the probability of spinning 7 or higher?

PRACTICE EXERCISES • For Extra Practice, see page 690.

A die is rolled 100 times with the following results.

Outcome	1	2	3	4	5	6
Frequency	15	18	22	9	16	20

What is the experimental probability of rolling each of the following results?

5. 2   6. 6   7. a number less than 4

8. What is the theoretical probability of rolling a number less than 4?

9. **CHAPTER INVESTIGATION** Working with a partner, prepare to make your own baseball simulation game. To begin, gather batting statistics on at least 18 players. You may use statistics from the most recent baseball season or statistics from prior years. For each player, you will need to know the number At Bats (AB), Hits (H), Doubles (2B), Triples (3B), Walks (BB) and Strike Outs (SO). This information is available in the newspaper, in sports magazines or on team websites.

List all the elements of the sample space for each of the following experiments.

10. You flip a dime and a quarter.

11. You spin each of these spinners once.

Find the probability of each of the following:

12. **CARD GAMES** drawing a black jack from a standard deck of cards

13. rolling a die and getting a multiple of 3

14. **EDUCATION** guessing correctly on one true-false question on a test

15. **EDUCATION** guessing incorrectly on a multiple-choice question with four choices

16. reaching into a drawer without looking and taking out a pair of black socks, when the drawer contains 3 pairs of black socks, 2 pairs of white socks, 1 pair of red socks, and 2 pairs of blue socks

17. **SPORTS** A basketball player has made 96 free-throws in his last 128 attempts. What is the probability he will be successful in his next attempt? How many successful free-throws do you predict this player will make in 500 attempts?

18. **WRITING MATH** You want to predict how many students in your school are right-handed. Describe how you would do it.

19. **WEATHER** The weather forecaster predicts a 25% chance of rain in your area tomorrow. Describe what this forecast means.

20. **TRANSPORTATION** A bus breaks down while traveling between two cities that are 200 mi apart. What is the probability the breakdown is within 25 mi of either city?

◣ EXTENDED PRACTICE EXERCISES

Write whether each of the following probabilities can be determined experimentally or theoretically.

21. The probability that Player A will win when Player A plays Player B in tennis.

22. The probability that a person will win the state's lottery.

23. The probability that a family with 4 children has all boys.

24. The probability that it will snow on January 9.

◣ MIXED REVIEW EXERCISES

Graph the image of triangle *ABC* with vertices at *A*(−2, 1) *B*(4, 2), and *C*(1, 4), under each transformation from the original position. (Lesson 8-11)

25. 3 units up   26. reflected across the *x*-axis

Graph the image of parallelogram *PQRS* with vertices at *P*(−1, 1) *Q*(2, 3), *R*(2, 6), and *S*(−1, 4), under each transformation from the original position. (Lesson 8-11)

27. 6 units down   28. reflected across the *x*-axis

Sometimes a probability problem is too difficult to solve theoretically or experimentally. One way to solve such a problem is to model it with a **simulation** to estimate the probability. Simulations often use **random numbers**; these can be readily generated and recorded by a computer. You can also find random numbers by rolling dice, flipping coins, using numbered slips of paper, or spinning a spinner.

Problem Solving Strategies

Guess and check

Look for a pattern

Solve a simpler problem

Make a table, chart or list

Use a picture, diagram or model

✔ Act it out

Work backwards

Eliminate possibilities

Use an equation or formula

Problem

MARKETING Each box of Batter-Up Pancake Mix contains one of 5 different classic baseball cards. Assuming that the company has evenly distributed the cards among the boxes, what is the probability that you will find all 5 cards if you buy 10 boxes of Batter-Up?

Solve the Problem

Work with a partner. Use 5 slips of paper numbered 1–5; each slip represents a box of cereal. Place the slips in a paper bag. Then draw one slip of paper from the bag, record its number, and place it back in the bag. Repeat the process until you have drawn and recorded 10 slips of paper. If you have drawn each of the 5 numbers at least once, consider the outcome of your experiment to be successful. If you have not drawn every number, the outcome is unsuccessful.

Repeat the experiment 50 times, recording all results in a table. Indicate which trials are successful. Then write a ratio comparing successful outcomes to the total number of outcomes. This ratio will be an estimate of the probability of getting every card in the set when you buy 10 boxes of cereal.

◤ TRY THESE EXERCISES

1. **COMPUTER SCIENCE** A computer generates a list of random 2-digit numbers. Zero cannot be the first digit. What is the probability that a randomly chosen number from the list contains the digit "1"?

2. **MARKETING** A candy company has placed 6 different prizes in its boxes. The prizes are uniformly distributed among the boxes of candy, only one per box. Describe a simulation you could do to estimate the probability of getting all 6 prizes in a 12-pack of candy.

3. **WRITING MATH** Describe a simulation you could do to find out how many cards you would expect to have to draw from a standard deck to get two kings.

4. **TEST TAKING** Suppose you are going to take a 10-question true-false test on the evolution of idiomatic phrases in Sri Lanka. You will need to guess each time, and you want to find out your chances of scoring 65% correct or better. Design a simulation to determine your chances. *Hint:* Use coin flipping.

5. A family wants to have 3 children. Do the following simulation to determine the probability that, if they do have 3 children, all 3 will be the same gender.

Five-step Plan

1 Read
2 Plan
3 Solve
4 Answer
5 Check

    a. Use 3 coins. Let heads = girl and tails = boy. Toss the coins and record the results. Repeat the coin tosses until you have recorded 50 sets of 3 tosses each.

    b. Count the successful outcomes—those with either 3 heads or 3 tails.

    c. Write a ratio comparing successful outcomes with total outcomes. What is your experimental probability of having 3 children, all of whom are of the same gender?

6. SPORTS One baseball player always arrives at the stadium between 5:30 P.M. and 6:30 P.M. for a night game. If batting practice always starts between 6:00 P.M. and 7:00 P.M., what is the probability on any given night that this player will arrive before batting practice begins? Design and do a simulation to find out.

7. PROGRAMMING A pitcher throws strikes about 60% of the time. If he throws 80 pitches, how many might be strikes? The following computer programming statements can be used to simulate 80 pitches.

1 S = 0	The experiment begins with no successes.
2 FOR I = 1 TO 80	80 pitches
3 X = RND(1)	Generate a random decimal.
4 IF X < .6 THEN S = S + 1	If the decimal < 0.6, increase S by 1.
5 NEXT I	Simulate the next pitch.
6 PRINT S	Total number of strikes.
7 END	

Use the program to simulate the problem. Then describe how you would adjust the program for a pitcher who throws strikes 40% of the time.

8. TALK ABOUT IT Petra is designing a simulation to determine the chance of guessing the correct answer on a multiple-choice test. Each item on the test has three choices. Petra plans to roll a 6-sided die to simulate random guesses. A roll of 1 or 2 will indicate choice A; a roll of 3 or 4, choice B; and a roll of 5 or 6, choice C. Will Petra's simulation work? Explain.

MIXED REVIEW EXERCISES

Find the slope of each line using the given information. (Lesson 6-1)

9. $A(-2, -1), B(5, 3)$
10. $C(7, 2), D(3, -2)$
11. $E(1, 8), F(-3, -4)$

12. $-3x + 2y = 9$
13. $4y + 2x = 8$
14. $-12 + x = 4y$

15. $G(-3, 5), H(-3, 9)$
16. $I(-3, 5), J(3, -5)$
17. $K(2, -1), L(-8, -1)$

Solve each proportion. (Lesson 7-1)

18. $\dfrac{5}{x} = \dfrac{15}{12}$
19. $\dfrac{9}{12} = \dfrac{x}{20}$
20. $\dfrac{4}{13} = \dfrac{16}{x}$
21. $\dfrac{7}{22} = \dfrac{x}{55}$

22. $\dfrac{14}{25} = \dfrac{x + 3}{10}$
23. $\dfrac{x + 1}{16} = \dfrac{x}{9}$
24. $\dfrac{x + 3}{12} = \dfrac{4}{8}$
25. $\dfrac{16}{x + 2} = \dfrac{8}{2 - x}$

Review and Practice Your Skills

A card is picked at random from a standard deck of 52 cards. Find each theoretical probability.

1. $P(\text{heart})$
2. $P(\text{jack})$
3. $P(\text{red card})$
4. $P(\text{black 2})$
5. $P(\text{2 or 3})$
6. $P(\text{7 of hearts})$
7. $P(3 \leq \text{card} \leq 8)$
8. $P(\text{king of clubs})$

You flip a coin four times. Find each theoretical probability.

9. $P(\text{no tails})$
10. $P(\text{exactly one head})$
11. $P(\text{2 tails, 2 heads})$
12. $P(\text{4 tails})$
13. $P(> 2 \text{ heads})$
14. $P(\text{0 or 1 head})$
15. $P(\text{3 tails})$
16. $P(\text{1, 2, 3 or 4 heads})$

You roll a pair of dice and calculate the sum. Find each theoretical probability.

17. $P(7)$
18. $P(11)$
19. $P(\text{even})$
20. $P(1)$
21. $P(12)$
22. $P(\text{4 or 5})$
23. $P(6)$
24. $P(< 6)$
25. $P(\text{10, 11, or 12})$
26. $P(9)$
27. $P(< 11)$
28. $P(2)$

29. A basketball player has made 48 free throws in her last 72 attempts. What is the probability she will be successful on her next attempt? How many free throws do you predict she will make in 600 attempts?

30. A car breaks down while traveling between two cities that are 300 mi apart. What is the probability that the breakdown is within 18 mi of either city?

31. A computer generates a list of random 3-digit numbers. Zero cannot be the first digit. What probability would you expect for a number in the list to contain the digit "4"?

32. Each box of Toasted Crunchies cereal contains a single prize. There are 4 different prizes uniformly distributed among all boxes of cereal at the production facility. Describe a simulation you could do to estimate the probability of getting all 4 prizes if you buy 10 boxes of Toasted Crunchies.

33. Perform and document a simulation to determine the probability that a family having 2 children will have 1 boy and 1 girl.

34. Perform and document a simulation to find the chances of scoring 50% or higher on a 5 question multiple choice test. Each question has four choices, and you guess on each question.

35. Agnes walks her dog each night outside the grounds of Tellco Corporation between 7:30 and 8:00 P.M. First shift workers leave the grounds between 7:30 and 8:30 P.M. each night. Describe a simulation using two spinners that you could use to calculate the probability of Agnes seeing first-shift workers leaving the grounds during her walk.

36. Describe a simulation you could do to find out how many cards you would have to draw from a standard deck to get a pair of hearts.

List all the elements of the sample space for each experiment. (Lesson 9-1)

37. Tossing a quarter and a nickel.

38. Spinning each of these two spinners:

39. Rolling a die and tossing a dime.

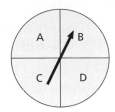 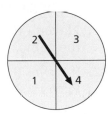

40. A computer randomly generates a list of 2-digit numbers. Zero cannot be the first digit. What is the probability that the next number generated is a multiple of 3? (Lesson 9-2)

MathWorks Career – Dietitian

Workplace Knowhow

Dietetics has applications in many different career fields. Clinical dietitians plan meals and nutritional plans for groups such as schools and hospitals. Community dietitians inform the public on nutritional habits to prevent disease and promote healthy lifestyles.

Consultant dietitians provide advice in the areas of sanitation and safety procedures. In the sports world, nutrition is important for maintaining muscle health and bone strength.

As the dietitian for a baseball team, you need to determine whether the team members are getting enough calcium in their diets. To find out, you separate the players into three categories: infielders, outfielders, and pitchers. During a buffet, you chart the food selections of the players.

1. Find the average amount of calcium consumed by players in each group.

A. Infielders	B. Outfielders	C. Pitchers
1B–300 mg	RF–113 mg	P1–233 mg
C–220 mg	CF–197 mg	P2–212 mg
2B–186 mg	LF–262 mg	P3–184 mg
3B–216 mg		P4–258 mg
SS–102 mg		

2. Your research shows that during lunch, calcium intake is $\frac{3}{4}$ of the amount consumed during the buffet. Breakfast amounts are $\frac{3}{5}$ of the buffet amount. How many milligrams of calcium is each player getting per day?

3. If 450 mg of calcium is recommended per player per day, which players are consuming too little calcium?

9-3 Compound Events

Goals ■ Find probabilities of compound events.

Applications Sports, Games, Business

Play this game with a partner. Use a pair of 6-sided dice.

1. Player A rolls first. If Player A rolls a 7, Player B wins the game. If not, Player B rolls.

2. If Player B rolls a 7 or an 11, Player A wins. If not, it is Player A's turn. Continue taking turns until there is a winner.

3. Play the game several times. Do you think one player has a better chance of winning?

◤ BUILD UNDERSTANDING

A **compound event** consists of two or more simple events. Compound events may involve finding the probability of one event *and* another event occurring. Or, the probability of one event *or* another event occurring. For example, when rolling a die, rolling a number that is even (event A) *and* greater than 2 (event B) is written $P(A \text{ and } B)$. Rolling a number that is even *or* greater than 2 is written $P(A \text{ or } B)$.

If two events *cannot* occur at the same time, they are called **mutually exclusive** events. For example, it is impossible to draw from a standard deck of playing cards a card that is both a heart and a club.

When two events A and B are mutually exclusive, the probability of the compound event A or B can be found using the formula $P(A \text{ or } B) = P(A) + P(B)$.

Example 1

SPORTS A standard deck of playing cards is used to simulate a baseball game. During the game, players draw a card at random. Spade number cards greater than 4 represent doubles. Home runs are represented by either a 3 or a queen.

a. Find P(spade and a number card greater than 4).

b. Find P(3 or queen).

Solution

There are 52 possible outcomes.

a. There are 6 outcomes in which spades are greater than 4: 5, 6, 7, 8, 9, and 10 of spades.

So, P(spade and number greater than 4) $= \frac{6}{52}$ or $\frac{3}{26}$.

The probability that the card will be a spade and a number greater than 4 is $\frac{3}{26}$.

b. A card cannot be both a 3 and a queen at the same time, so the events are mutually exclusive.

$$P(3 \text{ or queen}) = P(3) + P(\text{queen})$$
$$= \frac{4}{52} + \frac{4}{52}$$
$$= \frac{8}{52} \text{ or } \frac{2}{13}$$

The probability that the card will be a 3 or a queen is $\frac{2}{13}$.

Events that can happen at the same time are *not* mutually exclusive.

Example 2

GAMES You draw a card at random from a standard deck of playing cards. Find the probability that the card is a club or a jack.

Solution

These are not mutually exclusive events, because a card can be both a club and a jack.

There are 13 clubs, so $P(\text{club}) = \frac{13}{52}$.

There are 4 jacks, so $P(\text{jack}) = \frac{4}{52}$.

However, 1 club is a jack. $P(\text{club and jack}) = \frac{1}{52}$.

$$P(\text{club or jack}) = \frac{13}{52} + \frac{4}{52} - \frac{1}{52}$$
$$= \frac{16}{52} = \frac{4}{13}$$

The probability that the card is a club or a jack is $\frac{4}{13}$.

Example 2 illustrates that when two events A and B are not mutually exclusive, the probability of A or B can be found using the formula

$$P(A \text{ or } B) = P(A) + P(B) - P(A \text{ and } B)$$

Suppose you know the probability of event A. The set of outcomes in the sample space, but *not* in A, is called the **complement** of the event.

$$P(\text{not } A) = 1 - P(A)$$

Example 3

You select a marble from this jar without looking. You know $\frac{1}{5}$ of the marbles are red and $\frac{1}{5}$ are blue. What is the probability you will select *neither* a red *nor* a blue marble?

Solution

Find the probability of selecting red or blue.

$$P(\text{red or blue}) = \frac{1}{5} + \frac{1}{5} = \frac{2}{5}$$

Check Understanding

Classify each of the following pairs of events as *mutually exclusive* or *not mutually exclusive*.

1. drawing the 4 of clubs; drawing the 4 of spades

2. rolling two dice that show a sum of 8; rolling two dice and getting different numbers

3. tossing two coins and getting two tails; tossing two coins and getting two heads

Find the probability of *not* selecting red or blue.

$$P(\text{neither red nor blue}) = 1 - P(\text{red or blue})$$
$$= 1 - \frac{2}{5}$$
$$= \frac{3}{5}$$

The probability that you will not select a red or blue marble is $\frac{3}{5}$.

◤ TRY THESE EXERCISES

1. A die is rolled. Find the probability of rolling a 1 or a 2.

2. Two coins are tossed. Find the probability that the coins show two heads or one tail and one head.

3. A card is drawn at random from a standard deck of playing cards. Find the probability that it is a 7 or a black card.

4. Two dice are rolled. Find the probability that the sum of the numbers is not greater than 5.

5. Each student in your class writes his or her full name on a piece of paper. The pieces are put in a box and one is chosen without looking. What is the probability that your name will not be chosen?

◤ PRACTICE EXERCISES • For Extra Practice, see page 690.

GAMES A player rolls two 6-sided dice.

6. List the sample space for the rolls.

7. Find the probability that the sum of the numbers rolled is odd and greater than 5.

8. Find the probability that the sum of the numbers rolled is either 8 or 10.

9. Find $P(\text{not even})$.

10. Find $P(\text{neither odd nor sum of 6})$.

SPORTS Suppose you are on a team in the midst of a losing streak. The coach decides to "shake up" the line-up. He chooses the batting order by putting nine players' names into a hat and pulling them out one by one. The player whose name is drawn first bats first, the second bats second, and so on.

11. What is the probability you will bat second or fourth?

12. What is the probability you will bat fifth, or in the first third of the batting order?

13. What is the probability you will bat first, or in the first third of the batting order?

14. What is the probability you will bat in the last third of the batting order, or in an odd-numbered position?

15. **BUSINESS** Ms. Garrett plans to select a worker at random for a special training seminar. If there are 14 workers in sales, 6 in accounting and 5 in personnel, what is the probability that the worker will be from either sales or accounting?

GAMES You spin the spinner shown. Find each probability.

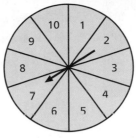

16. spinning a 2 or an odd number

17. spinning an odd or an even number

18. spinning a multiple of 2 or a multiple of 3

CARD GAMES To begin a game, the dealer draws a card at random from a standard deck of playing cards.

19. Find the probability that the card is a 2, a 5, or a face card.

20. Find the probability the card is a 7, an 8, or a red card.

A card is drawn at random from a standard deck of playing cards. For each event, estimate whether the probability is closer to 1, $\frac{1}{2}$, or 0.

21. P(red or face card)

22. P(2, 3, or 4)

23. P(black and face card)

24. P(black, heart, or 7)

PHOTOGRAPHY A team photo album contains photos of the players by themselves, the coaching staff by themselves, and the players and the coaches together. The players are in 15 of the photos and the coaches are in 12 of the photos. In 6 photos, the players and coaches appear together.

25. How many photos are in the album?

26. If you open the album at random to one of the team photos, what is the probability that the photo shows only coaches?

EXTENDED PRACTICE EXERCISES

27. **WRITING MATH** Suppose you roll a pair of dice. Why are rolling a multiple of 6 and a multiple of 4 not mutually exclusive events?

28. A pair of dice is rolled. What is the probability that the sum of the numbers is neither 5 nor a multiple of 2?

29. **SPORTS** Batting average is found by dividing hits by at-bats. In 1941, Ted Williams batted over .400 when he got 185 hits in 456 official at-bats for an average of .406. Suppose Ted Williams had 1 more at-bat in 1941. Based on his performance all season, what would you estimate the probability of his *not* getting a hit that time?

30. **CHAPTER INVESTIGATION** For each player you selected, the number of hits (H) is equal to the sum of the singles (S), doubles (2B), triples (3B), and home runs (HR). Since the number of singles is not usually reported as a separate statistic, calculate the number of singles (S) for each player using the formula: H − (2B + 3B + HR) = S. You may want to use a computer spreadsheet.

MIXED REVIEW EXERCISES

Evaluate each expression when $a = 5$ and $b = -4$. (Lesson 1-8)

31. a^2b^2

32. $a^2 + b^2$

33. ab^3

34. a^3b^2

35. $a(a^2b^2)$

36. $ab \cdot ab^2$

37. a^3b

38. $(a^2 - b^2)^2$

39. $(a^2)(b^2)(a^2)$

40. $a^2 - b(ab)^2$

41. $-(b^2)$

42. $(-b)^2$

Independent and Dependent Events

Goals ■ Find the probability of dependent and independent events.

Applications Sports, Surveys, Scheduling

Work with a partner. You will need a six-sided die and a coin.

1. Suppose one person rolls the die while the other tosses or flips the coin. What do you think the probability is of rolling a 3 and landing the coin heads up? Record your prediction.

2. Check your prediction by rolling the die and tossing the coin until you get both of these outcomes.

3. Share the results of your experiment with other groups.

◤ BUILD UNDERSTANDING

When the outcome of one event does not affect the outcome of another event, the events are said to be **independent**. To find the probability of both events occurring, multiply the probabilities of each event.

If A and B are independent events, then $P(A \text{ and } B) = P(A) \cdot P(B)$

To emphasize that A and B do not characterize a single event, sometimes $P(A$ and $B)$ is written $P(A, \text{then } B)$.

Example 1

A bag contains 3 white softballs, 2 yellow softballs, 3 green softballs, and 4 red softballs. You reach into the bag without looking and take out a ball. You replace it and then take out another ball at random. Find the probability that the first ball is red and the second ball is white.

Solution

Because the first ball is replaced before the second is taken, the sample space of 12 balls is the same for each event. The two events are independent. Multiply to find the probability that both will occur.

$$P(\text{red, then white}) = P(\text{red}) \cdot P(\text{white})$$

$$= \frac{\text{number of red balls}}{\text{total number of balls}} \cdot \frac{\text{number of white balls}}{\text{total number of balls}}$$

$$= \frac{1}{3} \cdot \frac{1}{4}$$

$$= \frac{1}{12}$$

The probability of picking red, then white, is $\frac{1}{12}$.

When the outcome of one event *is* affected by the outcome of another, the events are said to be **dependent**.

Check Understanding

Are these events independent or dependent?

1. tossing a coin twice

2. picking two marbles from a bag without replacing the first one

3. choosing a captain and then choosing a co-captain

4. rolling three dice

Example 2

SPORTS Six teams—the Panthers, Tigers, Lions, Bears, Cheetahs, and Elephants—are in the lottery round for this year's draft picks. The name of each team is written on a card and placed in a box.

To determine who gets the first lottery pick, one card will be drawn at random and *not* replaced. Then a second card will be drawn at random to determine the second pick. What is the probability that the Bears get the first draft choice and the Lions get the second draft choice?

Solution

Because the first card is not replaced, the sample space for the second drawing has been changed. The second event is dependent on the first event.

Probability of first event

$$P(\text{Bears}) = \frac{\text{number of Bears cards}}{\text{total number of cards}} = \frac{1}{6}$$

Probability of second event

$$P(\text{Lions, after Bears}) = \frac{\text{number of Lions cards}}{\text{new total number of cards}} = \frac{1}{5}$$

Multiply the probabilities.

$$P(\text{Lions, after Bears}) = \frac{1}{6} \cdot \frac{1}{5}$$
$$= \frac{1}{30}$$

The probability of drawing the Bears first and the Lions second is $\frac{1}{30}$.

Example 3

A bag contains 3 green marbles, 2 red marbles, 4 yellow marbles, and 1 black marble. Two are taken at random from the bag without replacement. Find P(green, then green).

Solution

$$P(\text{first green marble}) = \frac{\text{number of green marbles}}{\text{total number of marbles}} = \frac{3}{10}$$

$$P(\text{second green marble}) = \frac{\text{number of green marbles}}{\text{total number of marbles}} = \frac{3-1}{10-1} = \frac{2}{9}$$

Multiply the probabilities.

$$P(\text{green, then green}) = \frac{3}{10} \cdot \frac{2}{9}$$
$$= \frac{6}{90} = \frac{1}{15}$$

The probability of picking green, then green, is $\frac{1}{15}$.

A group of numbered cards contains three 3s, four 4s, and five 5s. Cards are picked at random, one at a time, and then replaced. Find each probability.

1. P(3, then 5)

2. P(4, then odd number)

3. P(even, then not 4)

SPORTS A baseball team has 10 pitchers, 3 catchers, 5 outfielders, and 7 infielders on its roster. Two players from this team will be chosen at random to represent the league in a tour of Japan. Find each probability.

4. P(pitcher, then catcher)

5. P(outfielder, then infielder)

SURVEYS A newspaper survey asked 100 men and 100 women whether they planned to vote for a proposed tax increase. Twenty men and 40 women said they are in support of the increase. A person from the survey is chosen at random. Find each probability.

6. What is the probability that the person chosen is in support of the tax increase?

7. What is the probability that the person is a woman in support of the increase?

8. What is the probability that the person is a man who is against the increase?

◥ **PRACTICE EXERCISES** • For Extra Practice, see page 691.

A box contains 3 red counters, 4 yellow counters, 2 green counters, and 1 blue counter. Counters are taken at random from the box, one at a time, and then replaced. Find each probability.

9. P(red, then yellow)

10. P(blue, then green)

11. P(red, then not red)

A drawer contains 2 pairs of black socks, 3 pairs of brown socks, a pair of beige socks, and 6 pairs of white socks. One sock at a time is taken at random from the drawer and not replaced. Find each probability.

12. P(black, then black)

13. P(white, then black)

14. P(beige, then white)

A billboard says "EAT HERE NOW." Two letters fall off, one after the other.

15. What is the probability that both letters are vowels?

16. What is the probability that the first letter is an E, and the second letter is not an E?

17. You are given one ticket each to two soccer games at a stadium with 48,000 seats. What is the probability you will sit in Section D in the first game, and then Section A in the second game, if Section D has 4000 seats and Section A has 3000 seats?

Maracana Stadium, Brazil

SCHEDULING Liu and Michi plan to sign up for a drawing class next term. Drawing is offered during the first 4 periods of the day, and students are assigned randomly to classes.

18. What is the probability that Liu and Michi will have drawing together?

19. What is the probability that both students will have drawing during first period?

Both spinners are spun. Find each probability.

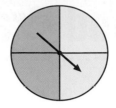 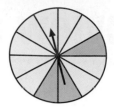

20. P(red, red)

21. P(red, yellow)

22. P(green, not green)

A golf bag pocket contains 4 yellow golf balls, 3 white balls, 1 green ball, and 4 red balls. You pull out one ball at a time, without replacing it. Find each probability.

23. P(red, then white, then yellow)   **24.** P(green, then red, then white)

25. WRITING MATH Explain the difference between events that are *mutually exclusive* and those that are *independent*.

EXTENDED PRACTICE EXERCISES

HISTORY Suppose that it is 1944, and the Homestead Grays of the Negro National League are playing the Birmingham Black Barons of the Negro American League in a "best two out of three" series. Then suppose the Grays are the favored team, and the probability they will win any individual game is $\frac{3}{4}$.

26. What is the probability that the Black Barons win a game?

27. What is the probability that the Grays win in two straight games?

28. What is the probability that the series goes for three games?

29. What is the probability that the Homestead Grays win the series?

30. DATA FILE Use the data on baseball on page 652. Suppose Davis had one more official at-bat in the 1943 season and Wagner had one more official at-bat in the 1948 season. What is the probability that both would have gone hitless?

31. CHAPTER INVESTIGATION When a baseball player comes to bat, the player can get a hit—either a single, a double, a triple or a home run—or the player can walk, strike out, or make an out some other way. For each player you have chosen, find the probability expressed as a percent that each event will occur. For example, to find the probability that a player will walk, divide the number of walks (BB) by the number of at bats (AB) and convert the decimal to a percent. Use a spreadsheet or calculator.

MIXED REVIEW EXERCISES

Copy quadrilateral *ABCD*. Then draw its dilation image.
(Lesson 8-3)

32. The center of dilation is the origin and the scale factor is 4.

33. The center of dilation is the origin and the scale factor is $\frac{3}{4}$.

34. The center of dilation is point *A* and the scale factor is 2.5.

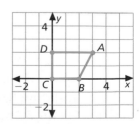

Review and Practice Your Skills

A card is picked at random from a standard deck of 52 cards. Find each probability.

1. P(heart and face card)   2. P(jack or queen)   3. P(red or black card)

4. P(black two)   5. P(2 or 3)   6. P(7 of hearts)

7. $P(2 \le \text{card} \le 5)$ 8. P(king of clubs)   9. P(club and (ten or king))

You flip a coin four times. Find each theoretical probability.

10. P(exactly one head)   11. P(2 tails, 2 heads)   12. P(3 or 4 tails)

13. P(more than 2 heads)   14. P(0 or 1 head)   15. P(3 tails)

You roll a pair of dice. Find each theoretical probability.

16. P(sum = 7)   17. P(sum = 11)   18. P(both even)

19. P(1 is rolled)   20. P(4 or 5 is rolled)   21. P(sum = 2)

22. P(sum is odd)   23. P(sum < 6)   24. P(sum = 10, 11, or 12)

25. P(sum is even and > 7)   26. P(sum is odd and < 11)   27. P(values are equal)

28. A spinner has 20 equal sectors numbered 1–20. Are spinning a multiple of 4 and multiple of 9 mutually exclusive events? Explain.

29. You spin a spinner with 8 equal sectors, numbered 1–8. What is the probability of spinning a number that is neither odd nor greater than 6?

A drawer contains 7 red shirts, 5 blue shirts, and 4 white shirts. One shirt at a time is taken at random from the drawer and not replaced. Find each probability.

30. P(red, then blue)   31. P(red, then not white)   32. P(white, then blue)

33. P(both white)   34. P(not blue, then not red)   35. P(both not blue)

A box contains 4 red cards, 5 black cards, 10 green cards, and 2 blue cards. Cards are taken at random from the box, one at a time, and then replaced. Find each probability.

36. P(red, then red)   37. P(red, then green, then blue)

38. P(not red, then green)   39. P(black, then black, then not green)

40. P(not green in each of three draws)   41. P(black, then not blue)

42. P(red, then red, then red, then red)   43. P(blue, then blue, then black)

44. You are given one ticket each to two hockey games in an arena with 18,000 seats. What is the probability that you will sit in Section B in the first game, and then Section C in the second game, if Section B has 4500 seats and Section C has 3000 seats?

List all the elements of the sample space for each of the following experiments.
(Lesson 9-1)

45. tossing a quarter and a nickel

46. choosing a month of the year

47. choosing the day of the week

48. choosing a letter of the alphabet

49. Conduct and document a simulation using a six-question multiple choice test. If each question has three choices for answers, and you guess on each question, what are your chances of getting 3 or more questions correct? (Lesson 9-2)

Three coins are tossed. Find each probability. (Lesson 9-3)

50. P(at least one heads)

51. P(no tails)

52. P(0 or 1 tails)

53. P(two tails)

54. P(all three coins the same)

55. P(1 or 2 heads)

A box contains 100 cards, numbered from 1–100. Cards are taken at random from the box, one at a time, and not replaced. Find each probability. (Lesson 9-4)

56. P(even number, then odd number)

57. P(multiple of 3, then 50)

58. P(45, then 45)

59. P(99, then 100)

60. P(number $<$ 40, then number $>$ 80)

61. P(prime number, then prime number)

Mid-Chapter Quiz

1. How many outcomes are there for an outfit chosen from three pairs of pants and five shirts? (Lesson 9-1)

2. A basketball player has made 60 out of 125 attempts. How many shots is he likely to make in 500 attempts? (Lesson 9-2)

3. A family has five children. What is P(three boys and two girls)? (Lesson 9-2)

A card is picked at random from a standard deck of 52 cards. (Lesson 9-3)

4. Find P(heart and less than 5)

5. Find P(heart or less than 5)

6. Find P(7 or king)

7. Find P(neither 5 nor diamond)

8. Find P(neither 3 nor red)

A bag contains 7 green marbles, 3 red marbles, and 5 blue marbles. A marble is picked and replaced. Then another marble is picked. (Lesson 9-4)

9. Find P(green, then red)

10. Find P(two red marbles)

From the same bag, a marble is picked and not replaced. Then another marble is picked. (Lesson 9-4)

11. Find P(green, then red)

12. Find P(two red marbles)

9-5 Permutations and Combinations

Goals ■ Find the number of permutations and combinations of a set.

Applications Travel, Sports, Office Work

Work with a partner to answer the questions.

You have just applied for your first set of license plates. Each license plate has 3 different letters followed by 3 different digits. The letters and digits are assigned randomly by the Department of Motor Vehicles.

1. How many arrangements of 3 different letters are possible?

2. How many arrangements of 3 different digits are possible? Remember, to include 0 as a digit.

3. Suppose you had hoped that your plate would read "ACE 123." What is the probability that you will receive this plate? How do you know?

◥ BUILD UNDERSTANDING

Thus far, you have used either a tree diagram or a set of ordered pairs to find the number of outcomes in a sample space. But sometimes the sample space is too large to use either of these methods.

Another way to find the total number of outcomes is to use the **fundamental counting principle**. This principle can be used to calculate the number of ways two or more events can happen in succession. It states that, if an event A can occur in m ways and an event B can occur in n ways, then events A and B can happen in $m \cdot n$ ways.

Example 1

TRAVEL Suppose the Carthage College Women's Basketball team will travel on a road trip to Grand Rapids, Peoria, and Battle Creek. They can go from Kenosha to Grand Rapids by car, train, or bus, then from Grand Rapids to Peoria by bus, train, or plane, and from there to Battle Creek by car, bus, train, or plane. Finally, from Battle Creek, they can either take the bus or the train back to Kenosha. How many different routes are possible on this road trip?

Solution

Use the fundamental counting principle.

The 3 possible routes for the first leg of the trip are car, train, or bus. Then they have 3 possible routes for the second leg of the trip, 4 for the third leg of the trip, and 2 for the return trip to Kenosha.

$$3 \cdot 3 \cdot 4 \cdot 2 = 72$$

Seventy-two different routes are possible.

An arrangement of items in a particular order is called a **permutation**. For the four letters M, A, T, H, there are 24 different 4-letter permutations.

MATH	MAHT	MHAT	MHTA	MTHA	MTAH
AMTH	AMHT	AHTM	AHMT	ATHM	ATMH
TAMH	TAHM	TMAH	TMHA	THAM	THMA
HAMT	HATM	HMAT	HMTA	HTAM	HTMA

You can use the fundamental counting principle to find the number of permutations of any group of items.

For each arrangement of M, A, T, H, there are 4 choices for the first letter, 3 choices for the second, 2 choices for the third, and 1 choice for the fourth.

$$4 \cdot 3 \cdot 2 \cdot 1 = 24.$$

$4 \cdot 3 \cdot 2 \cdot 1$ can be written in **factorial notation** as **4!**

In general, the number of permutations of n different items is written **n!** and read as **n factorial**.

Example 2

In how many different ways can you arrange your math, science, social studies, language arts, and literature anthology books in a row on a shelf?

Solution

There are five books. Find the number of permutations of five items.

number of permutations of five items = 5!

$$= 5 \cdot 4 \cdot 3 \cdot 2 \cdot 1 \text{ or } 120$$

There are 120 different ways to line up five books on a shelf.

Sometimes you need only part of a set of items, such as selecting two of nine players. The number of permutations of n different items, taken r items at a time, with no repetitions, is written $_nP_r$. Use the formula below to find the number of permutations when only part of a set is used.

$$_nP_r = \frac{n!}{(n-r)!}$$

Check Understanding

What is wrong with the notation $_4P_5$?

Example 3

SPORTS Eight teams enter a tournament. How many different arrangements of first-, second-, and third-place winners are possible?

Solution

Use the formula: $_nP_r = \dfrac{n!}{(n-r)!}$

$$_8P_3 = \frac{8!}{(8-3)!}$$

$$= \frac{8 \cdot 7 \cdot 6 \cdot \cancel{5} \cdot \cancel{4} \cdot \cancel{3} \cdot \cancel{2} \cdot \cancel{1}}{\cancel{5} \cdot \cancel{4} \cdot \cancel{3} \cdot \cancel{2} \cdot \cancel{1}} = 8 \cdot 7 \cdot 6 = 336$$ Cancel common factors to simplify the computation.

There are 336 ways for teams to finish first, second, and third.

For each situation described in the preceding Examples, the order of the items in consideration is important. A set of items in *no* particular order is called a **combination**. The number of combinations of n different items, taken r items at a time, where $0 \le r \le n$, is written $_nC_r$. You can use the formula below to find the number of combinations of a set of items.

$$_nC_r = \frac{n!}{(n-r)!r!}$$

Math: Who, Where, When?

Although several 16th- and 17th-century mathematicians, notably Pascal and Fermat, investigated probability, Jacques Bernoulli is considered by some to be the founder of probability theory. His book, *Ars Conjectandi*, published in 1713, is the first book devoted entirely to the subject of probability. This book contains a theory of combinations, essentially the same as we understand it today, as well as the first appearance, with today's meaning, of the word *permutation*.

Example 4

How many different four-person ensembles can be chosen from a pool of 10 musicians?

Solution

There are 10 people from which to pick, 4 at a time. So, $n = 10$ and $r = 4$. Use the formula:

$$_nC_r = \frac{n!}{(n-r)!r!}$$

$$_{10}C_4 = \frac{10!}{(10-4)!4!}$$

$$= \frac{10 \cdot 9 \cdot 8 \cdot 7 \cdot \cancel{6} \cdot \cancel{5} \cdot \cancel{4} \cdot \cancel{3} \cdot \cancel{2} \cdot \cancel{1}}{(\cancel{6} \cdot \cancel{5} \cdot \cancel{4} \cdot \cancel{3} \cdot \cancel{2} \cdot \cancel{1})(4 \cdot 3 \cdot 2 \cdot 1)}$$

$$= \frac{5040}{24} \quad \text{Cancel common factors to simplify the computation.}$$

$$= 210$$

There can be 210 different four-person ensembles.

◥ TRY THESE EXERCISES

1. **SPORTS** A manager is choosing her infield from among 4 third-base players, 3 shortstops, 2 second-base players, and 5 first-base players. How many different ways can an infield be chosen?

Tell whether each question involves a permutation or a combination. Then solve.

2. In how many different ways can you arrange the letters *a, c, e, g, i, k,* and *1*?

3. How many different selections of three tapes can be made by a consumer choosing from among a collection of six tapes?

4. In how many different ways can a starting lineup of 5 players be selected from a group of 12 basketball players?

5. In how many different ways can 4 winners be chosen from 15 contestants?

◥ PRACTICE EXERCISES • For Extra Practice, see page 691.

6. **OFFICE WORK** A secretary has to create ten new customer files. In how many different orders can he do this?

7. **HIRING** Six applicants apply for two jobs. How many different outcomes are possible?

8. Ralph is a tour guide. In how many ways can he choose 3 museums to visit from the 8 museums in a city?

9. In how many ways can a disk jockey select 5 of the 20 top hits?

10. **ERROR ALERT** On a test, students must choose 3 out of 5 essay questions to answer. Dale calculates that there are 60 ways to do this. What has Dale done wrong?

11. **SPORTS** Bill, Phil, and Jill are among 12 players competing for 3 spots on a table-tennis team. Every player has an equal chance of making the team. Find the probability that all three will make the team.

12. What is n, if $_nP_3 = 120$?

13. Suppose that license plates contain three different letters. What is the probability that Meg's plates will spell her name?

14. **SPORTS** How many ways can a batting order be made for 9 players if you know that one player has already been designated to bat first and another to bat fourth?

15. **WRITING MATH** Find the number of permutations of the letters in the word *shutout*. Explain how you did it.

16. Two students out of 8 will be chosen to speak at a school assembly. How many different outcomes are possible?

DATA FILE For Exercises 17–18, use the information about the All-American Girls Professional Baseball League on page 652.

17. Suppose you could interview all eight women who were batting champions of the All-American Girls Professional Baseball League to discover what playing in this league was like. How many orders for these interviews would be possible?

18. If you were to interview only four of the eight women, what is the probability that you would first interview a player from either Fort Wayne or Rockford?

Technology Note

Some calculators have a factorial key, marked $\boxed{x!}$. To find 7!, enter 7, then $\boxed{x!}$. On graphing calculators, you can choose the factorial function from a displayed mathematical menu. This menu may also include operations for finding permutations and combinations.

EXTENDED PRACTICE EXERCISES

19. Compare the values of $_8C_5$ and $_8C_3$. What do you notice?

20. Find the values of $_7C_4$ and $_7C_3$. What do you notice?

21. What can you say about the sum of the number of items taken at one time for each combination shown in Exercises 19 and 20?

22. **CRITICAL THINKING** Use what you have discovered to quickly find $_{67}C_{64}$.

MIXED REVIEW EXERCISES

Add. (Lesson 8-5)

23. $\begin{bmatrix} 2 & 4 \\ 3 & 0 \end{bmatrix} + \begin{bmatrix} 1 & 3 \\ -2 & 4 \end{bmatrix}$

24. $\begin{bmatrix} 6 & 0 & 3 \\ -5 & 2 & 1 \end{bmatrix} + \begin{bmatrix} 3 & -4 & -2 \\ 7 & 1 & 3 \end{bmatrix}$

Find the measure of each angle. (Lesson 3-2)

25. $\angle BZC$

26. $\angle CZD$

27. $\angle AZC$

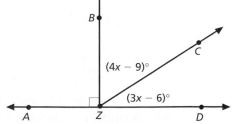

Scatter Plots and Box-and-Whisker Plots

Goals ■ Interpret and make scatter plots and box-and-whisker plots.

Applications Manufacturing, Sales, Sports

Work in groups of 3–4 students.

Find out your classmates' favorite music performer. Make a list of ten popular music groups or artists. Using a rating scale of 1–10, survey 25 students. Make a graph to display your findings. Compare findings and graphs with those of classmates.

◥ BUILD UNDERSTANDING

Data can be presented in many ways. Graphs are useful because they can help identify characteristics of data. Recall that a histogram shows frequencies of *intervals* of data. A stem-and-leaf plot shows *all* data ordered as in a frequency table, but also visually, as in a bar graph. It shows how data are clustered.

A **scatter plot** is another type of visual display used to explore the relationship between two sets of data, represented by unconnected points on a grid.

Example 1

MANUFACTURING The scatter plot shows the relationship between years of experience and hourly pay at one factory.

a. Why are the scales different?

b. What does each • represent?

c. Find the hourly pay of an employee with 8 years of experience.

d. Describe the relationship between experience and pay.

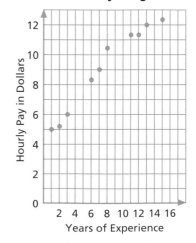

Factory Wages

Solution

a. There are two different sets of data—hourly pay and years of experience.

b. Each • shows the hourly pay given the years of experience.

c. $10.50

d. Hourly pay usually increases with years of work experience.

A pattern may emerge that shows a relationship between the two sets of data. If data clusters around a **line of best fit**, or **trend line**, from the bottom left upward to the top right of the graph, this shows a **positive correlation** between the sets of data. If the line slopes downward from left to right, it indicates a **negative correlation** between the data.

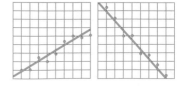

Example 2

SALES Use the scatter plot at the right for these questions.

a. What can you say about the correlation between the age of a car and its resale value?

b. Predict the resale value of an 8-year-old car with an original cost of $15,000.

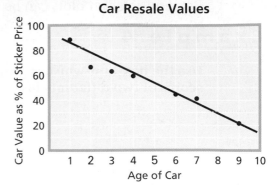

Car Resale Values

Solution

a. The trend line slopes downward from upper left to lower right, so there is a negative correlation between a car's age and its resale value.

b. Extend the pattern. A reasonable assumption would be for the resale value to be about 30% of the original cost for an 8-year-old car. So, a car that cost $15,000 originally might sell for about $4500 after 8 years.

Another way to display data is with a **box-and-whisker plot**, also known as **box plot**. This plot shows how data are dispersed around a median, but does not show each specific item in the data. By examining a box-and-whisker plot, you can tell if data are clustered closely together or spread far apart.

A box-and-whisker plot shows both the median and the **extremes** of a set of data. It also shows the median of the lower half of the data, called the **lower quartile**, and the median of the upper half of the data, called the **upper quartile**. Both quartiles include the median if the data contain an odd number of items.

Example 3

SPORTS Joe DiMaggio played center field for the New York Yankees for 13 years. During each year of his career, he hit the following number of home runs: 29, 46, 32, 30, 31, 30, 21, 25, 20, 39, 14, 32, and 12. Make a box-and-whisker plot for this data.

Solution

Write the data in numerical order. Find the least and greatest values, the median, the lower quartile, and the upper quartile.

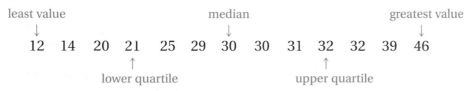

Use points to mark the values below a number line. Draw a box that starts and stops at the lower and upper quartiles, and a vertical line at the point for the median. Then draw *whiskers*, or line segments, from each end of the box to the least and greatest values. Finally, give your graph a title.

DiMaggio's Home Runs

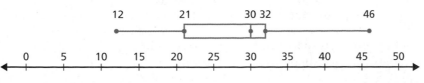

Check Understanding

In a box-and-whisker plot, what percent of a set of data is represented by the box? By the whisker to the right of the box?

Box-and-whisker plots can be used to compare sets of data.

Example 4

Use the box-and-whisker plots below to answer questions about the math test scores of two different classes.

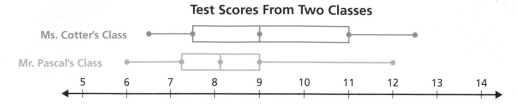

Test Scores From Two Classes

a. Which class had the higher median score?

b. What was the lower quartile in Mr. Pascal's class?

c. Which class had its scores grouped more closely around its median?

d. For which class were the lowest scores clustered more closely?

e. Which class, as a whole, scored better on the test?

Reading Math

The abbreviations Q_1, Q_2, and Q_3 are sometimes used for the lower quartile, the median, and the upper quartile. The **interquartile range** is the difference between the values of the upper and lower quartiles.

Solution

a. Ms. Cotter's

b. about 7.2

c. Mr. Pascal's; the range of the middle 50% of the scores is about 1.8. The range for the middle 50% in Ms. Cotter's class is about 3.5.

d. Ms. Cotter's; the range for the lowest quarter is about 1. In Mr. Pascal's class, it is about 1.2.

e. Ms. Cotter's

◣ TRY THESE EXERCISES

FITNESS Use the scatter plot at the right for Exercises 1–3.

1. What is the weight of the student who is 67 in. tall?

2. Does the scatter plot show a positive or negative correlation?

3. Give an estimate of the height of a student who weighs 145 lb.

4. **SPORTS** Make a box-and-whisker plot for the following data.

TOP PRICES OF TICKETS TO SPORTING EVENTS (IN DOLLARS)

45, 55, 40, 60, 15, 25, 35, 30, 10, 40

5. Use the box-and-whisker plot you made in Exercise 4. Are the data clustered more closely above or below the median?

Heights and Weights

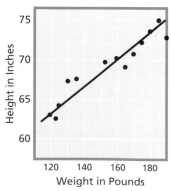

Height in Inches / Weight in Pounds

Technology Note

Use a graphing calculator to make a box-and-whisker plot.

1. Enter the data as a list using the STAT feature.

2. Select the STATPLOT menu and choose Plot1. Under Type, select the box-and-whisker plot diagram.

3. Adjust the window dimensions if necessary and press GRAPH.

4. Use the TRACE feature to find the median and upper and lower quartiles.

SPORTS This table shows how many points a basketball player scored during his career. Use this information for Exercises 6–8.

Age	Scoring Average
23	18
24	17.5
25	22.5
26	24
27	21.5
28	26
29	23.5
30	22.5
31	27.5
32	20.5

6. Make a scatter plot.

7. What is the range of this player's scoring average?

8. Does your scatter plot show a positive correlation, a negative correlation, or no correlation?

SPORTS These box-and-whisker plots show batting averages for 3 baseball teams.

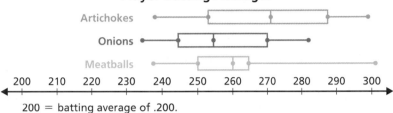

Player Batting Averages

200 = batting average of .200.

9. Which team has the highest median batting average?

10. Which team has the smallest range of batting averages?

11. **WRITING MATH** Why is the right whisker for the Meatballs longer than the left whisker?

12. **CHAPTER INVESTIGATION** Create a 10-by-10 table for each player. Number both the columns and rows from 1 to 10. The table represents all the possible outcomes for a player at bat. Using the percents you calculated, fill in the cells of the table with appropriate abbreviations. For example, in 1998 Gary Sheffield hit a home run in 5% of his at bats. To create a table for Sheffield, you would write HR in any 5 cells.

EXTENDED PRACTICE EXERCISES

Choose the graph you think works best to display the data described.

13. **MANUFACTURING** To show the relationship between the percent of polyester in an article of clothing and the price of the article of clothing

14. To show that the test scores in your class clustered around the middle-most score

15. **WRITING MATH** Is it possible for the mean of a set of data to fall outside the box part of a box-and-whisker plot? Explain.

MIXED REVIEW EXERCISES

Multiply. (Lesson 8-5)

16. $6 \cdot \begin{bmatrix} 6 & 3 \\ -4 & 2 \end{bmatrix}$

17. $4 \cdot \begin{bmatrix} 9 & 8 \\ 7 & 1 \end{bmatrix}$

18. $7 \cdot \begin{bmatrix} 7 & -3 & -5 \\ 4 & 6 & 0 \end{bmatrix}$

Review and Practice Your Skills

PRACTICE ◣ LESSON 9-5

Evaluate.

1. $_5P_3$
2. $_8P_7$
3. $_4P_1$
4. $_5P_5$
5. $_7P_4$
6. $_9P_0$
7. $_{14}P_7$
8. $_9P_6$
9. $_5C_3$
10. $_9C_6$
11. $_4C_0$
12. $_6C_1$
13. $_8C_7$
14. $_5C_4$
15. $_6C_2$
16. $_{12}C_4$

17. In how many ways can the positions of president, vice president, and secretary be chosen from a club containing 20 members?

18. In how many ways can a committee of three people be chosen from a club containing 20 members?

19. In how many ways can a volleyball coach choose 6 starters from a team of 14 players?

20. In how many ways can a disc jockey play 3 of the top ten hits?

21. In how many ways can first-place and runner-up winners be chosen from 15 entrants in a contest?

22. In how many ways can the numbers 1, 2, 3, 4, and 5 be arranged in a 5-digit password?

23. What is n, if $_nP_2 = 72$?

24. Find the values of $_{10}C_4$ and $_{10}C_6$. What do you notice?

PRACTICE ◣ LESSON 9-6

This table shows the appraised value of a house over time.

Age (years)	0	3	6	9	12	15	18
Value (thousands)	140	148	160	162	185	178	194

25. Make a scatter plot of the data.

26. What is the range of appraised values?

27. Does your scatter plot show a positive correlation, a negative correlation, or no correlation?

28. Draw a box-and-whisker plot that has the following attributes: (Lesson 9-6)
   a. range of 87
   b. median value of 135
   c. low value of 107
   d. upper quartile of 162
   e. range of middle 50% of data of 38

29. Draw a box-and-whisker plot that has the same value for its maximum value and its upper quartile. Define a collection of data points that would yield this type of plot. (Lesson 9-6)

A spinner with 8 equal sectors labeled A through H is spun. Find each probability. (Lesson 9-1)

30. P(spinning E)

31. P(spinning vowel)

32. P(spinning H)

33. P(spinning a letter before F)

34. P(spinning B or G)

35. P(spinning M)

36. Describe a simulation you could do to find out how many cards you would expect to have to draw from a standard deck to get three clubs. (Lesson 9-2)

For each situation, tell whether order does or does not matter. (Lesson 9-5)

37. You are selecting three-number combinations for school lockers.

38. You are selecting five books to check out from the library.

39. You are choosing the 9 starters on a baseball team.

40. You are choosing a 5-member committee from your leadership board.

Math*Works*
Workplace Knowhow

Career – Physical Therapist

Physical therapists work with people who have been injured. They improve mobility, relieve pain and prevent or limit permanent physical disabilities. To relieve pain and treat injuries, physical therapists use massages, electrical stimulation, hot and cold packs and traction. The sports world depends on physical therapists to help athletes who are injured during practices, exercise sessions or games. Some injuries require surgery, but many can be treated by rest followed by proper exercise. Physical therapists often travel with teams.

1. As a physical therapist, you have treated 236 injuries during the past year. Of the total number, 108 injuries were caused by an improper warm-up. What percent of the injuries resulted from an improper warm-up?

2. A team of 45 players suffered 15 ankle injuries during the season. If the squad is increased to 52 players, how many ankle injuries would you expect to see during a season?

3. A physical therapist employs 4 kinds of massage, 3 kinds of baths, 1 type of electrical stimulation, and 8 exercise programs. If each injury is treated with all four types of therapies, how many combinations of therapies does this therapist offer?

4. You are examining the player files for 30 players who have been injured during the season. For this sport, 1 out of 3 injuries are to the knee and 1 out of 2 of these cases require surgery. What is the probability that the first file you select will belong to a player requiring knee surgery?

Standard Deviation

Goals
- Find the variance, standard deviation and z-scores for a set of data.
- Use standard deviation to interpret data.

Applications Sports, Test-Taking, Education

Work in groups of 4–5 students.

1. Collect a set of data about students in class, such as heights, arm lengths, head circumference, lengths of thumbs and so on.

2. Study the data. Look for new ways to describe the data. Instead of focusing on central tendencies, study how the data are spread out, or *dispersed*.

3. Consider these questions: How much do individual values in your data differ from the greatest value? The least value? The mean, median or mode of the values?

4. Share your results with your classmates.

◥ BUILD UNDERSTANDING

Statistics that show how data is spread out are called **measures of dispersion.** For example, you know that the *range* of a set of data is the difference between the largest and smallest item.

Variance is another measure of dispersion. The variance of a set of numbers is the mean of the squared differences between *each* number in the set and the mean of *all* numbers in the set. For the set of numbers $x_1, x_2, \ldots x_n$, with a mean of m, use this formula.

$$v = \frac{(x_1 - m)^2 + (x_2 - m)^2 + \ldots + (x_n - m)^2}{n}$$

Example 1

SPORTS During a basketball tournament, the five starters for the Bulldogs made the following number of 3-pointers: Bowen, 3; White, 4; Fillmore, 5; Graham 6; and Bonilla, 7. Find the variance for the set of numbers.

Solution

1. Divide the sum of scores by 5 to find the mean, m. ($m = 5$)

2. Find the difference between each number and the mean. Then find the square of each difference.

3. Find the mean of all the squares in Step 2.

$$4 + 1 + 0 + 1 + 4 = 10 \qquad 10 \div 5 = 2$$

The variance is 2.

number	$x - m$	$(x - m)^2$
3	$3 - 5$	$(-2)^2 = 4$
4	$4 - 5$	$(-1)^2 = 1$
5	$5 - 5$	$0^2 = 0$
6	$6 - 5$	$1^2 = 1$
7	$7 - 5$	$2^2 = 4$

The **standard deviation**, s, of a set of numbers is the square root of the variance.

Example 2

Find the standard deviation for the set of numbers in Example 1.

Solution

Find the square root of the variance.

$\sqrt{2} \approx 1.4$ The standard deviation is 1.4

Example 3

	Test A	Test B
Molly's score	85	80
Mean score	65	60
Standard deviation	8	10

Molly took two tests. On which did she score better, compared with others in her class?

Solution

1. Compare both of her scores with the mean. She scored 20 points higher than the mean on both tests.

2. Use the standard deviation.

In Test A, Molly's score was $\frac{20}{8}$, or 2.5 standard deviations above the mean score.
In Test B, it was $\frac{20}{10}$, or 2 standard deviations above the mean score.
Relative to her classmates, Molly scored better on Test A.

The number of standard deviations between a score and the mean score is indicated by a **z-score**. Molly's z-score was 2.5 on Test A. A score below the mean would have a negative z-score.

◥ TRY THESE EXERCISES

Compute the variance and standard deviation for each set of data.

1. 4, 4, 4, 4, 4   **2.** 7, 3, 5, 9, 11   **3.** 2.7, 4.7, 6.7, 8.7, 10.7

◥ PRACTICE EXERCISES • For Extra Practice, see page 692.

Compute the variance and standard deviation for each set of data.

4. 6, 6, 6, 6

5. 1.5, 2.5, 3.5, 4.5, 5.5

6. 4.2, 9.2, 14.2, 19.2, 24.2

7. 8.9, 4, 9.4, 26.5, 14.9

8. Raymond took two tests. On the first test, his score was 45, while the mean score was 55 and the standard deviation was 5. On the second test, his score was 55, while the mean score was 65 and the standard deviation was 10. On which test did Raymond score better, relative to the scores of his classmates?

9. On a science test taken by 28 students, the mean score was 82.5. The standard deviation for the scores was 5.3. What was the sum of all the scores?

10. WRITING MATH What can you say about the relationship between the standard deviation of a set of scores and how spread out the scores are?

A visual display that shows the relative frequency of data is called a **frequency distribution**. A histogram is often used for this purpose.

11. Find the mean, median, mode, variance, and standard deviation to the nearest whole number for the data presented below.

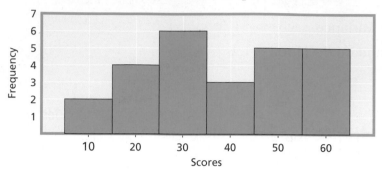

TEST TAKING Find out how well your classmates would score on a test on which they had to guess the answer to *every* question. Work in a small group of 3–4 students. Make up a 20-question multiple-choice test using the topic of obscure and unimportant sports data. Use an almanac, a sports encyclopedia, a book of records, or any other source to find facts unfamiliar to anyone in your class. Ask everybody to take the test. Then grade the test as a class. Record and analyze the results.

12. Find the range, mean, median and mode for the scores.

13. Find the standard deviation.

14. Make a visual display of the scores.

Have the class retake the test. Then analyze the results.

15. Find the range, mean, median and mode for the new scores.

16. Find the standard deviation.

17. Make a visual display of the scores.

Compare both sets of results.

18. On which test did the class perform better? Explain.

19. On which test did you perform better relative to your classmates? What was your *z*-score on that test?

20. CHAPTER INVESTIGATION Play the baseball simulation game. Draw a baseball diamond and use coins for markers. To play, two people choose nine baseball players each. Put the players' tables in batting order.

Use 10-sided polyhedral dice or a deck of standard playing cards with the kings, queens and jacks removed. Either roll two dice or draw two cards from the deck. The first die or card indicates the row on the player's table. The second die or card indicates the column. Find the cell at the intersection of the row and column to see what happens in the game.

If the baseball player gets a hit, place a marker in the appropriate place on the baseball diamond. Keep score as you would in a real baseball game. Play nine innings. Did your team do as well as you expected?

If you were to use a smooth curved line to connect the midpoints of the histogram on the previous page, you would form a frequency distribution known as a **bell curve**.

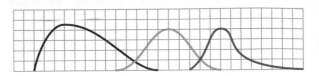

The **normal curve** is the best known frequency distribution. In a normal curve, the mean, median, and mode are the same.

Normal curves are determined by the mean and the standard deviation. In every normal curve, about 68% of the data are within one standard deviation unit of the mean. About 95% of the data are within two standard deviation units of the mean. Finally, about 99.7% of the data are within three standard deviation units.

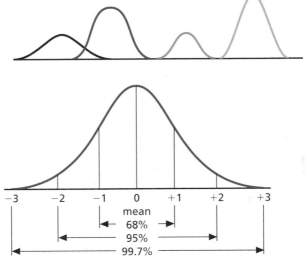

21. Suppose you drew two normal distributions on the same set of axes. Compare the appearances of these two curves if one has a greater mean than the other, but their variances are the same?

22. Suppose two normal curves drawn on the same set of axes have different variances but equal means. Compare the curves.

23. Which of these bell curves do you think might show the distribution of scores if your class were to take a third-grade spelling test?

24. As items in a set of data are dispersed more and more widely from the mean, what happens to the standard deviation?

■ MIXED REVIEW EXERCISES

Find each product. If not possible, write *NP*. (Lesson 8-6)

25. $[3 \quad 2 \quad 4] \cdot \begin{bmatrix} -1 \\ 6 \\ -5 \end{bmatrix}$

26. $\begin{bmatrix} 2 & 0 \\ 1 & 5 \end{bmatrix} \cdot \begin{bmatrix} 3 & 2 \\ 8 & 4 \end{bmatrix}$

27. $\begin{bmatrix} 5 & -4 \\ 3 & 1 \end{bmatrix} \cdot \begin{bmatrix} 0 & -5 & 6 \\ 2 & 8 & -4 \end{bmatrix}$

Solve. (Lesson 3-1)

28. On a number line, the coordinate of point *F* is −4. The length of $\overline{FG}$ is 13. Give 2 possible coordinates of point *G*.

29. On a number line, the coordinate of point *Q* is −18. The length of $\overline{QR}$ is 76. Give 2 possible coordinates of point *R*.

30. Point *S* is between points *R* and *T*. The length of $\overline{RS}$ is twice the length of $\overline{ST}$, and *RT* = 57. Find *RS* and *ST*.

Chapter 9 Review

VOCABULARY ◣

Choose the word from the list that best completes each statement.

1. The set of all possible outcomes of an experiment is the __?__.

2. A set of items in no particular order is called a(n) __?__.

3. If events cannot occur at the same time, they are __?__.

4. An upward sloping trend line on a scatter plot suggests a(n) __?__ correlation between the data.

5. The whiskers of a box-and-whisker plot show __?__ of the data.

6. The __?__ is a measure of dispersion that compares a number to the mean of a set of data.

7. If one event affects another event, they are __?__.

8. The __?__ uses multiplication to find the number of outcomes.

9. A(n) __?__ is an arrangement of items in a particular order.

10. __?__ is the number of favorable events divided by the total number of outcomes.

a. combination
b. dependent
c. extremes
d. fundamental counting principle
e. independent
f. mutually exclusive
g. permutation
h. positive
i. probability
j. sample space
k. simulation
l. standard deviation

LESSON 9-1 ◣ Review Percents and Probability, p. 384

▶ Divide the number of favorable observations by the number of total observations to find the **experimental probability** of an event. To find the **theoretical probability** of an event, divide the number of possible favorable outcomes by the number of possible outcomes.

A set of 30 cards is numbered 1, 2, 3, …, 30. Suppose you choose one card without looking. Find the probability of each event.

11. $P(12)$

12. $P(\text{odd})$

13. $P(\text{integer})$

14. $P(\text{less than 1})$

15. $P(\text{greater than 18})$

16. $P(\text{ends in 0})$

17. Find the probability of drawing a red 7 from a standard deck of playing cards.

LESSON 9-2 ◣ Problem Solving Skills: Simulations, p. 388

▶ One way to find a probability is to model the situation using a **simulation**. Simulations rely on **random numbers**.

18. A computer generates a list of random 2-digit numbers. What probability would you expect for a number in the list to contain the digit 2?

19. A fast food restaurant is putting 3 different toys in their children's meals. If the toys are placed in the meals at random, create a simulation to determine the experimental probability that a child will have all 3 toys after buying 5 meals.

20. Rodolfo must wear a tie when he works at the mall on Friday, Saturday, and Sunday. Each day, he picks one of his 6 ties at random. Create a simulation to find the experimental probability that he wears a different tie each day of the weekend.

LESSON 9-3 ◣ Compound Events, p. 392

▶ A **compound** event consists of two or more simple events. If A and B are **mutually exclusive** events, they cannot occur at the same time, and $P(A \text{ or } B) = P(A) + P(B)$. If A and B are not mutually exclusive events, they can occur at the same time, and $P(A \text{ or } B) = P(A) + P(B) - P(A \text{ and } B)$.

21. Two 1–6 spinners are spun. Find the probability that the sum of the numbers spun is 9 or less than 2.

22. A card is drawn at random from a standard deck. Find the probability that it is a black card or an 8.

23. There are 3 science books, 4 math books, and 2 history books on a shelf. If a book is randomly selected, what is the probability of selecting a science book or a history book?

24. In a drama club, 7 of the 20 girls are seniors, and 4 of the 14 boys are seniors. What is the probability of randomly selecting a boy or a senior to represent the drama club at an arts symposium?

LESSON 9-4 ◣ Independent and Dependent Events, p. 396

▶ Two events are **independent** if the outcome of one does not affect the outcome of the other. If A and B are independent events, $P(A, \text{ then } B) = P(A) \cdot P(B)$.

▶ Two events are **dependent** if the outcome of one affects the outcome of the other. If A and B are dependent events, $P(A, \text{ then } B) = P(A) \cdot P(B, \text{ after } A)$.

25. A die is rolled two times. Find $P(3, \text{ then even number})$.

26. A box contains 4 red marbles, 3 green marbles, 1 white marble, 2 yellow marbles, and 3 blue marbles. Two marbles are chosen at random and not replaced. Find $P(\text{red, then yellow})$.

27. A coin is tossed three times. What is the probability that all three times the coin shows heads?

28. Reiko has 3 quarters, 5 dimes, and 2 nickels in her pocket. She picks two coins at random without replacement. What is the probability that she chooses a quarter followed by a dime?

LESSON 9-5 ◣ Permutations and Combinations, p. 402

▶ The **fundamental counting principle** states that if an event A can occur in m ways and an event B can occur in n ways, then events A and B can occur in $m \cdot n$ ways.

▶ A **permutation** is a set of items arranged in a particular order. You can arrange a set of n items in $n!$ ways. To find the number of permutations of a set of n items taken r at a time, use the formula at the right. $\quad {_n}P_r = \dfrac{n!}{(n-r)!}$

▶ A set of items without consideration of order is called a **combination**. To find the number of combinations of a set of n items taken r at a time, use the formula at the right. $\quad {_n}C_r = \dfrac{n!}{(n-r)\,r!}$

29. In how many different ways can 6 pies be awarded first- through third-place prizes?

30. How many groups of 3 students can be chosen from a class of 20 students?

31. How many 3-digit whole numbers can you write using the digits 1, 3, 5, 7, and 9 if no digit can be used twice?

32. An ice cream store has 31 flavors of ice cream. Lamel wants to buy three pints of ice cream. If each pint of ice cream is a different flavor, how many different purchases can he make?

LESSON 9-6 ◼ Scatter Plots and Box-and-Whisker Plots, p. 406

▶ A **scatter plot** displays data as unconnected points. The trend line indicates whether the items being compared have a positive correlation, a negative correlation, or no correlation.

▶ A **box-and-whisker plot** shows extremes of data and how data are distributed.

33. The scatter plot shows the number of calories in different fruits compared to the number of milligrams of calcium they offer. Does there appear to be a positive, a negative, or no relationship between calories and calcium in fruit?

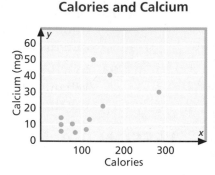

Calories and Calcium

34. Name a situation where a scatter plot would have a negative correlation. Sketch how it might look.

35. Some students at Johnson High rated the performance of their basketball team from 0 to 100, with 100 as the highest. These are the ratings: 67, 71, 58, 53, 65, 73, 64, 50, 52, 74, 48, 47, 53, 82, 63, 59, 67, 85, 45, 43, and 56. Make a box-and-whisker plot of this data.

36. Make a box-and-whisker plot for the following set of test scores.

77, 80, 75, 73, 77, 81, 62, 87, 99, 85, 82, 81, 77, 72, 78, 83, 86, 79, 80, 78

What does the plot tell you about the scores?

LESSON 9-7 ◼ Standard Deviation, p. 412

▶ The **variance** of a set of numbers is a measure of how the data are dispersed. To find the variance of a set of numbers $x_1, x_2, \ldots x_n$, with a mean of m, use the following formula.

$$\frac{(x_1 - m)^2 + (x_2 - m)^2 + \cdots + (x_n - m)^2}{n}$$

▶ The **standard deviation**, s, of a set of numbers is the square root of the variance.

37. Find the variance and standard deviation for the set of numbers 4, 7, 10, 13, and 16.

38. Compute the variance and standard deviation for 3, 5, 10, 7, 5.

39. Find the variance and standard deviation for the set of numbers 5, 2, 6, 8, and 19.

40. Kendra scored 95, 90 and 95 on three tests. The class mean and standard deviation for the first test were 75 and 10; for the second, 75 and 5; and for the third, 80 and 6. On which test did Kendra do best relative to her classmates?

CHAPTER INVESTIGATION

EXTENSION Present your game to your class. After listening to everyone's presentation, compare your game to those of your classmates. List the advantages and disadvantages to your game. Make improvements to your game based on your list of disadvantages.

Chapter 9 Assessment

1. A bowler got 32 strikes in her first 80 frames. What is the experimental probability that she will not get a strike in her next frame?

2. Find the probability of drawing a red king or a 5 from a standard deck of playing cards.

3. A pair of dice are rolled twice. Find *P*(sum is an even number, then sum is an odd number).

4. There are 3 red T-shirts, 2 white T-shirts, 2 green T-shirts, and 1 blue T-shirt in a drawer. You reach in without looking and take out two shirts. Find *P*(red, green).

5. How many different two-flavor ice cream cones can be chosen from a menu of 15 flavors?

6. Forty students are in the running for the science prize. In how many ways can a winner, a runner-up, and an alternate be chosen?

Use the scatter plot for Exercises 7 and 8.

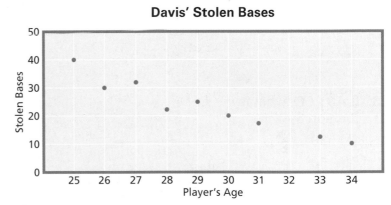

Davis' Stolen Bases

7. Does the scatter plot show a negative or a positive correlation?

8. Estimate the number of stolen bases Davis will have when he is 32 years old.

Groups of students and of community leaders rated the performance of the school superintendent on a scale from 0 to 100. The box-and-whisker plots below show the results.

Superintendent's Performance Rating

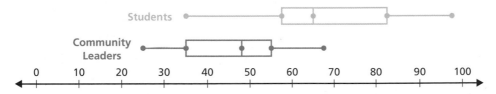

9. Which group gave the superintendent a higher median score?

10. For which group were the scores more widely spread?

11. Find the variance and standard deviation for the set of numbers: $-10, -5, 0, 5, 10$.

12. Patty scored 75 on a test in which the mean score in her class was 60 and the standard deviation was 10. Jamaal took the same test in his class. His score was 70, the class mean score was 50, and the standard deviation was 10. Who scored better, relative to his or her classmates?

Standardized Test Practice

Part 1 Multiple Choice

Record your answers on the answer sheet provided by your teacher or on a sheet of paper.

1. The first super computer, the Cray-1, was installed in 1976. It was able to perform 160 million different operations in a second. Which expression represents this number in scientific notation? (Lesson 1-8)

   (A) 1.6×10^6
   (B) 1.6×10^8
   (C) 160×10^6
   (D) 160×10^8

2. For which value of x is the y value in the equation $3x + 2y = 6$ the greatest? (Lesson 2-5)

   (A) $x = -2$
   (B) $x = 0$
   (C) $x = 2$
   (D) $x = 4$

3. If $\overleftrightarrow{AB} \parallel \overleftrightarrow{CD}$, what is $m\angle ECD$? (Lesson 3-4)

   (A) 38°
   (B) 56°
   (C) 124°
   (D) 153°

   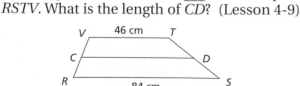

4. In the figure below, $\overline{CD}$ is a median of trapezoid $RSTV$. What is the length of $\overline{CD}$? (Lesson 4-9)

   (A) 60 cm   (B) 65 cm   (C) 70 cm
   (D) cannot be determined

5. What is the approximate area of the shaded region? (Lesson 5-2)

   (A) 14 ft^2
   (B) 34 ft^2
   (C) 50 ft^2
   (D) 114 ft^2

6. What is the y-intercept for the line with equation $3y - x = 6$? (Lesson 6-1)

   (A) $\frac{1}{3}$
   (B) 2
   (C) 3
   (D) 6

7. A 12-m flagpole casts a 9-m shadow. At the same time, the building next to it casts a 27-m shadow. How tall is the building? (Lesson 7-7)

   (A) 20.25 m
   (B) 36 m
   (C) 40 m
   (D) 84 m

8. Find the value of y. (Lesson 8-6)

   $$\begin{bmatrix} 5 & -4 \\ 3 & 1 \end{bmatrix} \cdot \begin{bmatrix} 0 & -5 & 6 \\ 2 & 8 & -4 \end{bmatrix} = \begin{bmatrix} -8 & -57 & y \\ 2 & -7 & 14 \end{bmatrix}$$

   (A) 14
   (B) 18
   (C) 22
   (D) 46

9. In the spinner, what color should the blank portion of the spinner be so that the probability of landing on this color is $\frac{3}{8}$? (Lesson 9-1)

   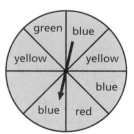

   (A) blue
   (B) green
   (C) red
   (D) yellow

10. The weather forecaster says there is a 35% chance of rain. What is the probability that it will *not* rain? (Lesson 9-3)

    (A) 50%
    (B) 65%
    (C) 70%
    (D) 75%

11. Use the box-and-whisker plot to determine the mean of the data. (Lesson 9-6)

    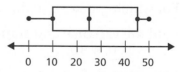

    (A) 0
    (B) 10
    (C) 25
    (D) 45

Part 2 Short Response/Grid In

Record your answers on the answer sheet provided by your teacher or on a sheet of paper.

12. Mrs. Hayashi made a tablecloth for her kitchen. She bought $4\frac{3}{4}$ yd of material. She used $3\frac{1}{8}$ yd of material to make the tablecloth. How much material was *not* used to make the tablecloth? (Lesson 1-4)

13. The Huang family had weekly grocery bills of $105, $115, $120, and $98 last month. What was their mean weekly grocery bill last month? (Lesson 2-7)

14. Find the value of x. (Lesson 3-2)

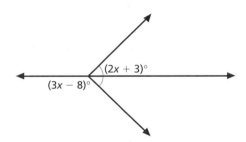

15. Change 0.08 kg to grams. (Lesson 5-1)

16. For a cleaning solution, bleach is mixed with water in the ratio of 1:8. How much bleach should be added to 12 qt of water to make the proper solution? (Lesson 7-1)

17. Suppose the segment shown below is translated 3 units to the left. What are the coordinates of the endpoints of the resulting segment? (Lesson 8-1)

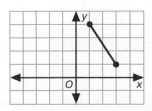

18. Only two of the five school newspaper editors can represent the school at the state awards banquet. How many different combinations of two editors can be selected to go to the banquet? (Lesson 9-5)

Part 3 Extended Response

Record your answers on a sheet of paper. Show your work.

19. Tiffany has a bag of 10 yellow marbles, 10 red marbles, and 10 green marbles. Tiffany picks two marbles at random and gives them to her sister. (Lesson 9-4)

    a. What is the probability of choosing 2 yellow marbles?

    b. Of the marbles left, what is the probability of choosing a green marble next?

    c. Of the marbles left, what color has a probability of $\frac{1}{3}$ of being picked? Explain.

20. Kenneth is recording the times it takes him to run various distances. The results are shown. (Lesson 9-6)

Distance (mi)	2	3	5	7	9
Time (mi)	13	20	35	53	72

    a. Make a scatter plot of the data.

    b. How many minutes do you think it will take Kenneth to run 4 mi? Explain.

Test-Taking Tip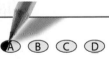

Question 19
Extended response questions often involve several parts. When one part of the question involves the answer to a previous part of the question, make sure you check your answer to the first part before moving on. Also, remember to show all of your work. You may be able to get partial credit for your answers, even if they are not entirely correct.

Right Triangles and Circles

THEME: Architecture

Imagine that you have the opportunity to design and build a new home for your family. How would you decide what design features to incorporate? Your new home must be functional to meet your family's needs. It must also be appealing to the eye to suit your family's personality and complement other buildings in the community.

Architects have been using geometric principles for centuries to design and build homes, schools and public buildings. In the search for pleasing and useful forms, they have explored and applied many important principles of geometry.

- **Building Inspectors** (page 435) ensure the safety of buildings before construction is complete. They make sure that safety guidelines and building codes are strictly followed. Inspectors may stop a project if safety standards are not met.

- **Landscape Architects** (page 453) design outdoor areas such as gardens, parks, and playgrounds. Landscape architects study soil, sunlight, topography, and climate when designing a landscaping plan.

mathmatters3.com/chapter_theme

Tall Buildings

Skyscraper	City	Built	Height	Stories
Central Towers	Hong Kong	1992	1227 ft	78
Petronas Towers	Kuala Lumpur	1997	1483 ft	88
Sears Tower	Chicago	1974	1450 ft	110
Jin Mao Building	Shanghai	1999	1379 ft	88
Empire State Building	New York	1931	1250 ft	102
Bank of China	Hong Kong	1989	1209 ft	70

Data Activity: Tall Buildings

Use the table for Questions 1–4.

1. Compare the number of stories to the height in feet for each building. Which building has the least amount of height per story? What has the greatest amount of height per story?

2. The framework for the Empire State Building rose at a rate of four-and-a-half stories per week. How many days were required to build the framework?

3. The Sears Tower consists of nine 75-ft square modules which rise to staggered levels. What is the combined area of the modules in square feet?

4. Create a bar graph of the buildings' heights with the building arranged according to the date they were built.

CHAPTER INVESTIGATION

Often, large cities do not have available space to build new buildings or the means to tear down older buildings and replace them with new ones. As a result, many cities are hiring architects to give older buildings a facelift or makeover. A new facade is attached to the front of the building to make the building more attractive and unify design elements.

Working Together

Make a sketch of the front of your school as it now looks. Research architectural styles and select appropriate design elements. Then create a facade that could fit over the front of your school to give it a new look. Draw plans for your new design. Use the Chapter Investigation icons to guide your group.

10 Are You Ready?

Refresh Your Math Skills for Chapter 10

The skills on these two pages are ones you have already learned. Review the examples and complete the exercises. For additional practice on these and more prerequisite skills, see pages 654–661.

PERCENT OF A NUMBER

Example Find 42% of 624.
Change the percent to a decimal: 42% → 0.42
Multiply:
$$\begin{array}{r} 624 \\ \times\ 0.42 \\ \hline 262.08 \end{array}$$

42% of 624 is 262.08.

Find the given percent of each number.

1. 15% of 500   2. 50% of 768   3. 24% of 496

4. 33% of 127   5. 64% of 481   6. 93% of 722

7. 81% of 3297   8. 29.6% of 17.84   9. 47% of 82.56

10. 23.5% of 1604   11. 106% of 300   12. 17.8% of 296

CIRCUMFERENCE AND AREA

The formula for the circumference of a circle is $C = 2\pi r$, where r is the radius of the circle. The formula for the area of a circle is $A = \pi r^2$, where r is the radius of the circle.

Find the circumference and area of each circle. Use 3.14 for π. Round answers to the nearest hundredth if necessary.

13.
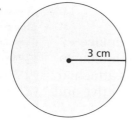
3 cm

14.
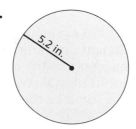
5.2 in.

15.
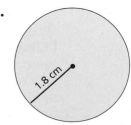
1.8 cm

16.
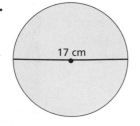
17 cm

17.
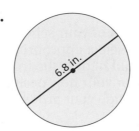
6.8 in.

18.
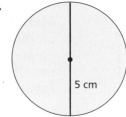
5 cm

SQUARES AND SQUARE ROOTS

Simplify each expression.

19. 16^2 **20.** $\sqrt{144}$ **21.** 24^2 **22.** $\sqrt{64}$

23. 9^2 **24.** $\sqrt{529}$ **25.** 15^2 **26.** $\sqrt{196}$

27. 13^2 **28.** $\sqrt{289}$ **29.** 30^2 **30.** $\sqrt{100}$

ANGLES

Find the measure of each indicated angle.

31. $\angle AEB$

32. $\angle AED$

33. $\angle DEC$

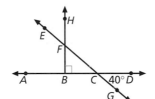

34. $\angle DCF$

35. $\angle EFB$

36. $\angle IIFC$

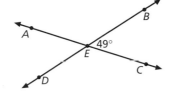

37. $\angle FEB$

38. $\angle ABG$

39. $\angle CBE$

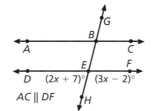

40. $\angle BAD$

41. $\angle FAE$

42. $\angle EAD$

43. $\angle CAE$

44. $\angle CAF$

45. $\angle DAF$

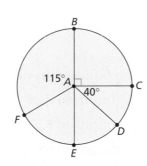

10-1 Irrational Numbers

Goals
- Find square roots.
- Simplify products and quotients containing radicals.

Applications Architecture, Small Business, Construction

Work with a partner to explore square roots.

A calculator gives the square root of 2 as 1.414213562. Copy and complete this chart, using a calculator.

$(1.4)^2 = $ __?__ $(1.4142)^2 = $ __?__ $(1.4142135)^2 = $ __?__

$(1.41)^2 = $ __?__ $(1.41421)^2 = $ __?__ $(1.41421356)^2 = $ __?__

$(1.414)^2 = $ __?__ $(1.414213)^2 = $ __?__ $(1.414213562)^2 = $ __?__

Do you think that eventually you will find a number, y, with enough decimal places so that $y^2 = 2$ exactly?

Technology Note

To find the square root of 2, the key sequence for most scientific calculators is 2 $\boxed{\sqrt{}}$.

For graphing calculators, the key sequence is usually $\boxed{\sqrt{}}$ 2 $\boxed{\text{ENTER}}$.

BUILD UNDERSTANDING

The symbol $\sqrt{x}$ means the **square root** of a number, x. It is the number that multiplied by itself equals x. For example, $\sqrt{4} = 2$, because $2^2 = 4$; and $\sqrt{144} = 12$, because $12^2 = 144$. These two examples have square roots that are rational numbers.

The number $\sqrt{2}$ is an **irrational number**, which means that it is not a rational number. Therefore, it is a number that cannot be written as a fraction, a terminating decimal, or a repeating decimal. It can be represented by a nonterminating, nonrepeating decimal, and it can be approximated to any decimal place.

Example 1

ARCHITECTURE An architect is drawing a landscaping plan which includes three square cement blocks with areas of the bases approximately 5, 31, and 98 ft^2. To find the length of a side of each block, she finds the square root of each area. Find the value of each to the nearest hundredth: $\sqrt{5}$, $\sqrt{31}$ and $\sqrt{98}$.

Solution

Use a calculator. Then round to the hundredths place.

$\sqrt{5} = 2.236067977 \ldots$ rounds to 2.24

$\sqrt{31} = 5.567764363 \ldots$ rounds to 5.57

$\sqrt{98} = 9.899494937 \ldots$ rounds to 9.90

Another way to read the expression $\sqrt{2}$ is "radical 2." The number under the radical sign is called the **radicand**. There are properties of radicals that allow you to simplify them.

Theorem	The square root of the product of two nonnegative numbers is the same as the product of their square roots. $$\sqrt{a \cdot b} = \sqrt{a} \cdot \sqrt{b}$$

Example 2

Simplify each expression.

    a. $\sqrt{32}$ **b.** $\sqrt{300}$ **c.** $\sqrt{125}$

Solution

Rewrite each radicand as a product of two numbers, so that one of them is a perfect square.

    a. $\sqrt{32} = \sqrt{16 \cdot 2} = \sqrt{16} \cdot \sqrt{2}$ or $4\sqrt{2}$

    b. $\sqrt{300} = \sqrt{3 \cdot 100} = \sqrt{3} \cdot \sqrt{100}$ or $10\sqrt{3}$

    c. $\sqrt{125} = \sqrt{25 \cdot 5} = \sqrt{25} \cdot \sqrt{5}$ or $5\sqrt{5}$

Numbers written in radical form can be multiplied together.

Example 3

Multiply $(4\sqrt{10})(-3.1\sqrt{6})$.

Check Understanding

What is the product, in simplest radical form, of $\sqrt{14} \cdot \sqrt{7}$?

Solution

Rewrite the product so that rational factors and irrational factors are grouped separately.

$$(4\sqrt{10})(-3.1\sqrt{6}) = (4 \cdot -3.1)(\sqrt{10} \cdot \sqrt{6})$$

$$= -12.4\sqrt{60} \qquad \sqrt{10} \cdot \sqrt{6} = \sqrt{10 \cdot 6}$$

The product can be simplified.

$$-12.4\sqrt{60} = -12.4(\sqrt{4 \cdot 15})$$

$$= -12.4(2\sqrt{15})$$

$$= -24.8\sqrt{15}$$

So, $(4\sqrt{10})(-3.1\sqrt{6}) = -24.8\sqrt{15}$.

A similar theorem about radicals applies to quotients.

Theorem	The square root of a quotient of two nonnegative numbers is equal to the quotient of their square roots. $$\sqrt{\frac{a}{b}} = \frac{\sqrt{a}}{\sqrt{b}} \text{ where } b \neq 0$$

Example 4

Express the quotient for $-5\sqrt{33} \div 3\sqrt{22}$ in simplest radical form.

Solution

$$-\frac{5\sqrt{33}}{3\sqrt{22}} = \frac{-5}{3} \cdot \sqrt{\frac{33}{22}} \quad \text{Simplify } \tfrac{33}{22}.$$

$$= \frac{-5}{3} \cdot \sqrt{\frac{3}{2}}$$

$$= \frac{-5\sqrt{3}}{3\sqrt{2}}$$

In simplest radical form, denominators cannot include radicals.

$$\frac{-5\sqrt{3}}{3\sqrt{2}} = \frac{-5\sqrt{3}(\sqrt{2})}{3\sqrt{2}(\sqrt{2})} \quad \text{Multiply numerator and denominator by } \sqrt{2}.$$

$$= \frac{-5\sqrt{6}}{3\sqrt{4}}$$

$$= \frac{-5\sqrt{6}}{3 \cdot 2}, \text{ or } \frac{-5\sqrt{6}}{6}$$

The process of rewriting a quotient to eliminate radicals from the denominator is called **rationalizing the denominator**.

◤ TRY THESE EXERCISES

CALCULATOR Find each value to the nearest hundredth.

1. $\sqrt{11}$ **2.** $\sqrt{56}$ **3.** $\sqrt{85}$ **4.** $\sqrt{196}$

Write each square root in simplest radical form.

5. $\sqrt{44}$ **6.** $\sqrt{242}$ **7.** $\sqrt{75}$ **8.** $\sqrt{48}$

Simplify.

9. $(2\sqrt{6})(7\sqrt{2})$ **10.** $(-3\sqrt{10})(5\sqrt{5})$ **11.** $(4\sqrt{2})^2$

12. $\dfrac{\sqrt{150}}{\sqrt{6}}$ **13.** $\dfrac{2}{\sqrt{3}}$ **14.** $\sqrt{\dfrac{21}{2}}$

◤ PRACTICE EXERCISES • For Extra Practice, see page 693.

CALCULATOR Find each value to the nearest hundredth.

15. $\sqrt{21}$ **16.** $\sqrt{47}$ **17.** $\sqrt{73}$ **18.** $\sqrt{200}$

Write each expression in simplest radical form.

19. $\sqrt{162}$ **20.** $\sqrt{500}$ **21.** $\sqrt{72}$

22. $\sqrt{40}$ **23.** $(2\sqrt{3})(4\sqrt{6})$ **24.** $(10\sqrt{2})(2\sqrt{6})$

25. $(3\sqrt{3})^2$ **26.** $\dfrac{\sqrt{45}}{\sqrt{5}}$ **27.** $\dfrac{7}{\sqrt{2}}$

28. $\sqrt{\dfrac{6}{5}}$ **29.** $(2\sqrt{5})^2$ **30.** $\dfrac{3\sqrt{12}}{4\sqrt{48}}$

31. **SMALL BUSINESS** Georgia is starting a pet-sitting service in her backyard. She wants to enclose an area of about 6 m², and has decided that the best shape for the enclosure is a square. What should be the length of a side of the enclosure, to the nearest tenth of a meter?

32. **WRITING MATH** Is the number $\sqrt{207.36}$ rational or irrational? Explain.

33. **DATA FILE** Use the data on rectangular architectural structures on page 645. For the Bakong Temple and the Wat Kukat Temple, express the length of a side as the square root of the area of the structure in meters.

34. **CONSTRUCTION** A mason is building a square pedestal from brick. The top of the pedestal must have an area of about 12 ft². To the nearest hundredth, find the length of a side of the pedestal.

The expression $\sqrt[n]{x}$ means the nth root of x. It means the number which, when raised to the nth power, equals x. Example: $\sqrt[3]{8} = 2$, because $2^3 = 8$.

Find each of the following.

35. $\sqrt[3]{27}$ 36. $\sqrt[3]{125}$ 37. $\sqrt[4]{16}$ 38. $\sqrt[4]{81}$

MATH HISTORY In the first century A.D., Heron of Alexandria discovered a formula for finding the area of a triangle when only the measures of its sides are known:

$$A = \sqrt{s(s-a)(s-b)(s-c)} \text{ where } s = 0.5(a + b + c).$$

Use Heron's formula to find the area of each triangle to the nearest tenth.

39. 3 cm, 5 cm, 6 cm   40. 7 in., 12 in., 15 in.

◼ EXTENDED PRACTICE EXERCISES

Decide if each statement is true or false. If it is false, give a counterexample.

41. The product of two even numbers is always an even number.

42. The square of a rational number is always a rational number.

43. The product of two irrational numbers is always an irrational number.

44. **CHAPTER INVESTIGATION** Make a detailed sketch of the front of the main building at your school. Include any current architectural elements. Estimate measurements for the elements in the sketch. If possible, consult building plans to find the length and width of standard doors and windows.

◼ MIXED REVIEW EXERCISES

Change each unit of measure as indicated. (Lesson 5-1)

45. 8 c = ___?___ qt   46. 7 yd = ___?___ in.   47. 10 km = ___?___ m

48. 6 gal = ___?___ pt   49. 14 m = ___?___ dm   50. 7 g = ___?___ kg

The Pythagorean Theorem

Goals ■ Use the Pythagorean Theorem to solve problems involving right triangles.

Applications Architecture, Construction, Home Repairs

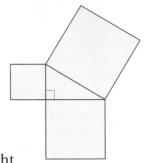

Draw any right triangle on graph paper. Make squares on each side of the triangle, as shown. Cut out the squares. Try to fit the two smaller squares on top of the larger square. You may cut up the smaller squares. What conclusion can you draw?

◥ BUILD UNDERSTANDING

In a right triangle, the shorter sides are **legs,** and the side opposite the right angle is the **hypotenuse.** The relationship between the squares of the lengths of the legs and the square of the length of the hypotenuse is called the **Pythagorean Theorem.**

Pythagorean Theorem	In a right triangle, the sum of the squares of the measures of the two legs is equal to the square of the measure of the hypotenuse. $a^2 + b^2 = c^2$	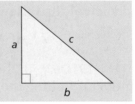

This proof of the Pythagorean Theorem uses similar triangles formed by the altitude to the hypotenuse.

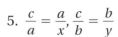

Given $\triangle ABC$, $\angle C$ is a right angle.
Prove $a^2 + b^2 = c^2$

Statements	Reasons
1. $\triangle ABC$ is a right triangle.	1. Given
2. $\overline{CD}$ is perpendicular to AB.	2. There is one and only one line through a point perpendicular to a given line.
3. $\overline{CD}$ is the altitude to the hypotenuse in $\triangle ABC$.	3. Definition of altitude.
4. $\triangle ABC \sim \triangle ACD \sim \triangle CBD$	4. The altitude to the hypotenuse forms two right triangles that are similar to each other and to the original triangle.
5. $\dfrac{c}{a} = \dfrac{a}{x}, \dfrac{c}{b} = \dfrac{b}{y}$	5. Corresponding sides of similar triangles are proportional.
6. $cx = a^2, cy = b^2$	6. Cross products are equal.
7. $cx + cy = a^2 + b^2$	7. Addition property of equality
8. $c(x + y) = a^2 + b^2$	8. Distributive property
9. $c = x + y$	9. Segment addition postulate
10. $c(c) = a^2 + b^2$	10. Substitution
11. $c^2 = a^2 + b^2$	11. Definition of c^2.

Math: Who, Where, When

The scarecrow in the classic movie The Wizard of Oz tried to recite the Pythagorean Theorem as proof of intelligence when he received his "brains." But he stated it incorrectly.

You can use the Pythagorean Theorem to find the length of a side of a right triangle when the lengths of the other two sides are known.

Example 1

ARCHITECTURE An architect draws a right triangle, $\triangle DEF$, on a blueprint. $DE = 3$ cm and $EF = 5$ cm.

Find DF to the nearest tenth.

Solution

Let $x = DF$. $3^2 + 5^2 = x^2$
$$9 + 25 = x^2$$
$$34 = x^2$$
$$\sqrt{34} = x$$
$5.830951895 \ldots = x$ Use a calculator or a square root table to find $\sqrt{34}$.

So, $DF = 5.8$ cm to the nearest tenth.

GEOMETRY SOFTWARE You can use geometry software to solve problems involving right triangles. Set the distance units to either inches, centimeters, or pixels. Choose pixels when the problem involves large numbers. Use the Show Grid option to make drawing figures easier. Draw the lengths given in the problem. Use the software to check the measurement of these segments. Draw and measure any remaining segments. Use the software to find unknown lengths and angle measures as required by the problem.

Sometimes, the missing measure in the right triangle is the length of one of the legs.

Example 2

Find x to the nearest hundredth of an inch.

Solution

$$x^2 + 9^2 = 11^2$$
$$x^2 + 81 = 121$$
$$x^2 = 40$$
$$x = \sqrt{40}$$
$x = 6.32455532 \ldots$ So, $x \approx 6.32$ in.

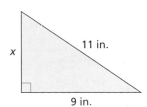

The converse of the Pythagorean Theorem is also true.

Converse of the Pythagorean Theorem	If the sum of the squares of the measures of two shorter sides of a triangle is equal to the square of the measure of the third side, then the triangle is a right triangle.

Example 3

Are triangles with the following side lengths right triangles?

a. 2 cm, $4\sqrt{2}$ cm, 6 cm   **b.** 4 in., 9 in., 10 in.

Solution

a. $2^2 + (4\sqrt{2})^2 \stackrel{?}{=} 6^2$

$4 + (16 \cdot 2) \stackrel{?}{=} 36$

$4 + 32 \stackrel{?}{=} 36$

$36 = 36$

b. $4^2 + 9^2 \stackrel{?}{=} 10^2$

$16 + 81 \stackrel{?}{=} 100$

$97 \neq 100$

Therefore, three sides with lengths 2 cm, $4\sqrt{2}$ cm, and 6 cm do form a right triangle, but sides with lengths 4 in., 9 in., and 10 in. do not.

◥ TRY THESE EXERCISES

Use the Pythagorean Theorem to find the unknown length. Round your answers to the nearest tenth.

1.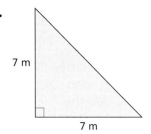

7 m, 7 m

2.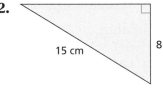

15 cm, 8

3.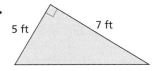

5 ft, 7 ft

4. CONSTRUCTION A triangular brace has sides that measure 6 cm, 8 cm, and 10 cm. Does the brace have a 90° angle?

5. HOME REPAIR Tim is cleaning out the rain gutters on his home. He has an 18-ft ladder. If the base of the ladder is placed 5 ft from the base of the building, how far up the wall will the ladder reach?

◥ PRACTICE EXERCISES • For Extra Practice, see page 693.

Use the Pythagorean Theorem to find the unknown length. Round your answers to the nearest tenth.

6.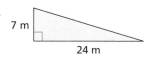

7 m, 24 m

7.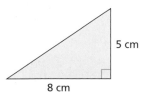

5 cm, 8 cm

8.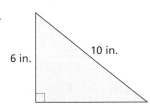

6 in., 10 in.

9.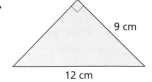

9 cm, 12 cm

10.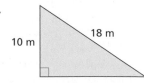

10 m, 18 m

11.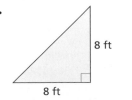

8 ft, 8 ft

Determine if the triangle is a right triangle. Write *yes* or *no*.

12.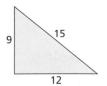

9, 15, 12

13.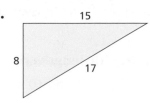

15, 8, 17

14.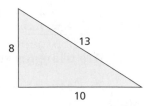

8, 13, 10

Solve. Round your answers to the nearest tenth.

15. **GEOMETRY SOFTWARE** Find the length of the diagonal of a rectangle with a length of 5 cm and a width of 4 cm.

16. A pole 4 m high is to be attached by a guy wire to a stake in the ground 1.2 m from the base of the pole. How long must the guy wire be?

17. A ramp 6 yd long reaches from the loading dock to a point on the ground 4 yd from the base of the dock. How high above ground is the loading dock?

Find x in each figure. Round your answer to the nearest tenth.

18.
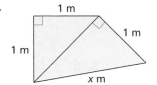
1 m
1 m
1 m
x m

19.
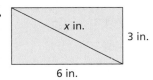
x in.
3 in.
6 in.

20.
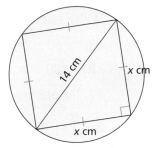
14 cm
x cm
x cm

21. **TALK ABOUT IT** Using *s* for the length of the side of a square and *d* for the length of the diagonal of the square, Michi has written the formula $d = s\sqrt{2}$. She says that the formula can be used to find the length of the diagonal of any square if the length of a side is known. Do you agree?

22. Figure *ABCDEFGH* is a rectangular prism. Its length *AB* is 8 cm, its width *BC* is 6 cm, and its height *BF* is 24 cm. Find the length of the diagonal $\overline{BH}$. (*Hint:* △*BDH* is a right triangle.)

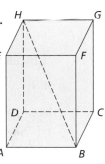

◼ EXTENDED PRACTICE EXERCISES

23. Prove that, in any rectangular prism with length *l*, width *w*, and height *h*, the length *d* of a diagonal is given by the formula, $d = \sqrt{l^2 + w^2 + h^2}$.

24. Three positive integers, *a*, *b*, and *c*, for which $a^2 + b^2 = c^2$ form a **Pythagorean triple**. Let *x* and *y* represent two positive integers such that $x > y$. Let $a = 2xy$, $b = x^2 - y^2$, and $c = x^2 + y^2$. Try three different pairs of values for *x* and *y*. What do you notice?

25. **WRITING MATH** In ancient Egypt, a pair of workers called rope stretchers would use a loop of rope divided by knots into 12 equal parts to mark off a perfect right angle in the sand. They would drive a stake through one knot. One worker would pull the rope taut at the third knot from the stake, while the other pulled the rope taut at the fourth knot on the other side of the stake. Write a paragraph and draw a diagram explaining why this system worked.

◼ MIXED REVIEW EXERCISES

A die is rolled 200 times with the following results: (Lesson 9-1)

Outcome	1	2	3	4	5	6
Frequency	31	40	30	29	34	36

What is the experimental probability of rolling each of the following results?

26. 3   27. 5   28. 6   29. 1   30. a number less than 5

Review and Practice Your Skills

PRACTICE ■ LESSON 10-1

Find each value to the nearest hundredth.

1. $\sqrt{53}$
2. $\sqrt{150}$
3. $2\sqrt{7}$
4. $\sqrt{624}$
5. $10\sqrt{10}$
6. $\sqrt{245}$
7. $\sqrt{1000}$
8. $\sqrt{37}$

Write each in simplest radical form.

9. $\sqrt{50}$
10. $\sqrt{48}$
11. $\sqrt{72}$
12. $\sqrt{600}$
13. $\sqrt{80}$
14. $\sqrt{192}$
15. $\sqrt{88}$
16. $\sqrt{147}$
17. $\sqrt{242}$
18. $3\sqrt{20}$
19. $5\sqrt{63}$
20. $10\sqrt{75}$
21. $\left(2\sqrt{3}\right)^2$
22. $\dfrac{\sqrt{12}}{\sqrt{3}}$
23. $\sqrt{\dfrac{3}{4}}$
24. $\left(3\sqrt{2}\right)\left(5\sqrt{10}\right)$
25. $\dfrac{\sqrt{10}}{\sqrt{2}}$
26. $\left(8\sqrt{18}\right)\left(4\sqrt{2}\right)$
27. $\sqrt{\dfrac{16}{7}}$
28. $\left(\sqrt{\dfrac{3}{8}}\right)\left(\dfrac{3}{\sqrt{2}}\right)\left(2\sqrt{27}\right)$

Find each of the following.

29. $\sqrt[3]{64}$
30. $\sqrt[4]{625}$
31. $\sqrt[5]{32}$
32. $\sqrt[5]{243}$
33. $\sqrt[3]{1000}$
34. $\sqrt[3]{\dfrac{1}{8}}$
35. $\sqrt[3]{\dfrac{1}{27}}$
36. $\sqrt[3]{-8}$

PRACTICE ■ LESSON 10-2

Use the Pythagorean Theorem to find the missing length. Round answers to the nearest tenth.

37.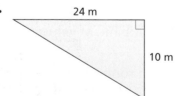
24 m
10 m

38.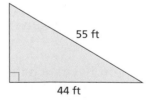
55 ft
44 ft

39.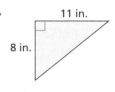
11 in.
8 in.

40.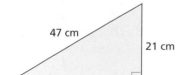
47 cm
21 cm

41.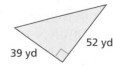
39 yd
52 yd

42.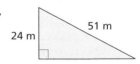
24 m
51 m

Solve. Round your answers to the nearest tenth.

43. Find the length of the diagonal of a rectangle with a length of 18 cm and a width of 24 cm.

44. Find the width of a rectangle with a length of 30 ft and a diagonal of 34 ft.

Write each in simplest radical form. (Lesson 10-1)

45. $\sqrt{24}$ **46.** $\sqrt{28}$ **47.** $\sqrt{40}$ **48.** $\sqrt{52}$

49. $\sqrt{\dfrac{1}{5}}$ **50.** $\dfrac{4}{\sqrt{2}}$ **51.** $\dfrac{\sqrt{28}}{\sqrt{7}}$ **52.** $\sqrt{\dfrac{6}{18}}$

53. $8\sqrt{63}$ **54.** $5\sqrt{200}$ **55.** $11\sqrt{96}$ **56.** $-2\sqrt{72}$

57. $\left(\sqrt{24}\right)\left(3\sqrt{6}\right)$ **58.** $\left(2\sqrt{5}\right)\left(5\sqrt{20}\right)$ **59.** $\left(7\sqrt{14}\right)\left(3\sqrt{2}\right)\left(8\sqrt{21}\right)$ **60.** $\sqrt{x^2}$

Determine if the triangle is a right triangle. Write *yes* or *no*. (Lesson 10-2)

61.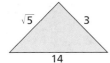

62.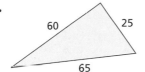

63.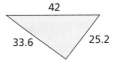

Career – Construction and Building Inspectors

MathWorks
Workplace Knowhow

Construction and building inspectors make sure new structures follow building codes and ordinances, zoning regulations, and contract specifications. They may also inspect alterations and repairs to existing structures.

Inspectors have the authority to stop projects if safety standards and building codes are not met. To do their work, inspectors use tape measures, survey instruments, metering devices, cameras, and calculators.

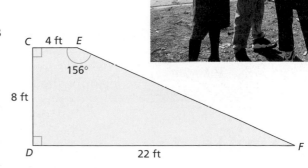

At a construction site, a temporary ramp has been built to assist workers in transporting materials up to the floor level. A diagram of the ramp is shown.

1. Use the Pythagorean Theorem to find the length of the ramp to the nearest tenth.

2. What is the measure of $\angle EFD$?

3. What is the slope of the ramp?

4. If *DF* is lengthened to 30 ft and *CE* remains the same, what will the new measure of *EF* be (to the nearest tenth)?

Special Right Triangles

Goals ■ Find the lengths of the sides of 30°–60°–90° and 45°–45°–90° triangles.

Applications Architecture, Road Planning, Plumbing

Work with a partner. You will need a compass, straightedge, and scissors.

1. Construct several different equilateral triangles.

2. Cut out each triangle and fold it so that one vertex matches another.

3. Draw a line segment along the fold.

4. Examine the figures formed by the triangle and the line segment you drew. Write as many different observations about the triangle and the line segment as you can.

◥ BUILD UNDERSTANDING

All right triangles with the same corresponding angle measures are similar. This fact and the Pythagorean Theorem lead to some theorems about the relationships among the sides in two types of right triangles.

In the equilateral triangles you examined above, you probably noticed that the segment that separated one side into two congruent segments was also perpendicular to that segment. In this way, a right triangle is formed that has a 60° angle and one leg that is half the measure of the hypotenuse.

In this 30°–60°–90° triangle, suppose the measure of the side opposite the 30° angle is s. Then the measure of the hypotenuse is $2s$. You can use the Pythagorean Theorem to find the measure of the longer leg of the triangle in terms of s.

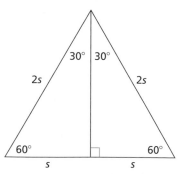

Let y represent the measure of the side opposite the 60° angle.

$$s^2 + y^2 = (2s)^2$$
$$s^2 + y^2 = 4s^2$$
$$y^2 = 3s^2$$
$$y = \sqrt{3s^2}$$
$$y = s\sqrt{3}$$

30°–60°–90° Triangle Theorem	In a 30°–60°–90° triangle, the measure of the hypotenuse is two times that of the leg opposite the 30° angle. The measure of the other leg is $\sqrt{3}$ times that of the leg opposite the 30° angle.

Example 1

ARCHITECTURE A triangular support shown on a blueprint forms a 30°–60°–90° triangle. On the plans, the leg opposite the 30° angle measures 4 cm. What are the measures of the other two sides?

Solution

In a 30°–60°–90° triangle, if the leg opposite the 30° angle is s, then the other leg is $s\sqrt{3}$, and the hypotenuse is $2s$.

So, in this triangle, the leg opposite the 30° angle is 4 cm, the other leg is $4\sqrt{3}$ cm, and the hypotenuse is 8 cm.

Example 2

In a 30°–60°–90° triangle, the hypotenuse measures 7 in. Find the measure of the other two sides to the nearest tenth.

Solution

$2s$	hypotenuse	$2s = 7$
s	side opposite 30° angle	$s = 3.5$
$s\sqrt{3}$	side opposite 60° angle	$s\sqrt{3} = (3.5)(\sqrt{3}) \approx 6.1$

So, the missing measures of this triangle are 3.5 in. and about 6.1 in.

> **Problem Solving Tip**
>
> When you are given information about a geometric figure, it is best to draw a quick sketch of the figure and label any given measurements. That way, you will be able to see what relationships can help you to find the missing measure.

When you are given the measure of the leg opposite the 60° angle, you may need to rationalize the denominator.

Example 3

In a 30°–60°–90° triangle, the measure of the leg opposite the 60° angle is 5 cm. Find the measures of the other two sides in simplest radical form.

Solution

You are given that $s\sqrt{3} = 5$.

$s = \dfrac{5}{\sqrt{3}}$ Rationalize the denominator. $s = \dfrac{5\sqrt{3}}{\sqrt{3} \cdot \sqrt{3}} = \dfrac{5\sqrt{3}}{3}$

The measure of the hypotenuse is $2s$.

Substitute the value you found for s. $2s = 2\left(\dfrac{5\sqrt{3}}{3}\right) = \dfrac{10\sqrt{3}}{3}$

So, the missing measures are $\dfrac{5\sqrt{3}}{3}$ cm and $\dfrac{10\sqrt{3}}{3}$ cm.

45°–45°–90° Triangle Theorem	In a 45°–45°–90° triangle, the measure of the hypotenuse is $\sqrt{2}$ times the measure of a leg of the triangle.

Example 4

Find the unknown measures.   a.

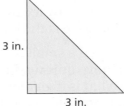

b.

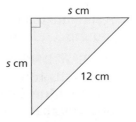

Solution

If s is the length of a side of a 45°–45°–90° triangle, then the hypotenuse is $s\sqrt{2}$.

a. Because $s = 3$, the measure of the hypotenuse is $3\sqrt{2}$ in.

b. The measure of the hypotenuse is given as 12 cm. You can use the equation $s\sqrt{2} = 12$ to solve for s.

$$s = \frac{12}{\sqrt{2}} = \frac{12\sqrt{2}}{\sqrt{2} \cdot \sqrt{2}} = \frac{12\sqrt{2}}{2} \qquad \text{Rationalize the denominator.}$$

$$s = 6\sqrt{2} \qquad \text{The unknown length is } 6\sqrt{2} \text{ cm.}$$

Right angles are common in our everyday life. You can find examples of right angles in and around any building. Many streets intersect at right angles, and trees form right angles with the ground. Therefore, right triangles and the Pythagorean Theorem have many applications.

Example 5

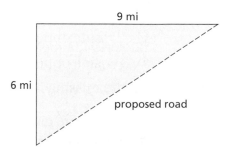

ROAD PLANNING Presently, to get from the ranger station to the park exit, you must drive 6 mi north and then 9 mi east. The park manager is considering having a new road constructed to provide a straight route. What will be the length of the new road to the nearest tenth?

Solution

The roads form a right triangle. Use the Pythagorean Theorem to find the hypotenuse, which will be the length of the new road.

Let x = the length of the new road.

$$6^2 + 9^2 = x^2$$

$$36 + 81 = x^2$$

$$117 = x^2$$

$$10.81665383 \approx x \qquad \text{The new road will be about 10.8 mi long.}$$

◤ TRY THESE EXERCISES

Find the unknown side measures. First find each in simplest radical form, and then find each to the nearest tenth.

1.
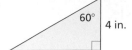
60° 4 in.

2.
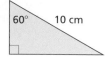
60° 10 cm

3.
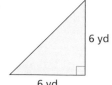
6 yd

6 yd

4.
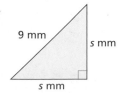
9 mm s mm

s mm

5. A satellite dish 5 ft high is attached by a cable from its top to a point on the ground 3 ft from its base. How long is the cable to the nearest tenth?

6. INVENTIONS Nefi has designed a small remote-controlled robot. He plans to test the robot's ability to move quickly on a rectangular piece of asphalt that measures 20 ft by 8 ft. What is the greatest distance to the nearest tenth of a foot that the robot can travel without turning?

Find the unknown measures. First find each in simplest radical form, and then find each to the nearest tenth.

7.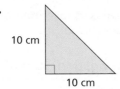
10 cm

10 cm

8.
60°

8 cm

9.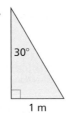
30°

1 m

10.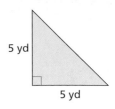
5 yd

5 yd

Solve. Round answers to the nearest tenth.

11. The diagonal of a square measures 15 cm. Find the length of a side of the square.

12. Find the measure of the altitude of an equilateral triangle with a side that measures 8 in.

13. Find the area of an equilateral triangle with a side that measures 10 cm.

14. **PLUMBING** A new pipe will be connected to two parallel pipes using 45° elbows. How long must the pipe be if the two parallel pipes are 2 ft apart?

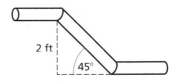
2 ft
45°

15. **CONSTRUCTION** A beam, 10 m in length, is propped up against a building. The beam has slipped so that its base is 3 m from the wall. How far up the building does the beam reach now?

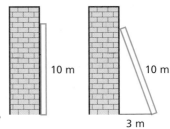
10 m   10 m

3 m

■ **EXTENDED PRACTICE EXERCISES**

16. Use algebra to show that, if *s* is the measure of a leg in a 45°–45°–90° triangle, then the hypotenuse measures $s\sqrt{2}$.

17. **WRITING MATH** Is this statement always, sometimes, or never true? Explain your choice.

If one acute angle of a right triangle is half the measure of the other acute angle, then the side opposite that angle measures half the length of the hypotenuse.

18. **CHAPTER INVESTIGATION** Research architectural styles that would complement your school grounds. Make a list of design elements that you could choose from in creating a new look for your school.

■ **MIXED REVIEW EXERCISES**

A card is drawn at random from a standard deck of cards. Find each probability. (Lesson 9-3)

19. The card is a 3, a 6, or a 9.

20. The card is a 5 or a face card.

21. The card is red and a 10.

22. The card is black and a face card.

23. The card is a club and a 7 or a jack.

24. The card is neither a spade nor a heart.

Solve each inequality and graph the solution set on a number line. (Lesson 2-6)

25. $2x + 1 > -3$

26. $2(x - 1) > 6$

27. $-2(x + 3) - 2 \geq 4$

28. $\frac{1}{2}(4x - 8) \leq 2$

10-4 Circles, Angles, and Arcs

Goals
- Find measures of central and inscribed angles.
- Find measures of angles formed by intersecting secants and tangents.

Applications   Landscape Architecture, Navigation, Surveying

Use a compass, straightedge and scissors to make these constructions:

1. Draw a circle, mark its center, and cut the circle out. Fold the circle in half and then in quarters. Draw line segments along the two creases to outline one-fourth of the circle. What kind of angle is formed at the center of the circle by the two line segments?

2. Draw another circle and cut it out. Fold the circle in half and draw along the crease to draw a diameter. Draw a line segment to connect one endpoint of the diameter with any point on the circle. Then draw another line segment to connect that point with the other endpoint of the diameter. What do you notice about the angle formed?

◥ BUILD UNDERSTANDING

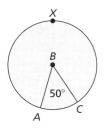

In a circle, a **central angle** is an angle that has its vertex at the center of the circle. The rays of the angle are said to **intercept** an arc. $\angle ABC$ intercepts **minor arc** $\overset{\frown}{AC}$ as well as **major arc** $\overset{\frown}{AXC}$. A minor arc is smaller than a semicircle, and a major arc is larger than a semicircle.

The degree measure of an arc is the same as the number of degrees of the corresponding central angle. Because $m\angle ABC = 50$, $m\overset{\frown}{AC} = 50$ and $m\overset{\frown}{AXC} = 360 - 50$, or 310.

Example 1

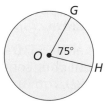

LANDSCAPE ARCHITECTURE A portion of the circumference of a circular pond will be tiled with handmade tiles donated by a local school. The portion to be tiled is represented on the plans as arc GH. Find $m\overset{\frown}{GH}$.

Solution

$\angle GOH$ is the central angle that intercepts $\overset{\frown}{GH}$. This angle measures 75°. Therefore, $m\overset{\frown}{GH} = 75°$.

A basic assumption about arcs is the *arc addition postulate*.

Arc Addition Postulate	If C is a point on an arc with endpoints A and B, then $$m\overset{\frown}{AC} + m\overset{\frown}{CB} = m\overset{\frown}{ACB}.$$	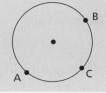

Example 2

Find the measure of each arc. a. $\overset{\frown}{PQR}$ b. $\overset{\frown}{SP}$

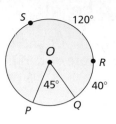

Solution

a. $m\overset{\frown}{PQR} = m\overset{\frown}{PQ} + m\overset{\frown}{QR}$

$\qquad = 45 + 40 = 85 \qquad$ $m\overset{\frown}{PQ} = 45$, because $m\angle POQ = 45$

b. $m\overset{\frown}{SP} = 360 - (m\overset{\frown}{SR} + m\overset{\frown}{RQ} + m\overset{\frown}{QP})$

$\qquad = 360 - (120 + 40 + 45) = 155$

Therefore, $m\overset{\frown}{PQR} = 85$ and $m\overset{\frown}{SP} = 155$.

In circle P, $\angle ABC$ is an **inscribed angle**. It has its vertex, B, on the circle, intersects the circle at two other points, and intercepts $\overset{\frown}{AC}$.

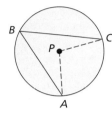

Theorem	The measure of an inscribed angle in a circle is one-half the measure of its intercepted arc.

Example 3

Find the measure of $\angle QRS$.

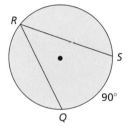

Solution

$\angle QRS$ intercepts $\overset{\frown}{QS}$. $m\overset{\frown}{QS} = 90$.

$\qquad m\angle QRS = \frac{1}{2}(90) = 45$

GEOMETRY SOFTWARE Use geometry software to explore the relationship between the measures of an inscribed angle and its intercepted arc.

1. Draw a circle and inscribed angle CBD.

2. Find $m\overset{\frown}{CD}$ and $m\angle CBD$.

3. Select and move point D around the circle, observing the changes to the measures of $\overset{\frown}{CD}$ and $\angle CBD$.

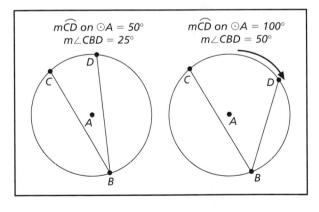

In circle O, $\overline{AB}$ is a line segment with both endpoints on the circle; it is called a **chord**. $\overleftrightarrow{CD}$ is a line that intersects the circle in two places; it is called a **secant**. $\overleftrightarrow{CF}$ is also a secant. $\overleftrightarrow{FG}$ intersects the circle in only one point; it is called a **tangent**.

The angles formed by secants and tangents are also related to the degree measures of their intercepted arcs.

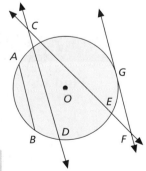

Theorem	If two secants intersect inside a circle, then the measure of each angle formed is equal to one-half the sum of the measures of the arcs intercepted by the angle and its vertical angle.

Example 4

Secants $\overleftrightarrow{AB}$ and $\overleftrightarrow{CD}$ intersect at point E inside a circle. Find the measure of $\angle AEC$.

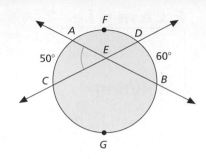

Solution

For $\angle AEC$, the two arcs intercepted by the secants are $\overparen{AC}$ and $\overparen{BD}$.

$$m\angle AEC = \frac{1}{2}(m\overparen{AC} + m\overparen{BD})$$

$$= \frac{1}{2}(50 + 60)$$

$$= \frac{1}{2}(110) = 55 \quad \text{So, } \angle AEC \text{ measures } 55°.$$

Theorem	If two secants, two tangents, or a tangent and a secant intersect outside a circle, then the measure of the angle formed is equal to one-half the positive difference of the measures of the intercepted arcs.

Example 5

Find $m \angle XZW$.

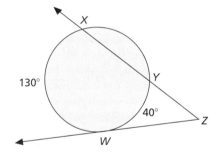

Solution

$\angle XZW$ is formed by secant $\overleftrightarrow{XZ}$ and tangent $\overleftrightarrow{WZ}$.

$$m\angle XZW = \frac{1}{2}(m\overparen{XW} - m\overparen{YW})$$

$$= \frac{1}{2}(130 - 40)$$

$$= \frac{1}{2}(90)$$

$$= 45$$

So, $m\angle XZW$ is 45°.

◤ TRY THESE EXERCISES

Find x.

1.
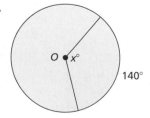

2.
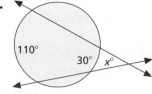

3.
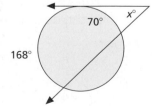

Find x.

4.

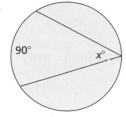

90°
x°

5.

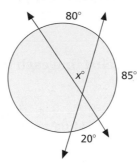

80°
x°
85°
20°

6.

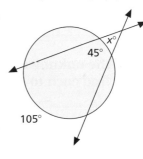

45°
x°
105°

7. **NAVIGATION** Explorers are searching for the site of an ancient shipwreck. From information found in a recently discovered diary, they draw a circle on a map with its center over a small island. They draw an inscribed angle with an intercepted arc of 80°. What is the measure of the inscribed angle?

8. **GEOMETRY SOFTWARE** $\triangle ABC$ is an isosceles triangle inscribed in circle O. Draw the triangle so that $m\angle BAC = 35°$. What is the measure of $\overset{\frown}{ADC}$?

A
B
C
O
D

9. **SURVEYING** On a surveyor's map, central angle COD intercepts minor arc CD which has a measure of 100°. What is the measure of inscribed angle CRD if it also intercepts $\overset{\frown}{CD}$?

10. **WRITING MATH** Write a paragraph that proves this statement: *If two inscribed angles intercept the same arc, the two angles are congruent.*

◼ EXTENDED PRACTICE EXERCISES

11. State the theorem that would apply to the measure of the angles formed by two chords intersecting inside a circle.

12. Describe the locus of all points x such that $\triangle XAB$ is a right triangle with hypotenuse $\overline{AB}$.

◼ MIXED REVIEW EXERCISES

A bag contains 5 white marbles, 8 red marbles, 7 green marbles, and 5 pink marbles. One marble at a time is taken from the bag and not replaced. Find each probability. (Lesson 9-4)

13. P(pink, then white)

14. P(red, then green)

15. P(white, then red, then pink)

16. P(red, then pink, then green)

Solve each equation. (Lesson 2-5)

17. $3(4a - 6) = -12$

18. $-7b + 6 = -b - 12$

19. $-3(h + 4) = 2(3h - 3)$

20. $5p + 3(p - 4) = -2$

21. $-2c + 6 = 2(5c - 3)$

22. $\frac{1}{2}m + 3(m - 1) = -10$

23. $12z + 3 - 6z + 9 = 3z$

24. $0.5x + 4.2 = 0.2(x - 3)$

25. $4(w - 3) + 2 = 5w - 8$

Review and Practice Your Skills

PRACTICE ◣ LESSON 10-3

Find the unknown side measures. First find each in simplest radical form and then find each to the nearest tenth.

1.
45°
20 in.

2.
7.5 cm
30°

3.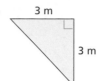
3 m
3 m

4.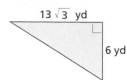
13 √3 yd
6 yd

For each 30°–60°–90° triangle, find the measures of the other two sides in simplest radical form.

5. side opposite 30° angle measures 11 cm

6. hypotenuse measures 48 ft

7. side opposite 60° angle measures $5\sqrt{3}$ in.

8. side opposite 30° angle measures $\sqrt{3}$ m

9. hypotenuse measures 37 ft

10. hypotenuse measures $6\sqrt{24}$ yd

11. side opposite 60° angle measures 30 km

12. hypotenuse measures $\frac{1}{3}$ cm

For each 45°–45°–90° triangle, find the measures of the other two sides in simplest radical form.

13. leg measures 44 in.

14. hypotenuse measures $15\sqrt{2}$ m

15. hypotenuse measures 8 ft

16. leg measures $\frac{3}{\sqrt{2}}$ cm

PRACTICE ◣ LESSON 10-4

Find *x*.

17.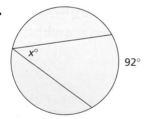
$x°$
92°

18.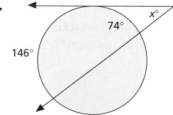
$x°$
74°
146°

19.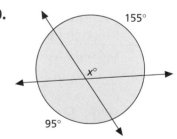
155°
$x°$
95°

20.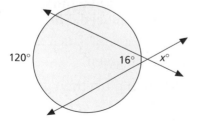
120°
16°
$x°$

21.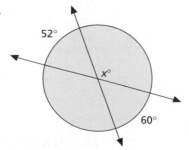
52°
$x°$
60°

22.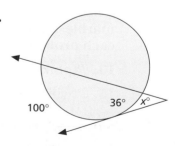
100°
36°
$x°$

23. *True* or *false*: Two tangents can intersect inside a circle.

24. *True* or *false*: A chord that passes through the center of a circle is called a diameter.

Write each in simplest radical form. (Lesson 10-1)

25. $\sqrt{\dfrac{3}{8}}$ **26.** $\dfrac{9}{\sqrt{3}}$ **27.** $(4\sqrt{22})(3\sqrt{33})(9\sqrt{8})$ **28.** $\sqrt{(4)(12)(18)}$

Use the Pythagorean Theorem to find the unknown length. Round to the nearest tenth. (Lesson 10-2)

29.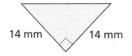

30.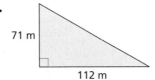

31.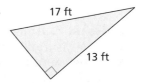

Mid-Chapter Quiz

Use your calculator to find the value to the nearest hundredth. (Lesson 10-1)

1. $\sqrt{44}$ **2.** $\sqrt{87}$

Write each in simplest radical form. (Lesson 10-1)

3. $\sqrt{270}$ **4.** $(4\sqrt{6})(-5\sqrt{3})$ **5.** $\dfrac{\sqrt{56}}{\sqrt{14}}$

Use the Pythagorean Theorem to find the unknown length. Round your answer to the nearest tenth. (Lesson 10-2)

6. legs: x, 15 in.
hypotenuse: 17 in.

7. legs: 7 cm, 9 cm
hypotenuse: x

Find the missing side lengths for 30°–60°–90° and 45°–45°–90° right triangles. Round your answers to the nearest tenth. (Lesson 10-3)

	Leg opposite 30° angle	Leg opposite 60° angle	Hypotenuse
8.			4 yd
9.	6 m		

	Leg opposite 45° angle	Hypotenuse
10.	9 ft	
11.		10 in.

Find the measure of each arc or angle. (Lesson 10-4)

12. arc *FG*

13. arc *FHG*

14. angle *FHG*

15. angle *FIG*

16. angle *LKF*

Chapter 10 **Review and Practice Your Skills** | **445**

Problem Solving Skills: Circle Graphs

A circle graph is a good way to compare data that are parts of a whole. Each part of the whole can be changed to a percent of the whole. Then the percents are used to divide a circle into sectors.

Problem Solving Strategies

Guess and check

Look for a pattern

Solve a simpler problem

✔ Make a table, chart or list

Use a picture, diagram or model

Act it out

Work backwards

Eliminate possibilities

Use an equation or formula

Problem

URBAN PLANNING A city planner is looking at the results of a survey of housing types. She decides to make a circle graph of the data to present to the next city council meeting.

Housing type	Number
Single-family home	9070
Two-family home	3023
Three-to-six family home	756
Seven or more units buildings	2267

Solve the Problem

Step 1: Add all of the data to find the total.

9070 + 3023 + 756 + 2267 = 15,116

Step 2: Find what percent each number is of the total. Use a calculator and round percents to the nearest whole percent.

Housing type	Number	Percent of total
Single-family home	9070	60%
Two-family home	3023	20%
Three-to-six family home	756	5%
Seven or more units buildings	2267	15%

Step 3: Find the central angles that correspond to each percent.

Since there are 360° in a circle, use the percents found in Step 2 to find corresponding central angle measures. For example, 60% of 360° is 216°.

Housing type	Percent of total	Central angle
Single-family home	60%	216°
Two-family home	20%	72°
Three-to-six family home	5%	18°
Seven or more units buildings	15%	54°

Step 4: Construct the graph using a compass, straightedge, and protractor.

Start with the smallest angle and work around to the largest angle. Draw a circle and one radius. Place the protractor so that the 18° mark aligns with the radius. Make a mark at 18° and draw a radius at that point.

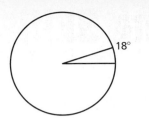 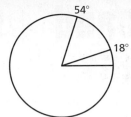 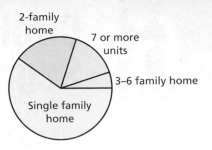

Ourtown—Housing Types

Place the protractor along the new radius. Make a mark at 54° and draw a radius at that point. Continue in this way around the circle.

Label each sector and write a title for the graph.

TRY THESE EXERCISES

Make a circle graph for each set of data.

1. **Rodriguez Family Monthly Budget**
   Mortgage, $750
   Food, $500
   Car payment, $125
   Utilities, $150
   Credit card payment, $250
   Transportation, $175
   Savings, $100
   Miscellaneous, $150

2. **Grace School Sports Budget**
   Football, $12,000
   Baseball, $9500
   Soccer, $2500
   Swimming, $5000
   Basketball, $6000

PRACTICE EXERCISES

Make a circle graph for each set of data.

3. **Town Population by Age**
   Under 5   3443
   5–13   4587
   14–18   2428
   19–25   4046
   26–39   7263
   40–64   1049
   Over 64   3125

4. **Window Types Sold, Fred's Building Supply**
   Single width, 520
   Double width, 241
   Bay, 183
   Round, 27
   Semicircular, 89
   Basement, 351

5. Use the data at the right to make a circle graph for family size.

Family Size*

Number of families	66,090
Two persons	27,606
Three persons	15,353
Four persons	14,026
Five persons	5938
Six persons	1997
Seven or more persons	1170
Total persons	209,515
Average per family	3.17

*Numbers in thousands except for averages

MIXED REVIEW EXERCISES

Solve for *x* and *y*. (Lesson 8-5)

6. $[x \quad 3y] = [y + 2 \quad 2x - 1]$

7. $[2x \quad y] = [y - 1 \quad 3x + 7]$

8. $\begin{bmatrix} x + 3 \\ y - 4 \end{bmatrix} = \begin{bmatrix} 3y \\ x - 5 \end{bmatrix}$

9. $\begin{bmatrix} x \\ 2y \end{bmatrix} = \begin{bmatrix} y - 2 \\ 3x + 1 \end{bmatrix}$

10. $\begin{bmatrix} 3x \\ 2y \end{bmatrix} = \begin{bmatrix} -2y + 1 \\ 6x - 2 \end{bmatrix}$

10-6 Circles and Segments

Goals ■ Find lengths of chord, secant and tangent segments.

Applications Architecture, Art, Surveying

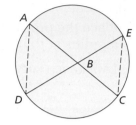

Work with a partner. You will need geometry software or a compass, straightedge, and protractor.

1. Draw a circle. Draw any two chords that intersect inside the circle and label them as shown at the right.

2. Draw $\overline{AD}$ and $\overline{EC}$ to form two triangles. Compare the angles of the triangles. Because all three pairs of angles are congruent, the triangles are similar.

3. Using the measures of the segments, write this proportion: $\dfrac{AB}{BD} = \dfrac{BE}{BC}$.

4. Use the cross-products rule. Discuss with your partner what conclusion you can make about the products of the lengths of segments of intersecting chords.

◣ BUILD UNDERSTANDING

As you can see from the above activity, any two intersecting chords determine two similar triangles. This leads to the following theorem.

Theorem	If two chords intersect inside a circle, then the product of the measures of the two segments of one chord is equal to the product of the measures of the two segments of the other chord.

This theorem can be used to find a missing measure.

Example 1

Find x.

Solution

Since there are two intersecting chords, the products of the lengths of the segments are equal.

$$3 \cdot x = 2 \cdot 12$$
$$3x = 24$$
$$x = 8$$

A similar type of relationship exists for intersecting secant segments. A **secant segment** intersects a circle in two points and has one endpoint on the circle and one endpoint outside the circle.

Theorem	If two secant segments have a common endpoint outside a circle, then the product of the measures of one secant segment and its external part is equal to the product of the other secant segment and its external part.

The following example will help you see what is meant by the length of a segment and the length of the external part of the segment.

Example 2

ARCHITECTURE An architect is redesigning a museum. One of the rooms will contain a circular platform to display the skeleton of a prehistoric mammal. Steel cables represented by the two secant segments shown in the drawing to the right will be used to brace the skeleton. Find x.

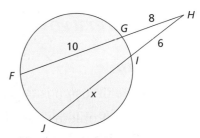

Solution

$\overline{HF}$ and $\overline{HJ}$ are secant segments. $\overline{HG}$ is the external part of $\overline{HF}$, and $\overline{HI}$ is the external part of $\overline{HJ}$. The theorem refers to the length of the entire secant segment and its external part.

$$HF \cdot HG = HJ \cdot HI$$
$$18 \cdot 8 = (6 + x)6 \qquad \text{The length of } \overline{HJ} \text{ is } 6 + x.$$
$$144 = 36 + 6x$$
$$108 = 6x$$
$$18 = x$$

Check Understanding

What is the length of $\overline{HJ}$?

A **tangent segment** of a circle is a segment that has one endpoint on a circle and one endpoint outside the circle, and the line containing the segment intersects the circle at exactly one point.

Theorem	If a tangent segment and a secant segment have a common endpoint outside a circle, then the square of the measure of the tangent segment is equal to the product of the measures of the secant segment and its external part.

Example 3

Find x.

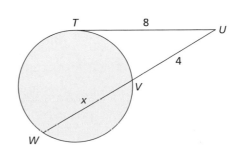

Solution

$\overline{TU}$ is a tangent segment, $\overline{UW}$ is a secant segment, and $\overline{UV}$ is its external segment.

$$(TU)^2 = UW \cdot UV$$
$$8^2 = (x + 4)4$$
$$64 = 4x + 16$$
$$48 = 4x$$
$$12 = x$$

Another interesting property of circles and chords arises from a radius perpendicular to a chord.

Theorem	If a radius of a circle is perpendicular to a chord of the circle, then that radius bisects the chord.

Example 4

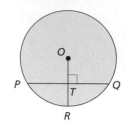

In circle *O*, radius $\overline{OR}$ is perpendicular to chord $\overline{PQ}$ at *T*. Find *PT* if *PQ* = 5 cm.

Solution

The first step is to make a diagram and label it using the given facts.

The problem states that $\overline{OR}$ is perpendicular to $\overline{PQ}$. Therefore, $\overline{OR}$ also bisects $\overline{PQ}$. If *PQ* = 5 cm, then *PT* = 2.5 cm.

◤ TRY THESE EXERCISES

Find *x*.

1.
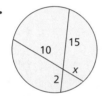

2.
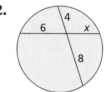

3.
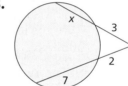

4.
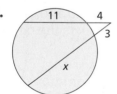

5.
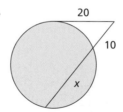

6.
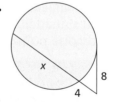

◤ PRACTICE EXERCISES • For Extra Practice, see page 695.

Find *x*.

7.
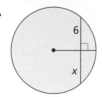

8.
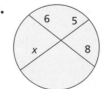

9.
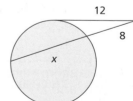

10.
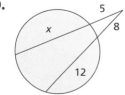

11.
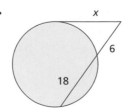

12.
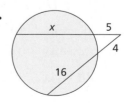

13. **ART** An artist is planning a circular mosaic. The design for the mosaic has two chords that intersect inside the circle. Chord $\overline{PR}$ consists of two segments 3 in. and 15 in. long. Chord $\overline{ST}$ consists of two segments, one of which is 5 cm in length. Find the length of the remaining segment.

14. **SURVEYING** On a map of a city, two chords of a circle, TR and KL, intersect at point X. $TX = 4$ in., $XR = 6$ in., and $KX = 3$ in. Find the measure of $\overline{KL}$.

Find *x* and *y*.

15.
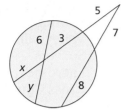

16.
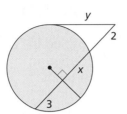

17.
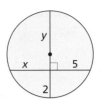

 18. **GEOMETRY SOFTWARE** The distance of a chord from the center of the circle is defined as the length of a segment from the center of the circle perpendicular to the chord. Draw a large circle. Then use a centimeter ruler to draw three different chords, all the same length, at different places in the circle. Find the distance from the center of the circle to each chord. What seems to be true?

 19. For a circle, $\overline{AB}$ is a secant segment 8 cm long. Its external part is 3 cm. $\overline{AC}$ is another secant segment with an external part of 4 cm. What is its length?

 20. **CHAPTER INVESTIGATION** Make a scale drawing of a new facade for a building on your school grounds. Display your original drawing and your new design.

■ EXTENDED PRACTICE EXERCISES

 21. **WRITING MATH** Look at the figure below. The theorem about the products of the lengths of secant segments with a common endpoint outside a circle can be proven using similar triangles. Which triangles are similar? Explain your answer.

22. **WRITING MATH** A tangent segment and a secant segment are drawn to a circle from a point outside the circle. The length of the secant segment is 27 m, and the length of its external part is 3 m. Is the length of the tangent segment a rational number? Explain your answer.

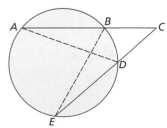

■ MIXED REVIEW EXERCISES

Find the surface area of each figure. (Lesson 5-6)

23.
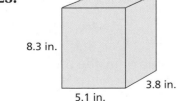
8.3 in.
3.8 in.
5.1 in.

24.
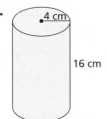
4 cm
16 cm

25.
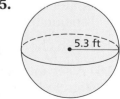
5.3 ft

Review and Practice Your Skills

PRACTICE ■ LESSON 10-5

Make a circle graph for each set of data.

1. Howe Family Budget
Mortgage	$860
Food	$645
Utilities	$322.50
Insurance	$107.50
Other	$215

2. Car Types Sold
Sport Coupe	50
2-Door Sedan	35
4-Door Sedan	45
Hatch Back	20

3. Fall Sports Athletes
Football	68
Volleyball	24
Soccer	36
Lacrosse	38
Cross Country	34

4. Heights of Freshman
<64 in.	24
64–66 in.	36
67–69 in.	72
70–72 in.	66
>72 in.	18

5. Technology Annual Budget
New equipment	$320,000
Repair/Upgrade	$80,000
Internet access	$40,000
Salaries	$180,000
Research	$100,000

6. Park Attendance by Age
<5	1714
5–12	2299
13–18	3617
19–55	18,196
>55	2991

PRACTICE ■ LESSON 10-6

Find x.

7.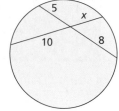

8.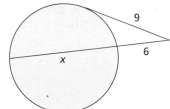

9.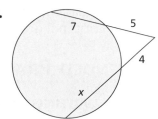

10.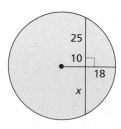

11.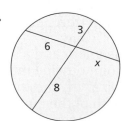

12.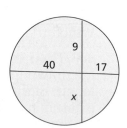

Classify each statement as *true* or *false*.

13. A radius of a circle bisects every chord of the circle.

14. If two secant segments have a common endpoint outside a circle and their external parts are equal in length, then the chords formed by each secant inside the circle will be equal in length.

15. Two chords of a circle will never intersect at the center of the circle.

16. A secant and a tangent to a circle can intersect either outside or inside the circle.

17. Chords of a circle which bisect each other are called diameters of the circle.

Determine if a triangle with the given sides is a right triangle. (Lesson 10-2)

18. 15 m, 36 m, 39 m

19. 10 ft, 22 ft, 30 ft

20. 16 cm, 30 cm, 34 cm

21. 18 in., 30 in., 24 in.

22. 4 yd, $4\sqrt{2}$ yd, $4\sqrt{3}$ yd

23. 21 m, $\sqrt{17}$ m, 21 m

Find the unknown side measures. First find each in simplest radical form, and then find each to the nearest tenth. (Lesson 10-3)

24.

25.

26.

27.

MathWorks Career – Landscape Architects

Workplace Knowhow

L andscape architects design outdoor areas such as public parks, playgrounds, shopping centers and industrial parks. They use knowledge of the natural environment to design areas that will complement the existing surroundings.

Landscape architects study soil, sunlight, vegetation and climate. They may work with government officials and environmentalists to find ways to build new structures and roads while preserving the natural beauty and wildlife in an area.

You are designing a circular fountain for a city park. A diagram of the fountain is shown below. You have the following measurements: DE = 4 ft 9 in., GE = 1 ft 4 in. and DF = 1 ft 6 in. Determine all measurements to the nearest tenth of a foot. Use 3.14 for pi.

1. A circular pedestal is at the center of the fountain. Find the circumference of the pedestal.

2. A wall is designed as a sitting area for park visitors. The outer edge of the wall will have a brass rim to reflect sunlight. Find the circumference of the outer edge of the wall.

3. The base of the pool, represented by the blue area of the diagram, will be tiled. Find the area of the base of the pool.

4. Find the total area of the fountain.

10-7 Constructions with Circles

Goals
- Construct regular polygons using circles.
- Inscribe a circle in a polygon and circumscribe a circle about a polygon.

Applications Architecture, Design, Art

Construct a polygon using a compass and a straightedge.

1. Use a compass to draw any circle.

2. Keeping the compass open to the same radius, place the compass point anywhere on the circle and draw a small arc that intersects the circle.

3. Place the compass point on the intersection you just made and draw another arc but not through the starting point.

4. Continue in this way around the circle. You will have drawn six points.

5. Connect each point to the one next to it with a line segment. Measure the sides and angles of the polygon you have drawn. How would you describe the figure?

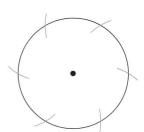

BUILD UNDERSTANDING

Several regular polygons can be constructed using a circle. For instance, you can construct an equilateral triangle using steps 1 through 4 above to draw six evenly spaced points. Then use a straightedge to connect every other point to form the triangle. The same basic construction can be adapted to construct a regular dodecagon, a 12-sided polygon.

Example 1

Construct a regular dodecagon.

Solution

Step 1: Begin with a circle and mark off 6 equal arcs, as you did above.

Step 2: Connect each point to the center of the circle. You now have 6 central angles that are all congruent.

Step 3: Bisect three consecutive central angles. Extend each bisector so that it intersects the circle on two sides. You should now have 12 equally spaced points on the circle.

Step 4: Connect each point to the one next to it with a straight line segment. The resulting figure is a regular dodecagon.

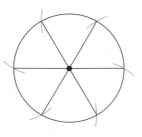
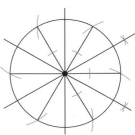

A **circumscribed polygon** of a circle has every side of the polygon tangent to the same circle. Any regular polygon can be circumscribed around a circle.

Example 2

Circumscribe a circle around this regular pentagon.

Solution

To circumscribe a circle around any regular polygon, construct the perpendicular bisector of any two of its sides. The point of intersection of the bisectors becomes the center of the circle. The radius of the circle is the distance from the center to any vertex of the polygon.

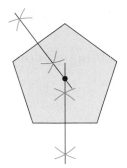
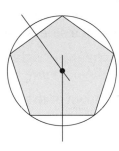

An **inscribed polygon** has every vertex of the polygon on the same circle. Any regular polygon can be inscribed in a circle.

Example 3

Inscribe a circle in a square.

Solution

Draw a square. To locate the point that will become the center of the circle, find the intersection of the perpendicular bisectors of any two sides. To find the radius of the circle, use the distance from the center of the circle along a perpendicular bisector to a side of the polygon. Draw the circle.

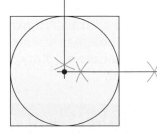

Many properties of circles and parts of circles can be used to solve real-life problems.

Example 4

ARCHITECTURE An architect is restoring an old house. She has found a part of a window that may have been used in the attic. How can she figure out the size of the original window from this fragment?

Problem Solving Tip

Use paper folding to construct a regular octagon using a circle.

1. Draw a circle and cut it out.

2. Fold the circle into halves, then fourths and finally eighths. Open the circle.

3. There are now 8 equally spaced points on the circle. Connect each point to the one next to it with a straight line segment. The resulting figure is a regular octagon.

Reading Math

The word *circumscribed* comes from Latin words which mean "drawn around." The word *inscribed* means "drawn in." The Example 3 text says that the circle has been inscribed in the square. It is also correct to say that the square has been circumscribed around the circle.

Solution

The perpendicular bisector of a chord passes through the center of a circle. To complete the circle, begin by drawing two chords. Construct the perpendicular bisector of each. The point where the bisectors intersect must be the center of the circle. Use the center of the circle and the radius to complete the circle.

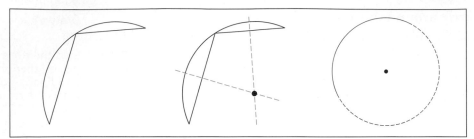

◥ TRY THESE EXERCISES

1. Construct a regular hexagon with a side measuring 4 cm.

2. Copy the regular octagon shown at the right. Circumscribe a circle around the octagon.

3. Copy the equilateral triangle shown at the right. Inscribe a circle in the triangle.

 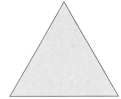

 TALK ABOUT IT Discuss the following statements with a partner. Decide whether each statement is true or false. Explain your reasoning.

4. In Example 3, the circle is inscribed in the square.

5. It is always possible to inscribe a rhombus with sides of given lengths in a circle.

6. It is impossible to inscribe a non-regular octagon in a circle.

7. **CONSTRUCTION** Circular saws come in different sizes. A builder bought a saw at a second-hand sale. The saw blade was broken as shown in the picture at the right. Copy the drawing. Use constructions to demonstrate how the builder can find the center and complete the circle in order to find out which size of replacement blade to buy.

 8. **WRITING MATH** An architect draws a circle and a diameter of the circle. She constructs the perpendicular bisector of the diameter and extends it so that it intersects the circle in two points. How could the architect use the drawing to construct a square?

◥ PRACTICE EXERCISES • For Extra Practice, see page 695.

9. Construct a regular octagon.

10. Copy this regular hexagon. Circumscribe a circle around the hexagon.

11. Copy this regular pentagon. Inscribe a circle in the pentagon.

12. Draw any regular pentagon. Describe how you can use a circle to help you construct a regular decagon from the pentagon.

13. A circle with a radius of 5 in. is inscribed in a square. What is the perimeter of the square?

14. A regular hexagon is inscribed in a circle. What is the measure of each of the six arcs of the circle?

15. Copy the regular hexagon in Exercise 10. Inscribe a circle around the hexagon.

16. ARCHITECTURE Ming Lee is an architect. She submitted the following plan for a new house to her clients, James and Odetta Williams. The Williamses ask Ms. Lee to alter the plans so that the sun room is circular rather than square. Copy Ming Lee's plans onto graph paper, using pencil, compass, and straightedge. Inscribe a circle within the square that represents the sun room. Then erase the outline of the square.

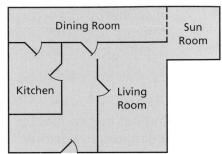

17. DESIGN Dave O'Brien makes and designs tiles. He has a client who wants a small tabletop covered in hexagonal tiles. Dave decides to make some sketches of different designs from which the client can choose. Construct a regular hexagon measuring 2 in. on all sides that Dave can use as a model.

18. ART Linda Soares is designing a logo for the new community center. She wants to take the pentagon at the right and circumscribe a circle around it while also inscribing a circle in the pentagon. Copy the pentagon and complete Linda's design.

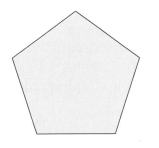

19. ARCHITECTURE Mike Whitehorse has a blueprint for a gazebo in the shape of a regular hexagon. He wants to change it so that it is about the same size, but has the shape of a regular dodecagon. How can he use a copy of the original blueprint to develop the other plan?

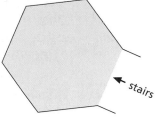

stairs

◢ EXTENDED PRACTICE EXERCISES

20. These regular polygons can be constructed using a compass and straightedge: square, octagon, 16-gon, and 32-gon. If n is the number of sides in a polygon, find an expression to represent the pattern.

◢ MIXED REVIEW EXERCISES

Compute the variance and standard deviation for each set of data. Round the final answer to the nearest tenth if necessary. (Lesson 9-7)

21. 8, 9, 10, 11, 12, 13

22. 28, 32, 31, 36, 29, 35

23. 2, 4, 6, 3, 5, 4

24. 12, 15, 16, 11, 15, 13

25. 74, 73, 82, 80, 77, 79

26. 52, 60, 65, 62, 58, 61

Chapter 10 Review

VOCABULARY ◤

Match the word from the list at the right with the description at the left.

1. longest side of a right triangle

2. angle that has its vertex on a circle and intersects the circle in 2 other points

3. segment with both endpoints on a circle

4. line that intersects a circle in two points

5. line that intersects a circle in only one point

6. number that cannot be written as a fraction, a terminating decimal, or a repeating decimal

7. number under the radical sign

8. angle that has its vertex at the center of a circle

9. polygon where every side is tangent to the same circle

10. arc with degree measure greater than 180°

a. central

b. chord

c. circumscribed

d. hypotenuse

e. inscribed

f. irrational

g. major

h. minor

i. radicand

j. secant

k. square root

l. tangent

LESSON 10-1 ◤ Irrational Numbers, p. 426

▶ An **irrational number** cannot be written as a fraction, terminating decimal, or repeating decimal. The square root of a number that is not a perfect square is always irrational.

▶ Many numbers can be written in simplest radical form using these theorems.

$$\sqrt{a \cdot b} = \sqrt{a} \cdot \sqrt{b} \qquad \sqrt{\frac{a}{b}} = \frac{\sqrt{a}}{\sqrt{b}}$$

Write each in simplest radical form.

11. $\sqrt{45}$

12. $\left(\sqrt{2}\right)\left(\sqrt{8}\right)$

13. $\left(7\sqrt{30}\right)\left(2\sqrt{6}\right)$

14. $\sqrt{\frac{3}{10}}$

15. $\frac{\sqrt{6}}{\sqrt{8}}$

16. $\left(2\sqrt{5}\right)^2$

LESSON 10-2 ◤ The Pythagorean Theorem, p. 430

▶ In a right triangle, the two shorter sides are called the legs, and the longest side is called the hypotenuse. In any right triangle, the Pythagorean Theorem is true: $a^2 + b^2 = c^2$, where a and b are the measures of the legs and c is the measure of the hypotenuse.

Use the Pythagorean Theorem to find the unknown length. Round to the nearest tenth.

17.
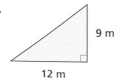
9 m

12 m

18.
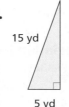
15 yd

5 yd

19.
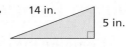
14 in.

5 in.

20. A pole that is 3 m high is connected by a guy wire from its top to a stake in the ground 1.5 m from its base. How long is the wire?

LESSON 10-3 ◼ Special Right Triangles, p. 436

▶ In a 30°–60°–90° triangle, the measure of the hypotenuse is two times that of the leg opposite the 30° angle. The measure of the longer leg is $\sqrt{3}$ times the leg opposite the 30° angle.

▶ In a 45°–45°–90° triangle, the measure of the hypotenuse is $\sqrt{2}$ times the measure of a leg of the triangle.

Find the unknown side measures. Give answers in simplest radical form.

21.

6 m
6 m

22.
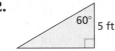
60° 5 ft

23.

60°
6 cm

24.
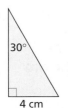
30°
4 cm

25.

45°
6 ft

26.
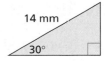
14 mm
30°

LESSON 10-4 ◼ Circles, Angles, and Arcs, p. 440

▶ The measure of a central angle of a circle is the same as the measure of its intercepted arc.

▶ The measure of an inscribed angle in a circle is one-half the measure of its intercepted arc.

▶ If two secants intersect inside a circle, the measure of each angle formed is equal to one-half the sum of the measures of the intercepted arcs.

▶ If two secants, two tangents, or a tangent and a secant intersect outside a circle, then the measure of the angle formed is equal to one-half the difference of the measures of the intercepted arcs.

Find x.

27.

x°
124°

28.
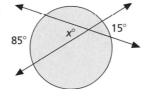
85° x° 15°

29.
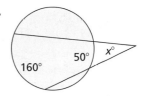
50° x°
160°

30.
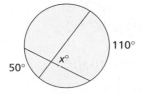
110°
x°
50°

31.
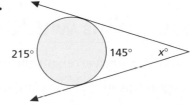
215° 145° x°

32.
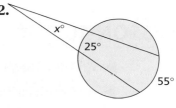
x°
25°
55°

LESSON 10-5 ◼ Problem Solving Skills: Circle Graphs, p. 446

▶ Circle graphs can represent data about how a quantity is subdivided.

33. Every Elm High School student takes one foreign language. This year, 251 take Italian, 478 take Spanish, 376 take French, and 50 take Japanese. Make a circle graph for this data.

34. The surface areas of the four oceans are given below. Make a circle graph for this data.

Ocean Surface Area	
Ocean	Area (square miles)
Pacific	64,186,300
Atlantic	33,420,000
Indian	28,350,500
Arctic	5,105,700

35. Students were asked to name their favorite flavor of ice cream. Eleven students said vanilla, 15 students said chocolate, 8 students said strawberry, 5 students said mint chip, and 3 said cookie dough. Make a circle graph for this data.

LESSON 10-6 ◥ Circles and Segments, p. 448

▶ If two chords intersect inside a circle, then the product of the measures of the two segments of one chord is equal to the product of the measures of the two segments of the other chord.

▶ If two secant segments have a common endpoint outside a circle, then the product of the measures of one secant segment and its external part is equal to the product of the measures of the other secant segment and its external part.

▶ If a tangent segment and a secant segment have a common endpoint outside a circle, then the square of the measure of the tangent segment is equal to the product of the measures of the secant segment and its external part.

Find *x*.

36.
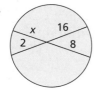

37.
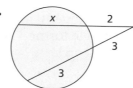

38.
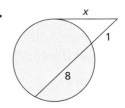

39.
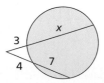

40.

41.
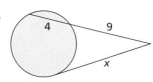

LESSON 10-7 ◥ Constructions with Circles, p. 454

▶ An inscribed polygon has every vertex on the same circle.

▶ A circumscribed polygon has every side tangent to the same circle.

42. Draw a square. Circumscribe a circle around the square.

43. Construct an equilateral triangle. Inscribe a circle in the triangle.

CHAPTER INVESTIGATION

EXTENSION Make a presentation to your class of your design. Explain why you chose the design you did and how it will make your school more attractive.

Chapter 10 Assessment

Simplify. Rationalize the denominator if necessary.

1. $\sqrt{700}$

2. $\dfrac{\sqrt{75}}{\sqrt{3}}$

3. $\dfrac{11}{\sqrt{3}}$

Find the unknown side measure in each right triangle. Round answers to the nearest tenth. (*a* and *b* are the measures of the legs; *c* is the measure of the hypotenuse.)

4. $a = 8$ in., $b = 15$ in.

5. $a = 5$ m $c = 7$ m

Find the missing measures. Give answers in simplest radical form.

6.

7.

8.

Find *x*.

9.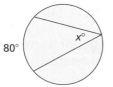

10.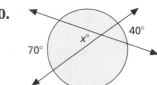

11.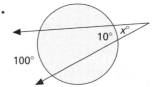

12.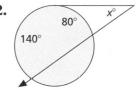

13.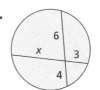

14.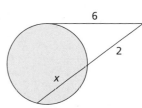

15.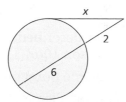

16.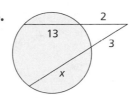

17.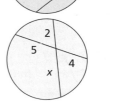

18. Construct a regular hexagon. Then inscribe a circle in it.

Solve.

19. A 20-ft ladder is placed against a building. The base of the ladder is 4 ft from the base of the building. How high up the building does the ladder reach, to the nearest tenth of a foot?

20. A survey of 200 department store customers showed that 24 had traveled more than 40 mi from home to the store, 52 traveled between 30 and 40 mi from home, 35 traveled between 20 and 30 mi, and 89 traveled less than 20 mi from home. Draw a circle graph that shows this data.

Standardized Test Practice

Record your answers on the answer sheet provided by your teacher or on a sheet of paper.

1. Write all the subsets of {*r, a, t, e*}. (Lesson 1-1)

   Ⓐ {*r, a*}, {*a, t*}, {*r, t*}, {*r, e*}, {*a, e*}, {*t, e*}

   Ⓑ {*r, a, t, e*}, {*r, a, t*}, {*a, t, e*}, {*r, a, e*}, {*r, t, e*}

   Ⓒ {*r*}, {*a*}, {*t*}, {*e*}, ∅

   Ⓓ all of these

2. Which of the following results in a negative number? (Lesson 1-8)

   Ⓐ $(-2)^5$

   Ⓑ 5^{-3}

   Ⓒ $-5 \cdot (-2)^5$

   Ⓓ $(-3)^{-2} \cdot 5$

3. Which graph shows a line with a slope of −2? (Lesson 6-1)

   Ⓐ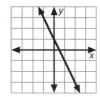

   Ⓑ

   Ⓒ

   Ⓓ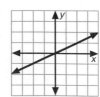

4. What is the location of the image of $A(-5, -2)$ if the point is translated three units to the right, translated four units up, and reflected over the y-axis? (Lesson 8-4)

   Ⓐ $(-2, 2)$

   Ⓑ $(2, -2)$

   Ⓒ $(1, 0)$

   Ⓓ $(2, 2)$

5. Which statement is correct concerning the probabilities of reaching into the jars without looking and pulling out a blue marble? (Lesson 9-1)

100 marbles	200 marbles
20 blue	50 blue
Jar 1	Jar 2

   Ⓐ greater for Jar 1 than Jar 2

   Ⓑ greater for Jar 2 than Jar 1

   Ⓒ equal for both jars

   Ⓓ cannot be determined

6. What is the approximate length of a diagonal of a rectangle that is 18 ft long and 12 ft wide? (Lesson 10-2)

   Ⓐ 6.0 ft   Ⓑ 13.4 ft

   Ⓒ 21.6 ft   Ⓓ 30.0 ft

7. What is the ratio of the measure of $\angle ACB$ to the measure of $\angle AOB$? (Lesson 10-4)

   Ⓐ 1:1   Ⓑ 2:1

   Ⓒ 1:2   Ⓓ cannot be determined

8. Which of the segments described could be a secant of a circle? (Lesson 10-6)

   Ⓐ intersects exactly one point on a circle

   Ⓑ has its endpoints on a circle

   Ⓒ has one endpoint at the center of a circle

   Ⓓ intersects exactly two points on a circle

Test-Taking Tip

Question 6
Be sure that you know and understand the Pythagorean Theorem. References to right angles, the diagonal of a rectangle, or the hypotenuse of a triangle indicate that you may need to use the Pythagorean Theorem to find the answer to the question.

Preparing for Standardized Tests
For test-taking strategies and more
practice, see pages 709–724.

Part 2 Short Response/Grid In

Record your answers on the answer sheet
provided by your teacher or on a sheet of paper.

9. In the figure below, $AD = 25$ and $AB = BC$.
Find BC. (Lesson 3-1)

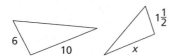

10. The angles of a
scalene triangle have
the measures shown
on the figure at the
right. What is the
value of x?
(Lesson 4-1)

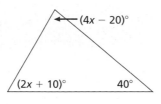

11. In the figure, what is
$m\angle PQR$? (Lesson 4-3)

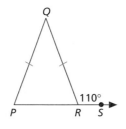

12. Write the ratio 18 ft to 9 yd in lowest terms.
(Lesson 5-1)

13. If $y = x + 2$ and $2y + 3x = 19$, what is the
value of $5y$? (Lesson 6-5)

14. The triangles below are similar. Find the value
of x. (Lesson 7-2)

15. On a blueprint, 1 in. represents 10 ft. Find the
actual length of a room that is $2\frac{1}{4}$ in. long on
the blueprint. (Lesson 7-3)

16. If $A = \begin{bmatrix} 8 & 2 & -1 \\ -9 & 0 & 2 \end{bmatrix}$ and $B = \begin{bmatrix} 0 & -5 & 0 \\ 2 & 4 & 1 \end{bmatrix}$,

find $A - 2B$. (Lesson 8-5)

17. A drawer has 4 black socks, 16 white socks,
and 2 blue socks. What is the probability of
reaching in the drawer without looking and
taking out two white socks? (Lesson 9-4)

18. How many ways can 4 students be selected
from a group of 7 students? (Lesson 9-5)

19. Simplify $\dfrac{\sqrt{6} \cdot \sqrt{8}}{\sqrt{3}}$. (Lesson 10-1)

20. A 15-ft ladder is propped against a shed. If
the top of the ladder rests against the shed
12 ft above ground, how far away in feet
from the shed is the base of the ladder?
(Lesson 10-2)

Part 3 Extended Response

Record your answers on a sheet of paper.
Show your work.

21. Haley hikes 3 mi north, 7 mi east, and then
6 mi north again. (Lesson 10-2)

   a. Draw a diagram showing the direction and
distance of each segment of Haley's hike.
Label Haley's starting point, her ending
point, and the distance, in miles, of each
segment of her hike.

   b. To the nearest tenth of a mile, how far
(in a straight line) is Haley from her
starting point?

   c. How did your diagram help you to find
Haley's distance from her starting point?

22. The following table shows how Jack uses his
time on a typical Saturday. Make a circle
graph of the data. (Lesson 10-5)

Saturday Time Use	
Activity	**Hours**
Jogging	1
Reading	2
Sleeping	9
Eating	2
Talking on the phone	1
Time with friends	4
Studying	5

Math Online mathmatters3.com/standardized_test

Polynomials

THEME: Consumerism

You would probably be surprised at the number of advertisments and commercials you see daily. Nearly one-fourth of every television hour is commercial time. Some radio stations devote 1 out of every 3 minutes to advertising.

How do companies decide which products to make and sell? Across America, businesses spend millions of dollars everyday to find out what consumers want and need. Marketing executives gather data about the spending habits and patterns of consumers in every age group. Product developers design new products for specific groups of consumers, and advertisers create exciting campaigns to convince the consumer to try the new product.

- **Brokerage Clerks** (page 477) assist in the buying and selling of stocks, bonds, commodities, and other types of investments. They monitor clients' accounts, make sure dividends are paid and check the accuracy of the paperwork used in making transactions.

- **Actuaries** (page 497) work for insurance companies to assemble and analyze statistical data about consumers in order to estimate the probabilities of death, sickness, injury, and property loss. This information helps insurance companies predict costs and charges for insurance coverage.

Math Online

mathmatters3.com/chapter_theme

American Spending Habits

Average Annual Expenses Per Household

Expense item	2000	2001	2002
Food at home	$3021	$3086	$3099
Food away from home	2137	2235	2276
Housing	12,319	13,011	13,283
Apparel and services	1856	1743	1749
Transportation	7417	7633	7759
Health care	2066	2182	2350
Entertainment	1863	1953	2079
Insurance and pensions	3365	3737	3899
Other	4001	3939	4182
Total average annual expenses	$38,045	$39,518	$40,677

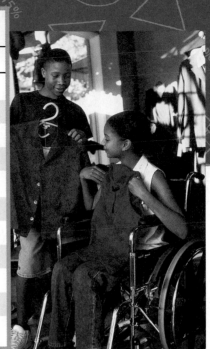

Data Activity: American Spending Habits

Use the table for Questions 1–4.

1. In which category was there the greatest percent increase from 2000 to 2002?

2. The government determined that there were 105,456,000 households in 2002. To the nearest million, how much was spent on apparel and services in 2002?

3. Which category demonstrated nearly a 14% increase from 2000 to 2002?

4. To the nearest tenth, what percent of a households' total expenses were housing costs in 2002?

CHAPTER INVESTIGATION

Demographics are the statistical characteristics of a particular population. Advertising decisions are often made based on the demographical profile of a market. For instance, car manufacturers generally buy commercial time during television programs that are watched by adult viewers.

Working Together

Conduct a survey to gather demographical information about your classmates. You will need to gather information about their viewing and listening preferences (television and radio), as well as their product preferences and brand loyalties. Discuss how the compiled results could be used by advertisers and manufacturers to sell products. Use the Chapter Investigation icons to guide your group.

11 Are You Ready?

Refresh Your Math Skills for Chapter 11

The skills on these two pages are ones you have already learned. Stretch your memory and complete the exercises. For additional practice on these and more prerequisite skills, see pages 654–661.

ORDER OF OPERATIONS

No matter what aspect of mathematics you study, the order of operations always applies.

Example Simplify: $3(4 + 6) - 3^2 + 9 \div 3 \cdot 8$

1. First, simplify anything in parentheses or involving exponents.

$3(\underline{4 + 6}) - 3^2 + 9 \div 3 \cdot 8$
$3(10) - \underline{3^2} + 9 \div 3 \cdot 8$

2. Then multiply and divide from left to right.

$\underline{3(10)} - 9 + 9 \div 3 \cdot 8$
$30 - 9 + \underline{9 \div 3} \cdot 8$
$30 - 9 + \underline{3 \cdot 8}$

3. Finally, add and subtract from left to right.

$\underline{30 - 9} + 24$
$\underline{21 + 24}$
45

Simplify each expression.

1. $25 \div 5 + 4 \cdot 2^2 - 15 \div 3$

2. $18 \div 3 + 6 - 9 + 3 \cdot 9 \div 6$

3. $45 \cdot 3 \div 9 + 8^2 - 6^2 + 3$

4. $15 \cdot 8 \div 40 - 3 + 16 - 2^2 + 18$

5. $108 \div 12 \cdot 3^2 - 8 + 16 \div 2$

6. $96 \div 4 + 3^2 - (5 + 3) + 11$

7. $64 \div (8 \div 2) \cdot 3 - 6^2 + 4$

8. $12 \cdot 9 \div 3 - 8^2 + 7^2 - (14 + 8)$

9. $3 \cdot 8 \cdot 4 \cdot 2^2 \div (8 \div 4) + 17$

10. $(9 \cdot 2) + (3 \cdot 4) - (4^2 \div 2) + 37$

SIMPLIFY EXPONENTS

Simplify each expression. Assume that $a \neq 0$, $b \neq 0$ and $c \neq 0$.

11. $(a^2)(a^3)(a^4)$

12. $(a^2 \cdot a^9)^2$

13. $(a^2 b^6 c^4)^3$

14. $[(a^2)^3]^5$

15. $\dfrac{a^9}{a^4}$

16. $\dfrac{a^7 b^6}{ab^4}$

17. $\dfrac{a^{12}}{a^9}$

18. $(a^4 b^6 c^7)^8$

19. $(a^2)^4 (a^3)^2 (a)^4$

20. $\dfrac{(a^3 \cdot a^4 \cdot a^2)}{a^5}$

21. $\dfrac{a^9 b^7 c^8}{a^6 b^2 c^5}$

22. $\dfrac{(a^5 b^9 c^4)^3}{(a^3 b^2 c^3)^2}$

PRIME FACTORIZATION

In this chapter you will learn to factor binomial and trinomial expressions. It may be helpful to practice this "un-multiplying" skill on simpler numbers.

Examples **Find the prime factorization of 36.** (Two methods are shown.)

You know that $6 \times 6 = 36$.   You know that $4 \times 9 = 36$.

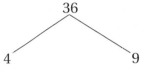

You know that $2 \times 3 = 6$.   You know that $2 \times 2 = 4$ and $3 \times 3 = 9$.

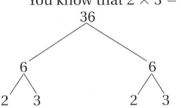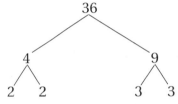

Both methods result in the same answer. Since 2 and 3 are both prime numbers, no more factoring is possible.

The prime factorization of 36 is $2 \cdot 2 \cdot 3 \cdot 3$ or $2^2 \cdot 3^2$.

Find the prime factorization of 156. (Two methods are shown.)

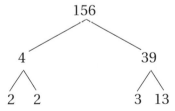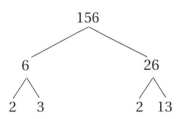

The prime factorization of 156 is $2 \cdot 2 \cdot 3 \cdot 13$ or $2^2 \cdot 3 \cdot 13$.

Name each prime factor of each number.

23. 81	**24.** 74	**25.** 100
26. 69	**27.** 58	**28.** 44
29. 29	**30.** 68	**31.** 75
32. 32	**33.** 99	**34.** 84

35–46. Write the prime factorization of each number in Exercises 23–34. Use exponents when possible.

Add and Subtract Polynomials

Goals
- Write polynomials in standard form.
- Add and subtract polynomials.

Applications Packaging, Transportation, Shipping

Work with a partner to answer the following questions.

Polynomials are expressions with several terms that follow patterns, such as $4x^3 + 3x^2 + 15x + 2$. Now consider the number 3946. As you know, the digits indicate 3 thousands, 9 hundreds, 4 tens, and 6 ones. Remember that one hundred is 10^2 and one thousand is 10^3. Can you see a connection between polynomials and our place value number system?

1. The number 3946 can be expressed as $3(10)^3 + 9(10)^2 + 4(10) + 6$. As you can see, this expression is similar to the polynomial pattern—the only difference is that a 10 is used instead of an x. Using this idea, write 62, then 832, and then 14,791 in polynomial form.

2. Write 1001 so that it looks like a polynomial. Omit the terms that are multiplied by zero.

3. Is it correct to say that $493 = 4(10)^2 + 9(10)^1 + 3(10)^0$?

4. Find the value of the polynomial $9x^3 + 7x^2 + 5x + 3$ if $x = 10$.

◣ BUILD UNDERSTANDING

Review the words used to discuss polynomials. A simple expression with only one term is called a **monomial**. A monomial is either a number or the product of a number and one or more variables. For example, $4x^3$ is a monomial. Other monomials are 15, m, ab, and $13p^2q$. If a monomial includes variables, the number part is called the **coefficient** of the term, and is written first. A number by itself is called a **constant**.

A polynomial is an expression that contains several monomial terms that are added or subtracted. If it has two terms, it is a **binomial**. With three terms, it is a **trinomial**. The expression $a^4 + 3b$ is a binomial; $6h^3 + 4gh + 39$ is a trinomial. Polynomials may have more than three terms. For example, $8s^4 - 5s^3t + s^2t^2 + 6st^3 - 7t^4$ is a polynomial with 5 terms.

Like terms are terms in which the variables or sets of variables are identical—though the coefficients may be different. Learn to recognize like terms, and do not be confused by unlike terms.

like terms:	$3b^2$	$15b^2$	$(-b^2)$
	$8x^3y$	$-14x^3y$	$25x^3y$

unlike terms:	$15a$	$15b$	
	$15b^2$	$12b$	
	$8x^3y$	$8xy^3$	$8x^3y^3$

You **simplify** a polynomial when you group and then combine all like terms.

$$4a^2 + 3bc - a^2 + 5c^2 + 9bc = (4a^2 - a^2) + (3bc + 9bc) + 5c^2$$
$$= 3a^2 + 12bc + 5c^2$$

A polynomial is in **standard form** if the terms are ordered from the greatest power of one of its variables to the least power of that variable.

$$15x + 13 - 9x^2 + 2x^3 = 2x^3 - 9x^2 + 15x + 13$$

To add polynomial expressions, place both expressions in parentheses with an addition sign between them, then simplify the combined expression and put it in standard form.

Problem Solving Tip

Rewriting polynomials in simplified and standard form will help you match the terms for adding and subtracting.

Example 1

Simplify $(8a^2b + 6ab^2) + (4a^2b - 3ab^2)$.

Solution

$$(8a^2b + 6ab^2) + (4a^2b - 3ab^2) = 8a^2b + 6ab^2 + 4a^2b - 3ab^2$$
$$= (8a^2b + 4a^2b) + (6ab^2 - 3ab^2)$$
$$= 12a^2b + 3ab^2$$

Another way to add polynomials is to set up the problem in vertical form with like terms aligned in columns.

Example 2

PACKAGING The cost of the materials for the inner packaging of a new product is determined by the expression $10x^2 + 8xy + y^2$. The cost of the outer packaging materials is $4x^2 - 3xy + 2$. Find the total cost of the packaging.

Solution

$$\begin{array}{l} 10x^2 + 8xy + y^2 \\ \underline{4x^2 - 3xy \qquad + 2} \\ 14x^2 + 5xy + y^2 + 2 \end{array}$$

To subtract polynomials, place the expressions in parentheses with a minus sign between them, then simplify and write the answer in standard form.

Think Back

Remember that to subtract an expression, you change all the signs and then add.

Example 3

Subtract:

a. $5m^2 - 2m$ from $8m^2 + m$

b. $s^2 + 3s - 4$ from $3s^2 - 5s - 3$

Solution

a. $(8m^2 + m) - (5m^2 - 2m) = 8m^2 + m + (-5m^2) + (2m)$
$$= 8m^2 + (-5m^2) + m + (2m)$$
$$= 3m^2 + 3m$$

b. $(3s^2 - 5s - 3) - (s^2 + 3s - 4)$

$\quad = 3s^2 - 5s - 3 + (-s^2) + (-3s) + 4$

$\quad = 3s^2 + (-s^2) + (-5s) + (-3s) + (-3) + 4$

$\quad = 2s^2 - 8s + 1$

You may also set up a subtraction problem vertically after you change the sign of each term.

$$\begin{array}{r} 3s^2 - 5s - 3 \\ -\ s^2 - 3s + 4 \\ \hline 2s^2 - 8s + 1 \end{array}$$ Change the signs of each term.

TRY THESE EXERCISES

Write each answer as a simplified polynomial in standard form for the variable x.

1. $x + 3x^3 - 4 + x^2 + 2x^3$

2. $4 + x^2 + 3 - 2x^2 + 4x^2$

3. Add $x^2 + 3$ to $3x^2 + 7$.

4. Add $7 - 2x^2$ to $5x^2 - 3$.

5. $(5x^2 - 7x) + (x^2 + 3x)$

6. $(4x^3 + 7) + (3x^3 - 4)$

7. Subtract $3x + 4$ from $5x - 3$.

8. Subtract $x - 4$ from $5 + 3x$.

9. $(2x + 14) - (x - 7)$

10. $(5x^2 - 5x) - (5x^2 - 5x)$

11. $(15x^3 + 12x^2 - 3xy) - (8x^3 - 3x^2 + 2xy)$

12. $(x + 6x^2y - 3x - 4x^3) + (x^2y + x^2 + 5x)$

13. Add $6x^3 + 2x^2 - 5x + 4$ to $2x^3 + 7x^2 + 2x - 1$, and then subtract $4x^2 - 3 + 9 - x^3$ from your answer.

14. WRITING MATH Explain how subtraction of polynomials is related to addition of polynomials.

PRACTICE EXERCISES • For Extra Practice, see page 696.

Simplify.

15. $(2a + 4) + (3a + 9)$

16. $(5p + q) + (2p + 2q)$

17. $(3x^2 + 2x) + (-x^2 + 5x)$

18. $(4h - 2g + k) + (h + 3j - 2k)$

19. $(5t + 7) - (3t + 2)$

20. $(3r + 2s) - (2r - s)$

21. $(4m^2 + 3n) - (-m^2 + 3n)$

22. $(y^2 - 2y + 3) - (-y^2 + y - 5)$

23. $(y^2 - 15x + 2x^2) - (7x - 2y^2 + x^2)$

24. $(12r^2 - 12rs + s^2) + (3r^2 - 4s)$

25. $(4v^2 - 9w^2) - (v^2 + 2vw + w^2)$

26. $(x^4 + 3x^2 + 2x) + (8x^3 + 4x)$

27. $(2b^2 - 15 + c) - (-c - 4b^2)$

28. $(-3f^2 + 4fg + g^2) - (4f^2 - g^2)$

29. INCOME Last week, Pedro worked 17 h at the pharmacy, where he earns p dollars an hour, and 12 h in the supermarket, where he earns s dollars an hour. This week, he worked 8 h at the pharmacy and 20 h in the supermarket. What were his earnings during the two weeks, expressed in terms of p and s?

30. TRANSPORTATION Airplane A uses $35d^2 + 3dr - 4r^2$ gal of fuel to make a trip. Airplane B uses $16d^2 + 45dr - 13r^2$ gal. How much less fuel does airplane B use than airplane A?

31. $(z^2 \quad 3z + 4) + (3z^2 + 2z - 2) - (4z^2 - z - 2)$

32. $-(122 + 7x + 2x^2) + (32 - 14x + 15x^2)$

33. $(4.2a^3 - 3.6b^3 + 8.8bc^2) - (4.2a^2b - 2.1a^3 + 3bc^2 - 1.9b^3)$

34. $[5(10)^3 + 6(10)^2 + 3(10)^0] - [2(10)^3 + 8(10)^2 + 4(10)^1 + 10^0]$

35. $2(8x^2 - 5x + 3) - (10x^2 - 16x + 2) + (13x^2 - 4)$

■ EXTENDED PRACTICE EXERCISES

36. ART The prism sculpture shown at the right is being shipped to a museum. The artist plans to build a wooden frame to protect the edges of the sculpture during shipping. The measure of the side of each triangle is equal to $(x^2 + y)$ ft and each long edge is $(2x^2 - 3y)$ ft. How many feet of wood will the artist need to protect the edges?

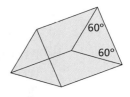

37. SHIPPING Janine's truck starts the day with a cargo of 54 large cubic boxes with each side measuring x feet. Each box contains z packages measuring 1 foot by x feet by y feet. In addition, 48 more of these packages are packed into the corners so the truck is full. At her first delivery, she drops off 12 large boxes—but she removes 3 packages from one of the boxes to keep on the truck for another customer. How much space is available on the truck after her first delivery, in terms of x and y?

38. The octal system of counting contains only eight digits. The number written 342, therefore, means only $3(8)^2 + 4(8) + 2$, not $3(10)^2 + 4(10) + 2$. Calculate $765 - 301$ in octal numbers, then convert answer to our own decimal system.

39. CHAPTER INVESTIGATION Suppose you have developed a new product targeted for consumers your own age. What do you know about the spending habits of people in your age group? Begin development of a survey to gather demographic information about your classmates. Working with your group, brainstorm a list of questions that can be used in a survey to find out information about your classmates' shopping interests and spending habits.

■ MIXED REVIEW EXERCISES

Find each square root to the nearest hundredth. (Lesson 10-1)

40. $\sqrt{52}$ **41.** $\sqrt{75}$ **42.** $\sqrt{83}$ **43.** $\sqrt{216}$

Write each expression in simplest radical form. (Lesson 10-1)

44. $\left(4\sqrt{2}\right)^2$ **45.** $\left(5\sqrt{7}\right)^2$ **46.** $\dfrac{\sqrt{50}}{\sqrt{3}}$ **47.** $\dfrac{\sqrt{96}}{\sqrt{5}}$

48. $\left(2\sqrt{3}\right)\left(3\sqrt{2}\right)$ **49.** $\left(4\sqrt{6}\right)\left(3\sqrt{3}\right)$ **50.** $\left(5\sqrt{2}\right)\left(2\sqrt{6}\right)$ **51.** $\left(4\sqrt{8}\right)\left(\sqrt{7}\right)$

Write each number in scientific notation. (Lesson 1-8)

52. 0.0000000743   **53.** 32,000,000,000   **54.** 0.000000904

Multiply by a Monomial

Goals   ■ Multiply polynomials by monomials.

Applications   Advertising, Landscaping, Payroll

Work with a partner to answer the following questions.

From your knowledge of geometry, you know that the area of a rectangle is calculated by multiplying width by length. Use the diagram shown at the right.

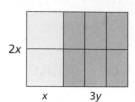

a. Express the area of the yellow section of the diagram, in terms of x. There is more than one possible answer.

b. Express the area of the orange section of the diagram, in terms of x and y.

c. Express the area of the whole diagram, in terms of x and y.

d. Trace the diagram and cut out the pieces. Use the pieces to form a different rectangle with the same area. Write expressions to represent the length and width of the new rectangle. How could you use the expressions to find the area?

◥ BUILD UNDERSTANDING

When you multiply a polynomial by a monomial, the answer always has the same number of terms as the original polynomial. To understand this, begin with the idea that a monomial is a product of constants and variables. If you multiply two monomial products, you will always get another product that is a monomial.

This is clear in (**a**) above: $(2x)(x) = (2)(x)(x) = 2x^2$. (Remember the associative property of multiplication.) It may be less clear in (**b**): $(2x)(3y)$, because of the two coefficients in the initial expression. But by the commutative property, the expression equals $(2)(3)(x)(y)$, or $6xy$. You can also see this in the diagram above.

Example 1

Simplify.

a. $(8a)(3b)$ **b.** $(3m)(-2n)$ **c.** $(-2x)(-5x^2)$

Solution

a. $(8a)(3b) = (8)(a)(3)(b) = (8)(3)(a)(b) = 24ab$

b. $(3m)(-2n) = (3)(m)(-2)(n) = (3)(-2)(m)(n) = -6mn$

c. $(-2x)(-5x^2) = (-2)(x)(-5)(x)(x) = (-2)(-5)(x)(x)(x) = 10x^3$

When you multiply a binomial by a monomial, the answer will be a binomial. This is because each term of the binomial must be multiplied by the monomial.

> **Problem Solving Tip**
>
> Remember that when you multiply two terms with negative coefficients together, the answer will be positive.

Example 2

TELEVISION To promote a new product, a company buys $2x$ minutes of airtime. The cost of one minute of airtime is $3x$ 4. Multiply to find an expression which represents the cost of advertising the new product on television.

Solution

$2x(3x - 4) = (2x)(3x) + (2x)(-4) = 6x^2 + (-8x) = 6x^2 - 8x$

When you multiply polynomials (including trinomials) by a monomial, the answer will have the same number of terms as the other polynomial.

Example 3

Simplify.

a. $3v^2(v^2 + v + 1)$

b. $12(a^2 + 3ab^2 - 3b^3 - 10)$

Solution

a. $3v^2(v^2 + v + 1) = (3v^2)(v^2) + (3v^2)(v) + (3v^2)(1)$
$= 3v^4 + 3v^3 + 3v^2$

b. $12(a^2 + 3ab^2 - 3b^3 - 10) = 12(a^2) + 12(3ab^2) + 12(-3b^3) + 12(-10)$
$= 12a^2 + 36ab^2 - 36b^3 - 120$

When you multiply $2x$ and $3y$, you first analyze each monomial into its simplest, **prime**, elements. Prime elements, including prime numbers, cannot be divided into smaller whole elements. To multiply $(2x)(3y)$, you thought $(2)(x)(3)(y)$, which was easily reorganized as $(2)(3)(x)(y)$, and then $6xy$. This type of analysis can also help you find **factors**, elements whose product is a given quantity.

Example 4

GEOMETRY List three possible dimensions for a rectangle with an area of $12x^2y$.

Solution

As you know, the area of a rectangle is the product of its length and width. To find a complete set of paired factors for the given area, start by analyzing its prime elements. Express the coefficient in prime numbers and separate the variables. The area $12x^2y$ is analyzed as $(2)(2)(3)(x)(x)(y)$.

Now use the analysis to find different factor pairs or sets of sides. Set up a table. The second factor contains all the elements not in the first factor.

First factor (length)	Second factor (width)
$(y) = y$	$(2)(2)(3)(x)(x) = 12x^2$
$(2)(2)(x) = 4x$	$(3)(x)(y) = 3xy$
$(2)(3)(x)(x) = 6x^2$	$(2)(y) = 2y$

There are many other possible sets of factors.

Simplify.

1. $(x)(3y)$

2. $(a^2)(2a)$

3. $(4p)(3q)$

4. $(3v^2)(2vw)$

5. $(-r)(-s^2)$

6. $(-5xy)^2$

7. $7(x^2 + x)$

8. $2y(y + z)$

9. $a^2(a^2 + a)$

10. $4pq(p - 2r)$

11. $-e^2f(e + f^2)$

12. $-13mn^3(2m^2 - n)$

13. $a(b^2 + b - 6)$

14. $3u(u^2 + uv + 2v^2)$

15. $-7x(x^2 - 2xy + y^2)$

16. $5ef^3(h + 3j + k^2)$

17. **MARKETING** A mailing list has x people from 14 to 18 years of age, y people from 19 to 25 years of age and z people from 26 to 40 years of age. A company decides to spend x dollars per person on the list to advertise its new product line. How much will the advertising cost the company?

◤ PRACTICE EXERCISES • For Extra Practice, see page 696.

Simplify.

18. $(2a)(3b)$

19. $(x^2)(3xy)$

20. $(-j)(-3jk)$

21. $(4x^3)(-3x^2y)$

22. $(6m^2n)(5mn^2)$

23. $(3a^2)^2$

24. $-7q(3q^2 - 5r)$

25. $2x^2[-(3x^2 + 2x)]$

26. $5rs(3r^4 + 5s^3)$

27. $-3mn^2(m^3n - m^4n^3)$

28. $2x^2y(4x^3z - 3xz^4)$

29. $8ef^2g(eg^3 - fg^3)$

30. $4abc(a^2b^3c + ab^4c^2)$

31. $-18lmn^4(l^2mn^3 - lm^5n)$

32. $3x(x^2 + 4x - 5)$

33. $ab(4e^2 - 2f + g)$

34. $-pq^2(3p^2 - pq + 10q^2)$

35. $4v^2w(3u + 2v + w^3)$

36. $-l^4[-(3l + 5m)]$

37. $7rs^3t^2(r^4st^3 - r^3s^2t^2 - r^2s^5t)$

Write and simplify an expression for the area of each rectangle.

38.

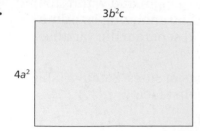

$3b^2c$

$4a^2$

39.

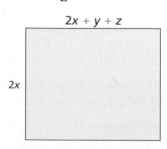

$2x + y + z$

$2x$

40. **PAYROLL** In 1990, a growing company employed c clerks, each of whom earned d dollars each week. The weekly pay rate increased by r dollars each year. Two years later, the number of clerks on staff had tripled. What was the total paid each week to the clerical staff in 1990? What was it in 1992? Simplify both answers if possible.

41. **CONSTRUCTION** A builder estimates that, for a typical office building, the height of each story is h ft from floor to floor, and the length of a building averages k ft per room. A company wants a structure that is 5 stories tall and has 12 rooms along the front; but each room is to be 3 ft longer than the standard. Estimate the area of the front wall of the building.

42. **WRITING MATH** How is algebraic multiplication of a monomial and a polynomial similar to arithmetic multiplication of a single-digit number and a multi-digit number?

43. Find the prime elements of $6ab^2$ and use them to list all factor pairs. (*Hint:* There are 11 pairs in all.)

44. **ERROR ALERT** A classmate says that $(-x^3)^2$ is equal to $(-x^5)$. Analyze the problem by writing the expression as the product of prime elements. What mistake has your classmate made?

Simplify.

45. $(x^2y)(xy^3)(xy^2)$

46. $(m^2n^4)(m^4n^2) - (m^3n^3)^2$

47. $(-a^3)^2 - (-a^2)^3$

48. $2pq(p + q) - p^2q(2 + q)$

49. $(5x^2)(3y)(x^2 - xy + y^2)$

50. $3r(2r - 5s + t) + 6s(3r - s + 2t)$

51. **TRANSPORTATION** Alva travelled for t hours at s miles per hour, then for twice that time at $(s + 10)$ miles per hour. How many miles did she travel in all? (Remember, distance = rate × time.)

52. **LANDSCAPING** A lawn has two flower gardens with the dimensions shown below. Write an expression for the area of grass left, then simplify.

EXTENDED PRACTICE EXERCISES

53. **ARCHAEOLOGY** An archaeologist finds a square-based pyramid rising in the Mexican jungle. From corner to corner, it is $60p$ (paces), and from each corner to the top is $50p$. What is the total surface area of its triangular sides, expressed in terms of p?

54. Using the diagram on the right, find factored expressions for three areas: the shaded area, the unshaded area, and the total area. Then simplify each expression.

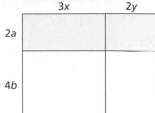

MIXED REVIEW EXERCISES

55. Three brothers, named Jarius, Keshawn, and Levon play football for the Cheetahs, the Gophers, and the Goats, not necessarily in that order. Jarius scored 2 touchdowns against the Cheetahs, but none against the Goats. Keshawn hasn't played against the Cheetahs yet. For which team does each brother play? (Lesson 3-8)

Review and Practice Your Skills

PRACTICE ▰ LESSON 11-1

Simplify.

1. $(8x + 3y) + (7y - 2x)$
2. $(13b + 6) + (7b - 14)$
3. $(4x^2 - 9x + 6) + (12x^2 + 5x - 13)$
4. $\left(\frac{1}{2}k + \frac{3}{4}g\right) + \left(\frac{3}{8}k - \frac{3}{4}h\right)$
5. $(4x - 6z) - (6x - 4z)$
6. $(-3m + 4n - p) - (6n - 7m + p)$
7. $(8x^3 - 5x^2 + 2x) - (6x^2 - 3x^3 + 10x)$
8. $(4a^2 + ab + 7b^2) - (8ab + 5b^2)$
9. $[y^2 - (-5y)] - (3y^2 + 6y + 1)$
10. $(14r^2 - 10rs + 15s^2) + (-8r^2 + 7s^2)$
11. $(x^2y + xy^2) + (3x^2y - 2xy - 4xy^2)$
12. $(m^2 - 15n + 4n^2) - (8n - 3m^2 + 2n^2)$
13. $(3x - 2y) - (4x - 3y) + (7y - 6x)$
14. $-9x + (11t - 2) + (5x - 4t) - 6$
15. $(20c^2 + 17cd) - (14d^2 + 3c^2) + 8d^2$
16. $(3d^2 + 8d - 1) - (-3d^2 + 8d - 1) + (5 - 5d^2)$

17. Notebooks cost n cents and pens cost p cents. Julia bought 5 notebooks and 6 pens. Her brother Tim bought 7 notebooks and 3 pens. How much did their mother pay for these purchases, expressed in terms of n and p?

18. A triangle has sides of $(x - 3y)$, $(6y - 5x)$, and $(4x + 2y)$. Write and simplify an expression for the perimeter of this triangle.

PRACTICE ▰ LESSON 11-2

Simplify.

19. $(3x)(-2x)$
20. $(8df)(2d^2)$
21. $(-6m)(7mn)$
22. $(5xy^2)(x^2y)$
23. $(-k)(-9k^5)$
24. $(8pqr)(3pr)$
25. $(7s^3t^2)(4s^2t)$
26. $(3x^2)^2$
27. $3x(4x - 10)$
28. $-2n(6n^2 - 5n)$
29. $11x^2(3x^2 + 2x - 1)$
30. $3c^2d(6d^2 - cd)$
31. $-pq(p^2q - 3pr + 7pq^3)$
32. $-2abc^2(a^2b^3c - a^2bc^2)$
33. $5x(3a + 2b - 4c)$
34. $7k^2[-(5 - 4k + 6k^2)]$
35. $x(3x + 4) + 2(x^2 - 5x + 8)$
36. $8(p^2 - 4pq + 5q^2) - 2(4p^2 + 20q^2)$
37. $-4pq(p^3q + 5pr - 3pq^2)$
38. $-6a^3bc^2(2a^3bc^2 - a^2bc)$
39. $2yz(4a + 3b - 10c)$
40. $8k^2[-(5k^3 - 13 + 9k^2)]$
41. $3x^{10}y^8z(x^5yz^9 + 2xy^2z^8 - xyz^{12})$
42. $-3(x + 2) - 3(2 - x) + 3(x - 2) - 3[x - (-2)]$

Write and simplify an expression for the area of each rectangle.

43.
$3x - 7$

$4x$

44.
$8p - 2q^2$

$6p^2q$

45.
$3x$

$x^2 - 6x + 7$

Simplify. (Lesson 11-1)

46. $(-5x + 2y) + (9y - 2x)$

47. $(15b - 6) + (-4b + 17)$

48. $(9x^2 + 4x - 6) + (13x^2 - 6x + 10)$

49. $\left(\dfrac{1}{2}h - \dfrac{3}{4}g\right) + \left(\dfrac{3}{8}g + \dfrac{3}{4}h\right)$

50. $(2x - 6z) - (4x + 6z)$

51. $(3m - 3n + 11p) - (-5n + 8m - p)$

52. $(5x^3 - 8x^2 - x) - (6x + 3x^2 - 8x^3)$

53. $(-4a^2 + 8ab + 12b^2) - (8ab - 12b^2)$

54. $[5y^2 - (-2y)] - (5y^2 + 6y - 21)$

55. $(6r^2 + 10rs - 13s^2) + (-8s^2 + 7r^2)$

56. $(x^2y + xy^2 - 2xy) + (4x^2y - 2xy - 3xy^2)$

57. $(-m^2 + 15n - 2n^2) - (-8n + 3m^2 - 2n^2)$

58. $(2x + 3y) - (3x - 2y) + (x + y)$

59. $(5x^4 - y^4) + (6x^3 + 2y^4) - (-7x^4 + 8x^3)$

Simplify. (Lesson 11-2)

60. $(k^2)(-3k^3)$

61. $(-8p^3qr)(2pr^2)$

62. $(-2x^3)^2$

63. $-2x(3x - 14)$

64. $2n^2(5n^2 - 4n)$

65. $-9c^2d(-4d^2 + 3cd)$

MathWorks
Workplace Knowhow

Career – Brokerage Clerks

Brokerage clerks work for financial institutions such as brokerages, insurance companies and banks. They perform many different tasks. Purchase and sale clerks make sure that orders to buy and sell are recorded accurately and balance. Dividend clerks pay dividends to customers from their investments. Margin clerks monitor the activity on clients' accounts, making sure clients make payments and abide by the laws covering stock purchases. Brokerage clerks often use computers to monitor all aspects of securities exchange. They use specialized software to enter transactions and check records for accuracy.

1. A client bought 60 shares of stock at x price per share and later sold 40 shares of the stock at y price. Write an expression that could be used to find the value of the client's stock after the sale.

2. A client wants to triple the number of gold certificates he owns. He has x certificates now, each worth y dollars today. Tomorrow the price of the certificates is expected to increase by z dollars. Write an expression to find the expected cost the client will pay tomorrow to triple his holdings.

3. A client wants to buy $(x + 3)$ shares of stock for $(x + 8)$ dollars. Write an expression for the total cost of the order.

4. A client bought $(x - 5)$ shares of stock A at a cost of $(x + 4)$ dollars. She also purchased $(x + 8)$ shares of stock B at a cost of $(x + 6)$ dollars. Write an expression to represent her total holdings of stocks A and B.

11-3 Divide and Find Factors

Goals ■ Factor polynomials into a monomial factor and a polynomial factor.

Applications Manufacturing, Sculpture, Landscaping

MODELING Did you realize that all monomials have factors? In fact, unless a monomial is a constant and also a prime number, it has more than one set of paired factors. What about polynomials? Can a binomial have a pair of factors? The answer is yes. The expression $4x + 2$ is equal to $1(4x + 2)$, because anything times 1 is equal to itself. Shown with Algeblocks or algebra tiles, the expression would look like this.

x	x	x	x	1	1

Are there any other paired factors of $4x + 2$? Use algebra tiles to see if you can multiply an expression by 2 and create the same area (it will be a different shape).

Now, use Algeblocks to arrange $4x^2 + 2$ into a rectangle with one side (factor) equal to 2.

◣ BUILD UNDERSTANDING

Using Algeblocks is not the only way to find the factors of a binomial or polynomial. Another technique, called **extracting factors**, begins by determining if a polynomial has a monomial factor other than 1. Check to see if any monomial will divide evenly into every term of the polynomial. If so, you can extract the monomial factor by dividing the polynomial by that monomial factor. The quotient from that division is the second factor of the original polynomial.

Example 1

Find factors of $4x + 2$.

Solution

2 will divide $4x$ evenly, and it will also divide 2 evenly. Therefore, 2 is a factor of the binomial. What is the other factor? You can find it by dividing each term of the binomial by 2.

$$\frac{4x + 2}{2} = \frac{(2)(2)(x)}{2} + \frac{2}{2}$$
$$= 2x + 1$$

The factors are the 2 that you extracted, and $(2x + 1)$, the quotient. So, $4x + 2 = 2(2x + 1)$.

As you may realize, a polynomial may have more than one monomial factor.

Example 2

Find the factors of $2x + 6x^2$.

Five-step Plan

1 Read
2 Plan
3 Solve
4 Answer
5 Check

Solution

You can see that 2 is a factor of both terms. You can also see that x is a factor of both terms. In addition, therefore, $(2)(x)$ or $2x$ is also a factor. In fact, $2x$ is the **greatest common factor**, or **GCF**, because it includes all the common factors. The paired factor is again found as follows.

$$\frac{2x + 6x^2}{2x} = \frac{(2)(x)}{(2)(x)} + \frac{(2)(3)(x)(x)}{(2)(x)}$$
$$= 1 + 3x$$

So, $2x + 6x^2 = 2x(1 + 3x)$.

Finding the monomial that is the GCF is very valuable for factoring a binomial.

Example 3

Find the greatest common factor of $15xy^3$ and $3x^2y^2$. Then write $15xy^3 - 3x^2y^2$ in factored form.

Solution

$$15xy^3 = (3)(5)(x) \quad (y)(y)(y)$$
$$3x^2y^2 = (3) \quad (x)(x)(y)(y)$$
$$ (3) \quad (x) \quad (y)(y) \quad \rightarrow \quad 3xy^2 \quad \text{Greatest Common Factor}$$

$$\frac{15xy^3 - 3x^2y^2}{3xy^2} = \frac{(3)(5)(x)(y)(y)(y)}{(3)(x)(y)(y)} - \frac{(3)(x)(x)(y)(y)}{(3)(x)(y)(y)}$$
$$= 5y - x$$

This technique lets you find the GCF by writing each monomial as a product of its prime elements. Thus,

$$15xy^3 - 3x^2y^2 = 3xy^2(5y - x).$$

Prime elements can help with division of monomials. Write the dividend and the divisor by using prime elements, then cancel each element they share.

Example 4

MANUFACTURING A company manufactures posters with inspirational sayings. Each poster has an area of $8mn^2$ in.^{2}. The length of each poster is $2mn$ in. Find the width.

Solution

$$8mn^2 \div 2mn = \frac{(2)(2)(2)(m)(n)(n)}{(2)(m)(n)} = (2)(2)(n) = 4n \text{ in.}$$

The width of the poster is $4n$ in.

Extract a monomial factor and find its paired binomial factor for the following.

1. $6x^2 + 9$ **2.** $2a + ab$ **3.** $5mn - n^2p$

Extract the GCF and indicate its paired binomial factor.

4. $16p - 20q$ **5.** $12x^2 + 18x$ **6.** $45a^2b - 27ab^2$

Extract a monomial factor and find the paired trinomial factor.

7. $7r^2 + 3rs + 2rt$ **8.** $h^2jk + jk^2l - 3klm$

9. SCULPTURE A sculptor has 2 columns of marble. One is 54 in. tall, the other is 90 in. tall. He wants to carve a set of identical figurines. He must use the full length of both columns and divide them into equal pieces. What is the maximum height of each figurine and how many will he make?

10. GEOMETRY A rectangle of area $9v^2w$ has a width of $3v$. What is its length?

Factor.

11. $6a + 8b$ **12.** $21x^2 - 35y^2$ **13.** $15p^3 - 35q$

14. $13e - 5ef$ **15.** $vw + wx$ **16.** $8gh - 3hj$

17. $5x^2y - 2y^2$ **18.** $18r^2s + 19st^2$ **19.** $13mn^2 + 25n$

Simplify.

20. $12x^3y^2 \div 6x^2y$ **21.** $45ef^2 \div 18ef$

22. Find the greatest common factor of $24u^3v^2$, $6u^2v^3$, and $8uv^4$.

Find the GCF and its paired factor for the following.

23. $14ab^2 + 35bc$ **24.** $45m^2n - 72mn$

25. $18r^3 + 27r^2$ **26.** $50u^4v^2 - 100u^3v^3$

27. $39j^7k^3l^4 - 65j^6k^5l^3 + 52j^5k^2l^6$ **28.** $6a^5b - 12a^4b^2 - 9a^3b^3$

29. $ax^3y^3 + bx^2y^2 + cxy$ **30.** $18r^5 + 45r^4s^2 - 63r^2s^4$

31. WRITING MATH The area of a trapezoid is $A = \frac{1}{2}th + \frac{1}{2}bh$, where t and b are the lengths of the bases, and h is the height. Factor this formula. Then find the area of a trapezoid with a top base of 6 in., a lower base of 5 in., and a height of 4 in. using the given formula and the factored formula. Which was easier to use? Explain.

32. LANDSCAPING Nguyen is calculating the price of a landscaping contract using her company's formula: $P = 4r^2 + 8rs - 4rt$. For this job, $r = 2.5$, $s = 5.4$, and $t = 3.3$. Hoping to avoid a lot of multiplication, Nguyen factors the formula, and finds the math is very simple. What is her factored version of the formula, and what price does she set for the contract?

Find the monomial and polynomial factors. Simplify first if necessary.

33. $6x^5y^2 + 8x^4y^3 + 6x^3y^4 + 14x^2y^5 + 2xy^6$ **34.** $3x(y^2 + 2z) - y(3xy - 6xz^2)$

Write, simplify, and factor an expression for each perimeter below.

35.

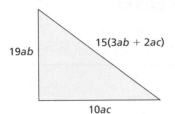

15(3ab + 2ac)

19ab

10ac

36.

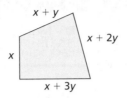

x + y

x + 2y

x

x + 3y

37.

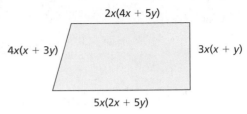

2x(4x + 5y)

4x(x + 3y)

3x(x + y)

5x(2x + 5y)

▰ EXTENDED PRACTICE EXERCISES

38. A snail usually travels $3a$ in. every hour. However, when it is climbing out of a slippery well, it also slides back $2b$ in. each hour. The distance it has climbed after x hours is found to be $3ax - 2bx$ in. Prove that this is exactly what you would expect by factoring this distance.

39. Factor $3x^n - 2x^{(n-1)}$.

40. The sum of a series of n positive even numbers starting with 2 is given by the formula $S = n^2 + n$. Test the formula on $(2 + 4)$, $(2 + 4 + 6)$, and $(2 + 4 + 6 + 8)$. Next, use the formula to calculate the sum of the first 14 even numbers. Then factor the formula, and use the factored version to sum the first 17 even numbers.
Note that the factored version saves a step.

41. CHAPTER INVESTIGATION Continue to work on your marketing survey. What advertising methods are most effective for your age group? Add questions to your survey to find out how much time each day your classmates spend in watching television, listening to the radio, reading newspapers and magazines and traveling by car or bus. Include questions to find out which television programs, radio stations and magazines are most popular.

▰ MIXED REVIEW EXERCISES

Find the unknown side lengths. First find each in simplest radical form, and then find each to the nearest hundredth. (Lesson 10-3)

42.

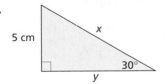

5 cm

x

30°

y

43.

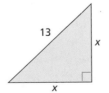

13

x

x

44.

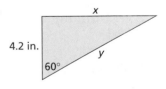

x

4.2 in.

y

60°

45. DATA FILE Use the data on money around the world on page 648. What is the value in United States dollars of 100 Indian rupees? (Lesson 7-1)

46. DATA FILE Mrs. Sanders is shopping for a coat. The original price of the coat at one store is $199. It is on sale for 25% off. A second store has a similar coat on sale for 40% off. The original price of this coat was $249. Use the data on page 649 on state sales tax to calculate the actual cost of each coat. Which is the better buy? (Prerequisite Skill)

11-4 Multiply Two Binomials

Goals ■ Multiply binomials.

Applications Packaging, Small Business, Product Development

Work with a partner to answer the following questions.

You have seen how a binomial can be multiplied and divided by a monomial. Binomials can also be multiplied (and divided) by other binomials. Look at the following diagram.

	a	b	
x	Area: ax	bx	$x(a + b) = ax + bx$
y	ay	by	$y(a + b) = ay + by$

As you can see, the whole diagram represents $(x + y)(a + b)$.

1. Express the large area as a polynomial by adding the areas of all four smaller rectangles.

2. Draw a diagram to show the expression

$(2p + 4q)(l + m)$.

3. Express your diagram as a polynomial by adding its parts.

◥ BUILD UNDERSTANDING

Multiplying a binomial by another binomial starts with the idea that a binomial is the sum of two monomials. To multiply two binomials, use the distributive property twice. Multiply the second binomial separately by each term in the first binomial. Then add the answers together. This is also called **expanding** the two binomials.

E x a m p l e   1

Find the product $(x + a)(2x + 3b)$.

Solution

$(x + a)(2x + 3b) = x(2x + 3b) + a(2x + 3b)$
$= 2x^2 + 3bx + 2ax + 3ab$

No further simplification is possible.

Sometimes simplification leads to a different-looking polynomial.

Example 2

PACKAGING The rectangular cover art for a new product has a length of $(x + 1)$ and a width of $(x + 5)$. Find the area of the cover art. Expand and simplify $(x + 1)(x + 5)$.

Solution

$$(x + 1)(x + 5) = x(x + 5) + 1(x + 5)$$
$$= x^2 + 5x + x + 5$$
$$= x^2 + (5 + 1)x + 5$$
$$= x^2 + 6x + 5$$

The area of the cover art is $x^2 + 6x + 5$.

Now that you have seen two examples, look for a pattern. The final solutions may seem quite different, but study the second line of each answer. In each case, the first term is the product of the binomials' first terms. Describe it as the **First product**. The second term is the product of the outer pair of terms in the binomials. It can be called the **Outer product**. The third term is the product of the inner terms—the **Inner product**. And the final term is the **Last product**, the product of the last terms of the two binomials. The whole multiplication process is often called the **FOIL** process—for First, Outer, Inner, and Last.

Notice that in Example 2 the inner and outer products can be simplified into a single term.

Reading About Math

The outer and inner products are also known as the **cross products**. If the binomials are placed one above the other, you can see why.

$(x + 1)$

$(x + 5)$

In each case, the first term of one binomial is multiplied by the last term of the other, making a cross.

Example 3

Expand and simplify $(y - 5)(y + 5)$.

Solution

$$\begin{array}{cccc} \text{F} & \text{O} & \text{I} & \text{L} \\ \downarrow & \downarrow & \downarrow & \downarrow \end{array}$$
$$(y - 5)(y + 5) = y^2 + 5y - 5y - 25$$
$$= y^2 - 25$$

This multiplication produces a polynomial pattern called the **difference of two squares**. The product of two binomials that differ only in their signs is always the square of the first binomial term minus the square of the second. The outer and inner products (the O and I terms) add to zero. In other words, $(a + b)(a - b) = a^2 - b^2$. This is true for any value of a and b.

TRY THESE EXERCISES

Multiply. Simplify if possible.

1. $(3a + 2b)(c + 5d)$

2. $(e - 6f)(2g - 3h)$

3. $(l - m)(l + n)$

4. $(3r + s)(2r - 3t)$

5. $(2x + 5)(3x + 3)$

6. $(y - 6)(y - 6)$

7. $(8x - y)(x + 2y)$

8. $(3u - 10v)(2u + v)$

9. $(p - q)(p + q)$

10. $(2x + 3y)(2x - 3y)$

11. **SMALL BUSINESS** As a summer project, Andre is making hand-painted ceramic plates. The material costs $10 for each plate, and 12 plates can be made comfortably each day. But if the work rate goes up, he uses up more materials because of mistakes. So the cost per item increases by $1 for each plate he makes over 12. To plan his work, he needs a formula. The cost of making 12 plates each day is $(12)(10)$. What is his daily cost when making $(12 + x)$ plates? Expand and simplify your answer.

12. **WRITING MATH** Can the product of two binomials ever have more than three terms? Explain your thinking.

PRACTICE EXERCISES • For Extra Practice, see page 697.

Simplify.

13. $(2p + 5q)(3r - 1)$

14. $(7k - l)(3m - n)$

15. $(4a + b)(a + 3c)$

16. $(8x - 3y)(3x - 8z)$

17. $(e - 3f)(2g + 5f)$

18. $(6w + 7x)(y - z)$

19. $(9p + 2q)(5p - 3r)$

20. $(7a - c)(3b + c)$

21. $(5m + 6n)(m + 9n)$

22. $(5 - 6n)(1 - 9n)$

23. $(3x - 4)(x - 2)$

24. $(3x + 4y)(x + 2y)$

25. $(j - 5k)(7j - 2k)$

26. $(8a + 1)(3a - 5)$

27. $(8b - c)(3b + 5c)$

28. $(l - 5)(7l + 2)$

29. $(w - 4z)(w - 4z)$

30. $(x - 4)(x - 4)$

31. $(x + 4)(x + 4)$

32. $(4w + x)(4w + x)$

33. $(a - 2)(a + 2)$

34. $(3b - 2)(3b + 2)$

35. $(2e + 5f)(2e - 5f)$

36. $(10x + 3y)(10x - 3y)$

37. **TRANSPORTATION** Four years ago, a $10 bill would buy x gallons of gas, and Jane's car averaged y mi/gal. Today, the car's gas mileage has decreased by 5 mi/gal, and a $10 bill buys 1 gal less. Find the difference between how far Jane could travel on $10 in those days, compared to now.

Expand and simplify.

38. $(4k + 1)(k + 3) - 4k^2$

39. $(3x - 4)(3x^2 + 6x - 2)$

40. $(7a + 3b)(6a^2 + 2ab - b^2)$

41. $(p - q)(p + 2q)(2p - q)$

42. $(a + b)(a + b)(a + b)$

43. $(a + b)^4$

Write, expand, and simplify expressions for the volumes of the two rectangular prisms shown below.

44.
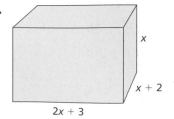

x

$x + 2$

$2x + 3$

45.
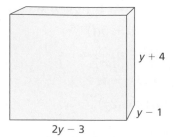

$y + 4$

$y - 1$

$2y - 3$

■ EXTENDED PRACTICE EXERCISES

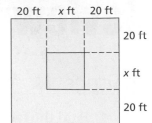

46. **CONSTRUCTION** A square fast-food restaurant building is surrounded by a square parking lot. The lot extends 20 ft beyond the restaurant in each direction, as shown on the map at the right. When the lot was paved, it took 4000 ft² of blacktop to cover it. How long is each wall of the restaurant?

47. **SEWING** The skateboard club, invited to enter a local parade, decided to have a flag. Their first idea was a beige pennant to represent a street ramp. It was a right triangle, twice as wide as it was high. For visibility, they then stitched a square lavender background around it. As shown in the picture, the background extended one foot above and below the triangle. The lavender area totaled 10 ft². About how much beige cloth did they use? (Don't worry about a seam allowance for your calculation.)

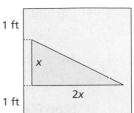

48. **PRODUCT DEVELOPMENT** A product engineer designs a new square handheld game. After field-testing the prototype, the engineer decides to change the shape of the game. She doubles the length and decreases the width by 4. Let s represent the length of a side on the original square. Write a polynomial to represent the area of the new rectangular game.

■ MIXED REVIEW EXERCISES

Find the volume of each figure. Round to the nearest whole number. (Lesson 5-7)

49.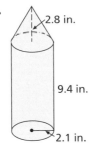

50.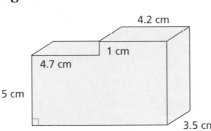

51.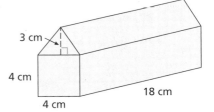

Add. (Lesson 8-5)

52. $\begin{bmatrix} 6 & 2 \\ 1 & 8 \\ -3 & 2 \end{bmatrix} + \begin{bmatrix} 5 & 4 \\ -3 & 1 \\ 0 & -8 \end{bmatrix}$

53. $\begin{bmatrix} 2 & -1 & 3 \\ -2 & 3 & 1 \\ 0 & 0 & 5 \end{bmatrix} + \begin{bmatrix} 2 & -2 & 0 \\ 3 & 4 & -1 \\ 2 & -3 & 2 \end{bmatrix}$

54. $\begin{bmatrix} 2 & 1 & 2 \\ 1 & 1 & 2 \\ 2 & 1 & 1 \end{bmatrix} + \begin{bmatrix} -2 & 0 & 1 \\ 0 & -2 & 0 \\ 1 & 0 & -2 \end{bmatrix}$

Multiply. (Lesson 8-5)

55. $5 \cdot \begin{bmatrix} 4 & 6 \\ 3 & 8 \\ -5 & 6 \end{bmatrix}$

56. $7 \cdot \begin{bmatrix} -1 & -3 \\ 4 & 2 \\ 3 & -7 \end{bmatrix}$

57. $4 \cdot \begin{bmatrix} -3 & -1 \\ 8 & 6 \\ 7 & -4 \end{bmatrix}$

Find the scale length for each of the following. Round to the nearest thousandth if necessary. (Lesson 7-3)

58. actual length: 7 mi
    scale is $\frac{1}{2}$ in.:3 mi

59. actual length: 12.4 yd
    scale is 1 in.:2 yd

60. actual length: 28.7 ft
    scale is $\frac{1}{4}$ in.:5 ft

Review and Practice Your Skills

Find the factors for the following.

1. $8x + 12y$

2. $6m^2 - 18n^2$

3. $7x^2 + 15x$

4. $5ab + 12b$

5. $2gh - ghk$

6. $12pq + 24rs$

7. $28abc - 11a^3$

8. $10mn^2 - 17m^2$

9. $17xy^2 + 24y^2z$

10. $2ab + 4bc - 8ac$

11. $5x^3 + 5x^2y^2$

12. $9r - 9r^5$

Find the GCF and its paired factor for the following.

13. $36a + 24b$

14. $17x + 34x^2$

15. $5ab + 10bc - 5b$

16. $8mn^2 - 12m^2$

17. $18p^2q - 36pr^2$

18. $14xy - 21xy^2$

19. $15s^2t^2 + 45s^3t$

20. $24a^3b^4 + 60a^2b^3$

21. $4x^3 - 2x^2 + 14x$

22. $x^2y + xy^2 + x^2y^2$

23. $3uv - 9u^2v^2 + 3u^3v^3$

24. $9mn - 3m^2 + 4mn^2$

25. $36m^3n^5 + 72m^2n^3 + 54m^5n^2$

26. $45x^2y^2 + 65u^3v - 35s^4t^2$

27. $6a^2bc + 2ab^2c - 4abc$

28. $15y^4z + 10y^2z^2 - 20yz$

29. $8mnp - 20m^2np^3 + 16mn^4p^2$

30. $32xy^3 + 100x^2y + 2xy$

Multiply. Simplify if possible.

31. $(x + 2)(x - 3)$

32. $(2x + 1)(3x - 5)$

33. $(x + 2y)(2x + 3y)$

34. $(3x + 2)(3x - 2)$

35. $(5x + 4)(5x + 4)$

36. $(7x - 4y)(8 + 3s)$

37. $(m - 5n)(4p + 5m)$

38. $(w + 3)(3 - w)$

39. $(a - 6b)(3a - 5b)$

40. $(2r - 7s)(5r - 3t)$

41. $(x + 6)(x - 6)$

42. $(8x - 3)(8x - 3)$

43. $(8x - 3)(8x + 3)$

44. $(a + b)(c + d)$

45. $(4y + 9z)(2y - 5z)$

46. $(5 - 2x)(11 + 5x)$

47. $(x + 1)(y + 2)$

48. $(10c - 13d)(2d + 3c)$

49. $(9x - 1)(9x + 1)$

50. $(9x + 1)(9x + 1)$

51. $(8p + 8q)(8p + 8q)$

52. $(x^2 + 1)(2x + 1)$

53. $(z^2 + 5)(z^2 - 5)$

54. $(x + 3)(3x^2 - 1)$

55. $(2r + 3s)(4r - 6s)$

56. $(4m + 13)(13m - 4)$

57. $(-7c + 3d)(6c - 5d)$

58. $2(m + 17)(m - 1)$

59. $x(x + 4)(x + 13)$

60. $(x + 5)(x - 5)(x + 5)$

61. The dimensions of a rectangle are $(7x - 5)$ ft and $(2x + 3)$ ft. Write and simplify an expression for the area of the rectangle.

62. Explain the difference between $(x + 4)(x - 4)$ and $(x - 4)(x - 4)$.

Simplify. (Lesson 11-1)

63. $(-8x + 7y) + (11y - 5x)$

64. $(21b - 16) - (-13b + 7)$

65. $(5x^2 + 9x - 10) + (-14x^2 + 11x + 10)$

66. $(gh - gh^2 + 3g^2h) + (g^2h + 5gh - gh^2)$

Multiply. Simplify if possible. (Lesson 11-2)

67. $-6x(5x - 11)$

68. $8n^2(n^2 - 7n)$

69. $5x^2(4x^2 - 3x + 1)$

70. $-9c^3d(-7d^3 + 2c)$

71. $-4pqr(2p^3q - 5pr - 3p^3q^2)$

72. $10a^3bc^2(4a^2b^2c - 3a^2bc^3)$

Find the GCF and its paired factor for the following. (Lesson 11-3)

73. $-30x + 54$

74. $18m - 30n$

75. $12g + 25g^2$

76. $4a^2 - 16a$

77. $45r^2st^3 + 75rs^2t^2$

78. $-26xyz + 52x^2yz^2$

79. $48a^3b^2 + 56a^2b^4 - 32a^4b^3$

80. $ab - abc + abcd - abcde$

81. $8x^3 + 6x^2 + 4x + 2$

Multiply. Simplify if possible. (Lesson 11-4)

82. $(3x + 2)(3y + 2z)$

83. $(7a - b)(7 + b)$

84. $(5m + 9n)(-2m + 3p)$

85. $(4x + 3)(4x + 3)$

86. $(4x + 3)(4x - 3)$

87. $(8p + 7q)(6p - 5q)$

88. $(8x + y)(4y - 7x)$

89. $(2x + 1)(1 - 2x)$

90. $(a - 11b)(5a - 13b)$

Mid-Chapter Quiz

1. Write $x^2y^2 + 3xy^3 + 4x^3y - 5$ in standard form for the variable x. (Lesson 11-1)

2. Write $2 - 4x^3 + 3x^2y^3 + y$ in standard form for the variable y. (Lesson 11-1)

Simplify. (Lesson 11-2)

3. $(5y - 2z) - (3y - 5z)$

4. $(3x^2 + 4x - 5) + (x^2 - 3x + 8)$

5. $(a^2 - 5ab + 2b^2) - (ab + b^2)$

6. $(-8p)(-2q)$

7. $-t^4(t^2 + u)$

8. $2v^2(3v^3 + 2v - 3)$

9. Write and simplify an expression for the area of a rectangle that has a length of $3x$ and a width of $(x^2 - y + 4)$.

Find factors for the following. (Lesson 11-3)

10. $6x - 9y$

11. $6a^3b - 4a^2b^2$

12. $3km^2n + 2mn^2 - 6k^2n$

Multiply. Simplify if possible. (Lesson 11-4)

13. $(c - d)(4g + 3h)$

14. $(12r + s)(3s - t)$

15. $(2k - 4)(2k + 4)$

16. $(z + 6)(z + 6)$

17. $(3b - c)(2b - 3c)$

18. $(x + 4)(x - 8)$

Find Binomial Factors in a Polynomial

Goals ■ Factor polynomials by grouping.

Applications Manufacturing, Design, Sales

Work in groups of 2 or 3 students.

As you know, multiplying a polynomial by a monomial does not change the number of terms. The answer has exactly as many terms as the polynomial you started with. But multiplying by a binomial is not so predictable.

1. Multiply each of the following pairs, and simplify each result.

   $(a + b)(c + d)$

   $(a + b)(a + b)$

   $(a + b)(a - b)$

2. The polynomials that result from these multiplications each have a different number of terms. Examine the three calculations and explain why there is a difference. Focus on what happens to the inner and outer products when you simplify each expression.

◥ BUILD UNDERSTANDING

You have seen that you can often extract a monomial factor from a polynomial. You may also be able to extract a binomial factor. Finding binomial factors is more complex, however, because of the greater variety of possible answers when you multiply by a binomial.

This lesson focuses on the $(a + b)(c + d)$ pattern you explored in the activity above. In this multiplication, the resulting polynomial has twice the terms of each polynomial that was multiplied.

When you factor a polynomial, the first step is always to look for a common monomial factor in all terms. If you find one (the GCF), extract it. The next step is to search for a binomial factor. If the number of terms in the polynomial is even, proceed as follows:

1. Group the terms in the polynomial into pairs with a common factor.

   $a^2 + ab + ab + b^2$

2. Extract the monomial factor from each pair.

   $(a^2 + ab) + (ab + b^2)$

3. If the binomials that remain for each pair are identical, this is a binomial factor of the expression.

   $a(a + b) + b(a + b)$

4. The monomials you extracted create a second polynomial.

   $(a + b)(a + b)$

Example 1

Find factors for $4x^3 + 4x^2y^2 + xy + y^3$.

Solution

1. Check for a monomial factor for the whole expression. There is none.

2. Within the polynomial, make pairs of terms that share monomial factors.

$$(4x^3 + 4x^2y^2) + (xy + y^3) \quad \text{or} \quad (4x^3 + xy) + (4x^2y^2 + y^3)$$

3. Extract the monomial factors in each pair.

$$4x^2(x + y^2) + y(x + y^2) \quad \text{or} \quad x(4x^2 + y) + y^2(4x^2 + y)$$

4. The binomials left in each pair are identical, so they are a factor of the whole polynomial. The binomial can be extracted; the monomials create a second factor as follows.

$$(x + y^2)(4x^2 + y) \quad \text{or} \quad (4x^2 + y)(x + y^2)$$

Note that these factorizations are the same, owing to the fact that multiplication is commutative.

Example 2

MANUFACTURING The volume of a box is $4pr - 6ps - 4qr + 6qs$. Find the possible dimensions of the box. (*Hint:* Volume is the product of three factors.)

Solution

Check for a monomial factor for the whole expression. The constant 2 can be extracted: $2(2pr - 3ps - 2qr + 3qs)$.

$$2[(2pr - 3ps) - (2qr - 3qs)] \quad \text{or} \quad 2[(2pr - 2qr) - (3ps - 3qs)]$$
$$= 2[p(2r - 3s) - q(2r - 3s)] \qquad\quad = 2[2r(p - q) - 3s(p - q)]$$
$$= 2(2r - 3s)(p - q) \qquad\qquad\qquad = 2(p - q)(2r - 3s)$$

Note that in the first step, the last sign had to be changed when the terms were grouped. Can you see why? There is a minus sign before the second group.

Example 3

Factor $2x^3 - 2x^2y - 3xy^2 + 3y^3 + xz^2 - yz^2$.

Solution

There is no shared monomial factor. Pair terms in the remaining polynomial, and factor if possible.

$$(2x^3 - 2x^2y) - (3xy^2 - 3y^3) + (xz^2 - yz^2)$$
$$= 2x^2(x - y) - 3y^2(x - y) + z^2(x - y)$$
$$= (x - y)(2x^2 - 3y^2 + z^2)$$

> ## Problem Solving Tip
>
> There may be more than one way to pair terms. You may need to try several approaches to find the one that works best.

Once again, note the sign changes during the grouping process.

Find factors for the following.

1. $9wx + 6wz + 6xy + 4yz$
2. $2e^2 + 14ef + 3eg + 21fg$
3. $18ab - 27ad - 8bc + 12cd$
4. $3x^3 - 12x^2y - xy + 4y^2$
5. $5rs - 40rt + 3s - 24t$
6. $24p^3 - 18p^2q + 4pq - 3q^2$
7. $kl + mn + ml + kn$
8. $8rs + 3tu + 2st + 12ru$
9. $3mr - 8ms + 5mt - 9nr + 24ns - 15nt$

LANDSCAPING In the exercises below, the areas of two rectangular lawns are expressed as polynomials. Find binomial expressions for the sides (one is given).

10.

Area:

$6ab + 4ac + 3bd + 2cd$

$2a + d$

11.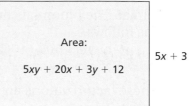

Area:

$5xy + 20x + 3y + 12$

$5x + 3$

12. **WRITING MATH** Suppose you are asked to factor $12pq + 8p - 3q - 2$. How would you decide the best way to group the terms? Explain your thinking.

▧ **PRACTICE EXERCISES** • For Extra Practice, see page 698.

Find factors for the following.

13. $4ab + 6ad + 6bc + 9cd$
14. $4a^2 + 6ab + 6ac + 9bc$
15. $4qr + 12qt + sr + 3st$
16. $4q^2 + 12qs + qr + 3rs$
17. $21ef - 12eh - 7fg + 4gh$
18. $21e^2 - 12e - 7ef + 4f$
19. $27w^2x - 18w^2z^2 - 3xy + 2yz^2$
20. $27 - 18z - 3y + 2yz$
21. $2k^2l^2 + 5k^2n - 6l^2m - 15mn$
22. $2kl + 5k - 6l - 15$
23. $15tu + 20t - 6vu - 8v$
24. $15tu + 20tv - 6u - 8v$
25. $3x^2y - x^2z + 24y - 8z$
26. $3x^2y - x^2 + 24y - 8$
27. $vy + 5vz + 3wy + 15wz + 2xy + 10xz$
28. $6j^2m^2 - 42j^2n + 5km^2 - 35kn - 3lm^2 + 21ln$
29. $10pr - 15ps + 20pt - 2qr + 3qs - 4qt$
30. $12a^2d + 4a^2e^2 - 6bd - 2be^2 - 15cd - 5ce^2$
31. $6df - 20eg - 35eh - 10ef + 12dg + 21dh$

Find factors for the following.

32. $8x^2 + 4xz + 4xy + 2yz$
33. $6j^3 - 12j^2l + 3j^2k - 6jkl$
34. $3abd - 3abe - 3acd + 3ace$
35. $3r^4 + 6r^3t - 6r^3s - 12r^2st$

Factoring can make calculations easier. For Exercises 36–37, calculate the value of each expression twice. First, calculate each term separately. Then factor the expressions before you calculate value.

36. DESIGN Changing the design of a computer monitor has decreased the cost of manufacturing the monitor. The change in cost is represented by the expression $8pr - 2qr - 20ps + 5qs$. Find the amount of change if $p = 2.1$, $q = 2.4$, $r = 0.5$ and $s = 1.2$.

37. SALES The number of units sold (in millions) of a new video game is represented by the expression $21x^2 - 14xz + 9xy - 6yz$. Find the number of sales if $x = 0.3$, $y = 0.9$ and $z = 0.2$.

38. ERROR ALERT When Monica attempts to factor $2a^2c^3 - 4a^2d - 4bc^3 + 8bd$, she gets $2a^2(c^3 - 2d) - 4b(c^3 + 2d)$. What mistake did Monica make?

■ EXTENDED PRACTICE EXERCISES

39. The area of the rectangle at the right is expressed as a polynomial. Find binomial expressions for the sides. (There are two possible answers.)

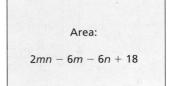

Area:

$2mn - 6m - 6n + 18$

SHIPPING The volumes of the boxes below are expressed as polynomials. Find expressions for the sides (one is given).

40.

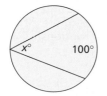

Volume:

$3wx^2 - 3wxz + 6wxy - 6wyz$

$3w$

41.

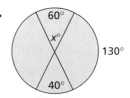

Volume:

$12a^3 + 18a^2c + 6a^2b + 9abc$

$3a$

42. Find the binomial expression for the base and height of this right triangle. (*Hint:* Remember the formula for the area of a triangle includes $\frac{1}{2}$.)

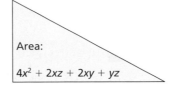

Area:

$4x^2 + 2xz + 2xy + yz$

■ MIXED REVIEW EXERCISES

Find x in each. (Lesson 10-4)

43.

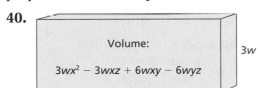

$x°$ $100°$

44.

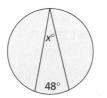

$x°$

$48°$

45.

$60°$

$x°$ $130°$

$40°$

46.

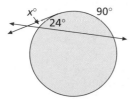

$x°$ $90°$

$24°$

47.

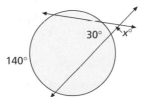

$30°$ $x°$

$140°$

48.

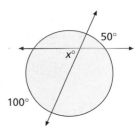

$50°$

$x°$

$100°$

Evaluate each product when $a = 4$, $b = -2$ and $c = -\frac{1}{2}$. (Lesson 1-7)

49. $7ab$

50. $3(abc)$

51. $2a + 3b$

52. $-4(a)(b)(c)$

53. $12bc$

54. $-5ac$

55. $3c - 4b$

56. $2(a)(c)$

11-6 Special Factoring Patterns

Goals
- ■ Factor perfect square trinomials and differences of perfect squares.
- ■ Use factoring to solve quadratic equations.

Applications Manufacturing, Landscaping, Art

Work with a partner to find patterns.

These two diagrams represent $(x + 2)^2$ and $(x + 5)^2$.

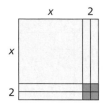 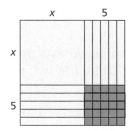

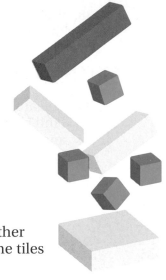

1. Express the area of each diagram as a trinomial. Do you see a pattern?

2. **MODELING** Use Algebra manipulatives such as Algeblocks to build other squared binomials; for example, $(x + 1)^2$ or $(x + 3)^2$. Find the sum of the tiles and express the areas as trinomials.

3. Discuss with your partner any patterns that you see. Apply the pattern to express $(x + 4)^2$ as a trinomial.

■ BUILD UNDERSTANDING

Finding binomial factors in polynomials with an even number of terms can be handled by pairing terms. Factoring a trinomial requires different strategies.

One strategy is to look for special patterns. You have already seen one such pattern—the difference of two squares. You can review this pattern by studying Example 3 on page 483. The activity above illustrates another pattern—the **perfect square trinomial**. Every binomial multiplied by itself fits this pattern.

Reading Math

Perfect square trinomials include squared negative binomials like $[-(a - b)]^2$, though this book does not explore all the negative options.

Pattern of a trinomial		How it relates to a binomial
First term	A perfect square	The square of the binomial's first term
Last term	A perfect square	The square of the binomial's last term
Middle term	The square roots of the two perfect squares multiplied together, and then doubled	The product of the binomial's terms, multiplied by two

If you spot this pattern in a trinomial, you can always find its binomial factors.

Example 1

Can you find binomial factors for the following?

a. $s^2 + 10s + 25$ **b.** $a^2 - 2ab + b^2$

Solution

a. The first term, s^2, is a perfect square. Therefore, the binomials' first terms would be s (or $-s$).

The last term, 25, is also a perfect square, so the binomials' last terms would be 5 or -5.

The middle term, $10s$, *does* equal $s \times 5 \times 2$. Therefore, the trinomial *is* a perfect square trinomial.

$s^2 + 10s + 25 = (s + 5)(s + 5)$

b. The first term, a^2, is a perfect square. Therefore, the binomials' first terms would be a (or $-a$).

The last term, b^2, is a perfect square. Therefore, the binomials' last terms would be b or $-b$.

The middle term, $(-2ab)$, is $a \times (-b) \times 2$, so the trinomial *is* a perfect square.

$a^2 - 2ab + b^2 = (a - b)(a - b)$

> ### Check Understanding
>
> Once again, negative options are not explored for the first quadratic terms. Is there a difference between $-(s)^2$, $(-s)^2$, and $-s^2$?

You may realize that the difference of two squares is also a special pattern that can be used for finding binomial factors. The difference of two squares is easy to recognize, because it is described fully by its name.

Example 2

MANUFACTURING Two rectangular metal covers have areas of $x^2 - 4$ and $25p^2 - 4q^2$. Both areas are examples of the difference of two squares. Find the dimensions of the metal covers by finding the binomial factors of each.

a. $x^2 - 4$ **b.** $25p^2 - 4q^2$

Solution

a. The first term, x^2, is a perfect square, so the first term of both binomials will be x. The second term, 4, is also a perfect square, so the binomials' second terms will be 2 and -2, respectively.

$x^2 - 4 = (x + 2)(x - 2)$

b. The first term, $25p^2$, is a perfect square, so the binomials' first terms would be $5p$ or $-5p$. The last term, $4q^2$, is a perfect square, so the binomials' second terms will be $2q$ and $-2q$.

$25p^2 - 4q^2 = (5p + 2q)(5p - 2q)$

We can use factoring to solve certain equations. Consider the equation $x^2 + 16 = 8x$. The variable x appears in an x^2-term. This type of equation is called a **quadratic equation**.

The logic used in solving quadratic equations is as follows. Start with the idea that if the product of two numbers or expressions is equal to zero, then at least one of the factors is equal to zero. (If $xy = 0$, then either $x = 0$ or $y = 0$.)

Example 3

Determine the possible solutions for $x^2 + 16 = 8x$.

Solution

Subtract $8x$ from both sides of the equation: $x^2 - 8x + 16 = 0$.

Then factor the expression on the left side: $(x - 4)(x - 4) = 0$.

One factor must equal 0. Since both factors are the same, both must be equal to zero. Solve the equation: $x - 4 = 0$, so $x = 4$.

This quadratic equation has a single solution because $x^2 - 8x + 16$ is a perfect square trinomial, and both factors are identical. When a quadratic equation has different factors, you may find more than one solution.

◥ TRY THESE EXERCISES

Find binomial factors for the following, if possible.

1. $s^2 + 10s + 25$
2. $4x^2 - 12xy + 9y^2$
3. $m^2 + 8mn + 16n^2$
4. $m^2 - 8mn + 16n^2$
5. $9r^2 - 36$
6. $25x^2 - 1$
7. $49a^2 - 28a + 2$
8. $81e^2 - 8f^2$
9. $64u^2 - 48uv + 9v^2$

10. A square is shown to have an area of $8w + 16 + w^2$. How long is each side?

 11. **WRITING MATH** Describe the special pattern shown by the polynomial $p^2 - 9$. Find the binomial factors.

◥ PRACTICE EXERCISES • For Extra Practice, see page 698.

Find binomial factors for the following, if possible.

12. $p^2 - 2p + 1$
13. $36a^2 + 24ab + 4b^2$
14. $9f^2 - 49g^2$
15. $4x^2 - 24xy + 27y^2$
16. $1 - 8x + 16x^2$
17. $100r^2 + 220r + 121$
18. $8v^2 - 25w^2$
19. $9m^2 - 6mn + 9n^2$
20. $h^2 - 14h + 49$
21. $9s^2 - 6st - t^2$
22. $y^2 + 2yz + z^2$
23. $36r^2 - s^2$
24. $4a^2 - 12b^2$
25. $9c^2d^2 - 64e$
26. $4c^2 + 20cd + 25d^2$

27. Find a monomial factor and two binomial factors for $4x^2 + 8x + 4$.

28. Find a monomial factor and two binomial factors for $16v^2 - 36w^2$.

29. Solve the equation $p^2 - 6p + 9 = 0$.
30. Solve the equation $m^2 + 25 = 10m$.

31. Solve the equation $(a - 3)(2a - 5) = 0$.

32. Solve the equation $k^2 - 16 = 0$ in two ways, one of which involves factoring. Your answers should be identical using either method.

33. LANDSCAPING A square garden with side length $8x$ is planted in the center of a square lawn with side length y. Write a polynomial to represent the area of the lawn. Then find two binomial factors.

34. ART A mosaic in the shape of a rectangle has an area of $49x^2 - 25y^2$. Find the possible length and width of the rectangle if $x = 9$ in. and $y = 2$ in.

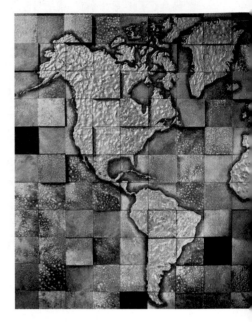

35. As a part of a problem, you have to calculate: $(8.35)^2 + (1.65)^2 + (8.35)(1.65)(2)$. Can you see a fast way to do this? What is the answer?

Find factors for the following.

36. $3c^2x + 18cdx + 27d^2x$

37. $5s^3 - 20s^2t + 20st^2$

38. $12a^2 - 12b^2$

39. $3x^3y - 12xy^3$

40. CHAPTER INVESTIGATION Distribute the final survey to your classmates and compile the data. Use the information to create a demographic profile of your class. Discuss with your group the best way to show your findings. Work together to prepare graphs and charts.

■ EXTENDED PRACTICE EXERCISES

41. Factor $x^3 - x^2y - 2x^2 + 2xy + x - y$.

42. The square floor of a shower with each side of x feet is covered with tiles measuring 1 ft^2 each. Some of these tiles are removed to insert a drain. 21 tiles are left on the shower floor. Find x and y, where y is the number of tiles removed.

■ MIXED REVIEW EXERCISES

These two spinners are spun. (Lesson 9-3)

43. List the sample spaces for the spinners.

44. Find the probability that the sum of the numbers is odd and greater than 6.

45. Find the probability that the sum of the numbers is either 6 or 10.

46. Find P(not an odd sum).

47. Find P(not an even sum or a sum of 9).

48. Find P(an odd sum or a sum of 8).

Trapezoids and their medians are shown. Find the length of each median. (Lesson 4-9)

49.

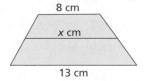

8 cm

x cm

13 cm

50.

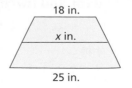

18 in.

x in.

25 in.

Review and Practice Your Skills

Find factors of the following.

1. $5(c + d) + b(c + d)$ **2.** $g(f^2 - 8) - 9(f^2 - 8)$ **3.** $a(b - 3) - c(b - 3)$

4. $xz + 10x + yz + 10y$ **5.** $2h - 2k + jh - jk$ **6.** $x^2 - x + xy - y$

7. $y^3 - 2y^2 + 3y - 6$ **8.** $3a - 3b + ab - a^2$ **9.** $2wz - w + 3 - 6z$

10. $xy + 5x + 2y + 10$ **11.** $mw - mx - nw + nx$

12. $gh + 3h^2 - 12h - 4g$ **13.** $2x^2y - 8x^2 + 3y - 12$

14. $3wz^2 + 12w - z^2 - 4$ **15.** $p^2r^3 - 2p^2s - qr^3 + 2qs$

16. $18w^2z - 3w^3 + 42wz^3 - 7w^2z^2$ **17.** $w - v + wv - v^2$

18. $8b^2 - 10b + 4b - 5$ **19.** $x - xy - 3ay^2 + 3ay$

20. $10m^2 + 15mp + 18mn + 27np$ **21.** $-9xy + 6xz - 6y + 4z$

22. $ax + bx + cx + 2a + 2b + 2c$ **23.** $xw + 2yw + 3zw - 4x - 8y - 12z$

24. $ap + aq - ar - bp - bq + br$ **25.** $x^2 - ax - bx + cx - ac - bc$

26. Find the possible dimensions of a rectangle whose area is $mn - 4m + 2n - 8$.

27. Find the possible dimensions of a rectangle whose area is $2g + 4f - 7ag - 14af$.

Find binomial factors of the following, if possible.

28. $x^2 + 10x + 25$ **29.** $x^2 - 20x + 100$ **30.** $m^2 + 16m + 64$

31. $z^2 - 6z + 36$ **32.** $16d^2 + 40d + 25$ **33.** $36b^2 - 12b + 1$

34. $64r^2 + 48r + 9$ **35.** $x^2 - 8xy + 16y^2$ **36.** $9g^2 + 12gh + 4h^2$

37. $w^2 - 144$ **38.** $121 - p^2$ **39.** $c^2 - 9d^2$

40. $x^2 + 25$ **41.** $16u^2 - 81v^2$ **42.** $1 - 4y^2$

43. $25s^2 - 70st + 49t^2$ **44.** $25x^2 - 49y^2$ **45.** $49p^2 + 28pq + 4q^2$

46. $49d^2 - 4f^2$ **47.** $64m^2 + 176mn + 121n^2$ **48.** $64x^2 - 121z^2$

49. $1 - 2a + a^2$ **50.** $1 - 8x + 64x^2$ **51.** $16 + 25v^2$

52. $25 - 4k^2$ **53.** $225x^2 + 330xy + 121y^2$ **54.** $625j^2 - 1$

Find a monomial factor and two binomial factors for each of the following.

55. $3x^2 - 12x + 12$ **56.** $5x^2 - 45$ **57.** $x^3 - 8x^2 + 16x$

58. $10x^2 - 140x + 490$ **59.** $3ax^2 - 12a$ **60.** $50by^2 - 18bx^2$

61. $x^4 - 25x^2$ **62.** $27y^3 - 36xy^2 + 12x^2y$ **63.** $4a^2 - 4b^2$

Simplify. (Lesson 11-1) and (Lesson 11-2)

64. $(-6n^2 + 7n - 11) + (17n^2 - 7n + 16)$ **65.** $(xy + 2x^2 - 8y) - (4x^2 + 8y - 3xy)$

66. $(7x + 15y) - (5x + 8y) + (2y - 4x)$ **67.** $2xyz(3xy - 7yz + 15xz)$

68. $x(4x^2 - 9) + 2(x^3 - 7x^2 + 4x)$ **69.** $3(x + 2y) - 4(2x - 5y) + 2(5x - 13y)$

Find the GCF and its paired factor for the following. (Lesson 11-3)

70. $22x + 55y$ **71.** $48x^2 + 32x$ **72.** $13x^2y^3 - 52x^3y^2$

73. $-4def - 8efg - 12ef$ **74.** $120a^2b^3 + 24a^3b - 72a^4b^2$ **75.** $2sk^2 + 58sq^2 + 34sy^2$

Find factors for the following. (Lesson 11-5)

76. $xy - 2x - 4y + 8$ **77.** $3xw + 7w - 12x - 28$

Find binomial factors of the following, if possible. (Lesson 11-6)

78. $121 + 22a + a^2$ **79.** $4x^2 - 28xy + 49y^2$ **80.** $81c^2d^2 - 25b^2$

MathWorks
Workplace Knowhow

Career – Actuaries

Actuaries assemble and analyze statistical data to estimate the probabilities of various types of loss. This information helps the insurance company determine how much to charge people in insurance premiums. For example, an actuary studies the effect of age on the number of driving accidents that occur. If a particular age group has more accidents than another, that group pays higher premiums.

The company must charge enough to pay all claims and still make a profit. However, if the company charges too much, customers will choose another company. Actuaries must have excellent math and statistics skills. They also need to understand economics, social trends, legislation and developments in health and medicine.

You are evaluating the risk factors involved in insuring the lives of firefighters over the course of their careers. You determine that the equation $y = -x^2 + 15x + 100$ can be used to predict risk where x equals the number of years a firefighter has been on the job and y equals risk.

1. What is the base risk at the start of a firefighter's career? Use 0 for x.

2. Find the amount of risk a firefighter faces at 2 years, 4 years and 6 years. (Remember, to evaluate $-x^2$, square x before multiplying by -1.)

3. Make a table to show the risk for the first 10 years. At what year(s) is the risk of insuring firefighters the highest?

4. At what year does the risk come back down to 100?

Factor Trinomials

Goals   ■ Factor trinomials with quadratic coefficients of one.

Applications   Product Development, Construction, Chemistry

Work with a partner to find factoring patterns.

A trinomial expression that does not fit a special pattern may still have binomial factors. Finding such factors requires a combination of logic and guess-and-check.

1. Start with the idea that finding factors of a trinomial is the reverse of multiplying binomials. Study these examples and look for patterns.

$$(x + 3)(x + 4) = x^2 + 7x + 12$$
$$(x - 3)(x - 4) = x^2 - 7x + 12$$
$$(x - 3)(x + 4) = x^2 + x - 12$$
$$(x + 3)(x - 4) = x^2 - x - 12$$

$$(y + 5)(y + 1) = y^2 + 6y + 5$$
$$(y - 5)(y - 1) = y^2 - 6y + 5$$
$$(y + 5)(y - 1) = y^2 + 4y - 5$$
$$(y - 5)(y + 1) = y^2 - 4y - 5$$

2. Look at the third term in each trinomial and the sign before it. How does each third term and its sign relate to the binomial factors?

3. Look at each second term and the sign before it. How does each second term and its sign relate to the binomial factors?

4. Set up an additional example using terms and signs similar to those in the examples above. Does your example follow the patterns you have found?

> **Reading About Math**
>
> Many trinomials have an x^2 term, an x term, and a constant. The x^2 term is called the **quadratic** term, from the Latin *quadrare*, which means "to make a square." Also, polynomials with a quadratic term as their highest power are called **quadratic polynomials**.

◤ BUILD UNDERSTANDING

In this lesson, you will study trinomials where the coefficient of the first term (the x^2, or **quadratic term**) is 1. This makes the pattern easier to see. From the activity above, you may have noticed the following.

a. The trinomial third term is always the product of the binomial second terms.

b. The coefficient of the trinomial second term is always the sum of the coefficients of the binomial second terms. (*Note:* When the signs in the binomials are different, this sum will *look* like a difference, because $a + (-b) = a - b$.)

c. If the sign of the trinomial third term is negative, the signs in the binomials are different. If it is positive, the signs in the binomials are the same.

d. The sign of the trinomial second term is always the same as the sign of the greater binomial second term.

> **Check Understanding**
>
> Before you count the terms, always be sure the trinomial is in standard form. Why would this be important?

With these four clues, you can find the factors of a standard form trinomial that begins with x^2.

Example 1

Find second-term constants or coefficients for the binomial factors of these polynomials.

a. $x^2 - 8x + 15$ **b.** $x^2 + 3xy - 18y^2$

Solution

a. The product of the binomial second terms is 15, and the sum is 8. So the binomial second-term constants are 5 and 3 (because $5 \times 3 = 15$ and $5 + 3 = 8$).

The binomials will be in the form $(x \quad 5)(x \quad 3)$.

b. The product of the binomial second terms is 18, and their sum is 3. Because the third term's sign is negative, the binomial signs differ, so the sum will look like a difference.

Think of factors of 18 that have a difference of 3.

Factors	18 1	9 2	6 3
Difference	17	7	③

Stop here; 3 is the difference you want.

> **Problem Solving Tip**
>
> Making an organized list is a good strategy when the third term in the trinomial has many pairs of factors.

The coefficients will be 6 and 3. The binomials will be in the form $(x \quad 6y)(x \quad 3y)$.

The next step in finding the factors involves determining the correct signs for the binomials.

Example 2

In the two expressions above, complete the binomial factors by determining the signs of the second terms.

Solution

a. The second trinomial term is negative, so the larger binomial second term has a negative sign. The third trinomial term is positive, so both binomial signs are the same—both negative.

The binomial factors are $(x - 5)(x - 3)$.

b. The second trinomial term is positive, so the larger binomial second term is also positive. But the third trinomial term is negative, so the two binomial signs are different.

The binomial factors of $x^2 + 3xy - 18y^2$ are $(x + 6y)(x - 3y)$.

> **Problem Solving Tip**
>
> As always in problem solving, you should check your solutions before you finally accept them. Whenever you identify a pair of factors, multiply them to be sure their product is the polynomial you started with.

You can handle the numbers and signs in a single step if you wish, though this takes a little more thought.

Example 3

PRODUCT DEVELOPMENT A software company determines that the cost of producing its new financial software is a product of the number of days spent working on the project and the number of programmers assigned to the project. The total cost is represented by $x^2 - 5x - 36$. Find the binomial factors.

Solution

The product of the binomial second terms is (-36) and the sum is (-5). So the two binomial constants are 4 and (-9).

The binomial factors of $x^2 - 5x - 36$ are $(x + 4)(x - 9)$.

◥ TRY THESE EXERCISES

Identify the binomial second terms when the following trinomials are factored.

1. $x^2 + 10x + 21$

2. $t^2 + 9t + 20$

3. $a^2 - 6ab + 8b^2$

4. $m^2 - mn - 2n^2$

5. $k^2 + 5k - 6$

6. $f^2 + 2fg - 15g^2$

Identify second-term signs for binomial factors of the following.

7. $v^2 + 18v + 77$

8. $x^2 - 19x + 90$

9. $b^2 - 15bc - 100c^2$

10. $n^2 + n - 42$

Factor the following trinomials.

11. $c^2 + 5c + 6$

12. $c^2 - 5c + 6$

13. $c^2 - 5c - 6$

14. $c^2 + 5c - 6$

15. **MODELING** What are the sides of the rectangle you can create with one "x^2" Algeblock piece, 21 "one" tiles, and 10 "x" tiles? *Do not* experiment. Use factoring—it will save time. Then use Algeblocks to check your answer.

◥ PRACTICE EXERCISES • For Extra Practice, see page 699.

Identify binomial second-term factors for the following.

16. $p^2 + 5p + 6$

17. $x^2 + 12xy + 35y^2$

18. $h^2 - 10h + 9$

19. $a^2 - 7ab + 10b^2$

20. $c^2 + 6cd - 16d^2$

21. $q^2 + 2q - 63$

22. $r^2 - 13r + 30$

23. $e^2 - 7ef - 30f^2$

Identify binomial second-term signs for the following.

24. $x^2 + x - 12$

25. $j^2 + 12j + 27$

26. $s^2 - 18st + 17t^2$

27. $b^2 - bc - 56c^2$

28. $l^2 + 5l - 36$

29. $v^2 - 10v + 24$

30. $j^2 + 12jk + 11k^2$

31. $z^2 - 3z - 18$

Factor the following trinomials.

32. $x^2 - 25x + 24$

33. $p^2 + 10pq + 24q^2$

34. $m^2 + 5mn - 24n^2$

35. $k^2 - 10k - 24$

36. $a^2 - 2a - 24$

37. $h^2 + 23h - 24$

38. $r^2 + 14r + 24$

39. $f^2 - 11fg + 24g^2$

40. $p^2 + 2p - 15$

41. $q^2 - 11q + 28$

42. $r^2 + 21r + 20$

43. $s^2 + 2st - 8t^2$

Technology Note

Computer spreadsheets allow businesses to explore decisions by using and varying data.

Coupled with a graphics program, spreadsheet formulas allow businesses to graph data as well.

Most spreadsheet applications use cell names in the data column as variables. The trinomial $x^2 + 10x + 21$ is entered as:

A2 * A2 + 10 * A2 + 21

The computer uses the value of cell A2 to calculate the expression.

44. **CONSTRUCTION** A rectangular trench x feet deep is being dug for the foundation of a wall. The area of the bottom is $x^2 + 34x - 35$ ft^2. Compare the depth of the trench to its width and to its length.

45. **WRITING MATH** Can a trinomial have different sets of binomial factors? Explain your thinking.

46. **CHAPTER INVESTIGATION** Work with your group to develop a strategy for marketing a new product aimed at people your own age. Use the demographic profile you developed in Lesson 11-6. Suppose you can afford to run one print advertisement, one radio spot and one television commercial. Determine when and where you would run your advertisements. Give an oral presentation of your marketing strategy to your classmates. Be ready to defend your choices using the demographic data.

Factor the following.

47. $1 - 5r + 6r^2$

48. $1 + 7x - 18x^2$

49. $24g^2 - 10g + 1$

50. $13a^2 - 12a - 1$

51. $5a^2x^2 - 15ax^2 + 10x^2$

52. $9 + 18x - 72x^2$

53. **CHEMISTRY** To dilute x pounds of a chemical, you need a water tank with a volume of $3x^3 - 12x^2 - 36x$. Indicate its dimensions, in terms of x.

■ EXTENDED PRACTICE EXERCISES

54. **SMALL BUSINESS** Andre receives a rush order for some hand-painted plates. But his budget for materials is limited to $255 per day. His cost formula indicates that if he works at a rate of $(12 + x)$ plates per day, the daily cost will be $\$(x^2 + 22x + 120)$. How many plates can he make each day—maximum—to fulfill the order? (*Hint:* Check Lesson 11–5. Make a quadratic equation about daily cost, adjust it so that one side equals zero, then factor and reject any negative answers. Remember, the final answer will be $12 + x$.)

■ MIXED REVIEW EXERCISES

Complete the chart in preparation for making a circle graph. *Do not make the graph.* (Lesson 10-5)

Budget Item	Percent of Total	Central Angle
Rent—$550	55.	63.
Food—$415	56.	64.
Car Payment—$260	57.	65.
Credit Card Payment—$150	58.	66.
Utilities—$210	59.	67.
Savings—$115	60.	68.
Insurance—$125	61.	69.
Misc.—$175	62.	70.

Write the equation for each line. (Lesson 6-3)

71. slope $= \dfrac{2}{3}$, y-intercept $= -2$

72. passes through $(-2, -3)$ and $(4, 5)$

73. slope $= -2$, y-intercept $= 3$

74. passes through $(-3, 4)$ and $(6, -4)$

Problem Solving Skills: The General Case

Drawing diagrams and looking at several examples are useful ways to find helpful patterns in mathematics. Another technique is to create a general case. Algebra is excellent for this. It allows you to use letters instead of numbers for an expression's coefficients. By searching for patterns formed by the letters and symbols, you can draw general conclusions that can be applied in specific situations.

Problem

Find a pattern to help discover factors of a polynomial with a quadratic (x^2) coefficient greater than 1.

Solve the Problem

Use letters instead of numbers to represent the coefficients and constants. (In this solution, a specific example is shown for comparison beside the general case.)

Step 1: Work forward from a pair of binomial factors. The letters a and b represent possible coefficients found in the first term of each monomial factor. The constants, or second term in each monomial, are represented by n_1 and n_2.

General: $(ax + n_1)(bx + n_2)$ Specific: $(2x - 5)(3x + 2)$

$$\begin{array}{cccc} \text{F} & \text{O} & \text{I} & \text{L} \\ \downarrow & \downarrow & \downarrow & \downarrow \end{array}$$

$= abx^2 + axn_2 + bxn_1 + n_1n_2$ $= 6x^2 + 4x - 15x - 10$

$= abx^2 + (an_2 + bn_1)x + n_1n_2$ $= 6x^2 + (4 - 15)x - 10$

                                       $= 6x^2 - 11x - 10$

Step 2: Study the pattern. Carefully compare the general case to the specific example.

Think about how this pattern differs from your work with trinomials in Lesson 11-7. The sum of the second terms of the binomial factors no longer equals the coefficient of the second term of the trinomial. This is only true if the quadratic coefficient is 1.

The product of the coefficients of the F and L terms (quadratic coefficient and constant) is abn_1n_2—identical to the product of the O and I coefficients. Call this product the **grand product**.

The cross product (O and I) coefficients *multiply* to give the grand product and add to give the trinomial's second term. Apply this general rule to the specific example above.

a. Multiply 6 and 10 to find the grand product: $6 \times 10 = 60$

b. Multiply the O and I coefficients: $4 \times 15 = 60$. The product equals the grand product.

c. Add the O and I coefficients: $(+4) + (-15) = -11$. The sum equals the coefficient of the trinomial's second term.

Five-step Plan

1 Read
2 Plan
3 Solve
4 Answer
5 Check

◥ TRY THESE EXERCISES

Suppose you have forgotten a useful pattern, or think you may have found a new one. Exploring a general case can be a useful strategy. As shown on the previous page, working a specific example beside the general case may help.

1. Explore the FOIL pattern for factoring a single-variable trinomial that has a first-term coefficient of 1. Work forward from $(x + n_1)(x + n_2)$ as the general case, and $(x - 6)(x + 3)$ as a specific example.

2. Explore the FOIL pattern for factoring a double-variable trinomial with first-term coefficient of 1. Work forward from $(x + ay)(x + by)$ as the general case, $(x - 9y)(x + 5y)$ as a specific example.

◥ PRACTICE EXERCISES

3. Using the same method, explore the pattern for perfect square trinomials. Use $(ax + by)^2$ for the general case, and select your own specific example.

4. Use the same method to explore the difference-of-two-squares pattern. (*Note*: This will *prove* that the pattern you first saw at the start of this lesson is correct for all expressions of its type.)

Study the following table of polynomial expansions. Notice that each expansion is a *difference of two cubes*.

Polynomial factors		Expansion
$(2x - 2)(4x^2 + 4x + 4)$	$=$	$8x^3 - 8$
$(2x - 1)(4x^2 + 2x + 1)$	$=$	$8x^3 - 1$
$(3x - 3y)(9x^2 + 9xy + 9y^2)$	$=$	$27x^3 - 27y^3$
$(3x - 2y)(9x^2 + 6xy + 4y^2)$	$=$	$27x^3 - 8y^3$
$(x - y)(x^2 + xy + y^2)$	$=$	$x^3 - y^3$

5. Work through the general case of $(ax - by)(a^2x^2 + abxy + b^2y^2)$.

6. Work through $(3x - 1)(9x^2 + 3x + 1)$. (*Note*: The second factor is a trinomial, so the FOIL technique will not apply. Use the original method for multiplying that you learned in Lesson 11-4.)

7. **WRITING MATH** Compare your work in Exercises 1 and 2. Decide whether the following statement is true or false, and explain your reasoning.

If you make the *y*-variable equal to 1, the single-variable pattern (Exercise 1) is a special case of the double-variable pattern (Exercise 2).

◥ MIXED REVIEW EXERCISES

Solve each proportion. (Lesson 7-1)

8. $\dfrac{n}{8} = \dfrac{5}{12}$

9. $\dfrac{3}{n} = \dfrac{15}{40}$

10. $\dfrac{2x}{5} = \dfrac{x + 1}{4}$

11. $\dfrac{49}{16} = \dfrac{x + 2}{12}$

12. $\dfrac{3x}{9} = \dfrac{x + 4}{6}$

13. $\dfrac{8}{x + 1} = \dfrac{10}{2x - 1}$

14. $\dfrac{3n}{15} = \dfrac{2n - 3}{8}$

15. $\dfrac{4}{3n - 1} = \dfrac{16}{9n + 8}$

16. $\dfrac{16}{x + 1} = \dfrac{58}{4x + 1}$

17. **DATA FILE** Use the data on size and depth of the oceans on page 646. What is the approximate volume in cubic miles of the Atlantic Ocean? Give your answer in scientific notation, rounded to the nearest tenth. (Lesson 5-7)

18. **DATA FILE** Use the data on the calorie count of food on page 650. Matthew had 2 c of spaghetti and meatballs for dinner with $1\frac{1}{2}$ c of lemonade. For dessert he had an apple and $\frac{1}{2}$ c of sherbet. How many kilocalories did he consume? (Prerequisite Skill)

Review and Practice Your Skills

Factor the following trinomials.

1. $x^2 + 7x + 6$

2. $m^2 + 11m + 28$

3. $d^2 + 13d + 42$

4. $b^2 + 17b + 42$

5. $x^2 + 16x + 28$

6. $p^2 + 12p + 11$

7. $x^2 - 9x + 20$

8. $g^2 - 8g + 12$

9. $w^2 - 10w + 21$

10. $f^2 - 30f + 200$

11. $x^2 - 12x + 32$

12. $n^2 - 18n + 32$

13. $m^2 + 3m - 54$

14. $b^2 + 6b - 7$

15. $c^2 + c - 20$

16. $h^2 + 5h - 24$

17. $t^2 + 3t - 10$

18. $x^2 + 4x - 45$

19. $a^2 - 2a - 48$

20. $k^2 - 8k - 48$

21. $p^2 - 5p - 36$

22. $z^2 - 6z - 40$

23. $d^2 - d - 56$

24. $x^2 - 4x - 32$

25. $m^2 + 11mn + 30n^2$

26. $g^2 + 2gh + h^2$

27. $p^2 + 17pq + 60q^2$

28. $x^2 - 9xy + 18y^2$

29. $r^2 - 3rs + 2s^2$

30. $c^2 - 8c + 15d^2$

31. $b^2 + 3bc - 4c^2$

32. $m^2 + 8mn - 9n^2$

33. $a^2 + 7ab - 18b^2$

34. $x^2 - 11xy - 26y^2$

35. $p^2 - 4pq - 77q^2$

36. $g^2 - 4gh - 60h^2$

37. $x^2 + 14x + 48$

38. $z^2 + 2z - 48$

39. $f^2 - 26f + 48$

40. $t^2 + 22t - 48$

41. $c^2 - 19cd + 48d^2$

42. $s^2 - 13st - 48t^2$

43. $48 - 49x + x^2$

44. $p^2 + 47pq - 48q^2$

45. $26x + x^2 + 48$

46. Explore the FOIL pattern for factoring a trinomial whose factors are of the form $(n_1 - x)(n_1 - x)$. Work forward from these factors as general case, and select your own specific example.

47. Explore the FOIL pattern for factoring a trinomial whose factors are of the form $(n_1 + x)(n_1 - x)$. Work forward from these factors as general case, and select your own specific example.

48. Explore the FOIL pattern for factoring a trinomial whose factors are of the form $(ax + y)(bx + y)$. Work forward from these factors as general case, and select your own specific example.

49. Explore the FOIL pattern for factoring a trinomial whose factors are of the form $(ax + y)(bx - y)$. Work forward from these factors as general case, and select your own specific example.

50. Explore the FOIL pattern for factoring a polynomial whose factors are of the form $(x + a)(x + a)(x + a)$. Work forward from these factors as general case, and select your own specific example from $a > 0$.

51. Repeat Exercises #50 for $a < 0$.

Simplify. (Lesson 11-1)

52. $(7x - 5y - 13z) + (-4y + 6x + z)$

53. $(-8n^2 + 9n - 13) + (13n^2 - 3n + 12)$

54. $(5xy + 7x^2 - 3y) - (-4x^2 - 8y + 3xy)$

55. $(15x + 8y) - (-5x + 8y) + (4y - 2x)$

Simplify. (Lesson 11-2)

56. $-5a(10 - 4a^2 - 5b)$

57. $6xyz(xy + 8yz - 2xz)$

58. $x^2(3x^2 + 5) + 3(2x^3 - 5x^2 + x)$

59. $5(x + 3y) - 2(2x - 3y) + 3(5x - 7y)$

Find the GCF and its paired factor for the following. (Lesson 11-3)

60. $78x + 39y$

61. $16x^2 + 60x$

62. $14x^3y - 42xy^2$

63. $-9def - 15efg - 12gde$

64. $48a^3b^2 - 24ab^5 + 72a^2b^4$

65. $7sm^2 + 28sw^2 + 63sy^2$

Simplify. (Lesson 11-4)

66. $(4r + 5y)(x - 2r)$

67. $(x + 9)(x + 11)$

68. $(8x - 5)(7x + 6)$

69. $(9 - 4x)(9 + 4x)$

70. $(13 - 5v)(13 - 5v)$

71. $(15f + 2)(9 - 2f)$

Find factors for the following. (Lesson 11-5)

72. $xy - 5x - 4y + 20$

73. $5xw - 4w + 20x - 16$

74. $24a^3 + 8a^3f + 12b + 4bf$

75. $8x^2z + 11x^2b - 40z - 55b$

76. $5x - 20y + 5z - 2ax + 8ay - 2az$

77. $6n - 21p + 42mp - 12mn$

78. $5x^2 - 2xz - 15xy + 6yz$

79. $ax - 2bx + 7x + 5a - 10b + 35$

80. $a^2c^2 + a^3b + bc^3 + ab^2c$

Find binomial factors of the following, if possible. (Lesson 11-6)

81. $169 + 26a + a^2$

82. $9x^2 - 42xy + 49y^2$

83. $x^2 + 28x + 196$

84. $1 - 100m^2$

85. $16a^2 - 49b^2$

86. $100c^2d^2 - b^2$

87. $x^2 - 144y^2$

88. $x^2 - 12xy + 144y^2$

89. $25m^2 + 110mn + 121n^2$

Factor the following trinomials. (Lesson 11-7)

90. $c^2 + 27c + 72$

91. $b^2 - 21b - 72$

92. $a^2 + ad - 72d^2$

93. $f^2 - 17fg + 72g^2$

94. $72 - 73x + x^2$

95. $72 - 71m - m^2$

96. $r^2 - 18r + 81$

97. $p^2 - 24pq + 81q^2$

98. $81x^2 + 30x + 1$

99. $a^2b^2 - 2ab - 3$

100. $3n^2 - 4mn + m^2$

101. $-20x + x^2 + 96$

Use the patterns explored in Lesson 11-8 to find all values of k which make each polynomial factorable. (Lesson 11-8)

102. $x^2 + kx + 24$

103. $x^2 + kx - 60$

104. $x^2 + 9x + k \ (k > 0)$

More on Factoring Trinomials

Goals ■ Factor trinomials of the form $ax^2 + bx + c$.

Applications Small Business, Packaging, Consumerism

Work with a partner to discuss the following questions.

1. Multiply each pair of binomials. Make sure that you show the FOIL multiplication step as part of your work.

    a. $(x + 4)(x - 5)$ b. $(3x + 4)(2x - 5)$ c. $(3x + 4y)(2x - 5y)$

2. Compare the multiplications and their products. Describe the ways in which the examples are similar.

3. Describe the ways in which the examples differ.

■ BUILD UNDERSTANDING

In the previous lessons, you have factored trinomials in the form $x^2 + bx + c$ or $x^2 + bxy + cy^2$.

In this lesson, you will learn to factor trinomials with a quadratic (x^2) term coefficient other than 1. Finding binomial factors for a trinomial that has a quadratic coefficient greater than 1 is a two-step process. First, you must identify the FOIL coefficients. Once these are found, you can use them to discover the binomial factors.

Step 1: Identify the FOIL coefficients. A standard-form trinomial already shows two possible FOIL coefficients. The coefficient of the quadratic (x^2) term will be the F-coefficient (ab in the previous lesson). The coefficient of the last trinomial term is the L-coefficient ($n_1 n_2$ in the previous lesson).

a. Multiply these coefficients together for the grand product coefficient.

b. Find two numbers whose *product* is the grand product coefficient and whose *sum* is the middle trinomial term. These two numbers are the cross-product (O- and I-) coefficients (an_2 and bn_1).

Step 2: Analyze the FOIL coefficients to find the four binomial coefficients (a, b, n_1, and n_2). (*Note:* Four is the maximum. There may appear to be fewer if some of the binomial coefficients are the same. For example, $(2x + 3)(3x + 1)$ has two coefficients of 3.)

a. List all possible paired factors for each FOIL coefficient.

b. Inspect the pairs, and select the pair for each coefficient that gives a total set including four or fewer individual factors. These will be the binomial coefficients.

c. Figure the signs as you did in Lesson 11-7; however, instead of focusing on which is the larger of the binomial second terms, you have to decide which is the larger of the two cross products.

Example 1

Find FOIL coefficients for the trinomial $6x^2 + 29x + 35$.

Solution

The F-coefficient is 6 (the coefficient of the quadratic term). The L-coefficient is 35 (the last term coefficient or the constant). The grand product coefficient is $(6)(35)$, or $(1)(2)(3)(5)(7)$, or 210. The cross-product (O- and I-) coefficients add to give 29, and multiply to give 210. The numbers 14 ($= 2 \times 7$) and 15 ($= 3 \times 5$) are the two coefficients you need. (*Note*: At this stage, you will not be able to tell which is the inner and which is the outer coefficient.)

Example 2

Given the four FOIL coefficients above, analyze their factor pairs to find the appropriate binomial coefficients for $6x^2 + 29x + 35$.

Solution

F-coefficient (ab):	$6 = (1)(6)$ or $(2)(3)$
O- and I-coefficients:	$14 = (1)(14)$ or $(2)(7)$
(an_2 and n_1b)	$15 = (1)(15)$ or $(3)(5)$
L-coefficient (n_1n_2):	$35 = (1)(35)$ or $(5)(7)$

Among these pairs, $(2)(3)$, $(2)(7)$, $(5)(3)$, and $(5)(7)$ share only four numbers. Therefore, they are the binomial coefficients. Thus:

a. 2 and 3 (the F pair) are the x coefficients $\qquad$ $(2x\ \ )(3x\ \ )$

b. 2 and 7 are a cross-product pair $\qquad$ $(2x\ \ )(3x\ \ 7)$

c. 3 and 5 are the other cross-product pair $\qquad$ $(2x\ \ 5)(3x\ \ 7)$

d. Trinomial signs are both positive, so signs are $\qquad$ $(2x + 5)(3x + 7)$.

Example 3

SMALL BUSINESS Ann designs and sells bracelets. Her gross profit is represented by the expression $2x^2 - 5x - 3$. The monomial factors represent the number of bracelets sold and the selling price per bracelet. Find the monomial factors.

Solution

The F-coefficient is 2, the L-coefficient is 3. The grand product coefficient is $(2)(3) = 6$. The L-coefficient sign is *negative*, so you need numbers with a product of 6 and an apparent *difference* of 5. The O- and I-coefficients must be 6 and 1.

F:	$2 = (1)(2)$	O and I:	$6 = (1)(6)$ or $(2)(3)$
L:	$3 = (1)(3)$		$1 = (1)(1)$

Binomial coefficients are $(2)(1)$, $(2)(3)$, $(1)(1)$, and $(1)(3)$.

Binomial factor values are $(2x\ \ 1)(x\ \ 3)$.

The second trinomial sign is negative, so the greater cross product (6) must be negative. Factors with signs are $(x - 3)(2x + 1)$.

Find FOIL coefficients/constants for the following.

1. $3x^2 + 19x + 6$

2. $10a^2 + 7a - 12$

Given the following FOIL coefficients, identify the binomial factor coefficients.

	3.	**4.**
F-coefficient	8	14
Cross-product coefficients	6	35
(O and I)	20	4
L-coefficient	15	10

Identify the correct signs for the binomial second terms.

5. $35v^2 + 11v - 6 = (7v \quad 2)(5v \quad 3)$

6. $15s^2 - 17s - 4 = (5s \quad 1)(3s \quad 4)$

7. $3a^2 - ab - 10b^2 = (3a \quad 5b)(a \quad 2b)$

Find binomial factors for the following.

8. $8m^2 - 26m + 15$

9. $7f^2 + 4fg - 3g^2$

10. $6r^2 - r - 35$

11. $6x^2 + 17x + 10$

12. PACKAGING The surface area of a rectangular package is represented by the trinomial $2x^2 - 30x + 108$. Find the possible dimensions of the package.

Find FOIL coefficients for the following trinomials.

13. $3p^2 - 11p - 4$

14. $5z^2 + 17z + 6$

15. $6d^2 + 13d - 5$

16. $21a^2 - 26ab + 8b^2$

17. $10x^2 - xy - 24y^2$

18. $4n^2 + 4n - 15$

For the following FOIL coefficients, identify the appropriate binomial factor coefficients.

	19.	**20.**	**21.**	**22.**
F-coefficient	3	21	4	27
Cross-product coefficients	15	35	24	21
(O and I)	2	6	3	18
L-coefficient	10	10	18	14

Place appropriate signs in these unsigned binomials.

23. $8q^2 + 22q + 15 = (2q \quad 3)(4q \quad 5)$

24. $15c^2 - 38cd + 24d^2 = (3c \quad 4d)(5c \quad 6d)$

25. $18m^2 - 9m - 20 = (3m \quad 4)(6m \quad 5)$

26. $10y^2 + 33y - 7 = (5y \quad 1)(2y \quad 7)$

27. $12j^2 - jk - k^2 = (3j \quad k)(4j \quad k)$

28. $22n^2 + 23n - 15 = (11n \quad 5)(2n \quad 3)$

Find binomial factors for the following trinomials.

29. $21x^2 - 22x - 8$

30. $6p^2 + 7p - 5$

31. $2z^2 + 11z + 12$

32. $3a^2 - 14ab + 8b^2$

33. $20r^2 - 20rs - 15s^2$

34. $20g^2 + 13gh - 15h^2$

35. $64m^2 - 16m - 15$

36. $49x^2 + 14xy - 24y^2$

Find factors for the following.

37. $18v^2x + 3vwx - 6w^2x$

38. $2e^2f^2 + 60d^2f^2 + 34def$

39. **TRAVEL** Goods are transported by train from City A to City B. The distance between the two cities is represented by the expression $2x^2 + 7x + 3$. Factor the expression to find binomials representing the time it took to transport the goods and the train's speed.

40. **WRITING MATH** What strategies do you use to determine the signs for the second terms of the binomials when factoring trinomials with quadratic coefficients larger than 1?

41. Find the binomial factors for the expression $5r^2 + r - 18$.

42. **CONSTRUCTION** The volume of a concrete block is $16x^2 - 20x + 6$. The height of the block is 2 ft. Find the possible remaining dimensions of the block.

■ EXTENDED PRACTICE EXERCISES

43. Solve the equation $3x^2 + 30 = 40 + x$ by writing it in standard-form equal to zero. Then factor the trinomial and state the positive and negative solutions.

44. **BOATING** For a sailboat to fit a particular design, its right triangle sail must be 2 ft shorter than the boat along its base, and 3 times taller than the boat's length plus an extra foot. To catch enough wind, the sail area must be 124 ft². How long must the boat be to fit these requirements? (*Hint:* Write a quadratic equation and solve it by factoring.)

■ MIXED REVIEW EXERCISES

Write each in simplest radical form. (Lesson 10-1)

45. $\sqrt{156}$

46. $\sqrt{300}$

47. $\sqrt{16 \cdot 9}$

48. $\sqrt{261}$

49. $(3\sqrt{5})(2\sqrt{7})$

50. $(4\sqrt{3})(2\sqrt{21})$

51. $(\sqrt{15})(2\sqrt{18})$

52. $(5\sqrt{5})(7\sqrt{5})$

53. $\left(4\sqrt{11}\right)^2$

54. $\dfrac{\sqrt{8}}{\sqrt{3}}$

55. $\dfrac{\sqrt{13}}{\sqrt{6}}$

56. $\sqrt{\dfrac{7}{3}}$

Given $f(x) = 3x - 2$, $g(x) = -2x + 2$, and $h(x) = 4x^2$, find each value. (Lesson 2-2)

57. $f(-2)$

58. $f(3)$

59. $f(-5)$

60. $f(8)$

61. $g(5)$

62. $h(-4)$

63. $g(3)$

64. $g(-1)$

65. $h(2)$

66. $h(-3)$

67. $h(4)$

68. $h(-5)$

Chapter 11 Review

VOCABULARY ◣

Choose the word from the list that completes each statement.

1. A __?__ cannot be divided into smaller whole elements.
2. A __?__ equation is an equation of the form $a^2 + bx + c = 0$.
3. A number itself is called a __?__.
4. A simple expression with only one term is called a __?__.
5. Terms in which the variables or sets of variables are identical even though the coefficient may be different are called __?__.
6. __?__ is a method used to multiply two binomials.
7. A __?__ has two terms.
8. The square of a binomial is a __?__.
9. A polynomial is in __?__ when its terms are in order from the greatest power of one of its variables to the least power of that variable.
10. In the term $4x^2y$, the 4 is the __?__.

a.	binomial
b.	coefficient
c.	constant
d.	FOIL
e.	grand product
f.	greatest common factor
g.	like terms
h.	monomial
i.	prime element
j.	quadratic
k.	standard form
l.	trinomial

LESSON 11-1 ◣ Add and Subtract Polynomials, p. 468

▶ A **polynomial** is an expression that involves only sums and differences of several monomial terms. It is a **binomial** if it has two terms and a **trinomial** if it has three terms.

▶ A polynomial is written in **standard form** when its terms are ordered from the greatest power to the least power of one of the variables.

▶ Simplify a polynomial by combining all like terms.

Simplify.

11. $(5x + 6y) + (2x + 8y)$

12. $(7n + 11m) - (4m + 2n)$

13. $(5a + 3b) + 8a$

14. $(13r + 9s) - 11s$

15. $(12x^2 - 5) + (3x^3 - 6x^2 + 2)$

16. $(4a - 3a^2 + 1) - (2a + a^2 - 5)$

17. $(n^2 + 5n + 3) + (2n^2 + 8n + 8)$

18. $(3 + 2a + a^2) - (5 + 8a + a^2)$

LESSON 11-2 ◣ Multiply By a Monomial, p. 472

▶ Use the distributive property and the rules for exponents to multiply a polynomial by a monomial.

Simplify.

19. $(3d)(4d^2f)$

20. $(-8a^3b^2)^2$

21. $-3(8k + 5)$

22. $3st(5s^2 + 2st)$

23. $2x(x + 3y - z)$

24. $4m^2(9m^2n + mn - 5n^2)$

25. $-8xy(4xy + 7x - 14y^2)$

26. $-5ab[a - (a^2b - 3b)]$

LESSON 11-3 ◣ Divide and Find Factors, p. 478

▶ To **extract a factor**, check to see if any monomial will divide exactly into every term of the polynomial.

▶ To **factor** an expression, use the GCF and the distributive property.

Find the GCF and its paired factors for the following.

27. $81x^2y - 27x^3y^2$

28. $3a^3b^2 - 6ab$

29. $11x + 44x^3y$

30. $25m^2n^2 + 30mn^3$

31. $12ax + 20bx + 32cx$

32. $28r^2s^2t^2 + 21r^2st^2 - 14rst$

33. $9a^3b + 18a^2b^2 - 6a^2b^3$

34. $5x^5y - 10x^4y^2 - 20x^3y^3$

LESSON 11-4 ◣ Multiply Two Binomials, p. 482

▶ To multiply two binomials, write the product of the first terms, the outer terms, the inner terms, and the last terms (FOIL), then simplify.

Simplify.

35. $(c + 2)(c + 8)$

36. $(y + 3)(y - 7)$

37. $(m - 2n)(m + 2n)$

38. $(4a - b)(4a + b)$

39. $(2r + 3s)(2r + 3s)$

40. $(2x - 5y)(3x + 8y)$

41. $(5v - 7w)(4v + 3w)$

42. $(5d - 3)(2d + 1)$

LESSON 11-5 ◣ Find Binomial Factors in a Polynomial, p. 488

▶ To factor a polynomial, group terms as pairs, extract the common monomial factor from each pair, and extract the identical binomial.

Find factors for the following.

43. $6a^2 + 9ab - 10ab - 15b^2$

44. $14x^2 + 15y^2 - 10xy - 21xy$

45. $5rt + 20ru + 2st + 8su$

46. $2v^2 + 3vx + 10vw + 15wx$

47. $2ax + 6cx + ab + 3bc$

48. $6mx - 4m + 3rx - 2r$

49. $a^3 + a^2b + ab^2 - b^3$

50. $2x^3 - 5xy^2 - 2x^2y + 5y^3$

LESSON 11-6 ◣ Special Factoring Patterns, p. 492

▶ Use these patterns to factor perfect square trinomials and polynomials that are differences of squares.

$$a^2 + 2ab + b^2 = (a + b)^2 \qquad a^2 - 2ab + b^2 = (a - b)^2 \qquad a^2 - b^2 = (a + b)(a - b)$$

Find the binomial factors for the following.

51. $d^2 + 16d + 64$

52. $4k^2 - 4k + 1$

53. $1 - 9y^2$

54. $49 - a^2b^2$

55. $81m^2 - 16n^2$

56. $4e^2 + 12e + 9$

57. $9x^2 - 30x + 25$

58. $25y^2 - 49z^4$

LESSON 11-7 ◼ Factor Trinomials, p. 498

▶ You can use four clues to factor a trinomial where the coefficient of the first term is one; the trinomial third term is always the product of the binomial second terms; the coefficient of the trinomial second term is always the sum of the coefficients of the binomial second terms; if the sign of the trinomial third term is negative (positive), the signs in the binomial are different (the same).

Factor the following trinomials.

59. $x^2 - xy - 6y^2$

60. $m^2 + 3mn - 40n^2$

61. $r^2 - 10r + 16$

62. $a^2 + 8a + 15$

63. $g^2 + 7g - 44$

64. $m^2 - 15mn + 36n^2$

65. $a^2 + 2ab - 3b^2$

66. $x^2 - 4xy - 5y^2$

LESSON 11-8 ◼ Problem Solving Skills: The General Case, p. 502

▶ Studying the general case can help you identify patterns and solve problems. Use letters instead of numbers to represent the coefficients and constants.

67. Explore the pattern for finding the square of a binomial. Work from $(ax + n)^2$ as the general case and $(3x + 5)^2$ as the specific example.

68. Explore the pattern for finding the product of the sum and the difference of two values. Work from $(ax + n)(ax - n)$ and the general case and $(3x + 7)(3x - 7)$ as the specific example.

LESSON 11-9 ◼ More on Factoring Trinomials, p. 506

▶ To find binomial factors for a trinomial that has quadratic coefficients greater than one is a two-step process. First, identify FOIL coefficients. Then analyze the coefficient to find the four binomials coefficient (a, b, n_1 and n_2).

Find binomial factors for the following trinomials.

69. $3s^2 - 10s - 8$

70. $2r^2 + 3r - 14$

71. $9k^2 + 30k + 25$

72. $15x^2 - 13x + 2$

73. $4s^2 - 4st - 15t^2$

74. $15a^2 + 2ab - 8b^2$

75. $28a^2 - ab - 2b^2$

76. $30p^2 - 57pq + 18q^2$

CHAPTER INVESTIGATION

EXTENSION Decide which of the three advertisements, print, radio, or television, would be most effective to sell a product to people your own age. Present your decision to the class and explain why that type of advertisement would be most effective.

Chapter 11 Assessment

Simplify.

1. $(x^2 - 5x + 4) - (3x^2 + 2x - 5)$
2. $(3x^4 - 3xy + 6y^2) + (y^2 - 2x^4)$
3. $2(-3a^3)^2$
4. $(-4a^2b)(5a - 3b + a^2b)$
5. $(5c + d)(3c - 2d)$
6. $(2x + 5)(2x - 5)$
7. $(6r + s)(r - 3s)$
8. $(x - y)(x - y)$
9. $-5 + 4m^2 - 3m - 8m^2 + 13 + m$
10. $(5r - 6s^2 + rs) - (rs - 6s^2 - 5r)$

Factor.

11. $25x^3yz^2 - 30xyz^3$
12. $13mn^5 - 52mn^4$
13. $8n^2 - 2mn - 3n^2 + 12mn$
14. $rs + 2r^2 - 10rs - 5s^2$
15. $25x^2 - 30xy + 9y^2$
16. $4a^2 - 49b^2$
17. $81x^2 + 16$
18. $9m^2 - 24mn + 16n^2$
19. $a^2 + 7ab - 18b^2$
20. $x^2 - 4xy - 5y^2$
21. $m^2 - 7m + 10$
22. $r^2 + 8rs + 7s^2$
23. $16e^2 + 2ef - 3f^2$
24. $15x^2 - 8xy - 12y^2$

Use the figure below for Exercises 25 and 26.

25. Write an expression for the perimeter of the rectangle.

26. Write an expression for the area of the rectangle.

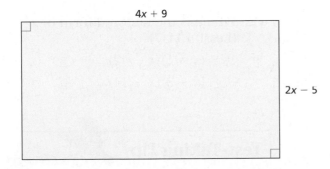

27. The width of a box is 9 in. more than its length. The height of the box is 1 in. less than its length. Write an expression for the volume of the box.

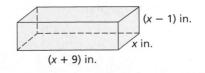

Standardized Test Practice

Record your answers on the answer sheet provided by your teacher or on a sheet of paper.

1. If $a + b = c$ and $a = b$, which of the following is *not* true? (Lesson 2-4)

   Ⓐ $a - c = b - c$ Ⓑ $a - b = 0$
   Ⓒ $2a + 2b = 2c$ Ⓓ $c - b = 2a$

2. Which inequality represents the shaded portion of the graph? (Lesson 2-6)

   Ⓐ $y \geq \dfrac{1}{3}x - 1$

   Ⓑ $y \leq \dfrac{1}{3}x - 1$

   Ⓒ $y \geq 3x - 1$

   Ⓓ $y \leq 3x - 1$

   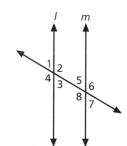

3. In the figure, $l \parallel m$. Choose two angles whose measures have a sum of 180°. (Lesson 3-4)

   Ⓐ $\angle 1$ and $\angle 5$
   Ⓑ $\angle 2$ and $\angle 8$
   Ⓒ $\angle 2$ and $\angle 5$
   Ⓓ $\angle 4$ and $\angle 8$

4. In $\triangle MNP$, $MP < NP$ and $MN < MP$. Which angle has the greatest measure? (Lesson 4-6)

   Ⓐ $\angle M$ Ⓑ $\angle N$ Ⓒ $\angle P$
   Ⓓ cannot be determined

5. If $3x + y = 5$ and $2x - 5y = 9$, what is the value of y? (Lessons 6-5, 6-6, and 6-7)

   Ⓐ -2 Ⓑ -1
   Ⓒ 1 Ⓓ 2

6. Solve $\dfrac{x+6}{5} = \dfrac{x}{4}$. (Lesson 7-1)

   Ⓐ -24 Ⓑ -12
   Ⓒ 12 Ⓓ 24

7. In the following equation, which value is the greatest? (Lesson 8-6)

   $$\begin{bmatrix} 2 & 3 \\ -1 & 5 \end{bmatrix}\begin{bmatrix} 0 & 1 \\ 4 & -2 \end{bmatrix} = \begin{bmatrix} a & b \\ c & d \end{bmatrix}.$$

   Ⓐ a Ⓑ b
   Ⓒ c Ⓓ d

8. *ABCD* is a rectangle. Find the approximate value of b. (Lesson 10-2)

   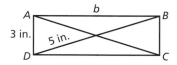

   Ⓐ 4.0 in.   Ⓑ 5.8 in.
   Ⓒ 9.5 in.   Ⓓ 10.4 in.

9. In the figure, what is the value of y? (Lesson 10-3)

   Ⓐ 5 m
   Ⓑ $5\sqrt{2}$ m
   Ⓒ $5\sqrt{3}$ m
   Ⓓ 10 m

   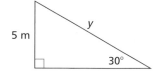

10. Which expression in simplest form is *not* a binomial? (Lesson 11-1)

    Ⓐ $x^2 - 1$ Ⓑ $7x + 2x$
    Ⓒ $m^3 + n^3$ Ⓓ $5t^2 + 3t - t$

11. If $s = t + 1$, which of the following is equal to $s^2 - t^2$? (Lesson 11-6)

    Ⓐ $(s - t)^2$ Ⓑ $t^2 - 1$
    Ⓒ $s^2 - 1$ Ⓓ $s + t$

12. Which is the factored form of $x^2 - 17x + 42$? (Lesson 11-7)

    Ⓐ $(x - 1)(x - 42)$ Ⓑ $(x - 2)(x - 21)$
    Ⓒ $(x - 3)(x - 14)$ Ⓓ $(x - 6)(x - 7)$

Test-Taking Tip   Ⓐ Ⓑ Ⓒ Ⓓ

Question 11
Sometimes you must use what you know about adding, subtracting, multiplying, and/or factoring polynomials to determine an answer. You can use the fact that $s^2 - t^2 = (s + t)(s - t)$ to solve this problem.

Preparing for Standardized Tests
For test-taking strategies and more
practice, see pages 709-724.

Part 2   Short Response/Grid In

Record your answers on the answer sheet provided by your teacher or on a sheet of paper.

13. If 0.00037 is expressed as 3.7×10^n, what is the value of n? (Lesson 1-8)

14. A plumber charges $70 for the first 30 minutes of each house call plus $4 for each additional minute that she works. The plumber charges Ke-Min $122 for her time. How much time did the plumber work? (Lesson 2-5)

15. The charge to enter a nature preserve depends on the number of people in each vehicle. The table shows some charges. Use this information to determine the charge for a vehicle with 8 people. (Lesson 3-5)

People	Charge
1	$1.50
2	$2.00
3	$2.50
4	$3.00

16. Brooke wants to fill her new aquarium two-thirds full of water. What volume of water is needed? (Lesson 5-7)

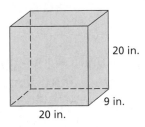

20 in.
20 in.
9 in.

17. Find the slope of $\overleftrightarrow{RS}$ containing points $R(-1, 5)$ and $S(4, 8)$? (Lesson 6-1)

18. What is the value of x in the figure? (Lesson 7-5)

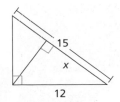

15
x
12

19. $\triangle ABC$ is rotated 180° clockwise about the origin. What are the coordinates of the image of $A(3, 4)$? (Lesson 8-2)

20. What is the probability of the spinner stopping on 5 or an even number? (Lesson 9-3)

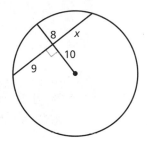

21. Use the figure below to find the value of x to the nearest tenth. (Lesson 10-6)

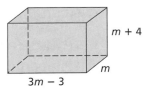

8
x
10
9

22. If $a^2 + b^2 = 40$ and $ab = 12$, find the value of $(a - b)^2$. (Lesson 11-4)

Part 3   Extended Responses

Record your answers on a sheet of paper. Show your work.

23. Use the rectangle prism below to solve the following problems. (Lessons 11-2 and 11-4)

$m + 4$
m
$3m - 3$

a. Write a polynomial expression that represents the surface area of the top of the prism.

b. Write a polynomial expression that represents the surface area of the front of the prism.

c. Write a polynomial expression that represents the volume of the prism.

d. If m represents 2 cm, what is the volume of the prism?

24. The polynomial $12ax^2 - 75ay^4$ can be factored as $3a(4x^2 - 25y^4)$. Can this expression be factored further? Explain. (Lesson 11-6)

Quadratic Functions

THEME: Gravity

Even very small children understand an important law of physics: When you drop something, it falls. But what makes the object fall? Scientists have named the force *gravity*. Gravity is measured by hanging the object on a spring scale. This measure is called the weight of the object.

By working to understand and measure gravity, scientists have succeeded in overcoming its effects.

- **Pilots** (page 529) command airplanes, jets, helicopters, and spacecraft. Pilots must understand how speed, altitude, temperature, and the weight of the plane, including its contents, affect air travel.

- **Air Traffic Controllers** (page 549) ensure safe air travel by monitoring the movement of aircraft. Controllers use radar and visual observation to monitor the progress of aircraft. They work together to make sure planes stay a safe distance apart.

Math
Online

mathmatters3.com/chapter_theme

How Does Gravity Affect Weight?

Location	Weight of 5-ton elephant on Earth
Mercury	2842 lb
Venus	9069 lb
Earth	10,000 lb
Moon	1656 lb
Mars	3803 lb
Jupiter	23,394 lb
Saturn	9253 lb
Uranus	7944 lb
Neptune	11,247 lb
Pluto	408 lb

Data Activity: How Does Gravity Affect Weight?

Use the table for Questions 1–4.

1. A tool weighs 2.5 lb on Earth. What is the weight of the tool on Mars? on Jupiter?

2. An astronaut has two oxygen tanks. On Earth, Tank A weighs twice as much as Tank B. If the two tanks are transported to Saturn, Tank A's weight will be how many times the weight of Tank B?

3. At which of the locations shown in the table would you weigh less than you do on Earth?

4. A bag of moon rocks weighs 225 lb on the moon. To the nearest tenth, what is the weight of the bag of rocks on Earth?

CHAPTER INVESTIGATION

A child's wagon has no engine or other visible means of moving itself forward. Yet, when the wagon is positioned at the top of a steep hill and begins to move down the hill, its speed increases as it goes. The wagon is propelled by gravity.

Working Together

Design a small gravity-driven vehicle weighing no more than 10 oz. Time the vehicle's descent down an incline, recording the angle of descent and the time. Explore how changing the shape or weight of the vehicle affects its speed. Use the Chapter Investigation icons to guide your group.

12 Are You Ready?

Refresh Your Math Skills for Chapter 12

The skills on these two pages are ones you have already learned. Review the examples and complete the exercises. For additional practice on these and more prerequisite skills, see pages 654–661.

MAKE A TABLE

Making a table of ordered pairs to help graph a linear equation is a skill that will be especially useful as you learn to graph quadratic functions.

Example   Graph the equation $2x + y = 8$.

Make a table of values. Generally, use -3 to 3 for the value of x.

x	y
−3	
−2	
−1	
0	
1	
2	
3	

Find the value of y for each value of x by substituting the value of x in the equation.
$$2(-3) + y = 8$$
$$2(-2) + y = 8$$
Solve each equation for y and fill in the table.

x	y
−3	14
−2	12
−1	10
0	8
1	6
2	4
3	2

With these ordered pairs you can graph the line of the equation.

Make a table of values for each equation. Round to the nearest thousandth if necessary.

1. $y = \frac{1}{2}x + 3$

2. $3y - 2x = 4$

3. $4x - 2y = 1$

4. $-2y + x = 3$

5. $y - 3x = 7$

6. $y + x = -6$

7. $2x - 3y = -6$

8. $4 + 3x = \frac{1}{2}y$

9. $5x + y = -8$

10. $4y + x = -7$

11. $8 - 3y = 2x$

12. $2x - 2y = -2$

GRAPHS OF FUNCTIONS

You can easily tell whether a graph is that of a function or not by the Vertical Line Test. When a vertical line is drawn through the graph of a relation, the relation is *not a function* if the vertical line intersects the graph in more than one point.

Graph of a function:

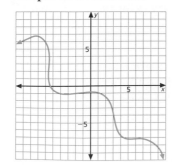

Not the graph of a function:

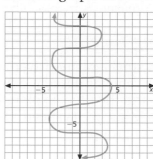

Use the Vertical Line Test to determine if each relation is a function.

13.

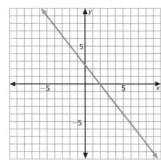

14.

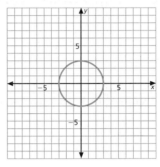

15.

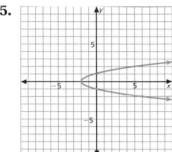

16.

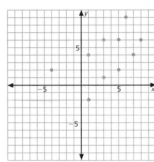

PYTHAGOREAN THEOREM

You have used the Pythagorean Theorem to find measures of the sides of right triangles. It is a very valuable formula to know, and one you will use in real life.

$$a^2 + b^2 = c^2$$

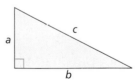

Find the missing side measures to the nearest hundredth.

17.

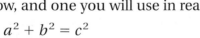

9
13

18.

7
8

19.

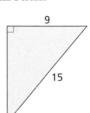

9
15

20.

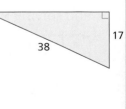

17
38

21.

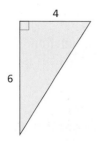

4
6

22.

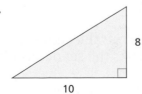

8
10

23.

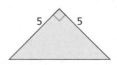

5 5

24.

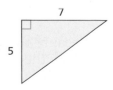

7
5

12-1 Graph Parabolas

Goals ■ Graph parabolas or second degree equations.

Applications Geology, Small Business, Physics

You will need a graphing calculator for this activity.

GRAPHING Use the ZOOM menu to make sure your display window is set on standard size. Then graph each of the following equations.

a. $y = -x$
b. $y = x^2 + 2x + 1$
c. $y = x + 3$
d. $y = -x^2 - 1$
e. $y = 2x - 2$
f. $y = 2x^2$

1. How can you tell by looking at an equation whether the graph will be a line or a curve?

2. Using the equations above, write two equations: one for a straight line and one for a curve.

◥ BUILD UNDERSTANDING

In this section you will learn to graph equations that contain second degree or **quadratic** terms.

A **quadratic equation** in x contains an x^2 term and involves no term with a higher power of x. The simplest quadratic equation is $y = x^2$.

Example 1

Graph $y = x^2$.

Solution

Find at least five ordered pairs by selecting x-values and solving the equation to find y-values.

x	-3	-1	0	1	3
y	9	1	0	1	9

Graph the ordered pairs.

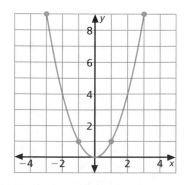

Draw a smooth curve through the points.

Because there is only one y-value for each x-value, this is the graph of a **function**. When the domain of a quadratic function is the set of real numbers, the graph is a **parabola**.

Notice that there is more than one x-value for each y-value. There are two x-values for each y-value except for point (0, 0), the lowest point on the parabola.

The parabola in Example 1 opens upward. For some functions, the parabolas open downward.

Example 2

Graph $y = -2x^2$.

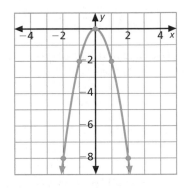

Solution

Make a table of ordered pairs.

x	-2	-1	0	1	2
y	-8	-2	0	-2	-8

Graph the points corresponding to the ordered pairs and draw a smooth curve through them.

The **vertex** is the lowest point on a parabola that opens upward, and the highest point on a parabola that opens downward. The graphs for $y = x^2$ and $y = -2x^2$ both have the point (0, 0) as their vertex.

Example 3

GEOLOGY The distribution of a trace element within a geologic sample can be modeled by the equation $y = 3x^2 - 2$. Graph and locate the vertex of the parabola.

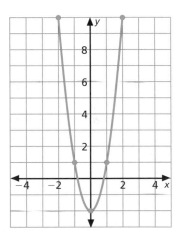

Solution

Make a table of ordered pairs.

x	-2	-1	0	1	2
y	10	1	-2	1	10

Graph the points and draw a smooth curve. Look for a y-value that has only one x-value. The vertex is (0, -2).

GRAPHING You can use a graphing calculator to find the coordinates of the vertex. Key in the quadratic equation and graph. You may have to use the zoom features to adjust the size of the display. Press the TRACE key. Your calculator will place a point at the y-intercept. If the y-axis is the line of symmetry, the coordinates for the y-intercept are also the coordinates for the vertex. Use a graphing calculator to locate the vertex of the parabola in Example 3.

You can use the arrow keys to move the trace point to locate the vertex or other points on the parabola. However, you must remember that there are limits to the display capabilities of a graphing calculator. The zoom feature allows you to see more detail.

The vertex lies on the line that divides the parabola in half. This line is called the **axis of symmetry** of the parabola.

In each of the three previous examples, the parabola is divided in half by the y-axis, which is the line $x = 0$. The axis of symmetry is not always $x = 0$. It is determined by the given equation.

Copy and complete each table. Then draw the graph.

1. $y = x^2 - 3$

x	-4	-2	0	2	4
y					

2. $y = -x^2$

x	-6	-3	0	3	6
y					

Graph each function for the domain of real numbers. For each graph, give the coordinates of the vertex.

3. $y = x^2 + 2$

4. $y = -x^2 - 3$

5. $y = -5x^2$

6. SMALL BUSINESS A study shows that the daily revenue from product sales can be modeled by the equation $y = -5x^2 + 12$, where y equals the revenue in hundreds of dollars and x equals possible increases and decreases in price. Graph the equation. What is the maximum revenue? (*Hint:* The y-coordinate of the vertex is the maximum revenue in hundreds of dollars.)

Graph each function for the domain of real numbers. For each graph, give the coordinates of the vertex.

7. $y = 4x^2$

8. $y = 2x^2 + 2$

9. $y = -x^2 + 1$

Determine if the graph of each equation below opens upward or downward.

10. $y = -3x^2 + 4$

11. $y = 7x^2$

12. $y = -x^2 - 10$

13. GRAPHING The equations below have the form $y = ax^2$. Graph each equation on a graphing calculator. How does the graph change as the value of a changes?

   a. $y = 10x^2$ **b.** $y = 4x^2$ **c.** $y = 0.5x^2$

   d. $y = -0.5x^2$ **e.** $y = -3x^2$ **f.** $y = -15x^2$

14. WRITING MATH What do you notice about the location of the vertex of a parabola that is the graph of an equation in the form $y = ax^2$? What do you notice about the location of the axis of symmetry of the graph?

15. GRAPHING The equations below are in the form $y = ax^2 + c$. Graph the equations on a graphing calculator. How does the graph change as the value of c decreases?

   a. $y = 2x^2 + 4$ **b.** $y = 2x^2 + 3$ **c.** $y = 2x^2 + 1$

   d. $y = 2x^2 - 1$ **e.** $y = 2x^2 - 3$ **f.** $y = 2x^2 - 5$

16. WRITING MATH What do you notice about the location of the vertex and axis of symmetry of the parabola you obtain when you graph an equation in the form $y = ax^2 + c$?

17. PHYSICS When an object is dropped and falls to the ground under the force of gravity, its height, y, in feet, x seconds after being dropped is given by $y = -16x^2 + 18$. Find the height from which it was dropped.

18. **PHYSICS** A projectile is shot vertically up in the air from ground level. Its distance d, in feet, after t seconds is given by $d = 96t - 16t^2$. Find the values of t when d is 96 ft.

EXTENDED PRACTICE EXERCISES

Each graph below is for an equation of the form $y = ax^2 + c$. The value of a is $+6$ or -6 for one equation, and $+1$ or -1 for the other. Write the equation for each graph.

19.
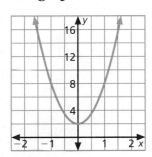

20.
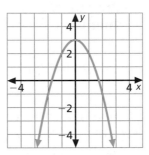

GRAPHING Use a graphing calculator. Graph each pair of equations in the same window. Find the vertex and axis of symmetry for each pair.

21. $y = \sqrt{x}$

$y = -\sqrt{x}$

22. $y = \sqrt{-x}$

$y = -\sqrt{-x}$

The graph of $y = ax^2$ has a maximum point or a minimum point.

23. What form of equation has a graph with a maximum point?

24. What form of equation has a graph with a minimum point?

25. **CHAPTER INVESTIGATION** Gravity can be used to power a vehicle on a ramp or incline. Work with your group to draw plans for a small vehicle weighing no more than 10 oz. As a first step, explore what types of materials to use to build the vehicle. Gather materials and weigh samples of each material in its raw form. Once you have made a final selection of materials, draw a design for a simple gravity-powered vehicle. Your vehicle should have wheels or employ some other technology to reduce friction.

MIXED REVIEW EXERCISES

Simplify. (Lesson 11-1)

26. $(6a - 3) + (4a + 6)$

27. $(2y^2 + y) - (5y + 8)$

28. $(4x^2 + 3x + 2) - (x + 4)$

29. $(3b^2 + 2b) - (4b - 3)$

30. $(6c + 2) + (5c - 8)$

31. $(-3d^2 + 8) - (4d - 9)$

In each triangle, $\overline{AB} \parallel \overline{CD}$. Find x to the nearest tenth. (Lesson 7-5)

32.
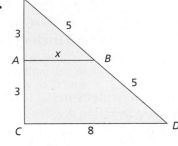

33.
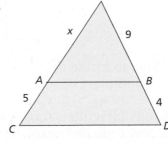

34.
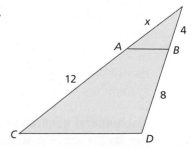

The General Quadratic Function

Goal ■ Graph functions defined by the general quadratic equation.

Applications Physics, Business, Astronomy

Work with a partner to answer the following questions.

1. **GRAPHING** The following quadratic equations have the form $y = ax^2 + bx$. Graph the equations on a graphing calculator.

$y = x^2 + 7x$ $y = x^2 + 5x$ $y = x^2 + 2x$

$y = x^2 - 7x$ $y = x^2 - 5x$ $y = x^2 - 2x$

$y = -x^2 + 7x$ $y = -x^2 + 5x$ $y = -x^2 + 2x$

$y = -x^2 - 7x$ $y = -x^2 - 5x$ $y = -x^2 - 2x$

2. Copy and complete each sentence. Write $>$ or $<$ in each blank.

   a. If a is __?__ 0 and b is __?__ 0, the vertex is in quadrant III.

   b. If a is __?__ 0 and b is __?__ 0, the vertex is in quadrant IV.

   c. If a is __?__ 0 and b is __?__ 0, the vertex is in quadrant I.

   d. If a is __?__ 0 and b is __?__ 0, the vertex is in quadrant II.

BUILD UNDERSTANDING

In Lesson 12-1, you investigated quadratic equations of the forms $y = ax^2$ and $y = ax^2 + c$. You learned that for quadratic equations, the value of y is determined by the value of x; y is a function of x. This relationship is expressed as $y = f(x)$.

As you have discovered, the graph of a quadratic equation is a parabola. For the parabolas you graphed in Lesson 12-1, the vertex was always located on the y-axis, and the y-axis was the axis of symmetry. Your exploration of equations in the form $y = ax^2 + bx$ showed that other locations are possible.

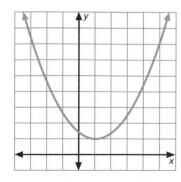

In this lesson, you will learn how to locate the vertex and axis of symmetry for any quadratic equation.

The **general quadratic function** may be written $f(x) = ax^2 + bx + c$ or $y = ax^2 + bx + c$, where a, b, and c are real numbers and $a \neq 0$.

Example 1

Graph $y = 3x^2 - 4x - 1$ on a graphing calculator. Estimate the coordinates of the vertex.

Technology Note

The *zoom* feature on a graphing calculator allows you to magnify a section of a graph. The *Zoom Box* defines the box to be enlarged.

To zoom in on the vertex:

1. Select *Box* from the *zoom* menu

2. Place the cursor on a corner of the area you want to magnify; press enter.

3. Move the cursor to the diagonally opposite corner of the box; press enter.

Solution

Enter the equation. Graph the function. Use the trace and zoom features to locate the coordinates of the vertex.

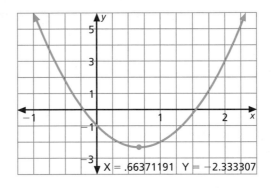

$X = .66371191 \quad Y = -2.333307$

The closer you zoom in, the closer the coordinates will be to the actual values of x and y. Eventually, you may be able to see a relationship between the decimal values on the screen and a common fraction or whole number. For example, you may have arrived at the coordinates $x = 0.6637119$ and $y = -2.333307$.

0.6637119 is about $\frac{2}{3}$.

2.333307 is about $2\frac{1}{3}$.

The vertex is approximately $\left(\frac{2}{3}, -2\frac{1}{3}\right)$.

For a parabola defined by the equation $y = ax^2 + bx + c$, the vertex is always at the point on the graph where the x-coordinate is $x = -\frac{b}{2a}$. The corresponding y-value can be found by substituting the x-value into the equation.

Example 2

Find the coordinates of the vertex for the graph of $y = 3x^2 - 4x - 1$.

Solution

$x = -\dfrac{b}{2a}$

$x = -\left(\dfrac{-4}{2(3)}\right)$ Substitute for a and b.

$x = \dfrac{4}{6}$ or $\dfrac{2}{3}$ Simplify.

$y = 3\left(\dfrac{2}{3}\right)^2 - 4\left(\dfrac{2}{3}\right) - 1$ Substitute the x-value into the equation.

$y = 3\left(\dfrac{4}{9}\right) - \dfrac{8}{3} - 1$ Simplify.

$y = -2\dfrac{1}{3}$

Check Understanding

Give the values of a, b, and c for each function.

1. $y = 6x^2 + 4x + 5$

2. $y = 7x^2 - 3x + 2$

3. $y = 9x^2 - 6x - 5$

4. $y = 7 - x^2$

The coordinates of the vertex for the graph of $y = 3x^2 - 4x - 1$ are $\left(\frac{2}{3}, -2\frac{1}{3}\right)$.

As you learned in Lesson 12-1, the axis of symmetry is a vertical line through the vertex of a parabola. The axis of symmetry for the graph of a quadratic function is $x = -\frac{b}{2a}$. For the graph of the function above, the axis of symmetry is $x = \frac{2}{3}$.

You can use the equation $x = -\dfrac{b}{2a}$ to graph a quadratic function.

Example 3

Graph $f(x) = -2x^2 - 3x + 1$.

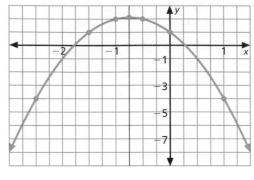

Solution

Locate the vertex.

$$x = -\frac{b}{2a} = -\left(\frac{-3}{2(-2)}\right) = -\frac{3}{4}$$

$$y = -2\left(-\frac{3}{4}\right)^2 - 3\left(-\frac{3}{4}\right) + 1$$

$$= -2\left(\frac{9}{16}\right) + \left(\frac{9}{4}\right) + 1 = 2\frac{1}{8}$$

The vertex is $\left(-\dfrac{3}{4}, 2\dfrac{1}{8}\right)$. The axis of symmetry is $x = -\dfrac{3}{4}$.

Because a is less than 0, the parabola opens downward. Substitute values in the equation to locate a few points.

x	-1	0	1
y	2	1	-4

Use the axis of symmetry to visually locate other points. Draw a smooth curve.

As you know, a function in x assigns only one y-value for each x-value. To tell whether a graph is of a function, check to see whether each vertical line in the coordinate plane contains at most one point of the graph.

◤ TRY THESE EXERCISES

GRAPHING Estimate the coordinates of the vertex for each parabola by graphing the equation on a graphing calculator. Then use $x = -\dfrac{b}{2a}$ to find the coordinates.

1. $y = x^2 - 4x + 3$ **2.** $y = 2x^2 + 3x - 1$

3. $5x^2 - 2x + 5 = y$ **4.** $y + 15 = x^2 - 2x$

Find the vertex and axis of symmetry. Then graph each equation.

5. $y = -3x^2 - 9x + 1$ **6.** $y = 2x^2 + 8x + 3$

7. $y = 3x^2 + 2x$ **8.** $-x^2 + y = 3 + x$

◤ PRACTICE EXERCISES • For Extra Practice, see page 700.

GRAPHING Estimate the coordinates of the vertex for each parabola by graphing the equation on a graphing calculator. Then use $x = -\dfrac{b}{2a}$ to find the coordinates.

9. $y = x^2 + 6x + 4$ **10.** $y - 3 = 2x^2 - 12x$ **11.** $-9 + y = -3x + 3x^2$

12. $y = x^2 + 4$ **13.** $-8x + 5 = -2x^2 + y$ **14.** $4x^2 - 16x + 1 = y$

Find the vertex and axis of symmetry.

15. $y = 5x^2 - 10x + 1$ **16.** $x^2 + x + 3 = y$ **17.** $y = -x^2 + 1$

18. BUSINESS The sales projections for a company can be represented by a quadratic equation in the form of $y = ax^2 + bx + c$ for which $c = 3$, and the axis of symmetry is $x = \frac{5}{6}$. Find the equation.

19. ASTRONOMY Observed movement in an object can be represented by a quadratic equation in the form $y = ax^2 + bx + c$ for which $c = -2$ and the vertex is $\left(-\frac{5}{8}, -\frac{7}{16}\right)$. Find the equation.

20. WRITING MATH Study graphs A and B shown at the right. Determine if each graph is the graph of a function. Explain your reasoning.

a. 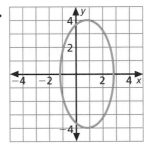   **b.**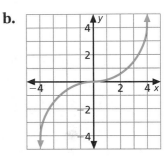

■ EXTENDED PRACTICE EXERCISES

PHYSICS A cannon is fired at several different angles. The paths of the cannonballs are shown on the graph.

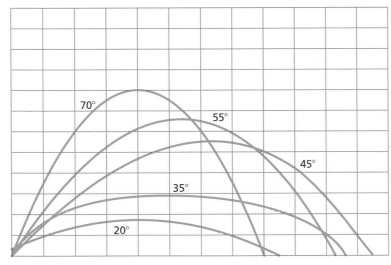

21. For what firing angle did the cannonball reach the highest point?

22. For what angles did the cannonball travel about the same distance?

23. WRITING MATH What conclusion can you draw about the maximum range of the cannon?

■ MIXED REVIEW EXERCISES

Simplify. (Lesson 11-2)

24. $(7bc)(3ab)$ **25.** $6s^2(s^2 + 3s + 2)$ **26.** $(5sz^4)(3s^4z^5)$

27. $(2a^2)(4ac - 3bd)$ **28.** $(4r^2s^3)(2r^3s^2 - 8rs^4)$ **29.** $(25x^4y^2)(4x^3y^2 - 2x^2y^5)$

Graph each function. (Lesson 2-3)

30. $y = \frac{2}{3}x + 8$ **31.** $y = 3x - 4$ **32.** $y = -2x + 3$

33. $y = \frac{1}{2}x - 2$ **34.** $y = -3x - 1$ **35.** $y = \frac{3}{4}x + 2$

Review and Practice Your Skills

Practice ◣ Lesson 12-1

1. The relation described by $y = x^2$ is a function. Explain why.

2. Describe the vertex of:

   a. a parabola that opens downward.

   b. a parabola that opens upward.

Copy and complete each table. Then draw the graph.

3. $y = x^2 + 4$

x	-3	-2	-1	0	1	2	3
y	13						

4. $y = 2x^2 - 1$

x	-3	-2	-1	0	1	2	3
y							

Determine if the graph of each equation opens upward or downward.

5. $y = -3x^2 + 4$ 6. $y = 3x^2 - 8$ 7. $y = 2x^2 + 4$ 8. $y = -(x^2 + 2)$

Use the graph of each function, where the domain is the real numbers, to name the coordinates of the vertex.

9. $y = x^2 - 6$

10. $y = 3x^2 + 1$

11. $y = -2x^2 + 1$

12. $y = -x^2 + 9$

Tell whether the vertex of the graph is at the origin.

13. $y = x^2 + 4$ 14. $y = x^2$ 15. $y = -x^2$ 16. $y = -x^2 + 3$

Tell whether the axis of symmetry of the graph is the y-axis.

17. $y = \pm \sqrt{2x}$

18. $y = 5x^2 - 3$

19. $y = \pm \sqrt{5x - 3}$

20. $y = -3x^2 + 1$

Practice ◣ Lesson 12-2

21. What is the general form of a quadratic equation?

22. If $f(x) = ax^2 + bx + c$ is a quadratic function, can $a = 0$?

Estimate the coordinates of the vertex for each parabola by graphing the equation on a graphing calculator. Then use $x = -\dfrac{b}{2a}$ to find the coordinates.

23. $y = 2x^2 - 3x + 4$

24. $y = -3x^2 + 4x - 5$

25. $y = x^2 - 2x + 3$

26. $y + 4x^2 = 2x - 6$

Find the vertex and axis of symmetry.

27. $y = -2x^2 + x - 4$ 28. $y = 3x^2 + 6x - 8$ 29. $y = 2x^2 - 3x + 5$

Use the graph of each function to name the coordinates of the vertex.
(Lesson 12-1)

30. $y = x^2 - 4x + 2$ **31.** $y = -2x^2 + x - 3$ **32.** $y = 3x^2 + 2x + 4$

Graph each equation on a graphing calculator. Give the quadrant of the vertex.
(Lesson 12-2)

33. $y = -3x^2 + 3x - 7$ **34.** $y = 2x^2 + 5x + 8$ **35.** $y = -x^2 + x + 5$

Find the vertex and axis of symmetry. (Lesson 12-2)

36. $y = x^2 + 4x - 3$ **37.** $y = -3x^2 - 2x + 4$ **38.** $y = 2x^2 - 3x - 5$

Math*Works* Career – Pilots
Workplace Knowhow

Pilots are the commanders of airplanes, jets, helicopters and space shuttles. Pilots transport passengers and goods, fight fires, test new aircraft, monitor traffic and crime, dust crops and conduct rescue missions. Pilots must make many calculations to fly safely. They must be able to read electronic instruments accurately. For example, they use math to calculate the speed necessary for take off. To do so, they must consider many variables such as altitude of the airport, outside temperature, weight of the plane and speed and direction of wind.

One plane left New York headed to Tokyo flying at an average speed of 375 mi/h. Another plane left New York 1 h later following the same route and flying at an average speed of 500 mi/h. If both planes followed the same course, how many hours after it left New York would the second plane catch up to the first plane?

1. Write an equation you could solve to answer the question.

2. Use a graphing utility to graph both sides of the equation and solve the problem.

3. How many miles will each plane have flown when they are equidistant from New York?

4. A plane gained altitude at a rate of 1000 ft/min, descended 1300 ft, then started to climb again for $2\frac{1}{4}$ min at a rate of 800 ft/min. The total gain in altitude was 6000 ft. How long did the plane climb at a rate of 1000 ft/min?

5. An airplane takes off and climbs at a steady 18° angle. After flying along a path of 2 mi, how much altitude has the plane gained in feet? Round to the nearest foot.

12-3 Factor and Graph

Goals ■ Use factoring to solve quadratic equations.

Applications Physics, Archery, Sports

GRAPHING Graph each equation below on a graphing calculator. How many x-intercepts, the points where the graph crosses the x-axis, does each graph have? Use the trace feature, if necessary, to locate the x-intercepts.

a. $y = x^2 - 25$ **b.** $y = 2x^2 + 3x - 1$

c. $y = 2x^2 + 3x + 5$ **d.** $y = x^2 - x$

e. $y = x^2 + 5$ **f.** $y = x^2 + 6x + 9$

◤ BUILD UNDERSTANDING

The x-intercepts are the solutions of the quadratic equation. They are the x-coordinates of points for which $y = 0$. As illustrated in the above activity, there may be 0, 1, or 2 x-intercepts.

You can determine the number of x-intercepts and estimate their values by graphing the quadratic equation. You can often find exact solutions by letting $y = 0$ and factoring the quadratic expression.

Example 1

Solve $3x^2 - 6x = 0$.

Solution

By graphing:

Graph the related quadratic function $y = 3x^2 - 6x$.

Locate the points where $y = 0$. Estimate the x-values for these points. Use the zoom feature to more closely estimate values. The x-values are the solutions.

The graph of the equation $y = 3x^2 - 6x$ intersects the x-axis at two points. They are located approximately at $x = 2$ and $x = 0$.

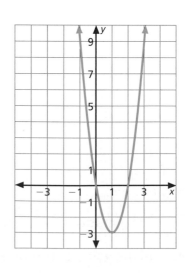

By factoring:

To solve by factoring, let $y = 0$.

$$3x^2 - 6x = 0$$
$$3x(x - 2) = 0 \qquad \text{Factor.}$$
$$3x = 0, \quad x - 2 = 0 \quad \text{Solve each equation.}$$
$$x = 0, \quad x = 2$$

The solutions for $y = 3x^2 - 6x = 0$ are 0 and 2.

If the parabola is tangent to the x-axis, there is only one solution to the quadratic equation.

Mental Math Tip

An expression in the form $ax^2 + bx + c$ in which a, b, and c are integers may be factored if ac has factors with a sum of b.

Example 2

Solve $x^2 + 8x + 16 = 0$.

Solution

Write the related quadratic function.

$$y = x^2 + 8x + 16$$

By graphing:

Graph the equation $y = x^2 + 8x + 16$ to determine the number of solutions. Locate the point where $y = 0$. The x-value is the solution.

The graph of the equation meets the x-axis at one point, approximately $x = -4$.

By factoring:

To solve by factoring, let $y = 0$.

$$x^2 + 8x + 16 = 0$$
$$(x + 4)(x + 4) = 0$$
$$x = -4$$

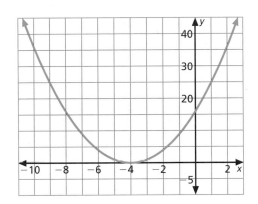

The solution for $x^2 + 8x + 16 = 0$ is -4.

If the parabola for the equation does not meet the x-axis, the equation has no solutions.

Example 3

Solve $-x^2 - x - 1 = 0$.

Solution

Graph the related function on a graphing calculator.

The graph of the equation does not cross the x-axis.

There are no solutions for the equation $y = -x^2 - x - 1$.

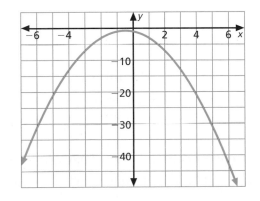

Technology Note

If you have access to charting and data analysis software, you may want to investigate methods for graphing functions. For many programs, you enter data and equations on a spreadsheet. The program then draws a graph of the function.

Suppose a projectile is lauched from ground level. If you know the velocity with which the projectile is launched, you can find the time between launch and landing using the equation $vt - 16t^2 = 0$, where v = velocity in feet per second (ft/sec) and t = time in seconds.

Example 4

PHYSICS A football is thrown with the initial velocity of 64 ft/sec. How long does it remain in the air?

Solution

Substitute 64 for v in the equation.	$64t - 16t^2 = 0$
Factor the equation.	$16t(4 - t) = 0$
Set each factor equal to 0 and solve for t.	$16t = 0 \qquad 4 - t = 0$
	$t = 0 \qquad\quad t = 4$

The equation has two solutions. The first solution represents the launch time; the second represents the landing time. The football stays in the air for 4 sec.

TRY THESE EXERCISES

GRAPHING Use a graphing calculator to determine the number of solutions for each equation. For equations with one or two solutions, find the exact· solutions by factoring.

1. $0 = x^2 + 10x + 21$

2. $0 = x^2 + 5x + 6$

3. $0 = x^2 + 25$

4. $0 = x^2 - 14x + 49$

5. $0 = x^2 + 18x + 81$

6. $0 = x^2 - 9$

PRACTICE EXERCISES • For Extra Practice, see page 701.

GRAPHING Use a graphing calculator to determine the number of solutions for each equation. For equations with one or two solutions, find the exact solutions by factoring.

7. $0 = x^2 - 100$

8. $0 = x^2 + 7x$

9. $x^2 - 9x + 14 = 0$

10. $0 = x^2 - 9x + 25$

11. $0 = -x^2 + 4$

12. $x^2 + x = 0$

13. $0 = x^2 + x + 1$

14. $0 = x^2 - 10x + 25$

15. $0 = x^2 + 8x - 48$

Write an equation for each problem. Then factor to solve.

16. The square of a positive integer is 20 less than 12 times the integer. Find the integer.

17. The square of a number exceeds the number by 30. Find the number.

18. The square of an integer is 5 more than 4 times the integer. Find the integer.

19. Seven times an integer plus 8 equals the square of the integer. Find the integer.

For Exercises 20–25, use the equation $vt - 16t^2 = 0$, **where** $v =$ **velocity in feet per second and** $t =$ **time in seconds.**

20. **PHYSICS** The initial velocity of a projectile is 128 ft/sec. In how many seconds will it return to the ground?

21. **ARCHERY** How long will an arrow shot from ground level with an initial velocity of 176 ft/sec remain in the air?

22. **SPORTS** How long will a football kicked with an initial velocity of 96 ft/sec remain in the air?

23. **PHYSICS** A rocket at a fireworks display was launched with the initial velocity of 208 ft/sec. How many seconds was it in the air before it splashed down in the lake?

24. **PHYSICS** A projectile is launched with the initial velocity of 896 ft/sec. How many seconds will it remain in the air?

 25. **WRITING MATH** The equation $vt - 16t^2 = 0$ is of the form $y = ax^2 + bx$. From Exercises 20–23, what generalization can you make about the solutions to quadratic equations in the form $0 = ax^2 + bx$?

Use your generalization from Exercise 25 to solve each of the following equations.

26. $x^2 - 3x = 0$

27. $x^2 + 2x = 0$

28. $5x^2 + 20x = 0$

29. $2x^2 - 8x = 0$

30. $9x^2 - 72x = 0$

31. $3x^2 + 33x = 0$

EXTENDED PRACTICE EXERCISES

The sum of the first n positive even integers is $S = n^2 + n$. How many integers must be added to give each sum?

32. 240

33. 2070

34. 1122

35. 40,200

 36. **CHAPTER INVESTIGATION** Working together, build your gravity-powered vehicle. Try to keep the weight as close to 10 oz as possible. If necessary add weight by adding new design elements. If your vehicle has wheels, make sure the wheels turn freely. You may want to use a drop of oil or powdered graphite to lubricate the axle.

MIXED REVIEW EXERCISES

Find factors for the following. (Lesson 11-3)

37. $9c - 27b$

38. $x^2y - x$

39. $3mn^2 - 9mn$

40. $8a^2b + 32ab^2$

41. $wx + xz$

42. $17a^2b - 68ab$

Find the GCF and its paired factor for the following. (Lesson 11-3)

43. $36a - 63b$

44. $12ab - 8a^2b$

45. $10x^3 - 15x^2$

46. $7a^3bc^2 - 14a^2bc$

47. $27x^3y^2 - 6x^2y$

48. $72p^2q^3r - 32p^2q^4r$

49. $6x^3y^3 - 9x^2y^4 + 6x^2y^3z$

50. $24a^3b^2c - 18ab^2c^3 + 12abc^2$

12-4

Complete the Square

Goals ■ Solve quadratic equations by completing the square.

Applications Science, Aeronautics, Sports

 MODELING Work with a partner to build equations using Algeblocks.

1. Use Algeblocks to illustrate each perfect square.

$$x^2 - 6x + 9 = (x - 3)^2 \qquad x^2 + 10x + 25 = (x + 5)^2$$

$$x^2 + 14x + 49 = (x + 7)^2 \qquad x^2 - 2x + 1 = (x - 1)^2$$

2. Each of the squares above is in the form $x^2 + bx + c = (x + h)^2$. Discuss the following:

    a. What is the relationship between c and h?

    b. What is the relationship between h and b?

    c. What is the relationship between c and b?

> **Problem Solving Tip**
>
> Use the relationships you discover for questions **a** and **b** to find **c**.

◥ BUILD UNDERSTANDING

Making a perfect square for an expression of the form $ax^2 + bx$ is called **completing the square**. Completing the square is another method for solving quadratic equations.

Example 1

Complete the square for $x^2 - 8x$.

Solution

 MODELING Use Algeblocks to illustrate $x^2 - 8x$. Add blocks to make a perfect square. Write the expression.

$$(x^2 - 8x + 16) - 16$$

In the activity at the top of the page, you discovered that for perfect squares in the form

$ax^2 + bx + c$, the constant $c = \left(\dfrac{b}{2}\right)^2$. Thus, by substitution, you know $ax^2 + bx + c = ax^2 + bx + \left(\dfrac{b}{2}\right)^2$.

You can use this relationship to complete the square for $x^2 - 8x$.

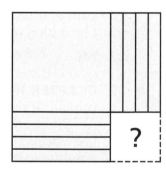

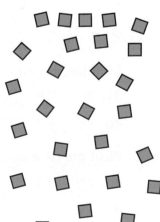

$$x^2 - 8x + \left(\dfrac{b}{2}\right)^2 \qquad \text{Add } \left(\dfrac{b}{2}\right)^2 \text{ to complete the square.}$$

$$x^2 - 8x + (-4)^2 \qquad \text{Find } \dfrac{b}{2}, -\dfrac{8}{2} = -4.$$

$$x^2 - 8x + 16 \qquad \text{Square } \dfrac{b}{2}, (-4)^2 = 16.$$

The expression is $(x^2 - 8x + 16) - 16$. This is equivalent to the original expression.

You can use the process of completing the square to solve quadratic equations.

Example 2

Solve by completing the square.

$$x^2 + 8x + 12 = 0$$

Solution

$x^2 + 8x + 12 = 0$	
$x^2 + 8x = -12$	Add -12 to each side.
$x^2 + 8x + 16 = -12 + 16$	Add $\left(\frac{b}{2}\right)^2$ to each side.
$(x + 4)^2 = 4$	Factor.
$x + 4 = \sqrt{4}$	Simplify.
$x + 4 = \pm 2$	
$x = -4 + 2, \; x = -4 + (-2)$	
$x = -2, \; x = -6$	

Problem Solving Tip

Always check your solutions by substituting them back into the original equation.

The solutions of the equation $x^2 + 8x + 12 = 0$ are -2 and -6.

To solve a quadratic equation by completing the square, the coefficient of the x^2 term must be 1.

Example 3

Mr. Bruno has a square garden in his yard. He wants to double the area by increasing the length 6 ft and the width by 4 ft. Find the original dimensions of the garden.

Solution

Let x be the length of each side of the garden. The original area of the garden is $A = x^2$. He wants this area to be doubled. Write this as a quadratic equation.

$$2x^2 = (x + 6)(x + 4)$$
$$2x^2 = x^2 + 6x + 4x + 24$$

$x^2 - 10x = 24$	Simplify.
$x^2 - 10x + 25 = 24 + 25$	Add $\left(\frac{b}{2}\right)^2$ to each side of the equation.
$(x - 5)^2 = 49$	Factor the left side of the equation.
$x - 5 = 7$	Take the square root of both sides.
$x = 12$	

The garden was 12 ft on each side.

TRY THESE EXERCISES

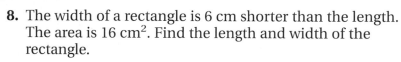

MODELING Complete the square. Use Algeblocks if you wish.

1. $x^2 + 4x$

2. $x^2 - 6x$

3. $x^2 - 2x$

Solve by completing the square. Check your solutions.

4. $x^2 - 2x - 8 = 0$

5. $x^2 + 6x = 7$

6. $2x^2 - 5x + 2 = 0$

7. $x^2 + 4x - 12 = 0$

Write an equation. Then complete the square to solve the problem.

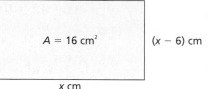

8. The width of a rectangle is 6 cm shorter than the length. The area is 16 cm². Find the length and width of the rectangle.

9. **WRITING MATH** Explain why you cannot use the negative solution to the equation to find the answer to the problem in Exercise 8.

10. Evan needs to solve the equation $x^2 - 2x = 15$. After completing the square, he writes $x^2 - 2x + 1 = 0$, factors the equation as $(x - 1)^2 = 0$ and solves for x. Evan is surprised when his solution, $x = 1$, doesn't check. What did he do wrong? Find the correct solutions.

PRACTICE EXERCISES • For Extra Practice, see page 701.

Complete the square.

11. $x^2 - 10x$

12. $x^2 + 20x$

13. $x^2 + x$

14. $x^2 - 14x$

15. $x^2 + 18x$

16. $x^2 - 30x$

17. $x^2 + 3x$

18. $x^2 - 16x$

19. $x^2 - x$

Solve by completing the square. Check your solutions.

20. $x^2 - 3x - 28 = 0$

21. $3x^2 - 2x - 5 = 0$

22. $9x^2 - 18x = 0$

23. $x^2 + 3x = 0$

24. $x^2 - 2x + 1 = 0$

25. $2x^2 - 9x = 5$

26. $x^2 - x - 12 = 0$

27. $2x^2 + 5x = 3$

28. $x^2 - 6x = 7$

29. $2x^2 - 4x = 0$

30. $6x^2 - 5x = -1$

31. $x^2 = 6x$

Find values for c and h to complete each perfect square.

32. $x^2 + 20x + c = (x + h)^2$

33. $x^2 - 4x + c = (x + h)^2$

34. $x^2 + x - c = (x + h)^2$

35. $x^2 - 3x + c = (x + h)^2$

36. $x^2 - 18x + c = (x + h)^2$

37. $x^2 + 22x + c = (x + h)^2$

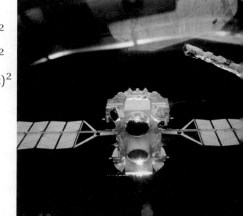

38. **AERONAUTICS** A length of a rectangular panel on a satellite is 4 cm greater than its width. Its area is 165 cm². Find its dimensions.

39. PHYSICS A ball is thrown from the ground up into the air at an initial velocity of 64 ft/sec. How long will it take the ball to reach the ground? Use the equation $64t - 16t^2 = 0$, where t equals time in seconds.

40. SPORTS A length of a rectangular playing field is twice the width plus 4 ft. Its area is 2310 ft². Find the length, width, and perimeter in feet of the field's boundary line.

■ EXTENDED PRACTICE EXERCISES

Solve each problem. If necessary, write an equation and then complete the square.

41. The width of a rectangle is 2 in. less than its length. Find its dimensions if its area is 360 in.².

42. If the width and length of a 4-in. by 2-in. rectangle are each increased by the same amount, the area of the rectangle will be 48 in.². Find the new length and new width.

43. A triangle with an area of 6 m² has a base that is 4 m longer than its height. What are the dimensions?

44. If the height and base of a triangle with a height of 5 cm and a base of 8 cm are each decreased by the same amount, the area of the triangle will be 14 cm². Find the new base and height.

45. ART A painting is 1 in. longer than it is wide. The painting and its frame have a total area of 156 in.². The frame is 1 in. wide on each side of the painting. What are the dimensions of the painting?

46. CHAPTER INVESTIGATION Work with your group to build a ramp from 24–36 in. in length. One possibility would be to attach a length of poster board to two yard sticks, using the sticks for stability. Set the ramp up so that the angle formed by the ramp and floor is 15°. Using a stopwatch, time your vehicle's descent from the top of the ramp to the bottom. Make sure you do not push the vehicle at the starting point. Increase the ramp's incline in 5° increments. How does the increase affect your vehicle's travel time? For each angle of descent, measure the distance the vehicle travels beyond the bottom of the ramp. How does changing the incline affect this distance?

■ MIXED REVIEW EXERCISES

Simplify. (Lesson 11-4)

47. $(3a - 2b)(2a - 4)$
48. $(6x + 2)(4k - 3)$
49. $(2a - 3c)(4b + 2d)$

50. $(5x - y)(3x - 1)$
51. $(3c - 2)(3c - 2)$
52. $(5a + 1)(5a - 1)$

53. $(4x + 2)(3x - 1)$
54. $(x + 9)(2x - 4)$
55. $(7a + 2b)(5a - 3b)$

56. $(2c - 5b)(4c - 3b)$
57. $(12x + 2y)(5x - 6y)$
58. $(4b - 7y)(4b + 7y)$

Write each in simplest radical form. (Lesson 10-1)

59. $\sqrt{77}$
60. $\sqrt{112}$
61. $\sqrt{56}$

62. $(2\sqrt{5})(3\sqrt{10})$
63. $(2\sqrt{17})(4\sqrt{22})$
64. $(5\sqrt{11})(4\sqrt{32})$

65. $\dfrac{\sqrt{18}}{\sqrt{6}}$
66. $\dfrac{\sqrt{40}}{\sqrt{5}}$
67. $\sqrt{\left(\dfrac{80}{12}\right)}$

Review and Practice Your Skills

PRACTICE ◤ LESSON 12-3

Determine if each statement is *true* or *false*.

1. There are always 2 solutions to every quadratic equation.

2. The x-coordinate of a point at which the graph of a quadratic function intersects the x-axis is a solution of the related quadratic equation.

3. For every quadratic equation, you can find exact solutions of an equation by factoring the equation, then substituting the value of x in the equation to find y.

4. If a parabola graphed on the coordinate plane does not meet the x-axis, how many real solutions of the related quadratic equation are there?

Use a graphing calculator to determine the number of solutions for each equation. For equations with one or two solutions, find the exact solutions by factoring.

5. $0 = x^2 - x - 12$
6. $0 = x^2 + 6x + 8$
7. $0 = x^2 + 4x + 4$
8. $0 = x^2 - x - 6$
9. $0 = x^2 - 9x + 25$
10. $0 = x^2 - 7x + 10$

Write an equation for each problem. Then factor to solve.

11. The square of a positive integer is 20 more than the integer.

12. The square of a positive integer is 9 less than 6 times the integer.

13. The square of a positive integer is 2 less than 3 times the integer.

14. The square of a positive integer is 6 less than 5 times the integer.

PRACTICE ◤ LESSON 12-4

15. For a perfect square in the form $ax^2 + bx + c$, what formula can be substituted for the constant c?

16. To solve a quadratic equation by completing the square, what must be true about the x^2 term?

Complete the square.

17. $x^2 - 4x$
18. $x^2 + 10x$
19. $x^2 - 18x$
20. $x^2 - 14x$
21. $x^2 + 5x$
22. $x^2 - 13x$

Solve by completing the square. Check your solutions.

23. $3x^2 - 5x - 2 = 0$
24. $x^2 - x = 2$
25. $x^2 - x = 0$
26. $2x^2 - 7x - 15 = 0$
27. $x^2 + 4x = 12$
28. $8x^2 - 6x = -1$

29. The length of a rectangle is 4 in. greater than its width. Find its dimensions if its area is 32 in.^{2}.

30. The length of a swimming pool is 3 times its width. Find its dimensions if its area is 432 ft^2.

Copy and complete each table. Then draw the graph. (Lesson 12-1)

31. $y = x^2 - 5$

x	-3	-2	-1	0	1	2	3
y							

32. $y = 3x^2 + 2$

x	-3	-2	-1	0	1	2	3
y							

Find the vertex and axis of symmetry. (Lesson 12-2)

33. $y = x^2 - 3x + 4$

34. $y = 3x^2 + x + 3$

35. $y = -2x^2 - 4x + 1$

36. $y = -x^2 + 5x - 3$

37. $y = 2x^2 - 3x + 4$

38. $y = -x^2 - 4x - 7$

Solve each equation by factoring or by completing the square.
(Lessons 12-3–12-4)

39. $y = x^2 + 12x - 1$

40. $y = x^2 - 4x + 3$

41. $y = x^2 + x - 12$

42. $y = x^2 - 5x - 14$

43. $y = x^2 + 10x - 4$

44. $y = x^2 - 2x - 15$

45. $y = x^2 + 8x + 12$

46. $y = x^2 + 7x - 2$

47. $y = x^2 - 6x - 7$

48. The length of a rectangle is 7 in. greater than its width. Find its dimensions if its area is 144 in.^{2}.

Mid-Chapter Quiz

Graph each function. Then name the vertex and the axis of symmetry, and state whether the graph opens upward or downward. (Lesson 12-1)

1. $y = 7x^2$

2. $y = -3x^2 + 1$

Use $x = -\dfrac{b}{2a}$ to find the coordinates of the vertex for each parabola. (Lesson 12-2)

3. $y = 2x^2 + x - 1$

4. $y = -3x^2 - 2x + 2$

Write the equation for the line of symmetry for the graph of each equation.
(Lesson 12-2)

5. $y + 2x = x^2 + 4$

6. $y - 3 = 3x^2 + 6x$

Find the solution or solutions of each equation by factoring. (Lesson 12-3)

7. $0 = x^2 + 2x$

8. $0 = -x^2 - 4x - 4$

9. $0 = -x^2 - 3x$

10. $0 = x^2 + 3x - 4$

Solve by completing the square. (Lesson 12-4)

11. $x^2 - 4x + 3 = 0$

12. $2x^2 + 16x - 18 = 0$

13. $x^2 - x = 3\dfrac{3}{4}$

14. $x^2 - 20x = 21$

Goals ■ Solve equations using the quadratic formula.

Applications Aeronautics, Skydiving, Physics

Work in groups of 2–3 students.

By looking for relationships among the variables, coefficients and constants in equations, we can write general rules or steps for solving all equations of the same form.

1. Work together to write a list of steps for solving any quadratic equation in the form $ax^2 + bx + c = 0$ by completing the square.

2. Check your steps by using them to solve the equation $x^2 - x - 3\frac{3}{4}$.

■ BUILD UNDERSTANDING

Any quadratic equation can be solved by completing the square; however, repeating the steps from the activity above for each equation you solve can be a lengthy process. Instead of repeating the steps, you can use the general quadratic equation $ax^2 + bx + c = 0$ to develop a formula for solving quadratic equations. The formula is found by solving the general quadratic equation for x.

$$ax^2 + bx + c = 0$$

$$x^2 + \frac{b}{a}x + \frac{c}{a} = 0 \qquad \text{Multiply each term by } \frac{1}{a}.$$

$$x^2 + \frac{b}{a}x = -\frac{c}{a} \qquad \text{Add } -\frac{c}{a} \text{ to each side of the equation.}$$

$$x^2 + \frac{b}{a}x + \frac{b^2}{4a^2} = \frac{b^2}{4a^2} - \frac{c}{a} \qquad \text{Half of } \frac{b}{a} = \frac{b}{2a}. \text{ Add } \left(\frac{b}{2a}\right)^2 \text{ or } \frac{b^2}{4a^2}$$
$$\text{to each side of the equation.}$$

$$\left(x + \frac{b}{2a}\right)^2 = \frac{b^2 - 4ac}{4a^2} \qquad \text{Factor the left side of the equation.}$$
$$\text{Combine terms on the right side of the equation.}$$

$$x + \frac{b}{2a} = \pm\frac{\sqrt{b^2 - 4ac}}{4a^2} \qquad \text{Find the square roots.}$$

$$x = -\frac{b}{2a} \pm \sqrt{\frac{b^2 - 4ac}{2a}} \qquad \text{Subtract } \frac{b}{2a} \text{ from each side of the equation.}$$

$$x = -\frac{b \pm \sqrt{b^2 - 4ac}}{2a} \qquad \text{Simplify.}$$

The formula for x in terms of a, b, and c is called the **quadratic formula**.

$$x = \frac{-b \pm \sqrt{b^2 - 4ac}}{2a}$$

The quadratic formula can be used to solve any quadratic equation of the form $ax^2 + bx + c = 0$, $a \neq 0$.

Example 1

Use the quadratic formula to solve $6x^2 - 5x + 1 = 0$.

Solution

$$x = \frac{-b \pm \sqrt{b^2 - 4ac}}{2a}$$

$$x = \frac{-(-5) \pm \sqrt{(-5)^2 - 4(6)(1)}}{2(6)} \quad \text{Substitute for } a, b, \text{ and } c.$$

$$x = \frac{5 \pm \sqrt{25 - 24}}{12} \quad \text{Simplify.}$$

$$x = \frac{5 \pm \sqrt{1}}{12} = \frac{5 \pm 1}{12}$$

$$x = \frac{5 + 1}{12} \quad \text{and} \quad x = \frac{5 - 1}{12}$$

$$x = \frac{6}{12} \text{ or } \frac{1}{2} \qquad x = \frac{4}{12} \text{ or } \frac{1}{3}$$

The equation $6x^2 - 5x + 1 = 0$ has two solutions, $\frac{1}{2}$ and $\frac{1}{3}$.

Problem Solving Tip

Remember the product property of square roots:

For any non-negative real numbers a and b,

$$\sqrt{a} \times \sqrt{b} = \sqrt{ab}$$

so, $\sqrt{12} = \sqrt{4 \times 3} =$

$$\sqrt{4} \times \sqrt{3} = 2\sqrt{3}$$

The radical part of the solutions is in simplified form if it contains no factors that are perfect squares.

Example 2

Use the quadratic formula to solve $x^2 = 6x + 3$.

Solution

$$x^2 - 6x - 3 = 0 \qquad \text{Write the equation in standard form.}$$

$$x = \frac{-b \pm \sqrt{b^2 - 4ac}}{2a} \qquad \text{Use the quadratic formula.}$$

$$x = \frac{-(-6) \pm \sqrt{(-6)^2 - 4(1)(-3)}}{2(1)} \qquad \text{Substitute for } a, b, \text{ and } c.$$

$$x = \frac{6 \pm \sqrt{36 + 12}}{2} \qquad \text{Simplify.}$$

$$x = \frac{6 \pm \sqrt{48}}{2}$$

$$x = \frac{6 \pm 4\sqrt{3}}{2} \text{ or } 3 \pm 2\sqrt{3}$$

The solutions for the quadratic equation $x^2 = 6x + 3$ are $3 + 2\sqrt{3}$ and $3 - 2\sqrt{3}$.

You can use your knowledge of quadratic equations to solve many problems involving distance. For example, gravity acts on a freely falling object according to the formula $h = 16t^2$. Using this formula, you can find the time it takes for an object to fall a certain distance.

Now consider the path of a projectile. Velocity, the force applied to the object, and gravity act upon the object. The path of a projectile is described by the formula $h = vt - 16t^2$, where h = height (ft), v = velocity (ft/sec) and t = time (sec).

Example 3

AERONAUTICS A model rocket is launched at a velocity of 80 ft/sec.

a. How long does it take the rocket to reach its maximum height?

b. What is the maximum height reached by the rocket?

c. How many seconds does it take the rocket to return to the ground?

Solution

The velocity is 80 ft/sec, so $h = 80t - 16t^2$.

a. At its maximum height, the rocket is at the vertex of the parabola formed by the height function. The time required to reach the maximum height is the x-coordinate of the vertex. Use $-\dfrac{b}{2a}$ to find the time.

$$-\frac{80}{2(-16)} = \frac{80}{32} = 2.5$$

It takes the rocket 2.5 sec to reach its maximum height.

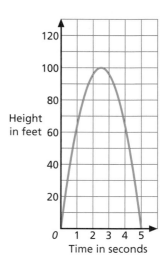

b. The maximum height is the y-coordinate of the vertex. Substitute 2.5 for t in the equation.

$$h = 80(2.5) - 16(2.5)^2$$

$$= 200 - 16(6.25)$$

$$= 200 - 100 = 100$$

The maximum height reached by the rocket is 100 ft.

c. At ground level, h is 0. To find when the rocket returns to the ground, solve the related quadratic equation.

$$80t - 16t^2 = 0$$

$$5t - t^2 = 0$$

$$t(5 - t) = 0 \quad t = 0, t = 5$$

The rocket returns to ground level 5 sec after launch.

◤ TRY THESE EXERCISES

Use the quadratic formula. Solve each equation.

1. $x^2 - 4x + 1 = 0$

2. $2x^2 + 15x = 8$

3. $4x^2 = 8x$

4. $-6x = -3x^2 - 3$

5. $x^2 + 2x = 6$

6. $7x - 15 = -2x^2$

◤ PRACTICE EXERCISES • For Extra Practice, see page 702.

Use the quadratic formula to solve each equation.

7. $x^2 - 9x = -14$

8. $2x^2 = 6 - x$

9. $4x^2 + 4x + 1 = 0$

10. $x^2 + 2x = 11$

11. $14x = 2x^2$

12. $3x^2 = 84 - 9x$

13. $2x^2 + 32 = 16x$

14. $x^2 = 50$

15. $x^2 + 4x + 2 = 0$

16. $x^2 + 7x = -12$

17. $45 = 2x^2 + x$

18. $x^2 - 8 = 0$

Choose factoring or the quadratic formula to solve each equation.

19. $x^2 + 2x = 35$

20. $2x - 4 = -x^2$

21. $x^2 = -3x$

22. $x^2 - 2x = 7$

23. $x - 21 = -2x^2$

24. $x = x^2$

25. $6x^2 = 2 - x$

26. $x^2 - 10x = 3$

27. $x^2 = 12x - 11$

28. WRITING MATH Suppose Galileo gathered the data shown. Explain how Galileo might use this data to show that free-fall distance is a function of the square of time.

29. SKYDIVING A parachutist free-falls for 20 sec. How far is the free-fall?

30. PHYSICS How many times longer does a projectile launched at 200 ft/sec stay in the air than one launched at 100 ft/sec?

Time (seconds)	Distance (feet)
0	0
1	16
2	64
3	144
4	256

■ EXTENDED PRACTICE EXERCISES

31. AERONAUTICS A model rocket is launched at an initial velocity of 96 ft/sec.

  a. How long will it take the rocket to reach its maximum height?

  b. What is the maximum height reached by the rocket?

  c. How many seconds does it take the rocket to return to the ground?

32. PHYSICS An object free-falls for 15 sec, and another free-falls for 30 sec. How many times farther does the second object fall than the first?

33. DATA FILE Use the data on the heights of some bridges on page 644. How long would it take a stone dropped from each bridge to reach the water below? Find your answer to the nearest tenth of a second.

34. PHYSICS A ball is thrown up into the air from the roof of a 128-ft-tall building. The initial velocity is 64 ft/sec. How long will it take the ball to reach the ground?

Math: Who, Where, When

Italian physicist and astronomer Galileo Galilei (1564–1642) probably did not drop cannonballs from the Leaning Tower of Pisa. However, from his research came the quadratic law of falling bodies.

■ MIXED REVIEW EXERCISES

Find factors for the following. (Lesson 11-5)

35. $2a^2 - 5ab + 3b^2$

36. $18x^2 - 3xy - 6y^2$

37. $16a^2 - 16ab + 4b^2$

38. $8x^2 + 18xy + 9y^2$

39. $6m^2 - 7mn - 20n^2$

40. $36a^2 - 12ab + 8b^2$

41. $24a^3 + 16a^2b - 8ab^2$

42. $10z^2 - 26xz - 12x^2$

43. $8m^2 - 10mn - 12n^2$

44. $49a^2 - 28ab + 4b^2$

45. $6x^2 + 4xy - 2y^2$

46. $3a^2 + 2ab - 5b^2$

Solve each system of equations. (Lesson 6-6)

47. $3x - 2y = -14$
$4x + 2y = 0$

48. $3y + x = 10$
$y - 2x = 1$

12-6 The Distance Formula

Goals ■ Use the Pythagorean Theorem, distance and midpoint formulas.

Applications Space Exploration, Sports, Archaeology

Work with a partner.

1. Using the coordinate plane shown below, count squares to find the length and midpoint of each leg of the right triangle.

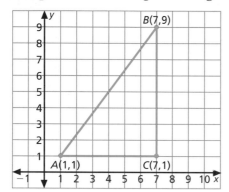

2. Note that $\overline{BC}$ is parallel to the y-axis, and $\overline{AC}$ is parallel to the x-axis. Answer the following questions.

   a. How can you calculate the length of $\overline{BC}$ given the y-coordinates of B and C?

   b. The midpoint of $\overline{BC}$ is halfway between the endpoints B and C. How can you calculate the y-coordinate of the midpoint, given the y-coordinates of B and C?

   c. How can you calculate the length of $\overline{AC}$, given the x-coordinates of A and C?

   d. How can you calculate the x-coordinate of the midpoint of $\overline{AC}$?

3. Use the Pythagorean Theorem to find the length of hypotenuse $\overline{AB}$.

4. The midpoint of $\overline{AB}$ is (4, 5). What is the relationship between this point and the midpoints of the two legs?

◥ BUILD UNDERSTANDING

A formula for calculating the distance, d, between any two points on the coordinate plane may be derived using the Pythagorean Theorem.

For any two points (x_1, y_1) and (x_2, y_2), a right triangle formed by drawing horizontal and vertical segments that intersect at (x_2, y_1) has legs with lengths $|x_2 - x_1|$ and $|y_2 - y_1|$.

These lengths may be substituted in the Pythagorean Theorem to find d.

$$d = \sqrt{(x_2 - x_1)^2 + (y_2 - y_1)^2}$$

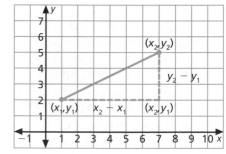

The distance, d, between any two points (x_1, y_1) and (x_2, y_2) may be found using the **distance formula**.

$$d = \sqrt{(x_2 - x_1)^2 + (y_2 - y_1)^2}$$

Example 1

SPACE EXPLORATION Suppose the grid shown was superimposed over a photograph taken by a space probe of the Martian landscape. Calculate the distance between points $P(-2, 4)$ and $Q(4, -1)$.

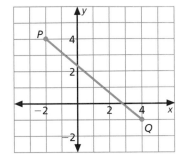

Solution

$$d = \sqrt{(x_2 - x_1)^2 + (y_2 - y_1)^2}$$

$$d = \sqrt{(-2 - 4)^2 + (4 - (-1))^2}$$

$$d = \sqrt{(-6)^2 + 5^2}$$

$$d = \sqrt{36 + 25} = \sqrt{61} \approx 7.8$$

The distance between points P and Q is $\sqrt{61}$ or about 7.8 units.

In the activity at the beginning of the lesson, you observed that for a segment parallel to the x-axis, the x-coordinate of the midpoint is $\dfrac{x_1 + x_2}{2}$. For a segment parallel to the y-axis, the y-coordinate of the midpoint is $\dfrac{y_1 + y_2}{2}$.

We can use these two facts to derive the **midpoint formula**. For a line segment with endpoints (x_1, y_1) and (x_2, y_2), the coordinates of the midpoint are $\left(\dfrac{x_1 + x_2}{2}, \dfrac{y_1 + y_2}{2}\right)$.

Example 2

Find the midpoint of the line segment with endpoints $M(-2, 5)$ and $N(6, -3)$.

Solution

$$\left(\frac{x_1 + x_2}{2}, \frac{y_1 + y_2}{2}\right)$$ Use the midpoint formula.

$$\left(\frac{-2 + 6}{2}, \frac{5 + (-3)}{2}\right)$$ Substitute coordinate values.

$$\left(\frac{4}{2}, \frac{2}{2}\right) = (2, 1)$$ Simplify.

The midpoint of line segment MN is $(2, 1)$.

Distances on aerial and satellite photos are often estimated using a coordinate grid. Even sporting events require these calculations. Suppose you are working for a television company broadcasting a major league baseball game. A blimp is transmitting overhead shots of the field at all times. Using a computer, you could instantly overlay each visual with a grid and calculate distances for the announcers to use in the broadcast.

Example 3

SPORTS On the grid shown at the right, one unit represents 30 ft. A batter hits a ball from home plate to point L. What is the distance?

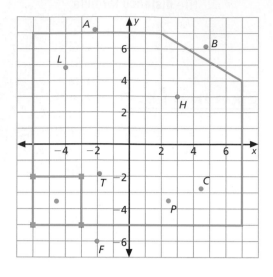

Solution

The coordinates of home plate are $(-6, -5)$. Point L is located at approximately $(-4, 5)$. Use the distance formula to find the distance between the points.

$$d = \sqrt{(x_2 - x_1)^2 + (y_2 - y_1)^2}$$

$$d = \sqrt{(-4 - (-6))^2 + (5 - (-5))^2}$$

$$d = \sqrt{2^2 + 10^2}$$

$$d = \sqrt{4 + 100} = \sqrt{104} \approx 10.2 \text{ units}$$

Since each unit equals 30 ft, multiply by 30: $10.2 \times 30 = 306$ ft. The ball traveled about 306 ft.

TRY THESE EXERCISES

Use the graph on the right. Round distances to the nearest tenth of a unit.

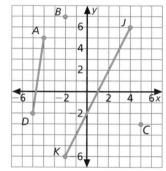

1. Find the length of segment AD.

2. Find the distance between B and C.

3. Locate the midpoint of segment JK.

Use the distance formula to calculate the distance between each pair of points. Round to the nearest tenth.

4. $G(-9, 4)$, $H(-5, -1)$

5. $P(3, -2)$, $Q(-2, 11)$

6. $M(0, 0)$, $N(6, -7)$

Use the midpoint formula to find the midpoint of the segment with the given endpoints.

7. $E(5, 5)$, $F(-3, -3)$

8. $L(-10, 4)$, $M(2, 6)$

9. $T(-9, 4)$, $S(1, -1)$

10. **SPORTS** Use the baseball field diagram from Example 3. One unit on the grid equals 30 ft. If a ball is caught at point H and thrown to second base at coordinates $(-3, -2)$, how long is the throw?

PRACTICE EXERCISES • For Extra Practice, see page 702.

Calculate the distance between each pair of points. Round to the nearest tenth.

11. $W(20, -1)$, $X(-6, 5)$

12. $C(8, 5)$, $D(-3, -6)$

13. $G(-1, 5)$, $H(2, 6)$

14. $P(5, 5)$, $Q(-6, 1)$

Find the midpoint of the segment with the given endpoints.

15. $E(16, -5)$, $F(4, 1)$

16. $Y(-10, -9)$, $Z(-2, -3)$

17. ARCHAEOLOGY The diagram of an archaeological site is drawn on a coordinate grid. An archaeologist notes that standing stones are placed at $A(-3, -1)$, $B(1, 3)$, and $C(7, -3)$. The stones are connected by a low wall to form a triangle. What are the coordinates of the midpoint of the segment connecting the midpoints of $\overline{AB}$ and $\overline{AC}$?

18. What type of triangle is formed by connecting the midpoints of line segments formed by $L(-1, -2)$, $M(-1, 4)$, $N(6, -2)$?

Mesa Verde National Park Cliff Dwellings, Colorado

SPORTS Use the baseball field diagram from Example 3 for Exercises 19–22. Estimate distances on the ground. One unit on the grid represents 30 ft.

19. A home run is hit from home plate to point B. What is the distance?

20. A foul ball is hit from home plate to point F. What is the distance?

21. The right fielder catches a ball at point C and throws the ball to first base at the point with coordinates $(-3, -5)$. How long is the throw?

22. What is the home-run distance from home plate to point A?

23. DATA FILE Use the diagram of a soccer field on page 653. Using only the lengths given and a grid overlay, estimate the following distances shown by colored arrows: green (Neyome to Young), red (Young to Lato), purple (Kasberczak to Carr), and black (Young to Correa).

24. YOU MAKE THE CALL The point $(3, 5)$ is the midpoint of a segment that has $(7, 11)$ as one endpoint. Janice says there isn't enough information to find the other endpoint of the segment. Do you agree? Explain your thinking.

25. WRITING MATHEMATICS When finding the distance between two points $(1, 4)$ and $(8, 3)$, explain why it makes no difference which point you use as (x_1, y_1) and (x_2, y_2).

◣ EXTENDED PRACTICE EXERCISES

26. Use the distance formula to find the equation for a circle with radius 5 and center at point $(0, 0)$.

27. Use the distance formula to find the equation for a circle with radius r and center at point $(0, 0)$.

28. Find an equation for the circle with center $(0, 2)$ and radius 8.

◣ MIXED REVIEW EXERCISES

Find binomial factors for the following, if possible. (Lesson 11-7)

29. $3x^2 + 10x - 8$

30. $2x^2 - 3x - 5$

31. $6x^2 - 26x + 24$

32. $x^2 - 4x + 4$

33. $2x^2 + 13x + 24$

34. $15x^2 + x - 2$

35. $12x^2 - 14xy - 6y^2$

36. $28x^2 + 30xy + 8y^2$

37. $10x^2 - 41x + 21$

38. $24x^2 + 4xy - 8y^2$

39. $16x^2 + 24xy + 9y^2$

40. $2x^2 - 16xy + 30y^2$

Review and Practice Your Skills

PRACTICE ◣ LESSON 12-5

Determine if each statement is *true* or *false*.

1. The quadratic formula was derived from the standard form of a quadratic equation.

2. The quadratic formula cannot be used if there is no x term in the quadratic equation.

3. The equation $0x^2 + 5x - 3 = 0$ is not a quadratic equation.

4. The quadratic formula cannot be used to solve $0x^2 - 8x + 4 = 0$. Explain why.

Use the quadratic formula to solve each equation.

5. $x^2 - 7x + 2 = 0$ **6.** $2x^2 + 5x - 3 = 0$ **7.** $x^2 - 4x + 3 = 0$

8. $3x^2 - 4x + 1 = 0$ **9.** $-2x^2 - 3x + 5 = 0$ **10.** $-x^2 + 4x + 8 = 0$

11. $2x^2 - 7x + 2 = 0$ **12.** $4x^2 + x - 2 = 0$ **13.** $-2x^2 + x + 5 = 0$

PRACTICE ◣ LESSON 12-6

14. When finding the distance on a coordinate plane, either endpoint may be designated as (x_1, y_1), or (x_2, y_2). Give an example that proves or disproves this theory.

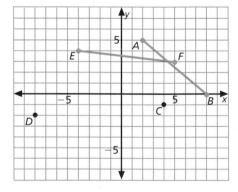

Use the graph on the right. Round distances to the nearest tenth of a unit.

15. Find the length of segment AB.

16. Find the distance between C and D.

17. Find the midpoint of segment EF.

Calculate the distance between each pair of points. Round to the nearest tenth.

18. $Y(1, 6), Z(-4, 3)$ **19.** $W(2, -5), X(4, 2)$ **20.** $U(1, -5), V(3, 7)$

21. $S(-4, -2), T(-5, 3)$ **22.** $Q(7, -2), R(-5, 3)$ **23.** $O(3, 3), P(-3, -3)$

Calculate the midpoint of the segment with the given endpoints.

24. $A(-4, 3), B(3, -4)$ **25.** $C(1, 5), D(3, -2)$ **26.** $E(4, -7), F(-2, 3)$

27. $G(5, 3), H(-5, -3)$ **28.** $I(6, -1), J(-2, 5)$ **29.** $K(4, 6), L(1, -3)$

Calculate the midpoint and distance between each pair of points.

30. $Z(-3, 5), Y(-2, -6)$ **31.** $X(4, -5), W(-2, 1)$ **32.** $V(0, 3), U(5, 4)$

33. $T(-1, 4), S(-2, -3)$ **34.** $R(2, 2), Q(-5, -5)$ **35.** $P(4, -8), O(-8, 4)$

36. Use the distance formula to find an equation for the circle with a radius of 4 and center at point $(0, 0)$.

For each equation, find the vertex and axis of symmetry of the graph. (Lessons 12-1–12-2)

37. $y = x^2 + 3x - 2$ **38.** $y = -2x^2 + x - 2$ **39.** $y = 3x^2 - 4x + 2$

40. $y = -x^2 - 5x + 4$ **41.** $y = 2x^2 - 5x - 1$ **42.** $y = x^2 + 7x - 3$

Solve each equation by graphing, factoring, or completing the square.
(Lessons 12-3–12-4)

43. $y = x^2 - 5$ **44.** $x^2 - x - 6 = 0$ **45.** $x^2 + 6x + 8 = 0$

46. $x^2 - 6x + 2 = 0$ **47.** $x^2 - 8x + 15 = 0$ **48.** $y = 2x^2 + 4$

49. $x^2 + 3x - 4 = 0$ **50.** $x^2 - 15x + 56 = 0$ **51.** $x^2 + 5x - 1 = 0$

52. The width of a rectangular mirror is 9 in. less than its length. Find its dimensions if the area of the mirror is 360 in.^{2}.

Solve each equation using the quadratic formula. (Lesson 12-5)

53. $x^2 - 4x - 3 = 0$ **54.** $-2x^2 - 5x + 2 = 0$ **55.** $2x^2 + 4x - 9 = 0$

MathWorks
Workplace Knowhow

Career – Air Traffic Controllers

Air traffic controllers manage the movement of air traffic through sections of air space. Controllers work together to monitor the movement of an aircraft from one section to another. They keep aircraft a safe distance apart and work to keep departures and arrivals on schedule. Controllers use radar and visual observation to monitor the progress of all aircraft. They also monitor the weather conditions for pilots. Air traffic controllers communicate directly with the pilot to direct the path of the flight. Together, controllers and flight crews make course corrections and respond to dangers caused by other aircraft, weather or emergency situations on the ground.

1. To determine the distance between two planes in the air, a three-coordinate system must be employed using ordered triplets (x, y, z). Each plane's position is measured in relation not only to a horizontal x-axis and a vertical y-axis, but also to a depth-measuring z-axis perpendicular to the other two axes. Place the origin $(0, 0, 0)$ of the three axes at the O'Hare Airport control tower and measure units in miles. Using the axes and units, describe a plane with coordinates $(8, 2, 5)$.

2. The space distance d between two points (x_1, y_1, z_1) and (x_2, y_2, z_2) is given by $d = \sqrt{(x_2 - x_1)^2 + (y_2 - y_1)^2 + (z_2 - z_1)^2}$. Two planes have position coordinates $(-3, 5, -4)$ and $(9, 2, 0)$. How far apart are they?

3. A passenger jet is midway between two private aircraft with position coordinates $(6, 2, -13)$ and $(1, -8, -5)$. Find the coordinates of the jet. Explain how you found your answer.

You have learned that the shape of the graph of a quadratic function is a parabola. You also know that quadratic functions are written in the form $y = ax^2 + bx + c$.

Suppose you were given the graph of a parabola. Can you use your knowledge to work backwards to discover the equation of the graph?

Problem Solving Strategies

Guess and check

Look for a pattern

Solve a simpler problem

Make a table, chart or list

Use a picture, diagram or model

Act it out

✔ Work backwards

Eliminate possibilities

Use an equation or formula

Problem

SCIENCE The results of an experiment are represented by the parabola shown at the right. What is the equation of the graph?

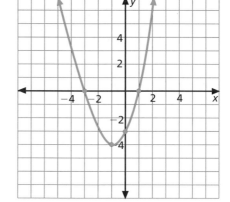

Solve the Problem

You can determine the equation of a parabola if you know three points on its graph.

Begin with the general form of a quadratic function: $y = ax^2 + bx + c$, where a, b and c represent coefficients and constants.

Locate three points on the graph. As you can see, the parabola passes through $(0, -3)$, $(-1, -4)$ and $(-3, 0)$.

Substitute the x-values and y-values for each point into the equation to create a system of three equations.

For $(0, -3)$
$ax^2 + bx + c = y$
$c = -3$

For $(-3, 0)$
$ax^2 + bx + c = y$
$9a - 3b + c = 0$

For $(-1, -4)$
$ax^2 + bx + c = y$
$a - b + c = -4$

Now use any of the methods you have learned for solving systems of equations.

$c = -3$
$9a - 3b + c = 0$
$a - b + c = -4$

Substitute -3 for c in the other two equations. Then solve for a.

$$9a - 3b - 3 = 0 \rightarrow \quad 9a - 3b - 3 = 0$$
$$a - b - 3 = -4 \rightarrow \quad \underline{3a - 3b - 9 = -12} \quad \text{Multiply by 3.}$$
$$6a \qquad + 6 = 12 \quad \text{Subtract}$$

Since $6a + 6 = 12$, $a = 1$. Substitute 1 for a and -3 for c in the second equation.

$$9a - 3b + c = 0$$
$$9 - 3b - 3 = 0$$
$$6 = 3b$$
$$b = 2$$

Thus, $a = 1$, $b = 2$ and $c = -3$. Check these values in the third equation. Then use these values in the form $y = ax^2 + bx + c$ to write an equation.

The equation of the parabola is $y = x^2 + 2x - 3$.

Five-step Plan

1 Read
2 Plan
3 Solve
4 Answer
5 Check

TRY THESE EXERCISES

1. The graph of a quadratic function contains the three points $(-5, 10)$, $(0, -5)$ and $(2, 3)$. Find the equation of the function.

2. The graph of a quadratic function contains the three points $(2, 7)$, $(0, -1)$ and $(-1, 1)$. Find the equation of the function.

PRACTICE EXERCISES

3. **WRITING MATH** In the example on page 550, will you obtain the same equation if you select three different points? Explain.

4. **PHYSICS** The relationship between two variables in an experiment can be represented by the parabola shown. Using three points from the graph, find the equation of the parabola.

5. A parabola contains the points $(1, -4)$, $(0, -3)$, $(2, -3)$, $(3, 0)$ and $(-2, 5)$. Choose three points and find the equation of the parabola.

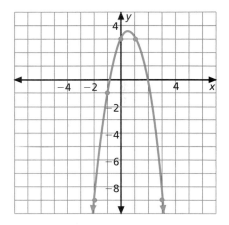

Find the equations for each parabola.

6.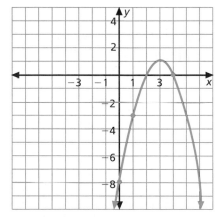

7.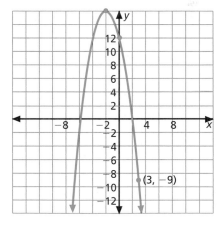

$(3, -9)$

8. **CHAPTER INVESTIGATON** Calculate your vehicle's average rate of speed in feet per second for each run of the ramp. Make a poster showing a diagram of the vehicle and the vehicle's speed for various inclines of the ramp. Is there a relationship between the increase in the incline and the vehicle's speed?

MIXED REVIEW EXERCISES

Factor the following trinomials. (Lesson 11-9)

9. $15a^2 - 14a - 8$

10. $a^2 + 5a - 14$

11. $a^2 + 11a + 24$

12. $12a^2 - 5a - 2$

13. $3a^2 - 14a - 24$

14. $4a^2 - 16a + 16$

Chapter 12 Review

VOCABULARY ◤

Match the letter of the word in the list at the right with the description on the left.

1. an equation that contains an x^2 term and involves no term with a higher power of x

2. the relationship in which each x-value has a unique y-value

3. the lowest point on a parabola that opens upward or the highest point on a parabola that opens downward

4. a function of the form $f(x) = ax^2 + bx + c$, where $a \neq 0$

5. the line that divides a parabola in half

6. the graph of a quadratic function

7. used to find the length of a segment given the coordinates of the endpoints

8. solving a quadratic function by adding $\left(\frac{b}{2}\right)^2$ to both sides of the related equation

9. express a polynomial as a product of monomials, binomials, or polynomials

10. formula used to solve quadratic functions in the form $ax^2 + bx + c = 0$

- **a.** axis of symmetry
- **b.** complete the square
- **c.** distance formula
- **d.** factoring
- **e.** function
- **f.** general quadratic function
- **g.** midpoint formula
- **h.** parabola
- **i.** quadratic equation
- **j.** quadratic formula
- **k.** vertex
- **l.** x-intercept

LESSONS 12-1 and 12-2 ◤ Quadratic Functions, p. 520

▶ To graph a quadratic function, find at least five ordered pairs by selecting x-values and finding the corresponding y-values. Locate and draw a smooth curve through the points.

▶ To locate the vertex, use $x = -\dfrac{b}{2a}$ to find the x-coordinate. Then substitute the x-value into the equation to find the y-coordinate.

▶ The axis of symmetry for the graph passes through the vertex and is parallel to the y-axis.

Graph each function. Give the coordinates of the vertex for each graph.

11. $y = -3x^2$
12. $y = 4x^2 - 4x$
13. $y = -2x^2 + 4x + 1$

Each graph below is in the form $y = ax^2 + c$. Write an equation for each parabola.

14.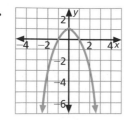

15.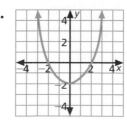

16.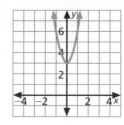

► A function of x assigns each x-value exactly one y-value. If each vertical line in the coordinate plane contains at most one point of the graph, the graph represents a function.

Determine if each graph is the graph of a function.

17. 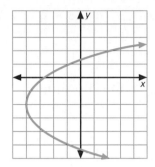 18. 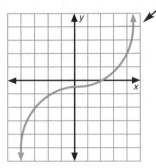 19.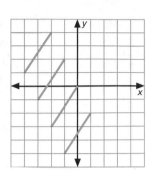

LESSON 12-3 ◣ Factor and Graph, p. 530

► The x-intercepts are the solutions of the related quadratic equation. The number of x-intercepts and approximate solutions may be found by graphing. To find the exact solutions, let $y = 0$, and factor the equation.

Use a graphing calculator to determine the number of solutions for each equation. Factor to solve each equation.

20. $x^2 - 14x = -49$ 21. $x^2 - 63 = -2x$ 22. $3x^2 + 9x - 84 = 0$

The length of a Rugby League field is 52 m longer than its width w.

23. Write an expression of the area of the field.

24. The area of a Rugby League field is 8160 m². What are the dimensions of the field?

LESSON 12-4 ◣ Complete the Square, p. 534

► To complete the square for an expression in the form $ax^2 + bx$, add $\left(\dfrac{b}{2}\right)^2$.

► To solve an equation by completing the square, rewrite the equation in the form $ax^2 + bx = c$, add $\left(\dfrac{b}{2}\right)^2$ to each side of the equation, factor, and simplify.

Solve each equation by completing the square. Round to the nearest tenth if necessary.

25. $x^2 - 2x - 8 = 0$ 26. $x^2 - 8x = -15$ 27. $x^2 + 14x = -49$

28. $x^2 - 4x - 12 = 0$ 29. $d^2 + 3d - 10 = 0$ 30. $y^2 - 19y + 4 = 70$

31. $d^2 + 20d + 11 = 200$ 32. $a^2 - 5a = -4$ 33. $p^2 - 4p = 21$

34. $x^2 + 4x + 3 = 0$ 35. $d^2 - 8d + 7 = 0$ 36. $r^2 - 10r = 23$

LESSON 12-5 ◣ The Quadratic Formula, p. 540

► The quadratic formula can be used to solve any quadratic equation of the form $ax^2 + bx + c = 0$, $a \neq 0$.

The quadratic formula is $x = \dfrac{-b \pm \sqrt{b^2 - 4ac}}{2a}$.

Use the quadratic formula to solve each equation.

37. $2x^2 - 5x = 3$

38. $6x^2 + x = 1$

39. $x^2 + 2x = 6$

40. $x^2 + 3x - 18 = 0$

41. $v^2 + 12v + 20 = 0$

42. $3t^2 - 7t - 20 = 0$

43. $5y^2 - y - 4 = 0$

44. $x^2 - 25 = 0$

45. $2x^2 + 98 = 28x$

46. $4r^2 + 100 = 40r$

47. $2t^2 + t - 14 = 0$

48. $2n^2 - 7n - 3 = 0$

LESSON 12-6 ◤ The Distance Formula, p. 544

▶ The distance, d, between any two points (x_1, y_1) and (x_2, y_2) on the coordinate plane may be found using the distance formula:

$$d = \sqrt{(x_2 - x_1)^2 + (y_2 - y_1)^2}$$

The coordinates of the midpoint of a line segment with endpoints (x_1, y_1) and (x_2, y_2) are $\left(\dfrac{x_1 + x_2}{2}, \dfrac{y_1 + y_2}{2}\right)$.

Find the midpoint of the segment with the given endpoints. Then, find the distance between each pair of points to the nearest tenth.

49. $L(-10, 5)$, $M(4, -9)$

50. $C(6, 4)$, $D(-8, -6)$

51. $P(0, -7)$, $Q(12, 2)$

Calculate the distance between each pair of points. Round to the nearest tenth.

52. $(12, 3)$, $(-8, 3)$

53. $(0, 0)$, $(5, 12)$

54. $(6, 8)$, $(3, 4)$

55. $(-4, 2)$, $(4, 17)$

56. $(-3, 8)$, $(5, 4)$

57. $(9, -2)$, $(3, -6)$

58. $(-8, -4)$, $(-3, -8)$

59. $(2, 7)$, $(10, -4)$

60. $(4, 2)$, $\left(6, -\dfrac{2}{3}\right)$

61. $\left(5, \dfrac{1}{4}\right)$, $(3, 4)$

62. $\left(\dfrac{4}{5}, -1\right)$, $\left(2, -\dfrac{1}{2}\right)$

63. $\left(3, \dfrac{3}{7}\right)$, $\left(4, -\dfrac{2}{7}\right)$

LESSON 12-7 ◤ Problem Solving Skills: Graphs to Equations, p. 550

▶ Work backwards to write the equation of a parabola. Use the general equation $y = ax^2 + bx + c$ and three points on the graph.

64. A parabola contains the points $(-4, 23)$, $(0, -1)$ and $(2, -49)$. Find an equation for the parabola.

Find the equation for each parabola.

65.

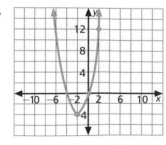

66.

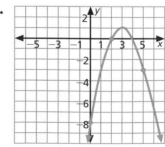

CHAPTER INVESTIGATION

EXTENSION Write a report about your experiment. Describe your vehicle and explain why you built it the way you did. Explain what happened when you used the ramp and changed its angle. Include all of the data you collected in your report.

Chapter 12 Assessment

Graph each function for the domain of real numbers. Give the coordinates of the vertex for each graph.

1. $y = -2x^2$

2. $y = 3x^2 - 4$

3. $y + 7 = -3x^2 + 2x$

Determine if each graph is a function.

4.

5.

6.

Factor to solve each equation.

7. $x^2 - 6x - 16 = 0$

8. $x^2 + 9 = 6x$

9. $4x^2 - 4x = -1$

Complete the square.

10. $x^2 + 14x$

11. $x^2 - 8x$

12. $x^2 + x$

Solve by completing the square.

13. $x^2 + 3x + 2 = 0$

14. $x^2 - 4x = 12$

15. $2x^2 + 2x = 40$

Use the quadratic formula to solve each equation.

16. $3x^2 - 4x = 15$

17. $x^2 + 8x = -3$

18. $24x = -x^2 - 136$

Calculate the distance between each pair of points to the nearest tenth of a unit.

19. $A(1, -5), B(-3, 7)$

20. $L(-6, 4), M(2, -2)$

21. $R(0, 8), S(-3, -12)$

Find the midpoint of the segment with the given endpoints.

22. $G(-13, 9), H(7, -11)$

23. $P(10, -6), Q(-4, -7)$

24. $C(-5, 3), D(12, 8)$

Solve.

25. How long will it take a ball thrown upwards with an initial velocity of 72 ft/sec to reach the ground?

Find the equation for each parabola.

26.

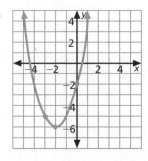

27.

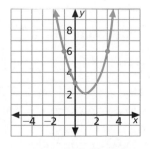

Standardized Test Practice

Part 1 Multiple Choice

Record your answers on the answer sheet provided by your teacher or on a sheet of paper.

1. Refer to the diagram below. Which pair of angles is supplementary? Assume the transversal is not perpendicular to the parallel lines. (Prerequisite Skill)

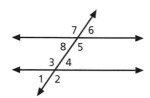

  (A) $\angle 7$ and $\angle 4$ (B) $\angle 8$ and $\angle 6$

  (C) $\angle 1$ and $\angle 6$ (D) $\angle 7$ and $\angle 2$

2. The figure below is an isosceles trapezoid with $\overline{AB} \parallel \overline{DC}$. What is $m\angle C$? (Lesson 4-9)

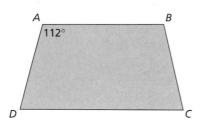

  (A) $22°$ (B) $68°$

  (C) $86°$ (D) $112°$

3. A 16-oz jar of spaghetti sauce costs \$3.59. If the cost per ounce is the same, how much would a 40-oz jar cost? (Lesson 5-1)

  (A)   \$7.95   (B)   \$8.98

  (C)   \$9.25   (D)   \$10.77

4. What is the solution to this system of equations? (Lesson 6-4)

$$3x + y = -11$$
$$2x + 2y = -10$$

  (A) $(4, -23)$ (B) $(-6, 7)$

  (C) $\left(1\frac{1}{2}, 3\right)$ (D) $(-3, -2)$

5. Solve the proportion: $\frac{2x}{x+1} = \frac{8}{9}$. (Lesson 7-1)

  (A) $\frac{4}{5}$ (B) $\frac{7}{8}$

  (C) $\frac{10}{11}$ (D) $\frac{9}{7}$

6. Determine whether the two triangles below are similar. If so, by which postulate? (Lesson 7-4)

  (A)   Angle Angle Postulate

  (B)   Side Angle Side Postulate

  (C)   Side Side Side Postulate

  (D)   The triangles are not similar.

7. Triangle RST is to be reflected over the y-axis. What will be the coordinates of S'? (Lesson 8-1)

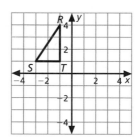

  (A) $(-3, -1)$ (B) $(3, 1)$

  (C) $(3, -1)$ (D) $(-3, 1)$

8. What is the value of z? (Lesson 8-5)

$$\begin{bmatrix} 1 & 2 \\ 0 & 4 \end{bmatrix} \cdot \begin{bmatrix} 3 & -3 \\ -2 & 1 \end{bmatrix} = \begin{bmatrix} w & x \\ y & z \end{bmatrix}$$

  (A) -12 (B) 0

  (C) 4 (D) 5

Test-Taking Tip

Question 8

In Question 8, you are only asked to solve for z. It is not necessary to find the product of the two matrices.

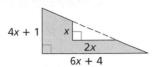

Preparing for Standardized Tests
For test-taking strategies and more practice, see pages 709–724.

Part 2 Short Response/Grid In

Record your answers on the answer sheet provided by your teacher or on a sheet of paper.

9. Mr. Russell purchased an antique tin for $24.50 at a flea market. After cleaning the piece, he sold it to a collector for $800. What is the percent increase of the cost of the tin? Round to the nearest whole percent. (Prerequisite skill)

10. Find the value of x for the quadrilateral. (Lesson 4-7)

11. If $ABCD$ is a trapezoid with median $\overline{EF}$, what is the length of $\overline{EF}$? (Lesson 4-9)

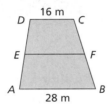

12. Find the volume in cubic inches of the three-dimensional figure. (Lesson 5-7)

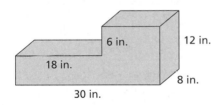

13. Jill and Gena went shopping and spent $122 altogether. Jill spent $25 less than twice as much as Gena. How much did Gena spend? (Lesson 6-4)

14. Find x. Round to the nearest tenth if necessary. (Lesson 7-5)

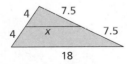

15. A card is drawn at random from a standard deck of cards. What is the probability that the card drawn is red or a 10? (Lesson 9-1)

16. Write an expression for the area of the figure. (Lesson 11-3)

$$4x + 1 \quad x \quad 2x \quad 6x + 4$$

17. The area of a rectangular playground is 750 m². The length of the playground is 5 m greater than its width. What are the length and width of the playground? (Lesson 12-5)

18. The function $h(t) = -16t^2 + v_0t + h_0$ describes the height in ft above the ground $h(t)$ of an object thrown vertically from a height of h_0 ft, with an initial velocity of v_0 ft/sec, if there is no air friction and t is the time in seconds that it takes for the ball to reach the ground. A ball is thrown upward from a 100-ft tower at an initial velocity of 60 ft/sec. How many seconds will it take for the ball to reach the ground? (Lesson 12-5)

Part 3 Extended Response

Record your answers on a sheet of paper. Show your work.

19. Paris's family is building a house on a lot that is 91 ft long and 158 ft wide. (Lesson 2-6)

   a. The town law states that the sides of a house cannot be closer than 10 ft from the edges of a lot. Write an inequality to represent the possible lengths of the house.

   b. They want their house to be at least 2800 ft² and no more than 3200 ft². They also want the house to have the maximum possible length. Write an inequality for the possible widths of the house. Round to the nearest whole number of feet.

Advanced Functions and Relations

THEME: Astronomy

Have you ever tried counting the stars at night? On a clear night, far away from bright city lights, you can see about 3000 stars using only your eyes. Today, astronomers use computers, telescopes, and satellites. Through color-spectrum analysis, they explore the history of our galaxy and solar system.

- **Astronauts** (page 571) are pilots and scientists, who travel in space. On a space shuttle, there are pilots, mission specialists, and payload specialists. They oversee and conduct experiments.

- **Astronomers** (page 589) use physics and mathematics to study the universe through observation and calculation. The knowledge gained through the science of astronomy helps in related fields of navigation and space flight. Astronomers work with engineers to design and launch space probes and satellites to gather data and transmit it back to Earth.

Math Online

mathmatters3.com/chapter_theme

The Solar System

Planet	Length of year	Mass in Earth masses	Diameter	Average orbital speed
Mercury	87.97 Earth days	0.055	3031 mi	29.8 mi/sec
Venus	224.7 Earth days	0.81	7521 mi	21.8 mi/sec
Earth	365.26 Earth days	1	7926 mi	18.5 mi/sec
Mars	1.88 Earth years	0.11	4217 mi	15 mi/sec
Jupiter	11.86 Earth years	318	88,850 mi	8.1 mi/sec
Saturn	29.46 Earth years	95.18	74,901 mi	6 mi/sec
Uranus	84.01 Earth years	14.5	31,765 mi	4.2 mi/sec
Neptune	164.79 Earth years	17.14	30,777 mi	3.4 mi/sec
Pluto	248.54 Earth years	0.0022	1429 mi	2.9 mi/sec

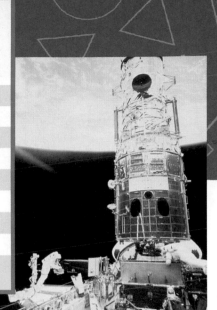

Data Activity: The Solar System

Use the table for Questions 1–4.

1. To the nearest million square miles, what is the surface area of the Earth? Assume the planet's shape is spherical.

2. Using the orbital speed and length of year in Earth days, calculate the length of Venus' orbit in miles.

3. An astronomer is 58 years old in Earth years. What is the astronomer's age in Martian years?

4. Find the circumference of Uranus to the nearest mile.

CHAPTER INVESTIGATION

How far is the Earth from the Sun? Because the Earth's orbit is elliptical, its distance from the Sun varies from 91.4 million miles to 94.5 million miles. The closest orbital point is called the *perihelion*. The farthest orbital point is called the *aphelion*. Astronomers have calculated each planet's aphelion and perihelion.

Working Together

Vast distances in space are difficult to imagine. Reducing these distances to a familiar scale can make them easier to visualize. Research the aphelion and perihelion for the nine planets of the Solar System. Then choose a location, such as your home or school, to represent the Sun. Using a system of maps, plot the aphelion and perihelion for each planet. Use the Chapter Investigation icons to guide your group.

13 Are You Ready?

Refresh Your Math Skills for Chapter 13

The skills on these two pages are ones you have already learned. Review the examples and complete the exercises. For additional practice on these and more prerequisite skills, see pages 654–661.

CIRCLES

Because you will be doing a lot of work with circles in this chapter, it may be helpful to review some of the basics.

Example   What is the diameter and the circumference of this circle?

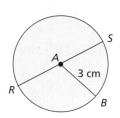

AB is a radius (r) of the circle.
RS is a diameter (d) of the circle.
A is the center point of the circle.

$$d = 2r$$
$$= 2(3)$$
$$= 6 \text{ cm}$$

$$C = 2\pi r$$
$$\approx 2(3.14)(3)$$
$$\approx 18.84 \text{ cm}$$

Find the diameter and circumference for each circle. Use 3.14 for π. Round answers to the nearest hundredth if necessary.

1.

7 cm

2.

1.5 in.

3.

8 ft

4.

3.9 dm

5.

4 mi

6.

17 km

7.

4.2 cm

8.

9.6 in.

9.

0.8 ft

10.

37.2 dm

11.

8.7 mm

12.

2.97 km

SOLVE PROPORTIONS

Example   Find x: $\dfrac{9}{12} = \dfrac{x}{30}$

Use cross-multiplication:

$$\dfrac{9}{12} = \dfrac{x}{30}$$
$$9 \cdot 30 = 12 \cdot x$$
$$270 = 12x$$
$$22.5 = x$$

Find the value of x in each proportion. Round answers to the nearest tenth if necessary.

13. $\dfrac{4}{12} = \dfrac{x}{28}$

14. $\dfrac{9}{6} = \dfrac{30}{x}$

15. $\dfrac{4}{5} = \dfrac{100}{x}$

16. $\dfrac{8}{5} = \dfrac{44}{x}$

17. $\dfrac{96}{8} = \dfrac{24}{x}$

18. $\dfrac{108}{4} = \dfrac{x}{6}$

19. $\dfrac{38}{9} = \dfrac{x}{14}$

20. $\dfrac{5}{20} = \dfrac{7}{x}$

21. $\dfrac{3}{17} = \dfrac{22}{x}$

22. $\dfrac{x}{4} = \dfrac{15}{56}$

23. $\dfrac{19}{13} = \dfrac{31}{x}$

24. $\dfrac{71}{86} = \dfrac{x}{50}$

SOLVE SYSTEMS OF EQUATIONS BY GRAPHING

Example   Solve: $2x + y = 1$

$\qquad\qquad\quad y - x = -5$

To solve, graph both equations on the same coordinate plane. The point at which the graphs intersect is the solution.

The solution is $(2, -3)$.

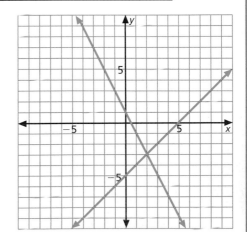

Solve each system of equations by graphing.

25. $3x - 2y = -11$

$\quad -2x + 3y = 9$

26. $x - y = 2$

$\quad 3y - x = 2$

27. $x - 2y = -2$

$\quad 2y - 2 = x$

28. $x - 6y = -1$

$\quad 4y + x = 1$

29. $2x + 2 = y$

$\quad 3y - 5x = 3$

30. $4y - 3x = 5$

$\quad x + 3 = 2y$

13-1 The Standard Equation of a Circle

Goals ■ Write equations for circles.

Applications Astronomy, Sports, Architecture

Work with a partner. You will need a compass or geometry software.

1. Draw a circle on a coordinate grid with a radius of 5 units.

2. Complete the table below by estimating the missing y-coordinates for points on the circle. Then find the values of the expressions in the third and fourth columns. Note that there will be two different y-coordinates for each x-coordinate.

x	y	$x^2 + y^2$	$\sqrt{x^2 + y^2}$
3			
3			
0			
0			
1.8			
1.8			

3. Explain the patterns that you see in the table.

Build Understanding

You can substitute into the distance formula to find the equation for a circle with its center located at any coordinate point (h, k).

$$d = \sqrt{(x_2 - x_1)^2 + (y_2 - y_1)^2}$$
$$r = \sqrt{(x - h)^2 + (y - k)^2}$$
$$r^2 = (x - h)^2 + (y - k)^2$$

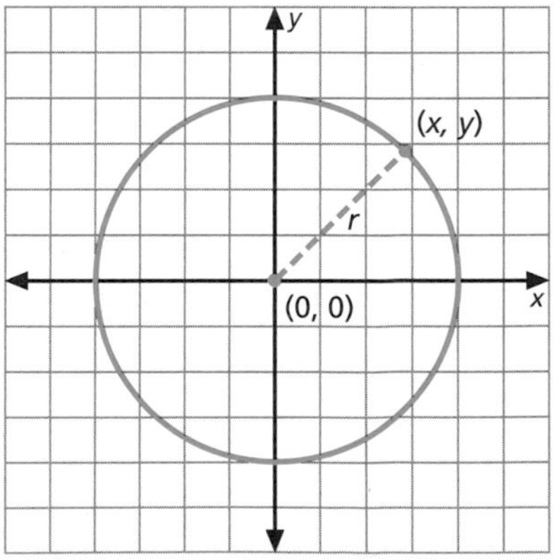

The **standard equation of a circle** is

$(x - h)^2 + (y - k)^2 = r^2, r \neq 0.$

If $h = 0$ and $k = 0$, then the standard equation simplifies to $x^2 + y^2 = r^2$.

Example 1

Write an equation for a circle with radius 6 units and center (0, 0).

Solution

Because the center is at the origin, substitute in the equation $x^2 + y^2 = r^2$.

$x^2 + y^2 = 6^2$ or

$x^2 + y^2 = 36$

Example 2

ASTRONOMY An astronomer is creating a computer model of a moon by entering the equation for a circle with a radius of 4 units and with the center located at point $(3, -2)$. Write the equation.

Solution

Substitute into the standard form for the equation of a circle.

$$(x - h)^2 + (y - k)^2 = r^2$$
$$(x - 3)^2 + (y - (-2))^2 = 4^2$$
$$(x - 3)^2 + (y + 2)^2 = 4^2 \text{ or}$$
$$(x - 3)^2 + (y + 2)^2 = 16$$

Example 3

GRAPHING Find the radius and center of the circle $x^2 + y^2 = 9$. Then graph the circle using a graphing calculator.

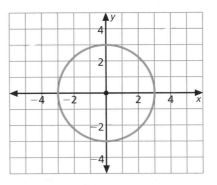

Solution

Because the equation is of the form $x^2 + y^2 = r^2$, the center is at the origin.

$$x^2 + y^2 = r^2$$
$$x^2 + y^2 = 9$$
$$r^2 = 9$$
$$r = 3$$

To graph a circle using a graphing calculator, rewrite the equation in terms of y.

$$x^2 + y^2 = 9$$
$$y^2 = 9 - x^2$$
$$y = \pm \sqrt{9 - x^2}$$

For most graphing calculators, you must enter separate formulas for the upper and lower portions of the circle. At the Y= screen, enter the functions $Y_1 = \sqrt{9 - x^2}$ and $Y_2 = -\sqrt{9 - x^2}$.

Use the ZSquare tool from the ZOOM menu to adjust the display so that each pixel represents a square. The graph is a circle with radius 3 and center at the origin.

You can also find the radius and center of a circle using the standard form for the equation of a circle.

Example 4

Find the radius and center of the circle $(x - 5)^2 + (y + 4)^2 = 18$.

Solution

Find what you would substitute into the standard form for the equation of a circle to get the given equation.

$$(x - h)^2 + (y - k)^2 = r^2$$

$$(x - 5)^2 + (y - (-4))^2 = (\sqrt{18})^2$$

You would substitute 5 for h, -4 for k, and $\sqrt{18}$ for r. So, the center (h, k) is $(5, -4)$. The radius is $\sqrt{18}$ or $3\sqrt{2}$.

◥ TRY THESE EXERCISES

Write an equation for each circle.

1. radius 7
center $(0, 0)$

2. radius 10
center $(-3, 0)$

3.

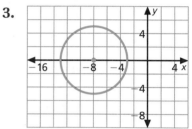

Find the radius and center for each circle.

4. $x^2 + y^2 = 100$

5. $x^2 + y^2 = 11$

6. $(x - 4)^2 + (y - 2)^2 = 49$

7. $(x + 8)^2 + (y - 4)^2 = 13$

8. WRITING MATH Does the equation for a circle describe a function? Explain your reasoning.

◥ PRACTICE EXERCISES • For Extra Practice, see page 703.

Write an equation for each circle.

9. radius 12
center $(0, 0)$

10. radius 9
center $(4, -6)$

11. radius 12
center $(-9, -4)$

12. radius 11
center $(0, 4)$

13. DESIGN The position of a circular knob on a design for a DVD player is shown on the graph below. Find the equation of the circle.

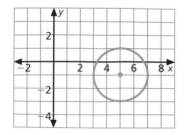

14. SCIENCE In a model, two metal balls are attached to the endpoints of a metal rod. The rod is attached to a machine at its center point. The machine spins the rod so that the balls move in a circle modeled by the equation $x^2 + y^2 = 144$. Find the length of the rod.

15. SPORTS A circular target is set up for hang glider landings. Write an equation to model a circle with a diameter of 5.2 meters and a center of (0, 0).

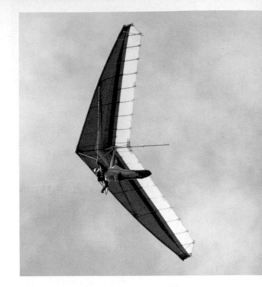

Find the radius and center for each circle.

16. $x^2 + y^2 = 121$

17. $x^2 + y^2 = 15$

18. $(x - 1)^2 + (y - 3)^2 = 9$

19. $(x + 4)^2 + (y - 2)^2 = 14$

20. $(x + 5)^2 + (y + 12)^2 = 17$

21. $(x - 12)^2 + (y + 6)^2 = 20$

22. $x^2 + (y - 4)^2 = 17$

23. $(x - 9)^2 + (x + 1)^2 = 400$

24. $(x + 3)^2 + y^2 = 50$

25. $(x + 1)^2 + (y - 9)^2 = 81$

26. ARCHITECTURE In the landscaping plan for a museum lawn, four circular paths are designed to intersect at the origin as shown. Write the equation for each circle.

■ EXTENDED PRACTICE EXERCISES

Write two equations for each circle, given the coordinates of the endpoints of the radius.

27. (0, 0), (9, 0)

28. (2, −3), (4, −3)

Find the radius and center point for each circle.

29. $x^2 + y^2 = 4x + 2y + 4$

30. $x^2 + y^2 = 16x$

■ MIXED REVIEW EXERCISES

For each function, name the coordinates of the vertex of the graph. (Lesson 12-1)

31. $y = 3x^2$

32. $y = x^2 - 4$

33. $y = -2x^2 + 3$

34. $y = -x^2 + 2$

35. $y = -4x^2 + 6$

36. $y = x^2 - 5$

Find the slope of the line containing the given points. (Lesson 6-1)

37. $A(-3, 4), B(6, -2)$

38. $C(3, 8), D(-2, -5)$

39. $E(-4, -5), F(5, 4)$

40. $G(7, -3), H(2, 5)$

41. $J(-2, 6), K(-2, -5)$

42. $L(4, -3), M(-4, 6)$

43. $N(5, 3), P(-4, -7)$

44. $Q(-5, 3), R(2, 3)$

45. $S(3, 0), T(-3, 3)$

46. $R(-4, -1), T(-3, -3)$

47. $P(-3, 3), W(1, 3)$

48. $J(-2, 1), K(-2, 3)$

49. $C(2, 3), D(9, 7)$

50. $A(5, 7), B(-2, -3)$

51. $M(-3, 6), N(2, 4)$

52. $T(-3, -4), W(5, -1)$

53. $R(2, -1), S(5, -3)$

54. $Y(2, 6), Z(-1, 3)$

13-2 More on Parabolas

Goals ■ Relate the equation of a parabola to its focus and directrix.

Applications Satellite Communications, Energy

Work with a partner to answer the following questions.

1. Study the graph shown at the right. From six randomly chosen points on the parabola, segments have been drawn to the parabola's **focus**, point F. From these same points perpendicular lines have been drawn to line ℓ, the **directrix** of the parabola.

2. Compare the lengths of the two segments from each point.

3. Describe the relationship between the points on the parabola and their distances from the focus and the directrix.

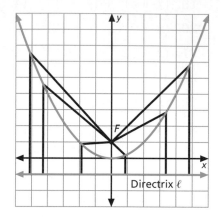

▼ BUILD UNDERSTANDING

A parabola is a set of points equidistant from a fixed point called the **focus** and a fixed line called the **directrix**.

Simple equations may be derived for parabolas that have a vertex at the origin and a directrix parallel to either the x-axis or the y-axis.

Let point $P(x, y)$ be any point on the parabola such that $FP = PD$. Then use the distance formula.

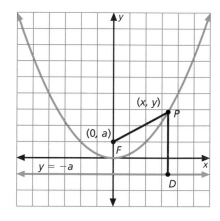

$$FP = PD$$
$$\sqrt{(x - 0)^2 + (y - a)^2} = \sqrt{(x - x)^2 + (y + a)^2}$$
$$(x - 0)^2 + (y - a)^2 = (y + a)^2 \quad \text{Square both sides. Simplify.}$$
$$x^2 + y^2 - 2ay + a^2 = y^2 + 2ay + a^2$$
$$x^2 - 2ay = 2ay$$
$$x^2 = 4ay$$

When the focus $(0, a)$ is on the y-axis and the directrix is $y = -a$, the equation for the parabola is $x^2 = 4ay$.

In the graph above, the variable $a = 1$. When $a > 0$, the parabola opens upward. When $a < 0$, the parabola opens downward.

GRAPHING Use a graphing calculator to graph $x^2 = 4ay$ when $a = 1$ and $a = -1$. You will need to write the equation $x^2 = 4ay$ in terms of y. Notice that the y-axis remains the axis of symmetry whether the parabola opens upwards or downwards.

Example 1

Find the focus and directrix of the parabola $x^2 = -6y$.

Solution

$$x^2 = -6y$$
$$x^2 = 4ay$$
$$4ay = -6y$$
$$4a = -6$$
$$a = -\frac{3}{2}$$
$$x^2 = 4\left(-\frac{3}{2}\right)y$$

Because a is negative, the parabola opens downward. The focus is located at $(0, a)$, so the focus is $\left(0, -\frac{3}{2}\right)$. The directrix is the line $y = -a$, so the directrix is $y = -\left(-\frac{3}{2}\right)$ or $y = \frac{3}{2}$.

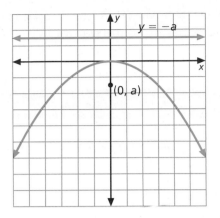

Example 2

SATELLITE COMMUNICATIONS A parabolic satellite dish directs all incoming signals to a receiver. The receiver is located at the vertex which is at the origin and the focus is at $(0, 5)$. Find the simple equation for the parabola.

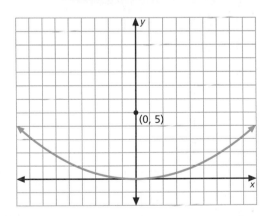

Solution

$$x^2 = 4ay$$
$$a = 5$$
$$x^2 = 4(5)y$$
$$x^2 = 20y$$

The equation is $x^2 = 20y$.

TRY THESE EXERCISES

Find the focus and directrix of each parabola.

1. $x^2 = 12y$

2. $x^2 = -28y$

3. $4x^2 - 32y = 0$

4. GRAPHING Use a graphing calculator to graph the parabolas for Exercises 1–3. Using the simple equation for a parabola $x^2 = 4ay$, where the focus is $(0, a)$ and the directrix is $y = -a$, graph the directrix. How does the shape of the parabola change as the value of $4a$ increases?

Find the equation for each parabola with vertex located at the origin.

5. Focus (0, 7)

6. Focus (0, −3)

7. Focus $\left(0, \dfrac{3}{4}\right)$

8. WRITING MATH A parabola has its vertex at the origin. Explain how to use the equation of the parabola to tell if the parabola opens up or down.

PRACTICE EXERCISES • For Extra Practice, see page 703.

Find the focus and directrix of each equation.

9. $x^2 = 16y$

10. $x^2 = -10y$

11. $x^2 - 20y = 0$

12. $2x^2 = -24y$

13. $16y - 4x^2 = 0$

14. $-5x^2 = -30y$

ENERGY Parabolic mirrors can be used to power steam turbines to generate electricity. Three mirrors have the following focus points. Find the simple equation for each. In each case, the vertex is located at the origin.

15. Focus (0, −2)

16. Focus (0, 9)

17. Focus $\left(0, -\dfrac{1}{2}\right)$

Write the equation for each parabola with vertex at the origin and focus and directrix as shown.

18.

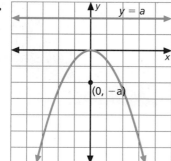

19.

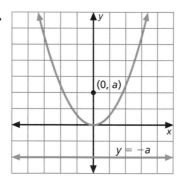

20.

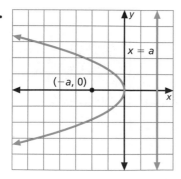

21.

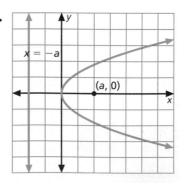

22. CHAPTER INVESTIGATION Research the perihelion (closest orbital point to the sun) and aphelion (farthest orbital point) for each planet of the Solar System. Make a rough sketch of the planets' orbits and label the perihelion and aphelion for each planet.

The standard form for the equation of a parabola with vertex (h, k) and axis parallel to the y-axis is $(x - h)^2 = 4a(y - k)$.

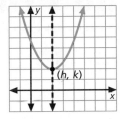

The standard form for the equation of a parabola with vertex (h, k) and axis parallel to the x-axis is $(y - k)^2 = 4a(x - h)$.

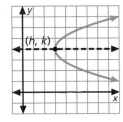

Find the equation of each parabola.

23. Focus $(2, 4)$
Vertex $(2, 2)$

24. Focus $(4, 1)$
Vertex $(2, 1)$

25. Focus $(2, 7)$
Directrix $y = -3$

26. WRITING MATH If a light is placed at the focus of a parabola, the rays will be reflected off the parabolic surface parallel to the axis as shown. How is this concept used in a flashlight?

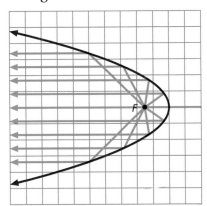

■ MIXED REVIEW EXERCISES

Use $x = -\dfrac{b}{2a}$ to find the vertex. (Lesson 12-2)

27. $y = x^2 - 3x + 6$

28. $y = 2x^2 - 5x + 8$

29. $y = x^2 + 3x - 4$

30. $-8 + y = 2x^2 + 2x$

31. $y = x^2 - 9$

32. $-x^2 = 3x - y + 5$

33. $y - 12 = x + 2x^2$

34. $3x^2 = y - 6x + 9$

35. $y = x^2 + 4x - 3$

36. $2x - y = -x^2 + 3$

37. $4x^2 = -y - 3x + 8$

38. $3x = -y + x^2 + 2$

Let $U = \{1, 2, 3, 4, 5, 6, 7, 8, 9\}$, $P = \{1, 3, 4, 6, 7\}$ and $Q = \{2, 5, 6, 7, 8\}$. Find each union or intersection. (Lesson 1-3)

39. P'

40. Q'

41. $P \cap Q$

42. $Q \cup P$

43. $(P \cup Q)'$

44. $P' \cap Q'$

45. $Q' \cup P$

46. $Q' \cap P$

47. $(P \cap Q)' \cup (P \cup Q)$

48. $(P \cup Q) \cap (P' \cup Q')$

49. $P' \cap (P \cup Q)$

Review and Practice Your Skills

1. In the standard equation for a circle, what do the variables h and k represent?

2. What is the radius of a circle?

3. When is the standard equation of a circle $x^2 + y^2 = r^2$, $r \neq 0$?

Write an equation for each circle.

4. radius 5
   center (0, 0)

5. radius 3
   center (2, 4)

6.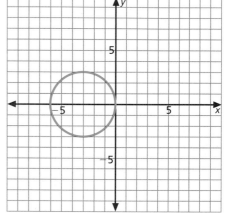

7. radius 4
   center (−1, 2)

8. radius 6
   center (−2, −3)

Find the radius and center for each circle.

9. $x^2 + y^2 = 13$

10. $(x − 2)^2 + (y + 1)^2 = 49$

11. $(x + 3)^2 + (y − 4)^2 = 25$

12. $x^2 + (y + 2)^2 = 64$

13. $(x − 3)^2 + y^2 = 36$

14. $(x − 4)^2 + (y + 3)^2 = 74$

15. $(x + 1)^2 + y^2 = 28$

16. $(x + 3)^2 + (y − 4)^2 = 32$

Determine if each statement is *true* or *false*.

17. When the focus is on the y-axis, and the directrix is $y = a$, the standard equation for the parabola is $x^2 = 4ay$.

18. The focus is equidistant from most points on the parabola.

19. When the y-coordinate of a focus is negative, the parabola opens downward.

Find the focus and directrix of each equation.

20. $x^2 = −4y$

21. $x^2 = 13y$

22. $x^2 = 7y$

23. $x^2 + 3y = 0$

24. $x^2 − 5y = 0$

25. $3x^2 + 27y = 0$

26. $x^2 = −8y$

27. $x^2 = 5y$

28. $x^2 = −3y$

29. $x^2 − 4y = 0$

30. $x^2 + 7y = 0$

31. $2x^2 + 12y = 0$

Find the standard equation for each parabola with vertex located at the origin.

32. Focus (0, 5)

33. Focus (0, −3)

34. Focus (0, 2)

35. Focus (0, −7)

36. Focus $\left(0, \dfrac{1}{2}\right)$

37. Focus (0, 2.5)

Without graphing, determine whether each equation is that of a circle or a parabola. Assume $r > 0$. (Lessons 13-1–13-2)

38. $x^2 + y^2 = r^2$
39. $x^2 = 4y$
40. $x^2 = r^2 - y^2$

41. $3x^2 = 9y$
42. $4x^2 = -8y$
43. $x^2 = r^2 - (y + 3)^2$

Find the radius and center for each circle. (Lesson 13-1)

44. $x^2 + y^2 = 81$
45. $(x + 3)^2 + (y - 4)^2 = 36$
46. $(x - 2)^2 + y^2 = 52$

Write an equation for each circle. (Lesson 13-1)

47. radius 7
center $(0, -3)$

48. radius 5
center $(4, 2)$

49. radius 6
center $(-3, -1)$

MathWorks — Career – Payload Specialist
Workplace Knowhow

Today astronauts are civilian and military specialists in scientific fields such as engineering. One particular type of astronaut is a payload specialist. A payload specialist is a professional in the physical or life sciences and is skilled in working with equipment developed specifically for the space shuttle. The payload specialist also oversees experiments.

1. In order to perform the experiment, the space shuttle must be kept ahead of the moon's orbit so that the ship, the Earth, and the moon form a right angle. The shuttle is orbiting the Earth at a distance of 325 km. The moon orbits the Earth at a distance of 384,403 km at its furthest distance (the current distance). Draw a diagram of the Earth, the moon and the ship forming a right angle.
The right angle will fall at the center of the Earth. The Earth's diameter is 12,756 km. The moon's diameter is 3476 km. (Make your diagram look as if the moon is quite a bit farther away from the Earth than your ship.

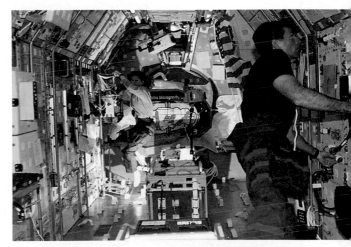

Payload specialists on space shuttle Discovery

2. How far is your ship from the center of the Earth? How far is the center of the Earth from the center of the moon?

3. Now that you have those two distances, draw the hypotenuse of the triangle between the Earth, the moon, and your ship. How far are you from the moon?

Problem Solving Skills: Visual Thinking

Think of a line that intersects the coordinate plane at the origin. The right circular cones formed by rotating the line about the y-axis are used to study **conics**.

A **conic section** is formed by a plane intersecting the right circular cones.

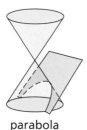 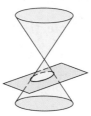 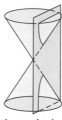

| parabola | circle | ellipse | hyperbola |

Problem

Draw and name the conic section formed by each plane.

a. The plane is parallel to a side of the cone and does not pass through the vertex.

b. The plane is parallel to the y-axis and does not pass through the vertex.

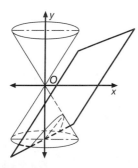 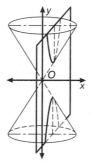

Solve the Problem

a.

The conic section is a parabola.

b.

The conic section is a hyperbola.

TRY THESE EXERCISES

Five-step Plan

1 Read
2 Plan
3 Solve
4 Answer
5 Check

Name the conic section formed by each plane.

1. The plane is parallel to the *x*-axis. It does not contain the vertex.

2. The plane does not contain a vertex or base. It is not parallel to the *x*-axis.

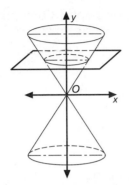

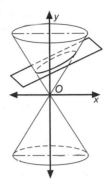

PRACTICE EXERCISES

Name the conic section or figure formed by each plane.

3. The plane intersects only the vertex.

4. The plane is perpendicular to the *x*-axis and passes through the vertex.

5. The plane is parallel to a side of the cone and does not pass through the vertex.

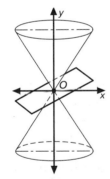

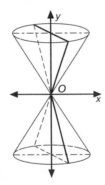

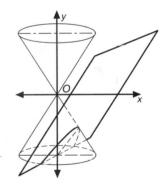

6. WRITING MATH Is it possible to intersect a double cone in such a way that two circles are formed? Explain.

MIXED REVIEW EXERCISES

Use a graphing calculator to determine the number of solutions for each equation. For equations with one or two solutions, find the exact solutions by factoring. (Lesson 12-3)

7. $0 = x^2 - 64$

8. $0 = x^2 + x - 12$

9. $0 = x^2 + 2x + 1$

10. $0 = x^2 + 2x - 8$

11. $0 = x^2 + 49$

12. $-x^2 + 5x + 6 = 0$

13. $x^2 = 5x + 6$

14. $0 = x^2 - 25$

15. $x^2 = -5x - 6$

16. DATA FILE Use the data on measuring earthquakes on page 647. An earthquake in southern California measured 6.0 on the Richter scale. Another earthquake in central California measured 3.0 on the Richter scale. How much more ground movement occured in southern California than in central California? How much more energy was released? (Lesson 1-8)

13-4 Ellipses and Hyperbolas

Goals ■ Graph equations of ellipses and hyperbolas.

Applications Astronomy, Oceanography, Communications

Work with a partner.

You will need a piece of cardboard, two thumbtacks, string, and scissors. Place two thumbtacks in a piece of cardboard. Label their positions F_1 and F_2. Let the distance between the two points be $2c$.

Tie the ends of a piece of string together to make a loop. Let the length of the string be $2a + 2c$ where a is any quantity greater than c. Place the loop over the tacks. With a pencil held upright, keep the string taut and draw the ellipse.

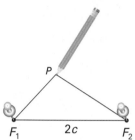

Use the same loop of string and increase the distance between F_1 and F_2. Draw another ellipse.

1. How does the shape of the ellipse change as F_1 and F_2 are farther apart?

2. For any point P on the ellipse, what is the sum of $F_1P + F_2P$?

3. If F_1 and F_2 were the same, what figure would you draw?

> **Check Understanding**
>
> Why must length a be greater than length c?

◤ BUILD UNDERSTANDING

As you discovered in the activity above, an **ellipse** is defined by a point moving about two fixed points. The two fixed points are called **foci**, the plural of focus. The sum of the distances from the two fixed points to any point on the ellipse is a constant, $F_1P + F_2P = 2a$. If F_1 and F_2 are on the x-axis, the major axis is horizontal. The major axis is vertical when F_1 and F_2 are on the y-axis.

An equation for the standard form of an ellipse can be derived by placing the ellipse on a coordinate grid. Locate $(0, 0)$ at the midpoint between F_1 and F_2. The distance from the center to each focus is c. When $x = 0$, $F_1P_1 = F_2P_1$ and $b^2 = a^2 - c^2$.

Use the distance formula to determine the **standard form** for the equation of an ellipse. By letting $y = 0$, x-intercepts are $(a, 0)$ and $(-a, 0)$. By letting $x = 0$, y-intercepts are $(0, b)$ and $(0, -b)$.

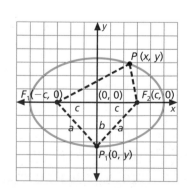

Equations of Ellipses with Centers at the Origin

Standard Form of Equation	$\dfrac{x^2}{a^2} + \dfrac{y^2}{b^2} = 1$	$\dfrac{y^2}{a^2} + \dfrac{x^2}{b^2} = 1$
Direction of Major Axis	horizontal	vertical
Foci	$(c, 0)$, $(-c, 0)$	$(0, c)$, $(0, -c)$

Example 1

Graph the equation $4x^2 + 9y^2 = 36$.

Solution

Divide both sides of the equation by 36 to change it to standard form.

$$4x^2 + 9y^2 = 36$$

$$\frac{x^2}{9} + \frac{y^2}{4} = 1$$

$$a^2 = 9, a = \pm 3$$

$$b^2 = 4, b = \pm 2$$

The x-intercepts are $(3, 0)$ and $(-3, 0)$. The y-intercepts are $(0, 2)$ and $(0, -2)$.

Locate the points and draw a smooth curve.

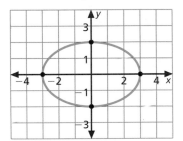

Technology Note

To graph an ellipse using a graphing calculator, rewrite the equation in terms of y.

$4x^2 + 9y^2 = 36$ becomes $y = \pm \sqrt{-\frac{4}{9}x^2 + 4}$.

1. Enter the positive form of the equation as Y_1 to graph the upper part of the ellipse.

2. To graph the lower portion, let $Y_2 = -Y_1$.

3. Graph the ellipse. You can use the trace feature to locate the x- and y-intercepts.

Example 2

ASTRONOMY Jae is using a computer to model the orbit of a moon. After placing a grid over a drawing of the orbit, he finds that the foci of the ellipse are $(4, 0)$ and $(-4, 0)$, and the x-intercepts are $(-5, 0)$ and $(5, 0)$. He needs to enter the equation of the ellipse into the computer to finish his work. Find the equation of the ellipse.

Solution

$$\frac{x^2}{a^2} + \frac{y^2}{b^2} = 1$$

$$a = \pm 5, c = \pm 4$$

$$b^2 = a^2 - c^2 \qquad \text{Use the Pythagorean Theorem to find } b^2.$$

$$b^2 = 25 - 16$$

$$b^2 = 9$$

$$\frac{x^2}{25} + \frac{y^2}{9} = 1 \qquad \text{Substitute in the standard form.}$$

$$9x^2 + 25y^2 = 225 \qquad \text{Multiply by 225 (9 × 25).}$$

The equation of the ellipse with foci $(4, 0)$ and $(-4, 0)$ and x-intercepts $(-5, 0)$ and $(5, 0)$ is $9x^2 + 25y^2 = 225$.

Check Understanding

Write *ellipse* or *hyperbola* for each equation.

1. $\frac{x^2}{10} + \frac{y^2}{4} = 1$

2. $\frac{x^2}{12} - \frac{y^2}{5} = 1$

3. $4x^2 - 7y^2 = 56$

4. $8x^2 + 13y^2 = 104$

A hyperbola has some similarities to an ellipse. The distance from the **center** to a vertex is a units. The distance from the center to a focus is c units. There are two axes of symmetry. The **transverse axis** is a segment of length $2a$ whose endpoints are the vertices of the hyperbola. The **conjugate axis** is a segment of length $2b$ units that is perpendicular to the transverse axis at the center. The values of a, b, and c are related differently for a hyperbola than for an ellipse. For a hyperbola, $c^2 = a^2 + b^2$.

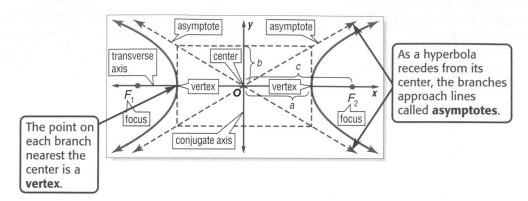

Equations of Hyperbolas with Centers at the Origin

Standard Form of Equation	$\dfrac{x^2}{a^2} - \dfrac{y^2}{b^2} = 1$	$\dfrac{y^2}{a^2} - \dfrac{x^2}{b^2} = 1$
Direction of Tranverse Axis	horizontal	vertical
Foci	$(c, 0), (-c, 0)$	$(0, c), (0, -c)$
Vertices	$(a, 0), (-a, 0)$	$(0, a), (0, -a)$
Length of Transverse Axis	$2a$ units	$2a$ units
Length of Conjugate Axis	$2b$ units	$2b$ units
Equations of Asymptotes	$y = \pm\dfrac{b}{a}x$	$y = \pm\dfrac{a}{b}x$

Example 3

ASTRONOMY Comets that pass by Earth only once may follow hyperbolic paths. Suppose a comet follows one branch of a hyperbola with center $(0, 0)$ and foci on the y-axis if $a = \pm15$ and $b = \pm20$. Find the equation of the hyperbola.

Solution

Write the standard form of the equation of the hyperbola.

$$\frac{y^2}{a^2} - \frac{x^2}{b^2} = 1$$

$$\frac{y^2}{15^2} - \frac{x^2}{20^2} = 1 \quad \text{Substitute in the standard equation.}$$

$$\frac{y^2}{225} - \frac{x^2}{400} = 1$$

$$400y^2 - 225x^2 = 90{,}000. \quad \text{Multiply by } 90{,}000 \ (400 \times 225).$$

The equation is $400y^2 - 225x^2 = 90{,}000$.

TRY THESE EXERCISES

Graph each equation.

1. $4x^2 + 16y^2 = 64$

2. $4x^2 - 25y^2 = 100$

3. Find the equation of the ellipse with foci $(12, 0)$ and $(-12, 0)$ and x-intercepts $(13, 0)$ and $(-13, 0)$.

4. Find the equation of the hyperbola with center $(0, 0)$ and foci on the x-axis if $a = \pm 9$ and $b = \pm 6$.

5. **WRITING MATH** Is a circle a type of ellipse? Explain your thinking.

PRACTICE EXERCISES • For Extra Practice, see page 704.

Graph each equation.

6. $9x^2 + 36y^2 = 36$

7. $9x^2 - 16y^2 = 144$

8. **COMMUNICATIONS** A communications satellite is launched into an elliptical orbit with foci $(8, 0)$ and $(-8, 0)$ and x-intercepts $(10, 0)$ and $(-10, 0)$. Find the equation of the ellipse.

9. **OCEANOGRAPHY** An object propelled through water travels along one branch of a hyperbola with an experimental submarine at its center $(0, 0)$ and in which $a = 7$ and $b = 6$ and the foci are on the x-axis. Find the equation for the hyperbola.

Graph each hyperbola.

10. $18y^2 - 8x^2 = 72$

11. $y^2 - x^2 = 144$

Graph each ellipse.

12. $4x^2 + y^2 = 4$

13. $25x^2 + 4y^2 = 100$

14. $2x^2 + y^2 = 8$

EXTENDED PRACTICE EXERCISES

The ellipse shown has center (h, k) and axes parallel to the coordinate axes.

15. Substitute in the standard form for the equation of an ellipse with its center at the origin and foci on the x-axis to find an equation for an ellipse with center (h, k).

16. Use your equation to find the center of the ellipse
$16(x + 2)^2 + 9(y - 1)^2 = 144$.

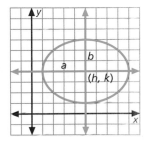

MIXED REVIEW EXERCISES

Solve by completing the square. (Lesson 12-4)

17. $x^2 + 10x = 0$

18. $x^2 + 3x = 0$

19. $x^2 - 8x = 0$

20. $x^2 + 4x - 7 = 0$

21. $x^2 - 6x + 2 = 0$

22. $x^2 + 12x - 3 = 0$

Review and Practice Your Skills

PRACTICE ■ LESSON 13-3

Draw and name the conic section formed by each plane.

1. The plane is parallel to a side of the cone and does not pass through the vertex.

2. The plane intersects the vertex and is parallel to the y-axis.

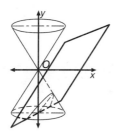

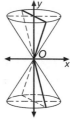

3. The plane is parallel to the x-axis and does not contain the vertex.

4. The plane does not contain a vertex or base, and is not parallel to the x-axis.

PRACTICE ■ LESSON 13-4

5. Demonstrate how to determine the x-intercepts of an ellipse using the standard form of the equation for an ellipse and letting $y = 0$.

6. Demonstrate how to determine the y-intercepts of an ellipse using the standard form of the equation for an ellipse and letting $x = 0$.

7. The standard form of the equation of an ellipse and for a hyperbola are similar. Explain how they are different.

Graph each equation.

8. $4x^2 - 16y^2 = 64$

9. $4x^2 + 16y^2 = 64$

10. $9x^2 + 9y^2 = 81$

11. $9x^2 - 9y^2 = 81$

12. $25x^2 + 4y^2 = 100$

13. $25x^2 - 4y^2 = 100$

Find the equation of each ellipse.

14. foci $(5, 0)$, $(-5, 0)$

    x-intercepts $(8, 0)$, $(-8, 0)$

15. foci $(3, 0)$ and $(-3, 0)$

    x-intercepts $(10, 0)$, $(-10, 0)$

16. foci $(4, 0)$ and $(-4, 0)$

    x-intercepts $(6, 0)$, $(-6, 0)$

Find the equation of each hyperbola. Assume all foci are on the x-axis.

17. center $(0, 0)$

    $a = 5, b = 3$

18. center $(0, 0)$

    $a = 9, b = 4$

19. center $(0, 0)$

    $a = 7, b = 4$

Name the conic section formed by the plane.
(Lesson 13-3)

20. The plane is parallel to a side of the cone and does not pass through the vertex.

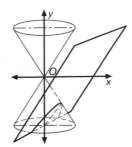

Graph each equation. (Lesson 13-4)

21. $4x^2 + 4y^2 = 16$

22. $4x^2 - 4y^2 = 16$

23. $9x^2 + 36y^2 = 324$

24. $9x^2 - 36y^2 = 324$

25. $49x^2 + 16y^2 = 784$

26. $49x^2 - 16y^2 = 784$

Mid-Chapter Quiz

Find the radius and center for each circle. (Lesson 13-1)

1. $(x - 3)^2 + (y + 1)^2 = 64$

2. $x^2 + (y + 6)^2 = 6$

3. $(x + 7)^2 + (y - 9)^2 = 20$

4. $(x + 2)^2 + y^2 = 18$

Find the standard equation for each parabola with the vertex located at the origin. (Lesson 13-2)

5. focus $(0, 2)$

6. focus $(0, -6)$

7. focus $\left(0, \dfrac{2}{3}\right)$

8. focus $\left(0, -\dfrac{1}{4}\right)$

9. focus $(6, 0)$

10. focus $(0.5, 0)$

Write the standard equation for each ellipse. (Lesson 13-4)

11. foci $(-2, 0)$ and $(2, 0)$; x-intercepts $(-5, 0)$ and $(5, 0)$

12. foci $(-3, 0)$ and $(3, 0)$; x-intercepts $(-6, 0)$ and $(6, 0)$

Write the standard equation for each hyperbola. (Lesson 13-4)

13. center $(0, 0)$; foci on x-axis; $a = 4$, $b = 7$

14. center $(0, 0)$; foci on x-axis; $a = 6$, $b = 8$

Name the figure or conic section formed by the plane. (Lesson 13-3)

15. The plane intersects the side of the cone and passes through the vertex.

16. The plane does not contain a vertex or base, and is not parallel to the x-axis.

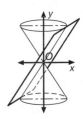

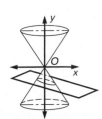

Direct Variation

Goals ■ Solve problems involving direct variation.

Applications Food Prices, Space Exploration, Physics

Work with a partner to answer the following questions:

The table below shows the diameter and approximate circumference of some circles.

Diameter (d)	3	6	9	12	24
Circumference (C)	9.42	18.84	28.26	37.68	75.36

a. Find the ratio $\frac{C}{d}$ for each pair of values.

b. When d doubles, what happens to C?

c. Write an equation for C as a function of d.

d. Graph the function. Describe the graph and find its slope.

◥ BUILD UNDERSTANDING

The circumference of a circle is a function of its diameter. This function can be written $C = 3.14d$. From the graph, you can see that the relationship between diameter and circumference is linear.

The relationship between diameter and circumference of a circle is an example of **direct variation**. The value of one variable increases as the other variable increases.

Direct variation can be represented by an equation in the form $y = kx$, where k is a nonzero constant and $x \neq 0$. The constant k is called the **constant of variation**. For the example above, circumference varies directly with diameter. The constant of variation is 3.14.

If y varies directly as x, the constant of variation can be found if one pair of values is known.

> **Reading Math**
>
> $y = kx$ is read "y is directly proportional to x" or "y varies directly as x."

Example 1

What is the equation for a direct variation when one pair of values is $x = 20$ and $y = 9$?

Solution

$y = kx$ Substitute in the equation for direct variation.

$9 = k(20)$

$\frac{9}{20} = k$

$0.45 = k$ Solve for k.

The equation is $y = 0.45x$.

Many examples of the use of direct variation may be found in everyday situations.

Example 2

FOOD PRICES The cost of apples varies directly with weight. If 9 lb of apples cost $4.32, how much will 17 lb of apples cost?

Solution

$$y = kx$$

$4.32 = k(9)$	Substitute.
$\dfrac{4.32}{9} = k$	
$0.48 = k$	Solve for k.
$y = 0.48x$	Write the equation.
$y = 0.48(17)$	Substitute 17 for x.
$y = 8.16$	Solve.

> **Check Understanding**
>
> k is sometimes called the constant of proportionality. How could this problem be solved as a proportion?

Seventeen pounds of apples will cost $8.16

The equation for the area of a circle is $A = \pi r^2$. The area varies directly as the square of the radius. This is an example of **direct square variation**. The equation represents a quadratic function.

Direct square variation may be expressed in the form $y = kx^2$ where k is a nonzero constant.

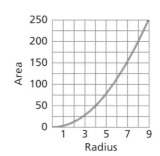

Example 3

SPACE EXPLORATION An air filter used in a space vehicle is in the shape of a cube. The surface area of a cube varies directly as the square of its sides. If the surface area of an air filter with sides 12 in. long is 864 in.^{2}, what is the surface area of an air filter in the shape of a cube with sides 11 in. long?

Solution

$$y = kx^2$$

$864 = k(12)^2$	Substitute in the equation.
$864 = 144k$	Solve for k.
$\dfrac{864}{144} = k$	
$6 = k$	
$y = 6x^2$	Write the equation.
$y = 6(11)^2$	Substitute 11 for x.
$y = 726$	Solve.

The surface area of a cube with sides 11 in. in length is 726 in.^{2}.

1. What is the equation of direct variation when one pair of values is $x = 72$ and $y = 18$?

2. If y varies directly as x and $y = 60$ when $x = 50$, find y when $x = 15$.

3. If y varies directly as x^2 and $y = 320$ when $x = 8$, find y when $x = 5$.

4. **POSTAGE** Fifteen stamps cost $5.55. How much will 26 stamps cost?

5. **EARNINGS** A person's income varies directly with the number of hours the person works. If the pay for 16 h is $200, what is the pay for working 40 h?

6. **CATERING** Three vegetable platters serve a party of 20 people. How many vegetable platters are needed for a party of 240 people?

7. The distance (d) a vehicle travels at a given speed is directly proportional to the time (t) it travels. If a vehicle travels 30 mi in 45 min, how far can it travel in 2 h?

8. **ASTRONOMY** The speed of a comet at perihelion (the closest orbital point to the sun) is 98,000 mi/h. How far will the comet travel in 30 sec?

9. **BIOLOGY** The expected increase (I) of a population of organisms is directly proportional to the current population (n). If a sample of 360 organisms increases by 18, by how many will a population of 9000 increase?

10. **PHYSICS** The distance (d) an object falls is directly proportional to the square of the time (t) it falls. If an object falls 256 ft in 4 sec, how far will it fall in 7 sec?

11. **SALES** The cost of ribbon is directly proportional to the length purchased. If 9.5 yd of ribbon cost $3.42, how much will 14.75 yd cost?

12. **SALES** A person is paid $153 for 9 baskets of dried flowers. To earn $450, how many baskets of dried flowers must the person produce?

13. **BAKED GOODS** A bakery earns $33.75 profit on 9 cakes. How many cakes must be sold for the bakery to make a profit of $150?

14. **WRITING MATH** A clothing store charges $1.75 in sales tax on an item that costs $25. The same store charges $2.80 in sales tax on a $40 item. Is sales tax an example of direct variation? Justify your answer.

15. **ERROR ALERT** The distance a spring will stretch, S, varies directly with the weight, W, added to the spring. A spring stretches 1.5 in. when 12 lb are added. Paige plans to add 2 more pounds to the spring. She concludes that the spring will stretch $1.5 + 2$, or 3.5 in. when the weight is added. Is Paige's thinking correct? Explain your reasoning.

The area of each regular polygon varies directly as the square of its sides. One pair of values is given for each. Find the area for each regular polygon. Then find the area of each polygon if one side is 9 units.

16. Pentagon
$s = 3$
$A = 15.48$

17. Hexagon
$s = 5$
$A = 64.95$

18. Octagon
$s = 7$
$A = 236.59$

ELECTRICITY The number of kilowatts of electricity used by an appliance varies directly as the time the appliance is used.

19. If you watch television for 4.5 h, about how many kilowatt hours of electricity do you use?

20. If you dry your hair for 10 min, how many kilowatt hours of electricity do you use?

21. If the refrigerator runs for 2 h/day, how many kilowatt hours of electricity does it use?

22. If the water heater runs 2.75 h/day, how many kilowatt hours of electricity does it use?

23. If electricity costs 12.3¢ per kilowatt hour, find the cost for each activity in Exercises 19–22 to the nearest cent.

Approximate Kilowatt Usage of Some Appliances

Appliance	Kilowatts per hour
Light bulbs	0.001 per watt
Electric blanket	0.07
Stereo	0.1
Color television	0.23
Hair dryer	1.5
Refrigerator	5.0
Iron	1.0
Freezer	3.0
Water heater	16.0

24. CHAPTER INVESTIGATION Choose a point within your school grounds or community to represent the Sun. Using a map of your school or community and an appropriate scale, plot the location of the aphelion and perihelion for each planet on the map. Make a rough sketch of the orbits of the planets.

◼ EXTENDED PRACTICE EXERCISES

Write *direct variation*, *direct square variation*, or *neither* to describe how P varies as V increases or decreases in each equation.

25. $P = 3V$

26. $P = \dfrac{K}{V}$

27. $MP = 2V^2$

28. $\dfrac{1}{8}V^2 = P$

29. $\dfrac{P}{V} = 1$

30. $(4V + 1) - P = -1$

◼ MIXED REVIEW EXERCISES

Use the quadratic formula to solve each equation. (Lesson 12-5)

31. $2x^2 + 3x - 6 = 0$

32. $x^2 - 4x + 3 = 0$

33. $x^2 - 4x - 8 = 0$

34. $-2x^2 - 4x + 1 = 0$

35. $-4x^2 - 6x + 1 = 0$

36. $2x^2 - 6 = 0$

Find each value to the nearest hundredth. (Basic Math Skills)

37. $\sqrt{72}$

38. $\sqrt{48}$

39. $\sqrt{175}$

40. $\sqrt{37}$

Write each in simplest radical form. (Lesson 10-1)

41. $(2\sqrt{12})(5\sqrt{27})$

42. $(2\sqrt{8})(3\sqrt{12})$

43. $\dfrac{\sqrt{18}}{\sqrt{8}}$

44. $\sqrt{\dfrac{28}{6}}$

13-6 Inverse Variation

Goals ■ Solve problems involving inverse variation and inverse square variation.

Applications Astronomy, Physics, Travel

Work with a partner to answer the following questions.

The table shows the cost per person of renting a vacation home.

Number of people (n)	1	2	4	5	8	10
Cost per person (c)	$800	$400	$200	$160	$100	$80

a. Find the product of nc for each pair of values.

b. When n doubles, what happens to c?

c. Write an equation for c as a function of n.

d. Graph the function. Explain how c varies as n increases.

◥ BUILD UNDERSTANDING

The cost per person for renting the vacation home is a function of the number of people. As the number of people increases, the cost per person decreases. The relationship between n and c is an example of **inverse variation**. The value of one variable decreases as the value of the other increases.

Inverse variation may be represented by an equation in the form $y = \dfrac{k}{x}$, where k is a nonzero constant and $x \neq 0$. For the example above, the cost per person of renting the vacation home varies inversely as the number of people sharing the cost. The constant of variation is $800. The graph you drew in the activity above illustrates the graph of an inverse variation.

Most applications involve only the part of the graph that lies in the first quadrant.

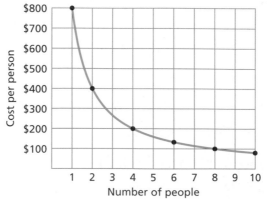

Example 1

Write an equation in which *y* varies inversely as *x* if one pair of values is *y* = 240 and *x* = 0.4.

Solution

$$y = \frac{k}{x}$$ Substitute in the equation for inverse variation.

$$240 = \frac{k}{0.4}$$

$$240 \times 0.4 = k$$

$$96 = k$$ Solve for *k*.

The equation is $y = \frac{96}{x}$.

Travel time varies inversely as travel speed. In other words, travel time decreases as speed increases.

Example 2

ASTRONOMY At its greatest distance from the sun, an asteroid travels a certain distance in 40 min while traveling at 250 mi/h. How long would it take the asteroid to travel the same distance, traveling at 400 mi/h?

Solution

$$y = \frac{k}{x}$$ Find an equation.

$$40 = \frac{k}{250}$$

$$40(250) = k$$

$$10,000 = k$$

$$y = \frac{10,000}{x}$$

$$y = \frac{10,000}{400}$$ Substitute 400 into the equation.

$$y = 25$$ Solve for *y*.

The trip will take 25 min travelling at 400 mi/h.

The table below shows how the brightness (in lumens) of a 60-watt light bulb varies with distance from the bulb.

Distance (in feet)	1	2	4	8
Brightness (in lumens)	880	220	55	13.75

This is an example of **inverse square variation**. The brightness of the light varies inversely as the square of the distance from its source.

Inverse square variation can be expressed in the form

$y = \frac{k}{x^2}$ or $x^2 y = k$, where *k* is a nonzero constant and $x \neq 0$.

For the example above, $l = \frac{880}{d^2}$. The constant of variation is 880.

Example 3

PHYSICS The brightness of a light bulb varies inversely as the square of the distance from the source. If a light bulb has a brightness of 400 lumens at 2 ft, what will be its brightness at 20 ft?

Solution

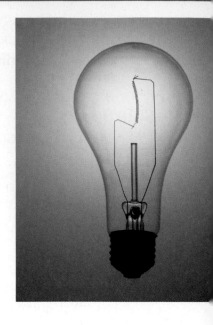

$$x^2y = k$$ Substitute known values into the equation for inverse square variation.

$$(2)^2(400) = k$$

$$4(400) = k$$

$$1600 = k$$ Solve.

$$x^2y = 1600$$ Write the equation.

$$(20)^2(y) = 1600$$ Substitute 20 for x.

$$400y = 1600$$

$$y = 4$$ Solve.

At 20 feet, the brightness will be 4 lumens.

◥ TRY THESE EXERCISES

1. Write an equation in which y varies inversely as x if one pair of values is $y = 85$ and $x = 0.8$.

2. If y varies inversely as x and one pair of values is $y = 44$ and $x = 5$, find y when $x = 8$.

3. In some cities, the amount paid by each person sharing a cab varies inversely as the number of people who share the cab. If 2 people pay $4.50 each for a ride, how much will the same ride cost 5 people?

4. If y varies inversely as the square of x and $y = 224$ when $x = 2$, find y when $x = 8$.

5. **PHYSICS** The brightness of a light bulb varies inversely as the square of the distance from the source. If a light bulb has a brightness of 300 lumens at 2 ft, what will be its brightness at 10 ft?

◥ PRACTICE EXERCISES • For Extra Practice, see page 705.

6. Write an equation in *which y varies* inversely as x if one pair of values is $y = 4550$ and $x = 0.05$.

7. If y varies inversely as x and one pair of values is $y = 39$ and $x = 3$, find y when $x = 39$.

8. If y varies inversely as x and one pair of values is $y = 12$ and $x = 10$, find y when $x = 20$.

9. **WRITING MATH** Think of a real-life example of inverse variation. Explain how you know the type of variation the example represents.

10. **TRAVEL** If it takes 30 min to drive from Ann's house to the museum traveling at 40 mi/h, how long will it take traveling at 50 mi/h?

11. **MAGNETISM** The force of attraction between two magnets varies inversely as the square of the distance between them. When two magnets are 2 cm apart, the force is 64 newtons. What will be the force when they are 8 cm apart?

12. If y varies inversely as the square of x and $y = 256$ when $x = 4$, find y when $x = 8$.

Write an equation of inverse variation for each.

13. Barometric pressure (p) is inversely proportional to the altitude (a).

14. The time (t) required to fill a swimming pool is inversely proportional to the square of the diameter (d) of the hose used to fill it.

15. The current (I) flowing in an electric circuit varies inversely as the resistance (R) in the circuit.

16. The intensity of the heat from a fireplace varies inversely as the square of the distance from the fireplace. Your friend is next to the fireplace. If you only feel $\frac{1}{16}$ the amount of heat that your friend feels, how much farther are you from the fireplace than your friend?

◣ EXTENDED PRACTICE EXERCISES

When a quantity varies directly as the product of two or more other quantities, the variation is called a **joint variation**. If y varies jointly as w and x, then $y = kwx$.

When a quantity varies directly as one quantity and inversely as another, the variation is called a **combined variation**. If y varies directly as w and inversely as x, then $y = \frac{kw}{x}$.

Write an equation of joint variation for each.

17. m varies directly as s and t.

18. a varies jointly as c and d.

19. r varies jointly as w, x, and y.

Write an equation of combined variation for each.

20. n varies directly as t and inversely as e.

21. v varies directly as r and inversely as the square of w.

22. d varies directly as the square of a and inversely as b.

◣ MIXED REVIEW EXERCISES

Calculate the distance between each pair of points. Round to the nearest hundredth if necessary. (Lesson 12-6)

23. $A(3, 2)$, $B(1, -8)$ 24. $C(-6, 2)$, $D(5, -9)$ 25. $E(3, -4)$, $F(8, -6)$

26. $G(-4, -3)$, $H(8, -1)$ 27. $J(7, -5)$, $K(3, 8)$ 28. $L(-4, -7)$, $M(-2, -3)$

In the figure, R is the midpoint of QS. Find each measure. (Lesson 3-3)

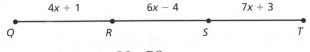

29. QR 30. RS 31. ST

Review and Practice Your Skills

1. In a direct variation, if y decreases, what must be true about x?

2. In the equation representing direct variation, $y = kx$, what must be true about k?

3. What is the difference between a graph showing a direct variation and a graph showing a direct square variation?

Write the equation of direct variation using the given values.

4. $x = 5, y = 20$ 5. $x = 12, y = 8$ 6. $x = 9, y = 10$

7. $x = 16, y = 4$ 8. $x = 24, y = 4$ 9. $x = 8, y = 18$

10. If y varies directly as x and $y = 2$ when $x = 8$, find y when $x = 20$.

11. If y varies directly as x and $y = 5$ when $x = 35$, find y when $x = 15$.

12. If y varies directly as x and $y = 8$ when $x = 4$, find y when $x = 7$.

13. If y varies directly as x^2, and $y = 12$ when $x = 4$, find y when $x = 8$.

14. If y varies directly as x^2, and $y = 9$ when $x = 2$, find y when $x = 12$.

15. If y varies directly as x^2, and $y = 45$ when $x = 3$, find y when $x = 9$.

16. A survey showed that 7 out of 10 people liked the taste of Shimmer toothpaste. At this rate, how many people out of 2400 would be expected to like the taste of Shimmer?

17. Refer to page 584. The example of an inverse variation was given as the cost of a vacation home per person. Why is the application only concerned with the part of the graph that lies in the first quadrant?

18. Write an equation in which y varies inversely as x if one pair of values is $y = 4$ and $x = 12$.

Find answers to the nearest hundredth.

19. Write an equation in which y varies inversely as x if one pair of values is $y = 7$ and $x = 0.9$.

20. If y varies inversely as x and $y = 4$ when $x = 2$, find y when $x = 7$.

21. If y varies inversely as x and $y = 7$ when $x = 4$, find y when $x = 9$.

22. If y varies inversely as the square of x and $y = 9$ when $x = 2$, find y when $x = 3$.

Write an equation of inverse variation for each.

23. The time (t) it takes to travel from one point to another is inversely proportional to the speed (s) of the travel.

24. The amount of oxygen (o) in the air is inversely proportional to the altitude (a).

Graph each equation. (Lessons 13-1–13-2)

25. $(x + 2)^2 + (y - 3)^2 = 49$ **26.** $x^2 = -5y$ **27.** $x^2 = 4y$

28. $(x + 2)^2 + y^2 = 25$ **29.** $(x - 4)^2 + (y + 3)^2 = 64$ **30.** $x^2 + (y + 5)^2 = 36$

Write the equation of direct variation using the given values. (Lesson 13-5)

31. $x = 10$, $y = 25$ **32.** $x = 9$, $y = 3$ **33.** $x = 5$, $y = 7$

34. One survey showed that 7 out of 12 eligible people in Greenville planned to vote in the next election. If the population of eligible voters in Greenville is 19,824, how many could be expected to vote?

MathWorks Career – Astronomer
Workplace Knowhow

Astronomers work in a sub-field of physics. Astronomers study things such as the birth of stars, the death of stars, natural satellites, the composition of planets, and the possibility of life on other planets. Astronomers work in observatories with very large telescopes, in universities teaching and in planetariums. Astronomers also work with engineers in the design, launch, and use of deep space probes and satellites that send astronomical data back to the Earth from planets too far away and too inhospitable for humans to visit. Precise calculations for the orbits of the satellites and probes are essential for their missions' success.

1. Suppose that you discovered a new solar system with an elliptical orbit of planets around two suns. What would be the mathematical term for the location of each sun?

The suns are observably 5 in. apart in scaled distance. You also observe the 4 planets' orbits for a period of six months for their measure. The point at which the planets stop moving away from the suns and start moving back toward them is the *x*-intercept of the ellipse when their orbits are graphed. These distances from one of the suns are as follows:

   a. Planet A is 1.2 in. from a sun   **b.** Planet B is 2.4 in. from a sun

   c. Planet C is 3.8 in. from a sun   **d.** Planet D is 5.2 in. from a sun

2. For graphing an ellipse around the origin on the coordinate plane, determine *x*-intercepts of each planet.

3. Calculate the *y*-intercept of each orbit.

4. Write the equation for each planet's elliptical orbit.

13-7 Quadratic Inequalities

Goals ■ Graph quadratic inequalities.

Applications Astronomy, Communications, Computer Design

Work with a partner. You may use a graphing calculator.

The graph of a quadratic function divides the coordinate plane into three sets of points. Graph $y = x^2 + 2x + 1$ on a coordinate plane.

a. Find 5 ordered pairs for which $y = x^2 + 2x + 1$.

b. Find 5 ordered pairs for which $y < x^2 + 2x + 1$.

c. Find 5 ordered pairs for which $y > x^2 + 2x + 1$.

d. Use the points you found for b to help you locate and shade the region of the graph where $y < x^2 + 2x + 1$.

e. Use the points you found for c to help you locate and draw horizontal lines through the region of the graph where $y > x^2 + 2x + 1$.

◣ BUILD UNDERSTANDING

Just as you used linear equations to graph linear inequalities, you can use quadratic equations to graph **quadratic inequalities**.

Example 1

Graph $-y^2 + 4x^2 > 16$.

Solution

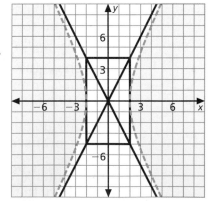

Graph the hyperbola $-y^2 + 4x^2 = 16$. Use the intercepts of $(2, 0)$, $(-2, 0)$, $(0, 4)$, and $(0, -4)$ to draw the rectangle. Then draw the asymptotes and sketch the hyperbola.

Because $-y^2 + 4x = 16$ is not part of the solution, the hyperbola is drawn with a dashed line.

To decide which points are part of the solution set, select points on the graph and substitute their coordinates into the equation.

Select $(0, 0)$: $-(0)^2 + 4(0)^2 \not> 16$ The point $(0, 0)$ and the region that
    $0 + 0 \not> 16$ contains it are not in the solution set.

Select $(-3, 1)$: $-(1)^2 + 4(-3)^2 > 16$ Select $(4, 0)$: $-(0)^2 + 4(4)^2 > 16$

    $-(1) + 4(9) > 16$ $0 + 4(16) > 16$

    $35 > 16$ $64 > 16$

These points and the regions they contain are in the solution set. The solution set is the shaded region shown on the graph.

Systems of inequalities can be solved by finding the intersections of their graphs.

Example 2

ASTRONOMY Radio commands may be sent to a space probe during a specific portion of its flight. If commands are sent too soon or too late, they will not be received by the probe. The solution set of the following system of inequalities is used to determine when commands may be sent. Solve the system of inequalities by graphing.

$$9x^2 + 4y^2 < 36$$

$$y \geq x - 2$$

Solution

Graph the ellipse $9x^2 + 4y = 36$. The center is at the origin. The x-intercepts are (2, 0) and (−2, 0). The y-intercepts are (0, 3) and (0, −3).

The points on the ellipse are not in the solution set, so the ellipse is drawn with a dashed line. Point (0, 0) is in the solution set, so points inside the ellipse are part of the solution set.

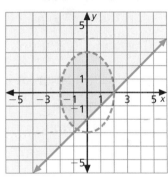

Graph $y = x - 2$. The solution set includes the line, so the line is solid. For the region above the line, $y > x - 2$, so part of this region is in the solution set.

The intersection of the two equations is shown by the green region of the graph.

Check Understanding

How can you check to be sure that the region inside the parabola is in the solution set?

Example 3

Solve this system of inequalities by graphing.

$$x^2 + y^2 \leq 49$$

$$y > 2x^2 + 2$$

Solution

Graph the circle with radius 7. The circle is in the solution set; draw the circle with a solid line. Point (0, 0) is in the solution set. The region inside the circle is in the solution set.

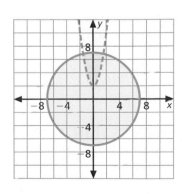

Graph parabola $y = 2x^2 + 2$. The parabola is not in the solution set; graph the parabola with a dashed line. Point (0, 0) is not in the solution set, so the region inside the parabola is in the solution set.

The region shaded green shows the intersection of the two equations.

Graph each inequality.

1. $x^2 + y^2 \leq 81$

2. $y < 2x^2 - 1$

3. $3x^2 + 12y^2 > 48$

4. $9x^2 - 25y^2 \geq 225$

Graph each system of inequalities.

5. $(x + 4)^2 + (y - 2)^2 > 4$

$y < x + 6$

6. $y \leq x^2 + 3x - 4$

$(x - 1)^2 + (y + 3)^2 \leq 4$

7. $4x^2 + 25y^2 \leq 100$

$x^2 + y^2 < 25$

8. $25x^2 - 4y^2 \geq 100$

$x \geq -1$

> **Problem Solving Tip**
>
> All hyperbolas and ellipses in these exercises have center (0, 0).

Graph each inequality.

9. $4x^2 + 16y^2 \geq 64$

10. $(x - 5)^2 + (y + 1)^2 \geq 4$

11. $y < x^2 + x + 1$

12. $9x^2 - 16y^2 \leq 144$

Graph each system of inequalities.

13. $x^2 - y^2 < 25$

$x^2 + y^2 < 100$

14. $y > x^2 - 2x$

$y \leq x + 3$

15. $(x - 1)^2 + (y - 3)^2 > 25$

$y < x^2 - 2x + 1$

16. $x^2 + 16y^2 \leq 16$

$x^2 + y^2 < 64$

Use the inequalities you have graphed in this lesson to help you complete each statement. Use <, >, or =.

17. If $x^2 + y^2$ __?__ r^2, points inside the circle are in the solution set.

18. If $\frac{x^2}{a^2} + \frac{y^2}{b^2}$ __?__ 1, points outside the ellipse are in the solution set.

19. If $ax^2 + bx + c$ __?__ y, the region inside the parabola is the solution set.

20. COMMUNICATIONS The limits of a transmitter can be modeled using the system of inequalities below. Graph the system and describe the solution set.

$x^2 + y^2 < 36$

$\frac{x^2}{100} + \frac{y^2}{36} < 1$

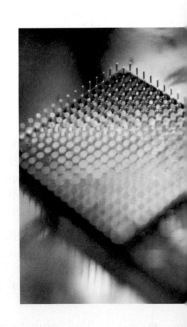

21. COMPUTER DESIGN The system of inequalities below defines the capabilities of a computer chip. Graph the system and describe the solution set.

$y > x^2 + 2x - 3$

$y < x^2 - x$

22. CHAPTER INVESTIGATION Select landmarks at your school or within your community to represent aphelion, or farthest orbital point, for each landmark. Mark these landmarks on your map. Share your findings with the class, and discuss how this activity has increased your understanding of the size of the solar system.

■ EXTENDED PRACTICE EXERCISES

The solution of the inequality $x^2 - 8x < -12$ is shown on the number lines below.

23. WRITING MATH Explain this method for solving quadratic inequalities.

24. Does the solution check? Try these points: 1, 3, 5, 8.

25. What is the solution to the given inequality?

26. Use this method to solve $x^2 - 8x \le -15$.

■ MIXED REVIEW EXERCISES

Calculate the midpoint of the segment with the given endpoints. (Lesson 12-6)

27. $A(-6, 3), B(4, -5)$ **28.** $C(3, -2), D(-8, 9)$ **29.** $E(1, 0), F(-3, -7)$

30. $G(3, 0), H(-8, 2)$ **31.** $J(2, 7), K(-7, 2)$ **32.** $L(4, -5), M(1, 3)$

33. $N(0, 7), P(3, -4)$ **34.** $Q(1, 6), R(-6, -1)$ **35.** $S(8, -3), T(5, -6)$

36. $U(-6, 4), V(-3, -5)$ **37.** $W(-2, 0), X(1, 6)$ **38.** $Y(3, 5), Z(8, 2)$

Each figure below is a parallelogram. Find a and b. (Lesson 4-8)

39.   **40.**   **41.**

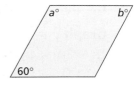

Find the volume of each figure to the nearest whole number. (Lesson 5-7)

42.   **43.**   **44.**

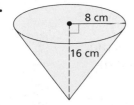
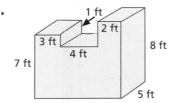
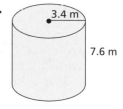

45. Solve by completing the square: $x^2 + 20x - 1 = 0$. (Lesson 12-4)

13-8 Exponential Functions

Goals
- Graph exponential functions.
- Solve problems involving exponential growth and decay.

Applications Population, Investments, Transportation

Work with a partner. You will need graphing paper.

1. Copy and complete the following table.

Expression	Exponent	Value of Expression
2^{-2}	-2	$\frac{1}{4}$
2^{-1}		
2^{0}		
2^{1}		
2^{2}		
2^{3}		

2. Use the table to write seven ordered pairs of the form (exponent, value of expression).

3. Locate the seven ordered pairs on a coordinate plane. Then draw a smooth curve through the points.

4. Describe the graph. Where does the curve cross the y-axis?

BUILD UNDERSTANDING

The graph you drew in the activity above is the graph of $y = 2^x$. This type of function, in which the variable is the exponent, is called an **exponential function**.

Example 1

Graph $y = 2^x - 3$. State the y-intercept.

Solution

You can use a graphing calculator to graph $y = 2^x - 3$. The y-intercept is -2.

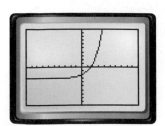

Example 2

Compare the graphs of $y = 3^x$ and $y = \left(\frac{1}{3}\right)^x$.

Solution

Make a table for $y = 3^x$. Graph the function.

x	−3	−2	−1	0	1	2	3
y	$\frac{1}{27}$	$\frac{1}{9}$	$\frac{1}{3}$	1	3	9	27

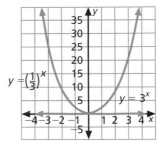

Make a table for $y = \left(\frac{1}{3}\right)^x$. Graph the function.

x	−3	−2	−1	0	1	2	3
y	27	9	3	1	$\frac{1}{3}$	$\frac{1}{9}$	$\frac{1}{27}$

The graphs are reflections of each other across the y-axis. Both graphs have a y-intercept of 1. The graph of $y = 3^x$ increases from left to right. The graph of $y = \left(\frac{1}{3}\right)^x$ decreases from left to right.

Two examples of exponential functions are exponential growth and exponential decay. **Exponential growth** happens when a quantity increases by a fixed rate each time period. **Exponential decay** happens when a quantity decreases by a fixed rate each time period. The general equations for exponential growth and decay are given below.

Exponential growth $y = C(1 + r)^t$

Exponential decay $y = C(1 - r)^t$

In these equations, y represents the final amount. C represents the initial amount, r represents the fixed rate or percent of change expressed as a decimal, and t represents time.

Example 3

POPULATION The country of Latvia has been experiencing a 0.6% annual decrease in population. In 2003, its population was 2,350,000. What would you predict the population to be in 2013 if the rate remained the same?

Solution

Since the population is decreasing by a fixed rate each year, this is an example of exponential decay.

$y = C(1 - r)^t$

$y = 2,350,000(1 - 0.006)^{10}$ Substitute in the equation.

$y \approx 2,212,747$ Use a calculator to solve for y.

In 2013, the population will be about 2,212,750.

Graph each function. State the *y*-intercept.

1. $y = 5^x$

2. $y = 2 \cdot 3^x$

3. $y = 2^x + 4$

4. **INVESTMENTS** A municipal bond pays 5% per year. If $2000 is invested in these bonds, find the value of the investment after 4 yr.

5. **FARMING** A farmer buys a tractor for $60,000. If the tractor depreciates 10% per year, what is the value of the tractor after 8 yr?

Graph each function. State the *y*-intercept.

6. $y = 4^x$

7. $y = 10^x$

8. $y = \left(\dfrac{1}{2}\right)^x$

9. $y = \left(\dfrac{1}{10}\right)^x$

10. $y = 5(2^x)$

11. $y = 3(5^x)$

12. $y = 3^x - 7$

13. $y = 3^x + 6$

14. $y = 2(3^x) - 1$

15. **TECHNOLOGY** Computer use around the world has risen 19% annually since 1980. If 18.9 million computers were in use in 1980, predict the number of computers that will be in use in 2015.

16. **POPULATION** The population of Mexico has been increasing at an annual rate of 1.7%. If the population of Mexico was 104,900,000 in the year 2003, predict its population in 2015.

17. **VEHICLE OWNERSHIP** A car sells for $24,000. If the annual rate of depreciation is 13%, what is the value of the car after 8 yr?

18. **NUTRITION** A cup of coffee contains 130 mg of caffeine. If caffeine is eliminated from the body at a rate of 11% per hour, how much caffeine will remain in the body after 3 h?

19. **INVESTMENTS** Determine the amount of an investment if $3000 is invested at an interest rate of 5.5% each year for 3 yr.

20. **REAL ESTATE** The Villa family bought a condominium for $115,000. Assuming that the value of the condo will appreciate 5% each year, how much will the condo be worth in 6 yr?

21. **BUSINESS** A piece of office equipment valued at $35,000 depreciates at a steady rate of 10% per year. What is the value of the equipment in 10 yr?

22. **VEHICLE OWNERSHIP** Carlos needs to replace his car. If he leases a car, he will have an option to buy the car after 2 yr for $14,458. The current price of the car is $17,369. If the car depreciates at 16% per year, how will the depreciated price compare with the buyout price of the lease?

23. ERROR ALERT Amanda graphed $y = \left(\frac{1}{4}\right)^x$ at the right. Is she correct? Explain.

CRITICAL THINKING For Exercises 24–26, describe the graph of each equation as a transformation of the graph of $y = 4^x$.

24. $y = \left(\frac{1}{4}\right)^x$ **25.** $y = 4^x + 2$ **26.** $y = 4^x - 6$

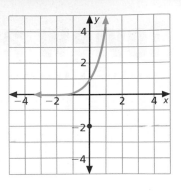

■ EXTENDED PRACTICE EXERCISES

In general, for any real number b and for any positive integer n, $b^{\frac{1}{n}} = \sqrt[n]{b}$, except when $b < 0$ and n is even.

Write each expression in radical form.

27. $6^{\frac{1}{5}}$ **28.** $26^{\frac{1}{4}}$ **29.** $m^{\frac{1}{6}}$

Write each radical using rational exponents.

30. $\sqrt{17}$ **31.** $\sqrt[3]{25}$ **32.** $\sqrt[8]{x}$

33. WRITING MATH Explain why $(-9)^{\frac{1}{2}}$ is not a real number.

■ MIXED REVIEW EXERCISES

Factor each trinomial. (Lesson 11-7)

34. $d^2 + 13d + 12$ **35.** $t^2 - 8t + 12$

36. $p^2 + 2p - 24$ **37.** $s^2 - 8s - 20$

38. $a^2 - ab - 2b^2$ **39.** $m^2 - 5mn + 6n^2$

Simplify. (Lesson 11-2)

40. $(x^3y^4)(xy^3)$ **41.** $(-3mn^2)(5m^3n^2)$

42. $3b(5b + 8)$ **43.** $\frac{1}{2}x(8x - 6)$

44. $5y(y^2 - 3y + 6)$ **45.** $-ab(3b^2 + 4ab - 6a^2)$

Use the Pythagorean Theorem to find the unknown length. Round your answers to the nearest tenth. (Lesson 10-2)

46.

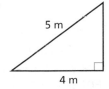

5 m

4 m

47.

6 cm

2 cm

48.

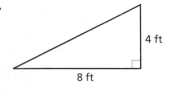

4 ft

8 ft

Review and Practice Your Skills

Graph each inequality.

1. $4x^2 + 2y^2 \geq 8$

2. $y < \frac{1}{8}x^2$

3. $x^2 - 4y^2 \leq 4$

4. $x^2 + y^2 > 25$

5. $(x + 2)^2 + (y - 3)^2 \leq 9$

6. $y \geq x^2 - 4x$

Graph each system of inequalities.

7. $x + 2y > 1$
   $x^2 + y^2 \leq 25$

8. $x^2 + y^2 \geq 4$
   $4y^2 - 9x^2 \leq 36$

9. $x + y < 4$
   $9x^2 - 4y^2 \geq 36$

10. $x^2 + y^2 < 25$
    $4x^2 - 9y^2 < 36$

11. $x + y \leq 2$
    $4x^2 - y^2 \geq 4$

12. $x^2 + y^2 < 36$
    $4x^2 + 9y^2 > 36$

13. $y^2 < x$
    $x^2 - 4y^2 < 16$

14. $x^2 \leq y$
    $y^2 - x^2 \geq 4$

15. $y < x$
    $y > x^2 - 4$

Graph each function. State the y-intercept.

16. $y = \left(\frac{1}{4}\right)^x$

17. $y = 9^x$

18. $y = 0.5(4^x)$

19. $y = 2\left(\frac{1}{2}\right)^x$

20. $y = \left(\frac{1}{3}\right)^x - 3$

21. $y = 2^x - 5$

22. **VEHICLE OWNERSHIP** A pick-up truck sells for $27,000. If the annual rate of depreciation is 12%, what is the value of the truck after 5 yr?

23. **INVESTMENTS** Determine the amount of an investment if $5000 is invested at an interest rate of 4.5% each year for 4 yr.

24. **POPULATION** In 2003, the population of Jamaica was 2,696,000. If the population increases at a rate of 1.2% per year, predict its population in 2015.

25. **COMPUTER COSTS** A computer package costs $1800. If it depreciates at a rate of 18% per year, find the value of the computer package after 3 yr.

26. **POPULATION** The population of Ukraine has been decreasing by an annual rate of 0.9%. The population of Ukraine was 48,055,000 in 2003. Predict its population in 2010.

27. **INCOME** The median income was $32,000 in 2003. If income increases at a rate of 0.5% per year, predict the median income in 2013.

28. **REAL ESTATE** A house is purchased for $180,000. If the value of the house increases 4.5% per year, what is its value after 8 yr?

Write an equation for each circle. (Lesson 13-1)

29. radius 12
center $(-1, 3)$

30. radius 6
center $(5, 4)$

31. radius 2
center $(-4, -1)$

Find the radius and center for each circle. (Lesson 13-1)

32. $x^2 + (y - 2)^2 = 16$

33. $(x + 1)^2 + (y + 5)^2 = 1$

34. $(x - 7)^2 + (y - 4)^2 = 49$

Find the focus and directrix of each equation. (Lesson 13-2)

35. $x^2 = 20y$

36. $3x^2 = -24y$

37. $28y - 7x^2 = 0$

38. Name the conic section formed by the plane.
The plane at the right is parallel to the x-axis
and does not contain the vertex.
(Lesson 13-3)

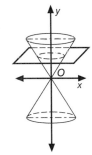

Graph each equation. (Lesson 13-4)

39. $x^2 - 36y^2 = 36$

40. $x^2 - 2y^2 = 2$

41. $x^2 - y^2 = 4$

42. $4x^2 + 8y^2 = 32$

43. $3x^2 + 9y^2 = 27$

44. $27x^2 + 9y^2 = 81$

45. If y varies directly as x and $y = 15$ when $x = 3$, find y when $x = 12$. (Lesson 13-5)

46. If y varies directly as x and $y = 8$ when $x = 6$, find y when $x = 15$. (Lesson 13-5)

47. If y varies directly as x and $y = 18$ when $x = 15$, find y when $x = 20$. (Lesson 13-5)

48. If y varies inversely as x and $y = 5$ when $x = 10$, find y when $x = 2$. (Lesson 13-6)

49. If y varies inversely as x and $y = 16$ when $x = 5$, find y when $x = 20$. (Lesson 13-6)

50. If y varies inversely as x and $y = 2$ when $x = 25$, find x when $y = 40$. (Lesson 13-6)

Graph each system of inequalities. (Lesson 13-7)

51. $x^2 + y^2 \leq 9$
$x^2 + 4y^2 \leq 16$

52. $x^2 + y^2 \geq 1$
$x^2 + y^2 \leq 16$

53. $y \geq x^2$
$y > -x + 2$

Graph each function. State the y-intercept. (Lesson 13-8)

54. $y = \left(\dfrac{1}{5}\right)^x$

55. $y = 5^x - 4$

56. $y = 4 \cdot 2^x$

57. **POPULATION** The population of Canada has been increasing by an annual
rate of 0.3%. The population of Canada was 32,207,000 in 2003. Predict its
population in 2018. (Lesson 13-8)

13-9 Logarithmic Functions

Goals
- Evaluate logarithmic expressions.
- Solve logarithmic equations.

Applications Sound, Chemistry, Earthquakes

Work with a partner.

1. In the function at the right, the variable x is an exponent of the base 2, and the value of 2^x is to be determined. Copy and complete the table.

x	$2^x = y$	y
−1	$2^{-1} = y$	$\frac{1}{2}$
0	$2^0 = y$	
1	$2^1 = y$	
2	$2^2 = y$	
3	$2^3 = y$	
4	$2^4 = y$	
5	$2^5 = y$	

2. In the function at the right, the variable x is given as the power of 2, and the exponent y is to be determined. Copy and complete the table.

x	$x = 2^y$	y
$\frac{1}{2}$	$\frac{1}{2} = 2^y$	−1
1	$1 = 2^y$	
2	$2 = 2^y$	
4	$4 = 2^y$	
8	$8 = 2^y$	
16	$16 = 2^y$	
32	$32 = 2^y$	

3. Compare the two functions.

BUILD UNDERSTANDING

In the second function, $x = 2^y$, the exponent y is called the **logarithm** base 2 of x. This function is written $\log_2 x = y$ and is read "the log base 2 of x is equal to y." The logarithm corresponds to the exponent.

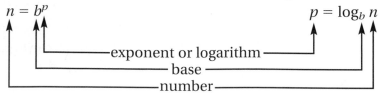

Exponential Function

$$n = b^p$$

Logarithmic Function

$$p = \log_b n$$

exponent or logarithm
base
number

Example 1

Write each equation in logarithmic form.

a. $8^0 = 1$

b. $2^{-4} = \frac{1}{16}$

Solution

a. $8^0 = 1 \longrightarrow \log_8 1 = 0$

b. $2^{-4} = \frac{1}{16} \longrightarrow \log_2 \frac{1}{16} = -4$

Example 2

Write each equation in exponential form.

a. $\log_9 81 = -2$ **b.** $\log_{10} 10{,}000 = 4$

Solution

a. $\log_9 \frac{1}{81} = -2 \longrightarrow \frac{1}{81} = 9^{-2}$ **b.** $\log_{10} 10{,}000 = 4 \longrightarrow 10{,}000 = 10^4$

You can use the definition of logarithm to find the value of a logarithmic expression.

Example 3

Evaluate $\log_5 125$.

Solution

$\log_5 125 = y$	Let the logarithm equal y.
$125 = 5^y$	Rewrite the equation using the definition of logarithm.
$5^3 = 5^y$	$5^3 = 125$

Since $5^3 = 5^y$, y must equal 3.

A **logarithmic equation** is an equation that contains one or more logarithms. You can use the definition of a logarithm to help you solve logarithmic equations.

Example 4

Solve $\log_4 m = -5$.

Solution

$\log_4 m = -5$	
$m = 4^{-5}$	Rewrite the equation using the definition of logarithm.
$m = \frac{1}{4^5}$	Use the definition of negative exponents.
$m = \frac{1}{1024}$	Simplify.

The **property of equality for logarithmic equations** states that if $\log_b x = \log_b y$, then $x = y$.

Example 5

Solve $\log_5 (x - 5) = \log_5 7$.

Solution

$\log_5 (x - 5) = \log_5 7$	
$x - 5 = 7$	Use the property of equality for logarithmic equations.
$x - 5 + 5 = 7 + 5$	Add 5 to each side.
$x = 12$	Simplify.

Write each equation in logarithmic form.

1. $5^4 = 625$

2. $7^{-2} = \dfrac{1}{49}$

3. $2^{-9} = \dfrac{1}{512}$

Write each equation in exponential form.

4. $\log_3 81 = 4$

5. $\log_6 216 = 3$

6. $\log_5 \dfrac{1}{25} = -2$

Evaluate each expression.

7. $\log_2 64$

8. $\log_3 \dfrac{1}{27}$

9. $\log_{10} 1{,}000{,}000$

Solve each equation.

10. $\log_3 k = 6$

11. $\log_5 (2a + 3) = \log_5 21$

12. $\log_2 (3d - 5) = \log_2 (d + 7)$

▗ PRACTICE EXERCISES • For Extra Practice, see page 707.

Write each equation in logarithmic form.

13. $8^5 = 32{,}768$

14. $\left(\dfrac{1}{3}\right)^4 = \dfrac{1}{81}$

15. $4^{-3} = \dfrac{1}{64}$

16. $\left(\dfrac{2}{5}\right)^3 = \dfrac{8}{125}$

17. $\left(\dfrac{1}{9}\right)^{-3} = 729$

18. $20^3 = 8000$

Write each equation in exponential form.

19. $\log_7 1 = 0$

20. $\log_4 64 = 3$

21. $\log_{10} \dfrac{1}{100} = -2$

22. $\log_3 729 = 6$

23. $\log_3 \dfrac{1}{243} = -5$

24. $\log_{\frac{1}{2}} 4 = -2$

Evaluate each expression.

25. $\log_2 8$

26. $\log_{12} 1$

27. $\log_{\frac{3}{4}} \dfrac{27}{64}$

28. $\log_4 \dfrac{1}{256}$

29. $\log_{\frac{1}{8}} 64$

30. $\log_{10} 0.00001$

Solve each equation.

31. $\log_7 s = 5$

32. $\log_2 t = -5$

33. $\log_{\frac{1}{6}} y = -6$

34. $\log_3 (4 + y) = \log_3 (2y)$

35. $\log_7 d = \log_7 (3d - 10)$

36. $\log_{10} (4w + 15) = \log_{10} 35$

37. $\log_3 (3x - 6) = \log_3 (2x + 1)$

38. $\log_6 (3r - 1) = \log_6 (2r + 4)$

39. $\log_9 (5p - 1) = \log_9 (3p + 7)$

40. YOU MAKE THE CALL Shane says that the value of $\log_4 2$ is 2. Do you agree? If not, why not?

41. SOUND An equation for loudness L, in decibels, is $L = 10 \log_{10} R$, where R is the relative intensity of the sound. Find the relative intensity of a fireworks display with a loudness of 150 decibels.

42. CHEMISTRY The pH of a solution is a measure of its acidity and is written as a logarithm with base 10. A low pH indicates an acidic solution. Neutral water has a pH of 7. A substance has a pH of 4. How many times as acidic is the substance as water?

43. EARTHQUAKES The magnitude of an earthquake is measured on a logarithmic scale called the Richter scale. The magnitude is given by $M = \log_{10} x$, where x represents the amplitude of the seismic wave. How many more times as great is the amplitude caused by an earthquake with a Richter scale rating of 6 as an aftershock with a Richter scale rating of 4?

◣ EXTENDED PRACTICE EXERCISES

For Exercises 44–46, use the same coordinate plane for each graph. Use the tables in the opening activity on page 600 to help graph the functions.

44. Graph seven ordered pairs that satisfy the function $y = 2^x$. Draw a smooth curve through the points. Label the graph $y = 2^x$.

45. Graph seven ordered pairs that satisfy the function $x = 2^y$. Draw a smooth curve through the points. Label the $y = \log_2 x$.

46. Graph $x = y$. Describe the relationship among the three graphs.

47. Compare the domain and range of the functions $y = 2^x$ and $y = \log_2 x$.

48. Graph the function $y = 3^x$. Without using a table of values, graph $y = \log_3 x$ on the same coordinate plane.

◣ MIXED REVIEW EXERCISES

Simplify. (Lesson 11-4)

49. $(2a - 4)(a + 1)$

50. $(3v - 1)(2v - 2)$

51. $(4g + 5)(4g - 5)$

There are 9 pennies, 7 dimes, and 5 nickels in an antique coin collection. Suppose two coins are to be selected at random from the collection without replacing the first one. Find the probability of each event. (Lesson 9-4)

52. P(a penny, then a dime)

53. P(two nickels)

54. P(two dimes)

Solve each system of equations. (Lesson 6-6)

55. $2r - 3s = 11$
$2r + 2s = 6$

56. $4c + 2d = 10$
$c + 3d = 10$

57. $4a + 2b = 15$
$2a + 2b = 7$

Given $f(x) = 3x - 10$, evaluate each function. (Lesson 2-2)

58. $f(2)$

59. $f(10)$

60. $f(-3)$

Chapter 13 Review

VOCABULARY ◥

Choose the word from the list at the right that completes each statement below.

1. The set of all points equidistant from a fixed point called a focus and a fixed line called a directrix is a(n) __?__.

2. A relationship in which one variable increases as the other variable increases is a(n) __?__.

3. A relationship in which one variable decreases as the other variable increases is a(n) __?__.

4. The equation $(x - h)^2 + (y - k)^2 = r$ where $r \neq 0$ is the standard form of a(n) __?__.

5. When a plane intersects right circular cones, a(n) __?__ is formed.

6. Exponential and __?__ functions are inverses of each other.

7. If a quantity decreases by a fixed rate each time period, there is exponential __?__.

8. A set of points such that the sum of the distances from two fixed points called foci is always the same is a(n) __?__.

9. A set of points such that the difference between the distances from two fixed points called foci is always the same is a(n) __?__.

10. A line that a graph approaches, but never meets is a(n) __?__.

a.	asymptote
b.	circle
c.	conic section
d.	decay
e.	directrix
f.	direct variation
g.	ellipse
h.	growth
i.	hyperbola
j.	inverse variation
k.	logarithmic
l.	parabola

LESSON 13-1 ◥ The Standard Equation of a Circle, p. 562

▶ The equation for a circle with its center at the origin and with radius r is $x^2 + y^2 = r^2$, $r \neq 0$.

▶ The standard equation for a circle with its center located at the point (h, k) with radius r is $(x - h)^2 + (y - k)^2 = r^2$, $r \neq 0$.

Write an equation for each circle.

11. radius 8
center (0, 0)

12. radius 4
center (2, 3)

13. radius 6
center (5, 0)

Find the radius and center of each circle.

14. $x^2 + y^2 = 25$

15. $x^2 + (y - 3)^2 = 9$

16. $(x + 9)^2 + (y + 4)^2 = 21$

LESSON 13-2 ◥ More on Parabolas, p. 566

▶ When the focus $(0, a)$ is on the y-axis and the directrix is $y = -a$, the standard equation for a parabola is $x^2 = 4ay$.

Find the focus and directrix of each parabola.

17. $x^2 = 20y$

18. $-40y = 5x^2$

19. $12x^2 - 48y = 0$

Find the standard equation for each parabola with vertex located at the origin.

20. Focus (0, 5)   **21.** Focus (0, −4)   **22.** Focus $\left(0, \frac{1}{2}\right)$

LESSON 13-3 ◣ Problem Solving Skills: Visual Thinking, p. 572

▶ You can visualize the conic section formed by a plane intersecting a cone or double cone.

23. Describe the intersection between a plane and a double cone that produces a circle.

24. Describe the intersection between a plane and a double cone that produces a hyperbola.

LESSON 13-4 ◣ Ellipses and Hyperbolas, p. 574

▶ The standard form for the equation of an ellipse with its center at the origin and foci on the x-axis is $\frac{x^2}{a^2} + \frac{y^2}{b^2} = 1$.

▶ The standard form for the equation of a hyperbola that is symmetric about the origin and has foci on the x-axis is $\frac{x^2}{a^2} - \frac{y^2}{b^2} = 1$.

Find an equation for each figure.

25. an ellipse with foci (8, 0) and (-8, 0) and x-intercepts (10, 0) and (−10, 0)

26. a hyperbola with center (0, 0) and foci on the x-axis if $a = 4$ and $b = 7$

LESSON 13-5 ◣ Direct Variation, p. 580

▶ Equations in which one variable increases as the other variable increases can be expressed as $y = kx$, where k is a positive constant and $x \neq 0$.

▶ Direct square variation is shown by the equation $y = kx^2$.

27. If y varies directly as x and $y = 75$ when $x = 7.5$, find y when $x = 5$.

28. If y varies directly as x^2 and $y = 51.2$ when $x = 4$, find y when $x = 9$.

29. Let y vary directly as the square of x. If $y = 45$ when $x = 3$, find y when $x = 8$.

LESSON 13-6 ◣ Inverse Variation, p. 584

▶ Equations in which one variable decreases as the other variable increases can be expressed as $y = \frac{k}{x}$, where k is a nonzero constant and $x \neq 0$.

▶ Inverse square variation is shown by the equation $y = \frac{k}{x^2}$, or $x^2y = k$.

30. Write an equation in which y varies inversely as x if one pair of values is $y = 90$ and $x = 0.7$.

31. If y varies inversely as the square of x and $y = 900$ when $x = 5$, find y when $x = 12$.

32. Let y vary inversely as x. If $y = 6.5$ when $x = 3$, find y when $x = 4$.

33. Let y vary inversely as the square of x. If $y = 40$ when $x = 9$, find y when $x = 6$.

LESSON 13-7 ◼ Quadratic Inequalities, p. 590

▶ Substitute coordinates into the equation for a quadratic inequality to locate regions in the solution set.

▶ Systems of inequalities can be solved by finding the intersections of their graphs.

Graph each inequality.

34. $x^2 + y^2 > 49$

35. $y \geq 2x^2 + x + 2$

36. $9x^2 + 36y^2 < 36$

LESSON 13-8 ◼ Exponential Functions, p. 594

▶ A function where the variable is an exponent is an **exponential function**.

▶ A quantity that increases or decreases by a fixed rate each time period is called **exponential growth** or **exponential decay**, respectively.

Graph each function. State the y-intercept.

37. $y = \left(\dfrac{1}{8}\right)^x$

38. $y = 4\left(\dfrac{1}{3}\right)^x$

39. $y = 5(2^x) + 4$

40. POPULATION In 2002, the population of South Carolina was about 4,107,000. If it continues to grow at a rate of 1.1% per year, predict the population in 2012.

LESSON 13-9 ◼ Logarithmic Functions, p. 600

▶ The exponential function $n = b^p$ can be written as the **logarithmic function** $p = \log_b n$.

▶ A **logarithmic equation** is an equation that contains one or more logarithms.

▶ The property of equality for logarithmic equations states that if $\log_b x = \log_b y$, then $x = y$.

Write each equation in logarithmic form.

41. $8^5 = 32{,}768$

42. $\left(\dfrac{2}{5}\right)^3 = \dfrac{8}{125}$

43. $17^2 = 289$

44. $22^2 = 484$

45. $\left(\dfrac{1}{5}\right)^4 = \dfrac{1}{625}$

46. $7^6 = 117{,}649$

Solve each equation.

47. $\log_5 1 = t$

48. $\log_4 a = 5$

49. $\log_7 m = -3$

50. $\log_6 216 = x$

51. $\log_y 25 = 2$

52. $\log_4 z = -4$

CHAPTER INVESTIGATION

EXTENSION Write a report to summarize your work. Be sure to include the results of your research, your map, and a description of anything you learned during your class discussion of your work.

Chapter 13 Assessment

Write an equation for each circle.

1. radius 6
   center (0, 0)

2. radius 10
   center (1, −5)

3. radius 13
   center (−1, 7)

Find the radius and center for each circle.

4. $x^2 + y^2 = 21$

5. $(x + 4)^2 + y^2 = 11$

6. $(x − 7)^2 + (y − 6)^2 = 225$

Find the focus and directrix for each equation.

7. $x^2 = 32y$

8. $x^2 = −24y$

9. $32y − 4x^2 = 0$

Find the standard equation for each parabola with vertex located at the origin.

10. Focus (0, 8)

11. Focus (0, −2)

12. Focus $\left(0, −\dfrac{1}{4}\right)$

Find an equation for each figure.

13. an ellipse with foci (4, 0) and (−4, 0) and x-intercepts (5, 0) and (−5, 0)

14. a hyperbola with center (0, 0) and foci on the x-axis if $a = 5$ and $b = 11$

Solve.

15. If y varies directly as x and $y = 36$ when $x = 15$, what is y when $x = 19$?

16. If y varies inversely as x and $y = 72$ when $x = 9$, what is y when $x = 6$?

17. If y varies inversely as x and $y = 144$ when $x = 6$, what is y when $x = 4$?

18. If y varies directly as x and $y = 360$ when $x = 12$, what is y when $x = 18$?

19. The total area of a picture and its frame is 456 in.^{2}. The picture is 21 in. long and 16 in. wide. What is the width of the frame?

Graph each inequality.

20. $(x + 1)^2 + (y − 6)^2 \geq 64$

21. $y > x^2 + 4x$

22. $9x^2 + 25y^2 > 225$

23. The amount of time a projectile is in the air after launch can be found by solving $vt − 16t^2 = 0$ (v = initial upward velocity, t = time). How long is a baseball in the air if it is thrown with an upward velocity of 64 ft/sec?

24. Write the equation of a circle having center at (2, −1) and radius of 3.

25. If y varies directly as x^2 and $y = 112$ when $x = 4$, find y when $x = 5$.

26. **INVESTMENTS** Determine the amount of an investment if $10,000 is invested at an interest rate of 4.5% each year for 6 yr.

27. Write $10^5 = 100,000$ in logarithmic form.

28. What is the value of $\log_3 243$?

Standardized Test Practice

Part 1 Multiple Choice

Record your answers on the answer sheet provided by your teacher or on a sheet of paper.

1. The sum of the measures of the complement of an angle and the measure of its supplement is 144°. Which is the measure of the angle? (Lesson 3-2)

(A) 18° (B) 27°

(C) 63° (D) 72°

2. Line p is parallel to line q. Which is the value of $c - a$? (Lesson 3-4)

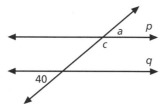

(A) 0° (B) 100°

(C) 120° (D) 150°

3. Which is the longest segment in the figure? (Lesson 4-6)

(A) $\overline{AB}$

(B) $\overline{BC}$

(C) $\overline{CD}$

(D) $\overline{DA}$

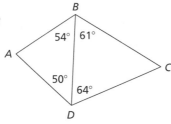

4. Choose the equation of a line parallel to the graph of $y = 3x + 4$. (Lesson 6-2)

(A) $y = -\frac{1}{3}x + 4$

(B) $y = -3x + 4$

(C) $y = -x + 1$

(D) $y = 3x + 5$

5. Drawing a card at random from a standard deck of cards, which is the probability that the card is a diamond or a face card? (Lesson 9-3)

(A) $\frac{3}{52}$ (B) $\frac{3}{13}$

(C) $\frac{11}{26}$ (D) $\frac{25}{52}$

6. Which is the simplest radical form of the product of ($7\sqrt{3}$ and ($2\sqrt{21}$)? (Lesson 10-1)

(A) $18\sqrt{6}$ (B) $42\sqrt{7}$

(C) $14\sqrt{63}$ (D) $9\sqrt{24}$

7. Find the value of x in the figure. (Lesson 10-6)

(A) $\frac{15}{7}$ (B) $\frac{21}{5}$

(C) 4 (D) $\frac{35}{3}$

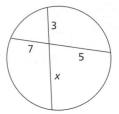

8. If $z^2 - 10z + 3 = 0$, which is the value of z? (Lesson 12-5)

(A) $-5 \pm 2\sqrt{22}$

(B) $-5 \pm \sqrt{22}$

(C) $5 \pm \sqrt{22}$

(D) $5 \pm 2\sqrt{22}$

9. Which equation represents a graph that is a parabola? (Lesson 13-2)

(A) $(x - 3)^2 + (y - 2)^2 = 100$

(B) $x^2 - 4y^2 = 4$

(C) $x^2 - 20y = 0$

(D) $x^2 + 25y^2 = 100$

10. Which graph represents the system of inequalities? (Lesson 13-7)

$$4x^2 + 9y^2 \geq 36$$
$$x - y + 1 \leq 0$$

(A) 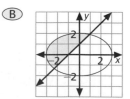 (B)

(C) 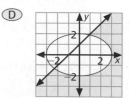 (D)

Preparing for Standardized Tests
For test-taking strategies and more
practice, see pages 709–724.

Part 2 Short Response/Grid In

**Record your answers on the answer sheet
provided by your teacher or on a sheet of paper.**

11. What is the value of $-\left|-(x+1)\right|$ if $x = -4$?
(Lesson 1-2)

12. Solve $9(x+4) - 2x = 19 - 3(x+6)$.
(Lesson 2-5)

13. What is the volume
of the pyramid?
(Lesson 5-7)

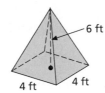
6 ft
4 ft   4 ft

14. What is the point of intersection of the graphs
of $x + 2y = 10$ and $2x - y = 5$? (Lesson 6-4)

15. Rectangle $ABCD$ is similar to rectangle
$EFGH$. If $AB = 6$, $BC = 7$, and $EF = 9$, what
is the perimeter of rectangle $EFGH$?
(Lesson 7-2)

16. In $\triangle XYZ$, $\overline{AB}$
and $\overline{CD}$ are
parallel to $\overline{XY}$.
If $YB = 2$,
$BD = 3$,
$DZ = 4$, and
$AC = 6$, find AZ.
(Lesson 7-6)

17. Point $A(2, 5)$ is rotated 90° counterclockwise
about the origin. What are the coordinates of
the image? (Lesson 8-2)

18. The measure of $\angle ABC$
is 56°. What is the
measure of $\angle AOC$?
(Lesson 10-4)

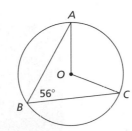

19. What is the product of $(2x - 3y)$ and $(4x + y)$?
(Lesson 11-4)

20. What number must be added to $a^2 - 14a$
to make it a perfect square trinomial?
(Lesson 12-4)

21. If y varies inversely as x, and one pair of
values is $y = 14$ and $x = 5$, find y when $x = 20$.
(Lesson 13-6)

22. Find $\log_2 \frac{1}{8}$. (Lesson 13-9)

Part 3 Extended Response

**Record your answers on a sheet of paper. Show
your work.**

23. Draw the graph of the equation
$16x^2 + 25y^2 = 400$. Plot and label the
x-intercepts, the y-intercepts, and the foci.
(Lesson 13-4)

24. A friend wants to enroll for cellular phone
service. Three different plans are available.
(Lesson 13-5)

Plan 1 charges $0.59 per minute.

Plan 2 charges a monthly fee of $10, plus
$0.39 per minute.

Plan 3 charges a monthly fee of $59.95.

a. For each plan, write an equation that
represents the monthly cost C for m
number of minutes per month.

b. Which plan(s) represent a direct
variation?

c. Your friend expects to use 100 min per
month. In which plan do you think that
your friend should enroll? Explain.

Test-Taking Tip

Question 21
You cannot write mixed numbers, such as $2\frac{1}{2}$, on an answer
grid. Answers such as these need to be written as improper
fractions, such as 5/2, or as decimals, such as 2.5. Choose
the method that you like best, so that you will avoid making
unnecessary mistakes.

Trigonometry

THEME: Navigation

Early explorers relied on the stars and simple tools such as sextants and quadrants to find their way. How did they do it? Long ago, people realized that the stars move in predictable patterns. By keeping careful records and taking angle measurements, they discovered a way to pinpoint their location on the Earth's surface with a reasonable degree of accuracy.

In the same way, modern navigators use information from satellites and guidance computers to find their way. Even automobiles are now equipped with global positioning systems which use data from satellites to determine an automobile's exact location in case of an emergency. These advances are made possible by a branch of mathematics called trigonometry. Trigonometry, which means "triangle measurement," is the study of relationships among the sides and angles of a triangle.

- **Commercial Fishers** (page 633) Aside from fishing duties, commercial fishers pilot small ships or boats and must be able to navigate to fishing areas. They use the stars as well as electronic equipment to pinpoint their location.

Math Online

mathmatters3.com/chapter_theme

U.S. Airport Traffic

Airport	Total passengers 2002	Total passengers 2003	Percent change 2002–2003
Atlanta, Hartsfield (ATL)	37,070,492	38,228,500	3.1%
Chicago, O'Hare (ORD)	28,356,224	30,797,513	8.6
Los Angeles (LAX)	20,320,299	20,913,455	2.9
Dallas/Ft. Worth (DFW)	24,072,162	24,502,273	1.8
San Francisco (SFO)	12,250,289	12,227,636	−0.2
Denver (DEN)	16,053,940	17,271,507	7.6
Miami (MIA)	11,125,611	11,049,687	−0.7
Newark (EWR)	13,113,997	13,087,544	−0.2
Memphis (MEM)	4,537,659	4,504,679	−0.7

Data Activity: U.S. Airport Traffic

Use the table for Questions 1–4.

1. On average, how many passengers arrive or depart from LAX each day?

2. If the passenger traffic for both Denver and Los Angeles continue to change at the same rate, in what year would you expect Denver to have surpassed Los Angeles' total passengers?

3. Of the airports shown on the table, which had the greatest decrease in actual numbers of passengers from 2002 to 2003?

4. If the total number of passengers traveling through Dallas/Ft. Worth was 30,343,500 in 2000, what was the percent change between 2000 and 2003? Round to the nearest tenth of a percent.

CHAPTER INVESTIGATION

In the Northern Hemisphere, the stars appear to move in a circular motion around a single star named *Polaris*, commonly known as the North Star. Explorers first navigated the globe using the star and an instrument called a *quadrant*.

Working Together

Build a quadrant using a photocopy of a protractor, heavy cardboard, string, and a small weight. Use the quadrant to find the angles of elevation for several tall objects. Use trigonometric relationships to find the height of the objects. Use the Chapter Investigation icons to guide your group.

The skills on these two pages are ones you have already learned. Review the examples and complete the exercises. For additional practice on these and more prerequisite skills, see pages 654-661.

NAMING SIDES OF TRIANGLES

In the study of trigonometry, it is very important to be able to shift your focus and see the triangles from different points of view.

Examples   Name the leg of $\triangle ABC$ that is adjacent to $\angle B$.

- Adjacent means "next to." The adjacent side of an angle is never the hypotenuse ($\overline{AB}$). Therefore, the adjacent side must be $\overline{BC}$.

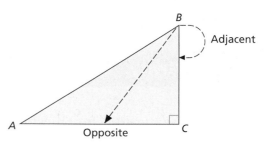

Name the side opposite $\angle A$.

- The side opposite an angle does not contain the vertex of the angle. Therefore, it must be $\overline{BC}$.

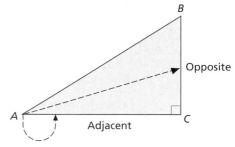

Name the sides in each triangle.

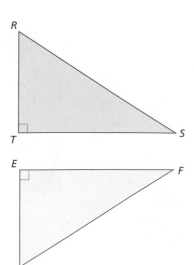

1. The side opposite $\angle S$.
2. The side adjacent to $\angle S$.
3. The side adjacent to $\angle R$.
4. The side opposite $\angle R$.

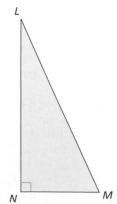

5. The side opposite $\angle F$.
6. The side adjacent to $\angle F$.
7. The side opposite $\angle G$.
8. The side adjacent to $\angle G$.

9. The side opposite $\angle M$.
10. The side opposite $\angle L$.
11. The side adjacent to $\angle L$.
12. The side adjacent to $\angle M$.

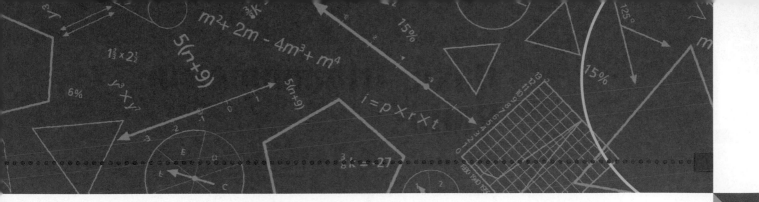

RATIONALIZING RADICALS

You will have to rationalize radicals much of the time while studying trigonometry.

Rationalize. Write in simplest radical form.

13. $\dfrac{\sqrt{8}}{\sqrt{3}}$

14. $\sqrt{\dfrac{32}{10}}$

15. $\dfrac{\sqrt{18}}{\sqrt{5}}$

16. $\sqrt{\dfrac{45}{3}}$

17. $\dfrac{\sqrt{96}}{\sqrt{24}}$

18. $\sqrt{\dfrac{36}{5}}$

19. $\dfrac{\sqrt{27}}{\sqrt{7}}$

20. $\sqrt{\dfrac{54}{8}}$

SPECIAL RIGHT TRIANGLES

- In a 30°-60°-90° triangle, the measure of the hypotenuse is two times that of the leg opposite the 30° angle. The measure of the other leg is $\sqrt{3}$ times that of the leg opposite the 30° angle.

- In a 45°-45°-90° triangle, the measure of the hypotenuse is $\sqrt{2}$ times the measure of a leg of the triangle.

Find the unknown measures.

21.
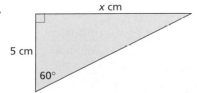

22.
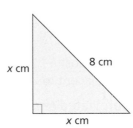

23.
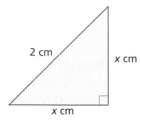

24.
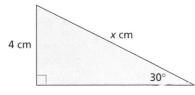

25.
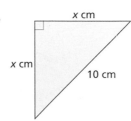

26.
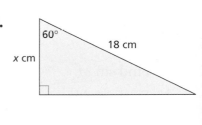

27.
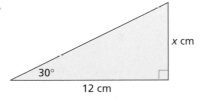

28.
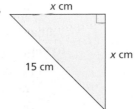

29.
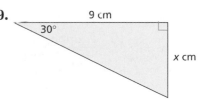

14-1 Basic Trigonometric Ratios

Goals
- Identify trigonometric ratios in a right triangle.
- Use trigonometric ratios to solve problems.

Applications Navigation, Construction, City Planning

Work with a partner.

Measure the segment lengths needed to calculate the following ratios. Use a calculator to evaluate the ratios.

a. $\dfrac{DE}{AD}$ **b.** $\dfrac{FG}{AF}$ **c.** $\dfrac{BC}{AB}$

What conclusions can you draw from your results?

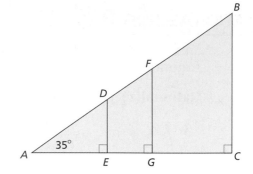

◣ BUILD UNDERSTANDING

Precise measurement shows that the ratio of the lengths of two sides of a right triangle depends on the angles of the triangle—but not the lengths of the sides. This fact forms the foundation for the study of trigonometry. In this lesson, you will study the most important **trigonometry ratios**: the **sine**, the **cosine**, and the **tangent**.

In a right triangle, angle A is an acute angle. Then,

$$\textbf{sine } A = \frac{\text{length of leg opposite } \angle A}{\text{length of hypotenuse}}$$

$$\textbf{cosine } A = \frac{\text{length of leg adjacent to } \angle A}{\text{length of hypotenuse}}$$

$$\textbf{tangent } A = \frac{\text{length of leg opposite } \angle A}{\text{length of leg adjacent to } \angle A}$$

Sine, cosine, and tangent are abbreviated sin, cos, and tan. Sin A means "the sine of $\angle A$."

Mental Math Tip

Use the memory device **SOH CAH TOA** (pronounced "sokatoah") to remember the trigonometric ratios.

SOH: **S**ine is **O**pposite over **H**ypotenuse.

CAH: **C**osine is **A**djacent over **H**ypotenuse.

TOA: **T**angent is **O**pposite over **A**djacent.

Example 1

a. Find sin M.

b. Find cos M.

c. Find tan N.

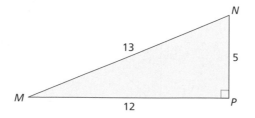

Solution

a. $\sin M = \dfrac{\text{opposite}}{\text{hypotenuse}}$

$= \dfrac{5}{13}$

b. $\cos M = \dfrac{\text{adjacent}}{\text{hypotenuse}}$

$= \dfrac{12}{13}$

c. $\tan N = \dfrac{\text{opposite}}{\text{adjacent}}$

$= \dfrac{12}{5}$

$= 2\dfrac{1}{5}$

Example 2

NAVIGATION A navigator at N sights a 37° angle between a buoy at B and a landmark at L. Find sin 37°.

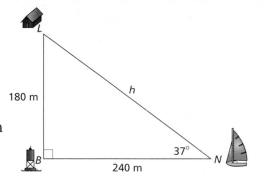

Check Understanding

In the figure at the top of page 614, which leg is opposite ∠A in △ADE? Which leg is adjacent to ∠AFG in △AFG?

Solution

You can use the Pythagorean Theorem to find the length of the hypotenuse.

$$180^2 + 240^2 = h^2$$
$$90,000 = h^2$$
$$300 = h$$

$$\sin 37° = \frac{180}{300}$$
$$= 0.6$$

You can use your calculator to find trigonometric ratios. Use the MODE key to set your calculator to compute in the "degree" mode. Your calculator may already be set for degree mode. Some calculators show the letters DEG in the display window. To check your calculator's mode, find sin 30°. The display should read 0.5. If your calculator displays −.988032, your calculator is in "radian" mode.

To find the angle that has a given trigonometric ratio, use the inverse function on your calculator.

Example 3

CALCULATOR An angle has a cosine of 0.55. Find the measure of the angle to the nearest degree.

Solution

On a graphing calculator, press 2nd [COS⁻¹] .55).

(Some calculators use INV COS rather than COS⁻¹.)

The display should read 56.6329. . ..
To the nearest degree, an angle with a cosine of 0.55 measures 57°.

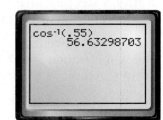

TRY THESE EXERCISES

Use the figure at the right to find each ratio. Express answers in lowest terms.

1. tan *K*   2. cos *H*   3. sin *K*

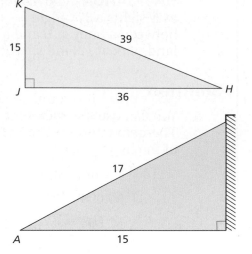

4. **CONSTRUCTION** A 17-ft wire is attached near the top of a wall. The wire is then anchored to the ground 15 ft from the base of the wall. Find tan *A* to the nearest hundredth.

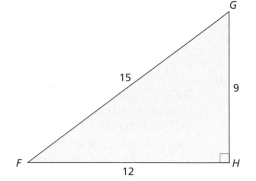

CALCULATOR Use a calculator to find each ratio. Round to the nearest ten-thousandth.

5. sin 22°   6. cos 81°   7. tan 52°

8. cos 40°   9. tan 12°   10. sin 64°

PRACTICE EXERCISES • For Extra Practice, see page 707.

Use the figure at the right to find each ratio.

11. tan *G*   12. sin *F*

13. cos *F*   14. sin *G*

15. tan *F*   16. cos *G*

17. In △*ABC*, ∠*C* is a right angle, *AB* = 29, and *AC* = 21. Find sin *A* and tan *B* to the nearest hundredth.

18. In right triangle *RST*, ∠*T* is the right angle and tan $R = \frac{9}{40}$. Write sin *R* and tan *S* as ratios.

19. **NAVIGATION** An airline navigator measures a 16° angle between the horizontal and an ocean liner. Find tan *A* to the nearest hundredth.

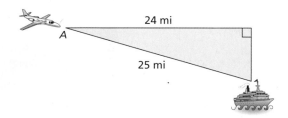

Find angles to the nearest tenth of a degree.

20. **CITY PLANNING** Filbert Street and 22nd Street in San Francisco are the nation's steepest streets. Each rises 1 ft for every 3.17 ft of horizontal distance. What angle do these streets form with the horizontal?

21. **NAVIGATION** From a boat at sea, the distance to the top of a 2325-ft cliff at the water's edge is 4370 ft. What angle does a line make with the horizontal from the boat to the top of the cliff?

22. A kite at the end of 545 ft of string is 130 ft above the ground. What angle does the kite string make with the ground?

NAVIGATION The table at the right shows measurements taken at six East Coast lighthouses. Use the table for Exercises 23–24.

Location of Lighthouse	Height (ft)
Annisquam, MA	41
Cape May, NJ	170
Fenwick Island, DE	87
McClellanville, SC	150
Millbridge, ME	123
Scituate, MA	25

23. The navigator of a ship standing on its bow 360 ft from the McClellanville lighthouse measures a 23° angle to the top of the lighthouse. Write cos 23° as a ratio.

24. A boat pilot standing on the bow 1120 ft from a lighthouse measures an 8° angle to the top of the lighthouse. If $\cos 8° = \frac{112}{113}$, where is the lighthouse located?

 25. **WRITING MATH** Suppose the top of a lighthouse measured 10° from your position offshore. If you knew the height of the lighthouse, how could you find your distance from shore?

26. **CHAPTER INVESTIGATION** Your task is to have each member of your group build a homemade quadrant and use them to find the heights of various objects. Gather the materials you will need: one plastic straw or unsharpened pencil per person, string, a weight such as a metal washer or bolt, and tape. Make a photocopy of a protractor or draw a protractor on heavy cardstock. If you choose to draw your own protractor, make and label markings for every ten degrees.

▰ EXTENDED PRACTICE EXERCISES

Find the exact value of each ratio.

27. tan 45°

28. cos 30°

29. cos 60°

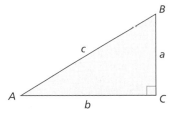

30. **WRITING MATH** True or false: In a right triangle, the sine of one acute angle equals the cosine of the other. Explain.

▰ MIXED REVIEW EXERCISES

Write an equation for each circle. (Lesson 13-1)

31. radius 5
    center (0, 0)

32. radius 8
    center (4, 3)

33. radius 3.5
    center (5, 1)

34. radius 10
    center (−2, 0)

35. radius 7
    center (3, −2)

36. radius 4.7
    center (−2, −2)

37. radius 6
    center (−3, 5)

38. radius 2
    center (2, −8)

39. radius 7.5
    center (−4, 3)

A bag contains 6 red marbles, 5 green marbles, 8 blue marbles, and 1 white marble. Marbles are taken from the bag at random one at a time and not replaced. Find each probability. (Lesson 9-4)

40. P(red, then blue)

41. P(green, then white)

42. P(green, then blue, then red)

43. P(white, then blue, then blue)

14-2 Solve Right Triangles

Goals ■ Find the lengths of sides and the measures of angles in right triangles.

Applications Surveying, Navigation, Safety

Work with a partner. Use the figures below to answer these questions.

a. How do you know the triangles are similar?

b. Write a proportion you could solve to find x.

c. Solve the proportion for x.

d. Explain how you could solve the equation $\sin 30° = \dfrac{x}{12}$ for x.

e. Solve for x: $\sin 40° = \dfrac{x}{16}$. Explain how you solved the equation.

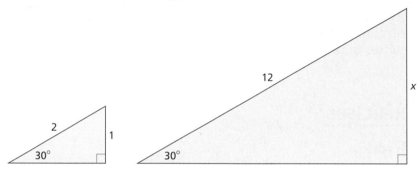

◣ BUILD UNDERSTANDING

You can find the measures of missing parts of a right triangle.

If you know the measure of one acute angle, you can find the measure of the other by subtracting the measure of the known angle from 90°.

If you know the lengths of two sides, you can use the Pythagorean Theorem to find the length of the third side.

If you know the measure of an angle and the length of a side, you can use trigonometric ratios and the first two rules to find the other parts of the triangle.

Example 1

Find the following in △PQR.

a. PR to the nearest tenth

b. $m\angle P$

c. RQ to the nearest tenth

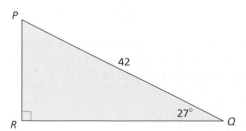

Solution

a. To decide which trigonometric ratio relates $\overline{PQ}$, $\overline{PR}$, and $\angle Q$, think: $\overline{PQ}$ is the *hypotenuse* and $\overline{PR}$ is *opposite* $\angle Q$. The ratio that relates the hypotenuse and the leg opposite an angle is the *sine*.

$$\sin Q = \frac{PR}{PQ}$$

$$\sin 27° = \frac{PR}{42}$$

$$0.4540 \approx \frac{PR}{42} \quad \text{calculator approximation of sin 27°}$$

$$PR \approx 42(0.4540) \approx 19.1$$

b. $m\angle P = 90° - 27° = 63°$

c. Use the Pythagorean Theorem to find RQ.

$$PR^2 + RQ^2 = PQ^2$$

$$19.1^2 + RQ^2 \approx 42^2$$

$$364.81 + RQ^2 \approx 1764$$

$$RQ^2 \approx 1399.19$$

$$RQ \approx 37.4$$

Math: Who, Where, When

Hipparchus, a Greek mathematician born about 160 B.C., is generally considered to be the creator of trigonometry. He was the first person to draw up a table of values for the sine, cosine, and tangent ratios.

Finding the measures of all parts of a right triangle is called **solving a right triangle**.

Trigonometry problems often involve angles of depression and elevation.

An **angle of depression** is formed by a horizontal line and a line slanting downward.

An **angle of elevation** is formed by a horizontal line and a line slanting upward.

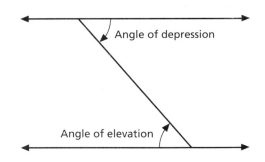

Example 2

SURVEYING A surveyor is standing 550 ft from the base of a redwood tree in Humboldt County, CA. The tree is 362 ft tall. Find the angle of elevation of the top of the tree from the spot where the surveyor is standing.

Solution

The angle of elevation is $\angle A$, formed by the horizontal line of the ground and the line slanting to the top of the tree. BC is opposite $\angle A$, and AC is adjacent to $\angle A$. The trigonometric ratio relating opposite and adjacent is the tangent.

$$\tan \angle A = \frac{362}{550}$$

$$\tan^{-1}\left(\frac{362}{550}\right) \approx 33.4°$$

The angle of elevation is approximately 33.4°.

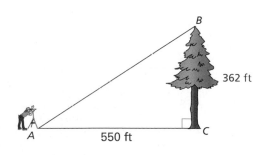

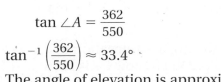

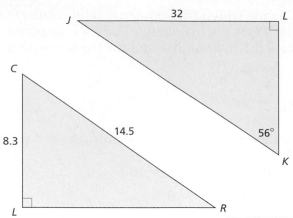

Find the following in △*JKL*.

1. *LK* to the nearest tenth

2. *JK* to the nearest tenth

3. *m∠J*

Find the following in △*CRL*.

4. *LR* to the nearest tenth

5. *m∠C* to the nearest degree

6. *m∠R* to the nearest degree

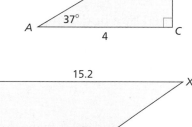

7. **NAVIGATION** A ship's sonar detects a submarine 880 ft below a point on the ocean's surface 1450 ft dead ahead of the ship. To the nearest degree, find the angle of depression to the submarine.

8. **SAFETY** Safety experts recommend that a ladder be placed at an angle of about 75° to the ground. Mr. Reese is using a 15-ft ladder. How far from the base of the wall should he place the foot of the ladder? Round the distance to the nearest tenth of a foot.

▜ **PRACTICE EXERCISES** • For Extra Practice, see page 708.

Round each length to the nearest tenth and each angle to the nearest degree.

CALCULATOR Use a calculator to find the following in △*ABC*.

9. *BC*

10. *m∠B*

Find the following in △*XYZ*.

11. *YZ*

12. *m∠Z*

13. *m∠X*

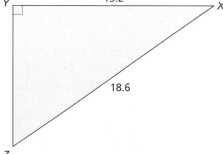

14. Find the angle of elevation to the top of the 1250-ft Empire State Building from a point 850 ft from the base.

15. The two legs of a right triangle measure 23.5 and 27.9. Solve the triangle.

16. In a right triangle, the leg adjacent to a 77° angle has a length of 87. Solve the triangle.

17. **AIR TRAFFIC CONTROL** From an airport runway, the angle of elevation to an approaching plane is 12.8°. If the plane's altitude is 2400 ft, how far is the plane from the runway?

18. **NAVIGATION** From the top of a cliff, the angle of depression of a ship at sea is 8.8°. If the direct-line distance from the clifftop to the ship is 2.3 mi, how high is the cliff?

19. **BOATING** The foot of a right-triangular sail is 64 in. long. The angle at the top of the sail measures 23°. Find the length of the luff of the sail.

20. **FOREST MANAGEMENT** A ranger in a fire tower spots a fire at an angle of depression of 4°. The tower is 36 m tall. How far from the base of the tower is the fire?

21. A silo casts a shadow 42 ft long. The angle of elevation from the sun to the ground is 38°. How tall is the silo?

22. **DATA FILE** Use the data on highest and lowest continental altitudes on page 647. A ship's navigator at sea level in Cook's Inlet, Alaska, sights the summit of Mount McKinley at a 6.8° angle of elevation. How many miles is it from the ship to the summit of the mountain?

23. **FLIGHT** A helicopter ascends 150 ft vertically, then flies horizontally 420 ft. Find the angle of elevation to the helicopter as seen by an observer at the takeoff point.

24. **COMMUNICATIONS** Orlando and Ryan are taking measurements related to the installation of a TV tower. Orlando measures a 62° angle of elevation to the top of the 950-ft TV tower. Find the angle of elevation for Ryan, standing 80 ft farther from the tower than Orlando.

25. **CHAPTER INVESTIGATION** Assemble the quadrant in the following manner: Tape the protractor image to the straw (or pencil), aligning the flat edge of the protractor with the length of the straw. Tie the weight to one end of a 8-in. length of string. Secure the other end of the string to the straw or pencil at the center point of the protractor base. The string should be able to move freely. As you sight an object through the straw, the string will indicate the angle of ascent or descent on the protractor scale.

▮ EXTENDED PRACTICE EXERCISES

Can you solve a right triangle from the given information? Answer *yes*, *no*, or *sometimes*.

26. two sides

27. three angles

28. one side

29. one leg and one angle

30. A pike is directly beneath a trout in a lake. The direct-line distance from an angler to the trout is 35 ft. The angle of depression to the trout is 20°. The angler's direct-line distance to the pike is 42 ft. The angle of depression to the pike is 24°. How far below the trout is the pike?

▮ MIXED REVIEW EXERCISES

Find the focus and directrix of each equation. (Lesson 13-2)

31. $x^2 = 4y$

32. $x^2 = -5y$

33. $x^2 = 8y$

34. $x^2 - 6y = 0$

35. $x^2 - 10y = 0$

36. $x^2 = 18y$

Review and Practice Your Skills

PRACTICE ◣ LESSON 14-1

1. For what type of triangle can you apply the trigonometric ratios?

Determine if each statement is true or false.

2. The sine of an acute angle is calculated by dividing the length of the opposite leg by the length of the hypotenuse.

3. The tangent of an acute angle is calculated by dividing the length of the hypotenuse by the length of the adjacent leg.

4. The cosine of an acute angle is calculated by dividing the length of the adjacent leg by the length of the hypotenuse.

Use the figure at the right to find each ratio. Express answers in lowest terms.

5. $\tan A$ 6. $\sin C$ 7. $\cos C$

8. $\tan C$ 9. $\sin A$ 10. $\cos A$

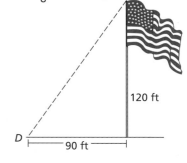

11. A 120-ft flagpole casts a shadow of 90 ft. Write $\sin D$ as a ratio.

12. In $\triangle ABC$, $\angle C$ is a right angle, $AB = 18$ and $AC = 12$. Find $\sin A$ and $\tan B$ to the nearest hundredth.

13. In $\triangle RST$, $\angle T$ is a right angle, $RS = 10$, and $ST = 5$. Find $\cos S$ and $\tan R$ to the nearest hundredth.

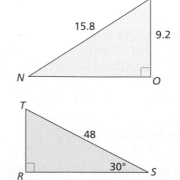

PRACTICE ◣ LESSON 14-2

14. Explain how you can use the Pythagorean Theorem to find the length of a side of a right triangle if you know the lengths of two sides.

15. Explain how to find the measure of an unknown acute angle in a right triangle.

Find the following in $\triangle NOP$.

16. NO to the nearest tenth

17. $m\angle N$ to the nearest degree

18. $m\angle P$ to the nearest degree

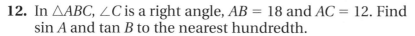

Find the following in $\triangle RST$.

19. RT to the nearest tenth

20. RS to the nearest tenth

21. $m\angle T$

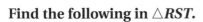

22. To the nearest tenth, find the angle of elevation to the top of the 1454-ft Sears Tower from a point 750 ft from the base.

In the figure at the right, use the Pythagorean Theorem to find the missing measure to the nearest hundredth. Then find each ratio to the nearest hundredth. (Lesson 14-1)

23. $\sin L$

24. $\cos L$

25. $\cos M$

26. $\tan L$

27. $\sin M$

28. $\tan M$

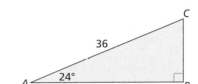

29. In $\triangle DEF$, $\angle F$ is a right angle, $DE = 30$, and $EF = 24$. Find $\sin D$ and $\tan E$ to the nearest hundredth.

30. In $\triangle LMN$, $\angle N$ is a right angle, $LM = 84$, $MN = 62$. Find $\cos L$ and $\tan M$ to the nearest hundredth.

Find the following in $\triangle ABC$. (Lesson 14-2)

31. $m\angle C$

32. BC to the nearest tenth

33. AB to the nearest tenth

34. Find the angle of elevation to the top of a 150-ft flagpole from a point 45 ft from the base of the pole.

35. The direct-line distance from the top of the slope to the ski lodge is 2500 ft. The top of the slope is 1050 ft above the level of the lodge. What is the angle of depression from the top of the slope to the lodge?

Mid-Chapter Quiz

Solve.

1. A ladder on a 10-ft tall fire truck is 75 ft long. If it makes a 45° angle with a building, what is the greatest height the ladder can reach up the side of the building? (Lesson 14-1)

2. When viewed from a horizontal distance of 32 ft, the top of a flagpole can be seen at an angle of 39°. What is the height of the flagpole? (Lesson 14-1)

3. Right triangle ABC has hypotenuse $\overline{AC}$. Right triangle DEF with hypotenuse $\overline{DF}$ is similar to ABC, and angle D corresponds to angle A. If $\overline{DE}$ measures 15 and $\overline{EF}$ measures 20, what is the value of $\tan A$? (Lesson 14-1)

4. An office worker on the fourteenth floor of a building sights a friend on the street. The angle of depression is 35°, and the fourteenth floor is 135 ft in the air. How far is the friend from the building? (Lesson 14-2)

5. The pilot of a plane flying east sights another plane ahead of him at an angle of elevation of 18°. The line of sight distance between the planes is 1850 m. At how much greater altitude is the lead plane than the trailing one? (Lesson 14-2)

14-3 Graph the Sine Function

Goals ■ Solve problems using trigonometry.

Applications Communications, Population, Flight

Work with a partner. You will need a calculator.

1. Choose several angles with measures in each of the given ranges below, and find the sine and cosine of each.

   a. 90°–180°   **b.** 180°–270°   **c.** 270°–360°

2. Draw a grid like the one shown below. Graph your results for parts **a**, **b**, and **c**.

   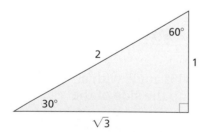

3. Study each range of angles. Describe any patterns you observe in the signs (+ and −) of the sine ratio and the cosine ratio.

◤ BUILD UNDERSTANDING

Until now, you have dealt only with acute angles in your work with trigonometric ratios. In this lesson, you will learn to find trigonometric ratios of angles with measures greater than or equal to 90°. To solve these problems, you will need to apply the relationships that hold in 30°-60°-90° and 45°-45°-90° right triangles.

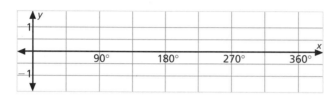

Example 1

Find sin 240°.

Solution

Sketch the angle on a coordinate plane. Use the positive *x*-axis as the **initial side**, and the ray resulting from a 240° counterclockwise rotation of the positive *x*-axis as the **terminal side**. The **reference angle** is the acute angle formed by the *x*-axis and the terminal side.

The reference angle measures $240° - 180° = 60°$. Complete a triangle with the terminal side as hypotenuse by drawing a segment perpendicular to the x-axis. The triangle is a 30°-60°-90° right triangle. In this example, the triangle is in the third quadrant because the legs were drawn by moving in negative directions from the x- and y-axes. The length of the terminal side is always considered to be positive.

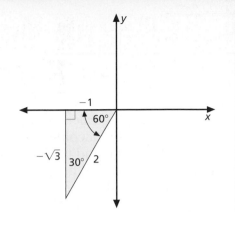

To find sin 240°, find the sine of the reference angle.

$$\sin 240° = \frac{\text{opposite}}{\text{hypotenuse}} = \frac{-\sqrt{3}}{2}$$

Example 2

Find sin 495°.

Solution

To form an angle of 495°, the initial side must complete a 360° rotation, then continue an additional 135°. The reference angle measures $180° - 135° = 45°$. The triangle formed is a 45°-45°-90° right triangle. The leg adjacent to the 45° angle measures -1 relative to the x-axis. The leg opposite the angle measures $+1$ relative to the y-axis. The length of the terminal side is always positive.

$$\sin 495° = \frac{1}{\sqrt{2}} = \frac{\sqrt{2}}{2}$$

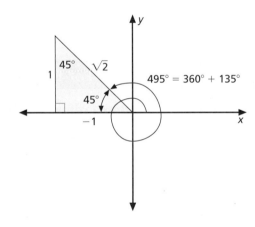

Example 3

COMMUNICATIONS A researcher is developing new technology to assist ships at sea to send urgent communications. Ships will use various tones and patterns to send messages. The electrical impulses produced by the tones are modeled by sine curves. Graph the sine curve $y = \sin x$ for $0° \le x \le 360°$.

Solution

Use your calculator to make a table of ordered pairs.

x	0	30	45	60	90	120	135	150	180	210	225	240	270	300	315	330	360
sin x	0	.50	.71	.87	1.0	.87	.71	.50	0	-.50	-.71	-.87	-1.0	-.87	-.71	-.50	0

Graph the points using sin x as the y-coordinate. Draw a smooth curve through the points. The graph is called a **sine curve**.

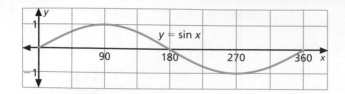

$y = \sin x$

GRAPHING If you have a graphing calculator, you can easily graph sine curves. To graph $y = \sin x$, make sure your calculator is in degree mode. To show the range of x-values, set the Xmin at 0 and the Xmax at 360. Use -1.5 (Ymin) and 1.5 (Ymax) as the range of y-values. In the graph above, the x-scale is set at 45 and the y-scale is at 1. You are ready to graph the function.

◥ TRY THESE EXERCISES

Find each ratio by drawing a reference angle.

1. $\sin 135°$ **2.** $\sin 300°$ **3.** $\sin 210°$

4. $\sin 405°$ **5.** $\sin 660°$ **6.** $\sin 855°$

7. POPULATION A sine curve models the population growth for wildlife in a wooded area. Graph $y = \sin x$ for $360° \leq x \leq 540°$.

8. WRITING MATH Explain how you would find the reference angle and draw a right triangle for a 1050° angle.

◥ PRACTICE EXERCISES • For Extra Practice, see page 708.

Find each ratio by drawing a reference angle.

9. $\sin 225°$ **10.** $\sin 330°$ **11.** $\sin 120°$ **12.** $\sin 150°$

13. $\sin 315°$ **14.** $\sin 480°$ **15.** $\sin 585°$ **16.** $\sin 690°$

17. $\sin 675°$ **18.** $\sin 930°$ **19.** $\sin 765°$ **20.** $\sin 1200°$

21. GRAPHING Use a graphing calculator to graph $y = \sin x$ for $360° \leq x \leq 720°$. Use -1.5 and 1.5 as your range of y-values.

Find each ratio by drawing a reference angle.

22. $\cos 120°$ **23.** $\tan 225°$ **24.** $\cos 330°$

25. $\tan 240°$ **26.** $\tan 660°$ **27.** $\cos 765°$

28. $\sin (-60°)$ **29.** $\sin (-45°)$ **30.** $\sin (-30°)$

31. $\cos (-315°)$ **32.** $\tan (-840°)$ **33.** $\cos (-1230°)$

34. Graph $y = \sin x$ for $-360° \leq x \leq 0°$.

35. Graph $y = \cos x$ for $0° \leq x \leq 360°$.

36. WRITING MATH Compare and contrast the sine curve and the cosine curve.

Solve for values of x with $0° \leq x \leq 360°$.

37. $\sin x = 0$ **38.** $\cos x = -1$ **39.** $\sin x = 1$ **40.** $\cos x = 0$

41. FLIGHT Use a graph of the equation $y = \sin x$ to estimate the value of sin 68°. Then use the value to find the length of a kite string pitched at a 68° angle to the ground if the kite is 450 ft above the ground.

42. CHAPTER INVESTIGATION Work in small groups to determine the height (or depth) of five objects on your school grounds or nearby community.

Follow these steps:

a. Stand at a particular point.

b. Measure the distance from the point to the base of the object.

c. Take an angle measure from the point using the quadrant.

d. Use your knowledge of trigonometry to determine the height of the object. Remember to take into account the distance from your eye level to the ground.

Share the results of your activity with the class. If more than one group measured the same object, compare measurements. Discuss how to account for any discrepancies.

◼ EXTENDED PRACTICE EXERCISES

43. a. Describe a method involving the graphs of $y = \sin x$ and $y = \cos x$ that you could use to solve the equation $\sin x = \cos x$.

b. Use your method to solve the equation for $0° \leq x \leq 360°$.

Solve for values of x with $0° \leq x \leq 360°$.

44. $\sin x = -\dfrac{1}{2}$

45. $\cos x = -\dfrac{\sqrt{3}}{2}$

46. $\tan x = -1$

47. $\tan x = \sqrt{3}$

48. The trademark on a 26-in. radius bicycle tire is 25.5 in. from the center of the wheel. The spoke touching the trademark is horizontal. Find the height of the trademark above the ground after the wheel turns through an angle of 495°.

← trademark

◼ MIXED REVIEW EXERCISES

Find the equation of each ellipse. (Lesson 13-4)

49. foci: (2, 0) and (−2, 0)
x-intercepts: (4, 0) and (−4, 0)

50. foci: (2, 0) and (−2, 0)
x-intercepts: (3, 0) and (−3, 0)

51. foci: (6, 0) and (−6, 0)
x-intercepts: (9, 0) and (−9, 0)

52. foci: (5, 0) and (−5, 0)
x-intercepts: (6, 0) and (−6, 0)

53. foci: (7, 0) and (−7, 0)
x-intercepts: (9, 0) and (−9, 0)

54. foci: (4, 0) and (−4, 0)
x-intercepts: (8, 0) and (−8, 0)

14-4

Experiment with the Sine Function

Goals ■ Determine the period, amplitude and position of sine curves.

Applications Communications, Music, Medicine

 Work with a partner to answer the following questions. You will need a graphing calculator.

1. Graph each equation on a graphing calculator. How does the coefficient of x^2 affect the shape of the graph?

   a. $y = x^2$ **b.** $y = 2x^2$ **c.** $y = 3x^2$

2. Graph each equation. How does the constant in parentheses affect the position of the graph?

   a. $y = x^2$ **b.** $y = (x + 1)^2$ **c.** $y = (x + 2)^2$

◤ BUILD UNDERSTANDING

Recall the sine curve you studied in the last lesson.

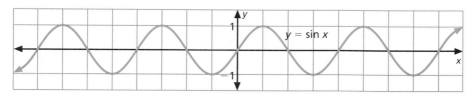

Notice that the shape of the curve repeats itself in every 360° interval along the x-axis. If you were to pick up a 360° section of the curve, you would find it congruent to the curve in each adjacent 360° section. Functions with repeating patterns like this are called **periodic functions**. The **period** of the function is the length of one complete cycle of the function. The period of the graph of $y = \sin x$ is 360°.

Example 1

COMMUNICATIONS A tone transmitted to a ship at sea produces a sound wave with the equation $y = \sin 2x$. State the period.

Solution

The effect of the coefficient 2 in the equation is to compress the sine curve horizontally. The period of $y = \sin 2x$ is $180°$, half the period of $y = \sin x$.

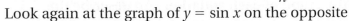

Example 1 suggests the following rule.

The period of the graph of $y = \sin nx$ is $\dfrac{360°}{n}$.

Look again at the graph of $y = \sin x$ on the opposite page. Notice that the maximum value of y is 1 and the minimum value of y is -1. The **amplitude** of a periodic function is half the difference between its maximum and minimum y-values. The amplitude of $y = \sin x$ is $\frac{1}{2}(1 - [-1]) = 1$.

Amplitude is a measure of the height and depth of a curve.

Example 2

Graph $y = 2 \sin x$. State the amplitude.

Solution

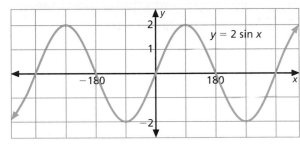

The graph of $y = 2 \sin x$ is twice as tall and twice as deep as the graph of $y = \sin x$. The amplitude is $\frac{1}{2}(2 - [-2]) = 2$.

Notice that, in Example 2, the amplitude is the same as the coefficient in the equation $y = 2 \sin x$. This suggests the following rule. The amplitude of the graph of $y = n \sin x$ is n.

 GRAPHING Graph $y = 2 \sin x$ using a graphing calculator to check your work. You may need to change the values in the display window in order to see the entire curve. Set the y-maximum equal to or greater than the amplitude (n), and the y-minimum equal to or less than $-n$.

Example 3

Graph $y = \sin x + 1$. Describe the position of the graph.

Solution

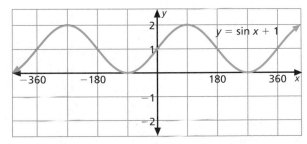

Check Understanding

Without drawing a graph, find the period and amplitude of the graph of $y = 3 \sin 4x - 2$ and describe the position of the graph.

The graph of $y = \sin x + 1$ is the graph of $y = \sin x$ raised 1 unit above its normal position.

Example 3 suggests the following rule.

The graph of $y = \sin x \pm n$ is the graph of $y = \sin x$ raised or lowered n units.

1. Graph $y = \sin 3x$. State the period.

2. Graph $y = 4 \sin x$. State the amplitude.

3. Graph $y = \sin x - 1$. Describe the position of the graph.

State the period and amplitude of the graph of each equation and describe the position of the graph.

4. $y = 3 \sin x + 2$

5. $y = 2 \sin 2x - 5$

6. **YOU MAKE THE CALL** To graph $y = \sin x + 3$, Cynthia enters $y = \sin (x + 3)$ on her graphing calculator. Has Cynthia made a mistake? Explain your thinking.

◥ PRACTICE EXERCISES • For Extra Practice, see page 708.

7. Graph $y = \sin 4x$. State the period.

8. Graph $y = 0.5 \sin x$. State the amplitude.

9. Graph $y = \sin x + 3$. Describe the position of the graph.

State the period and amplitude of the graph of each equation and describe the position of the graph.

10. $y = 2 \sin x - 1$

11. $y = 4 \sin 3x + 2.5$

Tell if the function is periodic. If it is, state the period.

12.
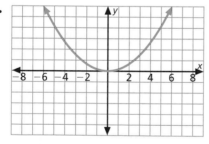

13.
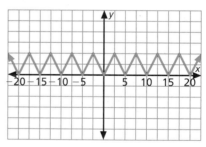

14.
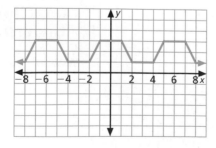

15.
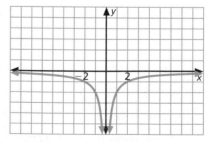

16. **MUSIC** A note played on a musical instrument produces a sound wave with the equation $y = 3 \sin 4x + 3$. State the period and amplitude, and describe the position of the graph.

17. **MEDICINE** An animal's heart rate can be modeled by a sine curve that has a period of 540° and an amplitude of $\frac{3}{2}$. Write the equation.

Find the period of the graph of each equation.

18. $y = \sin \frac{1}{2}x$

19. $y = \sin \frac{3}{5}x$

20. $y = \sin \frac{12}{7}$

21. Find an equation of a graph involving the sine function that has a period of 630° and an amplitude of 8.

State an equation for a sine function with the graph shown.

22.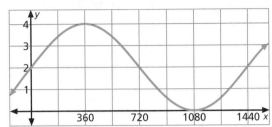

23.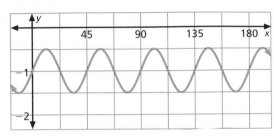

EXTENDED PRACTICE EXERCISES

Graph the equation.

24. $y = 2 \cos x$

25. $y = \cos 2x$

26. $y = \cos x + 1$

27. WRITING MATH Write rules you can use to find the period, amplitude, and position of a graph involving the cosine function.

Make a table of ordered pairs and graph the equation.

28. $y = -\sin x$

29. $y = \sin (x + 90°)$

30. WILDLIFE MANAGEMENT The equation $l = 50,000 + 48,000 \sin 90t$ approximates the number of lemmings on an arctic island on January 1 of a year t years after January 1, 1980.

a. Find the maximum number of lemmings.

b. What year was the maximum first reached?

c. Find the minimum number of lemmings.

d. What year was the minimum first reached?

MIXED REVIEW EXERCISES

Solve each variation. (Lesson 13-5)

31. If y varies directly as x, and $y = 9$ when $x = 6$, find y when $x = 27$.

32. If y varies directly as x, and $y = 8$ when $x = 3$, find y when $x = 45$.

33. If y varies directly as x, and $y = 7$ when $x = 2$, find y when $x = 38$.

Factor the following trinomials. (Lesson 11-7)

34. $x^2 + 3x - 10$

35. $2x^2 + 5x - 12$

36. $x^2 + 2x - 35$

37. $4x^2 - 8x - 5$

38. $x^2 - 6x + 9$

39. $3x^2 + 3x - 6$

40. $3x^2 + 14x - 5$

41. $x^2 - 64$

42. $2x^2 - 6x - 56$

Review and Practice Your Skills

PRACTICE ■ LESSON 14-3

Determine if each statement is *true* or *false*.

1. The initial side of an angle is always on the positive x-axis.

2. The terminal side of an angle is found in the second quadrant.

3. The reference angle is the acute angle formed by the y-axis and the terminal side of the angle.

Find each ratio by drawing a reference angle.

4. $\sin 270°$	**5.** $\sin 120°$	**6.** $\sin 150°$
7. $\sin 300°$	**8.** $\sin 315°$	**9.** $\sin 585°$

Find each ratio by drawing a reference angle.

10. $\tan 210°$	**11.** $\cos 240°$	**12.** $\tan 135°$
13. $\sin 330°$	**14.** $\tan 300°$	**15.** $\cos 225°$

Solve for values of x with $0° \leq x \leq 360°$.

16. $\sin x = -1$	**17.** $\cos x = \dfrac{1}{2}$	**18.** $\sin x = -\dfrac{1}{2}$
19. $\cos x = -\dfrac{1}{2}$	**20.** $\sin x = 1$	**21.** $\cos x = 1$

PRACTICE ■ LESSON 14-4

Define each term.

22. period   23. amplitude

24. Use the graph of $y = \sin 4x$ to state the period.

25. Use the graph of $y = 2 \sin x$ to state the amplitude.

Find the period of the graph of each equation.

26. $y = 2 \sin 5x$	**27.** $y = \sin 3x - 2$	**28.** $y = 3 \sin x + 4$
29. $y = 2 \sin 10x$	**30.** $y = \sin 6x - 1$	**31.** $y = 5 \sin x - 2$
32. $y = 3 \sin 4x - 1$	**33.** $y = \sin 40x + 3$	**34.** $y = 2 \sin 5x + 1$

State the period and amplitude of the graph of each equation and describe the position of the graph.

35. $y = 2 \sin x$	**36.** $y = \sin 4x$	**37.** $y = 2 \sin 3x + 1$
38. $y = 3 \sin 6x - 3$	**39.** $y = 5 \sin 2x + 2$	**40.** $y = 2 \sin 5x - 1$

Use the figure at the right to find each ratio. (Lesson 14-1)

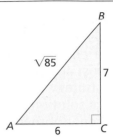

41. $\sin A$ **42.** $\cos B$

43. $\sin B$ **44.** $\tan A$

45. $\tan B$ **46.** $\cos A$

47. To the nearest tenth, find the angle of elevation to the top of a 90-ft tree from a point 40 ft from the base of the tree. (Lesson 14-2)

Find each ratio by drawing a reference angle. (Lesson 14-3)

48. $\sin 30°$ **49.** $\sin 135°$ **50.** $\sin 210°$

51. $\cos 270°$ **52.** $\tan 225°$ **53.** $\cos 45°$

Solve for values of x with $0° \leq x \leq 360°$. (Lesson 14-3)

54. $\sin x = \dfrac{-\sqrt{3}}{2}$ **55.** $\sin x = \dfrac{\sqrt{3}}{2}$ **56.** $\sin x = \dfrac{\sqrt{2}}{2}$

57. $\cos x = \dfrac{\sqrt{3}}{2}$ **58.** $\tan x = -1$ **59.** $\cos x = \dfrac{\sqrt{2}}{2}$

MathWorks
Workplace Knowhow

Career – Commercial Fisher

Commercial fishers are captains, deckhands, or boatswains (supervisor of the deckhands). Aside from fishing duties, the fishers aboard a fishing boat must be able to navigate to the fishing areas. Today this is mainly accomplished through the use of electronic equipment that pinpoints the ship's position on the surface of the Earth according to man-made satellites orbiting the planet. Before these devices were invented, sailors navigated according to the stars.

To avoid rocks near a shoreline, an experienced fisher uses the stars to know when to turn and make an arc to shore. When under the correct star, the fisher is 2.5 mi from shore. The arc begins there and ends 2.5 mi from the shore point where the arc began. His route is a quarter of a circle with a radius of 2.5 mi.

1. How many miles does the fisher travel in his arc?

2. On a coordinate plane, name the x, y coordinates of the beginning and ending points of the arc.

3. What is the equation of the whole circle of which the fisher traveled one quarter?

14-5 Problem Solving Skills: Choose a Strategy

In this book, you have studied a variety of problem solving strategies. Experience in applying these strategies will help you decide which will be most appropriate for solving a particular problem. Sometimes, only one strategy will work. In other cases, any one of several strategies will offer a solution. There may be times when you will want to use two different approaches to a problem to be sure the solution you found is correct. For certain problems, you will need to use more than one strategy to find the solution.

Problem Solving Strategies

Guess and check

Look for a pattern

Solve a simpler problem

Make a table, chart or list

Use a picture, diagram or model

Act it out

Work backwards

Eliminate possibilities

Use an equation or formula

◥ PRACTICE EXERCISES

Solve. Name the strategy you used to solve each problem. Find lengths to the nearest tenth and angles to the nearest degree.

1. **ARCHAEOLOGY** The Great Pyramid at El Giza, Egypt, measures 755 ft on a side. The faces stand at a 52° angle to the ground. The top 30 ft of the pyramid has been destroyed. How tall was the pyramid when it was first built?

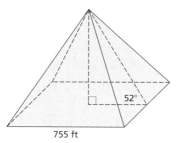
755 ft

2. The sine of a certain acute angle is equal to the cosine of 24°. Find the acute angle.

3. **SURVEYING** Two surveyors standing 2.8 mi apart each measure the angle of elevation to the top of a mountain. The surveyor nearer the mountain, standing 3.4 mi from the base of the peak, measures an angle of 26°. Find the angle of elevation measured by the other surveyor.

A   2.8 mi   D   3.4 mi   C

4. **GEOGRAPHY** The formula $l = 69.81 \cos d$ gives the length in miles, l, of one degree of longitude on the Earth's surface, where d is the latitude in degrees.

   a. Find the length of one degree of longitude at latitude 42° N.

   b. At what northern hemisphere latitude is the length of one degree of longitude 19.2 mi?

5. **SCIENCE** This sine wave appeared on a laboratory oscilloscope screen. The technician then generated a wave congruent to this one, but 3 units lower on the screen. Find the equation of the second wave.

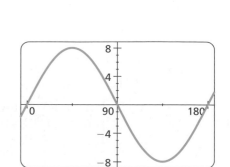

GEOGRAPHY The world's longest aerial ropeway ascends from the city of Merida, Venezuela (altitude: 5379 ft), to the summit of Pico Espejo (altitude: 15,629 ft). The ropeway is 42,240 ft in length.

Five-step Plan

1 Read
2 Plan
3 Solve
4 Answer
5 Check

6. Find the horizontal distance from the lower end of the ropeway to the point directly under the summit of Pico Espejo.

7. Find the angle of elevation of the ropeway.

8. PHYSICS A spring bounces up and down in such a way that the height in inches, h, of the weight at the end of the spring is given by $h = 6 \sin 360t$, where t is the number of seconds after the spring reaches the position shown for the first time.

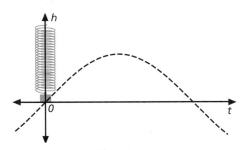

a. How high will the weight be after 5.1 sec?

b. What is the maximum height reached by the weight?

c. When is the maximum height first reached?

d. What is the minimum height reached by the weight?

e. When is the minimum height first reached?

◥ MIXED REVIEW EXERCISES

Round answers to the nearest hundredth if necessary. (Lesson 13-6)

9. If y varies inversely as x, and $y = 3$ when $x = 42$, find y when $x = 27$.

10. If y varies inversely as x, and $y = 9$ when $x = 8$, find y when $x = 13$.

11. If y varies inversely as x, and $y = 15$ when $x = 2$, find y when $x = 8$.

12. If y varies inversely as x, and $y = 30$ when $x = 9$, find y when $x = 14$.

13. If y varies inversely as x, and $y = 28$ when $x = 7$, find y when $x = 52$.

14. If y varies inversely as x, and $y = 18$ when $x = 3$, find y when $x = 38$.

Simplify. (Lesson 11-2)

15. $(x - 5)(x + 8)$

16. $(2x - 7)(3x + 4)$

17. $(-3x + 2)(-x + 8)$

18. $(6x - 4)(x + 4)$

19. $(2x - 9)(2x + 9)$

20. $4(x - 3)(x - 4)$

21. $3(2x - 1)(x + 6)$

22. $(4x + 7)(3x + 2)$

23. $(2x - 9)(3x + 8)$

Compute the variance and standard deviation for each set of data. Round answers to the nearest hundredth if necessary. (Lesson 9-7)

24. 7, 8, 8, 5, 6, 8, 9, 7

25. 12, 11, 13, 17, 15, 13, 12, 14

26. 21, 23, 20, 25, 25, 29, 27, 26

27. 15, 16, 19, 17, 18, 10, 4, 28

Chapter 14 Review

VOCABULARY

Choose the word or phrase from the list that best completes each statement.

1. __?__ is opposite over adjacent.

2. A(n) __?__ is formed by a horizontal line and a line slanting upward.

3. __?__ is adjacent over hypotenuse.

4. A(n) __?__ is formed by a horizontal line and a line slanting downward.

5. __?__ is opposite over hypotenuse.

6. A(n) __?__ is an acute angle formed by the x-axis and the terminal side of an angle.

7. In a periodic function, the __?__ is the length of one complete cycle of the function.

8. In a periodic function, the __?__ is half the difference between the maximum and minimum y-values.

9. The sine, cosine, and tangent are three __?__ ratios.

10. To __?__ a right triangle means to find the measures of all the parts of the triangle.

a. amplitude
b. angle of depression
c. angle of elevation
d. cosine
e. initial side
f. period
g. reference angle
h. sine
i. solve
j. tangent
k. terminal side
l. trigonometric

LESSON 14-1 ◢ Basic Trigonometric Ratios, p. 604

▶ The ratio of the lengths of two sides of a right triangle depends on the angles of the triangle.

▶ For a given angle, the ratios are always the same.

In a right triangle, $\angle A$ is an acute angle. Then,

$$\sin A = \frac{\text{length of leg opposite } \angle A}{\text{length of hypotenuse}},$$

$$\cos A = \frac{\text{length of leg adjacent to } \angle A}{\text{length of hypotenuse}}, \text{ and}$$

$$\tan A = \frac{\text{length of leg opposite } \angle A}{\text{length of leg adjacent to } \angle A}.$$

Use the figure at the right to find each ratio.

11. $\sin D$

12. $\cos F$

13. $\tan F$

14. $\cos D$

15. $\sin F$

16. $\tan D$

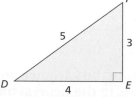

17. Visitors to Pittsburgh, Pennsylvania, can ride the Incline from the river valley up Mt. Washington. The Incline has a 403-ft rise and a 685-ft run. What is the angle made by the track and the horizontal?

LESSON 14-2 ◗ Solve Right Triangles, p. 608

▶ You can find the measures of the angles and sides of a right triangle.

    a. Given the measure of one acute angle, find the measure of the other by subtracting the measure of the known angle from 90°.

    b. Given the lengths of two sides, use the Pythagorean Theorem to find the length of the third side.

    c. Given an angle and length of a side, use trigonometric ratios to find other parts of the triangle.

▶ Trigonometry problems may involve angles of depression and elevation.

    a. An angle of depression is formed by a horizontal line and a line slanting downward.

    b. An angle of elevation is formed by a horizontal line and a line slanting upward.

Find the following in △*ABC*.

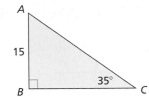

18. *BC*

19. *m∠A*

20. *AC*

21. A tower casts a shadow 55 ft long. Measuring from the end of the shadow, Brad determines that the angle of the sun is 43°. How tall is the tower?

22. A kite is flying at the end of a 300-ft string. Assuming the string is straight and forms an angle of 58° with the ground, how high is the kite?

23. From the top of an observation tower 50 m high, a forest ranger spots a deer at an angle of depression of 28°. How far is the deer from the base of the tower?

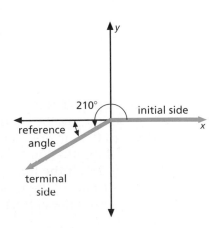

LESSON 14-3 ◗ Graph the Sine Function, p. 614

▶ The reference angle is the acute angle formed by the *x*-axis and the terminal side.

The reference angle measures $210° - 180° = 30°$. Draw a triangle with the terminal side as the hypotenuse by making a line perpendicular to the *x*-axis. The triangle is a 30°-60°-90° right triangle. In this case, the leg lengths are negative because they were drawn by moving in negative directions from the *x*- and *y*-axis. The length of the terminal side is always considered positive.

$$\sin 210° = \frac{\text{opposite}}{\text{hypotenuse}} = -\frac{1}{2}$$

Find each ratio by drawing a reference angle.

24. $\sin 120°$ **25.** $\tan 225°$ **26.** $\cos 960°$

27. $\tan 300$ **28.** $\cos 405°$ **29.** $\sin 390°$

LESSON 14-4 ■ Experiment with the Sine Function, p. 618

▶ Functions with repeating patterns are called periodic functions. The period of a function is the length of one complete cycle of the function.

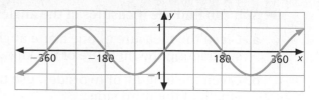

The period of $y = \sin x$ is 360°.

The period of $y = \sin nx$ is $\dfrac{360°}{n}$.

▶ The amplitude of a periodic function is half the difference between its maximum and minimum y-values. It is a measure of the height and depth of a curve.

30. Graph $y = 6 \sin x$. State the amplitude.

31. Graph $y = \sin \dfrac{1}{2}x$. State the period.

Tell if the function is periodic. If it is, state the period.

32.

33.

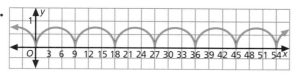

LESSON 14-5 ■ Problem Solving Skills: Choose a Strategy, p. 624

▶ Experience in applying strategies will help you solve a particular problem. Sometimes one strategy will work. There may be times that you will need to use more than one strategy to find a solution.

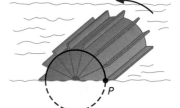

34. A steamboat paddlewheel with a radius of 40 in. makes one complete revolution every 4 sec. Half the wheel is submerged. Point P on the rim of the wheel is at the water line. Where in relation to the surface of the water will P be 17.2 sec from now?

35. Astronauts in a lunar lander see a large crater on the moon. The angle of depression to the far side of the crater is 18°. The angle of depression to the near side of the crater is 25°. If the lunar lander is 3 mi above the surface of the moon, what is the distance across the crater?

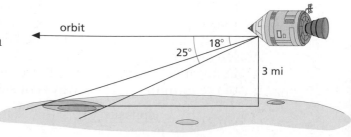

CHAPTER INVESTIGATION

EXTENSION Make a scale drawing to show how you determined the height or depth of one of the objects you measured. Label each part of your drawing and include all measurements.

Chapter 14 Assessment

Use the figure at the right to find each ratio.

1. sin *A*

2. cos *C*

3. sin *C*

4. tan *A*

5. cos *A*

6. tan *C*

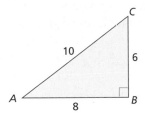

Find the following in △*CAT*.

7. *CT* to the nearest tenth.

8. *m*∠*C*

9. *CA* to the nearest tenth.

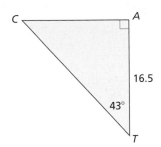

10. The General Sherman tree at Sequoia National Park is about 273 ft tall. If you're standing 64 ft from the base, what is the angle of elevation?

11. **GOLF** A golfer is standing at the tee, looking up to the green on a hill. If the tee is 36 yd lower than the green and the angle of elevation from the tee to the hole is 12°, find the distance from the tee to the hole.

Find each ratio by drawing a reference angle.

12. sin 120°

13. cos 240°

14. cos (−45°)

15. tan 300°

16. sin (−780°)

17. tan 405°

Graph. State the period and amplitude.

18. $y = \dfrac{3}{2} \sin x$

19. $y = \cos 3x$

20. $y = 3 \sin x - 2$

21. Sound is caused from continuous vibrations. You can think of a sine graph when describing sound. When the amplitude of sound vibrations is large, the sound is more intense. Suppose the graph $y = 3 \sin x$ describes a sound vibration. To make the sound more intense, the amplitude grows to 8. Write and graph the equation that describes the intense sound.

Standardized Test Practice

Part 1 Multiple Choice

Record your answers on the answer sheet provided by your teacher or on a sheet of paper.

1. Which is the simplest form of $x^7 \div x^{-11}$? (Lesson 1-8)

ⓐ $\dfrac{1}{x^{18}}$ ⓑ $\dfrac{1}{x^4}$ ⓒ x^4 ⓓ x^{18}

2. In the figure, find the value of x. (Lesson 4-1)

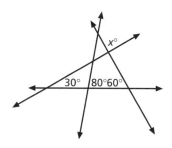

ⓐ 40   ⓑ 50   ⓒ 60   ⓓ 90

3. If you double the length and width of a rectangle, how does its perimeter change? (Lesson 5-2)

ⓐ It does not change.

ⓑ It is multiplied by $1\frac{1}{2}$.

ⓒ It doubles.

ⓓ It quadruples.

4. Which is the approximate volume of the cone? (Lesson 5-7)

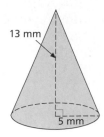

ⓐ 136.1 mm³

ⓑ 304.3 mm³

ⓒ 408.4 mm³

ⓓ 1021.0 mm³

5. Choose a proportion that can be used to find the distance across the lake (AB). (Lesson 7-7)

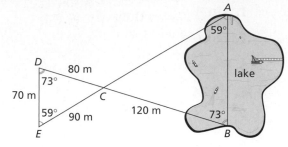

ⓐ $\dfrac{70}{AB} = \dfrac{90}{120}$ ⓑ $\dfrac{80}{70} = \dfrac{AB}{120}$

ⓒ $\dfrac{70}{AB} = \dfrac{80}{120}$ ⓓ $\dfrac{80}{90} = \dfrac{120}{AB}$

6. In the following equation, what is the value of c? (Lesson 8-5)

$$\begin{bmatrix} 7 & 6 \\ -3 & 8 \end{bmatrix} + 2\begin{bmatrix} 1 & -6 \\ -3 & -4 \end{bmatrix} = \begin{bmatrix} a & b \\ c & d \end{bmatrix}$$

ⓐ -9 ⓑ -6 ⓒ 0 ⓓ 3

7. Which are the solutions of the equation $x^2 + 7x - 18 = 0$? (Lesson 12-3)

ⓐ 2 or -9 ⓑ -2 or 9

ⓒ -2 or -9 ⓓ 2 or 9

8. Which value equals $\cos(-420°)$? (Lesson 14-3)

ⓐ $-\dfrac{\sqrt{3}}{2}$ ⓑ $-\dfrac{1}{2}$

ⓒ $\dfrac{1}{2}$ ⓓ $\dfrac{\sqrt{3}}{2}$

Test-Taking Tip

Question 2

The drawings provided for test items are often not drawn to scale. So remember what you can and cannot assume from a geometric figure.

You can assume the following from a drawing.
- When points appear on a line or line segment, they are collinear.
- Angles that appear to be adjacent or vertical are.
- When lines, line segments, or rays appear to intersect, they do.

*You **cannot** assume the following from a drawing.*
- Line segments or angles that appear to be congruent are.
- Lines or line segments that appear to be parallel or perpendicular are.
- A point that appears to be a midpoint of a segment is.

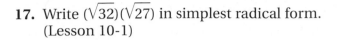

Preparing for Standardized Tests
For test-taking strategies and more
practice, see pages 709–724.

Part 2 Short Response/Grid In

**Record your answers on the answer sheet
provided by your teacher or on a sheet of paper.**

9. A competition diver receives a score from
each of 7 judges. The highest and lowest
scores are discarded and the remaining scores
are added together. The sum is multiplied by
the dive's degree of difficulty and then
multiplied by 0.6. What is the score of a dive
with a 1.5-degree of difficulty with the
following scores? (Lesson 1-5)

$$7.5, 7.0, 9.0, 8.5, 6.5, 8.0, 7.0$$

10. Solve $5(a + 3) = 2a$. (Lesson 2-5)

11. In the figure below, find $m\angle BEC$.
(Lesson 3-2)

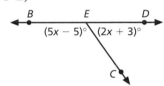

12. Find the midpoint of $\overline{MQ}$. (Lesson 3-3)

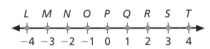

13. What is the slope of a line perpendicular
to the line represented by the equation
$3x - 6y = 12$? (Lesson 6-2)

14. If $-5x + y = 12$ and $5x + 8y = 6$, what is the
value of $9y$? (Lesson 6-6)

15. If parallelogram $ABCD$
is reflected over the
y-axis to become $A'B'C'D'$,
what are the coordinates
of C'? (Lesson 8-1)

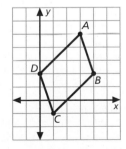

16. How many ways can 6 swimmers be arranged
on a 4-person relay team? (Lesson 9-5)

17. Write $(\sqrt{32})(\sqrt{27})$ in simplest radical form.
(Lesson 10-1)

18. If $\triangle ABC$ is an
equilateral
triangle, what
is the length of
$\overline{AD}$ in simplest
radical form?
(Lesson 10-3)

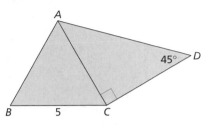

19. If $(4x + 2y)(3x - 6y) = ax^2 + bxy + cy^2$, what
is the value of b? (Lesson 11-4)

20. Factor $t^2 - 14t + 45$. (Lesson 11-7)

21. If y varies directly as x^2 and $y = 100$ when
$x = 5$, what is y when $x = 7$? (Lesson 13-5)

22. For the figure, find
$\cos R$. (Lesson 14-1)

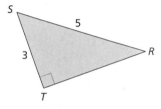

Part 3 Extended Response

**Record your answers on a sheet of paper.
Show your work.**

23. A drawbridge is normally 13 ft above the
water. Each section of the drawbridge is 210 ft
long. The angle of elevation of each section
when the bridge is up is 70°. (Lesson 14-2)
   a. Make a drawing of the situation.
   b. What trigonometric function would you
      use to find the distance from the top of a
      section of the drawbridge to the water
      when the bridge is up? Explain why you
      chose this function.
   c. To the nearest foot, what is the
      distance from the top of a section of
      the drawbridge to the water when the
      bridge is up?

24. Graph $y = 4 \sin 2x$. Describe the graph
including its period and amplitude.
(Lesson 14-4)

Student Handbook

Data File

Data File

Architecture

Bridges of the World

Bridge	Height (feet)
Firth of Forth Bridge, Scotland	148
Verrazano-Narrows Bridge, New York	213
Sydney Harbor Bridge, Australia	171
Tunkhannock Viaduct, Pennsylvania	240
Garabit Viaduct, France	480
Brooklyn Bridge, New York	135

Types of Structural Supports (Trusses) Used in Architecture

King-Post ← apex Queen-Post

Scissors

Fink

Fan-Fink

Sydney Harbor Bridge

Notable Tall Buildings of the World

Building	City	Year Completed	Stories	Height (meters)	Height (feet)
Petronas Tower I	Kuala Lumpur	1998	88	452	1483
Petronas Tower II	Kuala Lumpur	1998	88	452	1483
Sears Tower	Chicago	1974	110	443	1454
Empire State Building	New York	1931	102	381	1250
Central Plaza	Hong Kong	1992	78	374	1227
Bank of China Tower	Hong Kong	1988	72	368	1209
Amoco Building	Chicago	1973	80	346	1136
John Hancock Center	Chicago	1968	100	344	1127

Petronas Towers I and II

Noted Rectangular Structures

Structure	Country	Length (meters)	Width (meters)
Parthenon	Greece	69.5	30.9
Palace of the Governors	Mexico	96	11
Great Pyramid of Cheops	Egypt	230.6	230.6
Step Pyramid of Zosar	Egypt	125	109
Temple of Hathor	Egypt	290	280
Cleopatra's Needle (base)	England*	2.4	2.3
Ziggurat of Ur (base)	Middle East	62	43
Guanyin Pavilion of Dule Monastery	China	20	14
Izumo Shrine	Japan	10.9	10.9
Kibitsu Shrine (main)	Japan	14.5	17.9
Kongorinjo Hondo	Japan	21	20.7
Bakong Temple, Roluos	Cambodia	70	70
Ta Keo Temple	Cambodia	103	122
Wat Kukut Temple, Lampun	Thailand	23	23
Tsukiji Hotel	Japan	67	27

*Gift to England from Egypt

Parthenon

Housing Units-Summary of Characteristics and Equipment, by Tenure and Region: One Recent Year
(In thousands of units, except as indicated. Based on the American Housing Survey)

Item	Total Housing Units	Sea-sonal	Year-Round Units							Vacant
			Occupied							
			Total	Owner	Renter	Northeast	Midwest	South	West	
Total units	**99,931**	**3,182**	**88,425**	**56,145**	**32,280**	**18,729**	**22,142**	**30,064**	**17,490**	**8,324**
Percent distribution	100.0	3.2	88.5	56.2	32.3	21.2	25.0	34.0	19.8	8.3
Units in structure:										
Single family detached	60,607	1,834	55,076	46,703	8,373	9,368	14,958	19,984	10,766	3,697
Single family attached	4,514	64	4,102	2,211	1,890	1,431	814	1,212	645	349
2–4 units	11,655	134	10,217	1,996	8,221	3,324	2,515	2,426	1,952	1,304
5–9 units	5,134	73	4,372	344	4,029	903	967	1,408	1,094	689
10–19 units	4,558	99	3,760	261	3,500	818	766	1,360	817	699
20–49 units	3,530	146	2,913	287	2,627	904	603	651	756	470
50 or more units	3,839	135	3,230	438	2,792	1,540	661	583	446	474
Mobile home or trailer	6,094	698	4,754	3,906	848	440	860	2,440	1,014	642

Earth Science

Some Principal Rivers of the World

River	Length (miles)
Amazon	4000
Arkansas	1459
Columbia	1243
Danube	1776
Ganges	1560
Indus	1800
Mackenzie	2635
Mississippi	2340
Missouri	2540
Nile	4160
Ohio	1310
Orinoco	1600
Paraguay	1584
Red	1290
Rhine	820
Rio Grande	1900
St. Lawrence	800
Snake	1038
Thames	236
Tiber	252
Volga	2194
Zambezi	1700

Average Daily Temperatures (°F)

San Diego, CA

Legend: San Diego, CA; Milwaukee, WI

Months: January, February, March, April, May, June, July, August, September, October, November, December

Axis: 10° 20° 30° 40° 50° 60° 70° 80° 90°

Size and Depth of the Oceans

Ocean	Square Miles	Greatest Depth (feet)
Pacific	63,800,000	36,161
Atlantic	31,800,000	30,249
Indian	28,900,000	24,441
Arctic	5,400,000	17,881

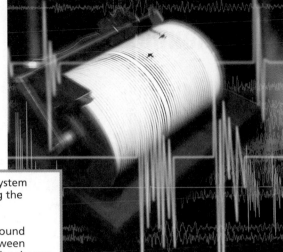
Richter Scale

Measuring Earthquakes

The energy of an earthquake is generally reported using the Richter scale, a system developed by American geologist Charles Richter in 1935, based on measuring the heights of wave measurements on a seismograph.

On the Richter scale, each single-integer increase represents 10 times more ground movement and 30 times more energy released. The change in magnitude between numbers on the scale can be represented by 10^x and 30^x, where x represents the change in the Richter scale measure. Therefore, a 3.0 earthquake has 100 times more ground movement and 900 times more energy released than a 1.0 earthquake.

Richter scale

2.5	Generally not felt, but recorded on seismometers.
3.5	Felt by many people.
4.5	Some local damage may occur.
6.0	A destructive earthquake that causes significant damage.
7.0	A major earthquake; about ten occur each year.
8.0 and above	Great earthquakes; these occur once every five to ten years.

Mount McKinley, Alaska

Highest and Lowest Continental Altitudes

Continent	Highest Point	Elevation (feet)	Lowest Point	Feet Below Sea Level
Asia	Mount Everest, Nepal-Tibet	29,028	Dead Sea, Israel-Jordan	1,312
South America	Mount Aconcagua, Argentina	22,834	Valdes Peninsula, Argentina	131
North America	Mount McKinley, Alaska	20,320	Death Valley, California	282
Africa	Kilmanjaro, Tanzania	19,340	Lake Assai, Djibouti	512
Europe	Mount El'brus, Russia	18,510	Caspian Sea, Russia	92
Antarctica	Vinson Massif	16,864	Unknown	–
Australia	Mount Kosciusko, New South Wales	7,310	Lake Eyre, South Australia	52

Economics

Money Around the World

Country	Basic Monetary Unit	Chief Fractional Unit	Exchange Rate 1 US dollar =
Australia	dollar	cent	1.44055
Canada	dollar	cent	1.34255
France	euro	cent	0.821272
India	rupee	paise	45.9797
Japan	yen	100 sen (not used)	108.9
Mexico	peso	centavo	11.5511
Sudan	dinar	piaster	259.502

In the United States the basic monetary unit is the dollar and the chief fractional unit is the penny. One dollar - 100 pennies. Unless noted otherwise, the basic monetary unit equals 100 chief fractional units for the countries listed above.

United States Foreign Trade
(millions of dollars)

Country	U.S. exports		U.S. imports	
	1996	2004	1996	2004
Canada	134,210	160,923	155,893	209,088
France	14,455	19,016	18,646	28,240
Japan	67,607	51,449	115,187	121,429
Mexico	56,791	97,470	74,297	134,616
Venezuela	4,749	4,430	13,173	15,094

State General Sales and Use Taxes, 2004

State	Percent Rate	State	Percent Rate	State	Percent Rate
Alabama	4	Kentucky	6	Ohio	6
Arizona	5.6	Louisiana	4	Oklahoma	4.5
Arkansas	5.125	Maine	5	Pennsylvania	6
California	7.25	Maryland	5	Rhode Island	7
Colorado	2.9	Massachusetts	5	South Carolina	5
Connecticut	6	Michigan	6	South Dakota	4
D.C.	5.75	Minnesota	6.5	Tennessee	7
Florida	6	Mississippi	7	Texas	6.25
Georgia	4	Missouri	4.225	Utah	4.75
Hawaii	4	Nebraska	5.5	Vermont	6
Idaho	6	Nevada	6.5	Virginia	4.5
Illinois	6.25	New Jersey	6	Washington	6.5
Indiana	6	New Mexico	5	West Virginia	6
Iowa	5	New York	4.25	Wisconsin	5
Kansas	5.3	North Carolina	4.5	Wyoming	4
		North Dakota	5		

NOTE: Alaska, Delaware, Montana, New Hampshire, and Oregon have no statewide sales and use taxes.

The Shrinking Value of the Dollar

Year	5 lb flour	1 lb round steak	1 qt milk	10 lb potatoes
1890	$0.15	$0.12	$0.07	$0.16
1910	0.18	0.17	0.08	0.17
1930	0.23	0.43	0.14	0.36
1950	0.49	0.94	0.21	0.46
1970	0.59	1.30	0.33	0.90
1975	0.98	1.89	0.45	0.99
1995	1.20	3.20	0.74	3.80
1999	1.70	2.93	0.83	2.82
2003	1.56	3.84	0.69	4.59

Health & Fitness

Calorie Count of Selected Dairy Products, Breads, Pastas, Snacks, Fruits, and Juices

Food	Approximate Amount	Food Energy (kcal)
Apple, raw	1	80
Apple juice	1 cup	120
Banana	1	100
Bread, white	1 slice	70
Butter or margarine	1 tbsp	100
Cheese, American	1 oz	105
Cheese, cottage	1 cup	235
Corn flakes	1 cup	95
Crackers, saltine	4	50
Lemonade	1 cup	105
Macaroni with cheese	1 cup	430
Milk, skim	1 cup	85
Milk, whole	1 cup	150
Oatmeal	1 cup	130
Orange	1	65
Orange juice	1 cup	120
Pizza, cheese	1 medium slice	145
Raisins	1/2 oz package	40
Sherbert	1 cup	270
Spaghetti with meatballs	1 cup	330

Height and Weight Tables

Men				Women			
Height ft in.	Small Frame	Medium Frame	Large Frame	Height ft in.	Small Frame	Medium Frame	Large Frame
5 2	128-134	131-141	138-150	4 10	102-111	109-121	118-131
5 3	130-136	133-143	140-153	4 11	103-113	111-123	120-134
5 4	132-138	135-145	142-156	5 0	104-115	113-126	122-137
5 5	134-140	137-148	144-160	5 1	106-118	115-129	125-140
5 6	136-142	139-151	146-164	5 2	108-121	118-132	128-143
5 7	138-145	142-154	149-168	5 3	111-124	121-135	131-147
5 8	140-148	145-157	152-172	5 4	114-127	124-138	134-152
5 9	142-151	148-160	155-176	5 5	117-130	127-141	137-155
5 10	144-154	151-163	158-180	5 6	120-133	130-144	140-159
5 11	146-157	154-166	161-184	5 7	123-136	133-147	143-163
6 0	149-160	157-170	164-188	5 8	126-139	136-150	146-167
6 1	152-164	160-174	168-192	5 9	129-142	139-153	149-170
6 2	155-168	164-178	172-197	5 10	132-145	142-156	152-173
6 3	158-172	167-182	176-202	5 11	135-148	145-159	155-176
6 4	162-176	171-187	181-207	6 0	138-151	148-162	158-179

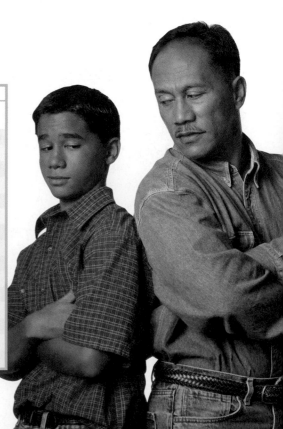

Patterns of Sleep

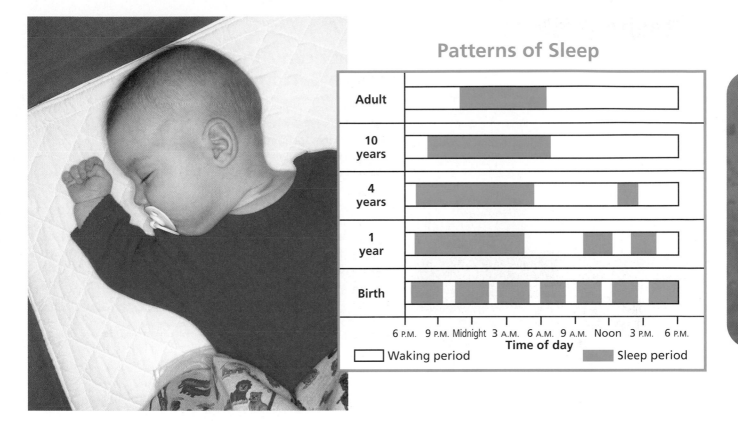

Adult		
10 years		
4 years		
1 year		
Birth		

6 P.M. 9 P.M. Midnight 3 A.M. 6 A.M. 9 A.M. Noon 3 P.M. 6 P.M.

Time of day

☐ Waking period ■ Sleep period

Number of Calories Burned Per Hour by People of Different Body Weights

Exercise	Calories Burned Per Hour		
	110 lb	154 lb	198 lb
Martial arts	620	790	960
Racquetball (2 people)	610	775	945
Basketball (full-court game)	585	750	910
Skiing–cross country (5 mi/h)	550	700	850
downhill	465	595	720
Running–8-min mile	550	700	850
12-min mile	515	655	795
Swimming–crawl, 45 yd/min	540	690	835
crawl, 30 yd/min	330	420	510
Stationary bicycle–15 mi/h	515	655	795
Aerobic dancing–intense	515	655	795
moderate	350	445	540
Walking–5 mi/h	435	555	675
3 mi/h	235	300	365
2 mi/h	145	185	225
Calisthenics–intense	435	555	675
moderate	350	445	540
Scuba diving	355	450	550
Hiking–20-lb pack, 4 mi/h	355	450	550
20-lb pack, 2 mi/h	235	300	365
Tennis–singles, recreational	335	425	520
doubles, recreational	235	300	365
Ice skating	275	350	425
Roller skating	275	350	425

Sports

All-American Girls Professional Baseball League Batting Champions, 1943-1954

Year	Player, Team	At-Bats	Average
1943	Gladys Davis, Rockford	349	.332
1944	Betsy Jochum, South Bend	433	.296
1945	Helen Callahan, Fort Wayne	408	.299
1946	Dorothy Kamenshek, Rockford	408	.316
1947	Dorothy Kamenshek, Rockford	366	.306
1948*	Audrey Wagner, Kenosha	417	.312
1949	Doris Sams, Muskegon	408	.279
1950	Betty Weaver Foss, Fort Wayne	361	.346
1951	Betty Weaver Foss, Fort Wayne	342	.368
1952	Joanne Weaver, Fort Wayne	314	.344
1953	Joanne Weaver. Fort Wayne	410	.346
1954	Joanne Weaver, Fort Wayne	333	.429

*First year overhand pitching was allowed.
Source: *A Whole New Ballgame*, Sue Macy, Henry Holt and Company, New York, 1993.

Olympic Gold Medal Winning Times for 400-m Freestyle Swimming, in Minutes

Year	1924	1928	1932	1936	1948	1952	1956	1960	1964	
Male	5:04.2	5:01.6	4:48.4	4:44.5	4:41.0	4:30.7	4:27.3	4:18.3	4:12.2	
Female	6:02.2	5:42.8	5:28.5	5:26.4	5:17.8	5:12.1	4:54.6	4:50.6	4:43.3	
Year	1968	1972	1976	1980	1984	1988	1992	1996	2000	2004
Male	4:09.0	4:00.27	3:51.93	3:51.31	3:51.23	3:46.95	3:45.00	3:47.97	3:40.59	3:43.10
Female	4:31.8	4:19.44	4:09.89	4:08.76	4:07.10	4:03.85	4:07.18	4:07.25	4:05.80	4:05.34

Sizes and Weights of Balls Used in Various Sports

Type	Diameter (centimeters)	Average Weight (grams)
Baseball	7.6	145
Basketball	24.0	596
Croquet ball	8.6	340
Field hockey ball	7.6	160
Golf ball	4.3	46
Handball	4.8	65
Soccer ball	22.0	425
Softball, large	13.0	279
Softball, small	9.8	187
Table tennis ball	3.7	2
Tennis ball	6.5	57
Volleyball	21.9	256

Soccer Playboard

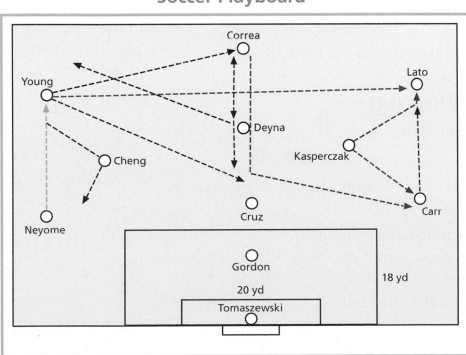

Prerequisite Skills

❶ Place Value and Order

Example 1

Write 2,345,678.9123 in words.

Solution

The place value chart shows the value of each digit. The value of each place is ten times the place to the right.

millions	hundred thousands	ten thousands	thousands	hundreds	tens	ones	.	tenths	hundredths	thousandths	ten thousandths
2	3	4	5	6	7	8	.	9	1	2	3

The number shown is *two million, three hundred forty-five thousand, six hundred seventy-eight and nine thousand one hundred twenty-three ten-thousandths.*

Example 2

Use < or > to make this sentence true.   **6 ▇ 2**

Solution

Remember, < means "less than" and > means "greater than." So, 6 > 2.

◣ Extra Practice Exercises

Write each number in words.

1. 3647

2. 6,004,300.002

3. 0.9001

Write each of the following as a number.

4. two million, one hundred fifty thousand, four hundred seventeen

5. five thousand, one hundred twenty and five hundred two thousandths

6. nine million, ninety thousand, nine hundred and ninety-nine ten-thousandths

Use < or > to make each sentence true.

7. 9 ▇ 8

8. 164 ▇ 246

9. 63,475 ▇ 6,435

10. 52 ▇ 50

11. 5.39 ▇ 9.02

12. 43.94 ▇ 53.69

❷ Multiply Whole Numbers and Decimals

To multiply whole numbers, find each partial product and then add.

When multiplying decimals, locate the decimal point in the product so that there are as many decimal places in the product as the total number of decimal places in the factors.

Example 1

Multiply 2.6394 by 3000.

Solution

$$\begin{array}{r} 2.6394 \\ \times\ \ \ 3000 \\ \hline 7918.2000 \end{array}\ \text{or } 7918.2$$

Zeros after the decimal point can be dropped because they are not significant digits.

Example 2

Multiply 3.92 by 0.023.

Solution

$$\begin{array}{r} 3.92 \\ \times\ 0.023 \\ \hline 1176 \\ +\ 7840 \\ \hline 0.09016 \end{array}$$

2 decimal places
+ 3 decimal places

5 decimal places

The zero is added before the nine so that the product will have five decimal places.

◤ EXTRA PRACTICE EXERCISES

Multiply.

1. 36×45

2. 500×30

3. $17,000 \times 230$

4. 6.2×8

5. 950×1.6

6. 3.652×20

7. 179×83

8. 257×320

9. 8560×275

10. 467×0.3

11. 2.63×183

12. 0.758×321.8

13. 49.3×1.6

14. 6.859×7.9

15. 794.4×321.8

16. 0.08×4

17. 0.062×0.5

18. 0.0135×0.003

19. 21.6×3.1

20. 8.76×0.005

21. 5.521×3.642

22. 5.749×3.008

23. 8.09×0.18

24. $89,946 \times 2.85$

25. 6.31×908

26. 391.05×25

27. $35,021 \times 76.34$

❸ Divide Whole Numbers and Decimals

Dividing whole numbers and decimals involves a repetitive process of estimating a quotient, multiplying and subtracting.

$$
\begin{array}{r}
34 \quad \leftarrow \text{quotient} \\
\text{divisor} \rightarrow 7\overline{)239} \quad \leftarrow \text{dividend} \\
21\downarrow \quad \leftarrow 3 \times 7 \\
\hline
29 \quad \leftarrow \text{Subtract. Bring down the 9.} \\
28 \quad \leftarrow 4 \times 7 \\
\hline
1 \quad \leftarrow \text{remainder}
\end{array}
$$

Example 1

Find: 283.86 ÷ 5.7

Solution

When dividing decimals, move the decimal point in the divisor to the right until it is a whole number. Move the decimal point the same number of places in the dividend. Then place the decimal point in the answer directly above the new location of the decimal point in the dividend.

$$
5.7.\overline{)283.8.6} \quad \rightarrow \quad
\begin{array}{r}
49.8 \\
57\overline{)2838.6} \\
228 \\
\hline
558 \\
513 \\
\hline
45\,6 \\
45\,6 \\
\hline
0
\end{array}
$$

If answers do not have a remainder of 0, you can add 0s after the last digit of the dividend and continue dividing.

◢ EXTRA PRACTICE EXERCISES

Divide.

1. $72 \div 6$
2. $6000 \div 20$
3. $26{,}568 \div 8$
4. $5.6 \div 7$
5. $120 \div 0.4$
6. $936 \div 12$
7. $3.28 \div 4$
8. $0.1960 \div 5$
9. $1968 \div 0.08$
10. $16 \div 0.04$
11. $1525 \div 0.05$
12. $109.94 \div 0.23$
13. $0.6 \div 24$
14. $7.924 \div 0.28$
15. $32.6417 \div 9.1$
16. $24 \div 0.6$
17. $1784.75 \div 29.5$
18. $0.01998 \div 0.37$
19. $7.8 \div 0.3$
20. $12{,}000 \div 0.04$
21. $820.94 \div 0.02$
22. $89{,}946 \div 28.5$
23. $15 \div 0.75$
24. $7.56 \div 2.25$
25. $0.19176 \div 68$
26. $0.168 \div 0.48$
27. $5.1 \div 0.006$
28. $55{,}673 \div 0.05$
29. $84.536 \div 4$
30. $261.18 \div 10$
31. $134{,}554 \div 0.14$
32. $90{,}294 \div 7.85$
33. $59{,}368 \div 47.3$
34. $11{,}633.5 \div 439$
35. $28.098 \div 14$
36. $16.309 \div 0.09$
37. $55.26 \div 1.8$
38. $8276 \div 0.627$
39. $10{,}693 \div 92.8$
40. $48.8 \div 1.6$
41. $27{,}268 \div 34$
42. $546.702 \div 0.078$

❹ Multiply and Divide Fractions

To multiply fractions, multiply the numerators and then multiply the denominators. Write the answer in simplest form.

Example 1

Multiply $\frac{2}{5}$ and $\frac{7}{8}$.

Solution

$$\frac{2}{5} \times \frac{7}{8} = \frac{2 \times 7}{5 \times 8} = \frac{14}{40} = \frac{7}{20}$$

To divide by a fraction, multiply by the reciprocal of that fraction. To find the reciprocal of a fraction, invert the fraction (turn it upside down). The product of a fraction and its reciprocal is 1. Since $\frac{2}{3} \times \frac{3}{2} = \frac{6}{6}$ or 1, $\frac{2}{3}$ and $\frac{3}{2}$ are reciprocals of each other.

Example 2

Divide $1\frac{1}{5}$ by $\frac{2}{3}$.

Solution

$$1\frac{1}{5} \div \frac{2}{3} = \frac{6}{5} \div \frac{2}{3} = \frac{6}{5} \times \frac{3}{2} = \frac{6 \times 3}{5 \times 2} = \frac{18}{10}, \text{ or } 1\frac{4}{5}$$

◤ EXTRA PRACTICE EXERCISES

Multiply or divide. Write each answer in simplest form.

1. $\frac{2}{3} \div \frac{5}{6}$

2. $\frac{3}{5} \times \frac{10}{12}$

3. $\frac{5}{8} \div \frac{1}{4}$

4. $\frac{1}{2} \times \frac{2}{3}$

5. $\frac{2}{3} \times \frac{1}{2}$

6. $\frac{3}{4} \times \frac{5}{8}$

7. $\frac{1}{2} \div \frac{2}{3}$

8. $\frac{2}{3} \div \frac{1}{2}$

9. $\frac{3}{4} \div \frac{5}{8}$

10. $2\frac{2}{3} \div 1\frac{3}{5}$

11. $1\frac{1}{5} \times 2\frac{1}{4}$

12. $3\frac{1}{3} \times 1\frac{1}{10}$

13. $5\frac{2}{5} \div 2\frac{4}{7}$

14. $2\frac{4}{7} \div 5\frac{2}{5}$

15. $2\frac{4}{7} \times 5\frac{2}{5}$

16. $1\frac{7}{8} \div 1\frac{7}{8}$

17. $\frac{3}{4} \times \frac{2}{3} \times 1\frac{5}{8} \times 2\frac{2}{3}$

18. $7\frac{1}{2} \div 2\frac{1}{4}$

19. $6\frac{2}{3} \times 4\frac{1}{2} \times 5\frac{3}{8}$

20. $11\frac{5}{9} \times 6\frac{1}{12}$

21. $\frac{25}{42} \div \frac{5}{21}$

22. $\frac{13}{18} \div \frac{8}{9}$

23. $\frac{3}{8} \times \frac{11}{12} \times \frac{16}{33}$

24. $\frac{51}{56} \div \frac{17}{24}$

❺ Add and Subtract Fractions

To add and subtract fractions, you need to find a common denominator and then add or subtract, renaming as necessary.

E x a m p l e   1

Add $\frac{3}{4}$ and $\frac{5}{6}$.

Solution

$$
\begin{array}{rcl}
\frac{3}{4} & = \frac{3}{4} \times \frac{3}{3} = & \frac{9}{12} \\[6pt]
+\frac{5}{6} & = \frac{5}{6} \times \frac{2}{2} = & +\frac{10}{12} \\[6pt]
& & \frac{19}{12}
\end{array}
$$

Add the numerators and use the common denominator.

Then simplify. $\frac{19}{12} = 1\frac{7}{12}$

E x a m p l e   2

Subtract $1\frac{3}{5}$ from $5\frac{1}{2}$.

Solution

$$
\begin{array}{rcccl}
5\frac{1}{2} & = & 5\frac{5}{10} & = & 4\frac{15}{10} \\[6pt]
-1\frac{3}{5} & = & -1\frac{6}{10} & = & -1\frac{6}{10} \\[6pt]
& & & & 3\frac{9}{10}
\end{array}
$$

You cannot subtract $\frac{6}{10}$ from $\frac{5}{10}$, so rename again.

◢ EXTRA PRACTICE EXERCISES

Add or subtract.

1. $\frac{1}{5} + \frac{1}{10}$

2. $\frac{2}{3} + \frac{1}{3}$

3. $\frac{5}{8} + \frac{3}{4}$

4. $\frac{6}{7} - \frac{2}{7}$

5. $\frac{3}{4} - \frac{1}{3}$

6. $\frac{5}{8} - \frac{1}{4}$

7. $2\frac{1}{2} + 3\frac{1}{2}$

8. $6\frac{5}{8} + 3\frac{7}{8}$

9. $3\frac{2}{3} + 4\frac{1}{2}$

10. $2\frac{3}{4} - 1\frac{1}{4}$

11. $5\frac{1}{8} - 3\frac{7}{8}$

12. $1\frac{1}{3} - \frac{2}{3}$

13. $6\frac{1}{2} + 5\frac{7}{9}$

14. $9\frac{2}{5} - 1\frac{1}{8}$

15. $7\frac{2}{3} + 6\frac{1}{5}$

16. $8\frac{1}{10} - 5\frac{2}{3}$

17. $6\frac{1}{2} - 5\frac{3}{5}$

18. $10\frac{5}{8} - 9\frac{3}{4}$

19. $1\frac{1}{5} + 2\frac{1}{3} + 5\frac{1}{4}$

20. $9\frac{2}{3} + 4\frac{3}{5} + 6\frac{1}{2}$

21. $10\frac{7}{8} + 3\frac{3}{4} + 6\frac{1}{2} + 2\frac{5}{8}$

6 Fractions, Decimals and Percents

Percent means per hundred. Therefore, 35% means 35 out of 100. Percents can be written as equivalent decimals and fractions.

$35\% = 0.35$ Move the decimal point two places to the left.

$35\% = \dfrac{35}{100}$ Write the fraction with a denominator of 100.

$= \dfrac{7}{20}$ Then simplify.

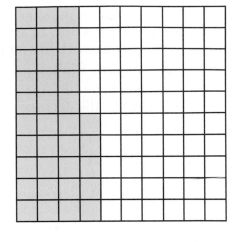

Example 1

Write $\dfrac{3}{8}$ as a decimal and as a percent.

Solution

$\dfrac{3}{8} = 0.375$ Divide to change a fraction to a decimal.

$0.375 = 37.5\%$ To change a decimal to a percent move the decimal point two places to the right and insert the percent symbol.

Percents greater than 100% represent whole numbers or mixed numbers.

$200\% = 2$ or 2.00 $350\% = 3.5$ or $3\dfrac{1}{2}$

◼ EXTRA PRACTICE EXERCISES

Write each fraction or mixed number as a decimal and as a percent.

1. $\dfrac{1}{2}$ 2. $\dfrac{1}{4}$ 3. $\dfrac{3}{4}$

4. $\dfrac{9}{10}$ 5. $\dfrac{3}{10}$ 6. $\dfrac{1}{25}$

7. $3\dfrac{7}{8}$ 8. $1\dfrac{1}{5}$ 9. $\dfrac{13}{25}$

Write each decimal or mixed number as a fraction and as a percent.

10. 0.63   11. 0.15   12. 0.4

13. 2.35   14. 10.125   15. 0.625

16. 0.05   17. 0.125   18. 0.3125

Write each percent as a decimal and as a fraction or mixed number.

19. 10%   20. 12%   21. 100%

22. 150%   23. 160%   24. 75%

25. 8%   26. 87.5%   27. 0.35%

❼ Multiply and Divide by Powers of Ten

To multiply a number by a power of 10, move the decimal point to the right. To multiply by 100 means to multiply by 10 two times. Each multiplication by 10 moves the decimal point one place to the right.

To divide a number by a power of 10, move the decimal point to the left. To divide by 1000 means to divide by 10 three times. Each division by 10 moves the decimal point one place to the right.

Example 1

Multiply 21 by 10,000.

Solution

$21 \times 10,000 = 210,000$ The decimal point moves four places to the right.

Example 2

Find 145 ÷ 500.

Solution

$145 \div 500 = 145 \div 5 \div 100$

$\qquad\qquad = 29 \div 100$

$\qquad\qquad = 0.29$ The decimal point moves two places to the left.

◣ EXTRA PRACTICE EXERCISES

Multiply or divide.

1. 15×100
2. $96 \times 10,000$
3. $1296 \div 100$
4. $9687.03 \div 1000$
5. $36 \times 20,000$
6. $7500 \div 3000$
7. 9×30
8. 94×6000
9. $561 \div 30$
10. $1505 \div 500$
11. $71 \times 90,000$
12. $9 \times 120,000$
13. $3159 \div 10,000$
14. $1,000,000 \times 0.79$
15. $601 \times 30,000$
16. $75 \div 300$
17. 4000×12
18. $14 \times 7,000,000$
19. $49,000 \div 7000$
20. $980 \div 10,000$
21. $216 \div 2000$
22. $108,000 \div 900$
23. $72 \times 10,000,000$
24. $953.16 \div 10,000$
25. $1472 \div 8000$
26. $490,000 \div 700$
27. $80 \times 90,000$
28. $8001 \div 90$
29. 50×6000
30. $950,000 \div 50,000$
31. $81,000 \times 5$
32. $1458 \times 30,000$
33. $452.3 \div 10$
34. $986,856.008 \div 10,000$
35. $316 \times 70,000$
36. $60 \div 1200$

⑧ Round and Order Decimals

To round a number, follow these rules:

1. Underline the digit in the specified place. This is the place digit. The digit to the immediate right of the place digit is the test digit.

2. If the test digit is 5 or larger, add 1 to the place digit and substitute zeros for all digits to its right.

3. If the test digit is less than 5, substitute zeros for it and all digits to the right.

Example 1

Round 4826 to the nearest hundred.

Solution

4826   Underline the place digit.

4800   Since the test digit is 2 and 2 is less than 5, substitute zeros for 2 and all digits to the right.

To place decimals in ascending order, write them in order from least to greatest.

Example 2

Place in ascending order: 0.34, 0.33, 0.39.

Solution

Compare the first decimal place, then compare the second decimal place.

0.33 (least), 0.34, 0.39 (greatest)

▰ EXTRA PRACTICE EXERCISES

Round each number to the place indicated.

1. 367 to the nearest ten
2. 961 to the nearest ten
3. 7200 to the nearest thousand
4. 3070 to the nearest hundred
5. 41,440 to the nearest hundred
6. 34,254 to the nearest thousand
7. 208,395 to the nearest thousand
8. 654,837 to the nearest ten thousand

Write the decimals in ascending order.

9. 0.29, 0.82, 0.35
10. 1.8, 1.4, 1.5
11. 0.567, 0.579, 0.505, 0.542
12. 0.54, 0.45, 4.5, 5.4
13. 0.0802, 0.0822, 0.00222
14. 6.204, 6.206, 6.205, 6.203
15. 88.2, 88.1, 8.80, 8.82
16. 0.007, 7.0, 0.7, 0.07

Extra Practice

Chapter 1

Extra Practice 1–1 • The Language of Mathematics • pages 6–9

Name each set using roster notation.

1. odd natural numbers greater than 6
2. months having 31 days
3. integers between 2 and 3
4. days beginning with the letter S

Determine whether each statement is *true* or *false*.

5. $7 \in \{x | x \text{ is a negative integer}\}$
6. $15 \in \{-3, 0, 3, 6, \ldots\}$
7. $\{a, h, t\} \subseteq \{m, a, t, h\}$
8. $\{-4\} \subseteq \{\text{natural numbers}\}$

Write all the subsets of each set.

9. $\{p\}$
10. $\{h, t\}$
11. $\{o, n, e\}$

Which of the given values is a solution of the equation?

12. $n - 8 = -3; 5, -5$
13. $d + 2 = -2; 4, -4$
14. $3a + 5 = 8; -1, 0, 1$
15. $\frac{4c}{3} = 4; -12, 3, 12$
16. $k + 5 = -5; 0, -5, -10$
17. $c - 7 = -10; 17, 3, -3$

Use mental math to solve each equation.

18. $x + 7 = 4$
19. $n - 6 = 3$
20. $7q = -28$
21. $\frac{c}{-6} = 6$
22. $n - 5 = -5$
23. $\frac{1}{4} + d = \frac{3}{4}$

24. Henry saved \$36 less than Alan. Henry saved \$57. Use the equation $57 = A - 36$ and the values $\{91, 93, 99\}$ for A. Find A, the amount of money Alan saved.

Extra Practice 1–2 • Real Numbers • pages 10–13

Determine whether each statement is *true* or *false*.

1. 3.16 is a rational number.
2. 0.121212... is an irrational number.
3. $\sqrt{8}$ is a real number.
4. $-\sqrt{16}$ is an integer.
5. $\sqrt{5\frac{5}{8}}$ is not a real number.
6. $-\frac{15}{16}$ is a rational number.

Graph each set of numbers on a number line.

7. $\{-3, -1, 1.5, 2\}$
8. $\left\{-1.5, -\frac{1}{2}, 0, \sqrt{9}\right\}$
9. $\left\{-\sqrt{3}, -0.3, 1\frac{3}{4}, 2\frac{1}{3}\right\}$
10. $\left\{-2\frac{3}{4}, -1\frac{3}{4}, 0.6, \sqrt{6}\right\}$
11. whole numbers less than 1
12. real numbers less than 3
13. real numbers from -3 to 2 inclusive.
14. real numbers greater than or equal to -2

Refer to the diagram. Find the sets by listing the members.

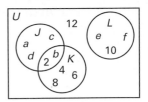

1. $J \cup K$
2. $J \cap L$
3. J'
4. K'
5. $(J \cap K)'$
6. $(J \cup K)'$
7. $K \cap L$
8. $K \cup L$

Let $U = \{g, r, a, p, h\}$, $B = \{g, r, a, p\}$, and $C = \{r, a, p\}$. Find each set, union, or intersection.

9. C'
10. $B \cup C$
11. $B \cap C$
12. B'
13. $(B \cup C)'$
14. $(B \cap C)'$
15. $B' \cup C$
16. $B' \cap C'$

17. Let $X = \{l, i, g, h, t\}$ and $Y = \{t, r, o, u, g, h\}$. Find $X \cap Y$.

18. Let $R = \{-4, -2, 0, 2, 4\}$ and $S = \{-2, 4, 10\}$. Find $R \cup S$.

19. Let $P = \{0, 6, 12\}$ and $Q = \{0, 3, 6, 9, 12\}$. Find $P \cap Q$.

Use the set of real numbers as the replacement set. Graph the solution set of each compound inequality.

20. $x \leq 0$ or $x \geq 1$
21. $x \leq 1$ and $x \geq 1$
22. $x > 4$ and $x \leq -1$
23. $x \geq 0$ and $x \geq 2$
24. $x < -4$ or $x < -1$
25. $x > 3$ or $x \leq 1$

26. Tondra's car stays in first gear until it reaches a speed of 12 mi/h. Graph the speeds at which her car is in first gear.

Add or subtract.

1. $-8 + (-37)$
2. $-46 + 17$
3. $-22 - 23$
4. $18 - (-18)$
5. $-16.4 + 9.3$
6. $-68.9 + 70$
7. $-2.1 + (-16.2)$
8. $-4.3 - 5.7$
9. $-2\frac{7}{8} - 1\frac{3}{8}$
10. $-6\frac{1}{4} + 5\frac{3}{4}$
11. $-3\frac{5}{8} - 2\frac{1}{6}$
12. $-7\frac{2}{3} + 3\frac{1}{4}$
13. $-9.5 + (-11.7) + 8.6 + 0.4$
14. $-19 + 21 + 16 + (-24)$
15. $-8\frac{4}{5} + \left(-6\frac{7}{10}\right) + \left(-3\frac{1}{5}\right)$
16. $-42 + 29 + (-16) + 39$

Evaluate each expression when $x = 24$ and $y = -18$.

17. $x + y$
18. $x - y$
19. $y - x$
20. $-x + y$

Evaluate each expression when $a = -3$ and $b = 1.8$.

21. $a - b$
22. $-a - b$
23. $-a + b$
24. $a + b$

25. Alfonse makes the following transactions to his savings account. Previous balance, $564.82; Withdrawal, $125; Deposit, $152.68; Deposit, $38.95; Withdrawal, $75. What is his new balance?

Perform the indicated operations.

1. $6.7(-2.8)$ **2.** $(-3.2)(-1.4)$ **3.** $\left(3\frac{1}{2}\right)\left(-2\frac{1}{3}\right)$ **4.** $\left(-6\frac{1}{4}\right)\left(-1\frac{1}{2}\right)$

5. $2\frac{2}{3} \div \left(-\frac{4}{9}\right)$ **6.** $\left(-3\frac{1}{3}\right) \div 2$ **7.** $(1.05) \div (0.35)$ **8.** $(2.25) \div (-15)$

9. $3\frac{7}{8} + (5)(-6)$ **10.** $-4\frac{5}{16} + (-3)(-4)$ **11.** $7.6 \div 1.9 - 4.1$

12. $(-9.1) \div (-7) - 1.3$ **13.** $5\frac{1}{3} \div (-4) + 6$ **14.** $3\frac{1}{8} \div \left(-\frac{5}{8}\right) - 1$

15. $-75 \div (-10) + \left(-2\frac{1}{2}\right)$ **16.** $17 \div (-2) + 2\frac{1}{2}$ **17.** $(-64) \div \left(-\frac{2}{3}\right) + (-96)$

Evaluate each expression when $r = -3$, $s = 1.5$, and $t = \frac{4}{5}$.

18. $r - s$ **19.** $r + t$ **20.** $r + s$ **21.** $r - t$

22. rs **23.** $t(r + s)$ **24.** $r + st$ **25.** $(r + s) \div t$

26. Nat earns \$6.40 per hour for each hour in his 32-h work week. For each hour over 32 h, he earns $1\frac{1}{2}$ times his hourly pay. How much will he earn if he works 42 h in one week?

27. Gloria earns \$7.50 per hour and $1\frac{1}{2}$ times that amount for each hour she works over 32 h in week. One week she earned \$273.75. How many hours of overtime did she work?

Use the distributive property to find each product.

1. $6.8 \cdot 7 + 6.8 \cdot 93$ **2.** $2.7 \cdot 8 + 2.7 \cdot 12$ **3.** $23 \cdot 16 - 23 \cdot 6$

4. $101 \cdot 27$ **5.** $35\left(2\frac{3}{7}\right)$ **6.** $24\left(20\frac{1}{8}\right)$

Evaluate each expression when $m = -2$ and $n = 5$.

7. m^2 **8.** $m^2 - n^2$ **9.** n^3

10. mn^2 **11.** m^3 **12.** $2mn^2$

13. $2m^2n$ **14.** $(m - n)^2$ **15.** $(n - m)^2$

16. $(m^2 - 1)^3$ **17.** $(-2mn)^2$ **18.** $-2mn^2$

Simplify.

19. $2^8 \cdot 2^6$ **20.** $\dfrac{x^7}{x^4}$ **21.** $y^3 \cdot y^3$ **22.** $(y^3)^3$

23. $\left(\dfrac{1}{n}\right)^8$ **24.** $m^{10} \cdot m^{15}$ **25.** $(x^3)(x^4)(x^5)$ **26.** $(x)(x^2)(x^2)$

Evaluate mentally each sum or product when $j = 4.5$, $k = 2$, and $l = 0$.

27. $10jk$ **28.** $67j^2l$ **29.** $-5jk$

30. $(5.5 - j)(11k)$ **31.** $(2j - l)k^2$ **32.** $(j + 0.5)(k + 2)$

33. $(jk)^2 - 80$ **34.** $(3.5 + k)(j - 5.5)$ **35.** $jk^3(2j - 9)$

Simplify.

1. $(-1)^{-4} + (-1)^{-5}$
2. $(-1)^{-4} + (-1)^{-6}$
3. $c^{-18} \div c^{-6}$
4. $n^{-4} \cdot n^3$
5. $x^{-4} \cdot x^{-3}$
6. $y^{-3} \cdot y^{-3}$

Evaluate each expression when $r = 3$ and $s = -3$.

7. r^{-3}
8. s^{-3}
9. $(rs)^{-2}$
10. $r^2 \cdot r^{-2}$
11. $s^{-3} \cdot s^2$
12. $r^{-2}r^{-2}r^4$
13. $s^3s^{-2}r^{-3}$
14. r^3rs^{-3}

Write each number in scientific notation.

15. 4700
16. 66,800
17. 1,410,000
18. 218,000
19. 0.0571
20. 0.00178
21. 0.00082
22. 0.971
23. 0.0000000505

Write each number in standard form.

24. $1.76 \cdot 10^5$
25. $2.6 \cdot 10^4$
26. $4.9 \cdot 10^{-2}$
27. $5.04 \cdot 10^{-6}$

Solve. Write your answer in scientific notation.

28. The distance from Earth to the Sun is about 93,000,000 mi. Write this distance in scientific notation.

29. The speed of light is $3.00 \cdot 10^{10}$ m/sec. How far does light travel in 1 h? Write the answer in scientific notation.

Chapter 2

Determine the next three terms in each sequence.

1. 1, 5, 9, 13, ____, ____, ____
2. 31, 26, 21, 16, ____, ____, ____
3. −5, −3, −1, 1, ____, ____, ____
4. 25, 18, 11, 4, ____, ____, ____
5. 9, 3, 1, $\frac{1}{3}$, ____, ____, ____
6. $\frac{1}{10}, \frac{1}{5}, \frac{2}{5}, \frac{4}{5}$, ____, ____, ____
7. 2, 8, 18, 32, ____, ____, ____
8. −8, −6, −4, −2, ____, ____, ____
9. 1.5, 3, 4.5, 6, ____, ____, ____
10. −1, −2, −4, −7, ____, ____, ____

Draw the iteration diagram for each sequence. Calculate the output for the first 7 iterations.

11. 128, 64, 32, 16, . . .
12. −10, −7, −4, −1, . . .
13. 1, 6, 36, 216, . . .
14. 12, 9.5, 7, 4.5, . . .

Graph each point on a coordinate plane.

1. $A(4, 3)$	**2.** $B(-3, -2)$	**3.** $C(5, -3)$	**4.** $D(-4, 4)$
5. $E(0, -3)$	**6.** $F(2, 0)$	**7.** $G(5, 3)$	**8.** $H(-3, 3)$
9. $J(-5, -4)$	**10.** $K(4, -3)$	**11.** $L(3, 1)$	**12.** $M(2, -2)$
13. $N(-1, 2)$	**14.** $P(-5, -2)$	**15.** $Q(0, 4)$	**16.** $R(-5, 0)$

Given $f(x) = 3x - 2$, evaluate each of the following.

17. $f(1)$	**18.** $f(0)$	**19.** $f(-1)$	**20.** $f(-3)$
21. $f(2)$	**22.** $f(-2)$	**23.** $f(-6)$	**24.** $f(6)$

Write each relation as a set of ordered pairs. Give the domain and range.

25.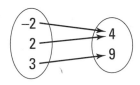

26.

x	3	5	7	9
y	2	4	6	8

27.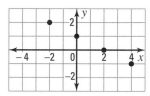

28. José charges \$3 for the first hour of baby-sitting and then \$5 per hour for each additional hour. The function that describes how he is paid is $f(x) = 3 + 5(x - 1)$ where x is the number of hours he works. How much does he earn if he works 7 h?

Graph each function.

1. $y = x + 3$	**2.** $y = x - 3$
3. $y = x$	**4.** $f(x) = x + 5$
5. $f(x) = x - 5$	**6.** $f(x) = -x$
7. $y = 2x + 2$	**8.** $f(x) = 2x - 2$
9. $f(x) = -4$	**10.** $y = 4$
11. $y = 0$	**12.** $f(x) = 3x$
13. $y = -x + 5$	**14.** $y = -3x + 2$
15. $f(x) = -2x - 3$	**16.** $f(x) = 4x - 6$
17. $f(x) = -3x + 3$	**18.** $y = -x - 1$

Given $f(x) = |3x + 4|$, find each value.

19. $f(1)$	**20.** $f(-1)$	**21.** $f(7)$
22. $f(0)$	**23.** $f(-6)$	**24.** $f(-2)$

Given $g(x) = |-5x - 3|$, find each value.

25. $g(0)$	**26.** $g(1)$	**27.** $g(2)$
28. $g(-3)$	**29.** $g(-4)$	**30.** $g(-2)$

Solve each equation.

1. $m + 17 = 45$
2. $9x = -54$
3. $17 + d = -5$
4. $-16 = j - 2$
5. $-24 = c + 9$
6. $16n = -12$
7. $-8x = 96$
8. $0.8a = 0.72$
9. $b + 0.8 = 0.72$
10. $36 = -\frac{4}{9}c$
11. $\frac{5}{8}x = 10$
12. $51 = \frac{3}{5}x$
13. $13.24 = x - 4.2$
14. $-8.6 = m + 2.15$
15. $j + \frac{3}{8} = 1$

Translate each sentence into an equation. Use n to represent the unknown number. Then solve the equation for n.

16. When n is increased by 18, the result is -12.

17. When a number is decreased by 7, the result is -4.

18. The quotient of a number and 7 is 0.6.

19. The product of 4 and a number is the same as the square of -6.

20. The difference between a number and 13 is 14.

21. One fourth of -64 is the same as the product of 2 and some number.

22. Liya decided to save $8 per week for the next 4 weeks so that her savings would total $100. Let n represent the amount she has before she begins saving. Write an equation that illustrates the situation. Then solve the equation.

Solve each equation and check the solution.

1. $6n + 5 = 23$
2. $4n - 3 = 17$
3. $-55 = 8x - 7$
4. $-36 = 5x + 4$
5. $-3j + 16 = -11$
6. $-2n - 17 = 17$
7. $2(3d - 4) = 10$
8. $-4(2x + 1) = 4$
9. $-3(2x - 3) = -9$
10. $8x - 7 = 2x + 5$
11. $-3x + 24 = 5x - 24$
12. $3c + 5 = 7c - 7$
13. $-3x - 1 = -2x - 1$
14. $2k + 3 + 3k = 1 + 7k$
15. $-4a + 7 - 2a = -11$
16. $\frac{1}{2}(16k + 10) = -11$
17. $4(1.5 - x) = 14$
18. $4(3c - 2) = -38 + 6$

Translate each sentence into an equation. Then solve.

19. Five more than 4 times a number is 33. Find the number.

20. Two less than 3 times a number is 13. Find the number.

21. When 20 is decreased by twice a number, the result is 8. Find the number.

22. Keisha bought 3 report binders that had the same price. The total cost came to $11.97, which included $0.57 sales tax. Write and solve an equation to find out how much each binder cost.

Solve each inequality and graph the solution on a number line.

1. $3a + 2 \leq -10$

2. $7n - 2 > 19$

3. $\frac{1}{2}n - 7 < -6$

4. $\frac{1}{3}c + 8 \geq 10$

5. $10 - 3r \leq 7$

6. $7 > 2a - 5$

7. $33 \leq -7n - 2$

8. $-19 < 14 - 11c$

9. $2 \geq -18 - 5t$

10. $\frac{2}{3}x + 8 \leq 10$

11. $2(3w + 4) < -28$

12. $3(4c - 2) \leq 18$

13. $2a + 5 \geq 8a - 7$

14. $2n - 13 > 11n + 14$

15. $2 < \frac{2}{3}(9 - 6a)$

16. $12 > \frac{4}{9}(18 - 9c)$

Graph each inequality on the coordinate plane.

17. $y < 2x + 5$

18. $y \leq -2x - 3$

19. $y > -x - 1$

20. $x + y \leq 5$

21. $x - y < 3$

22. $y < \frac{1}{2}x - 1$

23. $2x + 4y \geq 8$

24. $x - 2y < 10$

25. $1 \geq 2x - \frac{1}{2}y$

26. A pet store charges a minimum of $3 per hour to take care of a person's pet. The inequality that describes how the store charges is $y \geq 3x$ where x is the number of hours and y is the amount of money charged. Graph the inequality.

Thirty families were randomly sampled and surveyed as to the number of hours they watched television on a typical Friday. The results are listed below.

5	0	3	1	2
0	1	4	2	2
2	1	0	4	6
1	1	3	3	3
0	3	5	2	1
0	2	0	0	4

1. Construct a frequency table for these data.

2. Find the mean, median, and mode of the data.

As part of her research for a term paper on home entertainment, Lydia surveyed video stores to find the cost of renting a movie for one day. The results are listed below.

$2.50	$2.86	$1.99	$2.00	$3.10
$2.15	$1.55	$2.83	$3.49	$2.69
$1.85	$3.14	$2.62	$3.35	$3.32
$2.45	$2.12	$1.99	$2.05	$2.90
$2.49	$3.07	$1.68	$2.33	$3.00
$2.60	$2.00	$3.25	$2.25	$2.50

3. Construct a frequency table for these data. Group the data into intervals of $0.25.

4. Which interval contains the median of the data?

The weights in pounds of the 30 students who tried out for the Snyder High School football team were as follows:

145	160	172	129	149	202	183	176	170	169
157	146	177	200	162	164	168	165	150	161
145	171	173	162	164	164	166	175	181	179

1. Construct a stem-and-leaf plot to display the data.

2. Identify any outliers, clusters, and gaps in the data.

3. Find the mode of the data.   4. Find the median of the data.

On a test that measures reasoning aptitude on a scale of 0 to 100, a class of 30 students received the following scores.

59	38	48	75	78	81	52	45	55	62
91	56	39	47	80	55	72	60	58	60
63	70	65	52	42	72	70	50	47	55

5. Construct a stem-and-leaf plot to display the data.

6. Identify any outliers, clusters, and gaps in the data.

7. Find the mode of the data.   8. Find the median of the data.

9. A newspaper took a random survey of its readers about the number of miles they travel to and from work each day. The data are recorded in this frequency table. Construct a histogram of the data.

DISTANCES TRAVELED TO AND FROM WORK

Miles	Frequency
0 – 9	24
10 – 19	16
20 – 29	10
30 – 39	20
40 – 49	18
50 – 59	7
60 – 69	5

Chapter 3

Extra Practice 3–1 • Points, Lines, and Planes • pages 104–107

Use the figure at the right for Exercises 1–4. Which postulate justifies your answer?

1. Name two points that determine line ℓ.

2. Name three points that determine plane $\mathcal{A}$.

3. Name three lines that lie in plane $\mathcal{A}$.

4. Name the intersection of planes $\mathcal{A}$ and $\mathcal{B}$.

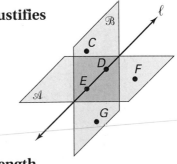

Use the number line at the right for Exercises 5–8. Find each length.

5. AD 6. EC

7. FB 8. EF

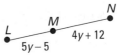

9. In the figure below, $RT = 85$. Find RS.   10. In the figure below, $LN = 79$. Find ML.

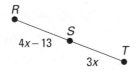

Exercises 1–4 refer to the protractor at the right.

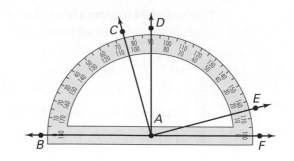

1. Name the straight angle.

2. Name the three right angles.

3. Name all the acute angles. Give the measure of each.

4. Name all the obtuse angles. Give the measure of each.

5. In the figure below, $m\angle QRS = x°$ and $m\angle SRT = 5x°$. Find $m\angle SRT$.

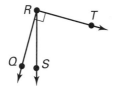

6. In the figure below, $m\angle MLO = (4x + 5)°$ and $m\angle KLO = (2x - 11)°$. Find $m\angle OLN$.

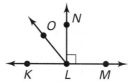

7. An angle measures 47°. What is the measure of its complement?

Exercises 1–4 refer to the figure.

$$G \quad H \quad I \quad J \quad K \quad L \quad M \quad N$$
$$-4 \quad -3 \quad -2 \quad -1 \quad 0 \quad 1 \quad 2 \quad 3$$

1. Name the midpoint of $\overline{GK}$.

2. Name the segment whose midpoint is point H.

3. Name all the segments whose midpoint is point J.

4. Assume that point O is the midpoint of $\overline{GN}$. What is its coordinate?

In the figure below, $\overleftrightarrow{JM}$ and $\overleftrightarrow{OL}$ intersect at point K, and $\overrightarrow{KN}$ bisects $\angle OKM$. Find the measure of each angle.

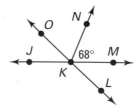

In the figure below, $\overleftrightarrow{UV}$, $\overleftrightarrow{WX}$, and $\overleftrightarrow{YZ}$ intersect at point O, and $\overrightarrow{OU}$ bisects $\angle XOZ$. Find the measure of each angle.

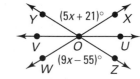

5. $\angle OKM$

6. $\angle LKJ$

7. $\angle MKL$

8. $\angle JKO$

9. $\angle ZOW$

10. $\angle WOY$

11. $\angle XOU$

12. $\angle WOV$

In the figure at the right, point F is the midpoint of $\overline{EG}$. Find the length of each segment.

13. $\overline{EF}$

14. $\overline{EG}$

15. $\overline{GH}$

16. $\overline{EH}$

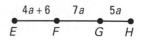

Extra Practice 3–4 • Constructions and Lines • pages 118–121

In the figure at the right, $\overleftrightarrow{AD} \parallel \overleftrightarrow{HE}$ and $\overleftrightarrow{BF} \perp \overleftrightarrow{GC}$. Find the measure of each angle.

1. $\angle AJB$
2. $\angle JKI$
3. $\angle CJD$
4. $\angle JIK$
5. $\angle GIK$
6. $\angle GJD$
7. $\angle BKE$
8. $\angle FKI$

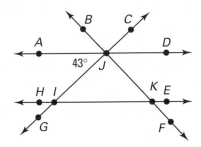

In the figure at the right, $\overrightarrow{XU} \parallel \overrightarrow{YV}$. Find the measure of each angle.

9. $\angle VYX$
10. $\angle VYZ$
11. $\angle UXW$
12. $\angle UXY$
13. Compare parallel and skew lines.

Extra Practice 3–5 • Inductive Reasoning • pages 124–127

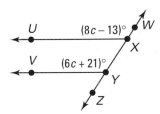

Draw the next figure in each pattern. Then describe the twelfth figure in the pattern.

1.

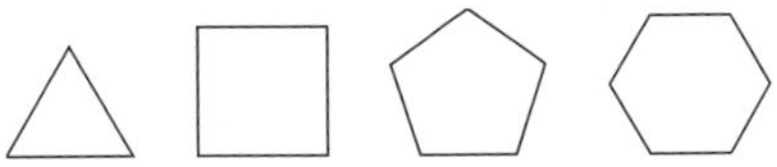

2.

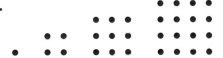

3.

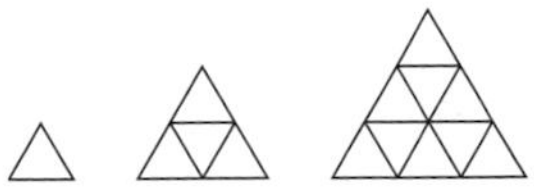

The figures below show one, two, three, and four segments drawn inside a triangle.

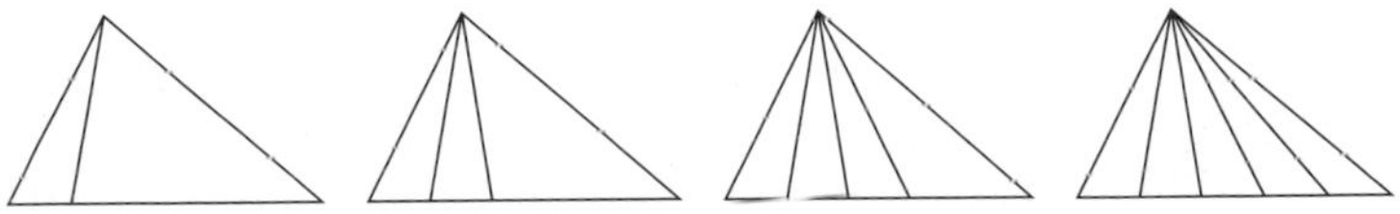

4. In each figure, the segments divide the interior of the triangle into regions. How many regions are formed in each of the figures shown?
5. Find the number of regions that would be formed when twelve segments are drawn through a triangle.

Sketch a counterexample that shows why each conditional is false.

1. If $m\angle XYZ + m\angle ZYW = 180°$, then $\overline{ZY} \perp \overline{XW}$.

2. If point B is between points A and C, then B is the midpoint of $\overline{AC}$.

3. If two lines are not parallel, then they intersect.

Write the converse of each statement. Then tell whether the given statement and its converse are true or false.

4. If the sum of the measures of two angles is 180°, then the angles are supplementary.

5. If two lines are parallel, then they intersect.

6. If $\overrightarrow{BC}$ and $\overrightarrow{BA}$ are opposite rays, then B is the midpoint of $\overline{AC}$.

Write each definition as two conditionals and as a single biconditional.

7. Parallel lines are coplanar lines that do not intersect.

8. Supplementary angles are two angles whose sum of their measures is 180°.

1. **Given:** $m\angle 1 = m\angle 4$
   $\angle 1$ and $\angle 2$ are complementary.
   $\angle 3$ and $\angle 4$ are complementary.
   Prove: $m\angle 2 = m\angle 3$

Statements	Reasons
1. $\angle 1$ and $\angle 2$ are __?__. $\angle 3$ and $\angle 4$ are __?__.	1. given
2. $m\angle 1 + m\angle 2 = 90°$ $m\angle 3 + m\angle 4 = $ __?__	2. definition of complementary angles
3. $m\angle 1 + m\angle 2 = m\angle 3 + m\angle 4$	3. __?__
4. $m\angle 1 = m\angle 4$	4. __?__
5. $m\angle 2 = m\angle 3$	5. __?__

2. **Given:** $m\angle TSW = m\angle TWS$
   Prove: $m\angle TSR = m\angle TWX$

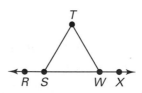

Statements	Reasons
1. $\angle TSR$ is supplementary to __?__. $\angle TWX$ is supplementary to __?__.	1. __?__
2. $m\angle TSW = m\angle TWS$	2. __?__
3. $m\angle TSR = $ __?__	3. __?__

Chapter 4

Extra Practice 4–1 • Triangles and Triangle Theorems • pages 150–153

Find the value of *x* in each figure.

1.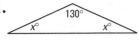

2.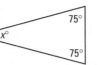

3.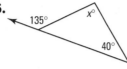

4.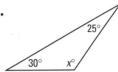

5.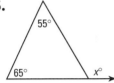

6.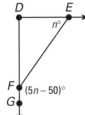

7.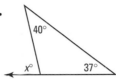

8.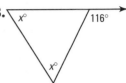

9. In the figure below, $\overline{ED} \perp \overline{DF}$. Find $m\angle DFE$.

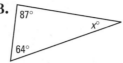

10. In the figure below, $\overleftrightarrow{AB} \parallel \overleftrightarrow{JK}$. Find $m\angle JCK$.

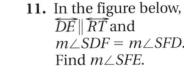

11. In the figure below, $\overleftrightarrow{DE} \parallel \overleftrightarrow{RT}$ and $m\angle SDF = m\angle SFD$. Find $m\angle SFE$.

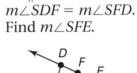

Extra Practice 4–2 • Congruent Triangles • pages 154–157

1. Copy and complete this proof.
Given: $\overline{AB} \cong \overline{CB}$; $\overline{DB}$ bisects $\angle ABC$.
Prove: $\triangle ABD \cong \triangle CBD$

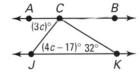

Statements	Reasons
1. ___?___	1. ___?___
2. $m\angle 1 = m\angle 2$ or $\angle 1 \cong \angle 2$	2. ___?___
3. ___?___	3. ___?___
4. $\triangle ABD \cong \triangle CBD$	4. ___?___

Write a two-column proof.

2. **Given:** $\overline{RS} \cong \overline{VT}$; $\overline{RV} \cong \overline{ST}$
Prove: $\triangle RSV \cong \triangle TVS$

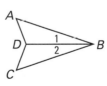

3. **Given:** $\overline{XV}$ and $\overline{WT}$ intersect at point Y; $\overline{XY} \cong \overline{VY}$; Y is the midpoint of $\overline{WT}$.
Prove: $\triangle WXY \cong \triangle TVY$

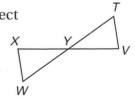

Find the value of *n* in each figure.

1.

2.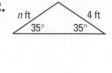

3.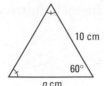

4.

Copy and complete the proof.

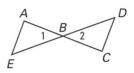

5. Given: Point *B* is the midpoint of *AC* and *ED*.
   Prove: $\angle E \cong \angle D$

Statements	Reasons
1. __?__	1. __?__
2. __?__	2. definition of midpoint
3. $\angle 1$ and $\angle 2$ are __?__	3. __?__
4. __?__ $\cong$ __?__	4. __?__
5. $\triangle ABE \cong \triangle CBD$	5. __?__
6. __?__	6. __?__

Trace each triangle onto a sheet of paper. Sketch all the altitudes and all the medians.

1.

2.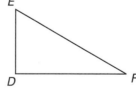

3.

Exercises 4–9 refer to $\triangle PQR$ with altitude $\overline{QT}$. Tell whether each statement is *true* or *false*.

4. $\overline{QT} \perp \overline{PR}$

5. $\overline{TQ} \cong \overline{PT}$

6. $\overline{PT} \cong \overline{TR}$

7. $m\angle PTQ = m\angle RTQ$

8. $m\angle QTP = 90°$

9. $\angle P \cong \angle R$

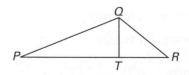

Can the given measures be the lengths of the sides of a triangle?

1. 3 m, 6 m, 8 m

2. 9 ft, 7 ft, 2 ft

3. 18 in., 13 in., 34 in.

4. 15 cm, 15 cm, 15 cm

5. 2.4 yd, 6.7 yd, 3.9 yd

6. $3\frac{1}{2}$ ft, $3\frac{1}{4}$ ft, $6\frac{1}{2}$ ft

7. 6 mm, 5 mm, 4 mm

8. 3 mi, 2 mi, 1 mi

9. 2 yd, 5 ft, 72 in.

In each figure, give the ranges of possible values for *x*.

10.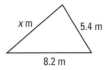

11.

12.

13. In $\triangle FGH$, $\overline{FG} > \overline{GH}$ and $\overline{HF} > \overline{FG}$. Which is the largest angle of the triangle?

14. In $\triangle ABC$, $\overline{BC} = 18$, $\overline{AB} = 16.5$, and $\overline{AC} = 14$. List the angles of the triangle in order from largest to smallest.

15. In $\triangle PQR$, $m\angle P = 73°$, $m\angle Q = 57°$, and $m\angle R = 50°$. List the sides of the triangle in order from longest to shortest.

Find the unknown angle measure or measures in each figure.

1.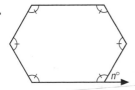

2.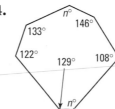

3.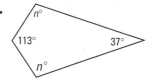

4.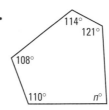

5. Find the measure of each interior angle of a regular heptagon.

6. Find the measure of each interior angle of a regular decagon.

7. Find the sum of the measures of the interior angles of a regular polygon with 16 sides.

8. Find the sum of the measures of the interior angles of a regular polygon with 20 sides.

In Exercises 1–6, the figure is a parallelogram. Find the values of *a, b, c,* and *d.*

1.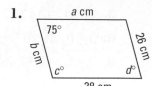

2.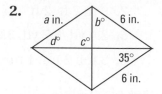

3.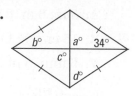

4. $AC = 9\frac{3}{4}$ ft; $AD = 4\frac{1}{2}$ ft

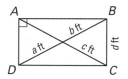

5. $VS = 18$ m; $RS = 14$ m; $RT = 12$ m

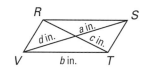

6. $HG = 5$ yd; $EG = 8.6$ yd; $FG = 5.4$ yd

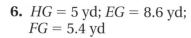

Tell whether each statement is *true* or *false.*

7. A rhombus is a parallelogram.

8. Every parallelogram is a quadrilateral.

9. A square is a rectangle.

10. Diagonals of a rectangle bisect each other.

11. Diagonals of a square are perpendicular.

12. Opposite sides of a square are parallel.

A trapezoid and its median are shown. Find the value of *n.*

1. 29 cm / *n* cm / 37 cm

2. 3.5 in. / *n* in. / 1.7

3. 152 mm / *n* mm / 104 mm

4. *n* ft / $6\frac{1}{2}$ ft / 5 ft

5. *n* yd / 27 yd / 42 yd

6. 13 cm / $(n-3)$ cm / 27 cm

7. *n* ft / 1.5 ft / 2.4 ft

8. $2\frac{1}{3}$ yd / 3 yd / *n* yd

The given figure is a trapezoid. Find all the unknown angle measures.

9.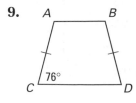

10.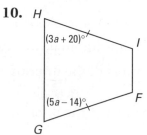

11.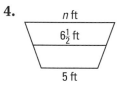

Chapter 5

Complete.

1. 12 qt = __?__ c
2. 312 in. = __?__ yd __?__ ft
3. 3 gal = __?__ fl oz
4. 1.8 T = __?__ oz
5. 0.7 cm = __?__ m
6. 500 mg = __?__ g
7. 0.003 kg = __?__ g
8. 5.9 mL = __?__ L
9. 3 gal = __?__ c
10. 6.4 L = __?__ mL
11. 31 ft = __?__ yd
12. 4.37 km = __?__ m

Name the best customary unit for expressing the measure of each.

13. weight of a computer
14. height of a seat
15. length of a room

Name the best metric unit for expressing the measure of each.

16. capacity of a cooler
17. mass of a box of cereal
18. length of a building

Write each ratio in lowest terms.

19. 27 m:45 m
20. 60 g to 420 g
21. 30 min/6 h

Find each unit rate.

22. 220 mi in 4 h
23. $16 for 320 prints
24. 15 L in 3 min
25. Which is the better buy, 6 grapefruit for $1.80, or 8 grapefruit for $2.56?
26. In 2 h 20 min Suzanne biked 14 mi. What was her biking rate?

1. What is the perimeter of a regular hexagon with 6-cm sides?
2. What is the circumference of a circle with a radius of 5.4 m?
3. Find the base of a triangle if area = 42 cm^2 and height = 8 cm.

Find the area of the shaded region of each figure.

4.
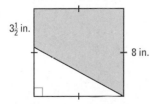
$3\frac{1}{2}$ in.   8 in.

5.
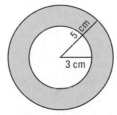
5 cm   3 cm

6.
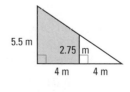
5.5 m   2.75 m   4 m   4 m

7. If you triple the length of the radius of a circle, how does the area change?

A standard deck of playing cards has 52 cards. A card is drawn at random from a shuffled deck. Find each probability.

1. P(king)　　　　**2.** P(black card)　　　　**3.** P(red face card)

Find the probability that a point selected at random in each figure is in the shaded region.

4.

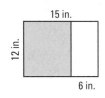

15 in.

12 in.

6 in.

5.

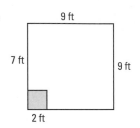

9 ft

7 ft

9 ft

2 ft

6.

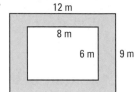

12 m

8 m

6 m

9 m

7.

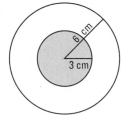

6 cm

3 cm

8.

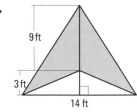

9 ft

3 ft

14 ft

9.

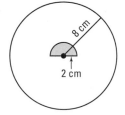

8 cm

2 cm

10. Suppose Mrs. O'Malley left her purse within her 1500 ft² apartment. What is the probability it is in the 15-ft by 12-ft kitchen?

Name the polyhedra shown below. Then state the number of faces, vertices, and edges each has.

1.

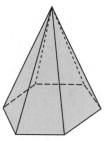

2.

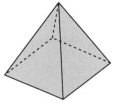

3.

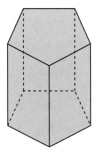

Draw the figure.

4. right rectangular prism

5. right cylinder

6. sphere

7. A figure has 5 triangular faces and 1 pentagonal face. What is the figure?

Extra Practice

Find the surface area of each figure. Assume that all pyramids are regular pyramids. Use 3.14 for π. Round answer to the nearest whole number.

1.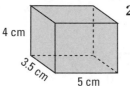

4 cm

3.5 cm

5 cm

2.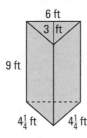

6 ft

3 ft

9 ft

$4\frac{1}{4}$ ft $4\frac{1}{4}$ ft

3.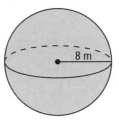

8 m

4.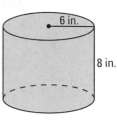

6 in.

8 in.

5. What is the surface area of a cone with a base that is 8 cm across and has a slant height of 5.6 cm?

Find the volume to the nearest whole number. Use 3.14 for π.

1.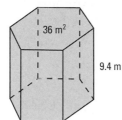

36 m²

9.4 m

2.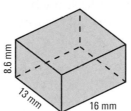

8.6 mm

13 mm

16 mm

3.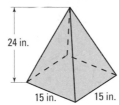

24 in.

15 in.   15 in.

4.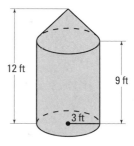

12 ft

9 ft

3 ft

5. How many cubic centimeters of water can a fish tank hold, if the tank is a rectangular prism 60 cm long, 40 cm wide, and 25 cm high?

Chapter 6

Find the slope of the line containing the given points.

1. $C(3, -1)$ and $D(0, -1)$ **2.** $M(-2, 4)$ and $N(5, 6)$ **3.** $S(-5, 0)$ and $T(4, -3)$

4. $X(-5, -3)$ and $Z(5, 5)$ **5.** $J(6, 2)$ and $K(0, 18)$ **6.** $P(-7, -3)$ and $Q(-2, 17)$

7. $Q(4, -1)$ and $R(-5, -3)$ **8.** $E(-3, -2)$ and $F(-4, 2)$ **9.** $J(6, -4)$ and $K(-4, -4)$

Graph the line that passes through the given point P and has the given slope.

10. $P(-1, 4)$, $m = \frac{1}{3}$ **11.** $P(5, -2)$, $m = -\frac{3}{4}$ **12.** $P(-2, -3)$, $m = \frac{3}{2}$

Find the slope of the line.

13. $4x - 6y = 12$ **14.** $4x - 5y = -15$

15. $8x - y = 2$ **16.** $-x - 2y = 8$

17. $5x = -2y + 7$ **18.** $2x = 20 - 6y$

19. Find the slope of a ramp that rises 8 ft for every 120 ft of horizontal run.

Find the slope and y-intercept for each line.

20. $y = \frac{2}{3}x - 5$

21. $y = -x + 9$

22. $y = 12x$

23. $2x + 8y = 16$

24. $-5x + 7y = 35$

25. $\frac{1}{2}x + 4y = -24$

Write an equation of the line with the given slope and y-intercept.

26. $m = 2, b = 6$

27. $m = \frac{3}{4}, b = 0$

28. $m = -3, b = -9$

29. $m = -\frac{1}{8}, b = \frac{1}{2}$

30. $m = 0, b = -7$

31. $m = -1, b = \frac{5}{9}$

Graph each equation.

32. $3x + 7y = 21$

33. $-2x + 8y = -32$

34. $y = -5x + 4$

35. Each week, the Weekly News prints 400 newspapers plus 20% of the total newspapers sold the previous week. The number of papers sold last week was 420. Write an equation to show how many newspapers will be printed this week. Solve. If 450 newspapers are sold this week, how many will be printed next week?

Extra Practice 6–2 • Parallel and Perpendicular Lines • pages 248–251

Find the slope of a line parallel to the given line and of a line perpendicular to the given line.

1. the line containing (2, 3) and (4, 9)

2. the line containing (−1, 7) and (2, −3)

3. the line containing (0, 9) and (3, −6)

4. the line containing (−4, −3) and (0, −7)

5. the line containing (−3, 0) and (−5, −3)

6. the line containing (4, −2) and (7, −6)

Determine whether each pair of lines is *parallel*, *perpendicular*, or *neither*.

7. the line containing points $C(-2, 5)$ and $D(5, 9)$
   the line containing points $E(2, 2)$ and $F(6, -5)$

8. the line containing points $M(-4, -2)$ and $N(3, -8)$
   the line containing points $O(-6, 3)$ and $P(1, -3)$

9. $7x - y = 4; -14x + 2y = 6$

10. $\frac{1}{2}x + 5y = 20; 2x + 10y = 15$

11. $x - y = 3; 3x - 4y = 9$

12. $4y = -x + 14; 8x - 2y = -10$

13. $4y + 10 = 6x; 3x - 2y = 10$

14. $6x - 10y = 20; 10x + 6y = 24$

15. Plot and connect the points $A(4, 5)$, $B(-4, 1)$, $C(2, -3)$, and $D(-1, 8)$. Determine whether $ABCD$ is a square.

Write an equation of the line with the given slope and y-intercept.

1. $m = 4, b = -1$
2. $m = -2, b = 7$
3. $m = -\dfrac{3}{4}, b = 0$

4. $m = \dfrac{1}{5}, b = -5$
5. $m = -8, b = \dfrac{1}{2}$
6. $m = -\dfrac{2}{3}, b = -\dfrac{4}{9}$

7. $m = \dfrac{3}{4}, b = -4$
8. $m = \dfrac{5}{3}, b = \dfrac{3}{4}$
9. $m = -5, b = 0$

Write an equation of the line that has the given slope and passes through the given point.

10. $m = 3, A(3, -7)$
11. $m = -1, B(-5, 2)$

12. $m = -\dfrac{2}{3}, C(3, 3)$
13. $m = \dfrac{1}{2}, D(-4, -6)$

14. $m = 5, E(8, -2)$
15. $m = -6, F(-3, -9)$

Write an equation for the line whose graph is shown.

16.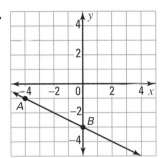

17.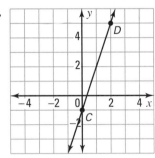

18.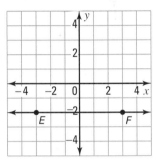

Determine the solution of each system of equations whose graph is shown.

1.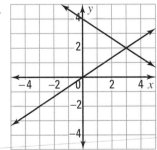

2.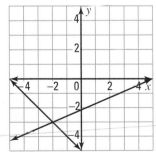

3.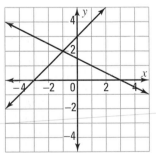

Solve each system of equations by graphing.

4. $y = 3x - 1$
   $y = -x + 3$

5. $\dfrac{1}{2}y = 2x + 2$
   $y = 3x + 2$

6. $x - y = -3$
   $2x - y = -4$

7. $2x + y = 3$
   $2x - 2y = -6$

8. $x + y = 3$
   $3x - y = 5$

9. $x = 3y + 6$
   $y = -2x + 5$

Solve and check each system of equations by the substitution method.

1. $x + 2y = 5$
$4x - 4y = 8$

2. $6x + y = 4$
$-x + 4y = 16$

3. $2x + 3y = 4$
$5x - 2y = -9$

4. $x + 3y = 5$
$4x + 8y = 16$

5. $-3x + 8y = -18$
$\frac{1}{2}x - y = 3$

6. $6x + 3y = 3$
$-x + 2y = -13$

7. $y = 3x + 8$
$x - 3y = 8$

8. $\frac{1}{2}x + y = 8$
$-3x + 6y = 0$

9. $y = 4x + 2$
$x + y = -3$

10. $y = 3x + 9$
$x = 8y - 3$

11. $4x + 3y = 3$
$-6y = 3 - 10x$

12. $y = -6x$
$3x + y = 2$

13. The perimeter of a rectangle is 96 in. If the length is three times the width, find the dimensions of the rectangle.

Solve each system of equations. Check the solutions.

1. $x + y = 5$
$-x + y = 1$

2. $6x + y = 13$
$4x - y = -3$

3. $x + 5y = 2$
$y = x + 4$

4. $6x + 3y = -15$
$-2x + 4y = 0$

5. $2x = 3y - 8$
$y = 6x$

6. $y = 3x - 14$
$2x + 3y = 2$

7. $y = -x + 1$
$x = y + 15$

8. $\frac{1}{3}x + y = 7$
$x - 2y = 1$

9. $-3x + 2y = -5$
$4y = -8 + 4x$

10. $4y = 3x + 3$
$\frac{1}{3}x = y - 2$

11. $5y = 10x - 5$
$3x + 2y = -9$

12. $x = 6y - 16$
$3y = x + 10$

13. Andrew has 25 coins with a total value of $3.05. The coins are all nickels and quarters. How many nickels and how many quarters does he have?

Determine whether the given ordered pair is a solution to the given system of inequalities.

1. $(2, 1)$; $2x + 5y \le 4$
$-x + 8y \le 4$

2. $(-3, 4)$; $3x + 3y > 2$
$2x - 6y \le 5$

3. $(1, 5)$; $4x - y < 5$
$-x + 3y > 0$

Write a system of linear inequalities for the given graph.

4.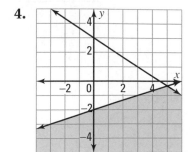

5.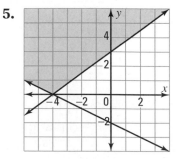

6.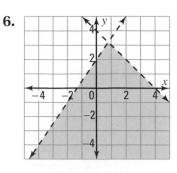

Graph the solution set of the system of linear inequalities.

7. $x \ge 2$
$y \le 3$

8. $y < 2x + 1$
$y > -x + 6$

9. $y - 3x \ge -5$
$y + \frac{2}{3}x < -1$

Determine the maximum value of $P = 2x - 3y$ for each feasible region.

1.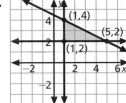

2.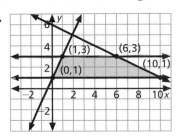

Determine the minimum value of $P = 15x + 12y$ for each feasible region.

3.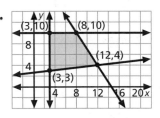

4.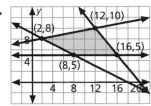

5. A receptionist for a veterinarian schedules appointments. He allots 20 min for a routine office visit and 40 min for a surgery. The veterinarian cannot do more than 6 surgeries per day. The office has 7 h available for appointments. If x represents the number of office visits and y represents the number of surgeries, the income for a day is $55x + 125y$. What is the maximum income for one day?

Chapter 7

Extra Practice 7–1 • Ratios and Proportions • pages 296–299

Is each pair of ratios equivalent? Write *yes* or *no*.

1. 4:8, 12:24   **2.** 14:18, 9:7   **3.** $\frac{2.7}{3.6}, \frac{6}{8}$ **4.** $\frac{1.5}{2.4}, \frac{10}{25}$

5. 10 to 7, 30 to 14   **6.** 12 to 8, 9 to 6   **7.** $\frac{18}{10}, \frac{54}{30}$ **8.** $\frac{4}{5}, \frac{2}{10}$

Solve each proportion.

9. $\frac{2}{9} = \frac{16}{x}$ **10.** $\frac{2}{6} = \frac{11}{x}$ **11.** $\frac{a}{16} = \frac{9}{1.8}$ **12.** $\frac{7.2}{6} = \frac{n}{5}$

13. $8.4:12 = 2.1:x$ **14.** $6:1.9 = n:7.6$ **15.** $\frac{15}{y} = \frac{4.8}{6.4}$ **16.** $\frac{7}{8} = \frac{k}{12}$

Use a calculator to solve these proportions.

17. $\frac{126}{21} = \frac{120}{x}$ **18.** $\frac{154}{231} = \frac{x}{99}$ **19.** $\frac{x}{325} = \frac{429}{165}$ **20.** $\frac{137}{x} = \frac{118}{354}$

21. A recipe for a sport drink calls for 3 parts cranberry juice to 8 parts lime juice. How much cranberry juice should be added to 20 pt of lime juice?

22. Cashew nuts cost $3 for 0.25 lb. How much will $1\frac{1}{2}$ lb cost?

23. Two college roommates share the cost of an apartment in a ratio of 5:6. The total monthly rent is $825. What is each person's share?

Extra Practice

Determine if the polygons are similar. Write *yes* or *no*.

1.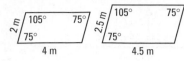

2.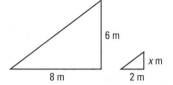

3.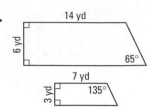

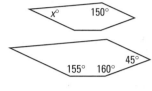

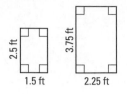

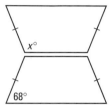

Find the value of *x* in each pair of similar figures.

4.

6 m

8 m   *x* m

2 m

5.

x°   150°

155° 160° 45°

6.

x°

68°

7. Draw any acute angle. Copy the angle using a straightedge and compass.

8. A photograph that measures 5 in. by 8 in. is enlarged so that the 8 in. side measures 10 in. How long is the 5-in. side in the enlargement?

Find the actual length of each of the following.

1. scale length is 5 cm
   scale is 2 cm:10 m

2. scale distance is 6.25 cm
   scale is 2.5 cm:10 m

3. scale length is $10\frac{1}{2}$ in.
   scale is $\frac{1}{4}$ in.:1 ft

Find the scale length for each of the following.

4. actual length is 15 ft
   scale is $\frac{1}{4}$ in.:1 ft

5. actual distance is 300 mi
   scale is 2 cm:50 mi

6. actual distance is 1.5 mi
   scale is 1 in.:0.6 mi

Find the actual distance using the map.

Hillsboro   Sanford

Franklin

Springvale   Lewiston

Scale: ▬▬ 6 mi

7. Franklin to Springvale

8. Sanford to Lewiston

9. Hillsboro to Franklin

10. Hillsboro to Lewiston

11. Springvale to Lewiston

12. Sanford to Franklin

Determine whether each pair of triangles is similar. If the triangles are similar, give a reason: write AA, SSS, or SAS.

1.

2.

3.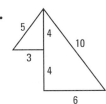

4.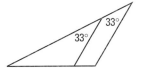

5.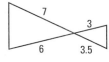

6.

7. The drawing at the right shows a smokestack and its shadow and a flagpole and its shadow. Explain why $\triangle PQR \sim \triangle MNO$.

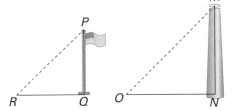

1. Copy and complete this proof.
   Given: $\triangle PQR \sim \triangle STV$; $\overline{PV} \cong \overline{VR}$; $\overline{SW} \cong \overline{WV}$
   Prove: $\dfrac{QV}{TW} = \dfrac{PR}{SV}$

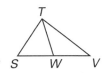

Statements	Reasons
1. $\triangle PQR \sim \triangle STV$; $\overline{PV} \cong \overline{VR}$; $\overline{SW} \cong \overline{WV}$	1. ___?___
2. V is the midpoint of $\overline{PR}$. W is the midpoint of $\overline{SV}$.	2. ___?___
3. $\overline{QV}$ is a median of $\triangle PQR$. $\overline{TW}$ is a median of $\triangle STV$.	3. ___?___
4. $\dfrac{QV}{TW} = \dfrac{PR}{SV}$	4. ___?___

Find x in each pair of similar triangles to the nearest tenth.

2.

3.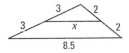

4.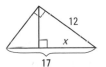

5. These cross sections of tents are similar triangles. If the support pole of the smaller tent is 4 ft, how tall is the support pole for the larger tent?

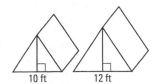

Extra Practice

In each figure, $\overline{AB} \parallel \overline{CD}$. Find the value of x to the nearest tenth.

1.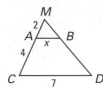

2.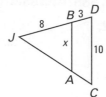

3.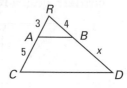

4.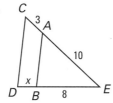

5.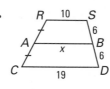

6.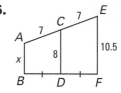

7. This map shows a vacant plot of land that is to be developed by creating four new equally-spaced north-south streets between Elm and Birch Streets. Copy the map and construct the points where the new streets would intersect Spruce Street.

Chapter 8

Extra Practice 8–1 • Translations and Reflections • pages 338–341

On a coordinate plane, graph $\triangle ABC$ with vertices $A(3, -2)$, $B(2, -7)$, and $C(9, -5)$. Then graph its image under each transformation from the original position.

1. 6 units up

2. reflected across the y-axis

3. Compare the slopes of all the sides of $\triangle ABC$ in both positions above.

On a coordinate plane, graph figure $WXYZ$ with vertices $W(-3, 9)$, $X(1, 7)$, $Y(-1, 2)$, and $Z(-5, 4)$. Then graph its image under each transformation from the original position.

4. 7 units right

5. reflected across the x-axis

6. Compare the slopes of $\overline{WX}$, $\overline{W'X'}$, and $\overline{W''X''}$.

7. On a coordinate plane, graph $\triangle RST$ with vertices $R(-4, 0)$, $S(1, -4)$, and $T(-6, -6)$. Graph its image under a reflection across the line with equation $y = -x$.

8. On a coordinate plane, graph figure $MNOP$ with vertices $M(-2, 1)$, $N(-4, -3)$, $O(-9, -1)$, and $P(-7, 2)$. Graph its image under a reflection across the line with equation $y = x$.

For each figure, draw the image after the given rotation about the origin. Then calculate the slope of each side before and after the rotation.

1. Use the rule $(-x, -y)$ for a 180° clockwise rotation.

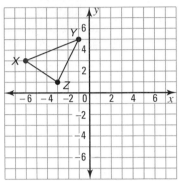

2. Use the rule $(-y, x)$ for a 90° counterclockwise rotation.

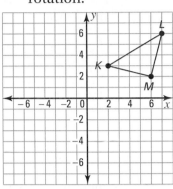

3. Use the rule $(y, -x)$ for a 90° clockwise rotation.

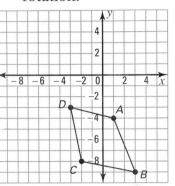

4. Triangle *XYZ* is rotated twice about the origin, as shown in the table below. Compare the slopes to determine how much of a rotation was completed each time. Each rotation is at most one full turn.

Original Position		After Rotation 1		After Rotation 2	
side	slope	side	slope	side	slope
YZ	-2	Y'Z'	$\frac{1}{2}$	Y''Z''	-2
XY	$\frac{1}{3}$	X'Y'	-3	X''Y''	$\frac{1}{3}$
XZ	5	X'Z'	$-\frac{1}{5}$	X''Z''	5

Copy each graph on graph paper. Then draw each dilation image.

1. Center of dilation: origin
   Scale factor: 2

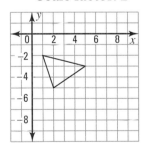

2. Center of dilation: point *A*
   Scale factor: 3

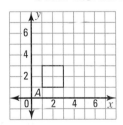

3. Center of dilation: origin Scale factor: $\frac{1}{2}$

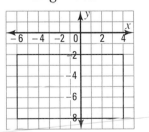

The following sets of points are the vertices of figures and their dilation images. For each two sets of points, give the scale factor.

4. $A(2, 0)$, $B(6, 0)$, $C(4, 4)$
   $A'(4, 0)$, $B'(12, 0)$, $C'(8, 8)$

5. $R(-2, 1)$, $S(-2, -7)$, $T(-10, 1)$
   $R'(-2, 1)$, $S'(-2, -3)$, $T'(-6, 1)$

6. $J(-8, -3)$, $K(-5, -3)$,
   $L(-5, -7)$, $M(-8, -7)$
   $J'(-8, -3)$, $K'(1, -3)$,
   $L'(1, -15)$, $M'(-8, -15)$

7. $D(2, -4)$, $E(8, -4)$, $F(8, -7)$, $G(2, -7)$
   $D'(6, -6)$, $E'(8, -6)$, $F'(8, -7)$, $G'(6, -7)$

For each exercise, draw the result of the first transformation as a dashed figure and the result of the second transformation in red.

1. a reflection over the x-axis followed by a translation 6 units to the left.

2. a clockwise rotation of 90° about the origin, followed by a reflection over the y-axis.

3. a counterclockwise rotation of 180° about the origin, followed by a dilation with center at the origin and a scale factor of 2.

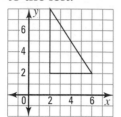

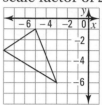

Determine the transformations necessary to create figure 2 from figure 1. There may be more than one possible answer.

4.

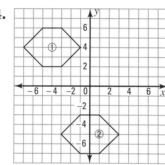

5.

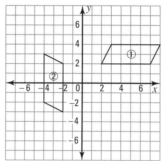

6.

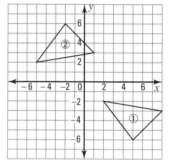

Find the dimensions of each matrix.

1. $\begin{bmatrix} 2 & 3 \\ 4 & 5 \end{bmatrix}$

2. $\begin{bmatrix} 3 \\ 6 \\ 9 \\ 5 \end{bmatrix}$

3. $\begin{bmatrix} 6 & 1 \\ 4 & 8 \\ 9 & 5 \end{bmatrix}$

Use the following matrices in Exercises 4–12.

$$J = \begin{bmatrix} 2 & 0 \\ 1 & 5 \\ -4 & -6 \end{bmatrix} \qquad K = \begin{bmatrix} -6 & 9 \\ 4 & -3 \\ 0 & 8 \end{bmatrix} \qquad L = \begin{bmatrix} 11 & -5 \\ -7 & 0 \\ -2 & -1 \end{bmatrix}$$

Find each of the following.

4. $K + L$

5. $J + K$

6. $J + L$

7. $-3L$

8. $\frac{1}{2}J$

9. $-\frac{1}{3}K$

10. $2J + K$

11. $L + (-3J)$

12. $\frac{1}{2}K + L$

13. Tyler Junior High School ordered school pennants. The seventh grade ordered 28 black, 24 white, and 16 green. The eighth grade ordered 30 black, 20 white, and 15 green. The ninth grade ordered 14 black, 25 white, and 27 green. Write two different 3 × 3 matrices to show this information.

Refer to the matrices below. Find the dimensions of each product, if possible. *Do not multiply.* If not possible to multiply, write NP.

$$A = \begin{bmatrix} 4 & 8 \\ 6 & 1 \\ 0 & 5 \end{bmatrix} \qquad B = [5 \quad 1 \quad 3] \qquad C = \begin{bmatrix} 4 & 5 & 3 \\ 8 & 9 & 6 \end{bmatrix} \qquad D = \begin{bmatrix} 3 & 9 \\ 5 & 7 \end{bmatrix}$$

1. AB **2.** AC **3.** AD **4.** BC

5. CD **6.** DC **7.** BA **8.** CA

Find each product. If not possible, write NP.

9. $\begin{bmatrix} 1 \\ 4 \\ -2 \end{bmatrix} [-5 \quad -1 \quad 3]$ **10.** $[4 \quad -1 \quad 0] \begin{bmatrix} 2 & 6 \\ 0 & -3 \\ 7 & 1 \end{bmatrix}$

11. $\begin{bmatrix} 3 \\ -2 \end{bmatrix} [-2 \quad 0]$ **12.** $[8 \quad -1] \begin{bmatrix} 3 & 2 \\ -2 & -1 \end{bmatrix}$

13. $\begin{bmatrix} -4 & 0 & 1 \\ 2 & 5 & -2 \\ 1 & -1 & 3 \end{bmatrix} \begin{bmatrix} 3 & 5 \\ -2 & 1 \\ 0 & -3 \end{bmatrix}$ **14.** $\begin{bmatrix} 0 & -2 & -5 \\ 1 & 0 & -3 \end{bmatrix} \begin{bmatrix} 3 & -2 & -5 \\ 4 & 3 & 0 \end{bmatrix}$

Represent each geometric figure with a matrix.

1. 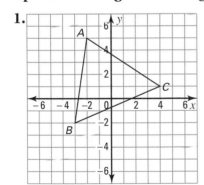   **2.** 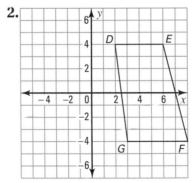   **3.**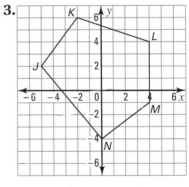

Find the reflection images of the triangle represented by $\begin{bmatrix} -2 & -5 & -7 \\ -4 & -1 & -5 \end{bmatrix}$.

4. over the y-axis   **5.** over the x-axis   **6.** over the line $y = -x$

Find the reflection images of the quadrilateral represented by $\begin{bmatrix} 2 & 4 & 7 & 8 \\ -5 & -2 & -5 & -9 \end{bmatrix}$.

7. over the line $y = x$ **8.** over the x-axis   **9.** over the y-axis

Interpret each equation as indicating: *The reflection image of point* ___?___ *over* ___?___ *is the point* ___?___ .

10. $\begin{bmatrix} -1 & 0 \\ 0 & 1 \end{bmatrix} \begin{bmatrix} -3 \\ 5 \end{bmatrix} = \begin{bmatrix} 3 \\ 5 \end{bmatrix}$ **11.** $\begin{bmatrix} 0 & -1 \\ -1 & 0 \end{bmatrix} \begin{bmatrix} -1 \\ -4 \end{bmatrix} = \begin{bmatrix} 4 \\ 1 \end{bmatrix}$

Chapter 9

Extra Practice 9–1 • Review Percents and Probability • pages 384–387

A spinner with 8 equal sectors labeled A through H is spun 100 times with the following results.

Outcome	A	B	C	D	E	F	G	H
Frequency	8	14	12	17	9	15	14	11

What is the experimental probability of spinning each of the following results?

1. B   **2.** E   **3.** H   **4.** C

5. a letter that comes before D   **6.** a letter that comes after D

List all the elements of the sample space for each of the following experiments.

7. You toss a penny and a dime.

8. You spin each of these spinners once.

Find the probability of each of the following.

9. Drawing a ten of hearts from a standard deck of cards.

10. Rolling a die and getting a prime number.

Extra Practice 9–3 • Compound Events • pages 392–395

Two dice are rolled.

1. Find the probability that the sum of the numbers rolled is either 4 or 5.

2. Find the probability that the sum of the numbers rolled is even and less than 7.

3. Find P(not a prime).   **4.** Find P(a sum of 8 or not prime).

Ashante's Little League coach chooses the line-up by placing the 9 names into a hat and then pulling them out one by one. Find each probability.

5. batting first or third   **6.** not batting in an even-numbered position

7. batting last or in the first two thirds of the batting order

8. batting second or in the first third of the order

You spin this spinner. Find each probability.

9. spinning 4 or an odd number

10. spinning a prime or an odd number

11. spinning a prime or an even number

12. spinning a prime or a number greater than 4

A bag contains marbles, all the same size. There are 5 red, 4 blue, 2 yellow, and 1 green. Marbles are drawn at random from the bag, one at a time, and then replaced. Find each probability.

1. *P*(red, then blue)
2. *P*(blue, then yellow)
3. *P*(green, then red)
4. *P*(blue, then not blue)
5. *P*(not green, then yellow)
6. *P*(green, then not red)

A box contains tennis balls. There are 4 white, 3 yellow, 1 green, and 2 pink. One ball at a time is taken at random from the box and not replaced. Find each probability.

7. *P*(green, then yellow)
8. *P*(white, then pink)
9. *P*(white, then not white)
10. *P*(yellow, then green)
11. *P*(green, then not white)
12. *P*(white, then not green)

A neon sign reading HOTEL CHELSEA has two of its letters go out.

13. What is the probability that both letters are vowels?

14. What is the probability that the first letter is an E and the second is also an E?

15. What is the probability that the first is L and the second is not L?

16. You are given tickets to two concerts at a theater with 3000 seats. What is the probability that you will sit in the orchestra section for the first concert, and then in the second balcony for the second concert, if the orchestra has 1800 seats and the second balcony has 600 seats?

For each situation, tell whether order does or does not matter.

1. You are recording the numbers and letters in an e-mail address.

2. You are at a video store and selecting 3 movies to rent for the weekend.

3. You are selecting candidates for president, vice president, and secretary.

4. You are selecting a 5-member committee from students in the class.

5. There are 5 different library books you would like to borrow, but the library allows you to borrow only 3 books at a time. How many ways can you select 3 of the books?

6. How many different ways can you arrange six videos in a row on a shelf?

7. A restaurant menu states that when you buy a dinner special you can select 3 side orders from 12 that are listed. How many ways can you do this?

8. Nine teams take part in an intramural volley ball tournament. How many different arrangements of first-, second-, and third-place winners are possible?

This table shows the average scores of students who participated in an annual math contest.

Year	1995	1996	1997	1998	1999	2000	2001	2002	2003	2004
Average Score	8.2	7.5	7.9	8.7	5.6	6.2	7.8	6.4	7.0	7.2

1. Make a scatter plot of the data.

2. What is the range of average scores?

3. Does your scatter plot show a positive correlation, a negative correlation, or no correlation?

These box-and-whisker plots show the weights of the members of three high school football teams.

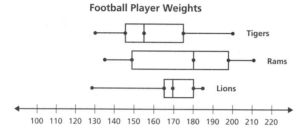

Football Player Weights

4. Which team has the highest median weight?

5. What is the lower quartile of the Tigers' weights?

6. What is the upper quartile of the Lions' weights?

7. What is the median weight of the Tigers?

8. Which team had the least range of weights? the greatest?

9. Which team had weights most closely clustered about its median?

Compute the variance and standard deviation for each set of data.

1. 3, 6, 9, 12, 15
2. 5, 5, 5, 5, 5
3. 0.5, 2.5, 4.5, 6.5, 8.5
4. 1, 3, 5, 7, 9
5. 4, 7, 10, 13, 16
6. 4, 8, 12, 16, 20
7. 2.3, 4.3, 6.3, 8.3, 10.3
8. 1.6, 5.6, 9.6, 13.6, 17.6
9. 7.6, 3.4, 6, 8.3, 5.7

Find the variance and standard deviation for each set of data.

10. The top five scores in an Olympics gymnastics trial were: 9.1, 8.5, 7.9, 8.2, and 8.5.

11. The prices of lunches in five country high schools are: $3.00, $3.50, $2.75, $3.25, and $3.75.

12. Ping took two tests. On the first test, his score was 78, while the mean score was 72 and the standard deviation was 3. On the second test, his score was 70, while the mean score was 68 and the standard deviation was 0.5. On which test did Ping score better, relative to the scores of his classmates?

13. Alicia took two tests. On Test A, her score was 82, while the mean score was 90 and the standard deviation was 10. On Test B, her score was 76, while the mean score was 82 and the standard deviation was 4. On which test did Alicia score better, relative to the scores of her classmates?

Chapter 10

Find the value to the nearest hundredth.

1. $\sqrt{13}$
2. $\sqrt{30}$
3. $\sqrt{62}$
4. $\sqrt{150}$

5. $\sqrt{189}$
6. $\sqrt{270}$
7. $\sqrt{666}$
8. $\sqrt{106}$

Write each in simplest radical form.

9. $\sqrt{45}$
10. $\sqrt{32}$
11. $\sqrt{147}$

12. $\sqrt{52}$
13. $\sqrt{28}$
14. $\sqrt{162}$

15. $\sqrt{125}$
16. $\sqrt{360}$
17. $(3\sqrt{5})(2\sqrt{10})$

18. $(4\sqrt{3})(2\sqrt{6})$
19. $(2\sqrt{5})^2$
20. $(2\sqrt{3})(-4\sqrt{7})$

21. $(-2\sqrt{2})(5\sqrt{8})$
22. $\dfrac{5}{\sqrt{7}}$
23. $\dfrac{\sqrt{64}}{\sqrt{2}}$

24. $\dfrac{\sqrt{45}}{\sqrt{9}}$
25. $\dfrac{4\sqrt{3}}{5\sqrt{6}}$
26. $-\dfrac{2\sqrt{6}}{5\sqrt{8}}$

27. $-\dfrac{3\sqrt{5}}{4\sqrt{15}}$
28. $\dfrac{4\sqrt{3}}{2\sqrt{8}}$
29. $-\sqrt{\dfrac{5}{6}}$

30. If the area of a square is 236 ft², find the length of each side to the nearest tenth of a foot.

Use the Pythagorean Theorem to find the unknown length. Round your answers to the nearest tenth.

1. 4 in. 5 in.
2. 7 cm / 7 cm
3. 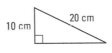 10 cm 20 cm
4. 15 yd / 13 yd

5. 3 ft 18 ft
6.  12 cm / 28 cm
7. 21 in. 35 in.
8. 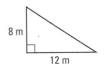 8 m / 12 m

Determine if each figure is a right triangle. Write _yes_ or _no_.

9. 15 ft 25 ft / 22 ft
10.  8 m 17 m / 15 m
11. 5 yd 13 yd / 12 yd
12. 18 in. 30 in. / 28 in.

Solve. Round your answers to the nearest tenth.

13. Find the length of a diagonal of a rectangle with a length of 24 ft and a width of 8 ft.

14. Find the width of a rectangle with a length of 9 m and a diagonal length of 11 m.

Find the unknown side measures. First find each in simplest radical form and then find each to the nearest tenth.

1.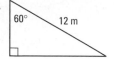

2.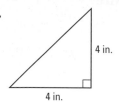

3.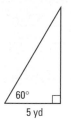

4.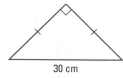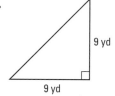

5.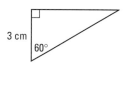

6.

7.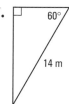

8.

9. The diagonal of a square measures 6 cm. Find the length of a side of the square to the nearest tenth.

10. The side of a square measures 10 in. Find the length of the diagonal of the square to the nearest tenth.

Find x.

1.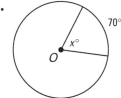

2.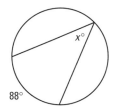

3.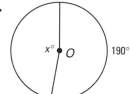

4.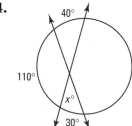

5.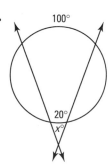

6.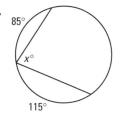

7.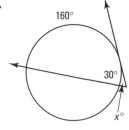

8.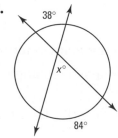

9.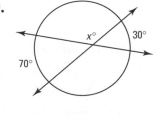

10. An inscribed angle intercepts an arc of 130°. What is the measure of the inscribed angle?

11. An inscribed angle measures 48°. What is the measure of the arc it intercepts?

12. An inscribed angle ∠*ABC*, measures 74°. What is the measure of the major arc *ABC*?

13. An inscribed angle ∠*JKL*, measures 50°. What is the measure of the central angle that contains the points *J* and *L*?

Extra Practice 10–6 • Circles and Segments • pages 448–451

Find *x*.

1. 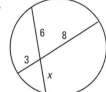 **2.** 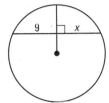 **3.** 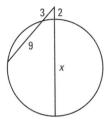 **4.**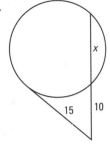

5. In circle *O*, two chords, $\overline{AB}$ and $\overline{CD}$ intersect at point *K*. $\overline{AK}$ = 12 cm, $\overline{KB}$ = 10 cm, and $\overline{KD}$ = 8 cm. Find the measure of $\overline{CK}$.

6. In circle *O*, radius $\overline{OM}$ is perpendicular to chord $\overline{JK}$ at point *L*. Find the measure of $\overline{JK}$ if $\overline{JL}$ = 18 cm.

Extra Practice 10–7 • Constructions with Circles • pages 454–457

1. Construct a regular hexagon.

2. Construct a square.

3. Copy this equilateral triangle. Inscribe the triangle in a circle.

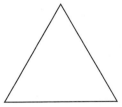

4. Copy this square. Inscribe the square in a circle.

5. Copy this regular pentagon. Inscribe the pentagon in a circle.

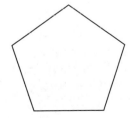

6. Copy this regular hexagon. Circumscribe the hexagon around a circle.

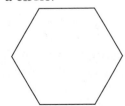

Chapter 11

Extra Practice 11–1 • Add and Subtract Polynomials • pages 468–471

Simplify.

1. $(3a + 7) + (4a + 5)$
2. $(6n + p) + (n + 7p)$
3. $(4x^2 + 3x) + (-x^2 + x)$
4. $(-3n^2 - 4n) + (3n^2 - 4n)$
5. $(3j + 4k - 2) + (2j - 2k + 5)$
6. $(4x^2 + 12x + 9) + (x^2 - 2x + 1)$
7. $(7a + 5) - (2a + 3)$
8. $(8x - y) - (5x + y)$
9. $(3x^2 + 4y) - (-x^2 + y)$
10. $(a^2 - 5a - 6) - (-a^2 + a + 12)$
11. $(ab - b + a) - (ab - b + a)$
12. $(4d^2 - 2de + e^2) - (3d^2 - de - 3e^2)$
13. $(m^2 - 16n + 3n^2) - (9n - 3m^2 + n^2)$
14. $(a^2 + 12b + b^2) + (4a^2 + 8b + 4b^2)$
15. $(5c^2 + 8cd + d^2) - (c^2 - 2c)$
16. $(7p^2 - 5q^2) - (p^2 - 6pq + q^2)$

17. Last week, Marisol worked 10 h at her part-time job, where she earns x dollars per hour, and 35 h at her full-time job, where she earns y dollars per hour. This week, she worked 15 h at her part-time job and 35 h at her full-time job. What were her earnings during the two weeks, expressed in terms of x and y?

Extra Practice 11–2 • Multiply by a Monomial • pages 472–475

Simplify.

1. $(3x)(4y)$
2. $(5a)(4)$
3. $(a^2)(2ab^3)$
4. $(m)(-3mn)$
5. $(-c)(-6cd)$
6. $(2p^2)(-3p^2q)$
7. $(7x^2y)(8xy^2)$
8. $(4c^2)^2$
9. $2x(7x^2 - 6y)$
10. $-8n(2n^2 - 5p)$
11. $2a^2[-(a^2 + a)]$
12. $3k^2[-(3k^2 - k)]$
13. $7cd(2c^2 + 3d^2)$
14. $-8c^2d(c^3d - c^2d)$
15. $3a^2b(5a^3c - 3ab^4)$
16. $9x^2yz(x^2y - y^3z)$
17. $3jkl(j^2k^2l - jkl^3)$
18. $-9abc^3(a^2bc^3 - ab^4c)$
19. $-15xyz(-xyz - x^2y^3z^2)$
20. $3d(d^3 + 2d^2 - d)$
21. $4c(c^2 + 5c - 6)$
22. $rs(5r^2 - 3rs + 4)$
23. $-xy(3a^2 - 2b + c)$
24. $-rs^2(4r^2 - rs + 3s^2)$
25. $2xy^4z^3(x^3yz^2 - x^2y^2z^3 - xy^5z^2)$

26. In 2001, a supermarket employed x clerks, each of whom earned y dollars per week. The weekly pay rate increased by d dollars each year. In 2004, the number of clerks quadrupled. What was the total paid each week to the staff of clerks in 2001? What was it in 2004?

Find the factors for the following.

1. $4x + 6y$

2. $8a^2 - 12b^2$

3. $24n^2 + 6$

4. $7xy + 21x$

5. $jk + jkl$

6. $7pq - 21q$

7. $7ab - 4bc$

8. $9d^2e - 5e^2$

9. $13x^2y + 15yz^2$

10. $v^2w + vw$

11. $3a^5 + 3a^5 + 3a^2b^2$

12. $4y - 4y^6$

Find the GCF and its paired factor for the following.

13. $48a + 56b$

14. $39x + 13x^2$

15. $18c^2 - 27cd^2$

16. $28xy^2 - 42yz^2$

17. $30m^4n^3 - 60m^3n^4$

18. $60x^3 + 45x^2$

19. $6a^3b + 12a^2b^2$

20. $8x^4b^3 + 12x^3b^2$

21. $r^2s^2 + r^2s + rs^2$

22. $xa^3b^3 + ya^2b^2 + 2ab$

23. $16d^5 + 40d^4e^2 - 56d^2e^4$

24. $63x^3y - 56w^3z + 28r^3t$

25. A carpenter has two planks of wood, one 8 in. long and the other 40 in. long. She wants to use all the wood in the two planks to cut a set of small pieces, each the same size and as long as possible. How long will each cut piece be, and how many can she cut?

Simplify.

1. $(3a + 4b)(5c - 2)$

2. $(6x - 5y)(z - 3)$

3. $(2r + 7s)(r + 2t)$

4. $(6a - 7p)(2a + 3q)$

5. $(m - 4n)(3p + 4n)$

6. $(2x + 5y)(2z - w)$

7. $(8r + 3s)(5r - 2t)$

8. $(8m - n)(2p + n)$

9. $(4x + 3y)(x + 5y)$

10. $(6 - 5n)(3 - 2n)$

11. $(5x - 3)(x - 3)$

12. $(x + 7y)(3x + y)$

13. $(c - 7d)(5c - 2d)$

14. $(7x + 1)(6x - 7)$

15. $(6x - y)(5x + 3y)$

16. $(m - 4)(3m + 5)$

17. $(a + 3b)(a + 3b)$

18. $(c - 9)(c - 9)$

19. $(4x + 5y)(4x + 5y)$

20. $(7c - e)(7c - e)$

21. $(d - 4)(d + 4)$

22. $(5a + 2)(5a - 2)$

23. $(y + 1)(y - 1)$

24. $(g - 6)(g + 6)$

25. $(4m + 1)(4m - 1)$

26. $(7x + 2y)(7x - 2y)$

27. The dimensions of a rectangle are $(3x + 2)$ ft and $(x + 5)$ ft. Write an expression for the area of the rectangle.

Find factors for the following.

1. $6wy + 9wz + 4xy + 6xz$

2. $10ac + 2bc + 15ad + 3bd$

3. $rt + rv + 3st + 3sv$

4. $x^2 + 7x + 6xy + 42y$

5. $8n^2 + 2np + 4nq + pq$

6. $3y^2 + xy - 12yz - 4xz$

7. $n^2 - 4n + 2mn - 8m$

8. $6k^2 - 8k + 3km - 4m$

9. $12a^2c - 8a^2d^2 - 3bc + 2bd^2$

10. $2r^3 - 4r^2t + 5rt - 10t^2$

11. $24w^2y - 16w^2z^2 + 9xy - 6xz^2$

12. $8xz + 12x - 6yz - 9y$

13. $4a^2b - a^2c + 28b - 7c$

14. $4a^2 - 24ab + 5a - 30b$

15. $2ac + 2ad - 3bc - 3bd$

16. $3ac^2 + 15bc^2 + 4ab + 20b^2$

17. $6ac + 9ad + 6ae + 2bc + 3+ 2be$

18. $16eg + 12eh + 8e^2 - 4fh - 2ef$

19. A rectangle has an area that can be expressed as $6a^2 + 2ab + 3ac + bc$. If the width can be expressed as $3a + b$, find an expression for the length.

20. The area of a rectangle can be expressed as $3s^2 + 2rs - 6st - 4rt$. Find expressions that might represent the dimensions of the rectangle.

Find binomial factors of the following, if possible. (Four do not have such factors.)

1. $y^2 + 2y + 1$

2. $x^2 + 18x + 81$

3. $x^2 + 22x + 121$

4. $a^2 - 14a + 49$

5. $n^2 - 12n + 36$

6. $y^2 - 7y + 49$

7. $d^2 - 64$

8. $r^2 - 1$

9. $4n^2 - 9$

10. $9j^2 + 6j + 1$

11. $25d^2 - 20d + 4$

12. $16 - 24c + 9c^2$

13. $4c^2 - 28cd + 49d^2$

14. $81r^2 - s^2$

15. $25p^2 - 144q^2$

16. $121k^2 - 66kl + 9l^2$

17. $64a^2 + 25b^2$

18. $64x^2 - 80xy + 25y^2$

19. $25c^2 - 64d^2$

20. $4x^2 + 20xy + 25y^2$

21. $81s^2 - 50t^2$

22. $49p^2 - 140pq + 100q^2$

23. $9x^2 + 36x + 64$

24. $64c^2 + 16c + 1$

Find a monomial factor and two binomial factors for each of the following.

25. $12x^2 - 27$

26. $36x^2 + 24x + 4$

27. $x^3 - 6x^2 + 9x$

28. The expression for the area of a certain square is $14x + 49 + x^2$. Find an expression for the length of a side of the square.

Identify binomial second-term factors for the following.

1. $x^2 + 6x + 8$

2. $m^2 - 11m + 24$

3. $x^2 - x - 30$

4. $y^2 - 11y + 30$

5. $d^2 + 6d - 27$

6. $p^2 - 15p + 44$

7. $a^2 + 17ab + 72b^2$

8. $x^2 + 17xy + 42y^2$

9. $r^2 - 15rt + 54t^2$

10. $j^2 + 3jk - 54k^2$

11. $t^2 - tr - 30r^2$

12. $l^2 - 19ln + 18n^2$

Identify binomial second-term signs for the following.

13. $x^2 + x - 20$

14. $t^2 + 14t + 33$

15. $a^2 - a - 56$

16. $c^2 - 24c + 23$

17. $a^2 + 2ab - 15b^2$

18. $j^2 - 3jk - 10k^2$

Factor the following trinomials.

19. $z^2 - 37z + 36$

20. $r^2 + 15r + 36$

21. $x^2 - 9x - 36$

22. $v^2 - 16vw - 36w^2$

23. $f^2 - 13f + 36$

24. $l^2 - 20lm + 36m^2$

25. $g^2 + 5g - 36$

26. $j^2 + 9jk - 36k^2$

27. $h^2 - 5h - 36$

28. A rectangular trench x feet deep is being dug for the foundation of a wall. The area of the bottom of the trench is $(x^2 + 22x - 48)$ ft^2. Compare the depth of the trench to its width. Then compare the depth of the trench to its length.

Find FOIL coefficients for the following trinomials.

1. $6x^2 + 19x + 10$

2. $8m^2 - 30m - 27$

3. $6c^2 - 11c + 3$

4. $21a^2 + 13a + 2$

5. $10x^2 - 23x + 12$

6. $18n^2 + 9n - 2$

Find binomial factors for the following trinomials.

7. $25x^2 - 25x + 4$

8. $6x^2 - 5x - 4$

9. $14n^2 + 5n - 24$

10. $81y^2 - 24y - 20$

11. $15a^2 + 38ab + 24b^2$

12. $36t^2 - 19t - 6$

13. $10x^2 - 23x - 14$

14. $16x^2 - 41x + 25$

15. $8a^2 - 14ab + 3b^2$

16. $56r^2 - 6rs - 2s^2$

17. $2m^2 + 9m - 81$

18. $9k^2 + 27k + 8$

19. A rectangle has an area of $12a^2 + a - 1$. Find expressions that might be the length and width of the rectangle.

Chapter 12

Graph each function for the domain of real numbers. For each graph, give the coordinates of the vertex.

1. $y = x^2$

2. $y = 3x^2 + 1$

3. $y = -2x^2 - 3$

4. $y = x^2 - 4$

5. $y = -x^2 + 3$

6. $y = 2x^2 - 1$

7. $y = 4x^2 - 1$

8. $y = -3x^2 + 1$

9. $y = -x^2 + 2$

Determine whether the graph of each equation below opens upward or downward.

10. $y = -3x^2 - 8$

11. $y = -x^2 - 4$

12. $y = 4x^2 - 5$

13. $y = -x^2 + 9$

14. $y = 2x^2 + 5$

15. $y = -9x^2 + 9$

16. $y = 6x^2 - 2$

17. $y = 3x^2 + 9$

18. $y = -5x^2 - 9$

19. $y = -5x^2 - 2$

20. $y = x^2 - 16$

21. $y = 5x^2 - 1$

22. Given the equations (a) $y = 2x^2 - 3$ and (b) $y = 2x^2 + 3$, explain the differences and similarities in the two graphs.

Estimate the coordinates of the vertex for each parabola by graphing the equation on a graphing calculator. Then use $x = \frac{-b}{2a}$ to find the coordinates.

1. $y = x^2 + 4x + 9$

2. $y = -2x^2 - 8x - 8$

3. $y = 2x^2 + 12x + 13$

4. $y - 5 = 2x^2$

5. $y = x^2 + 2x + 3$

6. $3x - 4 = x^2 + y$

7. $y = x^2 + 2x - 3$

8. $y = x^2 - \frac{1}{4}$

9. $y = x^2 - 4x - 5$

Find the vertex. Then graph each equation.

10. $y = 2x^2 + 12x + 18$

11. $y + 2 = x^2 - 4x$

12. $y + 3x^2 = 1 - 6x$

13. $y = x^2 - 6x + 4$

14. $y + 7 = x^2 + 5x$

15. $y = x^2 - 8x + 22$

16. $y = x^2 - 6x + 9$

17. $y = x^2 + 6x + 7$

18. $y = x^2 - 4x - 1$

19. Find a quadratic equation in the form $y = ax^2 + bx + c$ in which $c = -1$ and the vertex is $(2, -5)$.

Use a graphing calculator to determine the number of solutions for each equation. For equations with one or two solutions, find the exact solutions by factoring.

1. $0 = x^2 - 36$

2. $0 = x^2 - x - 2$

3. $0 = x^2 - x - 12$

4. $0 = 2x^2 + x + 1$

5. $\frac{1}{4} = x^2$

6. $0 = x^2 - 6x + 9$

7. $10 = x^2 - 3x$

8. $x^2 - 18 = 3x$

9. $0 = x^2 + 11x + 30$

10. $x^2 + 6x + 9 = 0$

11. $0 = x^2 - 0.5x - 3$

12. $0 = x^2 + 36$

13. $0 = x^2 - 6x - 16$

14. $0 = x^2 - x + 1$

15. $0 = x^2 - 4x + 5$

Write an equation for each problem. Then factor to solve.

16. The square of a number is 6 more than the number.

17. The square of a number is 4 more than 3 times the number.

18. The square of a number is 24 more than 2 times the number.

19. The square of a number is 27 more than 6 times the number.

Complete the square.

1. $x^2 + 12x$

2. $x^2 + 16x$

3. $x^2 - 4.2x$

4. $x^2 + \frac{2}{3}x$

5. $x^2 - 7x$

6. $x^2 + 8x$

7. $x^2 + 6x$

8. $x^2 + x$

9. $2x^2 - 5x$

Solve by completing the square. Check your solutions.

10. $x^2 + 12x + 11 = 0$

11. $x^2 - 2x - 3 = 0$

12. $x^2 - 8x - 9 = 0$

13. $2x^2 - 2x - 12 = 0$

14. $3x^2 + 11x - 4 = 0$

15. $5x^2 = 5$

16. $x^2 + 24x + 119 = 0$

17. $x^2 - 22x + 112 = 0$

18. $x^2 - 4x - 117 = 0$

19. The width of a rectangular pool is 5 m less than the length. The area is 24 m². Find the length and width.

20. The hypotenuse of a right triangle is 25 m. One leg is 17 m shorter than the other. Find the lengths of the legs.

21. The area of Harry's room is 132 ft². The length is 1 ft more than the width. Find the length and width.

Use the quadratic formula to solve each equation.

1. $x^2 - 5x = 0$

2. $x^2 - 7x + 6 = 0$

3. $x^2 - 6x = 0$

4. $2x^2 + 13x + 15 = 0$

5. $x^2 + 6x + 4 = 0$

6. $3x^2 + 8x + 5 = 0$

7. $x^2 - 2x - 15 = 0$

8. $x^2 - 7x - 30 = 0$

9. $5x^2 + 3x - 2 = 0$

10. $x^2 + x = 0$

11. $4x^2 = 20$

12. $3 = 5x^2 - 8x$

Choose factoring or the quadratic formula to solve each equation.

13. $0 = x^2 + 2x - 3$

14. $3x^2 - 8x = 3$

15. $x^2 - 2x = 24$

16. $0 = 3x^2 + 2x - 4$

17. $x^2 + 9 = 7x$

18. $x^2 = 2x + 15$

19. $x^2 + 6x - 16 = 0$

20. $0 = 2x^2 - 7x - 4$

21. $0 = x^2 + 3x - 1$

Write and solve an equation for each problem.

22. Five times an integer is 4 more than the integer squared.

23. The square of an integer minus three times the integer equals -2.

Calculate the distance between each pair of points. Round to the nearest tenth.

1. $A(9, 5)$, $B(6, 1)$

2. $X(0, -7)$, $Y(3, -4)$

3. $M(5, 6)$, $N(5, -2)$

4. $G(0, -3)$, $H(0, 6)$

5. $X(0, 0)$, $Y(3, 4)$

6. $A(1, 2)$, $B(4, 7)$

7. $K(2, 2)$, $L(-2, -2)$

8. $X(-2, 6)$, $Y(4, 6)$

9. $M(2, -2)$, $N(3, 5)$

10. $X(4, 3)$, $Y(6, 7)$

11. $M(9, 2)$, $N(5, 7)$

12. $A(9, 3)$, $B(4, 1)$

Calculate the midpoint between each pair of points.

13. $A(-4, 7)$, $B(2, 3)$

14. $M(2, -5)$, $N(-8, -9)$

15. $X(2, 2)$, $Y(6, 6)$

16. $D(4, 5)$, $E(4, -5)$

17. $A(3, 7)$, $B(3, 11)$

18. $K(2, -2)$, $L(-2, 5)$

19. $X(4, 5)$, $Y(6, -7)$

20. $X(8, -4)$, $Y(-3, 9)$

21. $A(3, 8)$, $B(6, 14)$

22. The vertices of a triangle are $A(1, 3)$, $B(8, 4)$, and $C(5, 0)$. What are the lengths of the sides to the nearest tenth?

23. A triangle ABC has vertex A at $(3, 1)$ and vertex B at $(1, 4)$. The length of side $\overline{BC}$ is 6 units, and the length of side $\overline{AC}$ is 5 units. Find the coordinates of vertex C.

24. Find the point on the x-axis that is the same distance from $(1, 3)$ as from $(8, 4)$.

25. Find the point on the y-axis that is the same distance from $(2, 2)$ as from $(2, 6)$.

Chapter 13

Extra Practice 13–1 • The Standard Equation of a Circle • pages 562–565

Write an equation for each circle.

1. radius 5, center (0, 0)   **2.** radius 3, center (−4, 2)   **3.** radius 10, center (5, −1)

4. 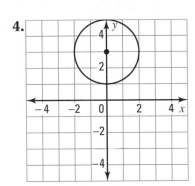   **5.** 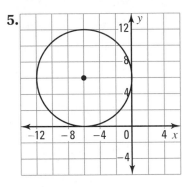   **6.**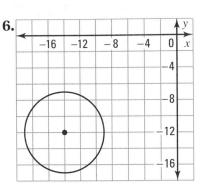

Find the radius and center of each circle.

7. $x^2 + y^2 = 64$

8. $x^2 + y^2 = 21$

9. $(x - 3)^2 + (y - 5)^2 = 49$

10. $(x - 4)^2 + (y + 2)^2 = 26$

11. $(x + 5)^2 + (y - 4)^2 = 41$

12. $(x + 4)^2 + y^2 = 70$

13. $x^2 + (y - 1)^2 = 6$

14. $(x + 5)^2 + (y + 5)^2 = 100$

15. The graph of $x^2 + y^2 = 25$ is translated 4 units to the right and 3 units up. Write the equation of the circle in the new position.

16. The graph of $x^2 + y^2 = 25$ is translated 6 units to the left and 5 units up. Write the equation of the circle in the new position.

Extra Practice 13–2 • More on Parabolas • pages 566–569

Find the focus and directrix of each parabola.

1. $x^2 = 8y$

2. $x^2 = 24y$

3. $x^2 = -8y$

4. $x^2 = -18y$

5. $x^2 - 15y = 0$

6. $x^2 - 28y = 0$

7. $2x^2 = -32y$

8. $4x^2 = -20y$

9. $8x^2 = 44y$

10. $54y - 18x^2 = 0$

11. $56y - 14x^2 = 0$

12. $-7x^2 = -42y$

13. $-8x^2 = 64y$

14. $6x^2 - 48y = 0$

15. $-9x^2 - 90y = 0$

Find the standard equation for each parabola with vertex located at the origin.

16. Focus $\left(0, -\dfrac{9}{4}\right)$

17. Focus $\left(0, \dfrac{15}{4}\right)$

18. Focus $\left(0, \dfrac{2}{5}\right)$

19. Focus (0, 12.5)

20. Focus (0, 1.5)

21. Focus (0, −7.5)

22. An elevated highway is supported by a parabolic arch that can be described by the equation $x^2 = -40y$. Find the value of a. Find the value of x when $y = -10$.

Graph each equation.

1. $x^2 + 4y^2 = 4$ 2. $4x^2 + y^2 = 4$ 3. $9x^2 + 4y^2 = 36$

4. $16x^2 + 9y^2 = 144$ 5. $25x^2 + 4y^2 = 100$ 6. $9x^2 + y^2 = 9$

7. $4x^2 - 9y^2 = 36$ 8. $9x^2 - y^2 = 9$ 9. $x^2 - 4y^2 = 16$

10. $25x^2 - 4y^2 = 100$ 11. $x^2 - 4y^2 = 4$ 12. $9x^2 + 16y^2 = 144$

13. $25x^2 + y^2 = 25$ 14. $25x^2 - 9y^2 = 225$ 15. $x^2 - 9y^2 = 9$

16. $x^2 + 25y^2 = 25$ 17. $9x^2 - 16y^2 = 144$ 18. $25x^2 + 9y^2 = 225$

19. Find the equation of the ellipse with foci $(2, 0)$ and $(-2, 0)$ and x-intercepts $(4, 0)$ and $(-4, 0)$.

20. Find the equation of the ellipse with foci $(6, 0)$ and $(-6, 0)$ and x-intercepts $(10, 0)$ and $(-10, 0)$.

21. Find the equation of the ellipse with foci $(3, 0)$ and $(-3, 0)$ and x-intercepts $(5, 0)$ and $(-5, 0)$.

22. Find the equation of the hyperbola with center $(0, 0)$ and foci on the x-axis if $a = \pm 6$ and $b = \pm 8$.

23. Find the equation of the hyperbola with center $(0, 0)$ and foci on the x-axis if $a = \pm 1$ and $b = \pm 4$.

24. Find the equation of the hyperbola with center $(0, 0)$ and foci on the x-axis if $a = \pm 4$ and $b = \pm 5$.

Extra Practice 13–5 • Direct Variation • pages 580–583

Find the equation of direct variation for each pair of values.

1. $x = 54$ and $y = 18$ 2. $x = 4$ and $y = 12$ 3. $x = 7$ and $y = 42$

4. $x = 15$ and $y = 10$ 5. $x = 3$ and $y = -12$ 6. $x = 1.5$ and $y = 10.5$

In each of the following, y varies directly as x.

7. If $y = 7$ when $x = 3$, find y when $x = 15$.

8. If $y = 30$ when $x = 40$, find y when $x = 32$.

9. If $y = 76$ when $x = 4$, find y when $x = 5$.

10. If $y = 2.4$ when $x = 8$, find y when $x = 0.3$.

In each of the following, y varies directly as x^2.

11. If $y = 176$ when $x = 4$, find y when $x = 5$.

12. If $y = 468$ when $x = 3$, find y when $x = 12$.

13. If $y = 180$ when $x = 6$, find y when $x = 4$.

14. If $y = 112$ when $x = 0.4$, find y when $x = 3$.

The distance (d) a vehicle travels at a given speed is directly proportional to the time (t) it travels.

15. If a vehicle travels 24 mi in 40 min, how far can it travel in 2 h?

16. If a vehicle travels 40 mi in 50 min, how far can it travel in 20 min?

17. In an electric circuit, the voltage varies directly as the current. If the voltage (v) is 90 volts when the current (c) is 15 amps, find the voltage when the current is 20 amps.

18. The distance (d) an object falls is directly proportional to the square of the time (t) it falls. If an object falls 400 ft in 5 sec, how far will it fall in 10 sec?

Extra Practice 13–6 • Inverse Variation • pages 584–587

For each pair of values, write an equation in which y varies inversely as x.

1. $x = 4$ and $y = 16$

2. $x = 6$ and $y = 60$

3. $x = 3.5$ and $y = 180$

4. $x = 0.6$ and $y = 32$

5. $x = 2.4$ and $y = 3.6$

6. $x = 3$ and $y = 1.8$

In each of the following, y varies inversely as x.

7. If $y = 13$ when $x = 2$, find y when $x = 13$.

8. If $y = 32$ when $x = 3$, find y when $x = 16$.

9. If $y = 4$ when $x = 2$, find y when $x = 6$.

10. If $y = 1.4$ when $x = 5$, find y when $x = 3.5$.

In each of the following, y varies inversely as the square of x.

11. If $y = 9$ when $x = 2$, find y when $x = 6$.

12. If $y = 2$ when $x = 10$, find y when $x = 5$.

13. If $y = 128$ when $x = 1$, find y when $x = 4$.

14. If $y = 800$ when $x = 20$, find y when $x = 25$.

In each of the following, travel time varies inversely as travel speed.

15. If it takes 40 min to travel a certain distance as at a speed of 25 mi/h, how long will it take to travel that distance at 40 mi/h?

16. If it takes 20 min for an airplane to travel a certain distance at a speed of 120 mi/h, how long will it take the plane to travel that distance at 100 mi/h?

The brightness of a light bulb varies inversely as the square of the distance from the source.

17. If a light bulb has a brightness of 600 lumens at 2 ft, what will be its brightness at 8 ft?

18. If a light bulb has a brightness of 1000 lumens at 3 ft, what will be its brightness at 15 ft?

Graph each inequality.

1. $y \geq 0.5x^2 + x - 1$

2. $y < -x^2 + x + 3$

3. $(x - 3)^2 + (y - 2)^2 \leq 9$

4. $9x^2 + 4y^2 > 36$

5. $x^2 - 4y^2 \geq 4$

6. $25x^2 - 9y^2 \leq 225$

Graph each system of inequalities.

7. $x^2 - y^2 < 16$
$x^2 + y^2 < 25$

8. $x^2 + y^2 > 4$
$x^2 + y^2 \leq 25$

9. $y \geq -0.5x^2 + 2$
$y \leq x + 3$

10. $y \leq -x^2 + x + 1$
$y \geq 0.5x^2 - 3$

11. $x^2 + (y - 2)^2 > 16$
$y < -x^2 - 2x + 1$

12. $9x^2 + 16y^2 < 144$
$x^2 - 9y^2 \geq 9$

13. $25x^2 + 4y^2 < 100$
$y \geq x^2 - 3$

14. $y \geq x^2 - 5$
$x^2 - 4y^2 \leq 4$

15. $(x + 1)^2 + (y - 1)^2 \leq 16$
$16x^2 + 9y^2 > 144$

Describe the part of the coordinate plane that is shaded.

16. $x^2 + y^2 > r^2$

17. $\dfrac{x^2}{a^2} + \dfrac{y^2}{b^2} \leq 1$

18. $ax^2 + bx + c < y \ (a > 0)$

Graph each function. State the y-intercept.

1. $y = 6^x$

2. $y = 8^x$

3. $y = \left(\dfrac{1}{8}\right)^x$

4. $y = 2\left(\dfrac{1}{3}\right)^x$

5. $y = 3(4)^x$

6. $y = 2(5)^x + 1$

7. **INVESTMENTS** Determine the amount of an investment if $10,000 is invested at an interest rate of 4% each year for 6 yr.

8. **REAL ESTATE** The Connor family bought a house for $185,000. Assuming that the value of the house will appreciate 4.5% each year, how much will the house be worth in 5 yr?

9. **VEHICLE OWNERSHIP** A pickup truck sells for $27,000. If the annual rate of depreciation is 12%, what is the value of the truck in 6 yr?

10. **BUSINESS** Mr. Rogers purchased a combine for $175,000 for his farm. It is expected to depreciate at a rate of 18% per year. Find the value of the combine in 3 yr.

11. **BIOLOGY** In a certain state, the population of black bears has been decreasing at the rate of 0.75% per year. In the year 2000, there were 400 black bears in the state. If the population continues to decline at the same rate, what will the population be in 2020?

12. **ENERGY** The Environmental Protection Agency (EPA) has called for businesses to find cleaner sources of energy. Coal is not considered to be a clean source of energy. In 1950, the use of coal by residential and commercial users was 114.6 million tons. Since then, the use of coal has decreased by 6.6% per year. Estimate the amount of coal that will be used in 2020.

Write each equation in logarithmic form.

1. $9^4 = 6561$

2. $\left(\frac{2}{3}\right)^4 = \frac{16}{81}$

3. $7^{-3} = \frac{1}{343}$

Write each equation in exponential form.

4. $\log_{\frac{1}{5}} \frac{1}{3125} = 5$

5. $\log_4 \frac{1}{1024} = -5$

6. $\log_6 7776 = 5$

Evaluate each expression.

7. $\log_5 1$

8. $\log_9 81$

9. $\log_{\frac{1}{5}} 125$

10. $\log_{18} 18$

11. $\log_{\frac{1}{2}} \frac{1}{32}$

12. $\log_{10} 0.01$

Solve each equation.

13. $\log_8 a = 3$

14. $\log_{\frac{3}{4}} m = -2$

15. $\log_4 y = \frac{1}{64}$

16. $\log_5 (2x + 1) = \log_5 (3x)$ **17.** $\log_9 (d + 5) = \log_9 (3d)$ **18.** $\log_{\frac{1}{2}} (t + 10) = \log_{\frac{1}{2}} (3t - 4)$

Chapter 14

Use the figures at the right to find each ratio.

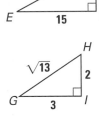

1. $\sin A$

2. $\cos A$

3. $\tan A$

4. $\sin B$

5. $\cos B$

6. $\tan B$

7. $\sin E$

8. $\tan E$

9. $\cos E$

10. $\sin D$

11. $\cos D$

12. $\tan D$

13. $\sin G$

14. $\cos G$

15. $\tan G$

16. $\sin H$

17. $\cos H$

18. $\tan H$

19. A tree is 75 ft tall. You stand x ft away from the tree, and the tangent of the angle to the top is about 1.5. How far are you from the tree?

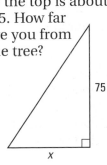

20. Suppose you stand 30 ft away from a building and measure the angle to the top to be 52°. Find the height of the building to the nearest tenth.

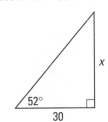

21. A ladder is leaning against a building. The foot of the ladder is 20 ft away from the building, and the top is 15 ft from the ground. How long is the ladder?

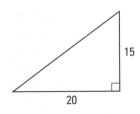

Find lengths to the nearest tenth and angle measures to the nearest degree.

Find the following in △LMN.

1. *LM*

2. *m∠N*

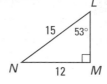

Find the following in △ABC.

3. *m∠B*

4. *AC*

5. *m∠A*

Find the following in △XYZ.

6. *XY*

7. *m∠X*

8. *m∠Y*

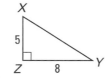

Find the following in △JKL.

9. *KL*

10. *JK*

11. *m∠K*

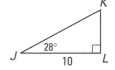

12. Find the angle of elevation to the top of the 1046-ft John Hancock Building in Chicago from a point 700 ft from the base.

Find each ratio by drawing a reference angle.

1. sin 300°

2. sin 210°

3. sin 135°

4. sin 450°

5. sin 585°

6. sin 660

7. sin 360°

8. sin 495°

9. sin 420°

10. sin 690°

11. sin 330°

12. sin 855°

1. Graph $y = \sin 6x$. State the period.

2. Graph $y = 3 \sin x$. State the amplitude.

3. Graph $y = \sin x + 2$. Describe the position of the graph.

4. Graph $y = \sin x - 2$. Describe the position of the graph.

State the period and amplitude of the graph of each equation and describe the position of the graph.

5. $y = 3 \sin x + 4$

6. $y = 1.5 \sin 2x$

7. $y = 8 \sin 2x - 3$

8. $y = 4.5 \sin 2x + 2.5$

9. $y = 1.5 \sin x + 2.5$

10. $y = 9 \sin 2x$

Preparing for Standardized Tests

Becoming a Better Test-Taker

At some time in your life, you will have to take a standardized test. Sometimes this test may determine if you go on to the next grade or course, or even if you will graduate from high school. This section of your textbook is dedicated to making you a better test-taker.

TYPES OF TEST QUESTIONS   In the following pages, you will see examples of four types of questions commonly seen on standardized tests. A description of each type of question is shown in the table below.

Type of Question	Description	See Pages
multiple choice	Four or five possible answer choices are given from which you choose the best answer.	710–711
gridded response	You solve the problem. Then you enter the answer in a special grid and color in the corresponding circles.	712–715
short response	You solve the problem, showing your work and/or explaining your reasoning.	716–719
extended response	You solve a multi-part problem, showing your work and/or explaining your reasoning.	720–724

PRACTICE   After being introduced to each type of question, you can practice that type of question. Each set of practice questions is divided into five sections that represent the categories most commonly assessed on standardized tests.

- Number and Operations
- Algebra
- Geometry
- Measurement
- Data Analysis and Probability

USING A CALCULATOR   On some tests, you are permitted to use a calculator. You should check with your teacher to determine if calculator use is permitted on the test you will be taking, and, if so, what type of calculator can be used.

TEST-TAKING TIPS   In addition to the Test-Taking Tips like the one shown at the right, here are some additional thoughts that might help you.

- Get a good night's rest before the test. Cramming the night before does not improve your results.
- Budget your time when taking a test. Don't dwell on problems that you cannot solve. Just make sure to leave that question blank on your answer sheet.
- Watch for key words like NOT and EXCEPT. Also look for order words like LEAST, GREATEST, FIRST, and LAST.

> **Test-Taking Tip**
>
> If you are allowed to use a calculator, make sure you are familiar with how it works so that you won't waste time trying to figure out the calculator when taking the test.

Multiple-Choice Questions

Multiple-choice questions are the most common type of question on standardized tests. These questions are sometimes called *selected-response questions*. You are asked to choose the best answer from four or five possible answers.

To record a multiple-choice answer, you may be asked to shade in a bubble that is a circle or an oval or just to write the letter of your choice. Always make sure that your shading is dark enough and completely covers the bubble.

Sometimes a question does not provide you with a figure that represents the problem. Drawing a diagram may help you to solve the problem. Once you draw the diagram, you may be able to eliminate some of the possibilities by using your knowledge of mathematics. Another answer choice might be that the correct answer is not given.

Incomplete Shading
Ⓐ Ⓑ Ⓒ Ⓓ

Too light shading
Ⓐ Ⓑ Ⓒ Ⓓ

Correct shading
Ⓐ Ⓑ Ⓒ Ⓓ

Example 1

A coordinate plane is superimposed on a map of a playground. Each side of a square represents 1 m. The slide is located at (5, −7), and the climbing pole is located at (−1, 2). What is the distance between the slide and the pole?

Ⓐ $\sqrt{15}$ m

Ⓑ 6 m

Ⓒ 9 m

Ⓓ $9\sqrt{13}$ m

Ⓔ none of these

Draw a diagram of the playground on a coordinate plane. Notice that the difference in the *x*-coordinates is 6 m and the difference in the *y*-coordinates is 9 m.

Since the two points are two vertices of a right triangle, the distance between the two points must be greater than either of these values. So we can eliminate Choices B and C.

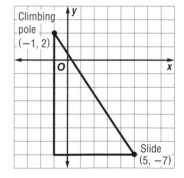

Use the Distance Formula or the Pythagorean Theorem to find the distance between the slide and the climbing pole. Let's use the Pythagorean Theorem.

$$a^2 + b^2 = c^2 \qquad \text{Pythagorean Theorem}$$
$$6^2 + 9^2 = c^2 \qquad \text{Substitution}$$
$$36 + 81 = c^2$$
$$117 = c^2$$
$$3\sqrt{13} = c \qquad \text{Take the square root of each side and simplify.}$$

So, the distance between the slide and pole is $3\sqrt{13}$ m. Since this is not listed as choice A, B, C, or D, the answer is Choice E.

If you are short on time, you can test each answer choice to find the correct answer. Sometimes you can make an educated guess about which answer choice to try first.

Multiple-Choice Practice

Choose the best answer.

Number and Operations

1. Carmen designed a rectangular banner that was 5 ft by 8 ft for a local business. The owner of the business asked her to make a larger banner measuring 10 ft by 20 ft. What was the percent increase in size from the first banner to the second banner?

  (A) 4%   (B) 20%

  (C) 80%   (D) 400%

2. A roller coaster casts a shadow 57 yd long. Next to the roller coaster is a 35-ft tree with a shadow that is 20 ft long at the same time of day. What is the height of the roller coaster to the nearest whole foot?

  (A) 98 ft   (B) 100 ft

  (C) 299 ft   (D) 388 ft

Algebra

3. At Speedy Car Rental, it costs $32 per day to rent a car and then $0.08 per mi. If y is the total cost of renting the car and x is the number of miles, which equation describes the relation between x and y?

  (A) $y = 32x + 0.08$ (B) $y = 32x - 0.08$

  (C) $y = 0.08x + 32$ (D) $y = 0.08x - 32$

4. Eric plotted his house, school, and the library on a coordinate plane. Each side of a square represents 1 mi. What is the distance from his house to the library?

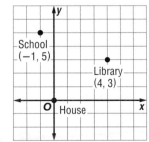

  (A) $\sqrt{24}$ mi

  (B) 5 mi

  (C) $\sqrt{26}$ mi

  (D) $\sqrt{29}$ mi

Geometry

5. The grounds outside of the Custer County Museum contain a garden shaped like a right triangle. One leg of the triangle measures 8 ft, and the area of the garden is 18 ft². What is the length of the other leg?

  (A) 2.25 in.   (B) 4.5 in.   (C) 13.5 in.

  (D) 27 in.   (E) 54 in.

6. The circumference of a circle is equal to the perimeter of a regular hexagon with sides that measure 22 in. What is the length of the radius of the circle to the nearest inch? Use 3.14 for π.

  (A) 7 in.   (B) 14 in.   (C) 21 in.

  (D) 24 in.   (E) 28 in.

Measurement

7. Eduardo is planning to install carpeting in a rectangular room that measures 12 ft 6 in. by 18 ft. How many square yards of carpet does he need for the project?

  (A) 25 yd²   (B) 50 yd²

  (C) 225 yd²   (D) 300 yd²

8. Marva is comparing two containers. One is a cylinder with diameter 14 cm and height 30 cm. The other is a cone with radius 15 cm and height 14 cm. What is the ratio of the volume of the cylinder to the volume of the cone?

  (A) 3 to 1   (B) 2 to 1

  (C) 7 to 5   (D) 7 to 10

Data Analysis and Probability

9. Refer to the table. Which statement is true about this set of data?

Country	Spending per Person
Japan	$8622
United States	$8098
Switzerland	$6827
Norway	$6563
Germany	$5841
Denmark	$5778

  (A) The median is less than the mean.

  (B) The mean is less than the median.

  (C) The range is 2844.

  (D) A and C are true.

  (E) B and C are true.

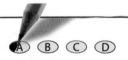

Test-Taking Tip

Questions 2, 5, and 7
The units of measure given in the question may not be the same as those given in the answer choices. Check that your solution is in the proper unit.

Gridded-Response Questions

Gridded-response questions are another type of question on standardized tests. These questions are sometimes called *student-produced response* or *grid-in*, because you must create the answer yourself, not just choose from four or five possible answers.

For gridded response, you must mark your answer on a grid printed on an answer sheet. The grid contains a row of four or five boxes at the top, two rows of ovals or circles with decimal and fraction symbols, and four or five columns of ovals, numbered 0–9. Since there is no negative symbol on the grid, answers are never negative. An example of a grid from an answer sheet is shown at the right.

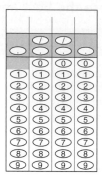

Example 1

In the diagram, $\triangle MPT \sim \triangle RPN$. Find PR.

What do you need to find?

You need to find the value of x so that you can substitute it into the expression $3x + 3$ to find PR. Since the triangles are similar, write a proportion to solve for x.

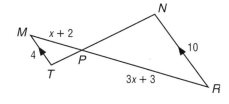

$\dfrac{MT}{RN} = \dfrac{PM}{PR}$	Definition of similar polygons
$\dfrac{4}{10} = \dfrac{x + 2}{3x + 3}$	Substitution
$4(3x + 3) = 10(x + 2)$	Cross products
$12x + 12 = 10x + 20$	Distributive Property
$2x = 8$	Subtract 12 and 10x from each side.
$x = 4$	Divide each side by 2.

Now find PR.
$PR = 3x + 3$
$\quad = 3(4) + 3$ or 15

How do you fill in the grid for the answer?
- Write your answer in the answer boxes.
- Write only one digit or symbol in each answer box.
- Do not write any digits or symbols outside the answer boxes.
- You may write your answer with the first digit in the left answer box, or with the last digit in the right answer box. You may leave blank any boxes you do not need on the right or the left side of your answer.
- Fill in only one bubble for every answer box that you have written in. Be sure not to fill in a bubble under a blank answer box.

Many gridded-response questions result in an answer that is a fraction or a decimal. These values can also be filled in on the grid.

Example 2

A triangle has a base of length 1 in. and a height of 1 in. What is the area of the triangle in square inches?

Use the formula $A = \frac{1}{2}bh$ to find the area of the triangle.

$A = \frac{1}{2}bh$ Area of a triangle

$\quad = \frac{1}{2}(1)(1)$ Substitution

$\quad = \frac{1}{2}$ or 0.5   Simplify.

How do you grid the answer?

You can either grid the fraction or the decimal. Be sure to write the decimal point or fraction bar in the answer box. The following are acceptable answer responses.

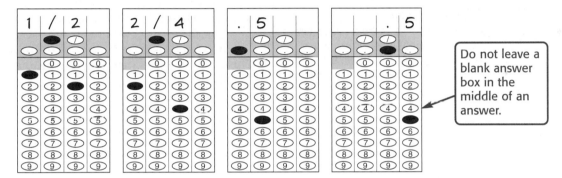

Do not leave a blank answer box in the middle of an answer.

Sometimes an answer is an improper fraction. Never change the improper fraction to a mixed number. Instead, grid either the improper fraction or the equivalent decimal.

Example 3

The shaded region of the rectangular garden will contain roses. What is the ratio of the area of the garden to the area of the shaded region?

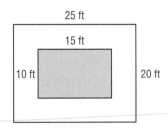

First, find the area of the garden.

$A = \ell w$

$\quad = 25(20)$ or 500

Then find the area of the shaded region.

$A = \ell w$

$\quad = 15(10)$ or 150

Write the ratio of the areas as a fraction.

$\dfrac{\text{area of garden}}{\text{area of shaded region}} = \dfrac{500}{150}$ or $\dfrac{10}{3}$

Leave the answer as the improper fraction $\frac{10}{3}$, as there is no way to correctly grid $3\frac{1}{3}$.

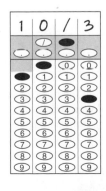

Gridded-Response Practice

Solve each problem and complete the grid.

Number and Operations

1. A large rectangular meeting room is being planned for a community center. Before building the center, the planning board decides to increase the area of the original room by 40%. When the room is finally built, budget cuts force the second plan to be reduced in area by 25%. What is the ratio of the area of the room that is built to the area of the original room?

2. Greenville has a spherical tank for the city's water supply. Due to increasing population, they plan to build another spherical water tank with a radius twice that of the current tank. How many times as great will the volume of the new tank be as the volume of the current tank?

3. In Earth's history, the Precambrian period was about 4600 million years ago. If this number of years is written in scientific notation, what is the exponent for the power of 10?

4. A virus is a type of microorganism so small it must be viewed with an electron microscope. The largest shape of virus has a length of about 0.0003 mm. To the nearest whole number, how many viruses would fit end to end on the head of a pin measuring 1 mm?

Algebra

5. Kaia has a painting that measures 10 in. by 14 in. She wants to make her own frame that has an equal width on all sides. She wants the total area of the painting and frame to be 285 in.^{2}. What will be the width of the frame in inches?

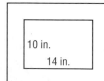

Test-Taking Tip

Question 1

Remember that you have to grid the decimal point or fraction bar in your answer. If your answer does not fit on the grid, convert to a fraction or decimal. If your answer still cannot be gridded, then check your computations.

6. The diagram shows a triangle graphed on a coordinate plane. If $\overline{AB}$ is extended, what is the value of the y-intercept?

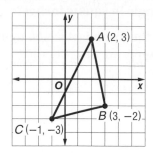

7. Tyree networks computers in homes and offices. In many cases, he needs to connect each computer to every other computer with a wire. The table shows the number of wires he needs to connect various numbers of computers. Use the table to determine how many wires are needed to connect 20 computers.

Computers	Wires	Computers	Wires
1	0	5	10
2	1	6	15
3	3	7	21
4	6	8	28

8. A line perpendicular to $9x - 10y = -10$ passes through $(-1, 4)$. Find the x-intercept of the line.

9. Find the positive solution of $6x^2 - 7x = 5$.

Geometry

10. The diagram shows $\triangle RST$ on the coordinate plane. The triangle is first rotated 90° counterclockwise about the origin and then reflected in the y-axis. What is the x-coordinate of the image of T after the two transformations?

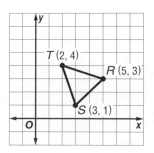

11. An octahedron is a solid with eight faces that are all equilateral triangles. How many edges does the octahedron have?

12. Find the measure of $\angle A$ to the nearest tenth of a degree.

Measurement

13. The Pep Club plans to decorate some large garbage barrels for Spirit Week. They will cover only the sides of the barrels with decorated paper. How many square feet of paper will they need to cover 8 barrels like the one in the diagram? Use 3.14 for π. Round to the nearest square foot.

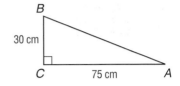

3 ft

14 in.

14. Kara makes decorative paperweights. One of her favorites is a hemisphere with a diameter of 4.5 cm. What is the surface area of the hemisphere including the bottom on which it rests? Use 3.14 for π. Round to the nearest tenth of a square centimeter.

|← 4.5 cm →|

15. The record for the fastest land speed of a car traveling for one mile is approximately 763 mi/h. The car was powered by two jet engines. What was the speed of the car in feet per second? Round to the nearest whole number.

16. On average, a B-777 aircraft uses 5335 gal of fuel on a 2.5-h flight. At this rate, how much fuel will be needed for a 45-min flight? Round to the nearest gallon.

Data Analysis and Probability

17. The table shows the heights of the tallest buildings in Kansas City, Missouri. To the nearest tenth, what is the positive difference between the median and the mean of the data?

Name	Height (m)
One Kansas City Place	193
Town Pavilion	180
Hyatt Regency	154
Power and Light Building	147
City Hall	135
1201 Walnut	130

18. A long-distance telephone service charges 40 cents per call and 5 cents per minute. If a function model is written for the graph, what is the slope of the function?

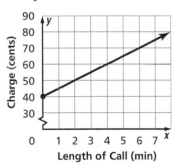

19. In a dart game, the dart must land within the innermost circle on the dartboard to win a prize. If a dart hits the board, what is the probability, as a percent, that it will hit the innermost circle?

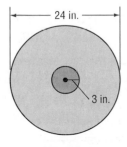

|← 24 in. →|

3 in.

Short-Response Questions

Short-response questions require you to provide a solution to the problem, as well as any method, explanation, and/or justification you used to arrive at the solution. These are sometimes called *constructed-response, open-response, open-ended, free-response,* or *student-produced questions.* The following is a sample rubric, or scoring guide, for scoring short-response questions.

Credit	Score	Criteria
Full	2	Full credit: The answer is correct and a full explanation is provided that shows each step in arriving at the final answer.
Partial	1	Partial credit: There are two different ways to receive partial credit. • The answer is correct, but the explanation provided is incomplete or incorrect. • The answer is incorrect, but the explanation and method of solving the problem is correct.
None	0	No credit: Either an answer is not provided or the answer does not make sense.

On some standardized tests, no credit is given for a correct answer if your work is not shown.

Example

Mr. Solberg wants to buy all the lawn fertilizer he will need for this season. His front yard is a rectangle measuring 55 ft by 32 ft. His back yard is a rectangle measuring 75 ft by 54 ft. Two sizes of fertilizer are available—one that covers 5000 ft^2 and another covering 15,000 ft^2. He needs to apply the fertilizer four times during the season. How many bags of each size should he buy to have the least amount of waste?

FULL CREDIT SOLUTION

Find the area of each part of the lawn and multiply by 4 since the fertilizer is to be applied 4 times. Each portion of the lawn is a rectangle, so $A = lw$.

$$4[(55 \times 32) + (75 \times 54)] = 23,240 \text{ ft}^2$$

If Mr. Solberg buys 2 bags that cover 15,000 ft^2, he will have too much fertilizer. If he buys 1 large bag, he will still need to cover $23,240 - 15,000$ or 8240 ft^2.

Find how many small bags it takes to cover 8240 ft^2.

$$8240 \div 5000 = 1.648$$

Since he cannot buy a fraction of a bag, he will need to buy 2 of the bags that cover 5000 ft^2 each.

Mr. Solberg needs to buy 1 bag that covers 15,000 ft^2 and 2 bags that cover 5000 ft^2 each.

The steps, calculations and reasoning are clearly stated.

The solution of the problem is clearly stated.

PARTIAL CREDIT SOLUTION

In this sample solution, the answer is correct. However, there is no justification for any of the calculations.

23,240

$23,240 - 15,000 = 8240$

$8240 \div 5000 = 1.648$

Mr. Solberg needs to buy 1 large bag and 2 small bags.

> There is not an explanation of how 23,240 was obtained.

PARTIAL CREDIT SOLUTION

In this sample solution, the answer is incorrect. However, after the first statement all of the calculations and reasoning are correct.

First find the total number of square feet of lawn. Find the area of each part of the yard.

$$(55 \times 32) + (75 \times 54) = 5810 \text{ ft}^2$$

The area of the lawn is greater than 5000 ft², which is the amount covered by the smaller bag, but buying the bag that covers 15,000 ft² would result in too much waste.

$$5810 \div 5000 = 1.162$$

Therefore, Mr. Solberg will need to buy 2 of the smaller bags of fertilizer.

> The first step of multiplying the area by 4 was left out.

NO CREDIT SOLUTION

In this sample solution, the response is incorrect and incomplete.

$55 + 75 = 130$
$32 + 54 = 86$
$130 \times 86 \times 4 = 44,720$
$44,720 \div 15,000 = 2.98$
Mr. Solberg will need 3 bags of fertilizer.

> The wrong operations are used, so the answer is incorrect. Also, there are no units of measure given with any of the calculations.

Short-Response Practice

Solve each problem. Show all your work.

Number and Operations

1. In 2000, approximately $191 billion in merchandise was sold by a popular retail chain store in the United States. The population at that time was 281,421,906. Estimate the average amount of merchandise bought from this store by each person in the U.S.

2. At a theme park, three educational movies run continuously all day long. At 9 A.M., the three shows begin. One runs for 15 min, the second for 18 min, and the third for 25 mi. At what time will the movies all begin at the same time again?

3. Ming found a sweater on sale for 20% off the original price. However, the store was offering a special promotion, where all sale items were discounted an additional 60%. What was the total percent discount for the sweater?

4. The serial number of a DVD player consists of three letters of the alphabet followed by five digits. The first two letters can be any letter, but the third letter cannot be O. The first digit cannot be zero. How many serial numbers are possible with this system?

Algebra

5. Solve and graph $2x - 9 \leq 5x + 4$.

6. Vance rents rafts for trips on the Jefferson River. You have to reserve the raft and provide a $15 deposit in advance. Then the charge is $7.50 per hour. Write an equation that can be used to find the charge for any amount of time, where y is the total charge in dollars and x is the amount of time in hours.

Test-Taking Tip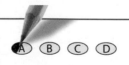
Ⓐ Ⓑ Ⓒ Ⓓ

Question 4
Be sure to completely and carefully read the problem before beginning any calculations. If you read too quickly, you may miss a key piece of information.

---

7. Hector is working on the design for the container shown below that consists of a cylinder with a hemisphere on top. He has written the expression $\pi r^2 + 2\pi rh + 2\pi r^2$ to represent the surface area of any size container of this shape. Explain the meaning of each term of the expression.

8. Find all solutions of the equation $6x^2 + 13x = 5$.

9. In 1999, there were 2,192,070 farms in the U.S., while in 2001, there were 2,157,780 farms. Let x represent years since 1999 and y represent the total number of farms in the U.S. Suppose the number of farms continues to decrease at the same rate as from 1999 to 2001. Write an equation that models the number of farms for any year after 1999.

Geometry

10. Refer to the diagram. What is the measure of $\angle 1$?

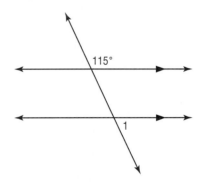

11. Quadrilateral *JKLM* is to be reflected in the line $y = x$. What are the coordinates of the vertices of the image?

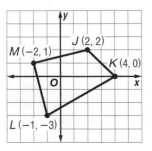

12. At what point does the graph of $y = -4x + 5$ intersect the x-axis?

13. In the Columbia Village subdivision, an unusually shaped lot, shown below, will be used for a small park. Find the exact perimeter of the lot.

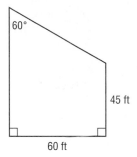

60°

45 ft

60 ft

Measurement

14. The Astronomical Unit (AU) is the distance from Earth to the Sun. It is usually rounded to 93,000,000 mi. The star Alpha Centauri is 25,556,250 million miles from Earth. What is this distance in AU?

15. Linesse handpaints unique designs on shirts and sells them. It takes her about 4.5 h to create a design. At this rate, how many shirts can she design if she works 22 days per month for an average of 6.5 h per day?

16. The world's largest pancake was made in England in 1994. To the nearest cubic foot, what was the volume of the pancake?

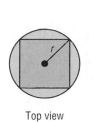

49 ft 3 in.   1 in.

17. Find the ratio of the volume of the cylinder to the volume of the pyramid.

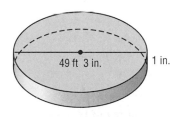

r

h

Top view   Front view

Data Analysis and Probability

18. The table shows the winning times for the Olympic men's 1000-m speed skating event. Make a scatter plot of the data and describe the pattern in the data. Times are rounded to the nearest second.

Men's 1000-m Speed Skating Event		
Year	Country	Time(s)
1976	U.S.	79
1980	U.S.	75
1984	Canada	76
1988	U.S.S.R.	73
1992	Germany	75
1994	U.S.	72
1998	Netherlands	71
2002	Netherlands	67

19. Bradley surveyed people about their favorite spectator sport. If a person is chosen at random from the people surveyed, what is the probability that the person's favorite spectator sport is basketball?

Favorite Spectator Sport	
Basketball	14
Football	18
Soccer	21
Golf	10
Other	7

20. The graph shows the altitude of a small airplane. Write a function to model the graph. Explain what the model means in terms of the altitude of the airplane.

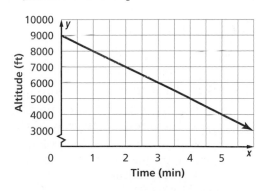

Extended-Response Questions

Extended-response questions are often called *open-ended* or *constructed-response questions*. Most extended-response questions have multiple parts. You must answer all parts to receive full credit.

Extended-response questions are similar to short-response questions in that you must show all of your work in solving the problem. A rubric is also used to determine whether you receive full, partial, or no credit. The following is a sample rubric for scoring extended-response questions.

Credit	Score	Criteria
Full	4	Full credit: A correct solution is given that is supported by well-developed, accurate explanations.
Partial	3, 2, 1	Partial credit: A generally correct solution is given that may contain minor flaws in reasoning or computation or an incomplete solution. The more correct the solution, the greater the score.
None	0	No credit: An incorrect solution is given indicating no mathematical understanding of the concept, or no solution is given.

> On some standardized tests, no credit is given for a correct answer if your work is not shown.

Make sure that when the problem says to *Show your work,* you show every part of your solution including figures, sketches of graphing calculator screens, or the reasoning behind your computations.

Example

> Polygon *WXYZ* with vertices $W(-3, 2)$, $X(4, 4)$, $Y(3, -1)$, and $Z(-2, -3)$ is a figure represented on a coordinate plane to be used in the graphics for a video game. Various transformations will be performed on the polygon to use for the game.

> **a.** Graph *WXYZ* and its image *W'X'Y'Z'* under a reflection in the *y*-axis. Be sure to label all of the vertices.

> **b.** Describe how the coordinates of the vertices of *WXYZ* relate to the coordinates of the vertices of *W'X'Y'Z'*.

> **c.** Another transformation is performed on *WXYZ*. This time, the vertices of the image *W'X'Y'Z'* are $W'(2, -3)$, $X'(4, 4)$, $Y'(-1, 3)$, and $Z'(-3, -2)$. Graph *WXYZ* and its image under this transformation. What transformation produced *W'X'Y'Z'*?

FULL CREDIT SOLUTION

Part a   A complete graph includes labels for the axes and origin and labels for the vertices, including letter names and coordinates.

- The vertices of the polygon should be correctly graphed and labeled.
- The vertices of the image should be located such that the transformation shows a reflection in the *y*-axis.
- The vertices of the polygons should be connected correctly. Optionally, the polygon and its image could be graphed in two contrasting colors.

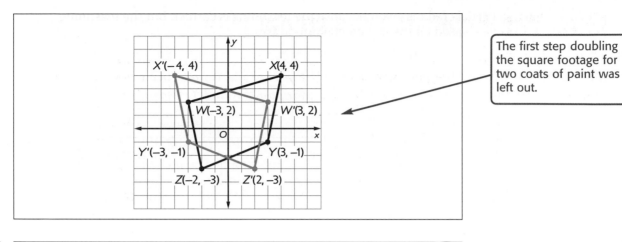

The first step doubling the square footage for two coats of paint was left out.

Part b

The coordinates of W and W' are (−3, 2) and (3, 2). The x-coordinates are the opposite of each other and the y-coordinates are the same.

Part c

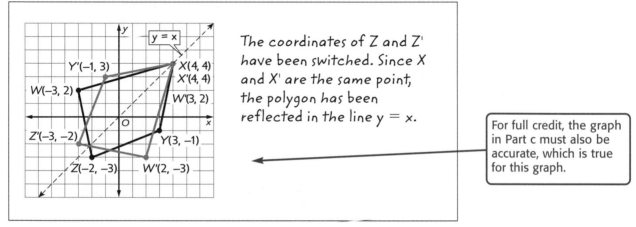

The coordinates of Z and Z' have been switched. Since X and X' are the same point, the polygon has been reflected in the line y = x.

For full credit, the graph in Part c must also be accurate, which is true for this graph.

PARTIAL CREDIT SOLUTION

Part a   This sample graph includes no labels for the axes and for the vertices of the polygon and its image. Two of the image points have been incorrectly graphed.

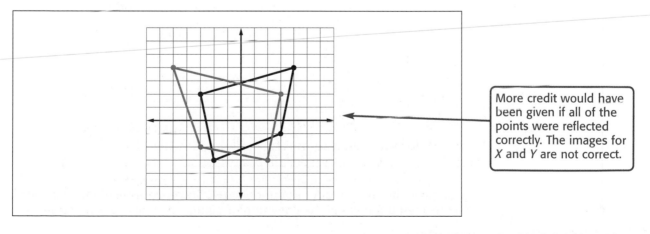

More credit would have been given if all of the points were reflected correctly. The images for *X* and *Y* are not correct.

Part b Partial credit is given because the reasoning is correct, but the reasoning was based on the incorrect graph in Part a.

> For two of the points, W and Z, the y-coordinates are the same and the x-coordinates are opposites. But, for points X and Y, there is no clear relationship.

Part c Full credit is given for Part c. The graph supplied by the student was identical to the graph shown for the full credit solution for Part c. The explanation below is correct, but slightly different from the previous answer for Part c.

> I noticed that point X and point X' were the same. I also guessed that this was a reflection, but not in either axis. I played around with my ruler until I found a line that was the line of reflection. The transformation from WXYZ to W'X'Y'Z' was a reflection in the line y = x.

This sample answer might have received a score of 2 or 1, depending on the judgment of the scorer. Had the student graphed all points correctly and gotten Part b correct, the score would probably have been a 3.

NO CREDIT SOLUTION

Part a The sample answer below includes no labels on the axes or the coordinates of the vertices of the polygon. The polygon *WXYZ* has three vertices graphed incorrectly. The polygon that was graphed is not reflected correctly either.

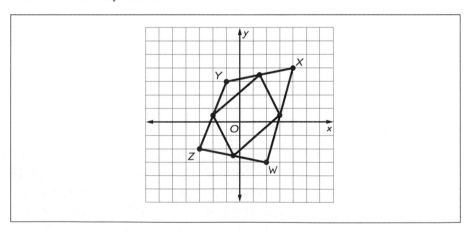

Part b

> I don't see any way that the coordinates relate.

Part c

> It is a reduction because it gets smaller.

In this sample answer, the student does not understand how to graph points on a coordinate plane and also does not understand the reflection of figures in an axis or other line.

Extended-Response Practice

Solve each problem. Show all your work.

Number and Operations

1. Refer to the table.

Population		
City	1990	2000
Phoenix, AZ	983,403	1,321,045
Austin, TX	465,622	656,562
Charlotte, NC	395,934	540,828
Mesa, AZ	288,091	396,375
Las Vegas, NV	258,295	478,434

a. For which city was the increase in population the greatest? What was the increase?

b. For which city was the percent of increase in population the greatest? What was the percent increase?

c. Suppose that the population increase of a city was 30%. If the population in 2000 was 346,668, find the population in 1990.

2. Molecules are the smallest units of a particular substance that still have the same properties as that substance. The diameter of a molecule is measured in angstroms (Å). Express each value in scientific notation.

a. An angstrom is exactly 10^{-8} cm. A centimeter is approximately equal to 0.3937 in. What is the approximate measure of an angstrom in inches?

b. How many angstroms are in 1 in.?

c. If a molecule has a diameter of 2 angstroms, how many of these molecules placed side by side would fit on an eraser measuring $\frac{1}{4}$ in.?

Algebra

3. The Marshalls are building a rectangular in-ground pool in their backyard. The pool will be 24 ft by 29 ft. They want to build a deck of equal width all around the pool. The final area of the pool and deck will be 1800 ft².

a. Draw and label a diagram.

b. Write an equation that can be used to find the width of the deck.

c. Find the width of the deck.

4. The depth of a reservoir was measured on the first day of each month. (Jan. = 1, Feb. = 2, and so on.)

Depth of the Reservoir

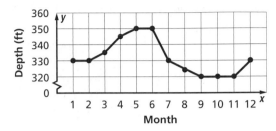

a. What is the slope of the line joining the points with x-coordinates 6 and 7? What does the slope represent?

b. Write an equation for the segment of the graph from 5 to 6. What is the slope of the line and what does this represent in terms of the reservoir?

c. What was the lowest depth of the reservoir? When was this depth first measured and recorded?

Geometry

5. The Silver City Marching Band is planning to create this formation with the members.

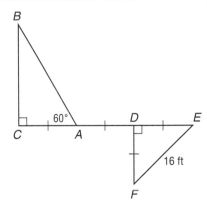

a. Find the missing side measures of △EDF. Explain.

b. Find the missing side measures of △ABC. Explain.

c. Find the total distance of the path: A to B to C to A to D to E to F to D.

d. The director wants to place one person at each point A, B, C, D, E, and F. He then wants to place other band members approximately one foot apart on all segments of the formation. How many people should he place on each segment of the formation? How many total people will he need?

Measurement

6. Rodrigo stands 50 yd away from the base of a building. From his eye level of 5 ft, Rodrigo sees the top of the building at an angle of elevation of 25°.

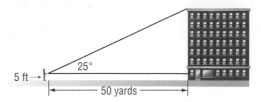

a. What is the height of the building? Round to the nearest tenth of a foot.

b. If Rodrigo moved back 15 more yards, how would the angle of elevation to the top of the building change?

c. If the building were taller, would the angle of elevation be greater than or less than 25°?

7. Kabrena is working on a project about the solar system. The table shows the maximum distances from Earth to the other planets in millions of miles.

Distance from Earth to Other Planets			
Planet	Distance	Planet	Distance
Mercury	138	Saturn	1031
Venus	162	Uranus	1962
Mars	249	Neptune	2913
Jupiter	602	Pluto	4681

a. The maximum speed of the Apollo moon missions spacecraft was about 25,000 mi/h. Make a table showing the time it would take a spacecraft traveling at this speed to reach each of the four closest planets.

b. Describe how to use scientific notation to calculate the time it takes to reach any planet.

c. Which planet would it take approximately 13.3 yr to reach? Explain.

Test-Taking Tip

Ⓐ Ⓑ Ⓒ Ⓓ

Question 6
While preparing to take a standardized test, familiarize yourself with the formulas for surface area and volume of common three-dimensional figures.

Data Analysis and Probability

8. The table shows the average monthly temperatures in Barrow, Alaska. The months are given numerical values from 1–12. (Jan. = 1, Feb. = 2, and so on.)

Average Monthly Temperature			
Month	°F	Month	°F
1	−14	7	40
2	−16	8	39
3	−14	9	31
4	−1	10	15
5	20	11	−1
6	35	12	−11

a. Make a scatter plot of the data. Let x be the numerical value assigned to the month and y be the temperature.

b. Describe any trends shown in the graph.

c. Find the mean of the temperature data.

d. Describe any relationship between the mean of the data and the scatter plot.

9. A dart game is played using the board shown. The inner circle is pink, the next ring is blue, the next red, and the largest ring is green. A dart must land on the board during each round of play.

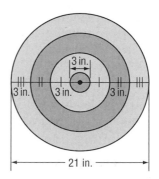

a. What is the probability that a dart landing on the board hits the pink circle?

b. What is the probability that the first dart thrown lands in the blue ring and the second dart lands in the green ring?

c. Suppose players throw a dart twice. For which outcome of two darts would you award the most expensive prize? Explain your reasoning.

Technology Reference Guide

Graphing Calculator Overview

This section summarizes some of the graphing calculator skills you might use in your mathematics classes using the TI-83 Plus or TI-84 Plus.

General Information

- Any yellow commands written above the calculator keys are accessed with the 2nd key, which is also yellow. Similarly, any green characters or commands above the keys are accessed with the ALPHA key, which is also green. In this text, commands that are accessed by the 2nd and ALPHA keys are shown in brackets. For example, 2nd [QUIT] means to press the 2nd key followed by the key below the yellow [QUIT] command.
- 2nd [ENTRY] copies the previous calculation so it can be edited or reused.
- 2nd [ANS] copies the previous answer so it can be used in another calculation.
- 2nd [QUIT] will return you to the home (or text) screen.
- 2nd [A-LOCK] allows you to use the green characters above the keys without pressing ALPHA before typing each letter.
- Negative numbers are entered using the (−) key, not the minus sign, −.
- The variable x can be entered using the X,T,θ,n key, rather than using ALPHA [X].
- 2nd [OFF] turns the calculator off.

Key Skills

Use this section as a reference for further instruction. For additional features, consult the TI-83 Plus or TI-84 Plus user's manual.

ENTERING AND GRAPHING EQUATIONS
Press Y=. Use the X,T,θ,n key to enter *any* variable for your equation. To see a graph of the equation, press GRAPH.

SETTING YOUR VIEWING WINDOW
Press WINDOW. Use the arrow or ENTER keys to move the cursor and edit the window settings. Xmin and Xmax represent the minimum and maximum values along the x-axis. Similarly, Ymin and Ymax represent the minimum and maximum values along the y-axis. Xscl and Yscl refer to the spacing between tick marks placed on the x- and y-axes. Suppose Xscl = 1. Then the numbers along the x-axis progress by 1 unit. Set Xres to 1.

THE STANDARD VIEWING WINDOW
A good window to start with to graph an equation is the **standard viewing window.** It appears in the WINDOW screen as follows.

To easily set the values for the standard viewing window, press ZOOM 6.

ZOOM FEATURES
To easily access a viewing window that shows only integer coordinates, press ZOOM 8 ENTER.

To easily access a viewing window for statistical graphs of data you have entered, press ZOOM 9.

USING THE TRACE FEATURE

To trace a graph, press TRACE. A flashing cursor appears on a point of your graph. At the bottom of the screen, x- and y-coordinates for the point are shown. At the top left of the screen, the equation of the graph is shown. Use the left and right arrow keys to move the cursor along the graph. Notice how the coordinates change as the cursor moves from one point to the next. If more than one equation is graphed, use the up and down arrow keys to move from one graph to another.

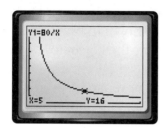

SETTING OR MAKING A TABLE

Press 2nd [TBLSET]. Use the arrow or ENTER keys to move the cursor and edit the table settings. Indpnt represents the x-variable in your equation. Set Indpnt to *Ask* so that you may enter any value for x into your table. Depend represents the y-variable in your equation. Set Depend to *Auto* so that the calculator will find y for any value of x.

USING THE TABLE

Before using the table, you must enter at least one equation in the Y= screen. Then press 2nd [TABLE]. Enter any value for x as shown at the bottom of the screen. The function entered as Y1 will be evaluated at this value for x. In the two columns labeled X and Y1, you will see the values for x that you entered and the resulting y-values.

PROGRAMMING ON THE TI–83 PLUS

When you press PRGM, you see three menus: EXEC, EDIT, and NEW. EXEC allows you to execute a stored program by selecting the name of the program from the menu. EDIT allows you to edit or change an existing program. NEW allows you to create a new program. For additional programming features, consult the TI–83 Plus or TI–84 Plus user's manual.

ENTERING INEQUALITIES

Press 2nd [TEST]. From this menu, you can enter the =, ≠, >, ≥, <, and ≤ symbols.

ENTERING AND DELETING LISTS

Press STAT ENTER. Under L1, enter your list of numerical data. To delete the data in the list, use your arrow keys to highlight L1. Press CLEAR ENTER. Remember to clear all lists before entering a new set of data.

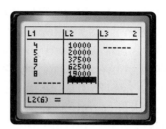

PLOTTING STATISTICAL DATA IN LISTS

Press Y=. If appropriate, clear equations. Use the arrow keys until Plot1 is highlighted. Plot1 represents a Stat Plot, which enables you to graph the numerical data in the lists. Press ENTER to turn the Stat Plot on and off. You may need to display different types of statistical graphs. To see the details of a Stat Plot, press 2nd [STAT PLOT] ENTER. A screen like the one below appears.

At the top of the screen, you can choose from one of three plots to store settings. The second line allows you to turn a Stat Plot on and off. Then you may select the type of plot: scatter plot, line plot, histogram, two types of box-and-whisker plots, or a normal probability plot. Next, choose which lists of data you would like to display along the x- and y-axes. Finally, choose the symbol that will represent each data point.

Cabri Jr. Overview

Cabri Junior for the TI-83 Plus and TI-84 Plus is a geometry application that is designed to reproduce the look and feel of a computer on a handheld device.

General Information

Starting Cabri Jr. To start Cabri Jr., press APPS and choose Cabri Jr. Press any key to continue. If you have not run the program on your calculator before, the F1 menu will be displayed. To leave the menu and obtain a blank screen, press CLEAR. If you have run the program before, the last screen that was in the program before it was turned off will appear. See Quitting Cabri Jr. for instructions on clearing this screen to obtain a blank screen.

In Cabri Jr., the four arrow keys (◀, ▶, ▼, ▲), along with ENTER, operate as a mouse would on a computer. The arrows simulate moving a mouse, and ENTER simulates a left click on a mouse. For example, when you are to select an item, use the arrow keys to point to the selected item and then press ENTER. You will know you are accurately pointing to the selected item when the item, such as a point or line, is blinking.

Quitting Cabri Jr. To quit Cabri Jr., press 2nd [QUIT], or [OFF] to completely shut off the calculator. Leaving the calculator unattended for approximately 4 minutes will trigger the automatic power down. After the calculator has been turned off, pressing ON will result in the calculator turning on, but not Cabri Jr. You will need to press APPS and choose Cabri Jr. Cabri Jr. will then restart with the most current figure in its most recent state.

As Cabri Jr. resembles a computer, it also has dropdown menus that simulate the menus in many computer programs. There are five menus, F1 through F5.

Navigating Menus To navigate each menu, press the appropriate key for F1 through F5. The arrow keys will then allow you to navigate within each menu. The ▲ and ▼ keys allow you to move within the menu items. The ▶ and ◀ keys allow you to access a submenu of an item. If an item has an arrow to the right, this indicates there is a submenu. Although not displayed, the menu items are numbered. You can also select a menu item by pressing the number that corresponds to each item. For example, to select the fourth menu item in the list, press [4]. If you press a number greater than the number of items in the list, the last item will be selected. If you press [0], you will leave the menu without selecting an item. This is the same as pressing CLEAR.

Key Skills

Use this section as a reference for further instruction. For additional features, consult the TI-84 Plus user's manual.

[F1] MANAGING FIGURES

The menu below provides the basic operations when working in Cabri Jr. These are commands normally found in menus within computer applications.

[F2] CREATING OBJECTS

This menu provides the basic tools for creating geometric figures. You can create points in three different ways, a line and line segment by selecting two points, a circle by defining the center and radius, a triangle by finding three vertices, and a quadrilateral by finding four vertices.

[F3] CONSTRUCTING OBJECTS

This menu provides the tools to construct new objects from existing objects. You can construct perpendicular and parallel lines, a perpendicular bisector of a segment, an angle bisector, a midpoint of a line segment, a circle using the center and a point on the circle, and a locus.

[F4] TRANSFORMING OBJECTS

This menu provides the tools to transform geometric figures. Using figures that are already created, you can access this menu to create figures that are symmetrical to other figures, reflect figures over a line of reflection, translate figures using a line segment or two points that define the translation, rotate figures by defining the center of rotation and angle of rotation, and dilate figures using the center of the dilation and a scale factor.

[F5] COMPUTING OBJECTS

This menu provides the tools for displaying, labeling, measuring, and computing. You can make an object visible or invisible, label points on figures, alter the way objects are displayed, measure length, area, and angle measures, display coordinates of points and equations of lines, make calculations, and delete objects from the screen.

While a graphing calculator cannot do everything, it can make some tasks easier. To prepare for whatever lies ahead, you should try to learn as much as you can about the technology. The future will definitely involve technology and the people who are comfortable with it will be successful. Using a graphing calculator is a good start toward becoming familiar with technology.

Technology Reference Guide

Glossary/Glosario

A mathematics multilingual glossary is available at **www.math.glencoe.com/multilingual_glossary.** The glossary includes the following languages.

Arabic	English	Korean	Tagalog
Bengali	Haitian Creole	Russian	Urdu
Cantonese	Hmong	Spanish	Vietnamese

English

Español

■ A ■

absolute value (p. 12) The distance of any number, *x*, from zero on the number line. Absolute value is represented by $|x|$.

valor absoluto (p. 12) La distancia a la que se encuentra un número, *x*, del cero en la recta numérica. Se representa: $|x|$.

acute triangle (p. 150) A triangle with three acute angles that measure less than 90°.

$$0° < m\angle A < 90°$$

triángulo acutángulo (p. 150) Un triángulo con tres ángulos que miden menos de 90°.

addition property of equality (p. 66) For all real numbers *a*, *b*, and *c*, if $a = b$, then $a + c = b + c$ and $c + a = c + b$.

propiedad aditiva de la igualdad (p. 66) Para todos los números reales *a*, *b*, y *c*, si $a = b$, entonces $a + c = b + c$ y $c + a = c + b$.

addition property of inequality (p. 76) For all real numbers *a*, *b*, and *c*, if $a < b$, then $a + c < b + c$ and $c + a < c + b$. If $a > b$, then $a + c > b + c$.

propiedad aditiva de la desigualdad (p. 76) Para todos los números reales *a*, *b*, y *c*, si $a < b$ entonces $a + c < b + c$ y $c + a < c + b$. Si $a > b$, entonces $a + c > b + c$.

addition property of opposites (p. 21) The sum of a number and its opposite equals 0. $a + (-a) = 0$.

propiedad aditiva de los opuestos (p. 21) La suma de un número y su opuesto es igual a 0. $a + (-a) = 0$

additive inverse (p. 21) The opposite of a number. The additive inverse of *a* is $-a$, and the additive inverse of $-a$ is *a*.

aditivo inverso (p. 21) El opuesto de un número. El aditivo inverso de *a* es $-a$ y el aditivo inverso de $-a$ es *a*.

adjacent angles (p. 109) Two angles in the same plane that have a common vertex and a common side but no common interior points.

ángulos adyacentes (p. 109) Dos ángulos en el mismo plano que tienen un vértice común y un lado común pero ningún punto interno común.

alternate exterior angles (p. 120) Two nonadjacent exterior angles on opposite sides of the transversal.

ángulos alternos externos (p. 120) Dos ángulos exteriores no adyacentes en lados opuestos a la transversal.

alternate interior angles (p. 120) Two nonadjacent interior angles on opposite sides of the transversal.

ángulos alternos internos (p. 120) Dos ángulos internos no adyacentes en lados opuestos a la transversal.

altitude of a triangle (p. 164) A perpendicular segment from a triangle's vertex to the line containing the opposite side.

altura de un triángulo (p. 164) Un segmento perpendicular desde el vértice de un triángulo al lado opuesto.

amplitude (p. 629) In a periodic function, half the difference between its maximum and minimum *y*-values.

amplitud (p. 629) En una función periódica, la mitad de la diferencia entre los valores máximo y mínimo de *y*.

angle (p. 108) The figure formed by two rays that have a common endpoint.

ángulo (p. 108) La figura formada por dos rayos que tienen un punto extremo en común.

English

Español

angle of depression (p. 619) The acute angle formed by a horizontal line and a line slanting downward.

ángulo de depresión (p. 619) El ángulo agudo formado por una línea horizontal y una línea que se inclina hacia abajo.

angle of elevation (p. 619) The acute angle formed by a horizontal line and a line slanting upward.

ángulo de elevación (p. 619) El ángulo agudo formado por una línea horizontal y una línea que se inclina hacia arriba.

angle of rotation (p. 342) In a rotation, the amount of turn expressed as a fractional part of a whole turn or as the angle of rotation in degrees.

ángulo de rotación (p. 342) En una rotación, la cantidad de una vuelta expresada como una parte fraccional de una vuelta completa.

associative property of addition (p. 21) For all real numbers a, b, and c, $a + (b + c) = (a + b) + c$.

propiedad asociativa de la adición (p. 21) Para todos los números reales a, b, y c, $(a + b) + c + a + (b + c)$.

associative property of multiplication (p. 27) For all real numbers a, b, and c, $a(bc) = (ab)c$.

propiedad asociativa de la multiplicación (p. 27) Para todos los números reales a, b, y c, $a(bc) = (ab)c$.

asymptotes (p. 576) The values that hyperbolic functions approach but never reach.

asíntotas (p. 576) Los valores a que las funciones hiperbólicas se acercan pero nunca alcanzan.

axis of a cylinder (p. 220) The segment joining the centers of the two bases.

eje de simetría de un cilindro (p. 220) El segmento que une los centros de las bases.

■ B ■

bell curve (p. 415) A frequency distribution that consists of a smooth curved line connecting the midpoints of a histogram. In a normal distribution of data, the curve is shaped like a bell.

curva de campana (p. 415) Una distribución de frecuencia que consiste de una línea curva lisa que conecta los puntos medios de un histograma. En una distribución normal de datos, la curva tiene la forma de una campana.

biconditional statement (p. 129) A statement in the *if-and-only-if* form. In the biconditional "*P* if and only if *Q*," *P* is both a necessary condition and a sufficient condition for *Q*.

proposición bicondicional (p. 129) Una proposición en la forma de *si y solamente si*. En la bicondicional "*P* si y solamente si *Q*", *P* es ambos una condición necesaria y suficiente para *Q*.

binomial (p. 468) A polynomial with two terms.

binomio (p. 468) Un polinomio con dos términos.

bisector of an angle (p. 115) A ray that divides an angle into two congruent adjacent angles.

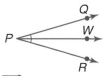

$\overrightarrow{PW}$ is the bisector of $\angle P$.
$\overrightarrow{PW}$ es la bisectriz del $\angle P$.

bisectriz de un ángulo (p. 115) Un rayo que divide un ángulo en dos ángulos adyacentes congruentes.

bisector of a segment (p. 114) Any line, segment, ray, or plane that intersects the segment at its midpoint.

bisectriz de un segmento (p. 114) Cualquier línea, segmento, rayo o plano que interseca el segmento en su punto medio.

boundary (p. 77) The line separating two half-planes in the coordinate plane.

frontera (p. 77) La línea que separa dos semiplanos en el plano coordenado.

box-and-whisker plot (p. 407) A means of visually displaying data that shows the median of a set of data, the median of each half of data, and the least and greatest value of the data.

diagrama de bloque (p. 407) Un medio visual de representar datos que muestra la mediana de un conjunto de datos, la mediana de cada mitad de datos y el valor menor y mayor de los datos.

English   # Español

cells   (p. 30) Areas that are formed by the vertical columns and horizontal rows on a spreadsheet.

celdas   (p. 30) Áreas que las forman las columnas verticales y las filas horizontales en una hoja de cálculos.

center of rotation   (p. 342) The point about which a figure is rotated in a rotation.

centro de rotación   (p. 342) El punto alrededor del cual se gira una figura.

chord   (p. 441) A segment with both endpoints on the circle.

cuerda   (p. 441) Un segmento con ambos puntos extremos en el mismo círculo.

circle   (p. 562) In a plane, the set of all points that are a given distance for a fixed point. That fixed point is the center of the circle.

P is the center of the circle.
P es el centro del círculo.

círculo   (p. 562) En un plano, el conjunto de todos los puntos a una distancia dada de un punto fijo. El punto fijo es el centro del círculo.

circle graph   (p. 446) A means of displaying data that represents items as parts of a whole circle; these parts are called sectors.

diagrama de círculo   (p. 446) Un medio de mostrar datos que representan artículos como partes de un círculo entero; estas partes se llaman sectores.

circumference   (p. 206) The distance around a circle.

circunferencia   (p. 206) La distancia alrededor un círculo.

circumscribed polygon   (p. 455) A polygon with all sides tangent to the same circle.

polígono circunscrito   (p. 455) Un polígono con todos los lados tangente al mismo círculo.

closed half-plane   (p. 77) The graph of either half-plane *and* the line that separates them.

semiplano cerrado   (p. 77) La gráfica de uno de los dos semiplanos y la línea que los separa.

clusters   (p. 87) Isolated groups of values on a stem-and-leaf plot.

conglomerados   (p. 87) Grupos aislados de valores en un diagrama de tallo y hoja.

cluster sampling   (p. 82) Statistical sampling in which the members of the population are randomly selected from particular parts of the population and then surveyed in groups, not individually.

muestra por conglomerado   (p. 82) Una muestra de estadística en que los miembros de la población se seleccionan al azar desde sectores particulares de la población y se inspeccionan en grupos, no individualmente.

coefficient   (p. 468) The numerical, nonvariable portion of a monomial.

coeficiente   (p. 468) La porción numérica, no variable de un monomio.

collinear points   (p. 104) Points that lie on the same line.

P, Q, and *R* are collinear.
P, Q, y R son colineales.

puntos colineales   (p. 104) Puntos que están en la misma línea.

combination   (p. 404) A set of items chosen from a larger set without regard to order.

combinación   (p. 404) Un conjunto de artículos seleccionados de un conjunto más grande sin considerar el orden.

combined variation   (p. 587) A variation in which a quantity varies directly as one quantity and inversely as another.

variación combinada   (p. 587) Una variación en que una cantidad varía directamente como una cantidad e inversamente como otra.

commutative property of addition   (p. 21) For all real numbers a, b, and c, $a + b = b + a$.

propiedad conmutativa de la adición   (p. 21) Para todos los números reales a, b, y c, $a + b = b + a$.

English

commutative property of multiplication (p. 27) For all real numbers a, b, and c, $ab = ba$.

complementary angles (p. 109) Two angles whose measures have a sum of 90°.

complement of an event (p. 393) The set of all outcomes in the sample space, not in A, when A is a subset of U. This is symbolized as A'.

completing the square (p. 534) Making a perfect square for an expression in the form $ax^2 + bx$.

compound event (p. 392) An event consisting of two or more simple events.

concave polygon (p. 178) A polygon with at least one diagonal that contains points in the exterior of the polygon.

conditional statement (p. 128) An *if-then* statement having two parts—a hypothesis and a conclusion.

cone (p. 221) A three-dimensional figure with a curved surface and one circular base.

congruent (p. 154) Term used to describe figures with the same size and shape.

congruent triangles (p. 154) Triangles whose vertices can be paired in such a way that all angles and sides of one triangle are congruent to corresponding angles and corresponding sides of the other.

conic section (p. 572) The section formed by a plane intersecting two circular cones whose vertices are at the origin.

conjecture (p. 124) A conclusion reached through inductive reasoning.

consecutive sides (p. 178) Two sides of a polygon that have a common vertex.

consecutive vertices (p. 178) The endpoints of any side of a polygon.

constant (p. 468) A monomial that contains no variables.

construction (p. 118) A precise drawing of a geometric figure made with the aid of only two tools: a compass and an unmarked straightedge.

convenience sampling (p. 82) Statistical sampling in which members of a population are selected because they are readily available, and all are surveyed.

Español

propiedad conmutativa de la multiplicación (p. 27) Para todos los números reales a, b, y c, $ab = ba$.

ángulos complementarios (p. 109) Dos ángulos cuyas medidas tienen una suma de 90°.

complemento de un evento (p. 393) El conjunto de todos los resultados en un espacio muestral, no en A, cuando A es un subconjunto de U. Esto se representa como A'.

completar el cuadrado (p. 534) Hacer un cuadrado perfecto para una expresión en la forma de $ax^2 + bx$.

suceso compuesto (p. 392) Un suceso que consiste de dos o más sucesos simples.

polígono cóncavo (p. 178) Un polígono con por lo menos una diagonal que contiene puntos en el exterior del polígono.

proposición condicional (p. 128) Una proposición *si - entonces* con dos partes, una hipótesis y una conclusión.

vertex
vértice

base
base

cono (p. 221) Una figura tridimensional con una superficie curva y una base circular.

congruente (p. 154) El término usado para describir figuras con el mismo tamaño y forma.

triángulos congruentes (p. 154) Triángulos con ángulos correspondientes congruentes y lados correspondientes congruentes.

$\triangle ABC \cong \triangle EDF$

sección cónica (p. 572) La sección formada por un plano que interseca dos conos circulares cuyos vértices están en el origen.

conjetura (p. 124) Una conclusión que se deriva mediante el razonamiento inductivo.

lados consecutivos (p. 178) Dos lados de un polígono que tienen un vértice común.

vértices consecutivos (p. 178) Los puntos extremos de cualquier lado de un polígono.

constante (p. 468) Un monomio que no tiene variables.

construcción (p. 118) Un dibujo de una figura geométrica hecha con la ayuda de solamente dos herramientas: un compás y una regla sin marcas.

muestra de conveniencia (p. 82) Una muestra de estadística en que los miembros de una población se seleccionan porque están fácilmente disponibles y todos son entrevistados.

English

converse (p. 129) A statement formed by interchanging the hypothesis and conclusion of a conditional statement.

converse of the Pythagorean Theorem (p. 431) If the sum of the squares of the measures of two sides of a triangle is equal to the square of the measure of the third side, then the triangle is a right triangle.

convex polygon (p. 178) A polygon with no diagonals that contain points in the exterior of the polygon.

coordinate of a point (p. 105) The real number that corresponds to a point. An ordered pair of numbers associated with a point on a grid are the coordinates of the point.

coordinate plane (p. 56) A two-dimensional mathematical grid system consisting of two perpendicular number lines, called the *x*-axis and the *y*-axis. The point where the axes intersect is the origin.

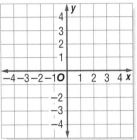

coplanar points (p. 104) Points that lie on the same plane.

corollary (p. 161) A statement that follows directly from a theorem.

corresponding angles (pp. 120, 154) Two angles in corresponding positions relative to two lines cut by a transversal. Also, angles in the same position in congruent or similar polygons.

corresponding sides (p. 154) Sides in the same position in congruent or similar polygons.

cosine (p. 614) In a right triangle, the cosine of acute $\angle A$ is equal to: $\dfrac{\text{length of side adjacent to } \angle A}{\text{length of hypotenuse}}$.

counterexample (p. 128) An instance that satisfies the hypothesis but not the conclusion of the conditional statement.

cross section (p. 223) The two-dimensional figure formed when a three-dimensional shape is cut with a plane.

customary units (p. 202) Units of measurement commonly used in the United States.

cylinder (p. 220) A three-dimensional shape made up of a curved region and two congruent circular bases that lie in parallel planes.

Español

converso (p. 129) Una declaración formada al intercambiar la hipótesis y la conclusión de una proposición condicional.

converso del teorema pitagórico (p. 431) Si la suma de los cuadrados de dos lados de un triángulo es igual al cuadrado del tercer lado, entonces el triángulo es un triángulo rectángulo.

polígono convexo (p. 178) Un polígono sin diagonales que contienen puntos en el exterior del polígono.

coordenada del punto (p. 105) El número real asociado con un punto en la recta numérica. Un par ordenado de números asociado con un punto en el plano.

plano coordenado (p. 56) Un sistema bidimensional que consiste de dos rectas numéricas perpendiculares, llamadas eje de *x* y eje de *y*.

puntos coplanares (p. 104) Puntos que están en el mismo plano.

corolario (p. 161) Una declaración que sigue directamente a un teorema.

ángulos correspondientes (pp. 120, 154) Dos ángulos en posiciones correspondientes relativos a dos líneas intersecadas por una transversal. También, ángulos en la misma posición en polígonos congruentes o similares.

lados correspondientes (p. 154) Lados en la misma posición en polígonos congruentes o similares.

coseno (p. 614) En un triángulo rectángulo, el coseno del ángulo agudo A es igual a: $\dfrac{\text{la longitud de lado adyacente a } \angle A}{\text{la longitud de la hipotenusa}}$.

contraejemplo (p. 128) Un ejemplo que satisface la hipótesis pero no la conclusión de una proposición condicional.

sección transversal (p. 223) La figura bidimensional formada cuando una figura tridimensional es cortada por un plano.

unidades inglesas (p. 202) Las unidades de medida usualmente usadas en los Estados Unidos.

base
base
base
base

cilindro (p. 220) Una figura tridimensional que consiste de una región curva y dos bases circulares congruentes que están en planos paralelos.

English

Español

■ D ■

data (p. 82) Factual information used as a basis for reasoning, discussion, or calculation.

deductive reasoning (p. 134) A process of reasoning in which the truth of the conclusion necessarily follows from the truth of the premises.

dependent events (p. 397) Events whose outcomes affect one another.

dependent system (p. 259) Two lines whose graphs coincide and thus have an infinite set of solutions.

dependent variables (p. 57) The elements of the range; also called the output values of a function.

determinant (p. 274) The difference of the products of the diagonal entries of a 2 x 2 square matrix.

diagonal (p. 178) A segment that joins two nonconsecutive vertices of a polygon.

datos (p. 82) Información objetiva usada como base para razonar, discusión o calcular.

razonamiento deductivo (p. 134) Un proceso de razonar en donde la verdad de la conclusión necesariamente sigue la verdad de las premisas.

sucesos dependientes (p. 397) Sucesos cuyos resultados se afectan uno a otro.

sistema dependiente (p. 259) Dos líneas cuyas gráficas coinciden y tienen un conjunto infinito de soluciones.

variables dependientes (p. 57) Los elementos del alcance; también llamados los valores de salida de una función.

determinante (p. 274) La diferencia entre los productos de los datos diagonales de una matriz 2 x 2.

$\overline{SQ}$ is a diagonal.
$\overline{SQ}$ es una diagonal.

diagonal (p. 178) Un segmento que une dos vértices no consecutivos de un polígono.

dilation (p. 348) A transformation that produces an image that is the same shape as the original figure but a different size.

directrix (p. 566) A line whose distance from a point on a parabola is equal to the distance from the same parabolic point to the focus.

direct square variation (p. 581) A function written in the form $y = kx^2$ where k is a nonzero constant.

direct variation (p. 580) A function that can be written in the form $y = kx$, where k is a nonzero constant. In a direct variation, the value of one variable increases as the other variable increases.

distance (p. 105) The absolute value of the difference between the coordinates of any two points.

distance formula (p. 544) For any points $P_1(x_1, y_1)$ and $P_2(x_2, y_2)$, the distance between P_1 and P_2 is given by: $P_1P_2 = \sqrt{(x_2 - x_1)^2 + (y_2 - y_1)^2}$.

distributive property (p. 34) Each factor outside parentheses can be used to multiply each term within the parentheses. $a(b + c) = ab + ac$

domain of a relation (p. 56) The set of all possible values of the x-coordinates for a relation.

dilatación (p. 348) Una transformación que produce una imagen que tiene la misma forma que la figura original pero un tamaño diferente.

directriz (p. 566) Una línea cuya distancia desde un punto en una parábola es igual a la distancia desde el mismo punto parabólico al foco.

variación cuadrada directa (p. 581) Una función escrita en la forma $y = kx^2$, donde k es una constante no igual a cero.

variación directa (p. 580) Una función que puede escribirse en la forma $y = kx$, donde k es una constante no igual a cero.

distancia (p. 105) El valor absoluto de la diferencia entre las coordenadas de dos puntos.

fórmula de distancia (p. 544) Para los puntos $P_1(x_1, y_1)$ y $P_2(x_2, y_2)$, la distancia entre P_1 y P_2 se da por: $P_1P_2 = \sqrt{(x_2 - x_1)^2 + (y_2 - y_1)^2}$.

propiedad distributiva (p. 34) Cada factor fuera del paréntesis puede usarse para multiplicar cada término dentro del paréntesis. Por ejemplo, $a(b + c) = ab + ac$.

dominio de una relación (p. 56) El conjunto de todos los valores posibles de las coordenadas de x para una relación.

English   Español

■ E ■

edge (p. 220) The set of linear points at which two faces of a polyhedron intersect.

arista (p. 220) El conjunto de puntos lineales en donde dos caras de un poliedro se cruzan.

ellipse (p. 574) In a plane, the figure created by a point moving about two fixed points, called foci. The sum of the distance from the foci to any point on the ellipse is a constant, $F_1P + F_2P = 2A$.

elipse (p. 574) En un plano, la figura creada por un punto que se mueve alrededor de dos puntos fijos, llamado los focos. La suma de la distancia desde los focos a cualquier punto sobre la elipse es una constante, $F_1P + F_2P = 2A$.

empty set (p. 6) A set containing no elements. The symbol for the empty set is $\varnothing$. This is also called the null set.

conjunto vacío (p. 6) Conjunto que carece de elementos. El símbolo para el conjunto vacío es $\varnothing$. También se llama conjunto nulo.

equation (p. 7) A mathematical statement that two numbers or expressions are equal.

ecuación (p. 7) Una declaración matemática en que dos números o expresiones son iguales.

equiangular triangle (p. 150) A triangle with three congruent angles.

equilateral triangle (p. 150) A triangle with three congruent sides.

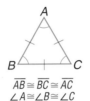

$\overline{AB} \cong \overline{BC} \cong \overline{AC}$
$\angle A \cong \angle B \cong \angle C$

triángulo equiangular (p. 150) Un triángulo con tres ángulos congruentes.

triángulo equilátero (p. 150) Un triángulo con tres lados congruentes.

equivalent ratios (p. 296) Two ratios that can both be named by the same fraction.

razones equivalentes (p. 296) Dos razones que pueden ambas ser nombradas por la misma fracción.

expanding binomials (p. 482) The multiplication and subsequent simplification of two binomials.

expansión del binomio (p. 482) La multiplicación y simplificación subsiguiente de dos binomios.

experiment (p. 384) An activity that is used to produce data that can be observed and recorded.

experimento (p. 384) Una actividad que se usa para producir datos que se pueden observar y registrar.

experimental probability (p. 384) The probability of an event determined by observation or measurement.

probabilidad experimental (p. 384) La probabilidad de un suceso determinado por una observación o medida.

exponent (p. 34) A superscripted number showing how many times the base is used as a factor. For example, in 2^4, 4 is the exponent.

exponente (p. 34) Un número superescrito que muestra cuántas veces la base se usa como un factor. Por ejemplo, en 2^4, 4 es el exponente.

exponential form (p. 34) A number written with a base and an exponent. For example, the exponential form of $(2)(2)(2)(2)$ is 2^4.

forma exponencial (p. 34) Un número escrito con una base y un exponente. Por ejemplo, la forma exponencial de $(2)(2)(2)(2)$ es 2^4.

exponential function (p. 594) A function that can be described by an equation of the form $y = a^x$, where $a > 0$ and $a \neq 1$.

función exponencial (p. 594) Función que puede describirse mediante una ecuación de la forma $y = a^x$, donde $a > 0$ y $a \neq 1$.

exterior angle of a polygon (p. 179) An angle both adjacent to and supplementary to an interior angle of a polygon.

∠1 is an exterior angle.
∠1 *es un ángulo externo.*

ángulo exterior de un polígono (p. 179) Un ángulo adyacente y suplementario al ángulo interno de un polígono.

exterior angles (p. 120) The angles formed by a transversal that are not between two coplanar lines.

ángulos exteriores (p. 120) Los ángulos formados por una transversal que no están entre dos líneas coplanares.

extremes (p. 296) The first and last terms of a proportion. (p. 407) In statistics, the data gathered that varies most from the median.

extremos (p. 296) Los primeros y últimos términos de una proporción. (p. 407) En estadísticas, los datos reunidos que más varían de la mediana.

English   Español

■ F ■

face (p. 220) The surface of a polyhedron.

factorial (p. 403) The product of all whole numbers from n to 1. Written as $n!$.

factors (p. 473) Elements whose product is a given quantity.

finite set (p. 6) A set whose elements can be counted or listed.

foci (p. 574) In an ellipse, the two fixed points whose combined distances to any point on the ellipse is constant.

focus (p. 566) The fixed point whose distance from a point on a parabola is equal to the distance from the same parabolic point to the directrix.

frequency distribution (p. 414) A visual display that shows the relative frequency of data.

frequency table (p. 82) A method of recording data that shows how often an item appears in a set of data.

function (p. 56) A set of ordered pairs in which each element of the domain is paired with exactly one element in the range.

function notation (p. 57) The notation that represents the rule associating the input value (independent variable) with the output value (dependent variable). The most commonly used function notation is the "f of x" notation, written $f(x)$.

fundamental counting principle (p. 402) The principle that states: If there are two or more stages of an activity, the total number of possible outcomes is the product of the number of possible outcomes for each stage of the activity.

cara (p. 220) La superficie de un poliedro.

factorial (p. 403) El producto de todos los números enteros desde n a 1. Escrito como $n!$.

factores (p. 473) Elementos cuyo producto es una cantidad determinada.

conjunto finito (p. 6) Un conjunto cuyos elementos pueden contarse o enumerarse.

focos (p. 574) En una elipse, los dos puntos fijos cuyas distancias combinadas a cualquier punto sobre la elipse es constante.

foco (p. 566) Los puntos fijos cuya distancia desde un punto en una parábola es igual a la distancia desde el mismo punto parabólico a la directriz.

distribución de frecuencia (p. 414) Una muestra visual que demuestra la frecuencia relativa de datos.

tabla de frecuencia (p. 82) Un método de registrar datos que muestra la frecuencia con que un artículo aparece en un conjunto de datos.

función (p. 56) Un conjunto de pares ordenados en donde cada elemento del dominio se aparea con exactamente un elemento del alcance.

notación de función (p. 57) La notación que representa la regla que asocia el valor de entrada (variable independiente) con el valor de salida (variable dependiente). La notación más comúnmente usada es la notación "f de x", escrita $f(x)$.

principio fundamental de conteo (p. 402) El principio que afirma: Si hay dos o más etapas de una actividad, el número total de resultados posibles es el producto del número de resultados posibles para cada etapa de la actividad.

■ G ■

gaps (p. 87) Large spaces between values on a stem-and-leaf plot.

general quadratic function (p. 524) A quadratic function written in the form $f(x) = ax^2 + bx + c$, where a, b, and c are real numbers, and $a \neq 0$.

greatest common factor (GCF) (p. 479) The greatest integer that is a factor of two or more integers. The GCF of two or more monomials is the greatest common numerical factor and the least power of the common variable factors.

greatest possible error (GPE) (p. 202) Half of the smallest unit used to make a measurement.

separacione (p. 87) Espacios grandes entre los valores en un diagrama de tallo y hoja.

función cuadrática general (p. 524) Una función cuadrática escrita en la forma $f(x) = ax^2 + bx + c$, donde a, b, y c son números reales, y $a \neq 0$.

máximo factor común (MFC) (p. 479) El entero más grande que es un factor de dos o más enteros. El MFC de dos o más monomios es el factor numérico común más grande y la potencia menor de los factores variables comunes.

máximo error posible (p. 202) La mitad de la unidad más pequeña usada para hacer una medida.

English

Español

half-plane (p. 77) The graphed region showing all solutions to a linear inequality.

histogram (p. 87) A type of bar graph used to visually display frequencies.

hyperbola (p. 572) A curve formed by the intersection of a double-right circular cone with a plane that cuts both halves of the cone.

hypotenuse (p. 175) The side opposite the right angle in a right triangle.

semiplano (p. 77) La región que muestra todas las soluciones de una desigualdad lineal.

histograma (p. 87) Un tipo del diagrama de barra usado para visualmente mostrar las frecuencias.

hipérbola (p. 572) Una curva formada por la intersección de un cono recto doble con un plano que corta ambas mitades del cono.

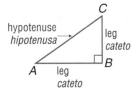

hipotenusa (p. 175) El lado opuesto al ángulo recto en un triángulo rectángulo.

identity property of addition (p. 21) The sum of any number and zero is that number. For example, $a + 0 = 0 + a = a$.

identity property of multiplication (p. 27) The product of any number and 1 is that number. For example, $a \times 1 = 1 \times a = a$.

image (p. 338) The new figure resulting from a translation.

included angle (p. 155) In a triangle, the term used to describe an angle's relative position to the two sides that form it.

included side (p. 155) In a triangle, the term used to describe a side's relative position to the two angles common to it.

inconsistent system (p. 258) Two lines that are parallel and thus have no solutions.

independent events (p. 396) Events whose outcomes are not affected by one another.

independent system (p. 258) Two linear equations that intersect at only one point.

independent variables (p. 57) The elements of the domain; also called the input values of a function.

indirect measurement (p. 326) The calculation of such a measurement that is difficult to measure directly. Using similar triangles is one method of indirect measurement.

indirect proof (p. 170) A proof in which one begins with the desired conclusion and assumes that it is not true. One then reasons logically until reaching a contradiction of the hypothesis or of a known fact.

propiedad aditiva de la identidad (p. 21) La suma de cualquier número y cero es ese número. Por ejemplo, $a + 0 = 0 + a = a$.

propiedad multiplicativa de identidad (p. 27) El producto de un número y 1 es ese número. Por ejemplo, $a \times 1 = 1 \times a = a$.

imagen (p. 338) La nueva imagen que resulta de una traslación.

ángulo incluido (p. 155) En un triángulo, el término usado para describir la posición relativa de un ángulo a los dos lados que lo forman.

lado incluido (p. 155) En un triángulo, el término usado para describir la posición relativa de un lado a los dos ángulos comunes a él.

sistema inconsistente (p. 258) Dos líneas que son paralelas y no tienen soluciones.

sucesos independientes (p. 396) Sucesos cuyos resultados no se afectan el uno al otro.

sistemas independientes (p. 258) Dos ecuaciones lineales que se intersecan en un solo punto.

variables independientes (p. 57) Los elementos del dominio; también se llaman los valores de entrada de la función.

medida indirecta (p. 326) El cálculo de una medida que es difícil de medir directamente. El usar triángulos similares es un método de medida indirecta.

prueba indirecta (p. 170) Una demostración en que se comienza con la conclusión deseada y se presume que no es cierta. Entonces se razona lógicamente hasta alcanzar una contradicción de la hipótesis.

English

inductive reasoning (p. 124) Logical reasoning where the premises of an argument provide some, but not absolute, support for the conclusion.

inequality (p. 11) A mathematical sentence that contains one of the symbols $<$, $>$, $\leq$, or $\geq$.

infinite set (p. 6) A set whose elements cannot be counted or listed.

initial side (p. 624) In the x and y coordinate plane, the side of an angle from which degree measurement begins.

inscribed polygon (p. 455) A polygon with all vertices lying on the same circle.

integers (p. 10) The set of whole numbers and their opposites.

interior angles (p. 120) The angles between coplanar lines that have been intersected by a transversal.

interior angles of a polygon (p. 178) The angles determined by the sides of a polygon.

interquartile range (p. 408) The difference between the values of the lower and upper quartiles.

intersection of geometric figures (p. 104) The set of points common to two or more figures.

inverse operation (p. 66) An operation that undoes what the previous operation did; used when simplifying and/or solving a math sentence. For example, addition and subtraction are inverse operations.

inverse square variation (p. 585) A function that can be written in the form $y = \frac{k}{x^2}$ or $x^2 y = k$, where k is a nonzero constant and $x \neq 0$.

inverse variation (p. 584) A function that can be written in the form $y = \frac{k}{x}$, where k is a nonzero constant and $x \neq 0$.

irrational numbers (p. 10) A number that cannot be written as a fraction, a terminating decimal, or a repeating decimal. Examples are π and $\sqrt{2}$.

isosceles triangle (p. 150) A triangle with at least two congruent sides.

iteration (p. 53) A process that is continually repeated. For example, the iterative process of multiplying by 2 is used to create the numerical sequence 1, 2, 4, 8, ...

Español

razonamiento inductivo (p. 124) Razonamiento lógico donde las premisas de un argumento proveen algunos apoyos para la conclusión.

desigualdad (p. 11) Una frase matemática que contiene uno de los símbolos $<$, $>$, $\leq$, o $\geq$.

conjunto infinito (p. 6) Un conjunto cuyos elementos no pueden contarse o enumerarse.

lado inicial (p. 624) En el plano coordenado de x e y, el lado de un ángulo desde donde comienza la medida en grados.

polígonos inscritos (p. 455) Polígonos cuyos vértices están en un mismo círculo.

enteros (p. 10) El conjunto de números enteros y sus opuestos.

ángulos interiores (p. 120) Los ángulos entre líneas coplanares que han sido intersecadas por una transversal.

ángulos interiores de un polígono (p. 178) Los ángulos determinados por los lados de un polígono.

amplitud intercuartílica (p. 408) La diferencia entre los valores de los cuartiles superior e inferior.

intersección de figuras geométricas (p. 104) El conjunto de puntos comunes a dos o más figuras.

operación inversa (p. 66) Operación que anula lo que ha hecho una operación anterior; se usa para reducir y/o resolver enunciados matemáticos. Por ejemplo, la adición y la sustracción son operaciones inversas.

variación cuadrada inversa (p. 585) Una función que puede escribirse en la forma $y = \frac{k}{x^2}$ ó $x^2 y = k$, donde k es una constante no igual a cero y $x \neq 0$.

variación inversa (p. 584) Una función que puede escribirse en la forma $y = \frac{k}{x}$, donde k es una constante no igual a cero y $x \neq 0$.

números irracionales (p. 10) Número que no puede escribirse como una fracción, un decimal terminal o un decimal periódico. Por ejemplo, π y $\sqrt{2}$.

triángulo isósceles (p. 150) Un triángulo con por lo menos dos lados congruentes.

iteración (p. 53) Proceso que se repite continuamente. Por ejemplo, el proceso iterativo de multiplicar por 2 se usa para crear la sucesión numérica 1, 2, 4, 8, ...

English

■ J ■

joint variation (p. 587) A variation in which a quantity varies directly as the product of two or more other quantities.

■ L ■

lateral edges (p. 220) The set of linear points at which lateral faces meet. See *prism*.

lateral faces (p. 220) The faces of prisms and pyramids that are not bases. See *prism* and *pyramid*.

line (p. 104) A set of points that extends infinitely in opposite directions.

linear function (p. 62) A function that can be represented by a linear equation. When graphed, a linear function yields a straight line.

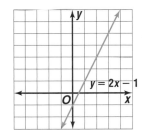

linear programming (p. 282) A method used to solve business-related problems involving linear inequalities. Also used to find the maximum or minimum of an expression involving a solution to the system of inequalities.

line of best fit (p. 406) The line that can be drawn near most of the points on a scatter plot that shows a relationship between two sets of data. This is also called a trend line.

line of reflection (p. 338) The line over which a figure is reflected or flipped.

line of symmetry (p. 338) A line on which a figure can be folded, so that when one part is reflected over that line it matches the other part exactly.

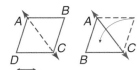

$\overleftrightarrow{AC}$ is a line of symmetry.
$\overleftrightarrow{AC}$ es un línea de simetría.

line segment (p. 105) A part of a line containing two endpoints and all points in between.

lower quartile (p. 407) In statistics, the median of the lower half of the data gathered.

logarithm (p. 600) In the function $x = b^y$, y is called the logarithm, base b, of x.

Español

■ J ■

variación conjunta (p. 587) Una variación en donde una cantidad varía directamente como el producto de dos o más otras cantidades.

■ L ■

aristas laterales (p. 220) El conjunto de puntos lineales en donde se encuentran las aristas laterales. Ver *prisma*.

caras laterales (p. 220) Las caras de los prismas y las pirámides que no son las bases. Ver *prisma* y *pirámide*.

línea (p. 104) Un conjunto de puntos que se extiende infinitamente en direcciones opuestas.

funciones lineales (p. 62) Una función que puede ser representada por una ecuación lineal. Su gráfica es un línea recta.

programación lineal (p. 282) Un método que se usa para resolver problemas relacionados con los negocios que contienen desigualdades lineales. También se usa para encontrar el máximo o el mínimo de una expresión que contiene una solución al sistema de desigualdades.

recta de óptimo ajuste (p. 406) La recta que se puede trazar cerca de la mayoría de los puntos en un diagrama de dispersión y que muestra una relación entre dos conjuntos de datos. También se conoce como la recta de tendencia.

línea de reflejo (p. 338) La línea sobre la cual una figura se refleja.

línea de simetría (p. 338) Una línea sobre la cual una figura puede doblarse, de tal manera que al reflejar una parte sobre esta línea coincide exactamente con la otra parte.

segmento de línea (p. 105) Parte de una línea que contiene dos puntos extremos y todos los puntos entre ellos.

cuartil inferior (p. 407) En estadísticas, la mediana de la mitad inferior de los datos reunidos.

logaritmo (p. 600) En la función $x = b^y$, y es el logaritmo en base b, de x.

English   Español

logarithmic equation (p. 601) An equation that contains one or more logarithms.

ecuación logarítmica (p. 601) Ecuación que contiene uno más logaritmos.

■ M ■

$m\overset{\frown}{PRG} > 180°$

major arc (p. 440) An arc that is larger than a semicircle.

arco mayor (p. 440) Un arco que es más grande que un semicírculo.

matrix (p. 275) (*plural: matrices*) A rectangular array of numbers arranged into rows and columns. Usually, square brackets enclose the numbers in a matrix.

matriz (p. 275) Arreglo rectangular de números acomodados en hileras y columnas. Los números en una matriz por lo general se encierran en corchetes.

mean (p. 83) The sum of the data divided by the number of data. Also known as the arithmetic average.

media (p. 83) La suma de los datos dividido entre el número de datos. También se conoce como el promedio aritmético.

measures of central tendency (p. 83) Statistics or measurements used to describe a set of data. Examples of these are the mean, the median, and the mode.

medidas de tendencia central (p. 83) Medidas usadas para describir un conjunto de datos. Ejemplos son la media, la mediana y el modo.

median (p. 83) The middle value of the data when the data are arranged in numerical order.

mediana (p. 83) El valor medio de los datos cuando los datos se arreglan en orden numérico.

median of a trapezoid (p. 188) The segment that joins the midpoints of a trapezoid's legs.

mediana de un trapezoide (p. 188) El segmento que une los puntos medios de los lados del trapezoide.

median of a triangle (p. 164) A segment with endpoints that are a vertex of a triangle and the midpoint of the opposite side.

mediana de un triángulo (p. 164) Un segmento cuyos puntos extremos son un vértice del triángulo y el punto medio del lado opuesto.

metric units (p. 202) Units of measurement that are based on multiples of 10. In the metric system, three basic units of measurement exist; the meter (to measure length), the gram (to measure mass), and the liter (to measure volume).

unidades métricas (p. 202) Unidades de medidas que se basan en los múltiplos de 10. Existen tres unidades básicas: el metro (mide longitud), el gramo (mide masa) y el litro (mide volumen).

midpoint (p. 114) The point that divides a segment into two congruent segments.

$SM = MT$

punto medio (p. 114) El punto que divide un segmento en dos segmentos congruentes.

minor arc (p. 440) An arc that is smaller than a semicircle. The degree measure of a minor arc is the same as the number of degrees in the corresponding central angle.

$m\overset{\frown}{PG} < 180°$

arco menor (p. 440) Arco que es más pequeño que un semicírculo. La medida angular de un arco menor es igual al número de grados en el ángulo central correspondiente.

mode (p. 83) The number that occurs most often in a set of data.

moda (p. 83) El número que ocurre más frecuentemente en un conjunto de datos.

monomial (p. 468) An expression that is either a single number, a variable, or the product of a number and one or more variables with whole-number exponents.

monomio (p. 468) Una expresión que es o un entero, una variable, o el producto de un número y una o más variables con exponentes enteros.

English

multiplication property of equality   (p. 67) For all real numbers *a, b,* and *c,* if $a = b$, then $ac = bc$.

multiplication property of zero   (p. 27) The product of any term and 0 is 0. For example, $a \times 0 = 0 \times a = 0$.

multiplicative inverses   (p. 27) Two numbers whose product is one; also called a reciprocal.

multiplicative property of inequality   (p. 76) For all real numbers *a, b,* and *c,* if $a < b$ and $c > 0$, then $ac < bc$; if $a < b$ and $c < 0$, then $ac > bc$; if $a > b$ and $c > 0$, then $ac > bc$; if $a > b$ and $c < 0$, then $ac < bc$.

mutually exclusive   (p. 392) Term used to describe events that cannot occur at the same time.

■ N ■

n factorial   (p. 403) The number of permutations of *n* different items; *n* factorial is written *n*!.

negative correlation   (p. 406) The inverse relationship between two sets of data. On a scatter plot, a negative correlation is evident if the trend line slopes downward from the top left to the bottom right corner of the graph.

negative reciprocals   (p. 248) Two fractions or ratios whose product is −1.

noncollinear points   (p. 104) Points that do not lie on the same line.

noncoplanar points   (p. 104) Points that do not lie on the same plane.

normal curve   (p. 415) The symmetrical bell-shaped curve resulting from a normal distribution of data. In a normal curve, the mean, median, and mode are the same.

null set   (p. 6) A set containing no elements. The symbol for the null set is $\varnothing$.

■ O ■

obtuse angle   (p. 109) An angle whose measure is greater than 90° but less than 180°.

$90° < m\angle A < 180°$

open half-plane   (p. 77) The region on either side of a line on a coordinate plane.

opposite angles   (p. 182) Two angles in a quadrilateral that do not share a common side.

opposite of the opposite property   (p. 12) The opposite of the opposite of any real number is the number. For example, $-(-n) = n$.

Español

propiedad multiplicativa de la igualdad   (p. 67) Para todos los números reales *a, b,* y *c,* si $a = b$, entonces $ac = bc$.

propiedad multiplicativa de cero   (p. 27) El producto de cualquier término y 0 es 0: $a \times 0 = 0 \times a = 0$.

inversos multiplicativos   (p. 27) Dos números cuyo producto es uno; también llamado recíproco.

propiedad multiplicativa de la desigualdad   (p. 76) Para todos los números reales *a, b,* y *c,* si $a < b$ y $c > 0$, entonces $ac < bc$; si $a < b$ y $c < 0$, entonces $ac > bc$; si $a > b$ y $c > 0$, entonces $ac > bc$; si $a > b$ y $c < 0$, entonces $ac < bc$.

mutuamente exclusivo   (p. 392) Término usado para describir sucesos que no pueden ocurrir a la misma vez.

n factorial   (p. 403) El número de permutaciones de *n* artículos diferentes; *n* factorial se escribe *n*!.

correlación negativa   (p. 406) La relación inversa entre dos conjuntos de datos. En un diagrama de dispersión, una correlación negativa es evidente si la recta de tendencia se inclina hacia abajo desde la esquina superior izquierda hasta la esquina inferior derecha de la gráfica.

recíprocos negativos   (p. 248) Dos fracciones o razones cuyo producto es −1.

puntos no colineales   (p. 104) Puntos que no están en la misma línea.

puntos no coplanares   (p. 104) Puntos que no están en el mismo plano.

curva normal   (p. 415) La curva simétrica acampanada que resulta de una distribución normal de datos. En una curva normal, la media, mediana y la moda son iguales.

conjunto nulo   (p. 6) Un conjunto que no contiene elementos. El símbolo para el conjunto nulo es $\varnothing$.

ángulo obtuso   (p. 109) Un ángulo cuya medida es mayor de 90° pero menor de 180°.

semiplano abierto   (p. 77) La región en cualquier lado de una línea en el plano coordenado.

ángulos opuestos   (p. 182) Dos ángulos en un cuadrilátero que no comparten un lado común.

propiedad del opuesto de la opuesto   (p. 12) Lo opuesto de lo opuesto de cualquier número real es el número. Por ejemplo, $-(-n) = n$.

English

Español

opposite sides   (p. 182) Two sides of a quadrilateral that do not share a common vertex.

lado opuesto   (p. 182) Dos lados de un cuadrilátero que no comparten un vértice común.

ordered pair   (p. 56) Two numbers named in a specific order.

par ordenado   (p. 56) Dos números nombrados en un orden específico.

origin   (p. 56) The point where the *x*-axis and *y*-axis intersect in the coordinate plane.

origen   (p. 56) El punto donde el eje de *x* y el eje de *y* se intersecan en el plano coordenado.

outcome   (p. 384) The result of each trial of an experiment.

resultado   (p. 384) El resultado de cada ensayo de un experimento.

outliers   (p. 87) Data values that are much greater or much less than most of the other values on a stem-and-leaf plot.

datos extremos   (p. 87) Valores de datos que son mucho mayores o menores que la mayoría de los otros valores en un digrama de tallo y hoja.

■ P ■

parabola   (p. 520) The locus of points whose distance from the focus is equal to the distance from a fixed line (the directrix).

parábola   (p. 520) Los lugares geométricos de puntos cuya distancia desde el foco es igual a la distancia desde una línea fija (la directriz).

parallel lines   (p. 119) Coplanar lines that do not intersect.

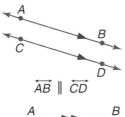

$\overrightarrow{AB} \parallel \overrightarrow{CD}$

líneas paralelas   (p. 119) Líneas coplanares que no se cruzan.

parallelogram   (p. 182) A quadrilateral whose two pairs of opposite sides are parallel.

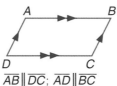

$\overline{AB} \parallel \overline{DC}$; $\overline{AD} \parallel \overline{BC}$

paralelogramo   (p. 182) Un cuadrilátero con dos pares de lados opuestos paralelos.

perfect square trinomial   (p. 492) A trinomial that results from squaring a binomial.

trinomio cuadrado perfecto   (p. 492) Un trinomio que resulta al elevar un binomio al cuadrado.

perimeter   (p. 206) The distance around a polygon.

perímetro   (p. 206) La distancia alrededor de un polígono.

period   (p. 628) The length of one complete cycle of a periodic function.

período   (p. 628) La longitud de un ciclo completo de una función periódica.

periodic function   (p. 628) A function that, when graphed, forms repeating patterns.

función periódica   (p. 628) Una función cuya gráfica forma patrones que se repiten.

permutation   (p. 403) An arrangement of items in a particular order.

permutación   (p. 403) Un arreglo de artículos en un orden particular.

perpendicular lines   (p. 119) Two lines that intersect to form adjacent right angles.

line $m \perp$ line n
recta $m \perp$ recta n

líneas perpendiculares   (p. 119) Dos líneas que se intersecan para formar ángulos rectos adyacentes.

plane   (p. 104) An infinite set of points extending in all directions along a flat surface.

plano   (p. 104) Un conjunto infinito de puntos que se extienden en cuatro direcciones a lo largo de una superficie plana.

English

Platonic solids (p. 221) The five polyhedrons studied by the Greek scholar, Plato. Each of the polyhedrons has faces that are congruent regular polygons.

point (p. 104) A specific location in space having no dimensions, represented by a dot, and named with a letter.

polygon (p. 178) A closed plane figure formed by joining three or more line segments at their endpoints. Each segment or side of the polygon intersects exactly two other segments, one at each endpoint.

polyhedron (p. 220) (*plural: polyhedra*) A closed three-dimensional figure made of only polygons.

polynomial (p. 468) An algebraic expression that is the sum of monomials. A polynomial is in standard form when its terms are ordered from the greatest to the least powers of one of the variables.

population (p. 82) The total number of people occupying a region or making up a whole.

positive correlation (p. 406) The direct relationship between two sets of data. On a scatter plot, a positive correlation is evident if the trend line slopes upward from the bottom left to the top right of the graph.

postulate (p. 105) A statement accepted as truth without proof.

precision (p. 202) The exactness to which measurement is made. Precision is relative to the unit of measurement used; the smaller the unit of measure, the more precise the measurement.

preimage (p. 338) The original figure of a translation.

prism (p. 220) A polyhedron that has two identical parallel bases and whose other faces are all parallelograms.

probability (p. 212) The chance or likelihood that an event will occur. The probability of an event can be expressed as a ratio:

$$P(\text{any event}) = \frac{\text{number of favorable outcomes}}{\text{number of possible outcomes}}.$$

An impossible event has a probability of zero. A certain event has a probability of one.

Español

sólidos platónicos (p. 221) Los cinco poliedros estudiados por el griego Platón, y cuyas caras son polígonos regulares congruentes.

punto (p. 104) Ubicación específica en el espacio que carece de dimensiones; se representa con una marca puntual y se denomina con una letra.

polígono (p. 178) Una figura cerrada plana formada al unir tres o mas segmentos en sus puntos extremos. Cada segmento o lado del polígono interseca exactamente dos otros segmentos, uno en cada extremo.

poliedro (p. 220) Cuerpo tridimensional cerrado compuesto solo de polígonos.

polinomio (p. 468) Expresión algebraica que es la suma de monomios. Un polinomio está en forma estándar cuando sus términos están ordenados según las potencias, de mayor a menor, de una de las variables.

población (p. 82) El número total de habitantes de una región que constituyen su totalidad.

correlación positiva (p. 406) La relación directa entre dos conjuntos de datos. En un diagrama de dispersión, una correlación positiva es evidente si la recta de tendencia se inclina hacia arriba desde la parte inferior izquierda hasta la parte superior derecha de la gráfica.

postulado (p. 105) Una declaración aceptada como verdadera sin prueba.

precisión (p. 202) El grado de exactitud al que se toma una medida. La precisión es relativa a la unidad de medida que se use; entre más pequeña es la unidad de medida, más precisa será la medida.

pre-imagen (p. 338) La figura original de una traslación.

prisma (p. 220) Un poliedro que tiene dos bases paralelas idénticas y cuyas otras caras son paralelogramos.

probabilidad (p. 212) La posibilidad de que un suceso ocurra.

$$P(\text{cualquier suceso}) = \frac{\text{número de resultados favorables}}{\text{número de resultados posibles}}.$$

Un suceso imposible tiene una probabilidad de cero. Un suceso seguro tiene una probabilidad de uno.

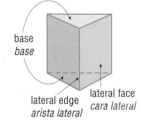

base
base

lateral face
cara lateral

lateral edge
arista lateral

triangular prism
prisma triangular

English

proportion (p. 296) An equation stating that two ratios are equivalent.

pyramid (p. 220) A polyhedron with only one base. The other faces are triangles that meet at a vertex.

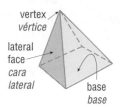

vertex
vértice

lateral
face
*cara
lateral*

base
base

rectangular pyramid
pirámide rectangular

Pythagorean Theorem (p. 430) In any right triangle, the square of the length of the hypotenuse c is equal to the sum of the squares of the lengths of the legs a and b. The Pythagorean Theorem is expressed as $c^2 = a^2 + b^2$.

Pythagorean triples (p. 433) Any three positive integers, a, b, and c, for which $a^2 + b^2 = c^2$.

Español

proporción (p. 296) Una ecuación que afirma que dos razones son equivalentes.

pirámide (p. 220) Un poliedro con una base y caras triangulares que se encuentran en un vértice.

teorema pitagórico (p. 430) En cualquier triángulo rectángulo, el cuadrado de la longitud de la hipotenusa c es igual a la suma de los cuadrados de las longitudes de los catetos a y b. El teorema pitagórico se expresa como $c^2 = a^2 + b^2$.

triples pitagóricos (p. 433) Enteros positivos a, b y c, para el cual $a^2 + b^2 = c^2$.

■ Q ■

quadrant (p. 56) One of the four regions formed by the axes of the coordinate plane.

quadratic equation (p. 493) An equation of the form $Ax^2 + Bx + C = 0$, where A, B, and C are real numbers and A is not zero.

quadratic term (p. 498) In a quadratic expression, the term that contains the squared variable.

quartiles (p. 407) The three values which divide an ordered set of data into four equal parts. The *lower quartile* is the median of the lower half of the data. The *upper quartile* is the median of the upper half. The *middle quartile* is the median of the entire set of data.

cuadrante (p. 56) Una de las cuatro regiones formadas por los ejes del plano coordenado.

ecuación quadrática (p. 493) Una ecuación de la forma $Ax^2 + Bx + C = 0$, donde A, B y C son números reales y A no es cero.

término cuadrático (p. 498) En una expresión cuadrática, el término que contiene la variable cuadrada.

cuartiles (p. 407) Los tres valores que dividen un conjunto ordenado de datos en cuatro partes iguales. El *cuartil inferior* es la mediana de la mitad inferior de los datos. El *cuartil superiores* la mediana de la mitad superior. El *cuartil central* es la mediana del conjunto de datos completo.

■ R ■

random sampling (p. 82) Statistical sampling in which each member of the population has an equal chance of being selected.

range (p. 49) The difference between the greatest and least values in a set of data.

range of relation (p. 56) The set of all possible y-coordinates for a relation.

rate (p. 204) A ratio that compares two different kinds of quantities.

muestra aleatoria (p. 82) Una muestra estadística en que cada miembro de la población tiene una oportunidad igual de ser seleccionado.

alcance (p. 49) La diferencia entre los valores mayores y menores en un conjunto de datos.

alcance de una relación (p. 56) El conjunto de todas las posibles coordenadas de y para una relación.

tasa (p. 204) Una razón que compara dos tipos diferentes de cantidades.

English

Español

ratio   (p. 202) A comparison of two numbers, a and b, represented in one of the following ways: $a{:}b$, $\frac{a}{b}$, or a to b.

razón   (p. 202) Una comparación de dos números, a y b. Se representan en una de las maneras siguientes: $a{:}b$, $\frac{a}{b}$ o a al b.

rationalizing the denominator   (p. 428) The process of rewriting a quotient to delete radicals from the denominator.

racionalizar el denominador   (p. 428) El proceso de escribir un cociente eliminando los radicales del denominador.

rational number   (p. 10) A number that can be expressed in the form $\frac{a}{b}$, where a and b are any integers and $b \neq 0$.

número racional   (p. 10) Un número que puede expresarse en la forma $\frac{a}{b}$, donde a y b son enteros y $b \neq 0$.

ray   (p. 105) Part of a line that starts at one endpoint and extends without end in one direction.

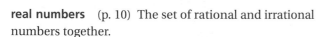

rayo   (p. 105) Parte de una línea que comienza en un punto y se extiende sin fin en una dirección.

real numbers   (p. 10) The set of rational and irrational numbers together.

números reales   (p. 10) El conjunto de números racionales e irracionales.

reciprocals   (p. 27) Two numbers that have a product of one.

recíprocos   (p. 27) Dos números cuyo producto es uno.

rectangle   (p. 183) A parallelogram that has four right angles.

rectángulo   (p. 183) Un paralelogramo que tiene cuatro ángulos rectos.

reduction   (p. 348) A dilated image that is smaller than the original figure.

reducción   (p. 348) Una imagen dilatada que es más pequeña que la figura original.

reference angle   (p. 624) In the coordinate plane, the acute angle formed by the x-axis and the terminal side.

ángulo de referencia   (p. 624) En el plano coordenado, el ángulo agudo formado por el eje de x y el lado terminal.

reflection   (p. 338) A transformation in which a figure is reflected, or flipped, over a line of reflection.

reflexión   (p. 338) Una transformación en que una figura se voltea sobre una línea de reflejo.

reflexive property   (p. 34) Any number is equal to itself. For example, $a = a$.

propiedad reflexiva   (p. 34) Para todos los números reales a, $a = a$.

regular polygon   (p. 179) A polygon that is both equilateral and equiangular.

polígono regular   (p. 179) Un polígono que es equilateral y equiangular.

rhombus   (p. 183) A parallelogram that has four congruent sides.

rombo   (p. 183) Un paralelogramo que tiene cuatro lados congruentes.

right triangle   (p. 150) A triangle that has one right angle.

triángulo rectángulo   (p. 150) Un triángulo que tiene un ángulo recto.

rotation   (p. 342) A transformation in which a figure is rotated, or turned, about a point.

rotación   (p. 342) Una transformación en que se gira una figura alrededor de un punto.

row-by-column multiplication   (p. 362) A method by which two matrices are multiplied together. Using this method, matrices can be multiplied together only when the number of columns in the first matrix is equal to the number of rows in the second matrix.

multiplicación por fila y columna   (p. 362) Un método en donde dos matrices se multiplican. Usando este método, las matrices se pueden multiplicar únicamente cuando el número de columnas en la primera matriz es igual al número de filas en la segunda matriz.

English

■ S ■

Español

sample   (p. 82) A representative portion of a population, often used for statistical study.

muestra   (p. 82) Una porción representativa de una población, frequentemente usada para el estudio estadístico.

sample space   (p. 385) The set of all possible outcomes of an event.

espacio muestral   (p. 385) El conjunto de todos los resultados posibles de un suceso.

scalar   (p. 359) The constant by which a matrix is multiplied.

escalar   (p. 359) La constante por la cual una matriz se multiplica.

scale factor   (p. 348) The number that is multiplied by the length of each side of a figure to create an altered image in a dilation.

factor de escala   (p. 348) El número que es multiplicado por la longitud de cada lado de una figura para crear una imagen alterada en una dilatación.

scalene triangle   (p. 150) A triangle with no congruent sides and no congruent angles.

triángulo escaleno   (p. 150) Un triángulo sin lados congruentes ni ángulos congruentes.

scatter plot   (p. 406) A method of visually displaying the relationship between two sets of data. The data are represented by unconnected points on a grid.

diagrama de dispersión   (p. 406) Método que presenta visualmente la relación entre dos conjuntos de datos. Los datos se representan mediante puntos conectados en una cuadrícula.

scientific notation   (p. 39) A system for writing a very large or very small number as the product of a factor that is greater than or equal to one and less than 10 and a second factor that is a power of 10. For example, 496,000,000 written in scientific notation is $4.96 \cdot 10^8$.

notación científica   (p. 39) Un sistema para escribir números grandes o pequeños como el producto de un factor mayor o igual a uno y menos de 10 y un segundo factor que es una potencia de 10. Por ejemplo, 496,000,000 escrito en notación científica es $4.96 \cdot 10^8$.

secant   (p. 441) A line that intersects a circle in two places.

secante   (p. 441) Una línea que interseca un círculo en dos lugares.

$\overleftrightarrow{CD}$ is a secant of $\odot P$.
$\overleftrightarrow{CD}$ es una secante de $\odot P$.

secant segment   (p. 448) A segment intersecting a circle in two points, having one endpoint on the circle and one endpoint outside the circle.

segmento secante   (p. 448) Segmento que interseca un círculo de dos puntos y que tiene un extremo en el círculo y el otro fuera del círculo.

sequence   (p. 52) An arrangement of numbers according to a pattern.

sucesión   (p. 52) Un arreglo de números según un patrón.

set   (p. 6) A well-defined collection of items. Each item is called an element, or member, of the set.

conjunto   (p. 6) Una colección bien definida de artículos que se le llama un elemento, o miembro.

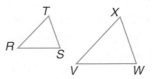

similar figures   (p. 300) Figures that have the same shape but not necessarily the same size.

figuras semejantes   (p. 300) Figuras que tienen la misma forma pero no el mismo tamaño.

$$\angle T \cong \angle X,\ \angle S \cong \angle W,\ \angle R \cong \angle V;\ \frac{RT}{VX} = \frac{ST}{WX} = \frac{RS}{VW}$$

English

simulation (p. 388) A model used to estimate the probability of an event.

sine (p. 614) In a right triangle, the sine of acute $\angle A$ is equal to: $\dfrac{\text{length of side opposite } \angle A}{\text{length of hypotenuse}}$.

skew lines (p. 119) Noncoplanar lines that do not intersect and are not parallel.

slope (p. 244) The ratio of the vertical change of a line (rise) to its horizontal change (run).

slope-intercept form (p. 245) A linear equation in the form $y = mx + b$, where m represents the slope, and b represents the y-intercept.

solution (p. 7) A replacement set for a variable that makes a mathematical sentence true.

sphere (p. 221) A three-dimensional figure consisting of the set of all points that are a given distance from a given point, called the center of the sphere.

C is the center of the sphere.
C es el centro de la esfera.

spreadsheet (p. 30) A computer application that simplifies preparation of tables.

square (p. 183) A parallelogram that has four right angles and four congruent sides.

square root (p. 426) One of two equal factors of a number. A number a is a square root of another number b if $a^2 = b$.

standard deviation (p. 417) The square root of the variance of a set of numbers.

standard equation of a circle (p. 562) The equation of a circle with its center at any coordinate point (h, k) is $(x - h)^2 + (y - k)^2 = r^2$, where r is the circle's radius. If the circle's center is at the origin, the standard equation reduces to $x^2 + y^2 = r^2$.

standard equation of an ellipse (p. 574) The equation of an ellipse with its center at the origin is: $\dfrac{x^2}{a^2} + \dfrac{y^2}{b^2} = 1$.

standard equation of a hyperbola (p. 576) The equation of a hyperbola with its center at the origin is: $\dfrac{x^2}{a^2} - \dfrac{y^2}{b^2} = 1$.

standard quadratic equation (p. 524) A quadratic equation written in the form $y = ax^2 + bx + c$, where a, b, and c are real numbers and $a \neq 0$.

statistics (p. 82) A branch of mathematics that involves the study of data, specifically the methods used to collect, organize, and interpret data.

Español

simulación (p. 388) Un modelo usado para estimar la probabilidad de un suceso.

seno (p. 614) En un triángulo rectángulo, el seno del ángulo agudo A es igual a: $\dfrac{\text{longitud del lado opuesto al } \angle A}{\text{longitud de la hipotesusa}}$.

líneas alabeadas (p. 119) Líneas no coplanares que no se intersecan ni son paralelas.

pendiente (p. 244) La razón del cambio vertical de una línea (subida) a su cambio horizontal (recorrido).

forma de pendiente e intersección (p. 245) Una ecuación lineal en la forma $y = mx + b$, donde m representa la pendiente y b representa el intercepto de y.

solución (p. 7) Un conjunto de reemplazo para una variable que hace que una frase matemática sea cierta.

esfera (p. 221) Una figura tridimensional que consiste del conjunto de todos los puntos equidistantes de un punto fijo, llamado el centro de la esfera.

hoja de cálculos (p. 30) Una aplicación de computadora que simplifica la preparación de tablas.

cuadrado (p. 183) Un paralelogramo que tiene cuatro ángulos rectos y cuatro lados congruentes.

raíz cuadrada (p. 426) Uno de dos factores iguales de un número. Un número a es una raíz cuadrada de otro número b si $a^2 = b$.

desviación estándar (p. 417) La raíz cuadrado de la varianza de un conjunto de números.

ecuación estándar de un círculo (p. 562) La ecuación de un círculo con su centro en cualquier punto coordenado (h, k) es $(x - h)^2 + (y - k)^2 = r^2$, donde r es el radio del círculo. Si el centro del círculo está en el origen, la ecuación estándar se reduce a $x^2 + y^2 = r^2$.

ecuación estándar de un elipse (p. 574) La ecuación de un elipse con su centro en el origen es: $\dfrac{x^2}{a^2} + \dfrac{y^2}{b^2} = 1$.

ecuación estándar de una hipérbola (p. 576) La ecuación de una hipérbola con su centro en el origen es: $\dfrac{x^2}{a^2} - \dfrac{y^2}{b^2} = 1$.

ecuación cuadrática estándar (p. 524) Una ecuación cuadrática en la forma $y = ax^2 + bx + c$; a, b, y c son números reales y $a \neq 0$.

estadisticas (p. 82) Una rama de las matemáticas que comprende el estudio de datos, específicamente los métodos usados para coleccionar, organizar e interpretar datos.

English

stem-and-leaf plot (p. 86) A method of displaying data in which certain digits are used as stems and the remaining digits are used as leaves.

subset (p. 6) If every element of set A is also an element of set B, then A is called a subset of B.

substitution property (p. 34) If expressions are equivalent, they may be substituted for one another in any statement. For example, if $a = b$, then b can be substituted for a or a can be substituted for b in any statement.

supplementary angles (p. 109) Two angles whose measures have a sum of 180°.

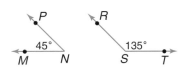

$m\angle MNP + m\angle RST = 180°$

surface area (p. 224) The sum of the areas of all the faces of a three-dimensional figure.

symmetric property (p. 34) The expressions on either side of an equals sign are equivalent and can thus be switched without affecting the equation.

system of equations (p. 258) Two or more linear equations with the same variables.

system of linear inequalities (p. 276) Two or more linear inequalities that can be solved by graphing.

Español

diagrama de tallo y hoja (p. 86) Un método para mostrar datos en que ciertos dígitos se usan como los tallos y los dígitos restantes se usan como las hojas.

subconjunto (p. 6) Si cada elemento de un conjunto A es también un elemento del conjunto B, entonces A se llama un subconjunto de B.

propiedad de sustitución (p. 34) Si son equivalentes, las expresiones se pueden reemplazar mutuamente en cualquier enunciado. Por ejemplo, si $a = b$, la b puede reemplazarse por a o la a puede reemplazarse por b en cualquier enunciado.

ángulos suplementarios (p. 109) Dos ángulos cuyas medidas tienen una suma de 180°.

área de superficie (p. 224) La suma de las áreas de todas las caras de una figura tridimensional.

propiedad de simetría (p. 34) Las expresiones en cualquier lado de un signo de igualdad son equivalentes y se pueden intercambiar sin afectar la ecuación.

sistema de ecuaciones (p. 258) Dos o más ecuaciones lineales con las mismas variables.

sistema de desigualdades lineales (p. 276) Dos o más desigualdades lineales que se pueden resolver usando gráficas.

■ T ■

tangent (p. 614) In a right triangle, the tangent of acute $\angle A$ is equal to:

$$\frac{\text{length of side opposite } \angle A}{\text{length of side adjacent to } \angle A} \text{ or } \frac{\text{sine } \angle A}{\text{cosine } \angle A}.$$

tangent of a circle (p. 441) A line that intersects a circle in only one point.

tangent segment (p. 449) A segment with one endpoint on a circle and one endpoint outside the same circle.

terminal side (p. 624) In the coordinate plane, the side of an angle that is not the initial side; the terminal side is the side to which one measures degrees from the initial side.

terms (p. 52) The parts of a variable expression that are separated by addition or subtraction signs.

theorem (p. 114) A statement whose truth can be proven.

tangente (p. 614) En un triángulo rectángulo, la tangente del ángulo agudo A es igual a:

$$\frac{\text{longitud del lado opuesto al } \angle A}{\text{longitud del lado adyacente al } \angle A} \text{ ó } \frac{\sin \angle A}{\cos \angle A}.$$

tangente de un círculo (p. 441) Una línea que interseca un círculo en un solo punto.

segmento tangente (p. 449) Un segmento con un extremo sobre un círculo y el otro extremo fuera del mismo círculo.

lado terminal (p. 624) En el plano coordenado, el lado de un ángulo que no es el lado inicial; el lado terminal es el lado en el que se miden los grados desde el lado inicial.

términos (p. 52) Las partes de una expresión variable que son separadas por signos de sustracción o adición.

teorema (p. 114) Una declaración cuya verdad puede probarse.

English

theoretical probability   (p. 385) The probability of an event, $P(E)$, assigned by determining the number of favorable outcomes and the number of possible outcomes in the sample space: $P(E) = \dfrac{\text{number of favorable outcomes}}{\text{number of possible outcomes}}$.

transformation   (p. 338) A way of moving or changing the size of a geometric figure in the coordinate plane.

transitive property of equality   (p. 34) If two expressions are equivalent, and a third expression is equivalent to the second expression, then the third expression is also equivalent to the first. For example, if $a = b$ and $b = c$, then $a = c$.

transitive property of inequality   (p. 77) The property that states: For real numbers a, b, and c, if $a < b$ and $b < c$, then $a < c$. Similarly, if $a > b$ and $b > c$, then $a > c$.

translation   (p. 338) A change in position of a figure such that all the points in the figure slide exactly the same distance and in the same direction at once.

transversal   (p. 120) A line that intersects at least two coplanar lines in different points, producing interior and exterior angles.

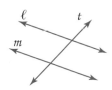

Line t is a transversal.
La recta t es una transversal.

trapezoid   (p. 188) A quadrilateral that has exactly one pair of parallel sides.

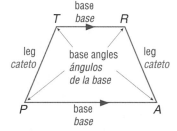

tree diagram   (p. 385) A diagram that shows all the possible outcomes in a sample space.

trend line   (p. 406) A line that can be drawn near most of the points on a scatter plot that shows the relationship between two sets of data; also called the line of best fit.

triangle   (p. 150) A polygon formed by three line segments joining three noncollinear points.

trinomial   (p. 468) A polynomial with three terms.

Español

probabilidad teórica   (p. 385) La probabilidad de un evento, $P(E)$, asignada al determinar el número de resultados favorables y el número de resultados posibles en un espacio muestral: $P(E) = \dfrac{\text{número de resultados favorables}}{\text{número de resultados posibles}}$.

transformación   (p. 338) Una manera de mover o cambiar el tamaño de una figura geométrica en el plano coordenado.

propiedad transitiva de la igualdad   (p. 34) Si dos expresiones son equivalentes y una tercera expresión es equivalente a la segunda expresión, entonces la tercera expresión también es equivalente a la primera. Por ejemplo, si $a = b$ y $b = c$, la $a = c$.

propiedad transitiva de la igualdad   (p. 77) Para números reales a, b, y c, si $a < b$ y $b < c$, entonces $a < c$. De la misma manera, si $a > b$ y $b > c$, entonces $a > c$.

traslación   (p. 338) Un cambio en la posición de una figura tal que todos los puntos en la figura se deslizan exactamente a la misma distancia y en la misma dirección.

transversal   (p. 120) Una línea que interseca por lo menos dos líneas coplanares en puntos diferentes produciendo ángulos interiores y exteriores.

trapezoide   (p. 188) Un cuadrilátero que tiene exactamente un par de lados paralelos.

diagrama de árbol   (p. 385) Un diagrama que muestra todos los resultados posibles en un espacio de muestra.

línea de tendencia   (p. 406) Una línea que puede dibujarse cerca de la mayoría de los puntos en un diagrama de dispersión que muestra la relación entre dos conjuntos de datos.

triángulo   (p. 150) Un polígono formado por tres segmentos que unen tres puntos no colineales.

trinomio   (p. 468) Un polinomio con tres términos.

English

Español

■ U ■

unit rate   (p. 204) A rate that has a denominator of one unit.

unlike terms   (p. 468) Terms in which the variables or sets of variables are not identical.

upper quartile   (p. 407) In statistics, the median of the upper half of the data.

tasa de unidad   (p. 204) Un valor que tiene un denominador de una unidad.

términos diferentes   (p. 468) Términos en que las variables o los conjuntos de variables no son idénticos.

cuartil superior   (p. 407) En estadísticas, la mediana de la mitad superior de los datos.

■ V ■

variable   (p. 7) A symbol, usually a letter, used to represent a number.

variance   (p. 412) For a set of numbers, the mean of the squared differences between *each* number in the set and the mean of *all* numbers in the set.

vertex angle   (p. 160) In an isosceles triangle, the angle opposite the base and adjacent to the two legs.

vertex of a polygon   (p. 178) The point at which two sides of a polygon meet.

vertex of a polyhedron   (p. 220) The point at which three or more edges of a polyhedron intersect.

variable   (p. 7) Un símbolo, comúnmente una letra, usado para representar un número.

varianza   (p. 412) Para un conjunto de números, la media de las diferencias cuadradas entre *cada* número en el conjunto y la media de *todos* los números en el conjunto.

ángulo de vértice   (p. 160) En un triángulo isósceles, el ángulo opuesto a la base y adyacente a los dos lados.

vértice de un polígono   (p. 178) El punto donde dos lados de un polígono se encuentran.

vértice de un poliedro   (p. 220) El punto donde tres o más aristas de un poliedro se intersecan.

vertical angles   (p. 115) The angles that are not adjacent to each other when two lines intersect. Vertical angles are congruent.

$\angle 1$ and $\angle 3$ are vertical angles.
$\angle 2$ and $\angle 4$ are vertical angles.
$\angle 1$ y $\angle 3$ son ángulos opuestos por el vértice.
$\angle 2$ y $\angle 4$ son ángulos opuestos por el vértice.

ángulos verticales   (p. 115) Los ángulos que no son adyacentes uno al otro cuando dos líneas se intersecan. Los ángulos verticales son congruentes.

vertical line test   (p. 57) A test used to determine whether or not a graph is a function. It states: When a vertical line is drawn through the graph of a relation, the graph is not a function if the vertical line intersects the graph in more than one point.

volume   (p. 230) A measure of the number of cubic units needed to fill a region of space.

prueba de verticalidad de línea   (p. 57) Una prueba que se usa para determinar si una gráfica es una función o no. Cuando una línea vertical se dibuja a través de la gráfica de una relación, la gráfica no es una función si la línea vertical cruza la gráfica en más de un punto.

volumen   (p. 230) Una medida del número de unidades cúbicas necesarias para llenar una región de espacio.

■ X ■

x-coordinate   (p. 56) The first number in an ordered pair. The x-coordinate determines the horizontal location of a point in a coordinate plane. Also called the abscissa.

x-intercept   (p. 245) The x-intercept of a line is the x-coordinate of the point where the line intersects the x-axis.

coordenada de x   (p. 56) El primer número en un par ordenado. La coordenada de x determina la ubicación horizontal de un punto en un plano coordenado. También se le llama la abscisa.

intercepto de x   (p. 245) El intercepto de x de una línea es la coordenada de x del punto donde la línea cruza el eje de x.

English   Español

■ Y ■

y-coordinate (p. 56) The second number in an ordered pair. The y-coordinate determines the vertical location of a point in a coordinate plane. Also called the ordinate.

coordenada de y (p. 56) El segundo número en un par ordenado. La coordenada de y determina la ubicación vertical de un punto en un plano coordenado. También se le llama la ordenada.

y-intercept (p. 245) The y-intercept of a line is the y-coordinate of the point where the line intersects the y-axis.

intercepto de y (p. 245) El intercepto de y de una línea es la coordenada de y del punto donde la línea cruza el eje de y.

■ Z ■

z-score (p. 413) The number of standard deviations between a score and the mean score.

calificación z (p. 413) El número de desviaciones estándares entre una calificación y la calificación media.

Selected Answers

Chapter 1: Essential Mathematics

Lesson 1-1, pages 6–9

1. $6 \notin \{1, 3, 5, 7, 9\}$ 3. -1 5. 11 7. No; there are eight subsets: three 1-element subsets, three 2-element subsets, one 3-element subset, and the empty set. 9. $\varnothing$ 11. true 13. $\{a\}, \varnothing$ 15. $\{t, e\}, \{t, n\}, \{e, n\}, \{t, e, n\}, \{t\}, \{e\}, \{n\}, \varnothing$ 17. 4 19. 1 21. -6 23. -3 25. false 27. $\{1, 2\} \subseteq \varnothing$ 29. $x = \{\ \}$ 31. $\frac{1}{12}$ 33. 6 35. 8 37. 32 39. $6, -6$ 41. $P = \{\ \}$ or $P = \varnothing$ 43. $8 \notin R$

Lesson 1-2, pages 10–13

1. false 3. false 5.

7.

9. 15 11. true 13. false

15.

17. 19. 9

21. $=$ 23. $>$ 25. $\{x \mid x$ is a real number, and $x < 4\}$ 27. $32 < 212$ 29. 284 million > 126 million 31. Volga $<$ Mackenzie or $2{,}194 < 2{,}635$.

33.

35. Answers will vary. 37. -10 39. -2 41. 3 43. 0

Review and Practice Your Skills, pages 14–15

1. $\varnothing, \{10\}, \{15\}, \{10, 15\}$ 3. -5 5. $8, -8$ 7. $6, -6$ 9. 33 nickels 11. 8 13. -5.5 15. $\varnothing \subseteq \{b, u, g\}$ 17. true 19. false 21. true 23. false 25. 8 27. 8

29.

31.

33. $<$ 35. $<$ 37. $>$ 39. $6, m$ and i 41. false 43. false 45. true 47. true 49. irrational, real 51. whole, integer, rational, real 53. irrational, real

Lesson 1-3, pages 16–19

1. $\{1, 2, 3, 4, a, b, c\}$ 3. $\{1, 2, 3, 4, a, b, c\}$

5.

7. {ban, can, fan, man, pan, ran, tan} 9. {ban, can, fan, man, ran, van, wan} 11. $\{c\}$ 13. $\{c\}$ 15. $\{c, h\}$ 17. $\{\ \}$ or $\varnothing$ 19. $\{x \mid x$ is a real number and $x \le 1$ or $x \ge 2\}$ 21. $\{1, 2, 4, 5, 7, 8\}$ 23. $\{0, 1, 2, 3, 4, 5, 6, 9\}$ 25. $\{1, 9\}$ 27. $\{0, 1, 3, 5, 7, 9\}$ 29. $\{3, 5\}$ 31. $\{1\}$ 33. 3 35. true 37. false, N 39. true

Lesson 1-4, pages 20–23

1. -27 3. $11\frac{1}{4}$ 5. -3.6 7. 3.6 9. No; 152 ft 11. Yes, round the amounts and add: $\$5 + \$8 + \$2 + \$9 = \$24$ 13. -3.9 15. -24.4 17. -8 19. 43 21. -11 23. $>$ 25. $<$ 27. 5.7 ft 29. 16 31. $\{1, 2, 5, 7, 8, 9\}$ 33. $\{0, 3, 4, 5, 6, 7\}$ 35. $\{1, 8\}$ 37. 3 39. -3 41. -3 43. -3 45. 20 47. 38 49. 38 51. $\frac{13}{2}$ or 6.5 53. 12 55. 0 57. -24

Review and Practice Your Skills, pages 24–25

1. $\{p, u, t, e, r\}$ 3. $\{e, r\}$ 5. $\varnothing$ 7. $\{p, u, t, e, r\}$ 9. $\{c, o, m, p, u, t, e, r\}$ 11.

13. 15.

17. 19. 15 21. 200 23. 5 25. -67 27. 455 29. -115 31. about 20 ft 33. No 35. false 37. false 39. false 41. true 43. -15 45. $\{c, l, e, a, n, y, r\}$ 47. $\{g, j, p, v, z\}$ 49. $\varnothing$ 51. 9.605 53. 48 55. 7.4

Lesson 1-5, pages 26–29

1. -4.93 3. -1 5. -32 7. 5.1 9. -1.2 11. 2 13. $8\frac{1}{8}$ 15. $-4°F$ 17. -21.84 19. 40 21. -500 23. 0.4 25. $4\frac{1}{3}$ 27. -6 29. -10.56 31. no; 22 33. 0.1 35. 8.4 37. 34.7 39. 193.75 calories 41. $+ \div$ 43. $\div + \div$ 45. $\frac{1}{1000} \times 16 = \frac{16}{1000} = \frac{4}{250} = \frac{2}{125}$ 47. -21 49. -4 51. -24 53. $-\frac{3}{2}$ 55. -18 57. 0 59. $\{0, 2, 3, 4, 5, 6, 8\}$ 61. $\{0, 2, 3, 4, 5, 8\}$ 63. $\{1, 3, 5, 7, 9\}$ 65. $\{0, 1, 7, 9\}$ 67. 27.5%

Lesson 1-6, pages 30–31

1. 2; 4; 7; 14 3. 8; 16; 19; 38 5. Sample answer: Store the value 4 for L. Enter the formulas for the perimeter, $2L + 2L * 0.5$ and area, $L * L * 0.5$. The perimeter is 12 ft and the area is 8 ft^2. 7. 133.65 in.^{2} 9. 1.65 yd^2 11. -1312 13. 7362 ft 15. -12.04 17. -30.91 19. 22.60 21. -2.98

Review and Practice Your Skills, pages 32–33

1. -21.84 3. $\frac{1}{10}$ 5. -1 7. -8.64 9. $\frac{9}{64}$ 11. -61.5 13. $-8\frac{8}{9}$ 15. $-\frac{1}{3}$ 17. $\frac{5}{8}$ 19. Letting $H =$ the number of hours worked, Ben's pay is equal to $8.82(40) + (H - 40)(1.5)(8.82)$. 21. $\$317.52$ 23. $\$273.42$ 25. $1, 6, -11$ 27. $2, 7, -12$ 29. $\varnothing, \left\{\frac{1}{2}\right\}, \left\{\frac{3}{5}\right\}, \left\{\frac{1}{2}, \frac{3}{5}\right\}$ 31. $\varnothing, \{0\}$ 33. $\{4, 8, 12, 16, 20, \ldots\}$ 35. $\{x \mid x$ is a natural number or a negative non-multiple of 4$\}$ 37. $\{0\}$ 39. Yes 41. 938.91 43. $\frac{1}{3}$ 45. $55\frac{2}{5}$

41.

43.

45.

Lesson 1-7, pages 34–37

1. 160 3. 284 5. −64 7. −576 9. c^4 11. c^{15}
13. $a^8 b^{12}$ 15. x^{11} 17. 4346 19. 9 21. 64 23. 96
25. −8 27. 2^{25} 29. r^9 31. $\frac{1}{d^9}$ 33. a^{25} 35. 35
37. 28 39. c^{18} 41. 1000^4 and $10,000^3$ both equal 10^{12}
43. 27 in.^{3}, 125 cm^3, e^3 cm^3, g^3 in.^{3} 45. true 47. 2000
49. $231.50 51. $292.60

Lesson 1-8, pages 38–41

1. $\frac{1}{16}$ 3. r^{-4} 5. m^{-21} or $\frac{1}{m^{21}}$ 7. d^{17} 9. $\frac{1}{9}$ 11. $\frac{1}{16}$
13. $5.9 \cdot 10^{10}$ 15. $6.052 \cdot 10^{-5}$ 17. 0.00043 19. $3.6 \cdot 10^7$
21. $\frac{1}{d^{12}}$ 23. $\frac{1}{64}$ 25. $-\frac{1}{64}$ 27. $9.3 \cdot 10^3$ 29. $2.15 \cdot 10^{-6}$
31. $3.9 \cdot 10^{-5}$, 0.000039 33. $2.592 \cdot 10^{15}$ m 35. y^{-2} or $\frac{1}{y^2}$
37. r^8 39. $9.00 41. 30.48 cm 43. $\frac{r^5}{25}$ 45. $\frac{1}{m^7}$
47. 25 49. −16 51. 225 53. −135 55. 34 57. 27
59. m^{10} 61. m^3 63. m^{37} 65. m^8 67. m^{16} 69. m^{14}

Chapter 1 Review, pages 42–44

1. j 3. c 5. d 7. l 9. e 11. false 13. true 15. −3
17.  19. $<$ 21. $=$ 23. {6, 12} 25. ∅
27. 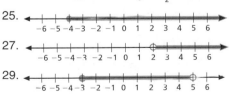 (all real numbers) 29. $\frac{5}{8}$
31. $\frac{1}{12}$ 33. $\frac{7}{12}$ 35. 12 37. 640 39. 1 41. $-\frac{27}{50}$ 43. 38
45. x^{11} 47. $\frac{a^4 b^6}{c^2}$ 49. $\frac{1}{16}$ 51. x^1 or $\frac{1}{x}$ 53. $-\frac{1}{8}$
55. $3.92 \cdot 10^{-4}$ 57. 46,000 59. 0.0000278

Chapter 2: Essential Algebra and Statistics

Lesson 2-1, pages 52–55

1. Add 4; 1, 5, 9. 3. Subtract 3; −7, −10, −13.
5. Multiply by 3; 162, 486, 1458. 7. 12, 17, 23
9. 21, 31, 43 11. 125, 216, 343 13. He added 6.2 rather than 5.2. 15. $300; $450; $1012.50; $1518.75
17. 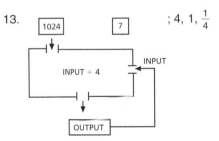 19. $1500 21. $637.50, $785.31, $944.21, $1115.03, $1298.66
23.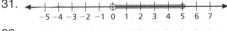
25.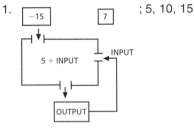
27.
29.
31.
33.

Lesson 2-2, pages 56–59

1–3. 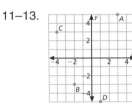 5. −2 7. −7 9. no; domain: {−3, −2, 0, 1}; range: {−3, −1, 0, 1}

11–13. 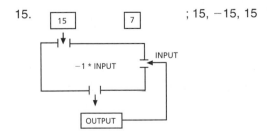 15. −17 17. 7
19. {(−2, −4), (0, −2), (2, 0), (4, 2)}; domain: {−2, 0, 2, 4}; range: {−4, −2, 0, 2} 21. 49
23. 17 25. 48 27. $1 + b$
29. c^3 31. No 33. Yes
35. $230 37. {0, 6, 8, 14, 20}
39. {0, 3, 5, 6, 10, 13, 14, 18, 20} 41. {0, 3, 6, 8, 14, 20}
43. {0, 3, 6, 8, 10, 14, 20}

Review and Practice Your Skills, pages 60–61

1. 80, −160, 320 3. 23, 33, 45 5. −7, −10, −13
7. $\frac{16}{7}, \frac{32}{7}, \frac{64}{7}$ 9. 0.125, 0.0625, 0.03125
11. 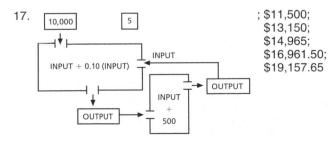 ; 5, 10, 15
13. ; 4, 1, $\frac{1}{4}$
15. ; 15, −15, 15
17. ; $11,500; $13,150; $14,965; $16,961.50; $19,157.65

19–25.

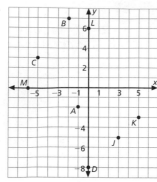

27. 25   29. −24
31. 0   33. 156
35. D: {0, 3, 4},
R: {0, 6, 12, 18};
not a function
37. (0, 3), (−1, 4),
(1, 4), (−2, 7),
(2, 7); D: {−2, −1,
0, 1, 2}, R: {3, 4, 7}
39. −136, 17,
170   41. *ZYXWV,*
ZYXWVU,
ZYXWVUT
43. 10^5, $−10^6$, 10^7

45. 12   47. $1\frac{1}{2}$ 49. II

Lesson 2-3, pages 62–65

1. $3 − x = 3$ 3. 1   5. $60,

7.   9.   11. 1

13.   15. 7   17.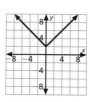

19. 8   21. 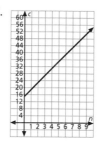   23.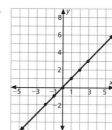

25. −16   27. 12   29. −3   31. 13,000   33. 16,000
35. $170.00   37. $170.00

Lesson 2-4, pages 66–69

1. 14   3. −9   5. −56   7. $\frac{1}{4}$ 9. −2.5
11. $n + 15 = \left(\frac{1}{2}\right)(72)$ 13. $n − 26 = −9$ 15. 22
17. −25   19. −0.4   21. −9   23. 9.34   25. $n + 25 = −15$, $n = −40$ 27. $\frac{n}{8} = 0.7$, $n = 5.6$ 29. −2   31. 25.5
33. −10   35. 6, −6   37. 0   39. 11   41. 40
43. Answers will vary; $|x| = −3$.   45. $0.6p = 101.25;
$168.75   47. −8   49. −23   51. 90   53. −13   55. 12

Review and Practice Your Skills, pages 70–71

1. 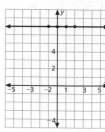   3.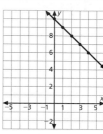

5. 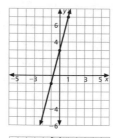   7.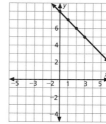

9. 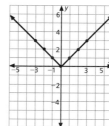   11. 13   13. −10
15. 7   17. $177
19. −3   21. 3
23. −6   25. $1\frac{1}{4}$
27. $\frac{1}{4}$ 29. 26   31. −102.3
33. −81   35. −70   37. $112\frac{2}{3}$
39. −36   41. 2   43. $−\frac{1}{4}$
45. −36   47. $−\frac{4}{3}$ 49. −2.2
51. −16   53. 3, −8, −19

55. −70, −44, −18

57.   59. $n + 13 = −29$, $n = −42$
61. $\frac{n}{−4} = 11$, $n = −44$
63. $2\frac{3}{8}$ 65. −4

Lesson 2-5, pages 72–75

1. 6   3. −6   5. −4   7. −11   9. $1.75   11. 3   13. −7
15. 4   17. −2   19. 2   21. −1   23. $3n + 4 = 31$, $n = 9$
25. 3   27. 1   29. −3.5   31. 11   33. $29.75   35. The
solution set is {all real numbers}. Any real number will
satisfy the equation.   37. Answers may vary.   39. 72
41. 13   43. 216   45. 27   47. 96   49. 100   51. 4.5 ·
$10^{−8}$ 53. 3.9 · $10^{−11}$ 55. 2.6 · $10^{−9}$

Lesson 2-6, pages 76–79

1. 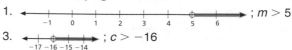   ; $m > 5$
3. 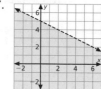   ; $c > −16$
5.   7.

9.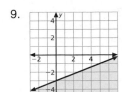

11.  ; $p > 6$

13.  ; $a > 7$

15. 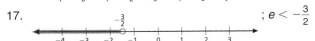 ; $z > 3$

17. 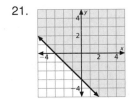 ; $e < -\dfrac{3}{2}$

19. 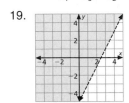   **21.**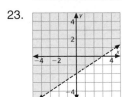

23.    **25.** $h < 18$ h

27.  ; $c \le -6.2$

29. 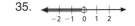 ; $d \le 0$

31. $y \le x$ **33.** $x + y \ge 4$ **35.** Students should discuss and compare the solutions as they appear on both a number line and a coordinate plane.   **37.** rule: $\times 4$; 1024; 4096; 16,384   **39.** rule: $+ 3$; 16, 19, 22   **41.** rule: $\times -3$; -243, 729, -2187

Review and Practice Your Skills, pages 80–81

1. 9   **3.** -3 **5.** 0   **7.** 20   **9.** 3.6 **11.** 84 **13.** -16 **15.** 7 **17.** 45 **19.** $-2\dfrac{4}{7}$ **21.** 4   **23.** 12.5   **25.** $6 + 5n = -29$, $n = -7$ **27.** $\dfrac{1}{3}n + \dfrac{1}{2} = 3\dfrac{1}{2}$, $n = 9$ **29.** $3(n + 2) = -27$, $n = -11$ **31.** ; $g < -12$

33.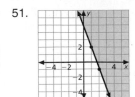

$n < -3$

35.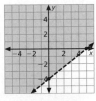

$d < 0$

37.

$c > \dfrac{2}{3}$

39.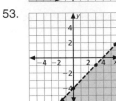

$-27 > k$

41.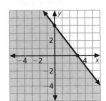

$-8.5 \ge h$

43.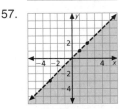

$x \ge \dfrac{5}{3}$

45.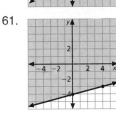

$m > 4$

47.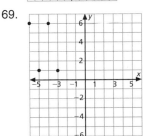

$x \le -2\dfrac{1}{6}$

49.

51.

53.

55.

57.

59.

61.

63. 2.4, 2.7, 3.0
65. -8.4, -10.8, -13.2
67. 1.21212, 1.212121, 1.2121212

69. ; $\{(-3, 1), (-4, 6), (-5, 1), (-6, 6)\}$; D: $\{-6, -5, -4, -3\}$; R: $\{1, 6\}$; Yes, it is a function.   **71.** 3   **73.** $-2\dfrac{2}{5}$ **75.** -98 **77.** -22 **79.** -24

Lesson 2-7, pages 82–85

1.

NUMBER OF ABSENCES PER STUDENT		
Absences	Tally	Frequency
0		2
1		5
2		4
3		3
4		3
5		1
6		0
7		0
8		1
9		1

3. median   **5.** 390–399
7. 1.9, 2, 1
9. $2.90–$2.99
11. 74   **13.** 100–109
15. greater than   **17.** 5
19. 26   **21.** 0   **23.** 16
25. -2

Lesson 2-8, pages 86–89

1.

Aptitude Test Scores	
1	6
2	7 8 8
3	2 3 4 8 8 8
4	0
5	0 2 5 6 6 7 9
6	0 0 1 2 5 6 9
7	1 1 4 5
8	6

1 | 6 represents a score of 16 on a scale of 0–100.

3. 38   **5.** 51.9

7. Time Spent on Homework

```
1 | 7
2 |
3 | 0 0 1 5 8
4 | 1 3 3 3 5 5 5 6 6 7
5 | 0 2 3 5 5 8 8
6 |
7 | 2 5 5 7 8
8 | 1        1 | 7 represents
9 | 8        17 min.
```

9. 43 and 45   **11.** 52.1

13.

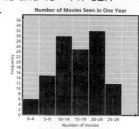

15. 21   **17.** 60.00–79.99   **19.** Answers will vary.

21.

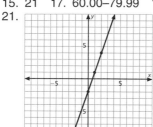

23.

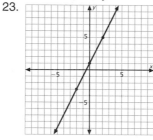

25.

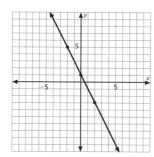

27. 41   **29.** −10   **31.** 8
33. 4.8   **35.** −15

25–31.

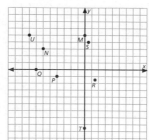

25. II
27. x-axis
29. I
31. II

33.

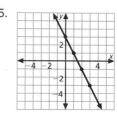

35.

37.

39. 23   **41.** $-4\frac{1}{2}$
43. −150
45. $2\frac{2}{3}$ **47.** −19
49. −6

51. −7   **53.** $\longleftarrow\!\!\circ\!\!\longrightarrow$; $x > 5\frac{3}{4}$

55. $\longleftarrow\!\!\!\!\!\!\!\!\!\!\!\!\!\!\!\!\!\!\!\bullet\!\!\!\!\!\longrightarrow$; $g \geq 6$

57.

59. about 29.1, 30, 22
61. mean; it is influenced by the salaries of president and group manager and will be higher than median salary

Review and Practice Your Skills, pages 90–91

1. 84, 81, 77   **3.** 240, 235, 235 and 210   **5.** 76.5, 75, 70

7.

Volumes	Frequency
201–210	12
211–220	7
221–230	5
231–240	3

201–210 : 12, 211–220 : 7, 221–230 : 5, 231–240 : 3
9. 211–220; 211–220

11.
```
20 | 1 2 5 6 7 8 8
21 | 0 0 0 0 0 5 5 5 8 8 8 8
22 | 4 5 8
23 | 0 0 4 5 8     20 | 1 represents 201mL.
```

13.

15. 23   **17.** 13–15   **19.** mean: increased by 3 h; median: increased by 3 h; mode: increased by 3 h   **21.** 67, 95, 123
23. Q, U, Y

Lesson 2-9, pages 92–93

1. Yes; the horizontal scale does not have uniform increments starting with zero—although one line is twice the length of the other, it does not show twice the number of hits.
3. different size intervals on the vertical scale
5. Graphs will vary.   **7.** 5   **9.** $2\frac{1}{3}$ **11.** $\frac{2}{3}$ **13.** 7   **15.** −2.5
17. 16   **19.** 10   **21.** 17   **23.** 0.1

Chapter 2 Review, pages 94–96

1. c   **3.** e   **5.** l   **7.** h   **9.** a   **11.** 1, $\frac{1}{4}$, $\frac{1}{16}$, divide by 4
13. 16, −32, 64, multiply by −2   **15.** function; domain: {−1, 0, 1, 2}, range: {2, 3, 4}   **17.** −4   **19.** −14

21.

23.

25. 7   **27.** 26
29. −1.1   **31.** $-\frac{7}{20}$
33. 12   **35.** −9
37 −2.5   **39.** 8   **41.** 2

43. $\longleftarrow\!\circ\!\!\!\!\!\!\!\!\!\longrightarrow$
    −3 −2 −1   0   1

45. 47. 49. 83; 85; 85
51. Outliers: 59 and 62;
clusters: 78-94; gaps:
between 62 and 78

53. Yes; both graphs show that 22 people preferred Brand X cookies and 30 people preferred Bill's cookies.

Chapter 3: Geometry and Reasoning

Lesson 3-1, pages 104–107

1. There are three possible answers: $\overrightarrow{XY}$ (or $\overrightarrow{YX}$), $\overrightarrow{XZ}$ (or $\overrightarrow{ZX}$), $\overrightarrow{RV}$ (or $\overrightarrow{VR}$), and $\overrightarrow{YZ}$ (or $\overrightarrow{ZY}$). Each answer is justified by Postulate 3.   3. 34   5. points R and S; Postulate 1.
7. $\overrightarrow{RS}$ (or $\overrightarrow{SR}$); Postulate 4.   9. 5   11. 6   13. 56
15. -25 and 9   17. $PN = 34$ and $NQ = 17$ 19. Think of the end of each leg as a point. When a table has four legs, the ends of the legs represent four points. If the legs are of different lengths, the four points are noncoplanar, and the table wobbles. When a table has three legs, the ends of the legs represent three points. Since Postulate 2 guarantees that any three points are coplanar, the length of the legs does not matter, and so the table does not wobble.

21. 23.

25. 27.

29. -2 31. -4 33. -4

Lesson 3-2, pages 108–111

1. 120°; obtuse   3. 120°; obtuse   5. 180°; straight
7. 79°   9. Answers will vary. Both postulates refer to the pairing of real numbers with geometric figures in a systematic way; both involve taking the absolute value of the difference of two real numbers.   11. $\angle MOR$
13. $\angle MOQ$, 125° and $\angle NOR$, 145°   15. 72°   23. 115°
25. 67°   27. never   29. 41   31. 1   33. -3200 35. -508 37. $-40{,}000$ 39. 1521   41. $2.4 \cdot 10^{13}$

Review and Practice Your Skills, pages 112–113

1. 5   3. 6   5. 60   7. $-18, 8$ 9. false   11. true
13. 12°   15. 84°   17. 1°   19. 48°   21. 97°   23. 54°
25. 149.5°   27. 162°   29. $m\angle UTR = 100°$; $m\angle QTU = 90°$; $m\angle QTR = 10°$; $m\angle QTS = 90°$; $m\angle RTS = 80°$; $m\angle UTS = 180°$ 31. $m\angle XWV = 90°$; $m\angle VWZ = 90°$; $m\angle YWZ = 90°$; $m\angle XWY = 90°$; $m\angle XWZ = 180°$; $m\angle VWY = 180°$;   33. 104°   35. U, V, Z; Post. 2
37. $\overline{UY}, \overline{UV}, \overline{UW}, \overline{YV}, \overline{YW}, \overline{VW}$; Post. 3   39. 176   41. 75°

Lesson 3-3, pages 114–117

1. point J 3. point K 5. 119°; 61°   7. $\overline{VX}$ 9. -0.5
11. 146°   13. 142°   15. 19°   17. 14   19. 49   21. 118°
23. 62°   25. 121°   27. 45°   29. E and S
31. Since $m\angle YXW + m\angle WXV + m\angle VXU = 176°$, $\angle AXB$ is an obtuse angle. Therefore, $\overrightarrow{XY}$ and $\overrightarrow{XU}$ are not opposite rays; false.   33. Since $m\angle TXV$ is a right angle, $\angle TXU$ and $\angle UXV$ are complementary, and $m\angle TXU = 90° - 43° = 47°$. Since $\angle TXU$ and $\angle UXV$ are not equal in measure, $\overrightarrow{XU}$ does not bisect $\angle TXV$.; false.   35. No information is given to indicate that $XZ = XV$, so it is not possible to identify X as the midpoint of $\overline{ZV}$; cannot tell.   37. 2.5
39. -14 41. rule: $\times -3$; $-243, 729, -2187$ 43. rule: -7.5; 62.5, 55, 47.5   45. rule: $+4, \div 2$; 19.5, 9.75, 13.75

Lesson 3-4, pages 118–121

1. 90°   3. 53°   5. 97°   7. Check students' drawings. Begin with the segment bisector construction shown in Example 1. Because point M is the midpoint of AB, $AN = MB$. Now use the segment bisector construction two more times to find the midpoint L of $\overline{AM}$ and the midpoint N of $\overline{MB}$. Points $L, M,$ and N divide $\overline{AB}$ into four segments of equal length: $AL = LM = MN = NB$.

9. 51°   11. 39°   13. 141°   15–17.
19. Answers will vary.   21. yes

Review and Practice Your Skills, pages 122–123

1. D 3. $\overline{BD}$ 5. 108°   7. 144°   9. 14   11. 28   13. false
19. 90°   21. 49°   23. 139°   25. 49°   27. false   29. 128
31. 54°   33. $\overline{CD}, \overline{EJ}, \overline{FH}$ 35. $\overline{AD}, \overline{BC}, \overline{DF}, \overline{CH}$

Lesson 3-5, pages 124–127

1. 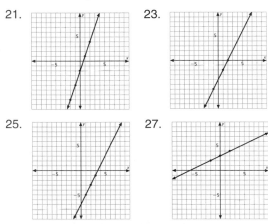 ; sixteenth figure: triangle with 16 dots on each side   3. 66

5. Next figure: 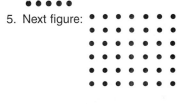 The fourteenth figure will be a rectangular arrangement of 210 points, with 14 points along one side of the rectangle and 15 points along the other.

7. 22   9. 153   11. $f(n) = n^2 + 1$ 13. Answers will vary.

15. 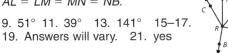 ; $a \le -\dfrac{10}{3}$

17. ; $c < -12$

19. ; 1. $x < \dfrac{1}{4}$

21.

Interval	Tally	Frequency
90-99	|||	3
80-89	ⅲ	6
70-79	ⅲ-ⅲ	8
60-69	|||	3
50-59			1
40-49			1

23.

Interval	Tally	Frequency
85-87	ⅲ |||	9
82-84	ⅲ	6
79-81	ⅲ	5
76-78	||	2
73-75			1
70-72	||	2

Lesson 3-6, pages 128–131

1. False; it is possible that the two lines are noncoplanar.
3. If an angle is bisected by a ray, then the two adjacent angles formed are equal in measure; if an angle is divided by a ray into two adjacent angles that are equal in measure, then the ray bisects the angle; an angle is bisected by a ray if and only if the two adjacent angles formed are equal in measure.   5. If an error is charged to the shortstop, then the shortstop made a bad throw to first base. False.

7. 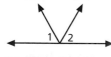   9.

11. Converse; If points *J, K,* and *L* are collinear, then they are coplanar. The given statement is false. Its converse is true.   13. Converse: If two angles are complementary, then the sum of their measures is 90°. Both the given statement and its converse are true.   15. Converse: If two lines do not intersect, then they are perpendicular. Both the given statement and its converse are false.
17. Conditionals: If a point is the midpoint of a segment, then it divides the segment into two segments of equal length; if a point divides a segment into two segments of equal length, then it is the midpoint of the segment. Biconditional: A point is the midpoint of a segment if and only if it divides the segment into two segments of equal length.   19. Conditionals: If a line is a transversal, then it intersects two or more coplanar lines in different points; if a line intersects two or more coplanar lines in different points, then it is a transversal. Biconditional: A line is a transversal if and only if it intersects two or more coplanar lines in different points.   21. *t, Q*   23. *P, X*   25. *X, Y*
27. Write the given definition as two conditional statements: *If two angles are vertical angles, then their sides form opposite rays* is true. However, *if the sides of two angles form opposite rays, then the angles are vertical angles* is false. Here is a counterexample in which the sides of ∠1 and ∠2 form a pair of opposite rays, but the angles are not vertical angles.

 For this reason, it is necessary to define vertical angles as two angles whose sides form two pairs of opposite rays.

29. Write the given definition as two conditionals: *If two angles are adjacent angles whose exterior sides form a right angle, then they are complementary* is true. However, *if two angles are complementary, then they are adjacent angles whose exterior sides form a right angle* is false. Complementary angles are not necessarily adjacent.
31. 72   33. 72

Review and Practice Your Skills, pages 132–133

1. 19 units horizontal; 10 units vertical   3. 10 units long
5. same orientation as second figure, but with ten lines in the interior   7. same as second figure
9. Sample answer:

11. Sample answer:

13. If $AB = 2(AC)$, then *C* is the midpoint of $\overline{AB}$. true, false
15. If two lines are perpendicular, then they intersect to form right angles. If two lines intersect to form right angles, then they are perpendicular. Two lines are perpendicular if and only if they intersect to form right angles.   17. −19, 7
19. 148°, 32°   21. 164°   23. 164°

Lesson 3-7, pages 134–137

1. given; parallel lines postulate; $m\angle 2 = m\angle 3$; transitive property of equality   3. Answers will vary.   5. Prove: $m\angle 1 = m\angle 3$; Statement: 1: ∠1 is comp. to ∠2 and ∠3 is comp. to ∠2; Reason 1: given; Reason 2: def. of comp. angles; Reason 3: trans. Prop. of inequality; Statement 4: $m\angle 1 = m\angle 3$.
7. Given: $k \parallel m, l \parallel m$
   Prove: $k \parallel l$

Statements	Reasons
1. $k \parallel m, l \parallel m$	1. given
2. $m\angle 1 = m\angle 2$; $m\angle 3 = m\angle 2$	2. parallel lines post.
3. $m\angle 1 = m\angle 3$	3. trans. property of equality
4. $k \parallel l$	4. corres. angles post.

9. Given: $l \parallel m, k \perp l$
   Prove: $k \perp m$

Statements	Reasons
1. $l \parallel m, k \perp l$	1. given
2. ∠1 is a right angle	2. def. of perpendicular lines
3. $m\angle 1 = 90°$	3. def. of right angle
4. $m\angle 1 = m\angle 2$	4. parallel lines postulate
5. $m\angle 2 = 90°$	5. substitution property
6. ∠2 is a right angle	6. def. of right angle
7. $k \perp m$	7. def. of perpendicular lines

11. 0.0000000146   13. 70,200,000   15. 0.0000000059
17. 21,000   19. 0.000397   21. 5,120,000,000   23. 33°, 123°   25. 15°, 105°   27. 7°, 97°   29. 29°, 119°

Lesson 3-8, pages 138–139

1.

	Cory	Srey	Molly	Mao
Des Moines	x	x	o	x
Pittsburgh	x	o	x	x
Santa Clara	o	x	x	x
Seattle	x	x	x	o

3. Ned—Miami; Carina—Dallas; Pedro—San Francisco; and Eva—San Diego

5–11. 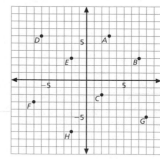   13. 1

Chapter 3 Review, pages 140–142

1. 1 3. e 5. f 7. g 9. j 11. 6 planes 13. no
15. 8 17. 13° 19. 110° 21. 68° 23. ∠2 and ∠5,
∠3 and ∠6, ∠4 and ∠7 25. ∠1 and ∠6 27. 140°
29. 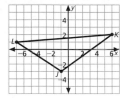 a shaded square 31. If ∠ABC and ∠DBE
are vertical angles, then m∠ABC = m∠DBE.
false, true 33. Marco: Wisconsin; Sue:
California; Stephanie: Florida

Chapter 4: Triangles, Quadrilaterals and Other Polygons

Lesson 4-1, pages 150–153

1. 45° 3. 26° 5. 153° 7. 71 9. 63 11. 38°, 76°, 66°
13. m∠F = 64°; m∠G = 26°; m∠H = 90°
15. 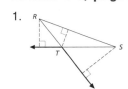 17.

scalene, obtuse   isosceles, right

19. never 21. sometimes 23. 52 25. mean: $7\frac{1}{6}$;
median: $7\frac{1}{2}$; mode: 8

Lesson 4-2, pages 154–157

1. $\overline{RQ} \cong \overline{RS}$; $\overline{RT}$ bisects ∠QRS; given; angle bisector;
$\overline{RT} \cong \overline{RT}$; reflexive; SAS postulate
3. Proofs may vary. A sample proof is given.

Statements	Reasons
1. $\overline{AB} \cong \overline{CB}$; $\overline{EB} \cong \overline{DB}$; AD and CE intersect at point B.	1. given
2. ∠ABE and ∠CBD are vertical angles.	2. definition of vertical angles
3. ∠ABE ≅ ∠CBD	3. vertical angle theorem
4. △ABE ≅ △CBD	4. SAS postulate

5. △ECB; ASA postulate 7. △MNK; SAS postulate
9. (−7, −11) or (−7, 15)
11. Proofs may vary. A sample proof is given.
Given: AB ≅ XY; BC ≅ YZ;
    ∠B and ∠Y are right angles.
Prove: △ABC ≅ △XYZ

Statements	Reasons
1. $\overline{AB} \cong \overline{XY}$; $\overline{BC} \cong \overline{YZ}$; ∠B and ∠Y are right angles.	1. given
2. m∠B = 90°; m∠Y = 90°	2. Definition of right angles
3. m∠B = m∠Y°	3. Transitive property of equality
4. △ABC ≅ △XYZ	4. SAS postulate

13. 104° 15. 71°

Review and Practice Your Skills, pages 158–159

1. 95 3. 23 5. 40 7. false 9. false
11.   13.

right isosceles   scalene acute

15. $\overline{RS} \cong \overline{VU}$, $\overline{RT} \cong \overline{VT}$, $\overline{ST} \cong \overline{UT}$, ∠SRT ≅ ∠UVT,
∠RTS ≅ ∠VTU, ∠TSR ≅ ∠TUV, △RST ≅ △VUT
17. always 19. sometimes 21. △NOM; SAS postulate

Lesson 4-3, pages 160–163

1. $\overline{FG} \cong \overline{HJ}$; $\overline{FG} \perp \overline{FH}$; $\overline{JH} \perp \overline{FH}$; given; ⊥ lines; right ∠;
m∠1 = m∠2, or ∠1 ≅ ∠2; $\overline{FH} \cong \overline{HF}$; SAS postulate;
∠J ≅ ∠G; CPCTC 3. 12 5. 3 7. 45 9. There are two
base angles. If each measures 70°, the vertex angle must
measure 40°. 11. ∠1 ≅ ∠2; ∠4 ≅ ∠5; ∠3 ≅ ∠6; ∠8 ≅
∠7; ∠XZY ≅ ∠XYZ

13.  15.   17. isosceles triangle
19. ∠ABC
21. ∠FBC = 115°, ∠ABD = 155°

Lesson 4-4, pages 164–167

1. 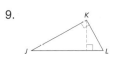 3. true 5. cannot be determined
7. The altitudes are concurrent lines.

9. 11. cannot be determined
13. true 15. true 17. false

19. The three distances are equal. 21. In a right triangle,
the two sides that are perpendicular (the legs) are two
altitudes of the triangle. It is never true that a side of a
triangle is also a median because a side always connects
two vertices of a figure. 23. Answers will vary. Possible
responses: △PQR is isosceles; QR ≅ PR; QS ≅ PS; ∠RQP
≅ ∠RPQ; ∠QRS ≅ ∠PRS; △QRS ≅ △PRS; $\overline{QT}$ is an
altitude of △PQR; $\overline{HS}$ is an altitude of △PQR; $\overline{RS}$ is a
median of △PQR; $\overline{RS}$ is a perpendicular bisector of $\overline{QP}$;
$\overline{RS}$ lies on the bisector of ∠QRP; ∠RSQ, ∠RSP, ∠QTP,
and ∠QTR are right angles. 25. D 27. $\overline{DF}$, $\overline{CG}$, $\overline{BH}$,
$\overline{AI}$ 29. −15 31. −21 33. 0 35. 10

Review and Practice Your Skills, pages 168–169

1. 8 cm 3. 46° 5. ∠3 ≅ ∠4, ∠2 ≅ ∠6, ∠1 ≅ ∠5,
∠3 ≅ ∠9, ∠2 ≅ ∠8, ∠1 ≅ ∠7, ∠3 ≅ ∠10, ∠2 ≅ ∠12,
∠1 ≅ ∠11, ∠4 ≅ ∠9, ∠6 ≅ ∠8, ∠5 ≅ ∠7, ∠4 ≅ ∠10,
∠6 ≅ ∠12, ∠5 ≅ ∠11, ∠9 ≅ ∠10, ∠8 ≅ ∠12, ∠7 ≅ ∠11
7. false (must be "included angle") 9. false 15. true
17. false 19. true 21. 71 23. 28 25. 112°

Lesson 4-5, pages 170–171

1. Assume that the triangle *can* be an obtuse triangle
3. two; *r; s; X; Y;* one; contradictory; false; true
5. *Step 1:* Assume that there are two lines through point *P* perpendicular to the given line. In particular, in the figure to the right, assume that *m* ⊥ *l* and *n* ⊥ *l*. *Step 2:* By the definition of right angles, *m*∠1 = 90° and *m*∠2 = 90°. By the transitive property of equality, *m*∠1 = *m*∠2. By the corresponding angles postulate, since *m*∠1 = *m*∠2, it follows that *m*‖*n*. By definition of intersecting lines, since *m* and *n* each pass through point *P*, *m* and *n* are intersecting lines. *Step 3:* The last two statements in *Step 2* are contradictory. Therefore, the assumption that there can be two lines perpendicular to a given line through a point outside the line is false. The given statement must be true. 7. 5 9. 9 11. 19

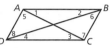

Lesson 4-6, pages 172–175

1. between 3 ft and 15 ft 3. between 5 ft and 9 ft
5. ∠*K*, ∠*M*, ∠*L* 7. *XY*, *YZ*, *XZ* 9. yes 11. no 13. yes
15. *DF*; *EF* 17. *VW*, *UV* and *UW* 19. 0 < *x* < 11
21. ∠*C* 23. Assume that *AB* + *AC* ≯ *BC*. Then, by the property of comparison, one of these two statements must be true: *AB* + *AC* = *BC* or *AB* + *AC* < *BC*; If *AB* + *AC* = *BC*, then there is a path connecting points *B* and *C* other than along *BC* that is equal in length to *BC*; this contradicts the shortest path postulate. Similarly, if *AB* + *AC* < *BC*, then there must be a path connecting points *B* and *C* that is shorter than *BC*; this also contradicts the shortest path postulate. Therefore, the assumption *AB* + *AC* ≯ *BC* must be false. It follows that the desired conclusion, *AB* + *AC* < *BC*, is true. 25. *BC*, *AB*, *AC*, *CD*, *AD* 27. 2 < *z* < 12
29. A right triangle can have only one right angle, and it cannot have an obtuse angle. Therefore, the one right angle is the angle with the greatest measure. By the unequal angles theorem, the side opposite that angle, the hypotenuse, is the longest side. 31. when the measure of the top angle = 60° 33. when the measure of the top angle < 60° 35. *f*(*n*) = *n*² − (*n* − 1) 37. 3.71 • 10¹¹
39. 2.56 • 10¹¹ 41. 8.9 • 10¹²

Review and Practice Your Skills, pages 176–177

1. Assume that a given triangle is not isosceles, but *is* equilateral. 3. Assume they are *not* equal in measure.
5. Assume they *do not* intersect. 7. Answers will vary.
9. Answers will vary. 11. yes 13. no 15. 0 < *x* < 14
17. *PM*, *PO*, *MO*, *MN*, *NO* 19. *XY*, *WY*, *YZ*, *WZ*, *XW*
21. false 23. true 25. false 27. 28 29. 7.5 m 31. 53

Lesson 4-7, pages 178–181

1. 101 3. 136 5. 36° 7. 129 9. 132 11. 72
13. 2880° 15. 15° 17. 30; 60; 60 19. 40 21. 20
23. Although the faces are all regular polygons, there are two different types of faces, pentagons and hexagons.
25. $\frac{(n-2)180}{n}$ 27. approaches 180° 29. right scalene
31. acute isosceles

Lesson 4-8, pages 182–185

1. 68; 112 3. 64° 5. 90° 7. 90° 9. 45; 135; 42; 28
11. $1\frac{5}{8}$; $1\frac{5}{8}$, $1\frac{1}{2}$; $1\frac{5}{8}$ 13. true 15. false 17. false

19. No; the angles that are equal in measure are not opposite angles. 21. Yes, the figure is a square, and every square is a parallelogram.
23. Given: *ABCD* is a parallelogram.
    Prove: *m*∠*B* = *m*∠*D*

Statements	Reasons
1. *ABCD* is a parallelogram.	1. given
2. *AB* ‖ *DC*; *AD* ‖ *BC*	2. definition of ‖-ogram
3. *m*∠5 = *m*∠7, or ∠5 ≅ ∠7; *m*∠6 = *m*∠8, or ∠6 ≅ ∠8	3. If two ‖ lines are cut by a trans., then alt. int. ∠*s* are = in measure.
4. *AC* ≅ *CA*	4. reflexive property
5. △*ABC* ≅ △*CDA*	5. ASA postulate
6. ∠*B* ≅ ∠*D*, or *m*∠*B* ≅ *m*∠*D*	6. CPCTC

25. Sample response: Draw parallelogram *ABCD* and both diagonals. The goal is to show that *BD* bisects *AC*, and that *AC* bisects *BD*. By the definition of parallelogram, *AB* ‖ *DC* and *AD* ‖ *BC*. When parallel lines are cut by a transversal, alternate interior angles are equal in measure, so there are four pairs of angles that are equal in measure: ∠1 and ∠3; ∠2 and ∠4; ∠5 and ∠7; and ∠6 and ∠8. Because the parallelogram-side theorem has been proved, it is known that *AB* = *CD* and *AD* = *CB*. Therefore, by the ASA postulate, there are two pairs of congruent triangles: △*BEA* and △*DEC*, and △*BEC* and △*DEA*. Because corresponding parts of congruent triangles are congruent, it follows that *AE* ≅ *CE*, or *AE* = *DE*, and that *DE* ≅ *BE*, or *DE* = *BE*. Therefore, point *E* is the midpoint of both *AC* and *BD*. It follows that *BD* bisects *AC* and *AC* bisects *BD*.
27. ∠*ABG* and ∠*FEH*, ∠*GBC* and ∠*DEH*
29. ∠*ABH* and ∠*GEF*, ∠*CBH* and ∠*GED* 31. yes; domain: {2, 3, 4, 5, 6}; range: {4.5, 6.5, 8.5, 10.5, 12.5}

Review and Practice Your Skills, pages 186–187

1. 143° 3. 60° 5. 18° 7. 360° 9. 9 11. 24 13. true
15. true 17. *a* = 115; *b* = 65; *c* = 37; *d* = 43
19. *a* = 8.8; *b* = 6.8; *c* = 133; *d* = 83 21. Yes, diagonals bisect each other. 23. 28.5° 25. 67° 27. 45
29. 60 31. 100 cm < *x* < 900 cm 33. 100 35. 43

Lesson 4-9, pages 188–191

1. 16 3. 3 5. *m*∠*C* = 75°; *m*∠*D* = 75°; *m*∠*E* = 105°; *m*∠*F* = 105° 7. 3.6 9. 9 11. 7 13. *m*∠*T* = 83°; *m*∠*V* = 97°; *m*∠*U* = 97°; *m*∠*W* = 83° 15. Answers may vary. *Possible likeness:* Each has an endpoint at the midpoint of a side of the figure. *Possible differences:* A median of a triangle has one endpoint that is also a vertex of the figure, whereas the median of a trapezoid does not; a triangle has three medians, whereas a trapezoid has only one. 17. quadrilateral 19. trapezoid, quadrilateral 21. isosceles trapezoid, quadrilateral 23. no; yes; yes; yes; yes; no; no 25. no; yes; yes; yes; yes; no; no 27. no; no; yes; no; yes; no; yes
29. Given: *ABCD* is a trapezoid with bases *AB* and *DC*.
    Prove: *ADC* and *DAB* are supplementary.

Statements	Reasons
1. $ABCD$ is a trapezoid with bases $\overline{AB}$ and $\overline{DC}$.	1. given
2. $\overline{AB} \parallel \overline{DC}$	2. definition of trapezoid
3. $m\angle EAB = m\angle ADC$	3. corr. $\angle s$ postulate
4. $m\angle EAB + m\angle DAB = 180°$	4. angle addition postulate
5. $m\angle ADC + m\angle DAB = 180°$	5. substitution property
6. $\angle ADC$ and $\angle DAB$ are supplementary	6. definition of supplementary angles

31. 9   33. 5   35. 8

Chapter 4 Review, pages 192–194

1. c   3. i   5. l   7. g   9. b   11. 18   13. 27   15. 51
17. $\triangle AES \cong \triangle BET$, SAS   19. $\triangle XYZ \cong \triangle QZY$, SSS
21. 35   23. 60   25. $\overline{AD}$ 27. Assume if a triangle is obtuse, then it can have a right angle.   29. Assume the angle bisector of the vertex angle of an isosceles triangle is not an altitude of the triangle.   31. no   33. 115°
35. 45°   37. 19   39. 110°   41. 38   43. 20.5 in.   45. 4 m

Chapter 5: Measurement

Lesson 5-1, pages 202–205

1. 32 c   3. 3500 mL   5. $\frac{2}{5}$ 7. 9:1   9. $1\frac{1}{4}$ ft per day
11. 32   13. 256   15. 0.004   17. 6   19. lb   21. kL
23. 2:5   25. 1:3   27. $0.06 per copy   29. 4-L vase
31. ±0.5 m; ± 0.05 m   33. 8000 billion qt   35. 2 mi/h
37. 48°; 70°; 62°

Lesson 5-2, pages 206–209

1. 10 m; 7 m²   3. 11 cm; 10 cm²   5. 40 cm   7. 4.8 cm
9. about 62.8 m²   11. It triples too.   13. about 144 ft
15. 3600 ft²   17. Luis entered 3.14 for π. Irene used the π key on her calculator.   21. Answers will vary.
23. Sample answers: Area of wall space, not including windows, doors, etc.; number and price of cans of paint needed based on the area one can of paint covers; painting speed and number of painters involved.   25. The circle ($r \approx 15.9$ ft and $A \approx 796$ ft²) has more area than the square (625 ft²).   27. yes   29. yes   31. yes   33. yes
35. 5   37. 6   39. 6

Review and Practice Your Skills, pages 210–211

1. 12   3. 0.3625   5. 7030   7. 40:9   9. 7:5   11. 5:6
13. b   15. c   17. 1 gal for $1.79   19. $P = 36$ in.
21. $C \approx 94.2$ in.   23. $A = 204$ ft²   25. 96 in.²
27. ≈ 27.93 mi²   29. 832   31. 22, 2   33. 0.0004
35. 1:10   37. 30:1   39. 1:1000   41. 200 L barrel
43. $P \approx 124.95$ yd; $A \approx 961.625$ yd²

Lesson 5-3, pages 212–215

1. $\frac{3}{16}$ 3. $\frac{1}{4}$ 5. $\frac{1}{13}$ 7. $\frac{3}{26}$ 9. $\frac{2}{3}$ 11. $\frac{3}{5}$ 13. about 0.785
15. Possible answer: about 0.009 or $\frac{9}{1000}$ 17. No. That actual probability is $\frac{1}{9}$, which is $\left(\frac{1}{3}\right)^2$ 19. likely   21. likely
23. about 0.56 or $\frac{14}{25}$ 25. $\frac{6}{31}$ 27. The map or actual

dimensions of the Palaestra and the total map or actual area of the ruins.

29. 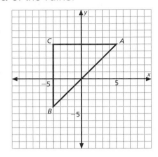   right isosceles

31. 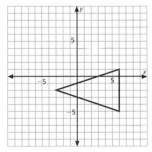   acute isosceles

33. −33   35. −48   37. 10.9   39. 39.5

Lesson 5-4, pages 216–217

1. 12,150 ft²   3. 396 ft²   5. 126.5 m²   7. at least 230 ft long, at least 150 ft wide   9. 7740°, 172°   11. 4680°, 167.14°   13. 5760°, 169.41°   15. 7200°, 171.43°

Review and Practice Your Skills, pages 218–219

1. $\frac{6}{25} = 0.24$ 3. $\frac{7.065}{27} \approx 0.262$ 5. $\frac{9}{28.26} \approx 0.318$
7. $\frac{1}{10} = 0.1$ 9. unlikely   11. Answers will vary. Most students will answer "likely."   13. 320 m²   15. 6300 in.²
17. 1:500   19. 1:2   21. $P \approx 67.1$ m; $A \approx 235.5$ m²
23. $\frac{2}{5} = 0.40$

Lesson 5-5, pages 220–223

1. Triangular prism; bases ABC and EFD; Remaining answers will vary. Sample answers include parallel edges $\overline{AB}$ and $\overline{EF}$; intersecting faces $BCDF$ and ABC; intersecting edges $\overline{EF}$ and $\overline{FD}$.   3. Triangular pyramid; base LNO; no parallel edges; Remaining answers will vary. Sample answers include intersecting faces LMO and LNO; intersecting edges $\overline{OM}$ and $\overline{OL}$.   5. a coplanar line halfway between them   7. hexagonal prism; 8; 12; 18
11. The sum of the faces and vertices is 2 more than the number of edges. Possible rule: $e = f + v - 2$.
13.   15. a cylindrical surface   19. a square
21. 45   23. 2.9   25. $\frac{2}{5}$ 27. 6   29. 3
31. $\frac{4}{5}$ 33. $\frac{23}{21}$

Lesson 5-6, page 224–227

1. 158 cm²   3. 336 m²   5. 222 in.²   7. 452 ft²
9. 166.4 m²   11. 476 in.²   13. 735 m²   15. the area of the base   17. about 24,727.5 mm²   19. a) SA is 4 times as great; b) SA is $\frac{1}{9}$ times as great.   21. $1\frac{2}{3}$ qt   23. 45

Review and Practice Your Skills, pages 228–229

1. rectangular prism; 6 faces, 8 vertices, 12 edges
3. oblique cone; faces and edges are undefined for cones, 1 vertex.   5. Answers will vary.   7.

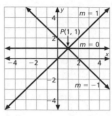

9.    11.   13. 240 cm²
15. 272 ft²
17. 1306.2 m²

19. 9 times as large   21. 65 $\frac{\text{mi}}{\text{h}}$ 23. 34 $\frac{\text{gal}}{\text{min}}$
25. $\frac{43}{200} \approx 0.215$ 27. 8478 in.²

Lesson 5-7, pages 230–233

1. 1309 mm³   3. 3561 cm³   5. 6 cm   7. 1570 cm³
9. 340 m³   11. about 254 m³   13. The volume of the pyramid is one-third the volume of the prism.
15. $V = 3840$ yd³   17. One answer: Area of base of Cheops is smaller than the area of the base of Cholula.
19. It is tripled.   21. Tetragonal system is combination of rectangular prism and two square pyramids; find volume of separate shapes and add.   23. the $0.40 can
25. Cutting 4-cm squares gives the greatest volume—16 cm × 16 cm × 4 cm = 1024 cm³. List all cuts and the resulting dimensions and volumes in a table.   27. 36.9 ft; 80 ft²   29. 14.5 in.; 8.22 in.²

Chapter 5 Review, pages 234–236

1. e   3. j   5. d   7. k   9. l   11. 2:5, $\frac{2}{5}$, 2 to 5
13. 0.005 m   15. $C \approx 15.42$ m; $A \approx 14.13$ m²   17. $\frac{3}{5}$
19. $\frac{1}{9}$ 21. 256 ft²   23. 300 in.²   25. rectangular prism; 6 faces; 8 vertices; 12 edges   27. triangular pyramid; 4 faces; 4 vertices; 6 edges
29. 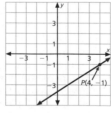   31. 164.9 cm²   33. 177.1 in.²
35. about 226 cm²   37. 135 cm³
39. 6188.9 mm³   41. 36 in.

Chapter 6: Linear Systems of Equations

Lesson 6-1, pages 244–F247

1. $-\frac{1}{3}$ 3. undefined   5.

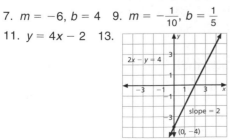

7. $m = -6$, $b = 4$ 9. $m = -\frac{1}{10}$, $b = \frac{1}{5}$
11. $y = 4x - 2$ 13.

15. -2 17. $\frac{3}{7}$ 19–21.

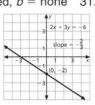

23. 2; yes   25. $m = 4$, $b = 0$
27. $m =$ undefined, $b =$ none
29. $m =$ undefined, $b =$ none   31. $y = -5x + 4$
33. $y = 6x$ 35.   37.

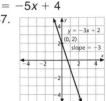

39.

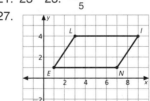

41. $3000   43. slope = a, y-intercept = $r - m$
45. $y = -\frac{A}{B}x - \frac{C}{B}$; $B \neq 0$
47. No, since there is no change in x, the line must be vertical and the slope is undefined.   49. 62 mi/h
51. $1.19 per lb   53. 4 mi/h   55. $1.67 per jar
57. 14, -14 59. 4, -4 61. 26, -26 63. 15, -15

Lesson 6-2, pages 248–251

1. $-\frac{2}{9}, \frac{9}{2}$ 3. 4, $-\frac{1}{4}$ 5. neither   7. neither   9. $-4, \frac{1}{4}$
11. $-\frac{5}{6}, \frac{6}{5}$ 13. $-7, \frac{1}{7}$ 15. neither   17. no   19. parallel
21. 23   23. $-\frac{2}{5}$
27.

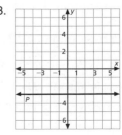

29. slope of $\overline{LN} = -\frac{3}{4}$ and slope of $\overline{EI} = \frac{3}{8}$; $\left(-\frac{3}{4}\right)\left(\frac{3}{8}\right) \neq -1$, so $\overline{LN}$ is not perpendicular to $\overline{EI}$.   31. 48 cm²   33. 5.25 cm²

Review and Practice Your Skills, pages 252–253

1. 1   3. $-\frac{5}{4}$ 5. $-\frac{4}{9}$ 7. 0   9. -6
11.   13.

15.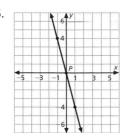

17. $m = -5$, $b = 9$

19. $m = \frac{2}{3}$, $b = -1$

21. $m = -\frac{2}{3}$, $b = 4$

23. $m = -\frac{1}{3}$, $b = \frac{1}{4}$

25. $y = 15$ **27.** $y = -\frac{7}{2}x - 3$

29. $y = -20x + 5$

31. 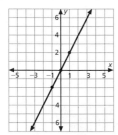 **33.**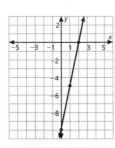

35.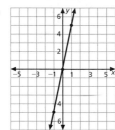

37. -2, $\frac{1}{2}$ **39.** 1, -1

41. $-\frac{5}{13}$, $\frac{13}{5}$ **43.** -3, $\frac{1}{3}$

45. $-\frac{12}{5}$, $\frac{5}{12}$ **47.** perpendicular

49. parallel **51.** neither

53. undefined **55.** $-\frac{3}{4}$

57. $m = -\frac{1}{4}$, $b = 7$

59. $m = 3$, $b = -4$ **61.** $m = \frac{1}{4}$, $b = 9$ **63.** $-\frac{19}{9}$, $\frac{9}{19}$

65. undefined, 0 **67.** $-\frac{1}{8}$, 8

Lesson 6-3, pages 254–257

1. $y = -3x - 2$ **3.** $y = 7x + 2$ **5.** $y = -5x - 1$

7. $y = 9$ **9.** $y = -\frac{3}{4}x + \frac{11}{4}$ **11.** $y = \frac{2}{3}x - 3$

13. $y = \frac{-1}{3}x - \frac{2}{3}$ **15.** $y = \frac{-1}{2}x - \frac{2}{5}$ **17.** $y = \frac{-3}{13}x + \frac{12}{13}$

19. $x = 2$ **21.** $d = 2t$ **23.** $2x + 9y = 32$ and $6x + y = 18$ **25.** $y = 0.01px + F$, where $x = $ sales, $y = $ pay

27. 96 cm² **29.** {2, 3, 5, 6, 7, 8, 11, 12} **31.** {4, 9, 13}

33. {1, 4, 9, 10, 13, 14}

Lesson 6-4, pages 258–261

1. yes **3.** 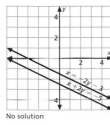 **5.**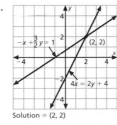

7. (3, 2) **9.** 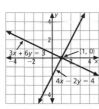 **11.**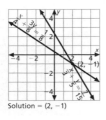

13. food, 7; personal care, 3 **15.** $10,092.50 **19.** no solution

21.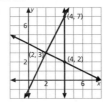

23. about 0.77

25. about 0.97

27. 190,000,000

29. 6,520,000,000,000

Review and Practice Your Skills, pages 262–263

1. $y = 4.5x - 7.5$ **3.** $y = -4$ **5.** $y = \frac{5}{4}x + 13$

7. $y = x + 0.05$ **9.** $y = 300x + 530$ **11.** $y = 2x + 7$

13. $y = \frac{7}{2}x + 2$ **15.** $y = -4x + 6$ **17.** $x = 13$

19. $y = \frac{1}{2}x + 7$ **21.** $x = 3$ **23.** $y = 3x + 2$

25. $y = x - 1$ **27.** $y = \frac{3}{4}x + 1$ **29.** $y = \frac{3}{2}x - 12$

31. $(-3, -4)$ **33.** $(8, 8)$ **35.** parallel lines (no solution)

37. $(9, 4)$ **39.** $(4, -2)$ **41.** 1 **43.** 5, -8 **45.** $m = 1$, $b = -3\frac{1}{2}$ **47.** $m = \frac{3}{2}$, $b = -7$ **49.** $m = 0$, $b = -12$

51. parallel **53.** perpendicular **55.** $y = -x - 8$

57. $y = 36$ **59.** $y = 3x + 15$ **61.** (45, 36)

Lesson 6-5, pages 264–267

1. $(-1, 2)$ **3.** $(-1, -2)$ **5.** length: 24 cm; width: 15 cm

7. $(4, 0)$ **9.** $(1, -2)$ **11.** $(5, -3)$ **13.** Choose the variable that will be easy to isolate in one of the equations.

15. $1.19, $0.79 **17.** $\left(\frac{4}{3}, \frac{6}{5}\right)$ **19.** 1975 **21.** $\left(\frac{1}{2}, -\frac{1}{3}\right)$

23. $(-1, 1, -2)$ **25.** 15 nickels, 9 dimes, 3 quarters

27. 12 **29.** 36 **31.** -72 **33.** -9408

Lesson 6-6, pages 268–271

1. (4, 5) **3.** $(-3, 1)$ **5.** 13 adults, 43 children

7. $7\frac{1}{2}$ months **9.** $(-1, 1)$ **11.** (4, 3) **13.** $\left(0, \frac{1}{9}\right)$

15. $(-2, 4)$ **17.** wheat, 950 acres; barley, 250 acres

19. (2, 3) **21.** (3, 9) **23.** $(-4, 5)$ **25.** 15 and 18

27. (10, 21) **29.** $(-1, 2, -1)$ **31.** 29,000,000

33. 35,500; 15,828,000,000 **35.** 15,300; 2,940,000,000

37.

39.

41.

43.

Review and Practice Your Skills, pages 272–273

1. (15, 3) **3.** (3, 2) **5.** $\left(\frac{1}{3}, 2\right)$ **7.** $(2, -7)$ **9.** $(2, -3)$

11. (0, 2) **13.** $(7, -1)$ **15.** $\left(-1, \frac{2}{7}\right)$ **17.** bagel, $0.55;

peach, $0.35 **19.** $(-1, 1)$ **21.** $(-2, 0)$ **23.** $(-2, 1)$

25. $(-3, -2)$ **27.** $(-1.5, 0.25)$ **29.** $\left(-\frac{1}{2}, 1\right)$

31. $\left(-11\frac{2}{3}, 5\right)$ **33.** $(0, -5)$ **35.** no solution

37.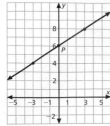

39.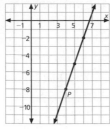

41.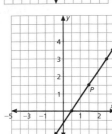

43. perpendicular
45. neither
47. $y = 3x - 19$
49. $y = 1.5x + 6$
51. $y = \frac{-1}{5}x + 1$
53. $(3, -1)$ **55.** $(-2, -1)$

Lesson 6-7, pages 274–275

1. $\begin{bmatrix} 5 & 1 \\ -3 & 4 \end{bmatrix}\begin{bmatrix} x \\ y \end{bmatrix} = \begin{bmatrix} 6 \\ 2 \end{bmatrix}$; det $A = 23$ Solution: $\left(\frac{22}{23}, \frac{28}{23}\right)$

3. $\begin{bmatrix} 4 & -7 \\ -2 & 1 \end{bmatrix}\begin{bmatrix} x \\ y \end{bmatrix} = \begin{bmatrix} 2 \\ -4 \end{bmatrix}$; det $A = -10$ Solution: $\left(\frac{13}{5}, \frac{6}{5}\right)$

5. $25D + 0.35M = 230$ $\begin{bmatrix} 25 & 0.35 \\ 35 & 0.25 \end{bmatrix}\begin{bmatrix} D \\ M \end{bmatrix} = \begin{bmatrix} 230 \\ 250 \end{bmatrix}$
  $35D + 0.25M = 250$
det $= -6$; $D = 5$ d; $M = 300$ mi
7. 41 in. **9.** 129 in. **11.** 58.4 in.

Lesson 6-8, pages 276–279

1. yes **3.** $y < 4$, $x \ge 3$

5. 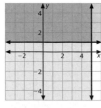 **7.**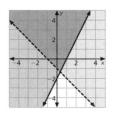

9. no **11.** $y \ge x - 3$; $y \ge \frac{-3}{4}x + 4$

13. 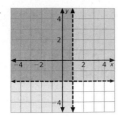 **15.**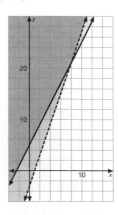

17. The solution set of $y < 2x + 3$ does not include solutions of $y = 2x + 3$. **19.** $x \ge 0$; $y \le x + 2$; $y < \frac{-3}{2}x + 5$ **21.** $y \ge x + 44$; $y \le 3x - 43$

23. 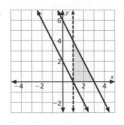 **25.**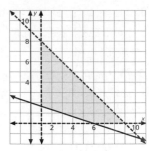

27. 78 **29.** 4800

Review and Practice Your Skills, pages 280–281

1. $(1, -5)$ **3.** $(1, 3)$ **5.** $(7, 8)$ **7.** $(-1, 2)$ **9.** $\left(0, -\frac{7}{3}\right)$
11. $(6, 9)$ **13.** $(2.8, 0.8)$ **15.** $(-2, 5)$ **17.** yes **19.** yes

21. 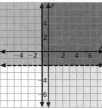 **23.**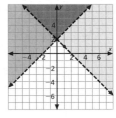

25. 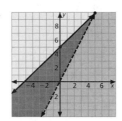 **27.**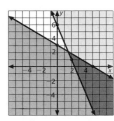

29.  **31.**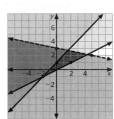

33.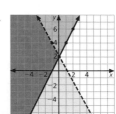

35. $m = \frac{5}{7}$, $b = -11$
37. $m = \frac{5}{7}$, $b = 11$
39. $m = -3$, $b = 13$
41. parallel **43.** perpendicular
45. $y = \frac{2}{3}x - 13$
47. $y = \frac{-2}{5}x + 7$

49. $y = 2.5x + 11$ **51.** $(5, 0)$ **53.** infinitely many solutions **55.** no solution **57.** $(4, 2)$ **59.** $(-5, 0)$
61. $(-8, 3)$ **63.** $(5, -4)$ **65.** $(-2, -4)$ **67.** $(2, -4)$
69. $(3, -1)$ **71.** $(-6, -43)$ **73.** $(1, 2)$

75.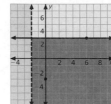

Lesson 6-9, pages 282–285

1. no 3. yes 5. maximum 225 at (15, 0)

7.

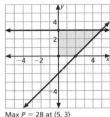

Max P = 28 at (5, 3)

9. maximum 140 at (5, 15); minimum 60 at (0, 10)
11. maximum 26.5 at (20, 2); minimum 3 at (0, 4)
13. These constraints insure that the feasible region will be confined to the first quadrant. This models constraints and feasible regions that can occur in the real world.
15. (0, 8), (4, 0), (0, 0) 17. minimum 12 at (0, 6)

19.

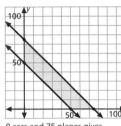

0 cars and 75 planes gives maximum profit of $2625

21. $x \geq 15$; $y \geq 10$; $x + y \leq 50$
23. They should sell 15 T-shirts and 10 sweatshirts.
25. Plant 1100 acres of Iceberg and 2500 acres of Romaine to yield a maximum profit of $845,000. 27. −142.272
29. 25.5 31. −12

Chapter 6 Review, pages 286–288

1. i 3. e 5. l 7. a 9. j 11. −11

13.

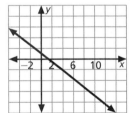

15. $y = -2x + 1$
17. neither 19. $-\frac{2}{5}$
21. $y = -x + 5$
23. $y = -2x - 4$
25. $y = -\frac{1}{4}x + 4$

27.

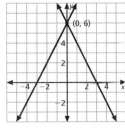

29. 13 m by 7 m
31. (−3, 4)
33. 78° and 102°
35. (3, −1) 37. $1.25
39. (2, −1) 41. Mario, 25; Danielle, 40

43.

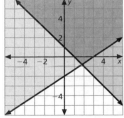

45. 15 47. 22

Chapter 7: Similar Triangles

Lesson 7-1, pages 296–299

1. yes 3. yes 5. 2 7. $900, $1200 9. no 11. no
13. 2.1 15. 988 17. 630 19. 1400 CDs 21. 7.5 lb
23. Sample answer: 100 : 10 = 30 : 3 25. 3 27. 5
29. about 8 h 31. 36°, 54° 33. The terms in the two ratios are not written in the same order. The correct proportion is either $\frac{3}{10} : \frac{48}{x}$ or $\frac{10}{3} : \frac{x}{48}$. 35. 6 37. $\frac{6}{7}$
39. $-\frac{5}{3}$ 41. $\frac{3}{2}$ 43. undefined 45. $\frac{7}{4}$ 47. $-\frac{5}{6}$
49. $1 < x < 15$ cm 51. $1 < x < 27$ cm
53. $6 < x < 12$ dm 55. $21 < x < 51$ in.

Lesson 7-2, pages 300–303

1. yes 3. 100 5. no 7. $6\frac{2}{3}$ ft 9. Corresponding angles are between corresponding sides. 11. 4.5 in.
13. 3 : 4 15. $7\frac{5}{9}$ in. by $9\frac{4}{9}$ in. 17. Answers will vary. One possible answer involves two rhombi with different angles. 19. −3 21. $-\frac{3}{2}$ 23. $\frac{2}{3}$ 25. −2 27. 5

Review and Practice Your Skills, pages 304–305

1. no 3. yes 5. yes 7. 24 9. 5 11. 100 13. 180
15. $\frac{4}{3}$ 17. 0.036 19. $1\frac{1}{4}$ c 21. 40.5° and 49.5°
23. yes 25. no 27. 5 29. 65

Lesson 7-3, pages 306–309

1. 12 ft 3. 1.25 cm 5. 40 mi 7. 32 m 9. 7.5 m
11. 0.5 m 13. 10 mi 15. 18 mi 17. St. Lawrence, 1 in.; Columbia, about $1\frac{1}{2}$ in. 21. 15 ha 23. $-\frac{5}{4}, \frac{4}{5}$

Lesson 7-4, pages 310–313

1. no 3. yes; AA 5. yes; AA 7. The rays of the sun form the same size angle with the ground for both the pole and tree. Therefore, $\angle A \cong \angle D$. Because $\angle B$ and $\angle E$ are both right angles, $\angle B \cong \angle E$. Therefore, the triangles are similar by the AA Similarity Postulate. 9. Yes; because $\angle 1 \cong \angle 2$, and $\angle D \cong \angle D$, there are 2 pairs of congruent angles. The AA Similarity Postulate applies.
11. always 13. sometimes 15. $y = \frac{1}{2}x - 1$
17. $y = x + \frac{1}{2}$ 19. $y = \frac{-9}{7}x + \frac{1}{7}$ 21. $y = 2x + 4$
23. $y = -2x - 2$ 25. 11

Review and Practice Your Skills, pages 314–315

1. 240 km 3. 0.02 ft 5. 52.5 mi 7. 5.55 yd 9. 1027 mi
11. 12.5 in. 13. 40 cm 15. 15 mm 17. 4 ft 19. 18 ft by 12 ft 21. yes, AA 23. no 25. no 27. −58.5 29. 6
31. 6 33. 960 cm 35. ≈ 40 mi 37. ≈ 56 mi

Lesson 7-5, pages 316–319

1. 4.5 3. 3.3 5. given; definition of altitude; Altitudes of similar triangles are in the same proportion as corresponding sides. 7. 3.2 9. The given triangles are similar, so $RW:AK = RY:AL$. Because $\overline{RX}$ is half the length of $\overline{RY}$, and $\overline{AB}$ is half the length of $\overline{AL}$, it follows that

$RX : AB = RY : AL = RW : AK$. $\angle R \cong \angle A$, because they are corresponding parts of similar triangles. You now have $\triangle WRX \sim \triangle KAB$ by SAS Similarity Postulate. Then, $WX:KB = WR:KA$, because they are corresponding parts of similar triangles.   **11.** 6.125 xy

13. ≈ 33.3 cm   **15.** $\left(\dfrac{-5}{3}, \dfrac{-4}{3}\right)$ **17.** 6   **19.** $\dfrac{3}{7}$

Lesson 7-6, pages 320–323

1. 2.5   **3.** 6   **7.** 3   **11.** 8.6   **13.** $YW : XY = ZW : XZ$
15. BE **17.** Congruent base angles mean the triangle is isosceles.   **19.** $\left(\dfrac{7}{5}, \dfrac{11}{5}\right)$ **21.** $\left(\dfrac{-1}{10}, \dfrac{43}{10}\right)$ **23.** $\left(\dfrac{27}{5}, -\dfrac{3}{5}\right)$
25. mean: 10; median: 9.5; modes: 2 and 10

Review and Practice Your Skills, pages 324–325

1. 24.5 m   **3.** 6 ft   **5.** 27 m   **7.** 20 in.   **9.** 46.5 m
11. 4 in.   **13.** 21 cm   **15.** 130°   **17.** true   **19.** $\dfrac{4}{3}$
21. 5.12   **23.** 33   **25.** 37.4 in.   **27.** 4.5

Lesson 7-7, pages 326–327

1. Answers may vary.   **3.** She can write a proportion, such as $\dfrac{3}{w} = \dfrac{10}{28}$ and solve for w, the width of the pond.
5. Yes; Ming formed a right triangle with a 45° angle, leaving 45° for the third angle. By the Base Angles Theorem, the triangle is isosceles, so both legs have length b. So the height of the tree is $a + b$.
7. (5.5, 8.5)   **9.** (1, 2)   **11.** $\left(\dfrac{39}{16}, \dfrac{-1}{16}\right)$ **13.** 20 : 36

Chapter 7 Review, pages 328–330

1. g   **3.** e   **5.** j   **7.** b   **9.** i   **11.** yes   **13.** no   **15.** 10.5
17. 12.5 c   **19.** yes   **21.** $21\dfrac{2}{3}$ **23.** 52.5 mi   **25.** $\dfrac{1}{4}$ in.
27. 10 cm   **29.** yes, AA   **31.** 3.8   **33.** 24.8   **35.** 10
37. 4.0   **39.** 30 ft

Chapter 8: Transformations

Lesson 8-1, pages 338–341

1.
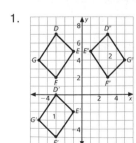

3. $-1, -1, 1, \dfrac{3}{2}, \dfrac{3}{2}, -\dfrac{3}{2}$

5. If part of the figure on one side of a line is the reflection image of the part on the opposite side, the line is a line of symmetry.

7.
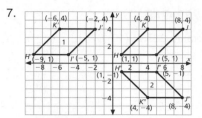

9.
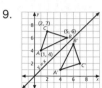

side	AB	BC	CA	A'B'	B'C'	C'A'
slope	$\dfrac{1}{2}$	$-\dfrac{1}{3}$	3	2	-3	$\dfrac{1}{3}$

11.   **13.**
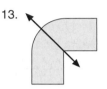

15.
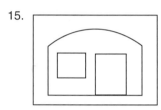

17. The slopes are equal.   **19.** 15   **21.** 18   **23.** 2
25. 4   **27.** -2 **29.** 5   **31.** l **33.** $\overline{DF}, \overline{CG}, \overline{BH}, \overline{AI}$

Lesson 8-2, pages 342–345

1
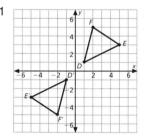
; the corresponding slopes are equal.

3. The first was a 90° rotation; the second was a 180° rotation.
5.
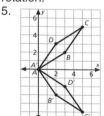
       7.

9. Triangle 6   **11.** Triangle 5   **13.** Triangle 8
15. The slopes of corresponding sides are equal; the product of corresponding slopes is -1; the product of corresponding slopes is -1.   **17.** 10   **19.** $\dfrac{1}{7}$ **21.** $-\dfrac{6}{7}$
23. $\dfrac{6}{5}$ **25.** $-\dfrac{2}{3}$ **27.** undefined slope   **29.** $-\dfrac{3}{4}$

Review and Practice Your Skills, pages 346–347

1–3.

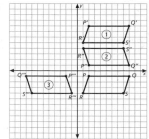

5–7.

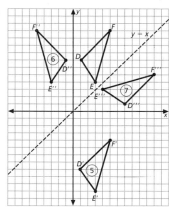

9–11.

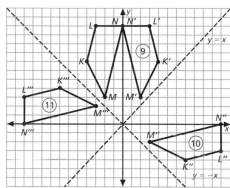

13–15.

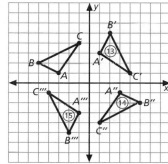

17.

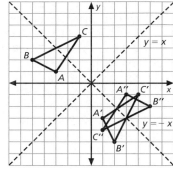

19–21.

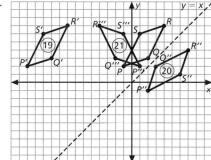

23–25.

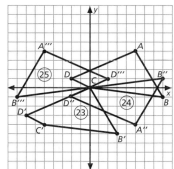

27–29.

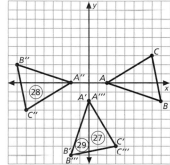

Lesson 8-3, pages 348–351

1. square $A'B'C'D'$ 3. 4 times as long

5.

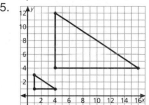

7.

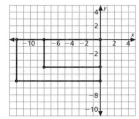

9.

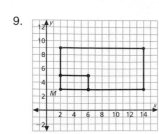

11. 1.25 **13.** 3 **17.** 24 square units; 96 square units; 6 square units; 4 times as large; $\frac{1}{4}$ as large **19.** 48 ft **21.** 60 km **23.** 54 m **25.** 81 m **27.** 102 km **29.** yes; domain: $\{-2, -1, 0, 1\}$; range: $\{-2, -1, 0, 1\}$

31. yes; domain: $\{-3, -2, 2, 3\}$; range: $\{-2, -1, 0, 1\}$

Lesson 8-4, pages 352–355

1.

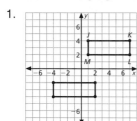

3. No; reflection over *x*-axis and a translation 8 units to the right.

5.

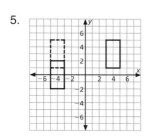

7.

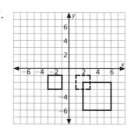

9. Answers will vary. **11.** Possible answers are given; reflection over *y*-axis and translation 5 units up **13.** yes **15.** yes **17.** yes **19.** rotation **21.** AA **23.** SAS

Review and Practice Your Skills, pages 356–357

1.

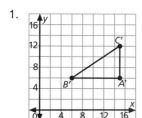

3.

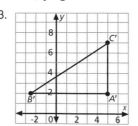

5.

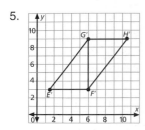

7.

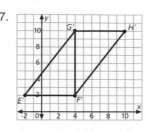

9. Scale factor: $\frac{1}{5}$; Center of dilation: $F(-5, 0)$

11.

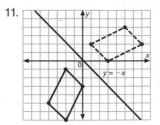

13.

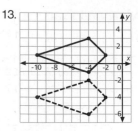

15. Answers will vary. Sample answers are given. reflection across *x*-axis; rotation 90° clockwise

17–19.

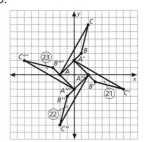

21–23.

25. Scale factor: 3; center of dilation: origin

Lesson 8-5, pages 358–361

1. 3×4 **3.** 10 **5.** $\begin{bmatrix} 0 & -3 \\ 5 & 0 \end{bmatrix}$ **7.** 2×3 **9.** 5×2

11. $\begin{bmatrix} 6 & -2 & -3 \\ -4 & 1 & 2 \end{bmatrix}$ **13.** $\begin{bmatrix} -10 & -1 & 7 \\ 11 & -1 & -11 \end{bmatrix}$

15. $\begin{bmatrix} 26 & -13 & -11 \\ -13 & 5 & 1 \end{bmatrix}$ **17.** $\begin{bmatrix} -4 & -3 & 4 \\ 7 & 0 & -9 \end{bmatrix}$

19.

	Serving	Calorie Count
White bread	1	70
Whole milk	1	150
Spaghetti	1	330

	White bread	Whole Milk	Spaghetti
Serving	1	1	1
Calorie Count	70	150	330

21. Answers will vary. **23.** $\begin{bmatrix} 4 & 4 \\ 3 & 3 \end{bmatrix}$ **25.** $\begin{bmatrix} 20 & 11 & 3 \end{bmatrix}$

27. $x = \frac{13}{2}, y = \frac{5}{2}$ **29.** $R + S = S + R = \begin{bmatrix} 10 & 11 \\ 4 & 3 \end{bmatrix}$; yes

31. $T = \begin{bmatrix} 894 & 930 & 1011 & 949 \\ 875 & 965 & 969 & 980 \end{bmatrix}$ **33.** 1 **35.** 11

37. 12 **39.** 75

Lesson 8-6, pages 362–365

1. 3×4 **3.** 2×4 **5.** NP **7.** NP **9.** $[163]$ **11.** NP **13.** 3×3 **15.** 1×1 **17.** NP **19.** 1×2 **21.** $[28]$

23. NP **25.** NP **27.** $\begin{bmatrix} 12 & 13 & 11 \\ 13 & 10 & 16 \end{bmatrix}$ **29.** $AB = \begin{bmatrix} 7 & 4 \\ 11 & 10 \end{bmatrix}$

$BA = \begin{bmatrix} 14 & 8 \\ 2 & 3 \end{bmatrix}$ **31.** $AI = \begin{bmatrix} 3 & 2 \\ -1 & 5 \end{bmatrix}$; $IA = \begin{bmatrix} 3 & 2 \\ -1 & 5 \end{bmatrix}$

33. $A(BC) = (AB)C = \begin{bmatrix} 9 & -4 \\ -45 & 33 \end{bmatrix}$

35. $\begin{bmatrix} 1 & 0 \\ 0 & 1 \end{bmatrix}$, $\begin{bmatrix} 0 & -1 \\ -1 & 0 \end{bmatrix}$, $\begin{bmatrix} 1 & 0 \\ 0 & 1 \end{bmatrix}$

37. $\begin{bmatrix} 90{,}200 & 70{,}000 & 119{,}500 \\ 233{,}400 & 181{,}200 & 309{,}000 \end{bmatrix}$ **39.** $x = 1, y = -1$

41. $(-1, 3)$ **43.** $(2, 3)$ **45.** $(6, 5)$

Review and Practice Your Skills, pages 366–367

1. $\begin{bmatrix} 12 & 0 & -15 & 21 \\ -9 & 24 & -6 & 3 \end{bmatrix}$ **3.** $\begin{bmatrix} -4 & -8 & 18 & -12 \\ 4 & -12 & 9.5 & 8 \end{bmatrix}$

5. $\begin{bmatrix} 4 & 8 & -18 & 12 \\ -4 & 12 & -9.5 & -8 \end{bmatrix}$ **7.** $\begin{bmatrix} -9 & 3 & -3.5 & 9.5 \\ 3.5 & 6.5 & -1 & -55 \end{bmatrix}$

9. $\begin{bmatrix} 15 & -11 & 9 & -4 \\ -7 & 1.5 & 5.5 & 16 \end{bmatrix}$ **11.** $\begin{bmatrix} 8 & -16 & 16 & 4 \\ -4 & 8 & 11 & 20 \end{bmatrix}$

13. $\begin{bmatrix} 4.4 & -1.2 & 0.4 & -2.4 \\ -2 & -1 & 0 & 2.4 \end{bmatrix}$ **15.** $\begin{bmatrix} -4 & 0 & 5 & -7 \\ 3 & -8 & 2 & -1 \end{bmatrix}$ **17.** 0

19. 2×4 **21.** $x = -5; y = 3$ **23.** $x = -1; y = 7$

25. $x = 0; y = -\frac{1}{2}$ **27.** $\begin{bmatrix} 15 & -4 \\ -5 & 8 \end{bmatrix}$ **29.** NP **31.** $\begin{bmatrix} 4 & 2 \\ -12 & 19 \end{bmatrix}$

33. $\begin{bmatrix} -28 & 26 \\ 22 & 6 \\ -2 & 14 \end{bmatrix}$ **35.** $\begin{bmatrix} -2 & 2 & 14 \\ -34 & 4 & -22 \end{bmatrix}$ **37.** $\begin{bmatrix} -20 \\ -70 \end{bmatrix}$

39. $\begin{bmatrix} -10 & -8 \\ -45 & 38 \\ -25 & 14 \end{bmatrix}$ **41.** $\begin{bmatrix} -78 & 62 \\ -12 & 42 \end{bmatrix}$ **43.** $\begin{bmatrix} -50 & 2 & -66 \\ -3 & 9 & 73 \\ -23 & 5 & 5 \end{bmatrix}$

45. $\begin{bmatrix} -68 \\ -84 \\ -68 \end{bmatrix}$ **47.** $\begin{bmatrix} -63 & 58 \\ -17 & 50 \end{bmatrix}$ **49.** $\begin{bmatrix} 10 & -15 \\ 10 & -5 \end{bmatrix}$ **51.** $\begin{bmatrix} 22 & 2 & 54 \\ -91 & 19 & 13 \end{bmatrix}$

53. $\begin{bmatrix} -276 & 122 \\ -168 & 166 \end{bmatrix}$ **55.** $\begin{bmatrix} 20 \\ -650 \\ -290 \end{bmatrix}$ **57.** $\begin{bmatrix} 34 \\ -68 \end{bmatrix}$

59. $x = -3; y = -5$

61–63.

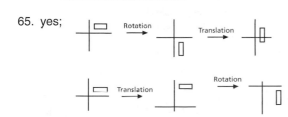

65. yes;

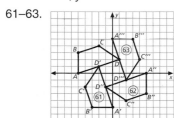

67. no affect

Lesson 8-7, pages 368–371

1. $\begin{bmatrix} -3 & 1 & -3 \\ 1 & 3 & -3 \end{bmatrix}$ **3.** $\begin{bmatrix} 1 & 5 & 7 & -4 \\ 3 & 3 & -3 & -3 \end{bmatrix}$ **5.** $\begin{bmatrix} 2 & 7 & 3 \\ -1 & 3 & 7 \end{bmatrix}$

7. $\begin{bmatrix} -2 & -7 & -3 \\ 1 & -3 & -7 \end{bmatrix}$ **9.** $\begin{bmatrix} -3 & 3 & 4 & -2 \\ 2 & 4 & 1 & -1 \end{bmatrix}$ **11.** $\begin{bmatrix} 7 & 3 & 4 & 7 \\ -2 & 1 & 7 & 4 \end{bmatrix}$

13. $\begin{bmatrix} -7 & -3 & -4 & -7 \\ 2 & -1 & -7 & -4 \end{bmatrix}$ **15.** $(4, 4)$, line $y = -x$, $(-4, -4)$
17. $(-2, 4)$, line $y = x$, $(4, -2)$

19. yes **21.** $y = -x$ **23.** y-axis **25.** dilation with center at origin and a scale factor of -2 **27.** c; b; a **29.** $\frac{1}{2}, -3$
31. $\frac{1}{2}, 3$ **33.** $\frac{4}{3}, \frac{7}{3}$

Lesson 8-8, pages 372–373

1. A: \$1602.50; B: \$1748.25; C: \$2344.50 **3.** A: \$465; B: \$571.25; C: \$586.25 **5.** 6300° **7.** 4860° **9.** 10,800° **11.** 10,080°

Chapter 8 Review, pages 374–376

1. k **3.** f **5.** e **7.** i **9.** g
11.

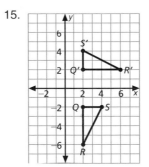

13.

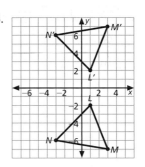

15.

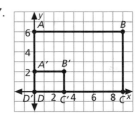

17.

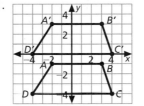

19. Possible answer: reflection over x-axis, then translation 4 units right **21.** 3×4

23. $\begin{bmatrix} -2 & -4 & -6 & -4 \\ 8 & -10 & 12 & -8 \\ -14 & 16 & -18 & -12 \end{bmatrix}$ **25.** $\begin{bmatrix} 9 & -6 & 11 & -15 \\ -14 & 20 & -3 & 5 \end{bmatrix}$

27. $\begin{bmatrix} -28 & -13 \\ 20 & 15 \end{bmatrix}$ **29.** $\begin{bmatrix} -4 & -6 & -5 \\ 2 & 1 & 4 \end{bmatrix}$

31. Advanced: \$326, Beginning: \$490

Chapter 9: Probability and Statistics

Lesson 9-1, pages 384–387

1. 0.7 **3.** 20 times **5.** 0.18 **7.** 0.55 **9.** Answers will vary. **11.** (A, 1), (A, 2), (A, 3), (B, 1), (B, 2), (B, 3), (C, 1), (C, 2), (C, 3) **13.** $\frac{1}{3}$ **15.** 0.75 **17.** 0.75; 375 **19.** Of past days when weather conditions were similar to those predicted for tomorrow, it rained 25% of the time.
21. experimentally **23.** theoretically

25. 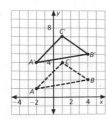 27.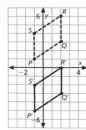

Lesson 9-2, pages 388–389

1. 1:5 (18 of the 90 possible numbers)   3. Answers will vary.   5. Answers will vary.   7. Change line 4 to: "IF $X < .4$ THEN $S = S + 1$"   9. $\frac{4}{7}$ 11. 3   13. $-\frac{1}{2}$
15. undefined   17. 0   19. 15   21. 17.5   23. $\frac{9}{7}$ 25. $\frac{2}{3}$

Review and Practice Your Skills, pages 390–391

1. $\frac{1}{4}$ 3. $\frac{1}{2}$ 5. $\frac{2}{13}$ 7. $\frac{6}{13}$ 9. $\frac{1}{16}$ 11. $\frac{3}{8}$ 13. $\frac{5}{16}$
15. $\frac{1}{4}$ 17. $\frac{1}{6}$ 19. $\frac{1}{2}$ 21. $\frac{1}{36}$ 23. $\frac{5}{36}$ 25. $\frac{1}{6}$ 27. $\frac{33}{36}$
29. $\frac{2}{3}$, 400   31. $\frac{7}{25}$ 33. Answers will vary.
35. Answers will vary.   37. (H, H) (T, H) (H, T) (T, T)
39. (1, H) (4, H); (1, T) (4, T); (2, H) (5, H); (2, T) (5, T); (3, H) (6, H); (3, T) (6, T)

Lesson 9-3, pages 392–395

1. $\frac{1}{3}$ 3. $\frac{7}{13}$ 5. $\frac{\text{number of pieces of paper} - 1}{\text{number of pieces of paper}}$ 7. $\frac{1}{3}$ 9. $\frac{1}{2}$
11. $\frac{2}{9}$ 13. $\frac{1}{3}$ 15. $\frac{4}{5}$ 17. 1   19. $\frac{5}{13}$ 21. $\frac{1}{2}$ 23. 0
25. 21   27. Sample Answer: 12 is a multiple of both numbers.   29. 0.594   31. 400   33. −320   35. 2000
37. −500   39. 10,000   41. −16

Lesson 9-4, pages 396–399

1. $\frac{5}{48}$ 3. $\frac{2}{9}$ 5. $\frac{7}{120}$ 7. $\frac{1}{5}$ 9. $\frac{3}{25}$ 11. $\frac{21}{100}$ 13. $\frac{2}{23}$
15. $\frac{2}{9}$ 17. $\frac{1}{192}$ 19. $\frac{1}{16}$ 21. $\frac{1}{6}$ 23. $\frac{2}{55}$ 25. Mutually exclusive events cannot occur at the same time. Independent events can occur at the same time, but neither event affects the other.   27. $\frac{9}{16}$ 29. $\frac{27}{32}$
33.

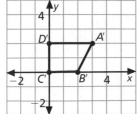

Review and Practice Your Skills, pages 400–401

1. $\frac{3}{52}$ 3. 1   5. $\frac{2}{13}$ 7. $\frac{4}{13}$ 9. $\frac{1}{26}$ 11. $\frac{3}{8}$ 13. $\frac{5}{16}$
15. $\frac{1}{4}$ 17. $\frac{1}{18}$ 19. $\frac{11}{36}$ 21. $\frac{1}{36}$ 23. $\frac{5}{18}$ 25. $\frac{1}{4}$ 27. $\frac{1}{6}$
29. $\frac{3}{8}$ 31. $\frac{77}{240}$ 33. $\frac{1}{20}$ 35. $\frac{11}{24}$ 37. $\frac{80}{9261}$ 39. $\frac{275}{9261}$

41. $\frac{95}{441}$ 43. $\frac{20}{9261}$ 45. {HH, HT, TH, TT}
47. {Sun., Mon., Tue., Wed., Thur., Fri., Sat.}   49. Answers will vary.   51. $\frac{1}{8}$ 53. $\frac{3}{8}$ 55. $\frac{3}{4}$ 57. $\frac{1}{300}$ 59. $\frac{1}{9900}$
61. $\frac{2}{33}$

Lesson 9-5, pages 402–405

1. 120   3. combination; 20   5. permutation; 32,760   7. 30
9. 15,504   11. $\frac{1}{220}$ 13. $\frac{1}{15,600}$ 15. 1260; Sample answer: Divide 7! by 2! · 2! · 1! · 1! · 1!   17. 8! or 40,320
19. They are the same.   21. They equal the total number of items.   23. $\begin{bmatrix} 3 & 7 \\ 1 & 4 \end{bmatrix}$ 25. 51°   27. 141°

Lesson 9-6, pages 406–409

1. about 130 lb   3. 67–68 in.   5. above   7. 10 points
9. Artichokes   11. There is a greater range of batting averages in the upper 25% of its players than in the lower 25%.   13. Answers will vary. Sample answers are given; scatterplot   15. Yes, when there are data values far from the median.
17. $\begin{bmatrix} 36 & 32 \\ 28 & 4 \end{bmatrix}$

Review and Practice Your Skills, pages 410–411

1. 60   3. 4   5. 840   7. 17,297,280   9. 10   11. 1
13. 8   15. 15   17. 6840   19. 3003   21. 210   23. 9
25.

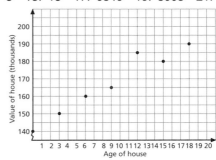

27. positive   29. Answers will vary.   31. $\frac{1}{4}$ 33. $\frac{5}{8}$ 35. 0
37. does   39. does

Lesson 9-7, pages 412–415

1. 0; 0   3. 8; $2\sqrt{2}$ 5. 2; $\sqrt{2}$ 7. ≈ 60; $2\sqrt{15}$ 9. 2310
11. mean: 38; median: 40; mode: 30; variance: 256; standard deviation: 16   13. Answers will vary.
15. Answers will vary.   17. Answers will vary.
19. Answers will vary.   21. The one with the higher mean is farther to the right.   23. The right one.
25. [−11]   27. $\begin{bmatrix} -8 & -57 & 46 \\ 2 & -7 & 14 \end{bmatrix}$ 29. 58, −94

Chapter 9 Review, pages 416–418

1. j   3. f   5. c   7. b   9. g   11. $\frac{1}{30}$ 13. 1   15. $\frac{2}{5}$
17. $\frac{1}{26}$ 19. Possible answer: Assign a number from 1 to 3 to each toy. Use the random number generator to randomly select numbers from 1 to 3 five at a time. Record whether or not all numbers are represented. Repeat the process numerous times.   21. $\frac{1}{9}$ 23. $\frac{5}{9}$ 25. $\frac{1}{12}$ 27. $\frac{1}{8}$ 29. 120

31. 60 whole numbers   33. positive

35.

Basketball Team Ratings

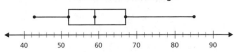

37. 18; $\sqrt{18}$ 39. 34; about 5.8

Chapter 10: Right Triangles and Circles

Lesson 10-1, pages 426–429

1. 3.32   3. 9.22   5. $2\sqrt{11}$ 7. $5\sqrt{3}$ 9. $28\sqrt{3}$ 11. 32
13. $\frac{2\sqrt{3}}{3}$ 15. 4.58   17. 8.54   19. $9\sqrt{2}$ 21. $6\sqrt{2}$
23. $24\sqrt{2}$ 25. 27   27. $\frac{7\sqrt{2}}{2}$ 29. 20   31. 2.4 m
33. Bakong Temple, $\sqrt{4900}$ m; Wat Kuhat Temple, $\sqrt{529}$ m
35. 3   37. 2   39. 7.5 cm^2 41. true   43. false; possible
counterexample $(\sqrt{2})(\sqrt{18}) = 6$ 45. 2   47. 10,000
49. 140

Lesson 10-2, pages 430–433

1. 9.9 m   3. 8.6 ft   5. about 17.3 ft   7. 9.4 cm
9. 7.9 cm   11. 11.3 ft   13. yes   15. 6.4 cm
17. about 4.5 yd   19. 6.7   21. Yes   23. See the figure
below. $d^2 = y^2 + h^2$ and $y^2 = l^2 + w^2$; Therefore, $d^2 = l^2 + w^2 + h^2$ and
$$d = \sqrt{l^2 + w^2 + h^2}$$

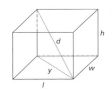

25. The system worked because the workers formed a
triangle that had sides measuring 3 rope lengths, 4 rope
lengths, and 5 rope lengths, $[12 - (3 + 4)] = 5$. Because
3, 4, and 5 form a Pythagorean triple, they formed a right
triangle.   27. 0.17   29. 0.155

Review and Practice Your Skills, pages 434–435

1. 7.28   3. 5.29   5. 31.62   7. 31.62   9. $5\sqrt{2}$ 11. $6\sqrt{2}$
13. $4\sqrt{5}$ 15. $2\sqrt{22}$ 17. $11\sqrt{2}$ 19. $15\sqrt{7}$ 21. 12
23. $\frac{\sqrt{3}}{2}$ 25. $\sqrt{5}$ 27. $\frac{4\sqrt{7}}{7}$ 29. 4   31. 2   33. 10
35. $\frac{1}{3}$ 37. 26 m   39. 13.6 in.   41. 65 yd   43. 30 cm
45. $2\sqrt{6}$ 47. $2\sqrt{10}$ 49. $\frac{\sqrt{5}}{5}$ 51. 2   53. $24\sqrt{7}$
44$\sqrt{6}$ 57. 36   59. $2352\sqrt{3}$ 61. no   63. yes

Lesson 10-3, pages 436–439

1. $4\sqrt{3}$ in., 8 in.; 6.9 in., 8.0 in.   3. $6\sqrt{2}$ yd; 8.5 yd
5. 5.8 ft   7. $10\sqrt{2}$ cm; 14.1 cm   9. $\sqrt{3}$ m, 2 m; 1.7 m,
2 m   11. 10.6 cm   13. $25\sqrt{3}$ cm^2 or about 43.3 cm^2

15. 9.5 m   17. Always; let x represent the measure of the
smaller angle. Then $90 - x$ can represent the larger angle.
You can solve this equation for x:

$x = 0.5(90 - x)$ Since the triangle is a 30°-60°-90°
$x = 45 - 0.5x$ triangle, the length of the side
$1.5x = 45$ opposite the 30° angle is half the
$x = 30$ length of the hypotenuse.

19. $\frac{3}{13}$ 21. $\frac{1}{26}$ 23. $\frac{1}{26}$ 25. ; $x > -2$

27. ; $x \le -6$

Lesson 10-4, pages 440–443

1. 140   3. 49   5. 50   7. 40°   9. 50°   11. If two
secants (or chords) intersect inside a circle, then the
measure of each angle formed is equal to one-half the
sum of the measures of the intercepted arcs.
13. $\frac{1}{24}$ 15. $\frac{1}{69}$ 17. $\frac{1}{2}$ 19. $-\frac{2}{3}$ 21. 1   23. -4
25. -2

Review and Practice Your Skills, pages 444–445

1. $10\sqrt{2}$ in., $10\sqrt{2}$ in.; 14.1 in., 14.1 in.   3. $3\sqrt{2}$ m; 4.2 m
5. $11\sqrt{3}$ cm, 22 cm   7. 5 in., 10 in.   9. 18.5 ft,
$18.5\sqrt{3}$ ft   11. $10\sqrt{3}$ km, $20\sqrt{3}$ km   13. 44 in.,
$44\sqrt{2}$ in.   15. $4\sqrt{2}$ ft, $4\sqrt{2}$ ft   17. 46   19. 125
21. 124   23. false   25. $\frac{\sqrt{6}}{4}$ 27. $4752\sqrt{3}$ 29. 19.8 mm
31. 11.0 ft

Lesson 10-5, pages 446–447

1.   3.

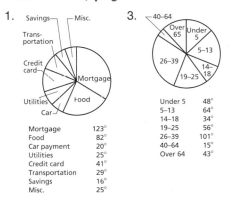

Mortgage	123°
Food	82°
Car payment	20°
Utilities	25°
Credit card	41°
Transportation	29°
Savings	16°
Misc.	25°

Under 5	48°
5–13	64°
14–18	34°
19–25	56°
26–39	101°
40–64	15°
Over 64	43°

5.

7. $(-6, -11)$ 9. $(3, 5)$

2 Persons	150°
3 Persons	84°
4 Persons	76°
5 Persons	32°
6 Persons	11°
7 or more	7°

Lesson 10-6, pages 448–451

1. 3   3. 3   5. 30   7. 6   9. 10   11. 12   13. 9 cm
15. $x = 13$; $y = 6.5$ 17. $x = 5$, $y = 12.5$ 19. 6 cm
21. $\triangle CEB \sim \triangle CAD$ by AA similarity. $\angle C \cong \angle C$ and
$\angle A \cong \angle E$, because both are inscribed angles that
intercept $\overset{\frown}{CD}$.   23. 186.5 in.^{2} 25. 352.99 ft^2

Review and Practice Your Skills, pages 452–453

1.
Howe Family Budget

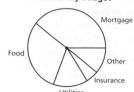

3.
Fall Sports Athletes

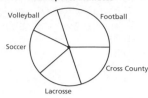

5.
Technology Annual Budget

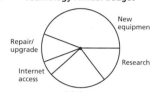

7. 4   9. 11   11. 4
13. false   15. false
17. true   19. no
21. yes   23. no
25. $7\sqrt{3}$ in., 21 in.;
12.1 in.,   21 in.

27. $\dfrac{39\sqrt{2}}{2}$ m, 27.6 m

Lesson 10-7, pages 454–457

5. false   7.

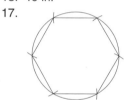

13. 40 in.

17.

Draw a circle with a radius of 2 in. Then inscribe a hexagon in the circle.

19. Mike could take the copy, find the perpendicular bisectors of two different sides of the hexagon, and use their intersections as the center of a circle. He could use a radius equal to the distance from the center of the circle to a vertex of the hexagon. Then he could construct the perpendicular bisectors of each of the other sides of the hexagon, and mark the point where these bisectors meet the circle. At this point, there should be 12 equally spaced points on the circle. He can connect these 12 points with a straightedge to draw a dodecagon.   21. 2.9, 1.7
23. 1.7, 1.3   25. 10.25, 3.2

Chapter 10 Review, pages 458–460

1. d   3. b   5. l   7. i   9. c   11. $3\sqrt{5}$ 13. $84\sqrt{5}$
15. $\dfrac{\sqrt{3}}{2}$ 17. 15 m   19. 13.1 in.   21. $6\sqrt{2}$ m
23. $2\sqrt{3}$ cm, $4\sqrt{3}$ cm   25. 6 ft, $6\sqrt{2}$ ft   27. 62
29. 55   31. 35   33. Check students' circle graphs. Percents should be Italian: 22%, Spanish: 41%, French: 33%, and Japanese: 4%.

35.
Favorite Flavor of Ice Cream

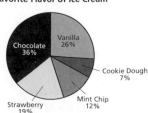

37. 2   39. $11\dfrac{2}{3}$
41. $\sqrt{117} \approx 10.8$
43.

Chapter 11: Polynomials

Lesson 11-1, pages 468–471

1. $5x^3 + x^2 + x - 4$ 3. $4x^2 + 10$ 5. $6x^2 - 4x$
7. $2x - 7$ 9. $x + 21$ 11. $7x^3 + 15x^2 - 5xy$
13. $9x^3 + 5x^2 - 3x - 3$ 15. $5a + 13$ 17. $2x^2 + 7x$
19. $2t + 5$ 21. $5m^2$ 23. $x^2 - 22x + 3y^2$
25. $3v^2 - 2vw - 10w^2$ 27. $6b^2 + 2c - 15$
29. $25p + 32s$ 31. 4   33. $6.3a^3 - 4.2a^2b +$
$5.8bc^2 - 1.7b^3$ 35. $19x^2 + 6x$ 37. $12x^3 - 3xy$
41. 8.66   43. 14.70   45. 175   47. $\dfrac{4\sqrt{30}}{5}$ 49. $36\sqrt{2}$
51. $8\sqrt{14}$ 53. 3.2×10^{10}

Lesson 11-2, pages 472–475

1. $3xy$ 3. $12pq$ 5. rs^2 7. $7x^2 + 7x$ 9. $a^4 + a^3$
11. $-e^3f - e^2f^3$ 13. $ab^2 + ab - 6a$ 15. $-7x^3 +$
$14x^2y - 7xy^2$ 17. $x^2 + xy + xz$ 19. $3x^3y$
21. $-12x^5y$ 23. $9a^4$ 25. $-6x^4 - 4x^3$
27. $-3m^4n^3 + 3m^5n^5$ 29. $8e^2f^2g^4 - 8ef^3g^4$
31. $-18l^3m^2n^7 + 18l^2m^6n^5$ 33. $4abe^2 - 2abf + abg$
35. $12uv^2w + 8v^3w + 4v^2w^4$ 37. $7r^5s^4t^5 - 7r^4s^5t^4 -$
$7r^3s^8t^3$ 39. $4x^2 + 2xy + 2xz$ 41. $60hk + 180h$ ft^2
43. $2b^2(3a)$ $2(3ab^2)$ $3(2ab^2)$ $6(ab^2)$ $a(6b^2)$ $2a(3b^2)$
$6a(b^2)$ $b(6ab)$ $2b(3ab)$ $3b(2ab)$ $6b(ab)$
45. x^4y^6 47. $2a^6$ 49. $15x^4y - 15x^3y^2 + 15x^2y^3$
51. $3ts + 20t$ 53. $4800p^2$ 55. Jarius plays for the Gophers, Keshawn, for the Goats, and Levon for the Cheetahs.

Review and Practice Your Skills, pages 476–477

1. $6x + 10y$ 3. $16x^2 - 4x - 7$ 5. $-2x - 2z$
7. $11x^3 - 11x^2 - 8x$ 9. $-2y^2 - y - 1$ 11. $4x^2y - 2xy$
$- 3xy^2$ 13. $-7x + 8y$ 15. $17c^2 + 17cd - 6d^2$
17. $12n + 9p$ 19. $-6x^2$ 21. $-42m^2n$ 23. $9k^6$
25. $28s^5t^3$ 27. $12x^2 - 30x$ 29. $33x^4 + 22x^3 - 11x^2$
31. $-p^3q^2 + 3p^2qr - 7p^2q^4$ 33. $15ax + 10bx - 20cx$
35. $5x^2 - 6x + 16$ 37. $-4p^4q^2 - 20p^2qr + 12p^2q^3$
39. $8ayz + 6byz - 20cyz$ 41. $3x^{15}y^9z^{10} + 6x^{11}y^{10}z^9 -$
$3x^{11}y^9z^{13}$ 43. $4x(3x - 7); 12x^2 - 28x$
45. $3x(x^2 - 6x + 7); 3x^3 - 18x^2 + 21x$ 47. $11b + 11$
49. $\dfrac{5}{4}h - \dfrac{3}{8}g$ 51. $-5m + 2n + 12p$ 53. $-4a^2 + 24b^2$
55. $13r^2 + 10rs - 21s^2$ 57. $-4m^2 + 23n$
59. $12x^4 - 2x^3 + y^4$ 61. $-16p^4qr^3$ 63. $-6x^2 + 28x$
65. $36c^2d^3 - 27c^3d^2$

Lesson 11-3, pages 478–481

1. $3(2x^2 + 3)$ 3. $n(5m - np)$ 5. $6x(2x + 3)$
7. $r(7r + 3s + 2t)$ 9. 18 in. tall, 8 figurines
11. $2(3a + 4b)$ 13. $5(3p^3 - 7q)$ 15. $w(v + x)$
17. $y(5x^2 - 2y)$ 19. $n(13mn + 25)$ 21. $\dfrac{5f}{2}$
23. $7b(2ab + 5c)$ 25. $9r^2(2r + 3)$
27. $13j^5k^2l^3(3j^2kl - 5jk^3 + 4l^3)$ 29. $xy(ax^2y^2 + bxy + c)$
31. $A = \dfrac{1}{2}h(t + b) = 22$ in.^{2} 33. $2xy^2(3x^4 + 4x^3y +$
$3x^2y^2 + 7xy^3 + y^4)$ 35. $8a(8b + 5c)$ 37. $25x(x + 2y)$
39. $x^{(n-1)}(3x - 2)$ 43. $x \approx \dfrac{13\sqrt{2}}{2} \approx 9.19$ 45. \$2.17

Lesson 11-4, pages 482–485

1. $3ac + 15ad + 2bc + 10bd$ 3. $l^2 + ln - lm - mn$
5. $6x^2 + 21x + 15$ 7. $8x^2 + 15xy - 2y^2$ 9. $p^2 - q^2$
11. $x^2 + 22x + 120$ 13. $6pr - 2p + 15qr - 5q$
15. $4a^2 + 12ac + ab + 3bc$ 17. $2eg + 5ef - 6fg - 15f^2$
19. $45p^2 - 27pr + 10pq - 6qr$ 21. $5m^2 + 51mn + 54n^2$
23. $3x^2 - 10x + 8$ 25. $7j^2 - 37jk + 10k^2$
27. $24b^2 + 37bc - 5c^2$ 29. $w^2 - 8wz + 16z^2$
31. $x^2 + 8x + 16$ 33. $a^2 - 4$ 35. $4e^2 - 25f^2$
37. $5x + y - 5$ 39. $9x^3 + 6x^2 - 30x + 8$
41. $2p^3 + p^2q - 5pq^2 + 2q^3$ 43. $a^4 + 4a^3b + 6a^2b^2 +$
$4ab^3 + b^4$ 45. $2y^3 + 3y^2 - 17y + 12$ 47. $2\frac{1}{4}$ ft^2

49. 143 in.^{3} 51. 396 cm^3 53. $\begin{bmatrix} 4 & -3 & 3 \\ 1 & 7 & 0 \\ 2 & -3 & 7 \end{bmatrix}$

55. $\begin{bmatrix} 20 & 30 \\ 15 & 40 \\ -25 & 30 \end{bmatrix}$ 57. $\begin{bmatrix} -12 & -4 \\ 32 & 24 \\ 28 & -16 \end{bmatrix}$ 59. 6.2 in.

Review and Practice Your Skills, pages 486–487

1. $4(2x + 3y)$ 3. $x(7x + 15)$ 5. $gh(2 - k)$
7. $a(28bc - 11a^2)$ 9. $y^2(17x + 24z)$ 11. $5x^2(x + y^2)$
13. $12(3a + 2b)$ 15. $5b(a + 2c - 1)$ 17. $18p(pq - 2r^2)$
19. $15s^2t(t + 3s)$ 21. $2x(2x^2 - x + 7)$
23. $3uv(1 - 3uv + u^2v^2)$ 25. $18m^2n^2(2mn^3 + 4n + 3m^3)$
27. $2abc(3a + b - 2)$ 29. $4mnp(2 - 5mp^2 + 4n^3p)$
31. $x^2 - x - 6$ 33. $2x^2 + 7xy + 6y^2$ 35. $25x^2 + 40x +$
16 37. $4mp + 5m^2 - 20np - 25mn$ 39. $3a^2 - 23ab +$
$30b^2$ 41. $x^2 - 36$ 43. $64x^2 - 9$ 45. $8y^2 - 2yz -$
$45z^2$ 47. $xy + 2x + y + 2$ 49. $81x^2 - 1$ 51. $64p^2 +$
$128pq + 64q^2$ 53. $z^4 - 25$ 55. $8r^2 - 18s^2$
57. $-42c^2 + 53cd - 15d^2$ 59. $x^3 + 17x^2 + 52x$
61. $(7x - 5)(2x + 3) = 14x^2 + 11x - 15$
63. $-13x + 18y$ 65. $-9x^2 + 20x$ 67. $-30x^2 + 66x$
69. $20x^4 - 15x^3 + 5x^2$ 71. $-8p^4q^2r + 20p^2qr^2 + 12p^4q^3r$
73. $6(-5x + 9)$ 75. $g(12 + 25g)$ 77. $15rst^2(3rt + 5s)$
79. $8a^2b^2(6a + 7b^2 - 4a^2b)$ 81. $2(4x^3 + 3x^2 + 2x + 1)$
83. $49a + 7ab - 7b - b^2$ 85. $16x^2 + 24x + 9$
87. $48p^2 + 2pq - 35q^2$ 89. $-4x^2 + 1$

Lesson 11-5, pages 488–491

1. $(3w + 2y)(3x + 2z)$ 3. $(9a - 4c)(2b - 3d)$
5. $(5r + 3)(s - 8t)$ 7. $(k + m)(\ell + n)$
9. $(m - 3n)(3r - 8s + 5t)$ 11. $(y + 4)$
13. $(2a + 3c)(2b + 3d)$ 15. $(4q + s)(r + 3t)$
17. $(3e - g)(7f - 4h)$ 19. $(9w^2 - y)(3x - 2z^2)$
21. $(k^2 - 3m)(2l^2 + 5n)$ 23. $(5t - 2v)(3u + 4)$
25. $(x^2 + 8)(3y - z)$ 27. $(y + 5z)(v + 3w + 2x)$
29. $(5p - q)(2r - 3s + 4t)$ 31. $(3d - 5e)(2f + 4g + 7h)$
33. $3j(2j + k)(j - 2l)$ 35. $3r^2(r - 2s)(r + 2t)$ 37. 2.4
39. $(2m - 6)$ and $(n - 3)$ or $(m - 3)$ and $(2n - 6)$
41. $3a(2a + b)(2a + 3c)$ 43. 50 45. 50 47. 55
49. -56 51. 2 53. 12 55. $6\frac{1}{2}$

Lesson 11-6, pages 492–495

1. $(s + 5)^2$ 3. $(m + 4n)^2$ 5. $(3r + 6)(3r - 6)$ 7. none
9. $(8u - 3v)(8u - 3v)$ 11. the difference of two squares;
$(p - 3)(p + 3)$ 13. $(6a + 2b)^2$ 15. $(2x - 9y)(2x - 3y)$
17. $(10r + 11)^2$ 19. none 21. none 23. $(6r - s)(6r + s)$
25. none 27. $4(x + 1)^2$ 29. $p = 3$ 31. $a = 3$ or $a =$
$2\frac{1}{2}$ 33. $y^2 - 64x^2$; $(y - 8x)(y + 8x)$ 35. It is a perfect

square trinomial; 100 37. $5s(s - 2t)^2$ 39. $3xy(x +$
$2y)(x - 2y)$ 41. $(x - y)(x - 1)^2$
43. 1, 1; 2, 1; 3, 1; 4, 1; 5, 1; 6, 1; 7, 1; 8, 1;
    1, 2; 2, 2; 3, 2; 4, 2; 5, 2; 6, 2; 7, 2; 8, 2;
    1, 3; 2, 3; 3, 3; 4, 3; 5, 3; 6, 3; 7, 3; 8, 3;
    1, 4; 2, 4; 3, 4; 4, 4; 5, 4; 6, 4; 7, 4; 8, 4;
    1, 5; 2, 5; 3, 5; 4, 5; 5, 5; 6, 5; 7, 5; 8, 5;
    1, 6; 2, 6; 3, 6; 4, 6; 5, 6; 6, 6; 7, 6; 8, 6;
    1, 7; 2, 7; 3, 7; 4, 7; 5, 7; 6, 7; 7, 7; 8, 7;
    1, 8; 2, 8; 3, 8; 4, 8; 5, 8; 6, 8; 7, 8; 8, 8;
45. 0.1875 47. 0.5 49. 10.5 cm

Review and Practice Your Skills, pages 496–497

1. $(5 + b)(c + d)$ 3. $(a - c)(b - 3)$ 5. $(2 + j)(h - k)$
7. $(y^2 + 3)(y - 2)$ 9. $(w - 3)(2z - 1)$
11. $(m - n)(w - x)$ 13. $(2x^2 + 3)(y - 4)$
15. $(p^2 - q)(r^3 - 2s)$ 17. $(1 + v)(w - v)$
19. $(x - 3ay)(1 - y)$ 21. $(3x + 2)(-3y + 2z)$
23. $(w - 4)(x + 2y + 3z)$ 25. $(x + c)(x - a - b)$
27. $(2 - 7a) \times (g + 2f)$ 29. $(x - 10)(x - 10)$
31. not possible 33. $(6b - 1)(6b - 1)$
35. $(x - 4y)(x - 4y)$ 37. $(w + 12)(w - 12)$
39. $(c + 3d)(c - 3d)$ 41. $(4u + 9v)(4u - 9v)$
43. $(5s - 7t)(5s - 7t)$ 45. $(7p + 2q)(7p + 2q)$
47. $(8m + 11n)(8m + 11n)$ 49. $(1 - a)(1 - a)$
51. not possible 53. $(15x + 11y)(15x + 11y)$
55. $3(x - 2)(x - 2)$ 57. $x(x - 4)(x - 4)$
59. $3a(x + 2)(x - 2)$ 61. $x^2(x + 5)(x - 5)$
63. $4(a + b)(a - b)$ 65. $4xy - 2x^2 - 16y$
67. $6x^2y^2z - 14xy^2z^2 + 30x^2yz^2$ 69. $5x$
71. $16x(3x + 2)$ 73. $-4ef(d + 2g + 3)$
75. $2s(k^2 + 29q^2 + 17y^2)$ 77. $(w - 4)(3x + 7)$
79. $(2x - 7y)(2x - 7y)$

Lesson 11-7, pages 498–501

1. 3 and 7 3. $4b$ and $2b$ 5. 6 and 1 7. $+$ and $+$
9. $-$ and $+$ 11. $(c + 2)(c + 3)$ 13. $(c - 6)(c + 1)$
15. $(x + 3)(x + 7)$ 17. $7y, 5y$ 19. $5b, 2b$ 21. 9, 7
23. $10f, 3f$ 25. $+, +$ 27. $+, -$ 29. $-, -$ 31. $-, +$
33. $(p + 6q)(p + 4q)$ 35. $(k - 12)(k + 2)$
37. $(h - 1)(h + 24)$ 39. $(f - 3g)(f - 8g)$
41. $(q - 4)(q - 7)$ 43. $(s - 2t)(s + 4r)$ 45. No. Only
one pair of factors will satisfy all conditions.
47. $(1 - 3r)(1 - 2r)$ 49. $(6g - 1)(4g - 1)$
51. $5x^2(a - 2)(a - 1)$ 53. $3x(x + 2)(x - 6)$
55. 27.5% 57. 13% 59. 10.5% 61. 6.25%
63. 99° 65. 46.8° 67. 37.8° 69. 22.5°

71. $y = \frac{2}{3}x - 2$ 73. $y = -2x + 3$

Lesson 11-8, pages 502–503

1a. $(x + n_1)(x + n_2) =$
  F O I L
  ↓ ↓ ↓ ↓
  $x^2 + xn_2 + xn_1 + n_1n_2 =$
  $x^2 + (n_2 + n_1)x + n_1n_2$

1b. $(x - 6)(x + 3) =$
  F O I L
  ↓ ↓ ↓ ↓
  $x^2 + 3x - 6x - 18 =$
  $x^2 + (3 - 6)x - 18 =$
  $x^2 - 3x - 18$

3a. $(ax + by)(ax + by) =$
  F O I L
  ↓ ↓ ↓ ↓
  $a^2x^2 + abxy + abxy + b^2y^2 =$
  $a^2x^2 + 2abxy + b^2y^2$

3b. Answers will vary.

5a. $(ax - by)(a^2x^2 + abxy + b^2y^2) = a^3x^3 + a^2bx^2y + ab^2xy^2$
    $- a^2bx^2y - ab^2xy^2 - b^3y^3 = a^3x^3 - b^3y^3$

5b. $(3x - 1)(9x^2 + 3x + 1) =$
$27x^3 + 9x^2 + 3x - 9x^2 - 3x - 1 = 27x^3 - 1$
7. True. In a single-variable trinomial, think of the constant as a y-coefficient multiplied by 1.   9. 8   11. 34.75   13. 3   15. 4   17. $1.8 \cdot 10^8$ mi^3

Review and Practice Your Skills, pages 504–505

1. $(x + 6)(x + 1)$ 3. $(d + 6)(d + 7)$ 5. $(x + 14)(x + 2)$
7. $(x - 4)(x - 5)$ 9. $(w - 3)(w - 7)$ 11. $(x - 8)(x - 4)$
13. $(m + 9)(m - 6)$ 15. $(c + 5)(c - 4)$ 17. $(t + 5)(t - 2)$
19. $(a - 8)(a + 6)$ 21. $(p - 9)(p + 4)$
23. $(d - 8)(d + 7)$ 25. $(m + 6n)(m + 5n)$
27. $(p + 12q)(p + 5q)$ 29. $(r - s)(r - 2s)$
31. $(b + 4c)(b - c)$ 33. $(a + 9b)(a - 2b)$
35. $(p - 11q)(p + 7q)$ 37. $(x + 6)(x + 8)$
39. $(f - 2)(f - 24)$ 41. $(c - 3d)(c - 16d)$
43. $(48 - x)(1 - x)$ 45. $(x + 2)(x + 24)$ 47. Specific examples will vary; $(n_1 + x)(n_1 - x) = n_1{}^2 - x^2$
49. Specific examples will vary; $(ax + y)(bx - y) = abx^2 - y^2$ 51. Answers will vary.   53. $5n^2 + 6n - 1$
55. $18x + 4y$ 57. $6x^2y^2z + 48xy^2z^2 - 12x^2yz^2$
59. $16x$ 61. $4x(4x + 15)$ 63. $-3e(3df + 5fg + 4gd)$
65. $7s(m^2 + 4w^2 + 9y^2)$ 67. $x^2 + 20x + 99$ 69. $81 - 16x^2$ 71. $-30f^2 + 131f + 18$ 73. $(w + 4)(5x - 4)$
75. $(x^2 - 5)(8z + 11b)$ 77. $(3 - 6m)(2n - 7p)$
79. $(x + 5)(a - 2b + 7)$ 81. $(13 + a)(13 + a)$
83. $(x + 14)(x + 14)$ 85. $(4a - 7b)(4a + 7b)$
87. $(x + 12y)(x - 12y)$ 89. $(5m + 11n)(5m + 11n)$
91. $(b - 24)(b + 3)$ 93. $(f - 8g)(f - 9g)$
95. $(72 + m)(1 - m)$ 97. $(p - 27q)(p + 3q)$
99. $(ab - 3)(ab + 1)$ 101. $(x - 8)(x - 12)$ 103. 59, 28, 17, 11, 7, 4, -4, -7, -11, -17, -28, -59

Lesson 11-9, pages 506–509

1. 3, 18, 1, 6   3. (2, 4); (2, 3); (5, 4); (5, 3)   5. $-$; $+$
7. $+$; $-$ 9. $(f + g)(7f - 3g)$ 11. $(6x + 5)(x + 2)$
13. 3, 12, 1, 4   15. 6, 2, 15, 5   17. 10, 15, 16, 24
19. (3, 1); (3, 5); (2, 1); (2, 5)   21. (4, 1); (4, 6); (3, 1); (3, 6)   23. $+$; $+$ 25. $-$; $+$ 27. $-$; $+$
29. $(3x - 4)(7x + 2)$ 31. $(z + 4)(2z + 3)$
33. $(2r - 3s)(10r + 5s)$ 35. $(8m + 3)(8m - 5)$
37. $3x(2v - w)(3v + 2w)$ 39. $(2x + 1)(x + 3)$
41. $(5r - 9)(r + 2)$ 43. $3x^2 - x - 10 = 0$; $x = 2$ or $-\frac{5}{3}$
45. $2\sqrt{39}$ 47. 12   49. $6\sqrt{35}$ 51. $6\sqrt{30}$ 53. 176
55. $\frac{\sqrt{78}}{6}$ 57. -8 59. -17 61. -8 63. -4
65. 16   67. 64

Chapter 11 Review, pages 510–512

1. i   3. c   5. g   7. a   9. k   11. $7x + 4y$
13. $13a + 3b$ 15. $3x^3 + 6x^2 - 3$ 17. $3n^2 + 13n + 11$
19. $12d^3f$ 21. $-24k - 15$ 23. $2x^2 + 6xy - 2xz$
25. $-32x^2y^2 - 56x^2y + 112xy^3$ 27. $27x^2y(3 - xy)$
29. $11x(1 + 4x^2y)$ 31. $4x(3a + 5b + 8c)$
33. $3a^2b(3a + 6b - 2b^2)$ 35. $c^2 + 10c + 16$
37. $m^2 - 4n^2$ 39. $4r^2 + 12rs + 9s^2$
41. $20v^2 - 13vw - 21w^2$ 43. $(3a - 5b)(2a + 3b)$
45. $(5r + 2s)(t + 4u)$ 47. $(2x + b)(a + 3c)$
49. $(a^2 + b^2)(a - b)$ 51. $(d + 8)(d + 8)$
53. $(1 + 3y)(1 - 3y)$ 55. $(9m + 4n)(9m - 4n)$

57. $(3x - 5)(3x - 5)$ 59. $(x - 3y)(x - 2y)$
61. $(r - 8)(r - 2)$ 63. $(g + 11)(g - 4)$
65. $(a + 3b)(a - b)$ 67. $(ax + n)(ax + n) = a^2x^2 + axn + axn + n^2 = a^2x^2 + 2axn + n^2$; $(3x + 5)(3x + 5) = 9x^2 + 15x + 15x + 25 = 9x^2 + 30x + 25$ 69. $(3s + 2)(s - 4)$
71. $(3k + 5)(3k + 5)$ 73. $(2s - 5t)(2s + 3t)$
75. $(4a + b)(7a - 2b)$

Chapter 12: Quadratic Functions

Lesson 12-1, pages 520–523

1. 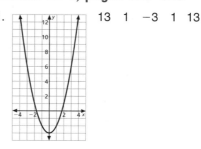   13   1 -3 1   13

3. 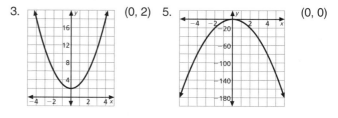 (0, 2)   5. 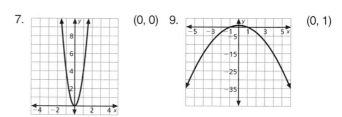 (0, 0)

7. (0, 0)   9. (0, 1)

11. upward   13. As the absolute value of a decreases the graph becomes wider.   15. The graph shifts downward.
17. 18 ft   19. $y = 6x^2 + 2$ 21. (0, 0) x-axis
23. An equation in which $a < 0$; the graph opens downward and has a maximum point.   27. $2y^2 - 4y - 8$
29. $3b^2 - 2b + 3$ 31. $-3d^2 - 4d + 17$ 33. 11.3

Lesson 12-2, pages 524–527

1. (2, 21)   3. $\left(\frac{1}{5}, 4\frac{4}{5}\right)$

5.  $\left(-\frac{3}{2}, 7\frac{3}{4}\right)$; $x = -\frac{3}{2}$

7. 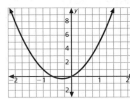 $\left(-\dfrac{1}{3}, -\dfrac{1}{3}\right); x = -\dfrac{1}{3}$

29. $\left(\dfrac{3}{4}, 3\dfrac{7}{8}\right); x = \dfrac{3}{4}$

31. $\left(\dfrac{1}{4}, -2\dfrac{7}{8}\right)$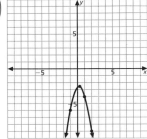

9. $(-3, -5)$ **11.** $\left(\dfrac{1}{2}, 8\dfrac{1}{4}\right)$ **13.** $(2, -3)$

15. $(1, -4); x = 1$ **17.** $(0, 1); y\text{-axis}$
19. $y = -4x^2 - 5x - 2$ **21.** $70°$
23. The maximum range is reached at about 45°.
25. $6s^4 + 18s^3 + 12s^2$ **27.** $8a^3c - 6a^2bd$
29. $100x^7y^4 - 50x^6y^7$

31. 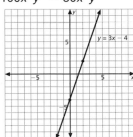   **33.**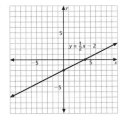

33. 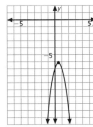 IV

35.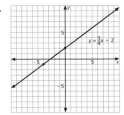

35. 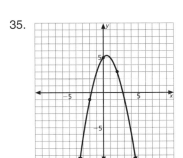 I   **37.** $\left(-\dfrac{1}{3}, 4\dfrac{1}{3}\right);$ $x = -\dfrac{1}{3}$

Review and Practice Your Skills, pages 528–529

1. There is only one y-value for each x-value.

3.

x	-3	-2	-1	0	1	2	3
y	13	8	5	4	5	8	13

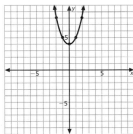 **5.** downward   **7.** upward

9. $(0, -6)$ **11.** $(0, 1)$

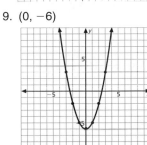 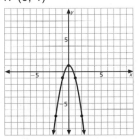

13. no   **15.** yes   **17.** no   **19.** no   **21.** $y = ax^2 + bx + c$
23. $\left(\dfrac{3}{4}, \dfrac{23}{8}\right)$ **25.** $(1, 2)$ **27.** $\left(\dfrac{1}{4}, -3\dfrac{7}{8}\right); x = \dfrac{1}{4}$

Lesson 12-3, pages 530–533

1. $x = -3, x = -7$ **3.** no solutions   **5.** $x = -9$
7. $x = 10, x = -10$ **9.** $x = 2, x = 7$ **11.** $x = 2, x = -2$
13. no solution   **15.** $x = 4, x = -12$ **17.** 6 or -5 **19.**
8 or -1 **21.** 11 sec   **23.** 13 sec   **25.** One solution is
always 0, the other is $-\dfrac{b}{a}$.   **27.** $x = 0, x = -2$
29. $x = 0, x = 4$ **31.** $x = 0, x = -11$ **33.** 45
35. 200   **37.** $9(c - 3b)$ **39.** $3mn(n - 3)$ **41.** $x(w + z)$
43. $9(4a - 7b)$ **45.** $5x^2(2x - 3)$ **47.** $3x^2y(9xy - 2)$
49. $3x^2y^3(2x - 3y + 2z)$

Lesson 12-4, pages 534–537

1. $x^2 + 4x + 4$ **3.** $x^2 - 2x + 1$ **5.** $x = -7, x = 1$
7. $x = -6, x = 2$ **9.** lengths cannot be negative
11. $(x^2 - 10x + 25) - 25$ **13.** $\left(x^2 + x + \dfrac{1}{4}\right) - \dfrac{1}{4}$
15. $(x^2 + 18x + 81) - 81$ **17.** $\left(x^2 + 3x + \dfrac{9}{4}\right) - \dfrac{9}{4}$
19. $\left(x^2 - x + \dfrac{1}{4}\right) - \dfrac{1}{4}$ **21.** $x = -1, x = \dfrac{5}{3}$
23. $x = -3, x = 0$ **25.** $x = -\dfrac{1}{2}, x = 5$
27. $x = -3, x = \dfrac{1}{2}$ **29.** $x = 0, x = 2$ **31.** $x = 0, x = 6$
33. $c = 4, h = -2$ **35.** $c = \dfrac{9}{4}, h = -\dfrac{3}{2}$

37. $c = 121$, $h = 11$ 39. 4 sec   41. 18 in. $\times$ 20 in.
43. $h = 2$ m, $b = 6$ m   45. 10 in. $\times$ 11 in.
47. $6a^2 - 12a - 4ab + 8b$ 49. $8ab + 4ad - 12bc - 6cd$ 51. $9c^2 - 12c + 4$ 53. $12x^2 + 2x - 2$
55. $35a^2 - 11ab - 6b^2$ 57. $60x^2 - 62xy - 12y^2$
59. $\sqrt{77}$ 61. $2\sqrt{14}$ 63. $8\sqrt{374}$ 65. $\sqrt{3}$ 67. $\frac{2\sqrt{15}}{3}$

Review and Practice Your Skills, pages 538–539

1. false   3. true   5. $x = -3, 4$ 7. $x = -2$ 9. no solutions   11. $y = x^2 - x - 20$; $x = 5$ 13. $y = x^2 - 3x + 2$; $x = 1, 2$ 15. $c = \left(\frac{b}{2}\right)^2$ 17. $(x^2 - 4x + 4) - 4$
19. $(x^2 - 18x + 81) - 81$ 21. $(x^2 + 5x + \frac{25}{4}) - \frac{25}{4}$
23. $-\frac{1}{3}, 2$ 25. 0, 8   27. $-6, 2$ 29. $w = 4$ in., $\ell = 8$ in.

31.

x	−3	−2	−1	0	1	2	3
y	4	−1	−4	−5	−4	−1	4

33. $\left(\frac{3}{2}, \frac{7}{4}\right)$; $x = \frac{3}{2}$

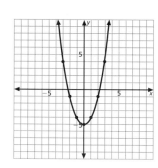

35. $(-1, 3)$; $x = -1$ 37. $\left(\frac{3}{4}, \frac{23}{8}\right)$; $x = \frac{3}{4}$
39. $x = \pm\sqrt{37} - 6$ 41. $x = -4, 3$ 43. $x = \pm\sqrt{29} - 5$
45. $x = -6, -2$ 47. $x = -1, 7$

Lesson 12-5, pages 540–543

1. $2 \pm \sqrt{3}$ 3. 0, 2   5. $1 \pm \sqrt{7}$ 7. 2, 7   9. $-\frac{1}{2}$ 11. 0, 7
13. 4   15. $-2 \pm \sqrt{2}$ 17. $4\frac{1}{2}, -5$ 19. 5, -7 21. 0, -3
23. $-\frac{7}{2}, 3$ 25. $-\frac{2}{3}, \frac{1}{2}$ 27. 1, 11   29. 6400 ft   31a. 3 sec
31b. 144 ft   31c. 6 sec   33. Firth of Forth Bridge, 3.0 sec; Verrazano Narrows Bridge, 3.6 sec; Sydney Harbor Bridge, 3.3 sec; Tunkhannock Viaduct 3.9 sec; Garabit Viaduct, 5.5 sec; Brooklyn Bridge, 2.9 sec.
35. $(a - b)(2a - 3b)$ 37. $4(2a - b)(2a - b)$
39. $(3m + 4n)(2m - 5n)$ 41. $8a(3a - b)(a - b)$
43. $2(m - 2n)(4m + 3n)$ 45. $2(3x - y)(x + y)$ 47. $(-2, 4)$

Lesson 12-6, pages 544–547

1. 7.1 units   3. (1, 0)   5. 13.9 units   7. (1, 1)
9. $(-4, 1.5)$ 11. 26.7 units   13. 3.2 units   15. $(10, -2)$
17. $(0.5, -0.5)$ 19. about 467 ft   21. about 234 ft
23. green: 18 yd, red: 58 yd, purple: 18 yd; black: 38 yd
25. Answers will vary. Since $(x_2 - x_1)^2 = (x_1 - x_2)^2$ and $(y_2 - y_1)^2 = (y_1 - y_2)^2$, the sum of the squared differences will be the same positive number.   27. $x^2 + y^2 = r^2$
29. $(x + 4)(3x - 2)$ 31. $2(x - 3)(3x - 4)$
33. not factorable   35. $2(3x + y)(2x - 3y)$
37. $(2x - 7)(5x - 3)$ 39. $(4x + 3y)(4x + 3y)$

Review and Practice Your Skills, pages 548–549

1. true   3. true   5. $x = \frac{7 \pm \sqrt{41}}{2}$ 7. $x = 3, 1$
9. $x = 1, -\frac{5}{2}$ 11. $x = \frac{7 \pm \sqrt{33}}{4}$ 13. $x = \frac{1 \pm \sqrt{41}}{4}$
15. 7.8   16. 12.0   17. $\left(\frac{1}{2}, \frac{7}{2}\right)$ 19. 7.3   21. 5.1
23. 8.5   25. $\left(2, \frac{3}{2}\right)$ 27. (0, 0)   29. $\left(\frac{5}{2}, \frac{3}{2}\right)$
31. $m = (1, -2)$, $d = 8.5$ 33. $m = \left(-\frac{3}{2}, \frac{1}{2}\right)$, $d = 7.1$
35. $m = (-2, -2)$, $d = 17.0$ 37. $\left(-\frac{3}{2}, -\frac{17}{4}\right)$; $x = -\frac{3}{2}$
39. $\left(\frac{2}{3}, \frac{2}{3}\right)$; $x = \frac{2}{3}$ 41. $\left(\frac{5}{4}, -\frac{33}{8}\right)$; $x = \frac{5}{4}$ 43. $x = \pm\sqrt{5}$
45. $x = 2, -4$ 47. $x = 3, 5$ 49. $x = 1, -4$
51. $x = \frac{-5 \pm \sqrt{29}}{2}$ 53. $x = 2 \pm \sqrt{7}$ 55. $x = \frac{-2 \pm \sqrt{22}}{2}$

Lesson 12-7, pages 550–551

1. $y = x^2 + 2x - 5$ 3. Yes. Any three non-collinear points located on the graph will work. However, you may find it easier to solve the system of equations if you use the y-intercepts as one of the points.   5. $y = x^2 - 2x - 3$
7. $y = x^2 - 4x + 12$ 9. $(5a - 2)(3a + 4)$
11. $(a + 8)(a + 3)$ 13. $(a - 6)(3a + 4)$

Chapter 12 Review, pages 552–554

1. i   3. k   5. a   7. c   9. d
11. (0, 0)   13. (1, 3)

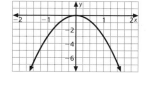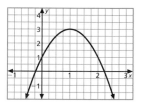

15. $y = \frac{1}{2}x^2 - 2$ 17. no   19. no   21. $-9, 7$
23. $[w(w + 52)]$ m^2 25. 4, -2 27. -7 29. $-5, 2$
31. $-27, 7$ 33. $-3, 7$ 35. 1, 7   37. $-0.5, 3$
39. $-1 \pm \sqrt{7}$ 41. $-10, -2$ 43. $-\frac{4}{5}, 1$ 45. 7
47. $-2.9, 2.4$ 49. $(-3, -2)$, 19.8 units   51. $(6, -2.5)$, 15 units   53. 13   55. 17   57. 7.2   59. 13.6   61. 4.3
63. 1.2   65. $y = x^2 + 4x$

Chapter 13: Advanced Functions and Relations

Lesson 13-1, pages 562–565

1. $x^2 + y^2 = 49$ 3. $(x + 8)^2 + y^2 = 25$ 5. $\sqrt{11}$, (0, 0)
7. $\sqrt{13}$, $(-8, 4)$ 9. $x^2 + y^2 = 144$
11. $(x + 9)^2 + (y + 4)^2 = 144$ 13. $(x - 5)^2 + (y + 1)^2 = 4$
15. $x^2 + y^2 = 6.76$ 17. $\sqrt{15}$, (0, 0)   19. $\sqrt{14}$, $(-4, 2)$

21. $2\sqrt{5}$, (12, −6) 23. 20, (9, −1) 25. 9, (−1, 9)
27. $x^2 + y^2 = 81$ or $(x − 9)^2 + y^2 = 81$ 29. 3, (2, 1)
31. (0, 0) 33. (0, 3) 35. (0, 6) 37. $\frac{-2}{3}$ 39. 1
41. undefined 43. $\frac{10}{9}$ 45. $-\frac{1}{2}$ 47. 0 49. $\frac{4}{7}$ 51. $-\frac{2}{5}$
53. $-\frac{2}{3}$

Lesson 13-2, pages 566–569

1. (0, 3), $y = −3$ 3. (0, 2), $y = −2$ 5. $x^2 = 28y$
7. $x^2 = 3y$ 9. (0, 4), $y = −4$ 11. (0, 5), $y = −5$
13. (0, 1), $y = −1$ 15. $x^2 = −8y$ 17. $x^2 = −2y$
19. $x^2 = 4ay$ 21. $y^2 = 4ax$ 23. $(x − 2)^2 = 8(y − 2)$
25. $(x − 2)^2 = 20(y − 2)$ 27. (1.5, 3.75)
29. (−1.5, −6.25) 31. (0, −9) 33. (−0.25, 11.875)
35. (−2, −7) 37. (−0.375, 8.5625) 39. {2, 5, 8, 9}
41. {6, 7} 43. {9} 45. {1, 3, 4, 6, 7, 9} 47. U
49. {2, 5, 8}

Review and Practice Your Skills, pages 570–571

1. h is the x-coordinate of the center, and k is the y-coordinate of the center. 3. when $h = 0$ and $k = 0$, that is, when the center is at the origin 5. $(x − 2)^2 + (y − 4)^2 − 9$ 7. $(x + 1)^2 + (y − 2)^2 = 16$ 9. $r = \sqrt{13}$, $c = (0, 0)$
11. $r = 5$, $c = (−3, 4)$ 13. $r = 6$, $c = (3, 0)$
15. $r = 2\sqrt{7}$, $c = (−1, 0)$ 17. false 19. false
21. focus = $\left(0, \frac{13}{4}\right)$; directrix is $y = −\frac{13}{4}$ 23. focus = $\left(0, −\frac{3}{4}\right)$; directrix is $y = \frac{3}{4}$
25. focus = $\left(0, −\frac{9}{4}\right)$; directrix is $y = \frac{9}{4}$
27. focus = $\left(0, \frac{5}{4}\right)$; directrix is $y = −\frac{5}{4}$
29. focus = (0, 1); directrix is $y = −1$
31. focus = $\left(0, −\frac{3}{2}\right)$; directrix is $y = \frac{3}{2}$ 33. $x^2 = −12y$
35. $x^2 = −28y$ 37. $x^2 = 10y$ 39. parabola
41. parabola 43. circle 45. $r = 6$, $c = (−3, 4)$
47. $x^2 + (y + 3)^2 = 49$ 49. $(x + 3)^2 + (y + 1)^2 = 36$

Lesson 13-3, pages 572–573

1. circle 3. point 5. parabola 7. 2; 8 and −8 9. 1; −1
11. no solution 13. 2; 6 and −1 15. 2; −2 and −3

Lesson 13-4, pages 574–577

1. 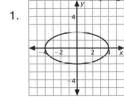 3. $25x^2 + 169y^2 = 4225$
5. No; a circle is not an ellipse. The foci of an ellipse are different points.

7. 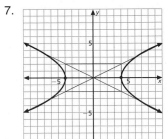 9. $36x^2 − 49y^2 = 1764$

11. 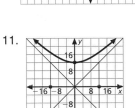 13.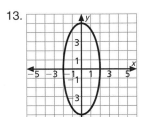

15. $\frac{(x − h)^2}{a^2} + \frac{(y − k)^2}{b^2} = 1$ 17. $x = 0$, $x = −10$ 19. $x = 0$, $x = 8$ 21. $x = 3 \pm \sqrt{7}$

Review and Practice Your Skills, pages 578–579

1. parabola 3. circle

5. The standard form is $\frac{x^2}{a^2} + \frac{y^2}{b^2} = 1$. If $y = 0$; $\frac{x^2}{a^2} + 0 = 1$; $\frac{x^2}{a^2} = 1$; $x^2 = a^2$; $x = \pm a$.

7. The standard equation for an ellipse uses the sum of the distances from a point on the ellipse to the foci; the equation for a hyperbola uses the differences of these distances.

9.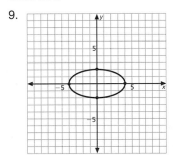

11.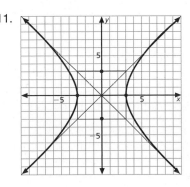

13.

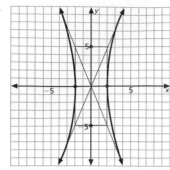

15. $91x^2 + 100y^2 = 9100$ 17. $9x^2 - 25y^2 = 225$
19. $16x^2 - 49y^2 = 784$

21.

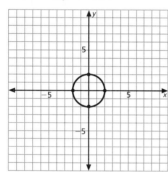

23.

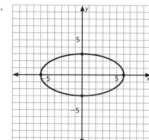

25.

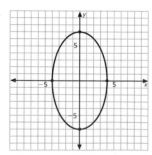

Lesson 13-5, pages 580–583

1. $y = 0.25x$ 3. $y = 125$ 5. \$500 7. 80 mi 9. 450
11. \$5.31 13. 40 15. No, Paige's conclusion is wrong.
Using the direction variation form, $y = kx$, if 14 lb are
added to the spring, it will stretch 1.75 in.
17. $A = 2.598s^2$, 210.438 19. 1035 21. 10
23. 19. \$0.13; 20. \$0.03; 21. \$1.23; 22. \$5.41
25. direct variation 27. direct square variation
29. direct variation 31. $x = \dfrac{-3 \pm \sqrt{57}}{4}$ 33. $x = 2 \pm 2\sqrt{3}$

35. $x = \dfrac{-3 \pm \sqrt{13}}{4}$ 37. 8.49 39. 13.23 41. 180 43. $\dfrac{3}{2}$

Lesson 13-6, pages 584–587

1. $y = \dfrac{68}{x}$ 3. \$1.80 each 5. 12 lumens 7. $y = 3$

9. Answers may vary. Students should mention that in an
inverse function, as one variable increases, the other
decreases. 11. 4 newtons 13. $p = \dfrac{k}{a}$ 15. $l = \dfrac{k}{R}$

17. $m = kst$ 19. $r = kwxy$ 21. $v = \dfrac{kr}{w^2}$ 23. 10.20
25. 5.39 27. 13.60 29. 11 31. 20.5

Review and Practice Your Skills, pages 588–589

1. x decreases if the constant of variation is positive.
3. A direct variation graph is linear; a direct square
variation graph is a quadratic function. 5. $y = \dfrac{2}{3}x$
7. $y = \dfrac{1}{4}x$ 9. $y = \dfrac{9}{4}x$ 11. $y = 2.14$ 13. $y = 48$
15. $y = 405$ 17. In real life, there are no negative people
and no negative costs. Quadrant I is the only quadrant
where both variables are positive.

19. $y = \dfrac{6.3}{x}$ 21. $y = 3.11$ 23. $t = \dfrac{k}{s}$

25.

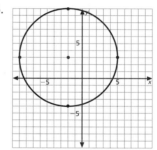

27.

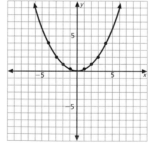

29.

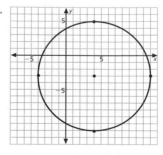

31. $y = \dfrac{5}{2}x$ 33. $y = \dfrac{7}{5}x$

Lesson 13-7, pages 590–593

1.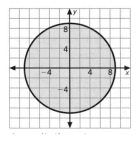

3.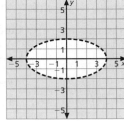

5.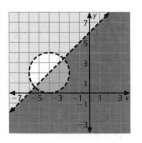

7.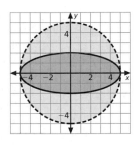

9.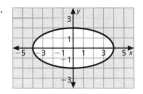

11.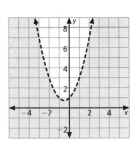

13.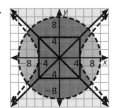

15.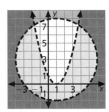

17. <   **19.** <   **21.** The graph of the solution set shows points above the first parabola and below the second parabola.   **23.** Answers will vary.   **25.** $2 < x < 6$
27. $(-1, -1)$ **29.** $\left(-1, -\frac{7}{2}\right)$ **31.** $\left(-\frac{5}{2}, -\frac{9}{2}\right)$ **33.** $\left(\frac{3}{2}, \frac{3}{2}\right)$
35. $\left(\frac{13}{2}, -\frac{9}{2}\right)$ **37.** $\left(-\frac{1}{2}, 3\right)$ **39.** $a = 105$, $b = 75$
41. $a = 120$; $b = 60$ **43.** 305 ft³   **45.** $x = -10 \pm \sqrt{101}$

Lesson 13-8, pages 594-597

1. 1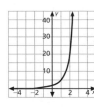

3. 5

5. about $25,828

7. 1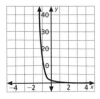

9. 1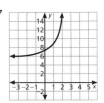

11. 3

13. 7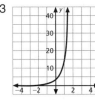

15. about 8329.2 million computers   **17.** about $7877
19. about $3522.72   **21.** about $12,204   **23.** No. The graph of $y = \left(\frac{1}{3}\right)^x$ decreases as x increases. This graph increases as x decreases.   **25.** translation 2 units up
27. $\sqrt[5]{6}$ **29.** $\sqrt[6]{m}$ **31.** $25^{\frac{1}{3}}$ **33.** In radical form, the expression would be $\sqrt{-16}$, which is not a real number because the radicand is negative.   **35.** $(t - 2)(t - 6)$
37. $(s + 2)(s - 10)$ **39.** $(m - 2n)(m - 3n)$ **41.** $-15m^4n^4$
43. $4x^2 - 3x$ **45.** $-3ab^3 - 4a^2b^2 + 6a^3b$ **47.** 6.3 cm

Review and Practice Your Skills, pages 598-599

1.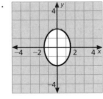

3.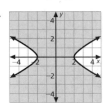

5.

7.

9.

11.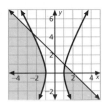

13.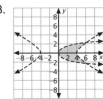

15.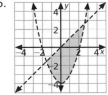

17. 1

19. 2

21. −4

23. about $5962.59
25. about $992
27. about $33,636
29. $(x + 1)^2 + (y − 3)^2 = 144$ 31. $(x + 4)^2 + (y + 1)^2 = 4$ 33. center $(−1, −5)$, radius 1 35. focus $(0, 5)$, directrix $y = −5$ 37. focus $(0, 1)$, directrix $y = −1$

39.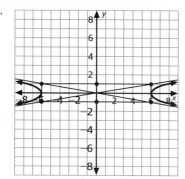

41.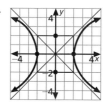

43.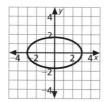

45. 60 47. 24 49. 4

51.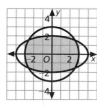

53.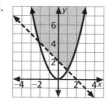

55.

57. about 33,687,150

Lesson 13-9, pages 600–603

1. $\log_5 625 = 4$ 3. $\log_2 \frac{1}{512} = −9$ 5. $216 = 6^3$ 7. 6
9. 6 11. 9 13. $\log_8 32{,}768 = 5$ 15. $\log_4 \frac{1}{64} = −3$
17. $\log_\frac{1}{9} 729 = −3$ 19. $1 = 7^0$ 21. $\frac{1}{100} = 10^{−2}$

---

23. $\frac{1}{243} = 3^{−5}$ 25. 3 27. 3 29. −2 31. 16,807
33. 46,656 35. 5 37. 7 39. 4 41. 10^{15} or 1,000,000,000,000,000 43. 10^2 or 100 times

45.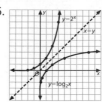

47. The domain of $y = 2^x$ is all real numbers and the range is all positive real numbers. The domain of $y = \log_2 x$ is all positive real numbers and the range is all real numbers. 49. $2a^2 − 2a − 4$

51. $16g^2 − 25$ 53. $\frac{1}{21}$

55. $(4, -1)$ 57. $\left(4, −\frac{1}{2}\right)$ 59. 20

Chapter 13 Review, pages 604–606

1. l 3. j 5. c 7. d 9. i 11. $x^2 + y^2 = 64$
13. $(x − 5)^2 + y^2 = 36$ 15. 3, $(0, 3)$ 17. $(0, 5)$, $y = −5$
19. $(0, 1)$, $y = −1$ 21. $x^2 = −16y$ 23. The plane is parallel to the x-axis and does not contain the vertex.
25. $36x^2 + 100y^2 = 3600$ 27. $y = 50$ 29. $y = 320$
31. $y = 156.25$ 33. $y = 90$

35. 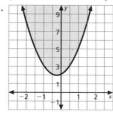 37.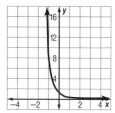

39.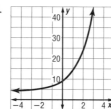

41. $\log_8 32{,}768 = 5$
43. $\log_{17} 289 = 2$
45. $\log_\frac{1}{5} \frac{1}{625} = 4$ 47. 0
49. $\frac{1}{343}$ 51. 5

Chapter 14: Trigonometry

Lesson 14-1, pages 614–617

1. $\frac{12}{5}$ 3. $\frac{12}{13}$ 5. 0.3746 7. 1.2799 9. 0.2126 11. $\frac{4}{3}$
13. $\frac{4}{5}$ 15. $\frac{3}{4}$ 17. 0.69; 1.05 19. 0.29 21. about 32.1°
23. $\frac{12}{13}$ 25. Divide the height of the lighthouse by the ratio of height to distance (or tan 10°). 27. 1 29. $\frac{1}{2}$
31. $x^2 + y^2 = 25$ 33. $(x − 5)^2 + (y − 1)^2 = 12.25$
35. $(x − 3)^2 + (y + 2)^2 = 49$ 37. $(x + 3)^2 + (y − 5)^2 = 36$
39. $(x + 4)^2 + (y − 3)^2 = 56.25$ 41. $\frac{1}{76}$ 43. $\frac{7}{855}$

Lesson 14-2, pages 618–621

1. 21.6 3. 34° 5. 55° 7. 31° 9. 3.0 11. 10.7
13. 35° 15. hypotenuse: 36.5; angles: 40°, 50°, 90°
17. 10,832.8 ft 19. 150.8 in. 21. 32.8 ft 23. 20°
27. no 29. Sometimes; the angle must be acute.
31. $(0, 1)$; $y = −1$ 33. $(0, 2)$; $y = −2$ 35. $\left(0, \frac{5}{2}\right)$; $y = −\frac{5}{2}$

Review and Practice Your Skills, pages 622–623

1. right triangle 3. false 5. $\frac{3}{4}$ 7. $\frac{3}{5}$ 9. $\frac{3}{5}$ 11. $\frac{4}{5}$

13. cos S = 0.50; tan R = 0.58   15. The sum of the angle measures of a triangle is 180°. If one angle is a right angle, you know its measure is 90°. Therefore, the sum of the other two angles must be 90°. If you know the measure of one of the acute angles, you can use this information to find the missing measure.   17. 36°   19. 24.0   21. 60°   23. 0.57   25. 0.57   27. 0.82   29. sin D = 0.8; tan E = 0.75   31. 66°   33. 32.9°   35. about 24.8

Lesson 14-3, pages 624–627

1. $\frac{\sqrt{2}}{2}$ 3. $-\frac{1}{2}$ 5. $\frac{-\sqrt{3}}{2}$

7. 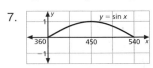   9. $-\frac{\sqrt{2}}{2}$ 11. $\frac{\sqrt{3}}{2}$
13. $-\frac{\sqrt{2}}{2}$ 15. $\frac{-\sqrt{2}}{2}$
17. $-\frac{\sqrt{2}}{2}$ 19. $\frac{\sqrt{2}}{2}$

21. 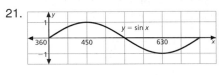   23. 1   25. $\sqrt{3}$
27. $\frac{\sqrt{2}}{2}$ 29. $-\frac{\sqrt{2}}{2}$
31. $\frac{\sqrt{2}}{2}$ 33. $-\frac{\sqrt{3}}{2}$

35.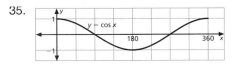

37. 0°, 180°, 360°   39. 90°   41. ≈ 485 ft
43a. Graph the equations on the same set of axes. Points of intersection of the curves represent solutions of the system of equations.   43b. x = 45° and x = 225°
45. 150°, 210°   47. 60°, 240°   49. $3x^2 + 4y^2 = 48$
51. $5x^2 + 9y^2 = 405$ 53. $32x^2 + 81y^2 = 2592$

Lesson 14-4, pages 628–631

1. period = 120°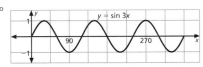

3.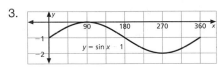

The graph is the graph of y = sin x lowered 1 unit from the origin.   5. period 180°; amplitude 2; the graph is the graph of y = 2 sin 2x lowered 5 units.

7. 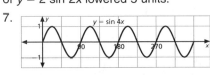   The period is 90°.

9.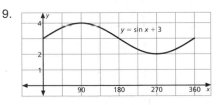

The graph is the graph of y = sin x raised 3 units.
11. period 120°; amplitude 4; the graph is the graph of y = 4 sin 3x raised 2.5 units.   13. yes; 5   15. no
17. Sample answer: $y = \frac{3}{2}\sin\frac{2}{3}x$ 19. 600°
21. Sample answer: $y = 8\sin\frac{4}{7}x$ 23. $y = \frac{1}{2}\sin 8x - 1$
25.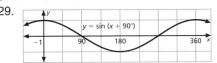

27. For the graph of the equation $y = m\cos nx \pm p$, the period is $\frac{360}{n}$, the amplitude is m, and the graph is the graph of $y = m\cos nx$ raised or lowered p units.

29.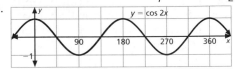

31. 40.5   33. 133   35. $(2x - 3)(x + 4)$
37. $(2x + 1)(2x - 5)$ 39. $3(x + 2)(x - 1)$
41. $(x - 8)(x + 8)$

Review and Practice Your Skills, pages 632–633

1. true   3. false   5. $\frac{\sqrt{3}}{2}$ 7. $\frac{-\sqrt{3}}{2}$ 9. $\frac{-\sqrt{2}}{2}$ 11. $-\frac{1}{2}$
13. $-\frac{1}{2}$ 15. $\frac{-\sqrt{2}}{2}$ 17. 60°, 300°   19. 120°, 240°
21. 0°, 360°   23. Half the difference between the maximum and minimum y-values.   25. 2   27. 120°   29. 36°
31. 360°   33. 9°   35. 360°, 2, centered on x-axis
37. 120°, 2, shifted 1 unit up   39. 180°, 5, shifted 2 units up
41. $\frac{7\sqrt{85}}{85}$ 43. $\frac{6\sqrt{85}}{85}$ 45. $\frac{6}{7}$ 47. 66.0°   49. $\frac{\sqrt{2}}{2}$ 51. 0
53. $\frac{\sqrt{2}}{2}$ 55. 60°, 120°   57. 30°, 330°   59. 45°, 315°

Lesson 14-5, pages 634–635

1. 483.2 ft   3. 15°   5. $y = 8\sin 2x - 3$ 7. 14°
9. 4.67   11. 3.75   13. 3.77   15. $x^2 + 3x - 40$
17. $3x^2 - 26x + 16$ 19. $4x^2 - 81$ 21. $6x^2 + 33x - 18$
23. $6x^2 - 11x - 72$ 25. 3.23; 1.80   27. 42.36; 6.51

Chapter 14 Review, pages 636–638

1. j   3. d   5. h   7. f   9. l   11. $\frac{3}{5}$ 13. $\frac{4}{3}$ 15. $\frac{4}{5}$
17. about 30.5°   19. 55°   21. ≈ 51.3 ft   23. about 94.0 m
25. 1   27. $-\frac{\sqrt{3}}{3}$ 29. $\frac{1}{2}$
31. period = 720°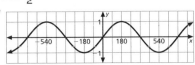

33. yes; 8   35. about 2.8 mi

Photo Credits

Index

Index

Index

Index

transversals, 120
Trapezoid-Median Theorem, 188
trapezoids, 188–191, 321
Triangle Inequality Theorem, 173
triangles, 147, 150–153, 157,
     172–175, 198, 316–319, 610
Triangle-Sum Theorem, 150
triangular pyramids, 220
trigonometry, 612–641
Unequal Angles Theorem, 173
Unequal Sides Theorem, 173
Unique Line Postulate, 105
Unique Plane Postulate, 105
vertex angles, 160
vertical angles, 100, 115
Vertical Angles Theorem, 115, 134
vertical lines, 245
vertices, 108, 150, 178, 220
volumes, 230–233
Geometry software, *see* Calculators,
     geometry software
Glide reflection, 353
Golden Rectangle, 205
GPE (greatest possible error), 202
Grand products, 502
Graphing
     ellipses, 575
     equations, 240
     factoring and, 530–533
     hyperbolas, 576
     inequalities, 50
     linear functions, 62–65
     parabolas, 520–523
     quadratic inequalities, 590–593
     real numbers, 11
     sine functions, 614–617
     solving systems of equations by,
          258–261, 561
Graphing calculators, *see* Calculators,
     graphing
Graphs, 86
     circle, 446–447
     of functions, 518–519
     misleading, 92–93
     of numbers, 11
     using, in writing equations,
          550–551
Gravity, center of, 166
Greater than or equal to symbol (≥), 11
Greater than symbol (>), 11
Greatest common factor (GCF), 479
Greatest possible error (GPE), 202
Grid In, 47, 199, 145, 197, 239, 291,
     333, 379, 421, 463, 515, 557,
     609, 641
Group work, 3, 10, 49, 86, 101, 104,
     147, 172, 212, 230, 241, 293, 306,
     335, 381, 406, 412, 423, 465, 471,
     488, 495, 501, 517, 540, 559, 611,
     617, 627

■ H ■

Half-plane, 77, 276
     closed, 77
     open, 77
Harriot, Thomas, 11
Hectare, 309
Hecto, prefix, 39
Heptagons, 178
Heron's formula, 429
Hexagonal prisms, 220
Hexagonal pyramids, 220
Hexagons, 178, 180
Hipparchus, 629
Histograms, 87, 381, 414
Horizontal line, 245
Hyperbolas
     ellipses and, 574–577
     standard equation of, 576
Hypotenuse, 175
Hypothesis, 128

■ I ■

Identifying angles, 102
Identity property
     of addition, 21
     of multiplication, 27
if-then statements, 128
Image, 338
Included angles, 155
Included sides, 155
Inconsistent systems, 258
Independent events
     defined, 396
     dependent events and, 396–399
Independent systems, 258
Independent variables, 57
Indirect measurement, 326–327
Indirect proofs
     defined, 170
     writing, 170–171
Inductive reasoning
     deductive reasoning versus, 135
     defined, 124
     in mathematics, 124–127
Inequality(ies), 11
     addition property of, 76
     compound, *see* Compound
          inequalities
     defined, 76
     graphing, 50
     multiplication property of, 76
     solving, 76–79
     systems of, *see* Systems of
          inequalities
     transitive property of, 77
     in triangles, 172–175
Infinite number of solutions, 259
Infinite sets, 6
Initial sides, 624

Inner product, 483
Input values, 57
Inscribed angles, 441
Inscribed polygons, 455
Inside calipers, 202
Integers, 10
     negative, 10
     positive, 10
Intercept, 440
Interior, of angles, 109
Interior angles, 120
     of polygons, 178
     of triangles, 147, 150
Internet, 419, 513
Interquartile range, 408
Intersection
     of figures, 104
     of sets, 16–19
     defined, 16
Inverse operations, 66
Inverse property
     of addition, 21
     of multiplication, 27
Inverses
     additive, 21
     multiplicative, 27
Inverse square variation, 585
Inverse variation, 584–587
     defined, 584
Irrational numbers, 10, 426–429
     defined, 426
Isosceles trapezoids, 189
Isosceles Trapezoid Theorem, 189
Isosceles triangles, 150
     legs of, 160
Isosceles Triangle Theorem, 161
Iteration, numeric, 53
Iterations
     defined, 53
     patterns and, 52–55

■ J ■

Joint variation, 587

■ K ■

Kilo, prefix, 39
Kites, 189

■ L ■

Last product, 483
Lateral edges, 220
Lateral faces, 220
Latitude, 102
Leaf, 86
Legs
     of isosceles triangle, 160
     of right triangle, 430
     of trapezoid, 188

Index

of inequalities, 276–279
defined, 276

■ **T** ■

30-60-90 Triangle Theorem, 436, 613
Tally system, 82
Tangent functions, 614
Tangent segments, 449
Tangents of circles, 441
Technology, 45, 289, 461
    calculators, 29, 30–31, 39, 40, 84,
        226, 247, 289, 298, 311, 405,
        428, 546, 615, 616, 625
    charting and data analysis
        software, 531
    computer program, 413
    geometric drawing software, 119,
        189
    geometric software, 106, 115, 179,
        316
    geometry software, 109, 121, 153,
        155, 157, 165, 166, 174, 320,
        351, 371, 431, 433, 441, 443,
        451
    graphing, 64–65, 245, 246, 248, 255,
        258, 261, 278, 283, 358, 360,
        363, 364, 370, 408, 426, 520,
        521, 522, 523, 524, 525, 526,
        530, 532, 563, 566, 567, 575,
        600, 626, 628, 629, 639
    spreadsheet program, 87
    spreadsheets, *see* Spreadsheets
    using, 30–31
Technology Note, 106, 115, 119, 179,
    189, 226, 245, 258, 278, 283, 311,
    316, 349, 353, 358, 360, 363, 405,
    408, 413, 426, 500, 525, 531, 575,
    625
Terminal sides, 624
Terminating decimals, 10
Terms, 52
    like, 73, 468
    in proportion, 296
    quadratic, 498
    undefined, 104
Theorems, 114, 134
    proofs of, 134
Theoretical probability, 385–386
Think Back, 82
Thinking, visual, 572–573
Three-dimensional figures
    defined, 220
    loci and, 220–223
    surface areas of, 224–227
    volumes of, 230–233
Transformations, 336–379
    composite of, 352
    with matrices, 369
    matrices and, 368–371
    multiple, 352–355

Transitive property, 34
    of equality, 34
    of inequality, 77
Translations
    defined, 338
    reflections and, 338–341
Transversals, 120
Trapezoid-Median Theorem, 188
Trapezoids, 188–191
    base angles of, 188
    defined, 188
    isosceles, 189
    legs of, 188
    medians of, 321
Tree diagrams, 385
Trend lines, 406
Triangle Inequality Theorem, 173
Triangles, 178
    acute, 150
    angles of, 149
    areas of, 198
    congruent, *see* Congruent triangles
    defined, 150
    equiangular, 150
    equilateral, 150
    exterior angles of, 151
    inequalities in, 172–175
    interior angles of, 147, 150
    isosceles, *see* Isosceles triangles
    medians of, 317
    naming sides of, 612
    obtuse, 150
    overlapping, 157
    perimeters of, 198
    proportional segments and,
        316–319
    quadrilaterals and other polygons
        and, 148–197
    right, *see* Right triangles
    scalene, 150
    similar, *see* Similar triangles
    triangle theorems and, 150–153
    vertex of, 150
Triangle-Sum Theorem, 150
Triangle theorems, triangles and,
    150–153
Triangular pyramids, 220
Trigonometric ratios, 614–617
Trigonometry, 612–641
Trinomials, 468
    factoring, 498–501, 506–509
    perfect square, 492
Triples, Pythagorean, 433
Turn, 342

■ **U** ■

Undefined slope, 245
Undefined terms, 104
Unequal Angles Theorem, 173

Unequal Sides Theorem, 173
Union of sets, 16–19
    defined, 16
Unique Line Postulate, 105
Unique Plane Postulate, 105
Unit price, 204
Unit rate, 204
Units of measure, 202–205
Universal set, 16
Universe, 16
Upper quartile, 407
Using
    graphs in writing equations,
        550–551
    logical reasoning, 138–139
    matrices, 372–373
    Pythagorean Theorem, 544–547
    technology, 30–31

■ **V** ■

Values
    absolute, 12
    input, 57
    output, 57
    range of, 49
Variables, 7
    dependent, 57
    independent, 57
Variance, 412
Variation
    combined, 587
    constant of, 580
    direct, *see* Direct variation
    direct square, 581
    inverse, *see* Inverse variation
    inverse square, 585
    joint, 587
Venn, John, 17
Venn diagrams, 16–19
Vertex
    of angle, 108
    of parabola, 521
    of polygon, 178
    of polyhedron, 220
    of triangle, 150
Vertex angles, 160
Vertical angles, 102, 115
Vertical Angles Theorem, 115, 134
Vertical line, 245
Vertical line test, 57, 518–519
Vertices, consecutive, 178
Visual thinking, 572–573
Vocabulary, 42, 94, 140, 192, 234, 286,
    328, 374, 416, 458, 510, 552, 604,
    636
Volumes
    defined, 230
    of three-dimensional figures,
        230–233

Formulas

Coordinate Geometry

Slope	$m = \dfrac{y_2 - y_1}{x_2 - x_1}$
Distance	on a coordinate plane: $d = \sqrt{(x_2 - x_1)^2 + (y_2 - y_1)^2}$
Midpoint	on a number line: $M = \dfrac{a + b}{2}$ on a coordinate plane: $M = \left(\dfrac{x_1 + x_2}{2}, \dfrac{y_1 + y_2}{2} \right)$

Perimeter and Circumference

square	$P = 4s$
rectangle	$P = 2\ell + 2w$
circle	$C = 2\pi r$ or $C = \pi d$

Area

square	$A = s^2$
rectangle	$A = \ell w$ or $A = bh$
parallelogram	$A = bh$
trapezoid	$A = \dfrac{1}{2}h(b_1 + b_2)$
triangle	$A = \dfrac{1}{2}bh$
circle	$A = \pi r^2$

Pythagorean Theorem	$a^2 + b^2 = c^2$
Quadratic Formula	$x = \dfrac{-b \pm \sqrt{b^2 - 4ac}}{2a}$

Total Surface Area

rectangular prism	$SA = 2(\ell w + \ell h + wh)$
cylinder	$SA = 2\pi rh + 2\pi r^2$
pyramid	$SA = \left(\dfrac{1}{2}bh\right)n + B$
cone	$SA = \pi rs + \pi r^2$
sphere	$SA = 4\pi r^2$

Volume

cube	$V = s^3$
rectangular prism	$V = \ell wh$
prism	$V = Bh$
cylinder	$V = \pi r^2 h$
pyramid	$V = \dfrac{1}{3}Bh$
cone	$V = \dfrac{1}{3}\pi r^2 h$
sphere	$V = \dfrac{4}{3}\pi r^3$

Equations for Figures on a Coordinate Plane

slope-intercept form of a line	$y = mx + b$
point-slope form of a line	$y - y_1 = m(x - x_1)$
circle	$(x - h)^2 + (y - k)^2 = r^2$

Addition and Multiplication Properties

Additive Identity	For any number a, $a + 0 = 0 + a = a$.
Multiplicative Identity	For any number a, $a \cdot 1 = 0 \cdot a = a$.
Additive Inverse	For any number a, there is exactly one number $-a$ such that $a + (-a) = 0$.
Multiplicative Inverse	For any number $\dfrac{a}{b}$, where $a, b \neq 0$, there is exactly one number $\dfrac{b}{a}$ such that $\dfrac{a}{b} \cdot \dfrac{b}{a} = 1$.
Commutative (+)	For any numbers a and b, $a + b = b + a$.
Commutative (×)	For any numbers a and b, $a \cdot b = b \cdot a$.
Associative (+)	For any numbers a, b, and c, $(a + b) + c = a + (b + c)$.
Associative (×)	For any numbers a, b, and c, $(a \cdot b) \cdot c = a \cdot (b \cdot c)$.
Distributive	For any numbers a, b, and c, $a(b + c) = ab + ac$ and $a(b - c) = ab - ac$.